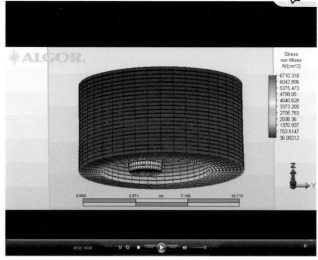

Figure 1–6 (b) The three-dimensional visual of the die as the elements in the plane are rotated through 360° around the *z*-axis of symmetry.

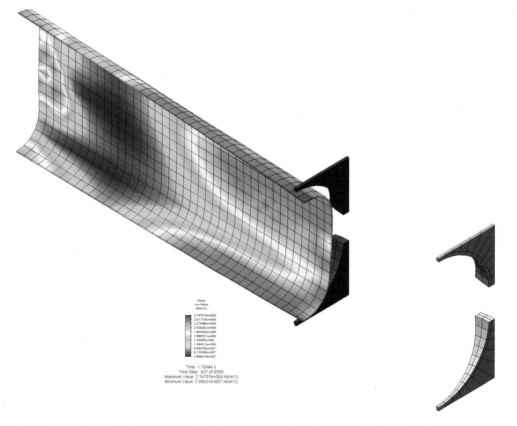

Figure 1–11 Finite element model of contour roll forming or cold roll forming process. (Courtesy of Valmont Coast Engineering Group)

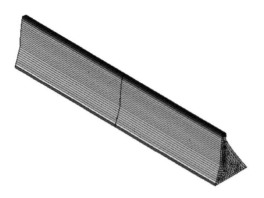

Figure 6-2 Plane strain problems: (a) dam subjected to horizontal loading. (Algor)

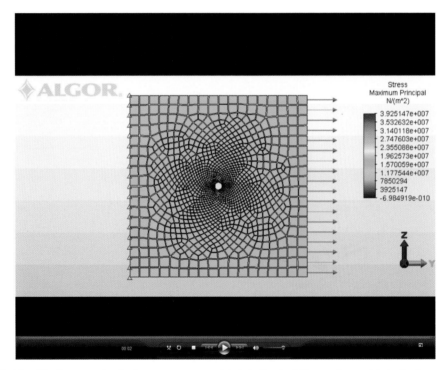

Figure 7-11 Maximum principal stress contour (shrink fit plot) for a plate with hole. Largest principal stresses of 29.48 MPa occur at the top and bottom of the hole, which indicates a stress concentration of 2.948. Stresses were obtained by using an average of the nodal values (called smoothing). (Same plate properties as in Figure 7-10.)

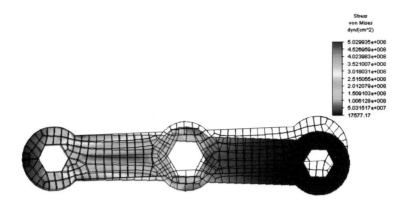

Figure 7–21 Bicycle wrench (a) Outline drawing of wrench, (b) meshed model of wrench, (c) boundary conditions and selecting surface where surface traction will be applied, (d) checked model showing the boundary conditions and surface traction, and (e) von Mises stress plot. (Compliments of Angela Moe)

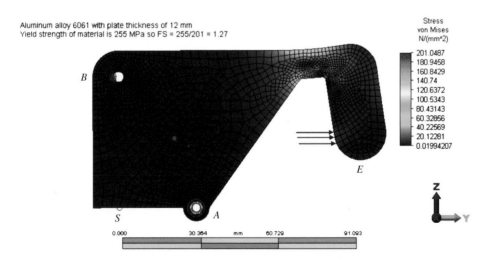

Figure 7–23 von Mises stress plot of overload protection device.

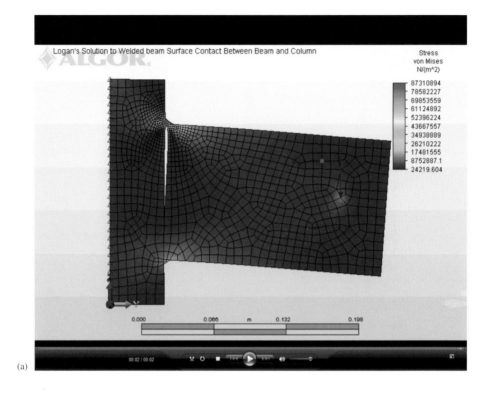

(a)

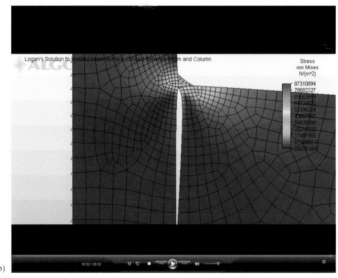

(b)

Figure 7–26 (a) von Mises stress plot of beam welded to column (largest von Mises stress of 87.3 MPa is located at toe of top fillet weld as shown by), (b) Zoomed-in view of top fillet weld (notice also that a surface contact was used between the beam and column that allowed for the gap to form where the beam separated from the column).

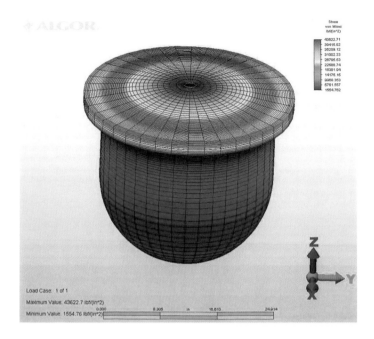

Figure 9–2 Examples of axisymmetric problems: (b) enclosed pressure vessel. (Courtesy of Algor, Inc.)

Figure 9–18   Three-dimensional visual of shaft of Figure 9–17 showing principal stress plot.

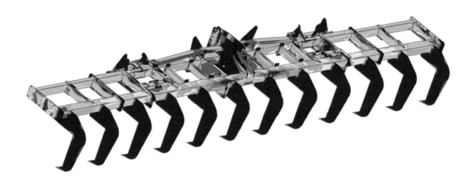

Figure 11–1   (c) *Subsoiler*—12-row subsoiler used in agricultural equipment. (Courtesy of Algor, Inc.)

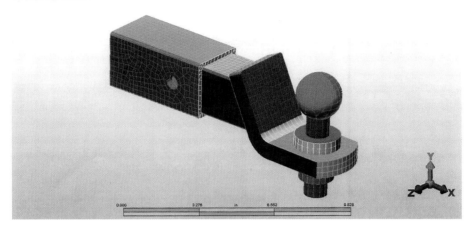

Figure 11–9 Meshed model of a trailer hitch. (Courtesy of David Anderson)

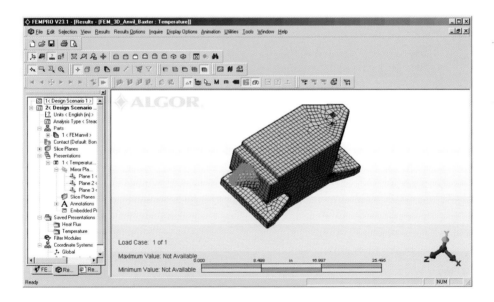

Figure P11–19 Anvil used for forging operation (showing dimensions in inch units) and typical finite element model. (Compliments of Dan Baxter)

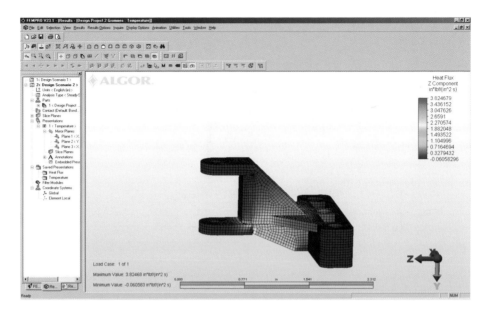

Figure P11–21 Radio-control front steering unit (all dimensions in inches) and finite element model. (Compliments of Phillip Grommes)

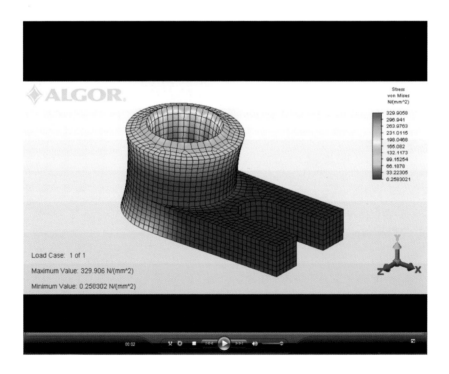

Figure P11–27 Locator part (dimensions in mm) with typical finite element model.

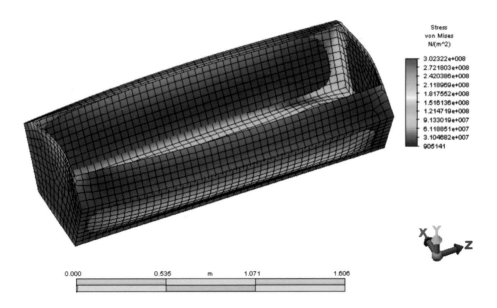

Figure 12–1 (b) Water tank.

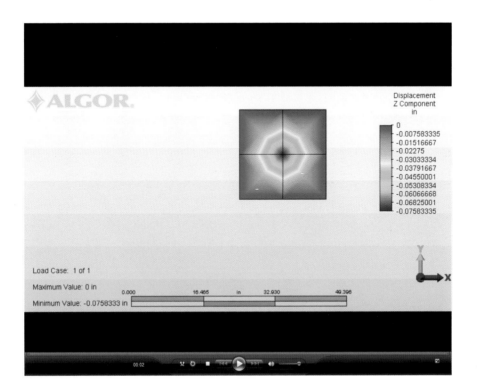

Figure 12–10 Displacement plot of the clamped plate of Example 12.1.

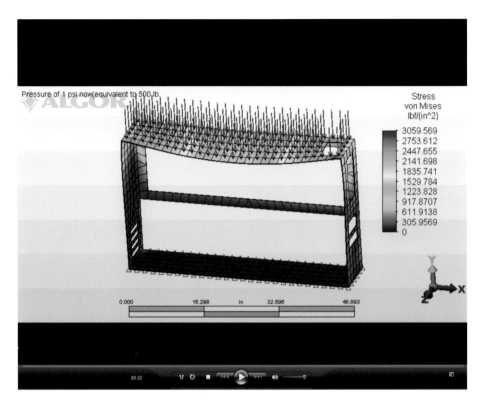

Figure 12–12 (b) The pressure load, boundary conditions, and resulting von Mises stress. (by Nicholas Dachniwskyj)

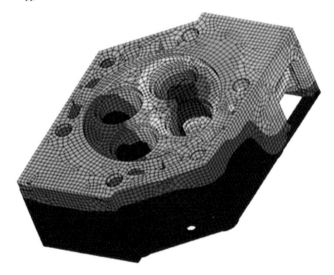

Figure 13–1 Finite element results of cylinder head showing temperature distribution (brick elements were used in the model). (Courtesy of Algor, Inc.)

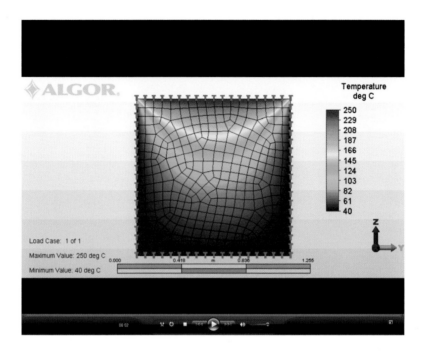

Figure 13–34 (b) Finite element model with resulting temperature variation throughout the plate. (Courtesy of David Walgrave)

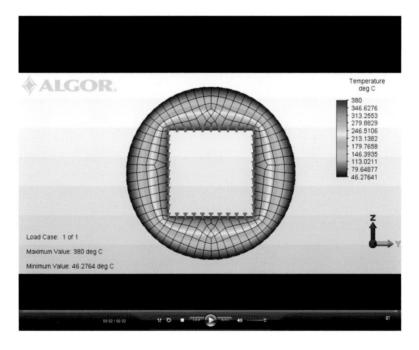

Figure 13–35 (b) The finite element model with resulting temperature variation through the insulation.

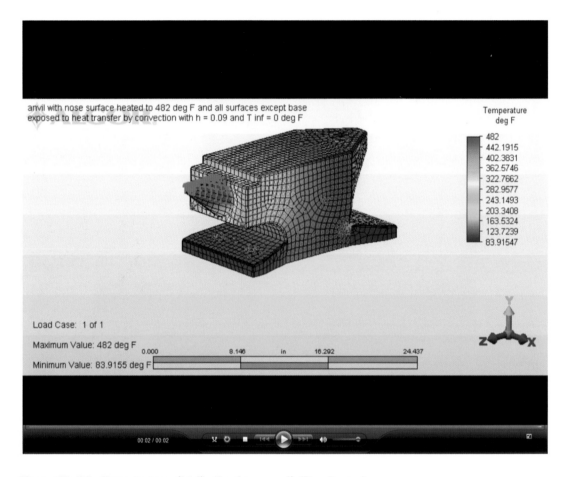

anvil with nose surface heated to 482 deg F and all surfaces except base
exposed to heat transfer by convection with h = 0.09 and T inf = 0 deg F

Temperature
deg F

482
442.1915
402.3831
362.5746
322.7662
282.9577
243.1493
203.3408
163.5324
123.7239
83.91547

Load Case: 1 of 1

Maximum Value: 482 deg F

Minimum Value: 83.9155 deg F

0.000   8.146   in   16.292   24.437

00 02 / 00 02

Figure 13–36 Temperature distribution in an anvil. (Dan Baxter)

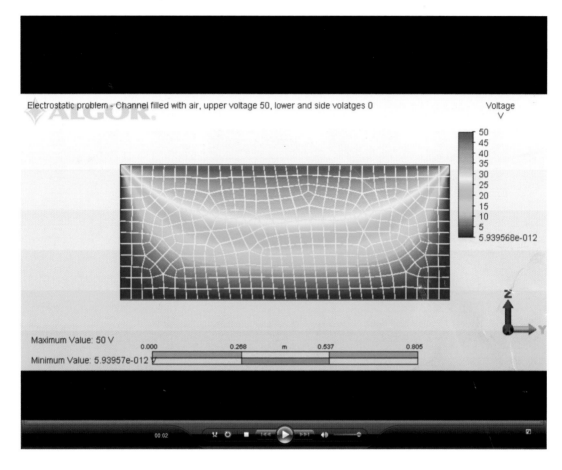

Figure 14–30 Voltage variation throughout channel.

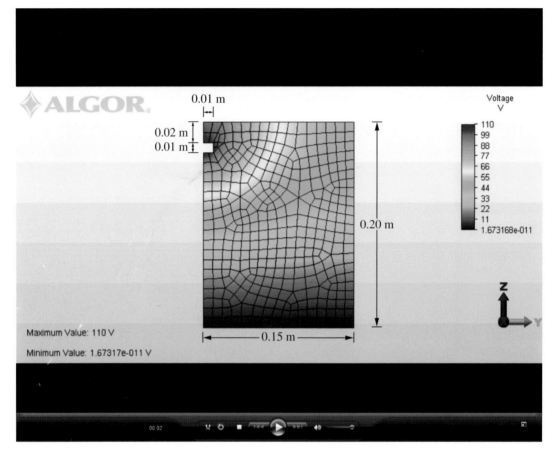

Figure 14–32 Busbar surrounded by air along with the finite element model and the resulting voltage distribution.

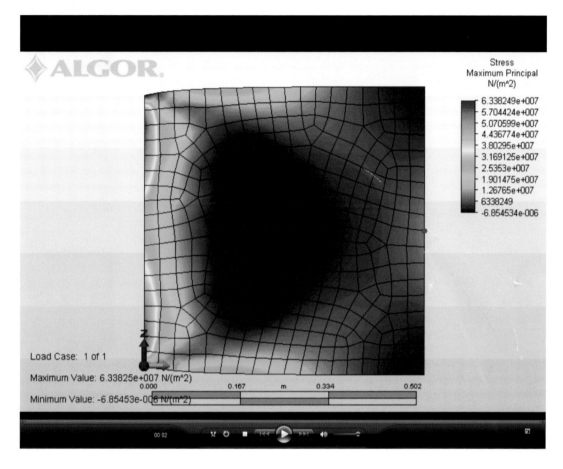

Figure 15-12 Discretized plate showing displaced plate superimposed with maximum principal stress plot in Pa.

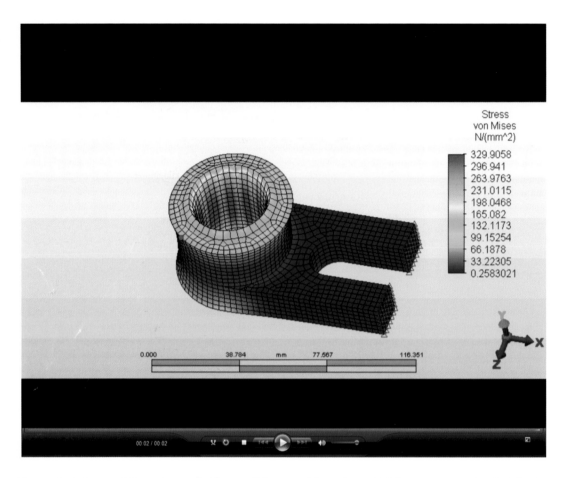

Figure 15–13 von Mises stress plot for a solid part subjected to 100 °C temperature rise inside the surface of the hole.

CONVERSIONS BETWEEN U.S. CUSTOMARY UNITS AND SI UNITS (Continued)

U.S. Customary unit		Times conversion factor		Equals SI unit	
		Accurate	Practical		
Moment of inertia (area)					
inch to fourth power	in.^{4}	416,231	416,000	millimeter to fourth power	mm^4
inch to fourth power	in.^{4}	0.416231×10^{-6}	0.416×10^{-6}	meter to fourth power	m^4
Moment of inertia (mass)					
slug foot squared	slug-ft^2	1.35582	1.36	kilogram meter squared	kg·m^2
Power					
foot-pound per second	ft-lb/s	1.35582	1.36	watt (J/s or N·m/s)	W
foot-pound per minute	ft-lb/min	0.0225970	0.0226	watt	W
horsepower (550 ft-lb/s)	hp	745.701	746	watt	W
Pressure; stress					
pound per square foot	psf	47.8803	47.9	pascal (N/m^2)	Pa
pound per square inch	psi	6894.76	6890	pascal	Pa
kip per square foot	ksf	47.8803	47.9	kilopascal	kPa
kip per square inch	ksi	6.89476	6.89	megapascal	MPa
Section modulus					
inch to third power	in.^{3}	16,387.1	16,400	millimeter to third power	mm^3
inch to third power	in.^{3}	16.3871×10^{-6}	16.4×10^{-6}	meter to third power	m^3
Velocity (linear)					
foot per second	ft/s	0.3048*	0.305	meter per second	m/s
inch per second	in./s	0.0254*	0.0254	meter per second	m/s
mile per hour	mph	0.44704*	0.447	meter per second	m/s
mile per hour	mph	1.609344*	1.61	kilometer per hour	km/h
Volume					
cubic foot	ft^3	0.0283168	0.0283	cubic meter	m^3
cubic inch	in.^{3}	16.3871×10^{-6}	16.4×10^{-6}	cubic meter	m^3
cubic inch	in.^{3}	16.3871	16.4	cubic centimeter (cc)	cm^3
gallon (231 in.^{3})	gal.	3.78541	3.79	liter	L
gallon (231 in.^{3})	gal.	0.00378541	0.00379	cubic meter	m^3

*An asterisk denotes an *exact* conversion factor

Note: To convert from SI units to USCS units, *divide* by the conversion factor

Temperature Conversion Formulas

$$T(°C) = \frac{5}{9}[T(°F) - 32] = T(K) - 273.15$$

$$T(K) = \frac{5}{9}[T(°F) - 32] + 273.15 = T(°C) + 273.15$$

$$T(°F) = \frac{9}{5}T(°C) + 32 = \frac{9}{5}T(K) - 459.67$$

A First Course in the Finite Element Method ▲

Fifth Edition, SI

Daryl L. Logan

University of Wisconsin–Platteville

SI Edition prepared by:

K. K. Chaudhry

ITM University, Gurgaon

CENGAGE
Learning™

Australia • Brazil • Japan • Korea • Mexico • Singapore • Spain • United Kingdom • United States

CENGAGE
Learning™

A First Course in the Finite Element Method,
Fifth Edition, SI
Daryl L. Logan
SI Edition prepared by: K. K. Chaudhry

Publisher, Global Engineering:
  Christopher M. Shortt

Acquisitions Editor, SI Edition:
  Swati Meherishi

Assistant Developmental Editor:
  Debarati Roy

Senior Acquisitions Editor: Randall Adams

Senior Developmental Editor:
  Hilda Gowans

Editorial Assistant: Tanya Altieri

Team Assistant: Carly Rizzo

Marketing Manager: Lauren Betsos

Media Editor: Chris Valentine

Content Project Manager: D. Jean Buttrom

Production Service:
  RPK Editorial Services, Inc.

Copyeditor: Shelly Gerger-Knechtl

Proofreader: Becky Taylor

Indexer: Shelly Gerger-Knechtl

Compositor: Glyph International

Senior Art Director: Michelle Kunkler

Cover Designer: Andrew Adams

Cover Image: Courtesy of Valmont
  Industries, Inc.

Internal Designer: Carmela Periera

Senior Rights, Acquisitions Specialist:
  Mardell Glinski-Schuttz

Text and Image Permissions Researcher:
  Kristiina Paul

First Print Buyer: Arethea L. Thomas

Library of Congress Control Number: 2011920717

ISBN-13: 978-0-495-66827-5

ISBN-10: 0-495-66827-3

Cengage Learning

200 First Stamford Place, Suite 400
Stamford, CT 06902
USA

Cengage Learning is a leading provider of customized learning solutions with office locations around the globe, including Singapore, the United Kingdom, Australia, Mexico, Brazil, and Japan. Locate your local office at: **international.cengage.com/region.**

Cengage Learning products are represented in Canada by Nelson Education Ltd.

For your course and learning solutions, visit
www.cengage.com/engineering

Purchase any of our products at your local college store or at our preferred online store **www.cengagebrain.com**

Printed in the United States of America
1 2 3 4 5 6 7 13 12 11

CONTENTS ▲

6 Development of the Plane Stress and Plane Strain Stiffness Equations

7 Practical Considerations in Modeling; Interpreting Results; and Examples of Plane Stress–Strain Analysis

The purpose of this fifth edition is again to provide an introductory approach to the finite element method that can be understood by both undergraduate and graduate students without the usual prerequisites (such as structural analysis and upper level calculus) required by many available texts in this area. The book is written primarily as a basic learning tool for the undergraduate student in civil and mechanical engineering whose main interest is in stress analysis and heat transfer, although new material on electrical networks and electrostatics has been included in this edition that should be of interest to the electrical engineer as well. The concepts are presented in sufficiently simple form with numerous example problems logically placed throughout the book, so that the book serves as a valuable learning aid for students with other backgrounds, as well as for practicing engineers. The text is geared toward those who want to apply the finite element method to solve practical physical problems.

General principles are presented for each topic, followed by traditional applications of these principles, which are in turn followed by computer applications where relevant. This approach is taken to illustrate concepts used for computer analysis of large-scale problems.

The book proceeds from basic to advanced topics and can be suitably used in a two-course sequence. Topics include basic treatments of (1) simple springs and bars, leading to two- and three-dimensional truss analysis; (2) beam bending, leading to plane frame, grid and space frame analysis; (3) elementary plane stress/strain elements, leading to more advanced plane stress/strain elements and applications to more complex plane stress/strain analysis; (4) axisymmetric stress analysis; (5) isoparametric formulation of the finite element method; (6) three-dimensional stress analysis; (7) plate bending analysis; (8) heat transfer and fluid mass transport; (9) basic fluid flow through porous media and around solid bodies, hydraulic networks, electrical networks, and electrostatics analysis; (10) thermal stress analysis; and (11) time-dependent stress and heat transfer.

Additional features include how to handle inclined or skewed supports, beam element with a nodal hinge, the concept of substructure analysis, the patch test, and practical considerations in modeling and interpreting results.

The direct approach, the principle of minimum potential energy, and Galerkin's residual method are introduced at various stages, as required, to develop the equations needed for analysis.

Appendices provide material on the following topics: (A) basic matrix algebra used throughout the text; (B) solution methods for simultaneous equations; (C) basic theory of elasticity; (D) work-equivalent nodal forces; (E) the principle of virtual work; and (F) properties of structural steel shapes.

More than 100 solved examples appear throughout the text. Most of these examples are solved "longhand" to illustrate the concepts. More than 560 end-of-chapter problems are provided to reinforce concepts. The answers to many problems are included in the back of the book to aid those wanting to verify their work. Those end-of-chapter problems to be solved using a computer program are marked with a computer symbol.

New features of this edition include updated standard notation used by most engineering instructors, chapter objectives at the start of each chapter to inform students about what is included in each chapter, summary equations for handy use at the end of each chapter, additional information on modeling, more comparisons of finite element solutions to analytical solutions, and numerous solid model- type examples and problems for solution. Also new to this edition is material on hydraulic networks, electrical networks, and electrostatics. Over 60 new problems for solution have been included, and additional design-type problems have been added to chapters 3, 5, 7, 11, and 12. Additional real-world applications from industry have been added. As in the 4th edition, this edition deliberately leaves out consideration of special purpose computer programs and suggests that instructors choose a program they are familiar with to integrate into their finite element course.

To access additional course materials, please visit www.cengagebrain.com. At the cengagebrain.com home page, search for the ISBN of your title (from the back cover of your book) using the search box at the top of the page. This will take you to the product page where an Instructor's Solutions Manual and PowerPoint slides can be found.

Following is an outline of suggested topics for a first course (approximately 44 lectures, 50 minutes each) in which this textbook is used.

Topic	Number of Lectures
Appendix A	1
Appendix B	1
Chapter 1	2
Chapter 2	3
Chapter 3, Sections 3.1–3.11, 3.14 and 3.15	5
Exam 1	1
Chapter 4, Sections 4.1–4.6	4
Chapter 5, Sections 5.1–5.3, 5.5	4
Chapter 6	4
Chapter 7	3
Exam 2	1
Chapter 9	2
Chapter 10	4
Chapter 11	3
Chapter 13, Sections 13.1–13.7	5
Chapter 15	3
Exam 3	1

This outline can be used in a one-semester course for undergraduate and graduate students in civil and mechanical engineering. (If a total stress analysis emphasis is desired, Chapter 13 can be replaced, for instance, with material from chapters 8, 12, and 16). The rest of the text can be finished in a second semester course with additional material provided by the instructor.

I express my deepest appreciation to the staff at Cengage Publishing Company, especially Chris Shortt, Publisher; Hilda Gowans, Senior Developmental Editor; Randall Adams, Senior Acquisitions Editor; and to Rose Kernan of RPK Editorial Services, for their assistance in producing this new edition.

I am grateful to Dr. Ted Belytschko for his excellent teaching of the finite element method, which aided me in writing this text. I want to thank Dr. Joseph Rencis for providing analytical solutions to structural dynamics problems for comparison to finite element solutions in Chapter 16. Also I want to thank the many students at the University for their suggestions on ways to make the topics easier to understand. These suggestions have been incorporated into this edition as well.

I thank many students at the University of Wisconsin-Platteville (UWP), whose names are credited throughout the book, for contributing various two-and-three dimensional models from the finite element course. Thank you also to UWP graduate students, Angela Moe and William Gobeli for Figure 7–19 and Table 11–2, respectively. Also, special thanks to Andrew Heckman, an alum of UWP and Design Engineer at Seagraves Fire Apparatus for permission to use Figure 11–10 and to Mr. Yousif Omer, Structural Engineer at John Deer Dubuque Works for allowing permission to use Figure 1–10. I want to thank the people at Autodesk (Algor) for their contribution of Figure 9–2b. Finally, I want to thank Ioan Giosan, Senior Design Engineer at Valmont West Coast Engineering for allowing permission to use Figures 1–11 and 1–12 and for allowing the wind mill model for the front cover.

Thank you also to the reviewers of this fifth edition: Andre Bernard, Michigan State University, Vincent C. Prantil, Milwaukee School of Engineering, Qinhua Qin, Australian National University, Robert Rizza, Milwaukee School of Engineering, Thomas J. Rudolphi, Iowa State University, and J. K. Spelt, University of Toronto, who made significant suggestions to make the book even more complete.

Special thanks to Joyce Clifton, program assistant at the university, for her invaluable assistance in scanning materials for this text.

A final special thanks to my wife Diane for her many sacrifices during the development of this fifth edition.

PREFACE TO THE SI EDITION ▲

This edition of *A First Course in the Finite Element Method* has been adapted to incorporate the International System of Units (Le Système International d'Unités or SI) throughout the book.

Le Système International d'Unités

The United States Customary System (USCS) of units uses FPS (foot–pound–second) units (also called English or Imperial units). SI units are primarily the units of the MKS (meter–kilogram–second) system. However, CGS (centimeter–gram–second) units are often accepted as SI units, especially in textbooks.

Using SI Units in this Book

In this book, we have used both MKS and CGS units. USCS units or FPS units used in the US Edition of the book have been converted to SI units throughout the text and problems. However, in case of data sourced from handbooks, government standards, and product manuals, it is not only extremely difficult to convert all values to SI, it also encroaches upon the intellectual property of the source. Some data in figures, tables, and references, therefore, remains in FPS units. For readers unfamiliar with the relationship between the FPS and the SI systems, a conversion table has been provided inside the front cover.

To solve problems that require the use of sourced data, the sourced values can be converted from FPS units to SI units just before they are to be used in a calculation. To obtain standardized quantities and manufacturers' data in SI units, the readers may contact the appropriate government agencies or authorities in their countries/regions.

Instructor Resources

The Instructors' Solution Manual in SI units is available through your Sales Representative or online through the book website at www.cengage.com/engineering.

The readers' feedback on this SI Edition will be highly appreciated and will go a long way in helping us improve subsequent editions.

The Publishers

NOTATION ▲

English Symbols

a_i	generalized coordinates (coefficients used to express displacement in general form)
A	cross-sectional area
$[B]$	matrix relating strains to nodal displacements or relating temperature gradient to nodal temperatures
c	specific heat of a material
$[C']$	matrix relating stresses to nodal displacements
C	direction cosine in two dimensions
C_x, C_y, C_z	direction cosines in three dimensions
$\{d\}$	element and structure nodal displacement matrix, both in global coordinates
$\{d'\}$	local-coordinate element nodal displacement matrix
D	bending rigidity of a plate
$[D]$	matrix relating stresses to strains
$[D']$	operator matrix given by Eq. (10.2.16)
e	exponential function
E	modulus of elasticity
$\{f\}$	global-coordinate nodal force matrix
$\{f'\}$	local-coordinate element nodal force matrix
$\{f_b\}$	body force matrix
$\{f_h\}$	heat transfer force matrix
$\{f_q\}$	heat flux force matrix
$\{f_Q\}$	heat source force matrix
$\{f_s\}$	surface force matrix
$\{F\}$	global-coordinate structure force matrix
$\{F_c\}$	condensed force matrix
$\{F_i\}$	global nodal forces
$\{F_0\}$	equivalent force matrix
$\{g\}$	temperature gradient matrix or hydraulic gradient matrix
G	shear modulus
h	heat-transfer (or convection) coefficient
i, j, m	nodes of a triangular element
I	principal moment of inertia
$[J]$	Jacobian matrix
k	spring stiffness
$[k]$	global-coordinate element stiffness or conduction matrix
$[k_c]$	condensed stiffness matrix, and conduction part of the stiffness matrix in heat-transfer problems
$[k']$	local-coordinate element stiffness matrix
$[k_n]$	convective part of the stiffness matrix in heat-transfer problems
$[K]$	global-coordinate structure stiffness matrix
K_{xx}, K_{yy}	thermal conductivities (or permeabilities, for fluid mechanics) in the x and y directions, respectively

L	length of a bar or beam element
m	maximum difference in node numbers in an element
$m(x)$	general moment expression
m_x, m_y, m_{xy}	moments in a plate
$[m'], [m]$	local element mass matrix
$[m_i']$	local nodal moments
$[M]$	global mass matrix
$[M^*]$	matrix used to relate displacements to generalized coordinates for a linear-strain triangle formulation
$[M']$	matrix used to relate strains to generalized coordinates for a linear-strain triangle formulation
n_b	bandwidth of a structure
n_d	number of degrees of freedom per node
$[N]$	shape (interpolation or basis) function matrix
N_i	shape functions
p	surface pressure (or nodal heads in fluid mechanics)
p_r, p_z	radial and axial (longitudinal) pressures, respectively
P	concentrated load
$[P']$	concentrated local force matrix
q	heat flow (flux) per unit area or distributed loading on a plate
$\bar{q}$	rate of heat flow
q^*	heat flow per unit area on a boundary surface
Q	heat source generated per unit volume or internal fluid source
Q^*	line or point heat source
Q_x, Q_y	transverse shear line loads on a plate
r, θ, z	radial, circumferential, and axial coordinates, respectively
R	residual in Galerkin's integral
R_b	body force in the radial direction
R_{ix}, R_{iy}	nodal reactions in x and y directions, respectively
s, t, z'	natural coordinates attached to isoparametric element
S	surface area
t	thickness of a plane element or a plate element
t_i, t_j, t_m	nodal temperatures of a triangular element
T	temperature function
T_∞	free-stream temperature
$[T]$	displacement, force, and stiffness transformation matrix
$[T_i]$	surface traction matrix in the i direction
u, v, w	displacement functions in the x, y, and z directions, respectively
u_i, v_i, w_i	x, y, and z displacements at node i, respectively
U	strain energy
ΔU	change in stored energy
v	velocity of fluid flow
V	shear force in a beam
w	distributed loading on a beam or along an edge of a plane element
W	work
x_i, y_i, z_i	nodal coordinates in the x, y, and z directions, respectively
x', y', z'	local element coordinate axes
x, y, z	structure global or reference coordinate axes

$[X]$	body force matrix
X_b, Y_b	body forces in the x and y directions, respectively
Z_b	body force in longitudinal direction (axisymmetric case) or in the z direction (three-dimensional case)

Greek Symbols

α	coefficient of thermal expansion
$\alpha_i, \beta_i, \gamma_i, \delta_i$	used to express the shape functions defined by Eq. (6.2.10) and Eqs. (11.2.5) through (11.2.8)
δ	spring or bar deformation
ε	normal strain
$\{\varepsilon_T\}$	thermal strain matrix
$\kappa_x, \kappa_y, \kappa_{xy}$	curvatures in plate bending
ν	Poisson's ratio
ϕ_i	nodal angle of rotation or slope in a beam element
π_h	functional for heat-transfer problem
π_p	total potential energy
ρ	mass density of a material
ρ_w	weight density of a material
ω	angular velocity and natural circular frequency
Ω	potential energy of forces
ϕ	fluid head or potential, or rotation or slope in a beam
σ	normal stress
$\{\sigma_T\}$	thermal stress matrix
τ	shear stress and period of vibration
θ	angle between the x axis and the local x' axis for two-dimensional problems
θ_p	principal angle
$\theta_x, \theta_y, \theta_z$	angles between the global x, y, and z axes and the local x' axis, respectively, or rotations about the x and y axes in a plate
$[\Psi]$	general displacement function matrix

Other Symbols

$\dfrac{d(\ )}{dx}$	derivative of a variable with respect to x
dt	time differential
$(\dot{\ })$	the dot over a variable denotes that the variable is being differentiated with respect to time
$[\ ]$	denotes a rectangular or a square matrix
$\{\ \}$	denotes a column matrix
$(\_)$	the underline of a variable denotes a matrix
$(')$	the prime next to a variable denotes that the variable is being described in a local coordinate system
$[\ ]^{-1}$	denotes the inverse of a matrix
$[\ ]^T$	denotes the transpose of a matrix
$\dfrac{\partial(\ )}{\partial x}$	partial derivative with respect to x
$\dfrac{\partial(\ )}{\partial \{d\}}$	partial derivative with respect to each variable in $\{d\}$
∎	denotes the end of the solution of an example problem

CHAPTER 1

INTRODUCTION ▲

CHAPTER OBJECTIVES

- To present an introduction to the finite element method.
- To provide a brief history of the finite element method.
- To introduce matrix notation.
- To describe the role of the computer in the development of the finite element method.
- To present the general steps used in the finite element method.
- To illustrate the various types of elements used in the finite element method.
- To show typical applications of the finite element method.
- To summarize some of the advantages of the finite element method.

Prologue

The finite element method is a numerical method for solving problems of engineering and mathematical physics. Typical problem areas of interest in engineering and mathematical physics that are solvable by use of the finite element method include structural analysis, heat transfer, fluid flow, mass transport, and electromagnetic potential.

For problems involving complicated geometries, loadings, and material properties, it is generally not possible to obtain analytical mathematical solutions. Analytical solutions are those given by a mathematical expression that yields the values of the desired unknown quantities at any location in a body (here total structure or physical system of interest) and are thus valid for an infinite number of locations in the body. These analytical solutions generally require the solution of ordinary or partial differential equations, which, because of the complicated geometries, loadings, and material properties, are not usually obtainable. Hence we need to rely on numerical methods, such as the finite element method, for acceptable solutions. The finite element formulation of the problem results in a system of simultaneous algebraic equations for solution, rather than requiring the solution of differential equations. These numerical methods yield approximate values of the unknowns at discrete numbers of points in the continuum. Hence this process of modeling a body by dividing it into an equivalent system of smaller bodies or units (finite elements) interconnected at points common to two or more elements (nodal points or nodes) and/or boundary lines and/or surfaces is called *discretization*. In the finite element method, instead of solving the problem for the entire body in one operation, we formulate the equations for each finite element and combine them to obtain the solution of the whole body.

Briefly, the solution for structural problems typically refers to determining the displacements at each node and the stresses within each element making up the structure that is subjected to applied loads. In nonstructural problems, the nodal unknowns may, for instance, be temperatures or fluid pressures due to thermal or fluid fluxes.

This chapter first presents a brief history of the development of the finite element method. You will see from this historical account that the method has become a practical one for solving engineering problems only in the past 55 years (paralleling the developments associated with the modern high-speed electronic digital computer). This historical account is followed by an introduction to matrix notation; then we describe the need for matrix methods (as made practical by the development of the modern digital computer) in formulating the equations for solution. This section discusses both the role of the digital computer in solving the large systems of simultaneous algebraic equations associated with complex problems and the development of numerous computer programs based on the finite element method. Next, a general description of the steps involved in obtaining a solution to a problem is provided. This description includes discussion of the types of elements available for a finite element method solution. Various representative applications are then presented to illustrate the capacity of the method to solve problems, such as those involving complicated geometries, several different materials, and irregular loadings. Chapter 1 also lists some of the advantages of the finite element method in solving problems of engineering and mathematical physics. Finally, we present numerous features of computer programs based on the finite element method.

▲ 1.1 Brief History ▲

This section presents a brief history of the finite element method as applied to both structural and nonstructural areas of engineering and to mathematical physics. References cited here are intended to augment this short introduction to the historical background.

The modern development of the finite element method began in the 1940s in the field of structural engineering with the work by Hrennikoff [1] in 1941 and McHenry [2] in 1943, who used a lattice of line (one-dimensional) elements (bars and beams) for the solution of stresses in continuous solids. In a paper published in 1943 but not widely recognized for many years, Courant [3] proposed setting up the solution of stresses in a variational form. Then he introduced piecewise interpolation (or shape) functions over triangular subregions making up the whole region as a method to obtain approximate numerical solutions. In 1947 Levy [4] developed the flexibility or force method, and in 1953 his work [5] suggested that another method (the stiffness or displacement method) could be a promising alternative for use in analyzing statically redundant aircraft structures. However, his equations were cumbersome to solve by hand, and thus the method became popular only with the advent of the high-speed digital computer.

In 1954 Argyris and Kelsey [6, 7] developed matrix structural analysis methods using energy principles. This development illustrated the important role that energy principles would play in the finite element method.

The first treatment of two-dimensional elements was by Turner et al. [8] in 1956. They derived stiffness matrices for truss elements, beam elements, and two-dimensional triangular and rectangular elements in plane stress and outlined the procedure commonly known as the *direct stiffness method* for obtaining the total structure stiffness matrix. Along with the development of the high-speed digital computer in the early 1950s, the work of Turner et al. [8] prompted further development of finite element stiffness equations expressed in matrix notation. The phrase *finite element* was introduced by Clough [9] in 1960 when both triangular and rectangular elements were used for plane stress analysis.

A flat, rectangular-plate bending-element stiffness matrix was developed by Melosh [10] in 1961. This was followed by development of the curved-shell bending-element stiffness matrix for axisymmetric shells and pressure vessels by Grafton and Strome [11] in 1963.

Extension of the finite element method to three-dimensional problems with the development of a tetrahedral stiffness matrix was done by Martin [12] in 1961, by Gallagher et al. [13] in 1962, and by Melosh [14] in 1963. Additional three-dimensional elements were studied by Argyris [15] in 1964. The special case of axisymmetric solids was considered by Clough and Rashid [16] and Wilson [17] in 1965.

Most of the finite element work up to the early 1960s dealt with small strains and small displacements, elastic material behavior, and static loadings. However, large deflection and thermal analysis were considered by Turner et al. [18] in 1960 and material nonlinearities by Gallagher et al. [13] in 1962, whereas buckling problems were initially treated by Gallagher and Padlog [19] in 1963. Zienkiewicz et al. [20] extended the method to visco-elasticity problems in 1968.

In 1965 Archer [21] considered dynamic analysis in the development of the consistent-mass matrix, which is applicable to analysis of distributed-mass systems such as bars and beams in structural analysis.

With Melosh's [14] realization in 1963 that the finite element method could be set up in terms of a variational formulation, it began to be used to solve nonstructural applications. Field problems, such as determination of the torsion of a shaft, fluid flow, and heat conduction, were solved by Zienkiewicz and Cheung [22] in 1965, Martin [23] in 1968, and Wilson and Nickel [24] in 1966.

Further extension of the method was made possible by the adaptation of weighted residual methods, first by Szabo and Lee [25] in 1969 to derive the previously known elasticity equations used in structural analysis and then by Zienkiewicz and Parekh [26] in 1970 for transient field problems. It was then recognized that when direct formulations and variational formulations are difficult or not possible to use, the method of weighted residuals may at times be appropriate. For example, in 1977 Lyness et al. [27] applied the method of weighted residuals to the determination of magnetic field.

In 1976, Belytschko [28, 29] considered problems associated with large-displacement nonlinear dynamic behavior, and improved numerical techniques for solving the resulting systems of equations. For more on these topics, consult the texts by Belytschko, Liu, Moran [58], and Crisfield [61, 62].

A relatively new field of application of the finite element method is that of bioengineering [30, 31]. This field is still troubled by such difficulties as nonlinear materials, geometric nonlinearities, and other complexities still being discovered.

From the early 1950s to the present, enormous advances have been made in the application of the finite element method to solve complicated engineering problems. Engineers, applied mathematicians, and other scientists will undoubtedly continue to develop new applications. For an extensive bibliography on the finite element method, consult the work of Kardestuncer [32], Clough [33], or Noor [57].

▲ 1.2 Introduction to Matrix Notation ▲

Matrix methods are a necessary tool used in the finite element method for purposes of simplifying the formulation of the element stiffness equations, for purposes of longhand solutions of various problems, and, most important, for use in programming the methods for high-speed electronic digital computers. Hence matrix notation represents a simple and easy-to-use notation for writing and solving sets of simultaneous algebraic equations.

Appendix A discusses the significant matrix concepts used throughout the text. We will present here only a brief summary of the notation used in this text.

A **matrix** *is a rectangular array of quantities arranged in rows and columns that is often used as an aid in expressing and solving a system of algebraic equations.* As examples of matrices that will be described in subsequent chapters, the force components $(F_{1x}, F_{1y}, F_{1z}, F_{2x}, F_{2y}, F_{2z}, \ldots, F_{nx}, F_{ny}, F_{nz})$ acting at the various nodes or points $(1, 2, \ldots, n)$ on a structure and the corresponding set of nodal displacements $(u_1, v_1, w_1, u_2, v_2, w_2, \ldots, u_n, v_n, w_n)$ can both be expressed as matrices:

$$\{F\} = \begin{Bmatrix} F_{1x} \\ F_{1y} \\ F_{1z} \\ F_{2x} \\ F_{2y} \\ F_{2z} \\ \vdots \\ F_{nx} \\ F_{ny} \\ F_{nz} \end{Bmatrix} \qquad \{d\} = \begin{Bmatrix} u_1 \\ v_1 \\ w_1 \\ u_2 \\ v_2 \\ w_2 \\ \vdots \\ u_n \\ v_n \\ w_n \end{Bmatrix} \tag{1.2.1}$$

The subscripts to the right of F identify the node and the direction of force, respectively. For instance, F_{1x} denotes the force at node 1 applied in the x direction. The x, y, and z displacements at a node are denoted by u, v, and w, respectively. The subscript next to u, v, and w denotes the node. For instance, u_1, v_1, and w_1 denote the displacement components in the x, y, and z directions, respectively, at node 1. The matrices in Eqs. (1.2.1) are called *column matrices* and have a size of $n \times 1$. The brace notation $\{\}$

will be used throughout the text to denote a column matrix. The whole set of force or displacement values in the column matrix is simply represented by $\{F\}$ or $\{d\}$.

The more general case of a known rectangular matrix will be indicated by use of the bracket notation [ ]. For instance, the element and global structure stiffness matrices $[k]$ and $[K]$, respectively, developed throughout the text for various element types (such as those in Figure 1–1 on page 10), are represented by square matrices given as

$$[k] = \begin{bmatrix} k_{11} & k_{12} & \cdots & k_{1n} \\ k_{21} & k_{22} & \cdots & k_{2n} \\ \vdots & \vdots & & \vdots \\ k_{n1} & k_{n2} & \cdots & k_{nn} \end{bmatrix} \tag{1.2.2}$$

and

$$[K] = \begin{bmatrix} K_{11} & K_{12} & \cdots & K_{1n} \\ K_{21} & K_{22} & \cdots & K_{2n} \\ \vdots & \vdots & & \vdots \\ K_{n1} & K_{n2} & \cdots & K_{nn} \end{bmatrix} \tag{1.2.3}$$

where, in structural theory, the elements k_{ij} and K_{ij} are often referred to as *stiffness influence coefficients*.

You will learn that the global nodal forces $\{F\}$ and the global nodal displacements $\{d\}$ are related through use of the global stiffness matrix $[K]$ by

$$\{F\} = [K]\{d\} \tag{1.2.4}$$

Equation (1.2.4) is called the *global stiffness equation* and represents a set of simultaneous equations. It is the basic equation formulated in the stiffness or displacement method of analysis.

To obtain a clearer understanding of elements K_{ij} in Eq. (1.2.3), we use Eq. (1.2.1) and write out the expanded form of Eq. (1.2.4) as

$$\begin{Bmatrix} F_{1x} \\ F_{1y} \\ \vdots \\ F_{nz} \end{Bmatrix} = \begin{bmatrix} K_{11} & K_{12} & \cdots & K_{1n} \\ K_{21} & K_{22} & \cdots & K_{2n} \\ \vdots & & & \\ K_{n1} & K_{n2} & \cdots & K_{nn} \end{bmatrix} \begin{Bmatrix} u_1 \\ v_1 \\ \vdots \\ w_n \end{Bmatrix} \tag{1.2.5}$$

Now assume a structure to be forced into a displaced configuration defined by $u_1 = 1, v_1 = w_1 = \cdots w_n = 0$. Then from Eq. (1.2.5), we have

$$F_{1x} = K_{11} \qquad F_{1y} = K_{21}, \ldots, F_{nz} = K_{n1} \tag{1.2.6}$$

Equations (1.2.6) contain all elements in the first column of $[K]$. In addition, they show that these elements, $K_{11}, K_{21}, \ldots, K_{n1}$, are the values of the full set of nodal forces required to maintain the imposed displacement state. In a similar manner, the second column in $[K]$ represents the values of forces required to maintain the displaced state $v_1 = 1$ and all other nodal displacement components equal to zero.

We should now have a better understanding of the meaning of stiffness influence coefficients.

Subsequent chapters will discuss the element stiffness matrices $[k]$ for various element types, such as bars, beams, plane stress and three-dimensional stress. They will also cover the procedure for obtaining the global stiffness matrices $[K]$ for various structures and for solving Eq. (1.2.4) for the unknown displacements in matrix $\{d\}$.

Using matrix concepts and operations will become routine with practice; they will be valuable tools for solving small problems longhand. And matrix methods are crucial to the use of the digital computers necessary for solving complicated problems with their associated large number of simultaneous equations.

▲ **1.3 Role of the Computer**   ▲

As we have said, until the early 1950s, matrix methods and the associated finite element method were not readily adaptable for solving complicated problems. Even though the finite element method was being used to describe complicated structures, the resulting large number of algebraic equations associated with the finite element method of structural analysis made the method extremely difficult and impractical to use. However, with the advent of the computer, the solution of thousands of equations in a matter of minutes became possible.

The first modern-day commercial computer appears to have been the Univac, IBM 701 which was developed in the 1950s. This computer was built based on vacuum-tube technology. Along with the UNIVAC came the punch-card technology whereby programs and data were created on punch cards. In the 1960s, transistor-based technology replaced the vacuum-tube technology due to the transistor's reduced cost, weight, and power consumption and its higher reliability. From 1969 to the late 1970s, integrated circuit-based technology was being developed, which greatly enhanced the processing speed of computers, thus making it possible to solve larger finite element problems with increased degrees of freedom. From the late 1970s into the 1980s, large-scale integration as well as workstations that introduced a windows-type graphical interface appeared along with the computer mouse. The first computer mouse received a patent on November 17, 1970. Personal computers had now become mass-market desktop computers. These developments came during the age of networked computing, which brought the Internet and the World Wide Web. In the 1990s the Windows operating system was released, making IBM and IBM-compatible PCs more user friendly by integrating a graphical user interface into the software.

The development of the computer resulted in the writing of computational programs. Numerous special-purpose and general-purpose programs have been written to handle various complicated structural (and nonstructural) problems. Programs such as [46–56] illustrate the elegance of the finite element method and reinforce understanding of it.

In fact, finite element computer programs now can be solved on single-processor machines, such as a single desktop or laptop personal computer (PC) or on a cluster of computer nodes. The powerful memories of the PC and the advances in solver programs have made it possible to solve problems with over a million unknowns.

To use the computer, the analyst, having defined the finite element model, inputs the information into the computer. This information may include the position of the element nodal coordinates, the manner in which elements are connected, the material properties of the elements, the applied loads, boundary conditions, or constraints, and the kind of analysis to be performed. The computer then uses this information to generate and solve the equations necessary to carry out the analysis.

▲ 1.4 General Steps of the Finite Element Method ▲

This section presents the general steps included in a finite element method formulation and solution to an engineering problem. We will use these steps as our guide in developing solutions for structural and nonstructural problems in subsequent chapters.

For simplicity's sake, for the presentation of the steps to follow, we will consider only the structural problem. The nonstructural heat-transfer, fluid mechanics and electrostatics problems and their analogies to the structural problem are considered in Chapters 13 and 14.

Typically, for the structural stress-analysis problem, the engineer seeks to determine displacements and stresses throughout the structure, which is in equilibrium and is subjected to applied loads. For many structures, it is difficult to determine the distribution of deformation using conventional methods, and thus the finite element method is necessarily used.

There are two general direct approaches traditionally associated with the finite element method as applied to structural mechanics problems. One approach, called the *force*, or *flexibility, method*, uses internal forces as the unknowns of the problem. To obtain the governing equations, first the equilibrium equations are used. Then necessary additional equations are found by introducing compatibility equations. The result is a set of algebraic equations for determining the redundant or unknown forces.

The second approach, called the *displacement*, or *stiffness, method*, assumes the displacements of the nodes as the unknowns of the problem. For instance, compatibility conditions requiring that elements connected at a common node, along a common edge, or on a common surface before loading remain connected at that node, edge, or surface after deformation takes place are initially satisfied. Then the governing equations are expressed in terms of nodal displacements using the equations of equilibrium and an applicable law relating forces to displacements.

These two direct approaches result in different unknowns (forces or displacements) in the analysis and different matrices associated with their formulations (flexibilities or stiffnesses). It has been shown [34] that, for computational purposes, the displacement (or stiffness) method is more desirable because its formulation is simpler for most structural analysis problems. Furthermore, a vast majority of general-purpose finite element programs have incorporated the displacement formulation for solving structural problems. Consequently, only the displacement method will be used throughout this text.

Another general method that can be used to develop the governing equations for both structural and nonstructural problems is the variational method. The variational method includes a number of principles. One of these principles, used extensively throughout this text because it is relatively easy to comprehend and is often introduced in basic mechanics courses, is the theorem of minimum potential energy that applies to materials behaving in a linear-elastic manner. This theorem is explained and used in various sections of the text, such as Section 2.6 for the spring element, Section 3.10 for the bar element, Section 4.7 for the beam element, Section 6.2 for the constant-strain triangle plane stress and plane strain element, Section 9.1 for the axisymmetric element, Section 11.2 for the three-dimensional solid tetrahedral element, and Section 12.2 for the plate bending element. A functional analogous to that used in the theorem of minimum potential energy is then employed to develop the finite element equations for the nonstructural problem of heat transfer presented in Chapter 13.

Another variational principle often used to derive the governing equations is the principle of virtual work. This principle applies more generally to materials that behave in a linear-elastic fashion, as well as those that behave in a nonlinear fashion. The principle of virtual work is described in Appendix E for those choosing to use it for developing the general governing finite element equations that can be applied specifically to bars, beams, and two- and three-dimensional solids in either static or dynamic systems.

The finite element method involves modeling the structure using small interconnected elements called *finite elements*. A displacement function is associated with each finite element. Every interconnected element is linked, directly or indirectly, to every other element through common (or shared) interfaces, including nodes and/or boundary lines and/or surfaces. By using known stress/strain properties for the material making up the structure, one can determine the behavior of a given node in terms of the properties of every other element in the structure. The total set of equations describing the behavior of each node results in a series of algebraic equations best expressed in matrix notation.

We now present the steps, along with explanations necessary at this time, used in the finite element method formulation and solution of a structural problem. The purpose of setting forth these general steps now is to expose you to the procedure generally followed in a finite element formulation of a problem. You will easily understand these steps when we illustrate them specifically for springs, bars, trusses, beams, plane frames, plane stress, axisymmetric stress, three-dimensional stress, plate bending, heat transfer, fluid flow and electrostatics in subsequent chapters. We suggest that you review this section periodically as we develop the specific element equations.

Keep in mind that the analyst must make decisions regarding dividing the structure or continuum into finite elements and selecting the element type or types to be used in the analysis (step 1), the kinds of loads to be applied, and the types of boundary conditions or supports to be applied. The other steps, 2 through 7, are carried out automatically by a computer program.

Step 1 Discretize and Select the Element Types

Step 1 involves dividing the body into an equivalent system of finite elements with associated nodes and choosing the most appropriate element type to model most closely the actual physical behavior. The total number of elements used and their variation in size and type within a given body are primarily matters of engineering judgment. The elements must be made small enough to give usable results and yet large enough to reduce computational effort. Small elements (and possibly higher-order elements) are generally desirable where the results are changing rapidly, such as where changes in geometry occur; large elements can be used where results are relatively constant. We will have more to say about discretization guidelines in later chapters, particularly in Chapter 7, where the concept becomes quite significant. The discretized body or mesh is often created with mesh-generation programs or preprocessor programs available to the user.

The choice of elements used in a finite element analysis depends on the physical makeup of the body under actual loading conditions and on how close to the actual behavior the analyst wants the results to be. Judgment concerning the appropriateness of one-, two-, or three-dimensional idealizations is necessary. Moreover, the choice of the most appropriate element for a particular problem is one of the major tasks that must be carried out by the designer/analyst. Elements that are commonly employed in practice—most of which are considered in this text—are shown in Figure 1–1.

The primary line elements [Figure 1–1(a)] consist of bar (or truss) and beam elements. They have a cross-sectional area but are usually represented by line segments. In general, the cross-sectional area within the element can vary, but throughout this text it will be considered to be constant. These elements are often used to model trusses and frame structures (see Figure 1–2 on page 16, for instance). The simplest line element (called a *linear element*) has two nodes, one at each end, although higher-order elements having three nodes [Figure 1–1(a)] or more (called *quadratic, cubic*, etc., *elements*) also exist. Chapter 10 includes discussion of higher-order line elements. The line elements are the simplest of elements to consider and will be discussed in Chapters 2 through 5 to illustrate many of the basic concepts of the finite element method.

The basic two-dimensional (or plane) elements [Figure 1–1(b)] are loaded by forces in their own plane (plane stress or plane strain conditions). They are triangular or quadrilateral elements. The simplest two-dimensional elements have corner nodes only (linear elements) with straight sides or boundaries (Chapter 6), although there are also higher-order elements, typically with midside nodes [Figure 1–1(b)] (called *quadratic elements*) and curved sides (Chapters 8 and 10). The elements can have variable thicknesses throughout or be constant. They are often used to model a wide range of engineering problems (see Figures 1–3 and 1–4 on pages 17 and 18).

The most common three-dimensional elements [Figure 1–1(c)] are tetrahedral and hexahedral (or brick) elements; they are used when it becomes necessary to perform a three-dimensional stress analysis. The basic three-dimensional elements (Chapter 11) have corner nodes only and straight sides, whereas higher-order elements with midedge nodes (and possible midface nodes) have curved surfaces for their sides [Figure 1–1(c)].

The axisymmetric element [Figure 1–1(d)] is developed by rotating a triangle or quadrilateral about a fixed axis located in the plane of the element through 360°. This element (described in Chapter 9) can be used when the geometry and loading of the problem are axisymmetric.

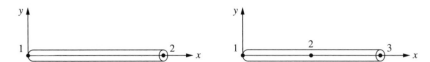

(a) Simple two-noded line element (typically used to represent a bar or beam element) and the higher-order line element

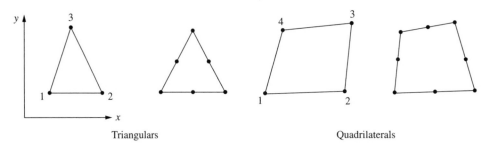

Triangulars   Quadrilaterals

(b) Simple two-dimensional elements with corner nodes (typically used to represent plane stress/ strain) and higher-order two-dimensional elements with intermediate nodes along the sides

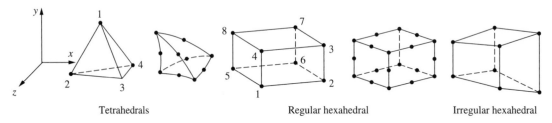

Tetrahedrals   Regular hexahedral   Irregular hexahedral

(c) Simple three-dimensional elements (typically used to represent three-dimensional stress state) and higher-order three-dimensional elements with intermediate nodes along edges

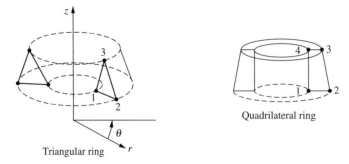

Triangular ring

Quadrilateral ring

(d) Simple axisymmetric triangular and quadrilateral elements used for axisymmetric problems

Figure 1–1 Various types of simple lowest-order finite elements with corner nodes only and higher-order elements with intermediate nodes

Step 2 Select a Displacement Function

Step 2 involves choosing a displacement function within each element. The function is defined within the element using the nodal values of the element. Linear, quadratic, and cubic polynomials are frequently used functions because they are simple to work with in finite element formulation. However, trigonometric series can also be used. For a two-dimensional element, the displacement function is a function of the coordinates in its plane (say, the x-y plane). The functions are expressed in terms of the nodal unknowns (in the two-dimensional problem, in terms of an x and a y component). The same general displacement function can be used repeatedly for each element. Hence the finite element method is one in which a continuous quantity, such as the displacement throughout the body, is approximated by a discrete model composed of a set of piecewise-continuous functions defined within each finite domain or finite element.

Step 3 Define the Strain/Displacement and Stress/Strain Relationships

Strain/displacement and stress/strain relationships are necessary for deriving the equations for each finite element. In the case of one-dimensional deformation, say, in the x direction, we have strain ε_x related to displacement u by

$$\varepsilon_x = \frac{du}{dx} \tag{1.4.1}$$

for small strains. In addition, the stresses must be related to the strains through the stress/strain law—generally called the *constitutive law*. The ability to define the material behavior accurately is most important in obtaining acceptable results. The simplest of stress/strain laws, Hooke's law, which is often used in stress analysis, is given by

$$\sigma_x = E\varepsilon_x \tag{1.4.2}$$

where σ_x = stress in the x direction and E = modulus of elasticity.

Step 4 Derive the Element Stiffness Matrix and Equations

Initially, the development of element stiffness matrices and element equations was based on the concept of stiffness influence coefficients, which presupposes a background in structural analysis. We now present alternative methods used in this text that do not require this special background.

Direct Equilibrium or Stiffness Method

According to this method, the stiffness matrix and element equations relating nodal forces to nodal displacements are obtained using force equilibrium conditions for a basic element, along with force/deformation relationships. Because this method is most easily adaptable to line or one-dimensional elements, Chapters 2, 3, and 4 illustrate this method for spring, bar, and beam elements, respectively.

Work or Energy Methods

To develop the stiffness matrix and equations for two- and three-dimensional elements, it is much easier to apply a work or energy method [35]. The principle of virtual work (using virtual displacements), the principle of minimum potential energy, and Castigliano's theorem are methods frequently used for the purpose of derivation of element equations.

The principle of virtual work outlined in Appendix E is applicable for any material behavior, whereas the principle of minimum potential energy and Castigliano's theorem are applicable only to elastic materials. Furthermore, the principle of virtual work can be used even when a potential function does not exist. However, all three principles yield identical element equations for linear-elastic materials; thus which method to use for this kind of material in structural analysis is largely a matter of convenience and personal preference. We will present the principle of minimum potential energy—probably the best known of the three energy methods mentioned here—in detail in Chapters 2 and 3, where it will be used to derive the spring and bar element equations. We will further generalize the principle and apply it to the beam element in Chapter 4 and to the plane stress/strain element in Chapter 6. Thereafter, the principle is routinely referred to as the basis for deriving all other stress-analysis stiffness matrices and element equations given in Chapters 8, 9, 11, and 12.

For the purpose of extending the finite element method outside the structural stress analysis field, a **functional**[1] (a function of another function or a function that takes functions as its argument) analogous to the one to be used with the principle of minimum potential energy is quite useful in deriving the element stiffness matrix and equations (see Chapters 13 and 14 on heat transfer and fluid flow, respectively). For instance, letting π denote the functional and $f(x,y)$ denote a function f of two variables x and y, we then have $\pi = \pi(f(x,y))$, where π is a function of the function f. A more general form of a functional depending on two independent variables $u(x,y)$ and $v(x,y)$, where independent variables are x and y in Cartesian coordinates, is given by

$$\pi = \int\int F(x,y,u,v,u_{,x},u_{,y},v_{,x},v_{,y},u_{,xx},\ldots v_{,yy})dx\,dy \qquad (1.4.3)$$

where the comma preceding the subscripts x and y denotes differentiation with respect to x or y, i.e., $u_{,x} = \frac{\partial u}{\partial x}$, etc.

Methods of Weighted Residuals

The methods of weighted residuals are useful for developing the element equations; particularly popular is Galerkin's method. These methods yield the same results as the energy methods wherever the energy methods are applicable. They are especially useful when a functional such as potential energy is not readily available. The weighted residual methods allow the finite element method to be applied directly to any differential equation.

---

[1] Another definition of a functional is as follows: A functional is an integral expression that implicitly contains differential equations that describe the problem. A typical functional is of the form $I(u) = \int F(x,u,u')\,dx$ where $u(x)$, x, and F are real so that $I(u)$ is also a real number. Here $u' = \partial u/\partial x$.

Galerkin's method, along with the collocation, the least squares, and the subdomain *weighted residual methods* are introduced in Chapter 3. To illustrate each method, they will all be used to solve a one-dimensional bar problem for which a known exact solution exists for comparison. As the more easily adapted residual method, Galerkin's method will also be used to derive the bar element equations in Chapter 3 and the beam element equations in Chapter 4 and to solve the combined heat-conduction/convection/mass transport problem in Chapter 13. For more information on the use of the methods of weighted residuals, see Reference [36]; for additional applications to the finite element method, consult References [37] and [38].

Using any of the methods just outlined will produce the equations to describe the behavior of an element. These equations are written conveniently in matrix form as

$$
\begin{Bmatrix} f_1 \\ f_2 \\ f_3 \\ \vdots \\ f_n \end{Bmatrix} = \begin{bmatrix} k_{11} & k_{12} & k_{13} & \dots & k_{1n} \\ k_{21} & k_{22} & k_{23} & \dots & k_{2n} \\ k_{31} & k_{32} & k_{33} & \dots & k_{3n} \\ \vdots & & & & \vdots \\ k_{n1} & & & \dots & k_{nn} \end{bmatrix} \begin{Bmatrix} d_1 \\ d_2 \\ d_3 \\ \vdots \\ d_n \end{Bmatrix}
\tag{1.4.4}
$$

or in compact matrix form as

$$
\{f\} = [k]\{d\}
\tag{1.4.5}
$$

where $\{f\}$ is the vector of element nodal forces, $[k]$ is the element stiffness matrix (normally square and symmetric), and $\{d\}$ is the vector of unknown element nodal degrees of freedom or generalized displacements, n. Here generalized displacements may include such quantities as actual displacements, slopes, or even curvatures. The matrices in Eq. (1.4.5) will be developed and described in detail in subsequent chapters for specific element types, such as those in Figure 1–1.

Step 5 Assemble the Element Equations to Obtain the Global or Total Equations and Introduce Boundary Conditions

In this step the individual element nodal equilibrium equations generated in step 4 are assembled into the global nodal equilibrium equations. Section 2.3 illustrates this concept for a two-spring assemblage. Another more direct method of superposition (called the *direct stiffness method*), whose basis is nodal force equilibrium, can be used to obtain the global equations for the whole structure. This direct method is illustrated in Section 2.4 for a spring assemblage. Implicit in the direct stiffness method is the concept of continuity, or compatibility, which requires that the structure remain together and that no tears occur anywhere within the structure.

The final assembled or global equation written in matrix form is

$$
\{F\} = [K]\{d\}
\tag{1.4.6}
$$

where $\{F\}$ is the vector of global nodal forces, $[K]$ is the structure global or total stiffness matrix, (for most problems, the global stiffness matrix is square and symmetric) and $\{d\}$ is now the vector of known and unknown structure nodal degrees of freedom or generalized displacements. It can be shown that at this stage, the global stiffness matrix $[K]$ is a singular matrix because its determinant is equal to zero. To remove this singularity problem, we must invoke certain boundary conditions (or constraints or supports) so that the structure remains in place instead of moving as a rigid body. Further details and methods of invoking boundary conditions are given in subsequent chapters. At this time it is sufficient to note that invoking boundary or support conditions results in a modification of the global Eq. (1.4.6). We also emphasize that the applied known loads have been accounted for in the global force matrix $\{F\}$.

Step 6   Solve for the Unknown Degrees of Freedom (or Generalized Displacements)

Equation (1.4.6), modified to account for the boundary conditions, is a set of simultaneous algebraic equations that can be written in expanded matrix form as

$$\begin{Bmatrix} F_1 \\ F_2 \\ \vdots \\ F_n \end{Bmatrix} = \begin{bmatrix} K_{11} & K_{12} & \cdots & K_{1n} \\ K_{21} & K_{22} & \cdots & K_{2n} \\ \vdots & & & \vdots \\ K_{n1} & K_{n2} & \cdots & K_{nn} \end{bmatrix} \begin{Bmatrix} d_1 \\ d_2 \\ \vdots \\ d_n \end{Bmatrix} \tag{1.4.7}$$

where now n is the structure total number of unknown nodal degrees of freedom. These equations can be solved for the ds by using an elimination method (such as Gauss's method) or an iterative method (such as the Gauss–Seidel method). These two methods are discussed in Appendix B. The ds are called the *primary unknowns*, because they are the first quantities determined using the stiffness (or displacement) finite element method.

Step 7   Solve for the Element Strains and Stresses

For the structural stress-analysis problem, important secondary quantities of strain and stress (or moment and shear force) can be obtained because they can be directly expressed in terms of the displacements determined in step 6. Typical relationships between strain and displacement and between stress and strain—such as Eqs. (1.4.1) and (1.4.2) for one-dimensional stress given in step 3—can be used.

Step 8   Interpret the Results

The final goal is to interpret and analyze the results for use in the design/analysis process. Determination of locations in the structure where large deformations and large stresses occur is generally important in making design/analysis decisions. Postprocessor computer programs help the user to interpret the results by displaying them in graphical form.

▲ 1.5 Applications of the Finite Element Method ▲

The finite element method can be used to analyze both structural and nonstructural problems. Typical structural areas include

1. Stress analysis, including truss and frame analysis (such as pedestrian walk bridges, high rise building frames, and windmill towers), and stress concentration problems, typically associated with holes, fillets, or other changes in geometry in a body (such as automotive parts, pressures vessels, medical devices, aircraft, and sports equipment)
2. Buckling, such as in columns, frames, and vessels
3. Vibration analysis, such as in vibratory equipment
4. Impact problems, including crash analysis of vehicles, projectile impact, and bodies falling and impacting objects

   Nonstructural problems include

1. Heat transfer, such as in electronic devices emitting heat as in a personal computer microprocessor chip, engines, and cooling fins in radiators
2. Fluid flow, including seepage through porous media (such as water seeping through earthen dams), cooling ponds, and in air ventilation systems as used in sports arenas, etc., air flow around racing cars, yachting boats, and surfboards, etc.
3. Distribution of electric or magnetic potential, such as in antennas and transistors

   Finally, some biomechanical engineering problems (which may include stress analysis) typically include analyses of human spine, skull, hip joints, jaw/gum tooth implants, heart, and eye.

   We now present some typical applications of the finite element method. These applications will illustrate the variety, size, and complexity of problems that can be solved using the method and the typical discretization process and kinds of elements used.

   Figure 1–2 illustrates a control tower for a railroad. The tower is a three-dimensional frame comprising a series of beam-type elements. The 48 elements are labeled by the circled numbers, whereas the 28 nodes are indicated by the uncircled numbers. Each node has three rotation and three displacement components associated with it. The rotations (θs) and displacements (ds) are called the *degrees of freedom*. Because of the loading conditions to which the tower structure is subjected, we have used a three-dimensional model.

   The finite element method used for this frame enables the designer/analyst quickly to obtain displacements and stresses in the tower for typical load cases, as required by design codes. Before the development of the finite element method and the computer, even this relatively simple problem took many hours to solve.

   The next illustration of the application of the finite element method to problem solving is the determination of displacements and stresses in an underground box culvert subjected to ground shock loading from a bomb explosion. Figure 1–3 shows the discretized model, which included a total of 369 nodes, 40 one-dimensional bar or truss elements used to model the steel reinforcement in the box

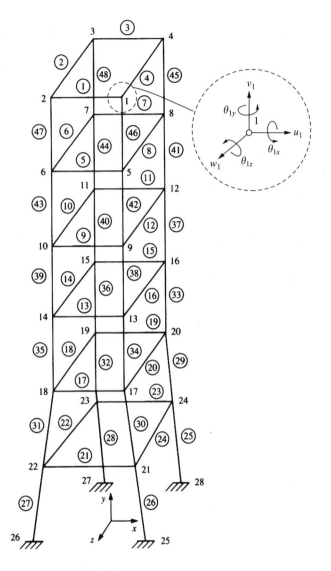

Figure 1–2 Discretized railroad control tower (28 nodes, 48 beam elements) with typical degrees of freedom shown at node 1, for example (By Daryl L. Logan)

culvert, and 333 plane strain two-dimensional triangular and rectangular elements used to model the surrounding soil and concrete box culvert. With an assumption of symmetry, only half of the box culvert need be analyzed. This problem requires the solution of nearly 700 unknown nodal displacements. It illustrates that different kinds of elements (here bar and plane strain) can often be used in one finite element model.

Another problem, that of the hydraulic cylinder rod end shown in Figure 1–4, was modeled by 120 nodes and 297 plane strain triangular elements. Symmetry was also applied to the whole rod end so that only half of the rod end had to be analyzed,

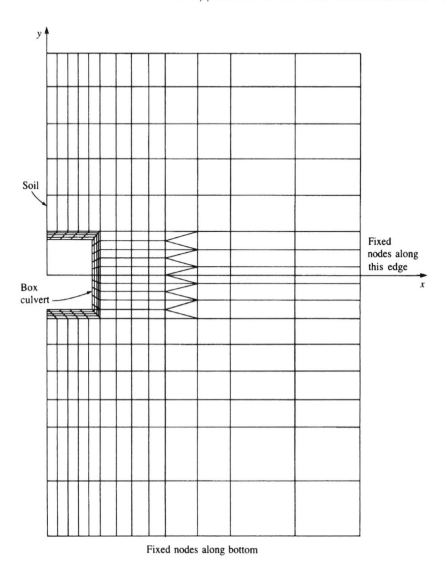

Figure 1–3 Discretized model of an underground box culvert (369 nodes, 40 bar elements, and 333 plane strain elements) [39]

as shown. The purpose of this analysis was to locate areas of high stress concentration in the rod end.

Figure 1–5 shows a chimney stack section that is four form heights high (or a total of 9.75 m high). In this illustration, 584 beam elements were used to model the vertical and horizontal stiffeners making up the formwork, and 252 flat-plate elements were used to model the inner wooden form and the concrete shell. Because of the irregular loading pattern on the structure, a three-dimensional model was necessary. Displacements and stresses in the concrete were of prime concern in this problem.

Figure 1–6 shows the finite element discretized model of a proposed steel die used in a plastic film-making process. The irregular geometry and associated

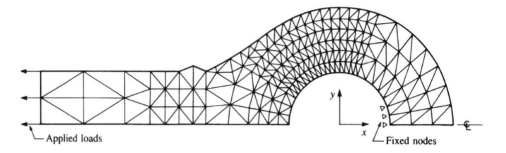

Applied loads

Fixed nodes

Figure 1–4 Two-dimensional analysis of a hydraulic cylinder rod end (120 nodes, 297 plane strain triangular elements)

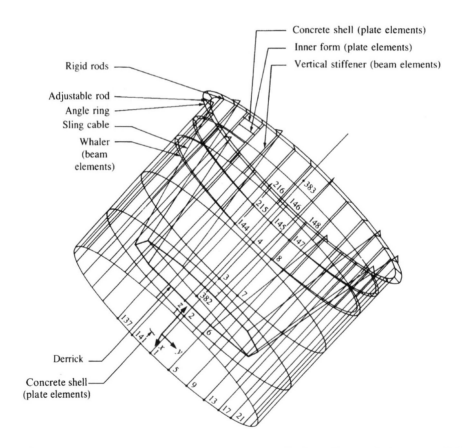

Concrete shell (plate elements)

Inner form (plate elements)

Vertical stiffener (beam elements)

Rigid rods

Adjustable rod
Angle ring
Sling cable
Whaler
(beam
elements)

Derrick

Concrete shell
(plate elements)

Figure 1–5 Finite element model of a chimney stack section (end view rotated 45°) (584 beam and 252 flat-plate elements) (By Daryl L. Logan)

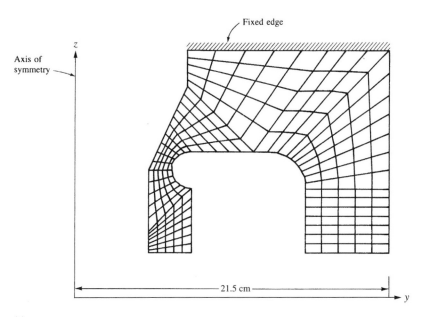

(a)

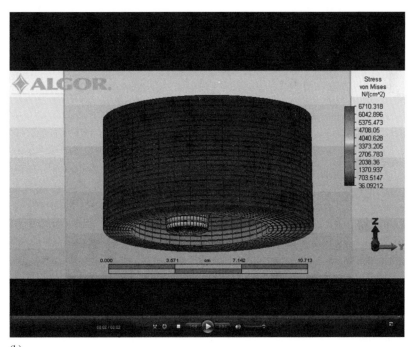

(b)

Figure 1–6 (a) Model of a high-strength steel die (240 axisymmetric elements) used in the plastic film industry (By Daryl L. Logan) and (b) the three-dimensional visual of the die as the elements in the plane are rotated through 360° around the z-axis of symmetry (See the full-color insert for a color version of this figure.)

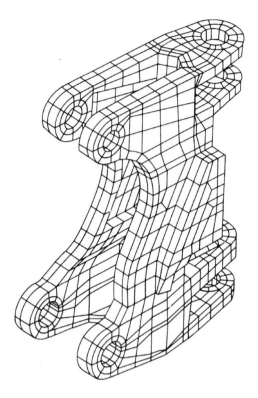

Figure 1–7 Three-dimensional solid element model of a swing casting for a backhoe frame

potential stress concentrations necessitated use of the finite element method to obtain a reasonable solution. Here 240 axisymmetric elements were used to model the three-dimensional die.

Figure 1–7 illustrates the use of a three-dimensional solid element to model a swing casting for a backhoe frame. The three-dimensional hexahedral elements are necessary to model the irregularly shaped three-dimensional casting. Two-dimensional models certainly would not yield accurate engineering solutions to this problem.

Figure 1–8 illustrates a two-dimensional heat-transfer model used to determine the temperature distribution in earth subjected to a heat source—a buried pipeline transporting a hot gas.

Figure 1–9 shows a three-dimensional model of human pelvis which can be used to study stresses in the bone and the cement layer between the bone and the implant.

More recently, mechanical event simulation (MES), including nonlinear behavior and contact, such as in roll forming processes, has been studied using finite element analysis [46], as shown in Figure 1–11 and wind mill generator stress analysis under various loading conditions, including wind, ice, and earthquake while the blades are rotating has been performed [46], as shown in Figure 1–12.

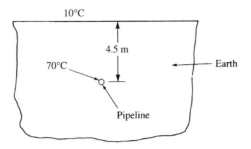

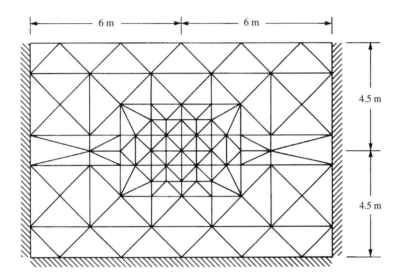

Figure 1–8 Finite element model for a two-dimensional temperature distribution in the earth

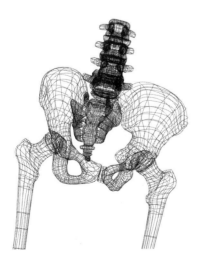

Figure 1–9 Finite element model of a human pelvis (Studio MacBeth/Science Photo Library)

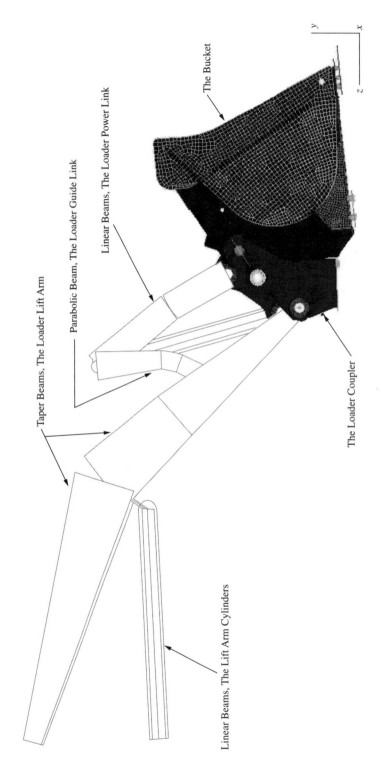

Figure 1–10 Finite element model of a 710G bucket with 169,595 elements and 185,026 nodes used (including 78,566 thin-shell linear quadrilateral elements for the bucket and coupler, 83,104 solid linear brick elements to model the bosses, and 212 beam elements to model lift arms, lift arm cylinders, and guide links) (Courtesy of Yousif Omer, Structural Design Engineer, Construction and Forestry Division, John Deere Dubuque Works)

Linear Beams, The Lift Arm Cylinders

Taper Beams, The Loader Lift Arm

Parabolic Beam, The Loader Guide Link

Linear Beams, The Loader Power Link

The Bucket

The Loader Coupler

y

x

z

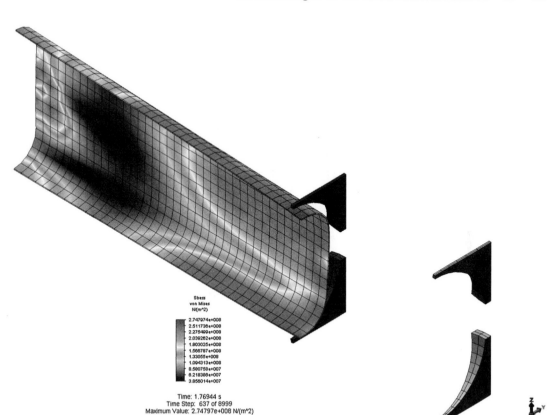

Figure 1-11 Finite element model of contour roll forming or cold roll forming process (Courtesy of Valmont West Coast Engineering) (See the full-color insert for a color version of this figure.)

Finally, the field of computational fluid dynamics (CFD) using finite element analysis has recently been used to design ventilation systems, such as in large sports arenas, and to study air flow around race cars and around golf balls when suddenly struck by a golf club [63].

These illustrations suggest the kinds of problems that can be solved by the finite element method. Additional guidelines concerning modeling techniques will be provided in Chapter 7.

▲ 1.6 Advantages of the Finite Element Method ▲

As previously indicated, the finite element method has been applied to numerous problems, both structural and nonstructural. This method has a number of advantages that have made it very popular. They include the ability to

1. Model irregularly shaped bodies quite easily
2. Handle general load conditions without difficulty

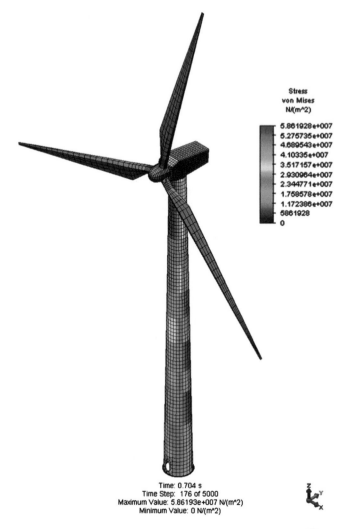

Figure 1-12   Finite element model showing the von Mises stress plot of a wind mill tower at a critical time step using a nonlinear finite element simulation (Courtesy of Valmont West Coast Engineering)

3. Model bodies composed of several different materials because the element equations are evaluated individually
4. Handle unlimited numbers and kinds of boundary conditions
5. Vary the size of the elements to make it possible to use small elements where necessary
6. Alter the finite element model relatively easily and cheaply
7. Include dynamic effects
8. Handle nonlinear behavior existing with large deformations and nonlinear materials

The finite element method of structural analysis enables the designer to detect stress, vibration, and thermal problems during the design process and to evaluate design changes *before* the construction of a possible prototype. Thus confidence in the acceptability of the prototype is enhanced. Moreover, if used properly, the method can reduce the number of prototypes that need to be built.

Even though the finite element method was initially used for structural analysis, it has since been adapted to many other disciplines in engineering and mathematical physics, such as fluid flow, heat transfer, electromagnetic potentials, soil mechanics, and acoustics [22–24, 27, 42–44].

▲ 1.7 Computer Programs for the Finite Element Method ▲

There are two general computer methods of approach to the solution of problems by the finite element method. One is to use large commercial programs, many of which have been configured to run on personal computers (PCs); these general-purpose programs are designed to solve many types of problems. The other is to develop many small, special-purpose programs to solve specific problems. In this section, we will discuss the advantages and disadvantages of both methods. We will then list some of the available general-purpose programs and discuss some of their standard capabilities.

Some advantages of general-purpose programs:

1. The input is well organized and is developed with user ease in mind. Users do not need special knowledge of computer software or hardware. Preprocessors are readily available to help create the finite element model.
2. The programs are large systems that often can solve many types of problems of large or small size with the same input format.
3. Many of the programs can be expanded by adding new modules for new kinds of problems or new technology. Thus they may be kept current with a minimum of effort.
4. With the increased storage capacity and computational efficiency of PCs, many general-purpose programs can now be run on PCs.
5. Many of the commercially available programs have become very attractive in price and can solve a wide range of problems [45–56].

Some disadvantages of general-purpose programs:

1. The initial cost of developing general-purpose programs is high.
2. General-purpose programs are less efficient than special-purpose programs because the computer must make many checks for each problem, some of which would not be necessary if a special-purpose program were used.
3. Many of the programs are proprietary. Hence the user has little access to the logic of the program. If a revision must be made, it often has to be done by the developers.

Some advantages of special-purpose programs:

1. The programs are usually relatively short, with low development costs.
2. Small computers are able to run the programs.
3. Additions can be made to the program quickly and at a low cost.
4. The programs are efficient in solving the problems they were designed to solve.

The major disadvantage of special-purpose programs is their inability to solve different classes of problems. Thus one must have as many programs as there are different classes of problems to be solved. A list of special-purpose, public-domain finite-element programs is given in the website [60].

There are numerous vendors supporting finite element programs, and the interested user should carefully consult the vendor before purchasing any software. However, to give you an idea about the various commercial personal computer programs now available for solving problems by the finite element method, we present a partial list of existing programs.

1. Algor [46]
2. Abaqus [47]
3. ANSYS [48]
4. COSMOS/M [49]
5. GT-STRUDL [50]
6. LS-DYNA [59]
7. MARC [51]
8. MSC/NASTRAN [52]
9. NISA [53]
10. Pro/MECHANICA [54]
11. SAP2000 [55]
12. STARDYNE [56]

Standard capabilities of many of the listed programs are provided in the preceding references and in Reference [45]. These capabilities include information on

1. Element types available, such as beam, plane stress, and three-dimensional solid
2. Type of analysis available, such as static and dynamic
3. Material behavior, such as linear-elastic and nonlinear
4. Load types, such as concentrated, distributed, thermal, and displacement (settlement)
5. Data generation, such as automatic generation of nodes, elements, and restraints (most programs have preprocessors to generate the mesh for the model)
6. Plotting, such as original and deformed geometry and stress and temperature contours (most programs have postprocessors to aid in interpreting results in graphical form)

7. Displacement behavior, such as small and large displacement and buckling
8. Selective output, such as at selected nodes, elements, and maximum or minimum values

All programs include at least the bar, beam, plane stress, plate-bending, and three-dimensional solid elements, and most now include heat-transfer analysis capabilities.

Complete capabilities of the programs and their cost are best obtained through program reference manuals and websites, such as References [46–56, 59].

▲ References

[1] Hrennikoff, A., "Solution of Problems in Elasticity by the Frame Work Method," *Journal of Applied Mechanics*, Vol. 8, No. 4, pp. 169–175, Dec. 1941.

[2] McHenry, D., "A Lattice Analogy for the Solution of Plane Stress Problems," *Journal of Institution of Civil Engineers*, Vol. 21, pp. 59–82, Dec. 1943.

[3] Courant, R., "Variational Methods for the Solution of Problems of Equilibrium and Vibrations," *Bulletin of the American Mathematical Society*, Vol. 49, pp. 1–23, 1943.

[4] Levy, S., "Computation of Influence Coefficients for Aircraft Structures with Discontinuities and Sweepback," *Journal of Aeronautical Sciences*, Vol. 14, No. 10, pp. 547–560, Oct. 1947.

[5] Levy, S., "Structural Analysis and Influence Coefficients for Delta Wings," *Journal of Aeronautical Sciences*, Vol. 20, No. 7, pp. 449–454, July 1953.

[6] Argyris, J. H., "Energy Theorems and Structural Analysis," *Aircraft Engineering*, Oct., Nov., Dec. 1954 and Feb., Mar., Apr., May 1955.

[7] Argyris, J. H., and Kelsey, S., *Energy Theorems and Structural Analysis*, Butterworths, London, 1960 (collection of papers published in *Aircraft Engineering* in 1954 and 1955).

[8] Turner, M. J., Clough, R. W., Martin, H. C., and Topp, L. J., "Stiffness and Deflection Analysis of Complex Structures," *Journal of Aeronautical Sciences*, Vol. 23, No. 9, pp. 805–824, Sept. 1956.

[9] Clough, R. W., "The Finite Element Method in Plane Stress Analysis," *Proceedings*, American Society of Civil Engineers, 2nd Conference on Electronic Computation, Pittsburgh, PA, pp. 345–378, Sept. 1960.

[10] Melosh, R. J., "A Stiffness Matrix for the Analysis of Thin Plates in Bending," *Journal of the Aerospace Sciences*, Vol. 28, No. 1, pp. 34–42, Jan. 1961.

[11] Grafton, P. E., and Strome, D. R., "Analysis of Axisymmetric Shells by the Direct Stiffness Method," *Journal of the American Institute of Aeronautics and Astronautics*, Vol. 1, No. 10, pp. 2342–2347, 1963.

[12] Martin, H. C., "Plane Elasticity Problems and the Direct Stiffness Method," *The Trend in Engineering*, Vol. 13, pp. 5–19, Jan. 1961.

[13] Gallagher, R. H., Padlog, J., and Bijlaard, P. P., "Stress Analysis of Heated Complex Shapes," *Journal of the American Rocket Society*, Vol. 32, pp. 700–707, May 1962.

[14] Melosh, R. J., "Structural Analysis of Solids," *Journal of the Structural Division*, Proceedings of the American Society of Civil Engineers, pp. 205–223, Aug. 1963.

[15] Argyris, J. H., "Recent Advances in Matrix Methods of Structural Analysis," *Progress in Aeronautical Science*, Vol. 4, Pergamon Press, New York, 1964.

[16] Clough, R. W., and Rashid, Y., "Finite Element Analysis of Axisymmetric Solids," *Journal of the Engineering Mechanics Division*, Proceedings of the American Society of Civil Engineers, Vol. 91, pp. 71–85, Feb. 1965.

[17] Wilson, E. L., "Structural Analysis of Axisymmetric Solids," *Journal of the American Institute of Aeronautics and Astronautics*, Vol. 3, No. 12, pp. 2269–2274, Dec. 1965.

[18] Turner, M. J., Dill, E. H., Martin, H. C., and Melosh, R. J., "Large Deflections of Structures Subjected to Heating and External Loads," *Journal of Aeronautical Sciences*, Vol. 27, No. 2, pp. 97–107, Feb. 1960.

[19] Gallagher, R. H., and Padlog, J., "Discrete Element Approach to Structural Stability Analysis," *Journal of the American Institute of Aeronautics and Astronautics*, Vol. 1, No. 6, pp. 1437–1439, 1963.

[20] Zienkiewicz, O. C., Watson, M., and King, I. P., "A Numerical Method of Visco-Elastic Stress Analysis," *International Journal of Mechanical Sciences*, Vol. 10, pp. 807–827, 1968.

[21] Archer, J. S., "Consistent Matrix Formulations for Structural Analysis Using Finite-Element Techniques," *Journal of the American Institute of Aeronautics and Astronautics*, Vol. 3, No. 10, pp. 1910–1918, 1965.

[22] Zienkiewicz, O. C., and Cheung, Y. K., "Finite Elements in the Solution of Field Problems," *The Engineer*, pp. 507–510, Sept. 24, 1965.

[23] Martin, H. C., "Finite Element Analysis of Fluid Flows," *Proceedings of the Second Conference on Matrix Methods in Structural Mechanics*, Wright-Patterson Air Force Base, Ohio, pp. 517–535, Oct. 1968. (AFFDL-TR-68-150, Dec. 1969; AD-703-685, N.T.I.S.)

[24] Wilson, E. L., and Nickel, R. E., "Application of the Finite Element Method to Heat Conduction Analysis," *Nuclear Engineering and Design*, Vol. 4, pp. 276–286, 1966.

[25] Szabo, B. A., and Lee, G. C., "Derivation of Stiffness Matrices for Problems in Plane Elasticity by Galerkin's Method," *International Journal of Numerical Methods in Engineering*, Vol. 1, pp. 301–310, 1969.

[26] Zienkiewicz, O. C., and Parekh, C. J., "Transient Field Problems: Two-Dimensional and Three-Dimensional Analysis by Isoparametric Finite Elements," *International Journal of Numerical Methods in Engineering*, Vol. 2, No. 1, pp. 61–71, 1970.

[27] Lyness, J. F., Owen, D. R. J., and Zienkiewicz, O. C., "Three-Dimensional Magnetic Field Determination Using a Scalar Potential. A Finite Element Solution," *Transactions on Magnetics*, Institute of Electrical and Electronics Engineers, pp. 1649–1656, 1977.

[28] Belytschko, T., "A Survey of Numerical Methods and Computer Programs for Dynamic Structural Analysis," *Nuclear Engineering and Design*, Vol. 37, No. 1, pp. 23–34, 1976.

[29] Belytschko, T., "Efficient Large-Scale Nonlinear Transient Analysis by Finite Elements," *International Journal of Numerical Methods in Engineering*, Vol. 10, No. 3, pp. 579–596, 1976.

[30] Huiskies, R., and Chao, E. Y. S., "A Survey of Finite Element Analysis in Orthopedic Biomechanics: The First Decade," *Journal of Biomechanics*, Vol. 16, No. 6, pp. 385–409, 1983.

[31] *Journal of Biomechanical Engineering*, Transactions of the American Society of Mechanical Engineers, (published quarterly) (1st issue published 1977).

[32] Kardestuncer, H., ed., *Finite Element Handbook*, McGraw-Hill, New York, 1987.

[33] Clough, R. W., "The Finite Element Method After Twenty-Five Years: A Personal View," *Computers and Structures*, Vol. 12, No. 4, pp. 361–370, 1980.

[34] Kardestuncer, H., *Elementary Matrix Analysis of Structures*, McGraw-Hill, New York, 1974.

[35] Oden, J. T., and Ripperger, E. A., *Mechanics of Elastic Structures*, 2nd ed., McGraw-Hill, New York, 1981.

[36] Finlayson, B. A., *The Method of Weighted Residuals and Variational Principles*, Academic Press, New York, 1972.

[37] Zienkiewicz, O. C., *The Finite Element Method*, 3rd ed., McGraw-Hill, London, 1977.

[38] Cook, R. D., Malkus, D. S., Plesha, M. E., and Witt, R. J., *Concepts and Applications of Finite Element Analysis*, 4th ed., Wiley, New York, 2002.

[39] Koswara, H., *A Finite Element Analysis of Underground Shelter Subjected to Ground Shock Load*, M.S. Thesis, Rose-Hulman Institute of Technology, 1983.

[40] Greer, R. D., "The Analysis of a Film Tower Die Utilizing the ANSYS Finite Element Package," M.S. Thesis, Rose-Hulman Institute of Technology, Terre Haute, Indiana, May 1989.

[41] Koeneman, J. B., Hansen, T. M., and Beres, K., "The Effect of Hip Stem Elastic Modulus and Cement/Stem Bond on Cement Stresses," 36th Annual Meeting, Orthopaedic Research Society, Feb. 5–8, 1990, New Orleans, Louisiana.

[42] Girijavallabham, C. V., and Reese, L. C., "Finite-Element Method for Problems in Soil Mechanics," *Journal of the Structural Division*, American Society of Civil Engineers, No. Sm2, pp. 473–497, Mar. 1968.

[43] Young, C., and Crocker, M., "Transmission Loss by Finite-Element Method," *Journal of the Acoustical Society of America*, Vol. 57, No. 1, pp. 144–148, Jan. 1975.

[44] Silvester, P. P., and Ferrari, R. L., *Finite Elements for Electrical Engineers*, Cambridge University Press, Cambridge, England, 1983.

[45] Falk, H., and Beardsley, C. W., "Finite Element Analysis Packages for Personal Computers," *Mechanical Engineering*, pp. 54–71, Jan. 1985.

[46] Algor Interactive Systems, 150 Beta Drive, Pittsburgh, PA 15238.

[47] Web site http://www.abaqus.com.

[48] Swanson, J. A., ANSYS-Engineering Analysis Systems User's Manual, Swanson Analysis Systems, Inc., Johnson Rd., P.O. Box 65, Houston, PA 15342.

[49] COSMOS/M, Structural Research & Analysis Corp., 12121 Wilshire Blvd., Los Angeles, CA 90025.

[50] web site http://ce6000.cegatech.edu.

[51] web site http://www.mscsoftware.com.

[52] MSC/NASTRAN, MacNeal-Schwendler Corp., 600 Suffolk St., Lowell, MA, 01854.

[53] web site http://emrc.com.

[54] Toogood, Roger, Pro/MECHANICA Structure Tutorial, SDC Publications, 2001.

[55] Computers & Structures, Inc., 1995 University Ave., Berkeley, CA 94704.

[56] STARDYNE, Research Engineers, Inc., 22700 Savi Ranch Pkwy, Yorba Linda, CA 92687.

[57] Noor, A. K., "Bibliography of Books and Monographs on Finite Element Technology," *Applied Mechanics Reviews*, Vol. 44, No. 6, pp. 307–317, June 1991.

[58] Belytschko, T., Liu W. K., and Moran, B., *Nonlinear Finite Elements For Continua and Structures*, John Wiley, 1996.

[59] Hallquist, J. O., LS-DYNA, Theoretical Manual, Livermore Software Technology Corp., 1998.

[60] web site http://homepage.usak.ca/~ijm451/finite/fe resources.

[61] Crisfield, M.A., *Non-linear Finite Element Analysis of Solids and Structures, Vol. 1: Essentials*, John Wiley & Sons, Chichester, UK, 1991.

[62] Crisfield, M.A., *Non-linear Finite Element Analysis of Solids and Structures, Vol. 2: Advanced Topics*, John Wiley & Sons, Chichester, UK, 1997.

[63] ANSYS Advantage, Vol. 1, Issue 1, 2007.

▲ Problems

1.1 Define the term *finite element*.

1.2 What does *discretization* mean in the finite element method?

1.3 In what year did the modern development of the finite element method begin?

1.4 In what year was the direct stiffness method introduced?

1.5 Define the term *matrix*.

1.6 What role did the computer play in the use of the finite element method?

1.7 List and briefly describe the general steps of the finite element method.

1.8 What is the displacement method?

1.9 List four common types of finite elements.

1.10 Name three commonly used methods for deriving the element stiffness matrix and element equations. Briefly describe each method.

1.11 To what does the term *degrees of freedom* refer?

1.12 List five typical areas of engineering where the finite element method is applied.

1.13 List five advantages of the finite element method.

CHAPTER 2

INTRODUCTION TO THE STIFFNESS (DISPLACEMENT) METHOD ▲

CHAPTER OBJECTIVES

- To define the stiffness matrix.
- To derive the stiffness matrix for a spring element.
- To demonstrate how to assemble stiffness matrices into a global stiffness matrix.
- To illustrate the concept of direct stiffness method to obtain the global stiffness matrix and solve a spring assemblage problem.
- To describe and apply the different kinds of boundary conditions relevant for spring assemblages.
- To show how the potential energy approach can be used to both derive the stiffness matrix for a spring and solve a spring assemblage problem.

Introduction

This chapter introduces some of the basic concepts on which the direct stiffness method is founded. The linear spring is introduced first because it provides a simple yet generally instructive tool to illustrate the basic concepts. We begin with a general definition of the stiffness matrix and then consider the derivation of the stiffness matrix for a linear-elastic spring element. We next illustrate how to assemble the total stiffness matrix for a structure comprising an assemblage of spring elements by using elementary concepts of equilibrium and compatibility. We then show how the total stiffness matrix for an assemblage can be obtained by superimposing the stiffness matrices of the individual elements in a direct manner. The term *direct stiffness method* evolved in reference to this technique.

After establishing the total structure stiffness matrix, we illustrate how to impose boundary conditions—both homogeneous and nonhomogeneous. A complete solution including the nodal displacements and reactions is thus obtained. (The determination of internal forces is discussed in Chapter 3 in connection with the bar element.)

We then introduce the principle of minimum potential energy, apply it to derive the spring element equations, and use it to solve a spring assemblage problem. We will illustrate this principle for the simplest of elements (those with small

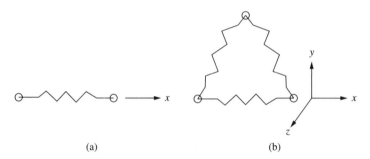

Figure 2–1 (a) Single spring element and (b) three-spring assemblage

numbers of degrees of freedom) so that it will be a more readily understood concept when applied, of necessity, to elements with large numbers of degrees of freedom in subsequent chapters.

▲ 2.1 Definition of the Stiffness Matrix ▲

Familiarity with the stiffness matrix is essential to understanding the stiffness method. We define the stiffness matrix as follows: For *an element, a* **stiffness matrix** $[k]$ is a matrix such that

$$\{f\} = [k]\{d\} \tag{2.1.1}$$

where $[k]$ relates nodal displacements $\{d\}$ to nodal forces $\{f\}$ of a single element, such as the spring shown in Figure 2–1a.

For a continuous medium or structure comprising a series of elements, such as shown for the spring assemblage in Figure 2–1b, stiffness matrix $[K]$ relates global-coordinate (x, y, z) nodal displacements $\{d\}$ to global forces $\{F\}$ of the whole medium or structure. such that

$$\{F\} = [K]\{d\} \tag{2.1.2}$$

where $[K]$ represents the stiffness matrix of the whole spring assemblage.

▲ 2.2 Derivation of the Stiffness Matrix for a Spring Element ▲

Using the direct equilibrium approach, we will now derive the stiffness matrix for a one-dimensional linear spring—that is, a spring that obeys Hooke's law and resists forces only in the direction of the spring. Consider the linear spring element shown in Figure 2–2. Reference points 1 and 2 are located at the ends of the element. These reference points are called the *nodes* of the spring element. The local nodal forces are f_{1x} and f_{2x} for the spring element associated with the local axis x. The local axis acts in the direction of the spring so that we can directly measure displacements and forces along the spring. The local nodal displacements are u_1 and u_2 for the spring element.

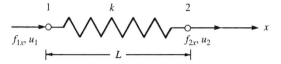

Figure 2–2 Linear spring element with positive nodal displacement and force conventions

These nodal displacements are called the *degrees of freedom* at each node. Positive directions for the forces and displacements at each node are taken in the positive x direction as shown from node 1 to node 2 in the figure. The symbol k is called the *spring constant* or *stiffness* of the spring.

Analogies to actual spring constants arise in numerous engineering problems. In Chapter 3, we see that a prismatic uniaxial bar has a spring constant $k = AE/L$, where A represents the cross-sectional area of the bar, E is the modulus of elasticity, and L is the bar length. Similarly, in Chapter 5, we show that a prismatic circular-cross-section bar in torsion has a spring constant $k = JG/L$, where J is the polar moment of inertia and G is the shear modulus of the material. For one-dimensional heat conduction (Chapter 13), $k = AK_{xx}/L$, where K_{xx} is the thermal conductivity of the material, and for one-dimensional fluid flow through a porous medium (Chapter 14), $k = AK_{xx}/L$, where K_{xx} is the permeability coefficient of the material.

We will then observe that the stiffness method can be applied to nonstructural problems, such as heat transfer, fluid flow, and electrical networks, as well as structural problems by simply applying the proper constitutive law (such as Hooke's law for structural problems, Fourier's law for heat transfer, Darcy's law for fluid flow and Ohm's law for electrical networks) and a conservation principle such as nodal equilibrium or conservation of energy.

We now want to develop a relationship between nodal forces and nodal displacements for a spring element. This relationship will be the stiffness matrix. Therefore, we want to relate the nodal force matrix to the nodal displacement matrix as follows:

$$\begin{Bmatrix} f_{1x} \\ f_{2x} \end{Bmatrix} = \begin{bmatrix} k_{11} & k_{12} \\ k_{21} & k_{22} \end{bmatrix} \begin{Bmatrix} u_1 \\ u_2 \end{Bmatrix} \tag{2.2.1}$$

where the element stiffness coefficients k_{ij} of the $[k]$ matrix in Eq. (2.2.1) are to be determined. Recall from Eqs. (1.2.5) and (1.2.6) that k_{ij} represent the force F_i in the ith degree of freedom due to a unit displacement d_j in the jth degree of freedom while all other displacements are zero. That is, when we let $d_j = 1$ and $d_k = 0$ for $k \neq j$, force $F_i = k_{ij}$.

We now use the general steps outlined in Section 1.4 to derive the stiffness matrix for the spring element in this section (while keeping in mind that these same steps will be applicable later in the derivation of stiffness matrices of more general elements) and then to illustrate a complete solution of a spring assemblage in Section 2.3. Because our approach throughout this text is to derive various element stiffness matrices and then to illustrate how to solve engineering problems with the elements, step 1 now involves only selecting the element type.

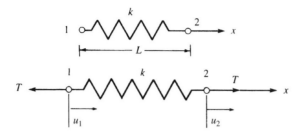

Figure 2–3 Linear spring subjected to tensile forces

Step 1 Select the Element Type

Consider the linear spring element (which can be an element in a system of springs) subjected to resulting nodal tensile forces T (which may result from the action of adjacent springs) directed along the spring axial direction x as shown in Figure 2–3, so as to be in equilibrium. The local x axis is directed from node 1 to node 2. We represent the spring by labeling nodes at each end and by labeling the element number. The original distance between nodes before deformation is denoted by L. The material property (spring constant) of the element is k.

Step 2 Select a Displacement Function

We must choose in advance the mathematical function to represent the deformed shape of the spring element under loading. Because it is difficult, if not impossible at times, to obtain a closed form or exact solution, we assume a solution shape or distribution of displacement within the element by using an appropriate mathematical function. The most common functions used are polynomials.

Because the spring element resists axial loading only with the local degrees of freedom for the element being displacements u_1 and u_2 along the x direction, we choose a displacement function u to represent the axial displacement throughout the element. Here a linear displacement variation along the x axis of the spring is assumed [Figure 2–4(b)], because a linear function with specified endpoints has a unique path. Therefore,

$$u = a_1 + a_2 x \qquad (2.2.2)$$

In general, the total number of coefficients a is equal to the total number of degrees of freedom associated with the element. Here the total number of degrees of freedom is two—an axial displacement at each of the two nodes of the element (we present further discussion regarding the choice of displacement functions in Section 3.2). In matrix form, Eq. (2.2.2) becomes

$$u = \begin{bmatrix} 1 & x \end{bmatrix} \begin{Bmatrix} a_1 \\ a_2 \end{Bmatrix} \qquad (2.2.3)$$

We now want to express u as a function of the nodal displacements u_1 and u_2, as this will allow us to apply the physical boundary conditions on nodal displacements directly as indicated in step 3 and to then relate the nodal displacements to the nodal

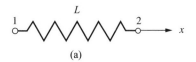

(a)

$u = a_1 + a_2 x$

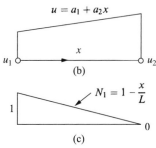

(b)

(c)

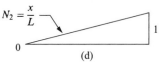

(d)

Figure 2–4 (a) Spring element showing plots of (b) displacement function u and shape functions, (c) N_1 and, (d) N_2 over domain of element

forces in step 4. We achieve this by evaluating u at each node and solving for a_1 and a_2 from Eq. (2.2.2) as follows:

$$u(0) = u_1 = a_1 \tag{2.2.4}$$

$$u(L) = u_2 = a_2 L + u_1 \tag{2.2.5}$$

or, solving Eq. (2.2.5) for a_2,

$$a_2 = \frac{u_2 - u_1}{L} \tag{2.2.6}$$

Upon substituting Eqs. (2.2.4) and (2.2.6) into Eq. (2.2.2), we have

$$u = \left(\frac{u_2 - u_1}{L}\right) x + u_1 \tag{2.2.7}$$

In matrix form, we express Eq. (2.2.7) as

$$u = \begin{bmatrix} 1 - \dfrac{x}{L} & \dfrac{x}{L} \end{bmatrix} \begin{Bmatrix} u_1 \\ u_2 \end{Bmatrix} \tag{2.2.8}$$

or

$$u = \begin{bmatrix} N_1 & N_2 \end{bmatrix} \begin{Bmatrix} u_1 \\ u_2 \end{Bmatrix} \tag{2.2.9}$$

Here

$$N_1 = 1 - \frac{x}{L} \quad \text{and} \quad N_2 = \frac{x}{L} \tag{2.2.10}$$

are called the *shape functions* because the N_i's express the shape of the assumed displacement function over the domain (x coordinate) of the element when the ith element degree of freedom has unit value and all other degrees of freedom are zero. In this case, N_1 and N_2 are linear functions that have the properties that

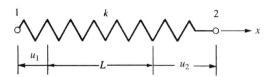

Figure 2–5 Deformed spring

$N_1 = 1$ at node 1 and $N_1 = 0$ at node 2, whereas $N_2 = 1$ at node 2 and $N_2 = 0$ at node 1. See Figure 2–4(c) and (d) for plots of these shape functions over the domain of the spring element. Also, $N_1 + N_2 = 1$ for any axial coordinate along the bar. (Section 3.2 further explores this important relationship.) In addition, the N_i's are often called *interpolation functions* because we are interpolating to find the value of a function between given nodal values. The interpolation function may be different from the actual function except at the endpoints or nodes, where the interpolation function and actual function must be equal to specified nodal values.

Step 3 Define the Strain/Displacement and Stress/Strain Relationships

The tensile forces T produce a total elongation (deformation) δ of the spring. The typical total elongation of the spring is shown in Figure 2–5. Here u_1 is a negative value because the direction of displacement is opposite the positive x direction, whereas u_2 is a positive value.

The deformation of the spring is then represented by

$$\delta = u(L) - u(0) = u_2 - u_1 \tag{2.2.11}$$

From Eq. (2.2.11), we observe that the total deformation is the difference of the nodal displacements in the x direction.

For a spring element, we can relate the force in the spring directly to the deformation. Therefore, the strain/displacement relationship is not necessary here.

The stress/strain relationship can be expressed in terms of the force/deformation relationship instead as

$$T = k\delta \tag{2.2.12}$$

Now, using Eq. (2.2.11) in Eq. (2.2.12), we obtain

$$T = k(u_2 - u_1) \tag{2.2.13}$$

Step 4 Derive the Element Stiffness Matrix and Equations

We now derive the spring element stiffness matrix. By the sign convention for nodal forces and equilibrium, (see Figures 2–2 and 2–3) we have

$$f_{1x} = -T \qquad f_{2x} = T \tag{2.2.14}$$

Using Eqs. (2.2.13) and (2.2.14), we have

$$T = -f_{1x} = k(u_2 - u_1)$$
$$T = f_{2x} = k(u_2 - u_1) \tag{2.2.15}$$

Rewriting Eqs. (2.2.15), we obtain

$$f_{1x} = k(u_1 - u_2)$$
$$f_{2x} = k(u_2 - u_1)$$

(2.2.16)

Now expressing Eqs. (2.2.16) in a single matrix equation yields

$$\begin{Bmatrix} f_{1x} \\ f_{2x} \end{Bmatrix} = \begin{bmatrix} k & -k \\ -k & k \end{bmatrix} \begin{Bmatrix} u_1 \\ u_2 \end{Bmatrix}$$

(2.2.17)

This relationship holds for the spring along the x axis. From our basic definition of a stiffness matrix and application of Eq. (2.2.1) to Eq. (2.2.17), we obtain

$$[k] = \begin{bmatrix} k & -k \\ -k & k \end{bmatrix}$$

(2.2.18)

as the stiffness matrix for a linear spring element. Here $[k]$ is called the *local stiffness matrix* for the element. We observe from Eq. (2.2.18) that $[k]$ is a symmetric (that is, $k_{ij} = k_{ji}$) square matrix (the number of rows equals the number of columns in $[k]$). Appendix A gives more description and numerical examples of symmetric and square matrices.

**Step 5   Assemble the Element Equations to Obtain
the Global Equations and Introduce Boundary Conditions**

The global stiffness matrix and global force matrix are assembled using nodal force equilibrium equations, force/deformation and compatibility equations from Section 2.3, and the direct stiffness method described in Section 2.4. This step applies for structures composed of more than one element such that

$$[K] = \sum_{e=1}^{N} [k^{(e)}] \quad \text{and} \quad \{F\} = \sum_{e=1}^{N} \{f^{(e)}\}$$

(2.2.19)

where $[k^{(e)}]$ and $\{f^{(e)}\}$ are now element stiffness and force matrices expressed in a global reference frame. This concept becomes relevant for instance when considering truss structures in Chapter 3. (Throughout this text, the $\sum$ sign used in this context does not imply a simple summation of element matrices but rather denotes that these element matrices must be assembled properly according to the direct stiffness method described in Section 2.4.)

Step 6   Solve for the Nodal Displacements

The displacements are then determined by imposing boundary conditions, such as support conditions, and solving a system of equations simultaneously as

$$\{F\} = [K]\{d\}$$

(2.2.20)

Step 7   Solve for the Element Forces

Finally, the element forces are determined by back-substitution, applied to each element, into equations similar to Eqs. (2.2.16).

▲ 2.3 Example of a Spring Assemblage   ▲

Structures such as trusses, building frames, and bridges comprise basic structural components connected together to form the overall structures. To analyze these structures, we must determine the total structure stiffness matrix for an interconnected system of elements. Before considering the truss and frame, we will determine the total structure stiffness matrix for a spring assemblage by using the force/displacement matrix relationships derived in Section 2.2 for the spring element, along with fundamental concepts of nodal equilibrium and compatibility. Step 5 will then have been illustrated.

We will consider the specific example of the two-spring assemblage shown in Figure 2–6.* This example is general enough to illustrate the direct equilibrium approach for obtaining the total stiffness matrix of the spring assemblage. Here we fix node 1 and apply axial forces for F_{3x} at node 3 and F_{2x} at node 2. The stiffnesses of spring elements 1 and 2 are k_1 and k_2, respectively. The nodes of the assemblage have been numbered 1, 3, and 2 for further generalization because sequential numbering between elements generally does not occur in large problems.

The x axis is the global axis of the assemblage. The local x axis of each element coincides with the global axis of the assemblage.

For element 1, using Eq. (2.2.17), we have

$$\left\{ \begin{array}{c} f_{1x}^{(1)} \\ f_{3x}^{(1)} \end{array} \right\} = \left[\begin{array}{cc} k_1 & -k_1 \\ -k_1 & k_1 \end{array} \right] \left\{ \begin{array}{c} u_1^{(1)} \\ u_3^{(1)} \end{array} \right\} \tag{2.3.1}$$

and for element 2, we have

$$\left\{ \begin{array}{c} f_{3x}^{(2)} \\ f_{2x}^{(2)} \end{array} \right\} = \left[\begin{array}{cc} k_2 & -k_2 \\ -k_2 & k_2 \end{array} \right] \left\{ \begin{array}{c} u_3^{(2)} \\ u_2^{(2)} \end{array} \right\} \tag{2.3.2}$$

Furthermore, elements 1 and 2 must remain connected at common node 3 throughout the displacement. This is called the *continuity* or *compatibility requirement*. The compatibility requirement yields

$$u_3^{(1)} = u_3^{(2)} = u_3 \tag{2.3.3}$$

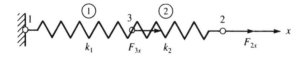

Figure 2–6   Two-spring assemblage

---

* Throughout this text, element numbers in figures are shown with circles around them.

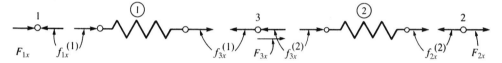

Figure 2–7 Nodal forces consistent with element force sign convention

where, throughout this text, the superscripts in parentheses above u refers to the element number to which they are related. Recall that the subscript to the right identifies the node of displacement and that u_3 is the node 3 displacement of the total or global spring assemblage.

Free-body diagrams of each element and node (using the established sign conventions for element nodal forces in Figure 2–2) are shown in Figure 2–7.

Based on the free-body diagrams of each node shown in Figure 2–7 and the fact that external forces must equal internal forces at each node, we can write nodal equilibrium equations at nodes 3, 2, and 1 as

$$F_{3x} = f_{3x}^{(1)} + f_{3x}^{(2)} \tag{2.3.4}$$

$$F_{2x} = f_{2x}^{(2)} \tag{2.3.5}$$

$$F_{1x} = f_{1x}^{(1)} \tag{2.3.6}$$

where F_{1x} results from the external applied reaction at the fixed support.

Here Newton's third law, of equal but opposite forces, is applied in moving from a node to an element associated with the node. Using Eqs. (2.3.1) through (2.3.3) in Eqs. (2.3.4) through (2.3.6), we obtain

$$F_{3x} = (-k_1 u_1 + k_1 u_3) + (k_2 u_3 - k_2 u_2)$$
$$F_{2x} = -k_2 u_3 + k_2 u_2 \tag{2.3.7}$$
$$F_{1x} = k_1 u_1 - k_1 u_3$$

In matrix form, Eqs. (2.3.7) are expressed by

$$\begin{Bmatrix} F_{3x} \\ F_{2x} \\ F_{1x} \end{Bmatrix} = \begin{bmatrix} k_1 + k_2 & -k_2 & -k_1 \\ -k_2 & k_2 & 0 \\ -k_1 & 0 & k_1 \end{bmatrix} \begin{Bmatrix} u_3 \\ u_2 \\ u_1 \end{Bmatrix} \tag{2.3.8}$$

Rearranging Eq. (2.3.8) in numerically increasing order of the nodal degrees of freedom, we have

$$\begin{Bmatrix} F_{1x} \\ F_{2x} \\ F_{3x} \end{Bmatrix} = \begin{bmatrix} k_1 & 0 & -k_1 \\ 0 & k_2 & -k_2 \\ -k_1 & -k_2 & k_1 + k_2 \end{bmatrix} \begin{Bmatrix} u_1 \\ u_2 \\ u_3 \end{Bmatrix} \tag{2.3.9}$$

Equation (2.3.9) is now written as the single matrix equation

$$\{F\} = [K]\{d\} \tag{2.3.10}$$

where $\{F\} = \begin{Bmatrix} F_{1x} \\ F_{2x} \\ F_{3x} \end{Bmatrix}$ is called the *global nodal force matrix*, $\{d\} = \begin{Bmatrix} u_1 \\ u_2 \\ u_3 \end{Bmatrix}$ is called the *global nodal displacement matrix*, and

$$[K] = \begin{bmatrix} k_1 & 0 & -k_1 \\ 0 & k_2 & -k_2 \\ -k_1 & -k_2 & k_1+k_2 \end{bmatrix} \tag{2.3.11}$$

is called the *total* or *global* or *system stiffness matrix*.

In summary, to establish the stiffness equations and stiffness matrix, Eqs. (2.3.9) and (2.3.11), for a spring assemblage, we have used force/deformation relationships Eqs. (2.3.1) and (2.3.2), compatibility relationship Eq. (2.3.3), and nodal force equilibrium Eqs. (2.3.4) through (2.3.6). We will consider the complete solution to this example problem after considering a more practical method of assembling the total stiffness matrix in Section 2.4 and discussing the support boundary conditions in Section 2.5.

▲ 2.4 Assembling the Total Stiffness Matrix by Superposition (Direct Stiffness Method) ▲

We will now consider a more convenient method for constructing the total stiffness matrix. This method is based on proper superposition of the individual element stiffness matrices making up a structure (also see References [1] and [2]).

Referring to the two-spring assemblage of Section 2.3, the element stiffness matrices are given in Eqs. (2.3.1) and (2.3.2) as

$$[k^{(1)}] = \begin{matrix} u_1 & u_3 \\ \begin{bmatrix} k_1 & -k_1 \\ -k_1 & k_1 \end{bmatrix} & \begin{matrix} u_1 \\ u_3 \end{matrix} \end{matrix} \qquad [k^{(2)}] = \begin{matrix} u_3 & u_2 \\ \begin{bmatrix} k_2 & -k_2 \\ -k_2 & k_2 \end{bmatrix} & \begin{matrix} u_3 \\ u_2 \end{matrix} \end{matrix} \tag{2.4.1}$$

Here the u_i's written above the columns and next to the rows in the $[k]$'s indicate the degrees of freedom associated with each element row and column.

The two element stiffness matrices, Eqs. (2.4.1), are not associated with the same degrees of freedom; that is, element 1 is associated with axial displacements at nodes 1 and 3, whereas element 2 is associated with axial displacements at nodes 2 and 3. Therefore, the element stiffness matrices cannot be added together (superimposed) in their present form. To superimpose the element matrices, we must expand them to the order (size) of the total structure (spring assemblage) stiffness matrix so that each element stiffness matrix is associated with all the degrees of freedom of the structure. To expand each element stiffness matrix to the order of the total stiffness matrix, we simply add rows and columns of zeros for those displacements not associated with that particular element.

For element 1, we rewrite the stiffness matrix in expanded form so that Eq. (2.3.1) becomes

$$
k_1
\begin{array}{c}
\begin{array}{ccc} u_1 & u_2 & u_3 \end{array} \\
\begin{bmatrix} 1 & 0 & -1 \\ 0 & 0 & 0 \\ -1 & 0 & 1 \end{bmatrix}
\end{array}
\begin{Bmatrix} u_1^{(1)} \\ u_2^{(1)} \\ u_3^{(1)} \end{Bmatrix}
=
\begin{Bmatrix} f_{1x}^{(1)} \\ f_{2x}^{(1)} \\ f_{3x}^{(1)} \end{Bmatrix}
\tag{2.4.2}
$$

where, from Eq. (2.4.2), we see that $u_2^{(1)}$ and $f_{2x}^{(1)}$ are not associated with $[k^{(1)}]$. Similarly, for element 2, we have

$$
k_2
\begin{array}{c}
\begin{array}{ccc} u_1 & u_2 & u_3 \end{array} \\
\begin{bmatrix} 0 & 0 & 0 \\ 0 & 1 & -1 \\ 0 & -1 & 1 \end{bmatrix}
\end{array}
\begin{Bmatrix} u_1^{(2)} \\ u_2^{(2)} \\ u_3^{(2)} \end{Bmatrix}
=
\begin{Bmatrix} f_{1x}^{(2)} \\ f_{2x}^{(2)} \\ f_{3x}^{(2)} \end{Bmatrix}
\tag{2.4.3}
$$

Now, considering force equilibrium at each node results in

$$
\begin{Bmatrix} f_{1x}^{(1)} \\ 0 \\ f_{3x}^{(1)} \end{Bmatrix}
+
\begin{Bmatrix} 0 \\ f_{2x}^{(2)} \\ f_{3x}^{(2)} \end{Bmatrix}
=
\begin{Bmatrix} F_{1x} \\ F_{2x} \\ F_{3x} \end{Bmatrix}
\tag{2.4.4}
$$

where Eq. (2.4.4) is really Eqs. (2.3.4) through (2.3.6) expressed in matrix form. Using Eqs. (2.4.2) and (2.4.3) in Eq. (2.4.4), we obtain

$$
k_1
\begin{bmatrix} 1 & 0 & -1 \\ 0 & 0 & 0 \\ -1 & 0 & 1 \end{bmatrix}
\begin{Bmatrix} u_1^{(1)} \\ u_2^{(1)} \\ u_3^{(1)} \end{Bmatrix}
+ k_2
\begin{bmatrix} 0 & 0 & 0 \\ 0 & 1 & -1 \\ 0 & -1 & 1 \end{bmatrix}
\begin{Bmatrix} u_1^{(2)} \\ u_2^{(2)} \\ u_3^{(2)} \end{Bmatrix}
=
\begin{Bmatrix} F_{1x} \\ F_{2x} \\ F_{3x} \end{Bmatrix}
\tag{2.4.5}
$$

where, again, the superscripts on the u's indicate the element numbers. Simplifying Eq. (2.4.5) results in

$$
\begin{bmatrix} k_1 & 0 & -k_1 \\ 0 & k_2 & -k_2 \\ -k_1 & -k_2 & k_1+k_2 \end{bmatrix}
\begin{Bmatrix} u_1 \\ u_2 \\ u_3 \end{Bmatrix}
=
\begin{Bmatrix} F_{1x} \\ F_{2x} \\ F_{3x} \end{Bmatrix}
\tag{2.4.6}
$$

Here the superscripts indicating the element numbers associated with the nodal displacements have been dropped because $u_1^{(1)}$ is really u_1, $u_2^{(2)}$ is really u_2, and, by Eq. (2.3.3), $u_3^{(1)} = u_3^{(2)} = u_3$, the node 3 displacement of the total assemblage. Equation (2.4.6), obtained through superposition, is identical to Eq. (2.3.9).

The expanded element stiffness matrices in Eqs. (2.4.2) and (2.4.3) could have been added directly to obtain the total stiffness matrix of the structure, given in Eq. (2.4.6). This reliable method of directly assembling individual element stiffness matrices to form the total structure stiffness matrix and the total set of stiffness equations is called the *direct stiffness method*. It is the most important step in the finite element method.

For this simple example, it is easy to expand the element stiffness matrices and then superimpose them to arrive at the total stiffness matrix. However, for problems involving a large number of degrees of freedom, it will become tedious to expand each element stiffness matrix to the order of the total stiffness matrix. To avoid this expansion of each element stiffness matrix, we suggest a direct, or short-cut, form of the direct stiffness method to obtain the total stiffness matrix. For the spring assemblage example, the rows and columns of each element stiffness matrix are labeled according to the degrees of freedom associated with them as follows:

$$[k^{(1)}] = \begin{matrix} u_1 & u_3 \\ \begin{bmatrix} k_1 & -k_1 \\ -k_1 & k_1 \end{bmatrix} & \begin{matrix} u_1 \\ u_3 \end{matrix} \end{matrix} \qquad [k^{(2)}] = \begin{matrix} u_3 & u_2 \\ \begin{bmatrix} k_2 & -k_2 \\ -k_2 & k_2 \end{bmatrix} & \begin{matrix} u_3 \\ u_2 \end{matrix} \end{matrix} \qquad (2.4.7)$$

$[K]$ is then constructed simply by directly adding terms associated with degrees of freedom in $[k^{(1)}]$ and $[k^{(2)}]$ into their corresponding identical degree-of-freedom locations in $[K]$ as follows. The u_1 row, u_1 column term of $[K]$ is contributed only by element 1, as only element 1 has degree of freedom u_1 [Eq. (2.4.7)], that is, $k_{11} = k_1$. The u_3 row, u_3 column of $[K]$ has contributions from both elements 1 and 2, as the u_3 degree of freedom is associated with both elements. Therefore, $k_{33} = k_1 + k_2$. Similar reasoning results in $[K]$ as

$$[K] = \begin{matrix} u_1 & u_2 & u_3 \\ \begin{bmatrix} k_1 & 0 & -k_1 \\ 0 & k_2 & -k_2 \\ -k_1 & -k_2 & k_1 + k_2 \end{bmatrix} & \begin{matrix} u_1 \\ u_2 \\ u_3 \end{matrix} \end{matrix} \qquad (2.4.8)$$

Here elements in $[K]$ are located on the basis that degrees of freedom are ordered in increasing node numerical order for the total structure. Section 2.5 addresses the complete solution to the two-spring assemblage in conjunction with discussion of the support boundary conditions.

▲ 2.5 Boundary Conditions ▲

We must specify boundary (or support) conditions for structure models such as the spring assemblage of Figure 2–6, or $[K]$ will be singular; that is, the determinant of $[K]$ will be zero, and its inverse will not exist. This means the structural system is unstable. Without our specifying adequate kinematic constraints or support conditions, the structure will be free to move as a rigid body and not resist any applied loads. In general, the number of boundary conditions necessary to make $[K]$ nonsingular is equal to the number of possible rigid body modes.

Boundary conditions relevant for spring assemblages are associated with nodal displacements. These conditions are of two types. Homogeneous boundary conditions—the more common—occur at locations that are completely prevented from movement; nonhomogeneous boundary conditions occur where finite nonzero values of displacement are specified, such as the settlement of a support.

In the mathematical sense in regard to solving boundary value problems, we encounter two general classifications of boundary conditions when imposed on an ordinary or partial differential equation or derived upon taking the first variation of a functional as shown in References [4, 5, 8], but these are avoided in this more basic textbook.

The first type—primary, essential, or Dirichlet—boundary condition (named after Johann Dirichlet (1805–1859)), specifies the values a solution, such as the displacement, must satisfy on the boundary of the domain.

The second type—natural or Neumann—boundary condition (named after Carl Neumann (1832–1925)), specifies the values that the derivatives of a solution must satisfy on the boundary of the domain.

To illustrate the two general displacement types of boundary conditions, let us consider Eq. (2.4.6), derived for the spring assemblage of Figure 2–6. which has a single rigid body mode in the direction of motion along the spring assemblage.

Homogeneous Boundary Conditions

We first consider the case of homogeneous boundary conditions. Hence all boundary conditions are such that the displacements are zero at certain nodes. Here we have $u_1 = 0$ because node 1 is fixed. Therefore, Eq. (2.4.6) can be written as

$$\begin{bmatrix} k_1 & 0 & -k_1 \\ 0 & k_2 & -k_2 \\ -k_1 & -k_2 & k_1 + k_2 \end{bmatrix} \begin{Bmatrix} 0 \\ u_2 \\ u_3 \end{Bmatrix} = \begin{Bmatrix} F_{1x} \\ F_{2x} \\ F_{3x} \end{Bmatrix} \tag{2.5.1}$$

Equation (2.5.1), written in expanded form, becomes

$$k_1(0) + (0)u_2 - k_1 u_3 = F_{1x}$$
$$0(0) + k_2 u_2 - k_2 u_3 = F_{2x} \tag{2.5.2}$$
$$-k_1(0) - k_2 u_2 + (k_1 + k_2)u_3 = F_{3x}$$

where F_{1x} is the unknown reaction and F_{2x} and F_{3x} are known applied loads.

Writing the second and third of Eqs. (2.5.2) in matrix form, we have

$$\begin{bmatrix} k_2 & -k_2 \\ -k_2 & k_1 + k_2 \end{bmatrix} \begin{Bmatrix} u_2 \\ u_3 \end{Bmatrix} = \begin{Bmatrix} F_{2x} \\ F_{3x} \end{Bmatrix} \tag{2.5.3}$$

We have now effectively partitioned off the first column and row of $[K]$ and the first row of $\{d\}$ and $\{F\}$ to arrive at Eq. (2.5.3).

For homogeneous boundary conditions, Eq. (2.5.3) could have been obtained directly by deleting the row and column of Eq. (2.5.1) corresponding to the zero-displacement degrees of freedom. Here row 1 and column 1 are deleted because one is really multiplying column 1 of $[K]$ by $u_1 = 0$. However, F_{1x} is not necessarily zero and can be determined once u_2 and u_3 are solved for.

After solving Eq. (2.5.3) for u_2 and u_3, we have

$$\begin{Bmatrix} u_2 \\ u_3 \end{Bmatrix} = \begin{bmatrix} k_2 & -k_2 \\ -k_2 & k_1 + k_2 \end{bmatrix}^{-1} \begin{Bmatrix} F_{2x} \\ F_{3x} \end{Bmatrix} = \begin{bmatrix} \dfrac{1}{k_2} + \dfrac{1}{k_1} & \dfrac{1}{k_1} \\ \dfrac{1}{k_1} & \dfrac{1}{k_1} \end{bmatrix} \begin{Bmatrix} F_{2x} \\ F_{3x} \end{Bmatrix} \tag{2.5.4}$$

Now that u_2 and u_3 are known from Eq. (2.5.4), we substitute them in the first of Eqs. (2.5.2) to obtain the reaction F_{1x} as

$$F_{1x} = -k_1 u_3 \qquad (2.5.5)$$

We can express the unknown nodal force at node 1 (also called the *reaction*) in terms of the applied nodal forces F_{2x} and F_{3x} by using Eq. (2.5.4) for u_3 substituted into Eq. (2.5.5). The result is

$$F_{1x} = -F_{2x} - F_{3x} \qquad (2.5.6)$$

Therefore, for all homogeneous boundary conditions, we can delete the rows and columns corresponding to the zero-displacement degrees of freedom from the original set of equations and then solve for the unknown displacements. This procedure is useful for hand calculations. (However, Appendix B.4 presents a more practical, computer-assisted scheme for solving the system of simultaneous equations.)

Nonhomogeneous Boundary Conditions

We now consider the case of nonhomogeneous boundary conditions. Hence one or more of the specified displacements are nonzero. For simplicity's sake, let $u_1 = \delta$, where δ is a known displacement (Figure 2–8), in Eq. (2.4.6). We now have

$$\begin{bmatrix} k_1 & 0 & -k_1 \\ 0 & k_2 & -k_2 \\ -k_1 & -k_2 & k_1 + k_2 \end{bmatrix} \begin{Bmatrix} \delta \\ u_2 \\ u_3 \end{Bmatrix} = \begin{Bmatrix} F_{1x} \\ F_{2x} \\ F_{3x} \end{Bmatrix} \qquad (2.5.7)$$

Equation (2.5.7) written in expanded form becomes

$$k_1 \delta + 0u_2 - k_1 u_3 = F_{1x}$$
$$0\delta + k_2 u_2 - k_2 u_3 = F_{2x} \qquad (2.5.8)$$
$$-k_1 \delta - k_2 u_2 + (k_1 + k_2)u_3 = F_{3x}$$

where F_{1x} is now a reaction from the support that has moved an amount δ. Considering the second and third of Eqs. (2.5.8) because they have known right-side nodal forces F_{2x} and F_{3x}, we obtain

$$0\delta + k_2 u_2 - k_2 u_3 = F_{2x}$$
$$-k_1 \delta - k_2 u_2 + (k_1 + k_2)u_3 = F_{3x} \qquad (2.5.9)$$

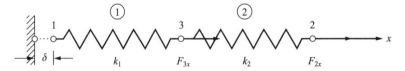

Figure 2–8 Two-spring assemblage with known displacement δ at node 1

Transforming the known δ terms to the right side of Eqs. (2.5.9) yields

$$k_2 u_2 - k_2 u_3 = F_{2x}$$
$$-k_2 u_2 + (k_1 + k_2)u_3 = +k_1\delta + F_{3x}$$

(2.5.10)

Rewriting Eqs. (2.5.10) in matrix form, we have

$$\begin{bmatrix} k_2 & -k_2 \\ -k_2 & k_1 + k_2 \end{bmatrix} \begin{Bmatrix} u_2 \\ u_3 \end{Bmatrix} = \begin{Bmatrix} F_{2x} \\ k_1\delta + F_{3x} \end{Bmatrix}$$

(2.5.11)

Therefore, when dealing with nonhomogeneous boundary conditions, we cannot initially delete row 1 and column 1 of Eq. (2.5.7), corresponding to the nonhomogeneous boundary condition, as indicated by the resulting Eq. (2.5.11) because we are multiplying each element by a nonzero number. Had we done so, the $k_1\delta$ term in Eq. (2.5.11) would have been neglected, resulting in an error in the solution for the displacements. For nonhomogeneous boundary conditions, we must, in general, transform the terms associated with the known displacements to the right-side force matrix before solving for the unknown nodal displacements. This was illustrated by transforming the $k_1\delta$ term of the second of Eqs. (2.5.9) to the right side of the second of Eqs. (2.5.10).

We could now solve for the displacements in Eq. (2.5.11) in a manner similar to that used to solve Eq. (2.5.3). However, we will not further pursue the solution of Eq. (2.5.11) because no new information is to be gained.

However, on substituting the displacement back into Eq. (2.5.7), the reaction now becomes

$$F_{1x} = k_1\delta - k_1 u_3$$

(2.5.12)

which is different than Eq. (2.5.5) for F_{1x}.

Notice that if the displacement is known at a node (say $u_1 = \delta$), then the force F_{1x} at the node in the same direction as the displacement is not initially known and is determined using the global equation of Eq. (2.5.7) after solving for the unknown nodal displacements.

At this point, we summarize some properties of the stiffness matrix in Eq. (2.5.7) that are also applicable to the generalization of the finite element method.

1. $[K]$ is square, as it relates the same number of forces and displacements.
2. $[K]$ is symmetric, as is each of the element stiffness matrices. If you are familiar with structural mechanics, you will not find this symmetry property surprising. It can be proved by using the reciprocal laws described in such References as [3] and [4].
3. $[K]$ is singular (its determinant is equal to zero), and thus, no inverse exists until sufficient boundary conditions are imposed to remove the singularity and prevent rigid body motion.
4. The main diagonal terms of $[K]$ are always positive. Otherwise, a positive nodal force F_i could produce a negative displacement d_i— a behavior contrary to the physical behavior of any actual structure.
5. $[K]$ is positive semidefinite (that is $\{x\}^T[K]\{x\} > 0$ for all non-zero vector $\{x\}$ with real numbers). (For more about positive semidefinite matrices, see Appendix A.)

In general, specified support conditions are treated mathematically by partitioning the global equilibrium equations as follows:

$$\begin{bmatrix} [K_{11}] & | [K_{12}] \\ [K]_{21} & | [K_{22}] \end{bmatrix} \begin{Bmatrix} \{d_1\} \\ \{d_2\} \end{Bmatrix} = \begin{Bmatrix} \{F_1\} \\ \{F_2\} \end{Bmatrix} \tag{2.5.13}$$

where we let $\{d_1\}$ be the unconstrained or free displacements and $\{d_2\}$ be the specified displacements. From Eq. (2.5.13), we have

$$[K_{11}]\{d_1\} = \{F_1\} - [K]_{12}\{d_2\} \tag{2.5.14}$$

and

$$\{F_2\} = [K_{21}]\{d_1\} + [K_{22}]\{d_2\} \tag{2.5.15}$$

where $\{F_1\}$ are the known nodal forces and $\{F_2\}$ are the unknown nodal forces at the specified displacement nodes. $\{F_2\}$ is found from Eq. (2.5.15) after $\{d_1\}$ is determined from Eq. (2.5.14). In Eq. (2.5.14), we assume that $[K_{11}]$ is no longer singular, thus allowing for the determination of $\{d_1\}$.

To illustrate the stiffness method for the solution of spring assemblages we now present the following examples.

Example 2.1

For the spring assemblage with arbitrarily numbered nodes shown in Figure 2–9, obtain (a) the global stiffness matrix, (b) the displacements of nodes 3 and 4, (c) the reaction forces at nodes 1 and 2, and (d) the forces in each spring. A force of 25 kN is applied at node 4 in the x direction. The spring constants are given in the figure. Nodes 1 and 2 are fixed.

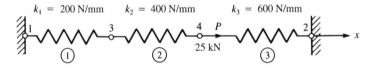

Figure 2–9 Spring assemblage for solution

SOLUTION:

(a) We begin by making use of Eq. (2.2.18) to express each element stiffness matrix as follows:

$$[k^{(1)}] = \begin{matrix} & 1 & 3 \\ & \begin{bmatrix} 200 & -200 \\ -200 & 200 \end{bmatrix} & \begin{matrix} 1 \\ 3 \end{matrix} \end{matrix} \qquad [k^{(2)}] = \begin{matrix} & 3 & 4 \\ & \begin{bmatrix} 400 & -400 \\ -400 & 400 \end{bmatrix} & \begin{matrix} 3 \\ 4 \end{matrix} \end{matrix}$$

$$[k^{(3)}] = \begin{matrix} & 4 & 2 \\ & \begin{bmatrix} 600 & -600 \\ -600 & 600 \end{bmatrix} & \begin{matrix} 4 \\ 2 \end{matrix} \end{matrix} \tag{2.5.16}$$

where the numbers above the columns and next to each row indicate the nodal degrees of freedom associated with each element. For instance, element 1 is associated with degrees of freedom u_1 and u_3. Also, the local element x axis coincides with the global x axis for each element.

Using the concept of superposition (the direct stiffness method), we obtain the global stiffness matrix as

$$[K] = [k^{(1)}] + [k^{(2)}] + [k^{(3)}]$$

or $$[K] = \begin{array}{c} \begin{array}{cccc} u_1 & u_2 & u_3 & u_4 \end{array} \\ \begin{bmatrix} 200 & 0 & -200 & 0 \\ 0 & 600 & 0 & -600 \\ -200 & 0 & 200+400 & -400 \\ 0 & -600 & -400 & 400+600 \end{bmatrix} \begin{array}{c} u_1 \\ u_2 \\ u_3 \\ u_4 \end{array} \begin{array}{c} \\ \\ \dfrac{\text{N}}{\text{mm}} \\ \\ \end{array} \end{array} \qquad (2.5.17)$$

(b) The global stiffness matrix, Eq. (2.5.17), relates global forces to global displacements as follows:

$$\begin{Bmatrix} F_{1x} \\ F_{2x} \\ F_{3x} \\ F_{4x} \end{Bmatrix} = \begin{bmatrix} 200 & 0 & -200 & 0 \\ 0 & 600 & 0 & -600 \\ -200 & 0 & 600 & -400 \\ 0 & -600 & -400 & 1000 \end{bmatrix} \begin{Bmatrix} u_1 \\ u_2 \\ u_3 \\ u_4 \end{Bmatrix} \qquad (2.5.18)$$

Applying the homogeneous boundary conditions $u_1 = 0$ and $u_2 = 0$ to Eq. (2.5.18), substituting applied nodal forces, and partitioning the first two equations of Eq. (2.5.18) (or deleting the first two rows of $\{F\}$ and $\{d\}$ and the first two rows and columns of $[K]$ corresponding to the zero-displacement boundary conditions), we obtain

$$\begin{Bmatrix} 0 \\ 2500 \end{Bmatrix} = \begin{bmatrix} 600 & -400 \\ -400 & 1000 \end{bmatrix} \begin{Bmatrix} u_3 \\ u_4 \end{Bmatrix} \qquad (2.5.19)$$

Solving Eq. (2.5.19), we obtain the global nodal displacements

$$u_3 = \frac{250}{11}\ \text{mm} \qquad u_4 = \frac{375}{11}\ \text{mm} \qquad (2.5.20)$$

(c) To obtain the global nodal forces (which include the reactions at nodes 1 and 2), we back-substitute Eqs. (2.5.20) and the boundary conditions $u_1 = 0$ and $u_2 = 0$ into Eq. (2.5.18). This substitution yields

$$\begin{Bmatrix} F_{1x} \\ F_{2x} \\ F_{3x} \\ F_{4x} \end{Bmatrix} = \begin{bmatrix} 200 & 0 & -200 & 0 \\ 0 & 600 & 0 & -600 \\ -200 & 0 & 600 & -400 \\ 0 & -600 & -400 & 1000 \end{bmatrix} \begin{Bmatrix} 0 \\ 0 \\ \frac{250}{11} \\ \frac{375}{11} \end{Bmatrix} \qquad (2.5.21)$$

Multiplying matrices in Eq. (2.5.21) and simplifying, we obtain the forces at each node

$$F_{1x} = \frac{-50{,}000}{11} \text{ N} \qquad F_{2x} = \frac{-225{,}000}{11} \text{ N} \qquad F_{3x} = 0$$

$$F_{4x} = \frac{275{,}000}{11} \text{ N}$$

(2.5.22)

From these results, we observe that the sum of the reactions F_{1x} and F_{2x} is equal in magnitude but opposite in direction to the applied force F_{4x}. This result verifies equilibrium of the whole spring assemblage.

(d) Next we use local element Eq. (2.2.17) to obtain the forces in each element.

Element 1

$$\left\{ \begin{array}{c} f_{1x}^{(1)} \\ f_{3x}^{(1)} \end{array} \right\} = \left[\begin{array}{cc} 200 & -200 \\ -200 & 200 \end{array} \right] \left\{ \begin{array}{c} 0 \\ \frac{250}{11} \end{array} \right\}$$

(2.5.23)

Simplifying Eq. (2.5.23), we obtain

$$f_{1x}^{(1)} = \frac{-50{,}000}{11} \text{ N} \qquad f_{3x}^{(1)} = \frac{50{,}000}{11} \text{ N}$$

(2.5.24)

A free-body diagram of spring element 1 is shown in Figure 2–10(a). The spring is subjected to tensile forces given by Eqs. (2.5.24). Also, $f_{1x}^{(1)}$ is equal to the reaction force F_{1x} given in Eq. (2.5.22). A free-body diagram of node 1 [Figure 2–10(b)] shows this result.

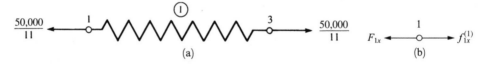

(a)

(b)

Figure 2–10 (a) Free-body diagram of element 1 and (b) free-body diagram of node 1

Element 2

$$\left\{ \begin{array}{c} f_{3x}^{(2)} \\ f_{4x}^{(2)} \end{array} \right\} = \left[\begin{array}{cc} 400 & -400 \\ -400 & 400 \end{array} \right] \left\{ \begin{array}{c} \frac{250}{11} \\ \frac{375}{11} \end{array} \right\}$$

(2.5.25)

Simplifying Eq. (2.5.25), we obtain

$$f_{3x}^{(2)} = \frac{-50{,}000}{11} \text{ N} \qquad f_{4x}^{2} = \frac{50{,}000}{11} \text{ N}$$

(2.5.26)

A free-body diagram of spring element 2 is shown in Figure 2–11. The spring is subjected to tensile forces given by Eqs. (2.5.26).

Figure 2–11 Free-body diagram of element 2

Element 3

$$\begin{Bmatrix} f_{4x}^{(3)} \\ f_{2x}^{(3)} \end{Bmatrix} = \begin{bmatrix} 600 & -600 \\ -600 & 600 \end{bmatrix} \begin{Bmatrix} \frac{375}{11} \\ 0 \end{Bmatrix} \tag{2.5.27}$$

Simplifying Eq. (2.5.27) yields

$$f_{4x}^{(3)} = \frac{225,000}{11} \text{ N} \qquad f_{2x}^{(3)} = \frac{-225,000}{11} \text{ N} \tag{2.5.28}$$

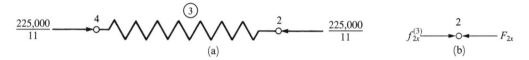

(a)   (b)

Figure 2–12 (a) Free-body diagram of element 3 and (b) free-body diagram of node 2

A free-body diagram of spring element 3 is shown in Figure 2–12(a). The spring is subjected to compressive forces given by Eqs. (2.5.28). Also, f_{2x} is equal to the reaction force F_{2x} given in Eq. (2.5.22). A free-body diagram of node 2 (Figure 2–12b) shows this result. ■

Example 2.2

For the spring assemblage shown in Figure 2–13, obtain (a) the global stiffness matrix, (b) the displacements of nodes 2–4, (c) the global nodal forces, and (d) the local element forces. Node 1 is fixed while node 5 is given a fixed, known displacement $\delta = 20.0$ mm. The spring constants are all equal to $k = 200$ kN/m.

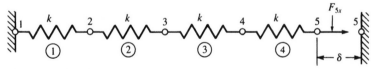

Figure 2–13 Spring assemblage for solution

SOLUTION:

(a) We use Eq. (2.2.18) to express each element stiffness matrix as

$$[k^{(1)}] = [k^{(2)}] = [k^{(3)}] = [k^{(4)}] = \begin{bmatrix} 200 & -200 \\ -200 & 200 \end{bmatrix} \qquad (2.5.29)$$

Again using superposition, we obtain the global stiffness matrix as

$$[K] = \begin{bmatrix} 200 & -200 & 0 & 0 & 0 \\ -200 & 400 & -200 & 0 & 0 \\ 0 & -200 & 400 & -200 & 0 \\ 0 & 0 & -200 & 400 & -200 \\ 0 & 0 & 0 & -200 & 200 \end{bmatrix} \frac{\text{kN}}{\text{m}} \qquad (2.5.30)$$

(b) The global stiffness matrix, Eq. (2.5.30), relates the global forces to the global displacements as follows:

$$\begin{Bmatrix} F_{1x} \\ F_{2x} \\ F_{3x} \\ F_{4x} \\ F_{5x} \end{Bmatrix} = \begin{bmatrix} 200 & -200 & 0 & 0 & 0 \\ -200 & 400 & -200 & 0 & 0 \\ 0 & -200 & 400 & -200 & 0 \\ 0 & 0 & -200 & 400 & -200 \\ 0 & 0 & 0 & -200 & 200 \end{bmatrix} \begin{Bmatrix} u_1 \\ u_2 \\ u_3 \\ u_4 \\ u_5 \end{Bmatrix} \qquad (2.5.31)$$

Applying the boundary conditions $u_1 = 0$ and $u_5 = 20$ mm ($= 0.02$ m), substituting known global forces $F_{2x} = 0$, $F_{3x} = 0$, and $F_{4x} = 0$, and partitioning the first and fifth equations of Eq. (2.5.31) corresponding to these boundary conditions, we obtain

$$\begin{Bmatrix} 0 \\ 0 \\ 0 \end{Bmatrix} = \begin{bmatrix} -200 & 400 & -200 & 0 & 0 \\ 0 & -200 & 400 & -200 & 0 \\ 0 & 0 & -200 & 400 & -200 \end{bmatrix} \begin{Bmatrix} 0 \\ u_2 \\ u_3 \\ u_4 \\ 0.02 \text{ m} \end{Bmatrix} \qquad (2.5.32)$$

We now rewrite Eq. (2.5.32), transposing the product of the appropriate stiffness coefficient (-200) multiplied by the known displacement (0.02 m) to the left side.

$$\begin{Bmatrix} 0 \\ 0 \\ 4 \text{ kN} \end{Bmatrix} = \begin{bmatrix} 400 & -200 & 0 \\ -200 & 400 & -200 \\ 0 & -200 & 400 \end{bmatrix} \begin{Bmatrix} u_2 \\ u_3 \\ u_4 \end{Bmatrix} \qquad (2.5.33)$$

Solving Eq. (2.5.33), we obtain

$$u_2 = 0.005 \text{ m} \qquad u_3 = 0.01 \text{ m} \qquad u_4 = 0.015 \text{ m} \qquad (2.5.34)$$

(c) The global nodal forces are obtained by back-substituting the boundary condition displacements and Eqs. (2.5.34) into Eq. (2.5.31). This substitution yields

$$F_{1x} = (-200)(0.005) = -1.0 \text{ kN}$$

$$F_{2x} = (400)(0.005) - (200)(0.01) = 0$$

$$F_{3x} = (-200)(0.005) + (400)(0.01) - (200)(0.015) = 0 \qquad (2.5.35)$$

$$F_{4x} = (-200)(0.01) + (400)(0.015) - (200)(0.02) = 0$$

$$F_{5x} = (-200)(0.015) + (200)(0.02) = 1.0 \text{ kN}$$

The results of Eqs. (2.5.35) yield the reaction F_{1x} opposite that of the nodal force F_{5x} required to displace node 5 by $\delta = 20.0$ mm. This result verifies equilibrium of the whole spring assemblage.

Remember if the displacement is known at a node in a given direction (in this example, $u_5 = 20$ mm) then the force F_{5x} at that same node and in that same direction is not initially known. The force is determined after solving for the unknown nodal displacements.

(d) Next, we make use of local element Eq. (2.2.17) to obtain the forces in each element.

Element 1

$$\begin{Bmatrix} f_{1x}^{(1)} \\ f_{2x}^{(1)} \end{Bmatrix} = \begin{bmatrix} 200 & -200 \\ -200 & 200 \end{bmatrix} \begin{Bmatrix} 0 \\ 0.005 \end{Bmatrix} \qquad (2.5.36)$$

Simplifying Eq. (2.5.36) yields

$$f_{1x}^{(1)} = -1.0 \text{ kN} \qquad f_{2x}^{(1)} = 1.0 \text{ kN} \qquad (2.5.37)$$

Element 2

$$\begin{Bmatrix} f_{2x}^{(2)} \\ f_{3x}^{(2)} \end{Bmatrix} = \begin{bmatrix} 200 & -200 \\ -200 & 200 \end{bmatrix} \begin{Bmatrix} 0.005 \\ 0.01 \end{Bmatrix} \qquad (2.5.38)$$

Simplifying Eq. (2.5.38) yields

$$f_{2x}^{(2)} = -1 \text{ kN} \qquad f_{3x}^{(2)} = 1 \text{ kN} \qquad (2.5.39)$$

Element 3

$$\begin{Bmatrix} f_{3x}^{(3)} \\ f_{4x}^{(3)} \end{Bmatrix} = \begin{bmatrix} 200 & -200 \\ -200 & 200 \end{bmatrix} \begin{Bmatrix} 0.01 \\ 0.015 \end{Bmatrix} \qquad (2.5.40)$$

Simplifying Eq. (2.5.40), we have

$$f_{3x}^{(3)} = -1 \text{ kN} \qquad f_{4x}^{(3)} = 1 \text{ kN} \qquad (2.5.41)$$

Element 4

$$\left\{ \begin{matrix} f_{4x}^{(4)} \\ f_{5x}^{(4)} \end{matrix} \right\} = \begin{bmatrix} 200 & -200 \\ -200 & 200 \end{bmatrix} \left\{ \begin{matrix} 0.015 \\ 0.02 \end{matrix} \right\} \tag{2.5.42}$$

Simplifying Eq. (2.5.42), we obtain

$$f_{4x}^{(4)} = -1 \text{ kN} \qquad f_{5x}^{(4)} = 1 \text{ kN} \tag{2.5.43}$$

You should draw free-body diagrams of each node and element and use the results of Eqs. (2.5.35) through (2.5.43) to verify both node and element equilibria. ■

Finally, to review the major concepts presented in this chapter, we solve the following example problem.

Example 2.3

(a) Using the ideas presented in Section 2.3 for the system of linear elastic springs shown in Figure 2–14, express the boundary conditions, the compatibility or continuity condition similar to Eq. (2.3.3), and the nodal equilibrium conditions similar to Eqs. (2.3.4) through (2.3.6). Then formulate the global stiffness matrix and equations for solution of the unknown global displacement and forces. The spring constants for the elements are $k_1, k_2,$ and k_3; P is an applied force at node 2.

(b) Using the direct stiffness method, formulate the same global stiffness matrix and equation as in part (a).

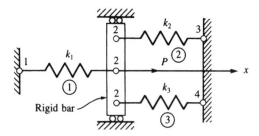

Figure 2–14 Spring assemblage for solution

SOLUTION:

(a) The boundary conditions are

$$u_1 = 0 \qquad u_3 = 0 \qquad u_4 = 0 \tag{2.5.44}$$

The compatibility condition at node 2 is

$$u_2^{(1)} = u_2^{(2)} = u_2^{(3)} = u_2 \tag{2.5.45}$$

The nodal equilibrium conditions are

$$F_{1x} = f_{1x}^{(1)}$$
$$P = f_{2x}^{(1)} + f_{2x}^{(2)} + f_{2x}^{(3)}$$
$$F_{3x} = f_{3x}^{(2)}$$
$$F_{4x} = f_{4x}^{(3)}$$

(2.5.46)

where the sign convention for positive element nodal forces given by Figure 2–2 was used in writing Eqs. (2.5.46). Figure 2–15 shows the element and nodal force free-body diagrams.

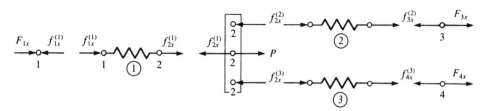

Figure 2–15 Free-body diagrams of elements and nodes of spring assemblage of Figure 2–14

Using the local stiffness matrix Eq. (2.2.17) applied to each element and compatibility condition Eq. (2.5.45), we obtain the total or global equilibrium equations as

$$F_{1x} = k_1 u_1 - k_1 u_2$$
$$P = -k_1 u_1 + k_1 u_2 + k_2 u_2 - k_2 u_3 + k_3 u_3 - k_3 u_4$$
$$F_{3x} = -k_2 u_2 + k_2 u_3$$
$$F_{4x} = -k_3 u_2 + k_3 u_4$$

(2.5.47)

In matrix form, we express Eqs. (2.5.47) as

$$\begin{Bmatrix} F_{1x} \\ P \\ F_{3x} \\ F_{4x} \end{Bmatrix} = \begin{bmatrix} k_1 & -k_1 & 0 & 0 \\ -k_1 & k_1 + k_2 + k_3 & -k_2 & -k_3 \\ 0 & -k_2 & k_2 & 0 \\ 0 & -k_3 & 0 & k_3 \end{bmatrix} \begin{Bmatrix} u_1 \\ u_2 \\ u_3 \\ u_4 \end{Bmatrix}$$

(2.5.48)

Therefore, the global stiffness matrix is the square, symmetric matrix on the right side of Eq. (2.5.48). Making use of the boundary conditions, Eqs. (2.5.44), and then considering the second equation of Eqs. (2.5.47) or (2.5.48), we solve for u_2 as

$$u_2 = \frac{P}{k_1 + k_2 + k_3}$$

(2.5.49)

We could have obtained this same result by deleting rows 1, 3, and 4 in the $\{F\}$ and $\{d\}$ matrices and rows and columns 1, 3, and 4 in $[K]$, corresponding to zero displacement, as previously described in Section 2.4, and then solving for u_2.

Using Eqs. (2.5.47), we now solve for the global forces as

$$F_{1x} = -k_1 u_2 \qquad F_{3x} = -k_2 u_2 \qquad F_{4x} = -k_3 u_2 \qquad (2.5.50)$$

The forces given by Eqs. (2.5.50) can be interpreted as the global reactions in this example. The negative signs in front of these forces indicate that they are directed to the left (opposite the x axis).

   (b) Using the direct stiffness method, we formulate the global stiffness matrix. First, using Eq. (2.2.18), we express each element stiffness matrix as

$$[k^{(1)}] = \begin{matrix} u_1 & u_2 \\ \begin{bmatrix} k_1 & -k_1 \\ -k_1 & k_1 \end{bmatrix} \end{matrix} \quad [k^{(2)}] = \begin{matrix} u_2 & u_3 \\ \begin{bmatrix} k_2 & -k_2 \\ -k_2 & k_2 \end{bmatrix} \end{matrix} \quad [k^{(3)}] = \begin{matrix} u_2 & u_4 \\ \begin{bmatrix} k_3 & -k_3 \\ -k_3 & k_3 \end{bmatrix} \end{matrix} \quad (2.5.51)$$

where the particular degrees of freedom associated with each element are listed in the columns above each matrix. Using the direct stiffness method as outlined in Section 2.4, we add terms from each element stiffness matrix into the appropriate corresponding row and column in the global stiffness matrix to obtain

$$[K] = \begin{matrix} u_1 & u_2 & u_3 & u_4 \\ \begin{bmatrix} k_1 & -k_1 & 0 & 0 \\ -k_1 & k_1 + k_2 + k_3 & -k_2 & -k_3 \\ 0 & -k_2 & k_2 & 0 \\ 0 & -k_3 & 0 & k_3 \end{bmatrix} \end{matrix} \qquad (2.5.52)$$

We observe that each element stiffness matrix $[k]$ has been added into the location in the global $[K]$ corresponding to the identical degree of freedom associated with the element $[k]$. For instance, element 3 is associated with degrees of freedom u_2 and u_4; hence its contributions to $[K]$ are in the 2–2, 2–4, 4–2, and 4–4 locations of $[K]$, as indicated in Eq. (2.5.52) by the k_3 terms.

   Having assembled the global $[K]$ by the direct stiffness method, we then formulate the global equations in the usual manner by making use of the general Eq. (2.3.10), $\{F\} = [K]\{d\}$. These equations have been previously obtained by Eq. (2.5.48) and therefore are not repeated. ∎

   Another method for handling imposed boundary conditions that allows for either homogeneous (zero) or nonhomogeneous (nonzero) prescribed degrees of freedom is called the *penalty method*. This method is easy to implement in a computer program.

   Consider the simple spring assemblage in Figure 2–16 subjected to applied forces F_{1x} and F_{2x} as shown. Assume the horizontal displacement at node 1 to be forced to be $u_1 = \delta$.

   We add another spring (often called a boundary element) with a large stiffness k_b to the assemblage in the direction of the nodal displacement $u_1 = \delta$ as shown in Figure 2–17. This spring stiffness should have a magnitude about 10^6 times that of the largest k_{ii} term.

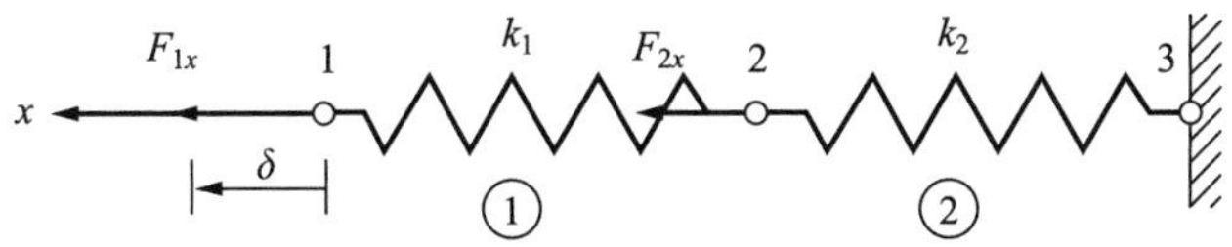

Figure 2–16 Spring assemblage used to illustrate the penalty method

Figure 2–17 Spring assemblage with a boundary spring element added at node 1

Now we add the force $k_b\delta$ in the direction of u_1 and solve the problem in the usual manner as follows.

The element stiffness matrices are

$$[k^{(1)}] = \begin{bmatrix} k_1 & -k_1 \\ -k_1 & k_1 \end{bmatrix} \qquad [k^{(2)}] = \begin{bmatrix} k_2 & -k_2 \\ -k_2 & k_2 \end{bmatrix} \tag{2.5.53}$$

Assembling the element stiffness matrices using the direct stiffness method, we obtain the global stiffness matrix as

$$[K] = \begin{bmatrix} k_1 + k_b & -k_1 & 0 \\ -k_1 & k_1 + k_2 & -k_2 \\ 0 & -k_2 & k_2 \end{bmatrix} \tag{2.5.54}$$

Assembling the global $\{F\} = [K]\{d\}$ equations and invoking the boundary condition $u_3 = 0$, we obtain

$$\begin{Bmatrix} F_{1x} + k_b\delta \\ F_{2x} \\ F_{3x} \end{Bmatrix} = \begin{bmatrix} k_1 + k_b & -k_1 & 0 \\ -k_1 & k_1 + k_2 & -k_2 \\ 0 & -k_2 & k_2 \end{bmatrix} \begin{Bmatrix} u_1 \\ u_2 \\ u_3 = 0 \end{Bmatrix} \tag{2.5.55}$$

Solving the first and second of Eqs. (2.5.55), we obtain

$$u_1 = \frac{F_{2x} - (k_1 + k_2)u_2}{-k_1} \tag{2.5.56}$$

and

$$u_2 = \frac{(k_1 + k_b)F_{2x} + F_{1x}k_1 + k_b\delta k_1}{k_bk_1 + k_bk_2 + k_1k_2} \tag{2.5.57}$$

Now as k_b approaches infinity, Eq. (2.5.57) simplifies to

$$u_2 = \frac{F_{2x} + \delta k_1}{k_1 + k_2} \tag{2.5.58}$$

and Eq. (2.5.56) simplifies to

$$u_1 = \delta \qquad\qquad (2.5.59)$$

These results match those obtained by setting $u_1 = \delta$ initially.

In using the penalty method, a very large element stiffness should be parallel to a degree of freedom as is the case in the preceding example. If k_b were inclined, or were placed within a structure, it would contribute to both diagonal and off-diagonal coefficients in the global stiffness matrix $[K]$. This condition can lead to numerical difficulties in solving the equations $\{F\} = [K]\{d\}$. To avoid this condition, we transform the displacements at the inclined support to local ones as described in Section 3.9.

▲ 2.6 Potential Energy Approach to Derive Spring Element Equations ▲

One of the alternative methods often used to derive the element equations and the stiffness matrix for an element is based on the principle of *minimum potential energy*. (The use of this principle in structural mechanics is fully described in Reference [4].) This method has the advantage of being more general than the method given in Section 2.2, which involves nodal and element equilibrium equations along with the stress/strain law for the element. Thus the principle of minimum potential energy is more adaptable to the determination of element equations for complicated elements (those with large numbers of degrees of freedom) such as the plane stress/strain element, the axisymmetric stress element, the plate bending element, and the three-dimensional solid stress element.

Again, we state that the principle of virtual work (Appendix E) is applicable for any material behavior, whereas the principle of minimum potential energy is applicable only for elastic materials. However, both principles yield the same element equations for linear-elastic materials, which are the only kind considered in this text. Moreover, the principle of minimum potential energy, being included in the general category of *variational methods* (as is the principle of virtual work), leads to other variational functions (or functionals) similar to potential energy that can be formulated for other classes of problems, primarily of the nonstructural type. These other problems are generally classified as *field problems* and include, among others, torsion of a bar, heat transfer (Chapter 13), fluid flow (Chapter 14), and electric potential (Chapter 14).

Still other classes of problems, for which a variational formulation is not clearly definable, can be formulated by *weighted residual methods*. We will describe Galerkin's method in Section 3.12, along with collocation, least squares, and the subdomain weighted residual methods in Section 3.13. In Section 3.13, we will also demonstrate these methods by solving a one-dimensional bar problem using each of the four residual methods and comparing each result to an exact solution. (For more information on weighted residual methods, also consult References [5–7].)

Here we present the principle of minimum potential energy as used to derive the spring element equations. We will illustrate this concept by applying it to the simplest of elements in hopes that the reader will then be more comfortable when applying it to handle more complicated element types in subsequent chapters.

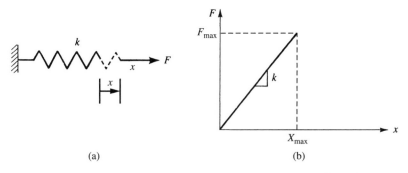

Figure 2–18   (a) Spring subjected to gradually increasing force *F*
(b) Force/deformation curve for linear spring

The total potential energy π_p of a structure is expressed in terms of displacements. In the finite element formulation, these will generally be nodal displacements such that $\pi_p = \pi_p(d_1, d_2, \ldots, d_n)$. When π_p is minimized with respect to these displacements, equilibrium equations result. For the spring element, we will show that the same nodal equilibrium equations $[k]\{d\} = \{f\}$ result as previously derived in Section 2.2.

We first state the principle of minimum potential energy as follows:

Of all the geometrically possible shapes that a body can assume, the true one, corresponding to the satisfaction of stable equilibrium of the body, is identified by a minimum value of the total potential energy.

To explain this principle, we must first explain the concepts of potential energy and of a stationary value of a function. We will now discuss these two concepts.

Total potential energy is defined as *the sum of the internal strain energy U and the potential energy of the external forces Ω*; that is,

$$\pi_p = U + \Omega \qquad (2.6.1)$$

Strain energy is the capacity of internal forces (or stresses) to do work through deformations (strains) in the structure; Ω is the capacity of forces such as body forces, surface traction forces, and applied nodal forces to do work through deformation of the structure.

To understand the concept of internal strain energy, we first describe the concept of external work. In this section, we consider only the external work due to an applied nodal force. In Chapter 3, Section 10, we consider work due to body forces (typically self weight) and surface tractions (distributed forces). External work is done on a linear-elastic behaving member (here we consider an elastic spring shown in Figure 2–18(a)) by applying a gradually increasing magnitude force F to the end of the spring up to some maximum value F_{max} less than that which would cause permanent deformation in the spring. The maximum deformation X_{max} occurs when the maximum force occurs as shown in Figure 2–18(b). The external work is given by the area under the force-deformation curve shown in Figure 2–18(b), where the slope of the straight line is equal to the spring constant k. The external work W_e is then given from basic mechanics principles as the integral of the dot product of vector force **F** with the differential displacement *dx*. This expression is represented by Eq. (2.6.2) as

$$W_e = \int \mathbf{F} \cdot d\mathbf{x} = \int_0^{x_{max}} F_{max}\left(\frac{x}{x_{max}}\right) dx = F_{max}x_{max}/2 \qquad (2.6.2)$$

where *F* in Eq. (2.6.2) is given by

$$F = F_{max}(x/x_{max}) \qquad (2.6.3)$$

In Eq. (2.6.2), we note that **F** and *dx* are in the same direction when expressing the second integral on the right side of Eq. (2.6.2).

By the conservation of mechanical energy principle, the external work due to the applied force F is transformed into the internal strain energy U of the spring. This strain energy is then given by

$$W_e = U = F_{max}x_{max}/2 \qquad (2.6.4)$$

Upon gradual reduction of the force to zero, the spring returns to its original undeformed state. This returned energy that is stored in the deformed elastic spring is called *internal strain energy* or just *strain energy*. Also

$$F_{max} = kx_{max} \qquad (2.6.5)$$

By substituting Eq. (2.6.5) into Eq. (2.6.4), we can express the strain energy as

$$U = kx_{max}^2/2 \qquad (2.6.6)$$

The potential energy of the external force, being opposite in sign from the external work expression because the potential energy of the external force is lost when the work is done by the external force, is given by

$$\Omega = -F_{max}x_{max} \qquad (2.6.7)$$

Therefore, substituting Eqs. (2.6.6) and (2.6.7) into (2.6.1), yields the total potential energy as

$$\pi_p = \frac{1}{2}kx_{max}^2 - F_{max}x_{max} \qquad (2.6.8)$$

In general for any deformation *x* of the spring corresponding to force *F*, we replace x_{max} with x and F_{max} with F and express U and Ω as

$$U(x) = kx^2/2 \qquad (2.6.8a)$$

$$\Omega(x) = -Fx \qquad (2.6.8b)$$

Substituting Eq. (2.6.8a) and (2.6.8b) into Eq. (2.6.1), we express the total potential energy as

$$\pi_p(x) = \frac{1}{2}kx^2 - Fx \qquad (2.6.9)$$

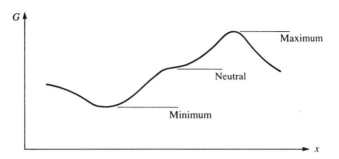

Figure 2–19 Stationary values of a function

The concept of a *stationary value* of a function G (used in the definition of the principle of minimum potential energy) is shown in Figure 2–19. Here G is expressed as a function of the variable x. The stationary value can be a maximum, a minimum, or a neutral point of $G(x)$. To find a value of x yielding a stationary value of $G(x)$, we use differential calculus to differentiate G with respect to x and set the expression equal to zero, as follows:

$$\frac{dG}{dx} = 0 \qquad (2.6.10)$$

An analogous process will subsequently be used to replace G with π_p and x with discrete values (nodal displacements) d_i. With an understanding of variational calculus (see Reference [8]), we could use the first variation of π_p (denoted by $\delta\pi_p$, where δ denotes arbitrary change or variation) to minimize π_p. However, we will avoid the details of variational calculus and show that we can really use the familiar differential calculus to perform the minimization of π_p. To apply the principle of minimum potential energy—that is, to minimize π_p—we take the *variation* of π_p, which is a function of nodal displacements d_i defined in general as

$$\delta\pi_p = \frac{\partial\pi_p}{\partial d_1}\delta d_1 + \frac{\partial\pi_p}{\partial d_2}\delta d_2 + \cdots + \frac{\partial\pi_p}{\partial d_n}\delta d_n \qquad (2.6.11)$$

The principle states that equilibrium exists when the d_i define a structure state such that $\delta\pi_p = 0$ (change in potential energy $= 0$) for arbitrary admissible variations in displacement δd_i from the equilibrium state. An *admissible variation* is one in which the displacement field still satisfies the boundary conditions and interelement continuity. Figure 2–20(a) shows the hypothetical actual axial displacement and an admissible one for a spring with specified boundary displacements u_1 and u_2. Figure 2–20(b) shows inadmissible functions due to slope discontinuity between endpoints 1 and 2 and due to failure to satisfy the right end boundary condition of $u(L) = u_2$. Here δu represents the variation in u. In the general finite element formulation, δu would be replaced by δd_i. This implies that any of the δd_i might be nonzero. Hence, to satisfy $\delta\pi_p = 0$, all coefficients associated with the δd_i must be zero independently. Thus,

$$\frac{\partial\pi_p}{\partial d_i} = 0 \quad (i = 1, 2, 3, \ldots, n) \qquad \text{or} \qquad \frac{\partial\pi_p}{\partial\{d\}} = 0 \qquad (2.6.12)$$

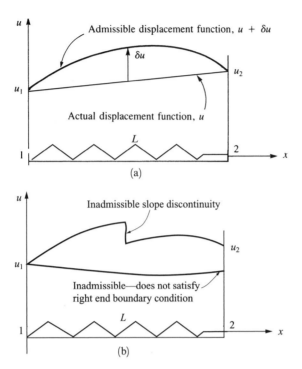

Figure 2–20 (a) Actual and admissible displacement functions and (b) inadmissible displacement functions

where n equations must be solved for the n values of d_i that define the static equilibrium state of the structure. Equation (2.6.12) shows that for our purposes throughout this text, we can interpret *the variation of* π_p as a compact notation equivalent to differentiation of π_p with respect to the unknown nodal displacements for which π_p is expressed. For linear-elastic materials in equilibrium, the fact that π_p is a minimum is shown, for instance, in Reference [4].

Before discussing the formulation of the spring element equations, we now illustrate the concept of the principle of minimum potential energy by analyzing a single-degree-of-freedom spring subjected to an applied force, as given in Example 2.4. In this example, we will show that the equilibrium position of the spring corresponds to the minimum potential energy.

Example 2.4

For the linear-elastic spring subjected to a force of 5000 N shown in Figure 2–21, evaluate the potential energy for various displacement values and show that the minimum potential energy also corresponds to the equilibrium position of the spring.

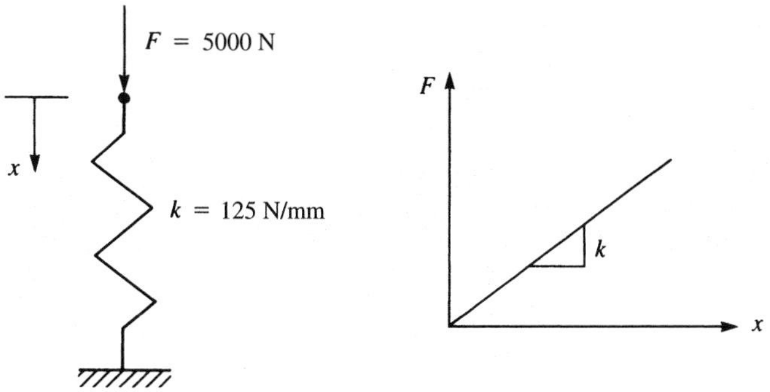

Figure 2–21 Spring subjected to force; load/displacement curve

SOLUTION:

We evaluate the total potential energy as

$$\pi_p = U + \Omega$$

where $$U = \tfrac{1}{2}(kx)x \quad \text{and} \quad \Omega = -Fx$$

We now illustrate the minimization of π_p through standard mathematics. Taking the variation of π_p with respect to x, or, equivalently, taking the derivative of π_p with respect to x (as π_p is a function of only one displacement x), as in Eqs. (2.6.11) and (2.6.12), we have

$$\delta\pi_p = \frac{\partial\pi_p}{\partial x}\delta x = 0$$

or, because δx is arbitrary and might not be zero,

$$\frac{\partial\pi_p}{\partial x} = 0$$

Using our previous expression for π_p, we obtain

$$\frac{\partial\pi_p}{\partial x} = 125x - 5000 = 0$$

or $$x = 40 \text{ mm}$$

This value for x is then back-substituted into π_p to yield

$$\pi_p = 62.5(40)^2 - 5000(40) = -100{,}000 \text{ N-mm}$$

which corresponds to the minimum potential energy obtained in Table 2–1 by the following searching technique. Here $U = \tfrac{1}{2}(kx)x$ is the strain energy or the area under the load/displacement curve shown in Figure 2–21, and $\Omega = -Fx$ is the potential energy of load F. For the given values of F and k, we then have

$$\pi_p = \frac{1}{2}(125)x^2 - 5000x = 62.5x^2 - 5000x$$

We now search for the minimum value of π_p for various values of spring deformation x. The results are shown in Table 2–1. A plot of π_p versus x is shown in

Table 2–1 Total potential energy for
various spring deformations

Deformation x, mm	Total Potential Energy π_p, N-mm $\times 10^3$ (N-m)
−80	800
−60	52.5
−40	300
−20	125
0.0	0
20	−75
40	−100
60	−75
80	0
100	125

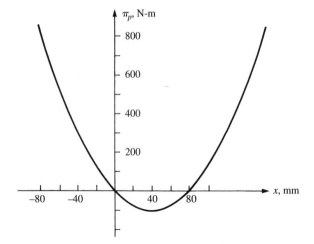

Figure 2–22 Variation of potential energy with spring deformation

Figure 2–22, where we observe that π_p has a minimum value at $x = 40$ mm. This deformed position also corresponds to the equilibrium position because $(\partial \pi_p / \partial x) = 125(40) - 5000 = 0$. ∎

We now derive the spring element equations and stiffness matrix using the principle of minimum potential energy. Consider the linear spring subjected to nodal forces shown in Figure 2–23. Using Eq. (2.6.9) reveals that the total potential energy is

$$\pi_p = \frac{1}{2}k(u_2 - u_1)^2 - f_{1x}u_1 - f_{2x}u_2 \qquad (2.6.13)$$

Figure 2–23 Linear spring subjected to nodal forces

where $u_2 - u_1$ is the deformation of the spring in Eq. (2.6.9). The first term on the right in Eq. (2.6.13) is the strain energy in the spring. Simplifying Eq. (2.6.13), we obtain

$$\pi_p = \frac{1}{2}k(u_2^2 - 2u_2u_1 + u_1^2) - f_{1x}u_1 - f_{2x}u_2 \tag{2.6.14}$$

The minimization of π_p with respect to each nodal displacement requires taking partial derivatives of π_p with respect to each nodal displacement such that

$$\frac{\partial \pi_p}{\partial u_1} = \frac{1}{2}k(-2u_2 + 2u_1) - f_{1x} = 0$$

$$\frac{\partial \pi_p}{\partial u_2} = \frac{1}{2}k(2u_2 - 2u_1) - f_{2x} = 0 \tag{2.6.15}$$

Simplifying Eqs. (2.6.15), we have

$$k(-u_2 + u_1) = f_{1x}$$

$$k(u_2 - u_1) = f_{2x} \tag{2.6.16}$$

In matrix form, we express Eq. (2.6.16) as

$$\begin{bmatrix} k & -k \\ -k & k \end{bmatrix} \begin{Bmatrix} u_1 \\ u_2 \end{Bmatrix} = \begin{Bmatrix} f_{1x} \\ f_{2x} \end{Bmatrix} \tag{2.6.17}$$

Because $\{f\} = [k]\{d\}$, we have the stiffness matrix for the spring element obtained from Eq. (2.6.17):

$$[k] = \begin{bmatrix} k & -k \\ -k & k \end{bmatrix} \tag{2.6.18}$$

As expected, Eq. (2.6.18) is identical to the stiffness matrix obtained in Section 2.2, Eq. (2.2.18).

We considered the equilibrium of a single spring element by minimizing the total potential energy with respect to the nodal displacements (see Example 2.4). We also developed the finite element spring element equations by minimizing the total potential energy with respect to the nodal displacements. We now show that the total potential energy of an entire structure (here an assemblage of spring elements) can be minimized with respect to each nodal degree of freedom and that this minimization results in the same finite element equations used for the solution as those obtained by the direct stiffness method.

Example 2.5

Obtain the total potential energy of the spring assemblage (Figure 2–24) for Example 2.1 and find its minimum value. The procedure of assembling element equations can then be seen to be obtained from the minimization of the total potential energy.

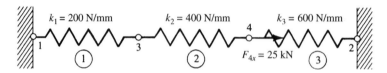

Figure 2–24 Spring assemblage

SOLUTION:

Using Eq. (2.6.8a), the strain energy stored in spring 1 is given by

$$U^{(1)} = k_1(u_3 - u_1)^2/2 \tag{2.6.19}$$

where the difference in nodal displacements $u_3 - u_1$ is the deformation x in spring 1. Eq. (2.6.19) can be written in matrix form as

$$U^{(1)} = \frac{1}{2}[u_3 \ u_1] \begin{bmatrix} k_1 & -k_1 \\ -k_1 & k_1 \end{bmatrix} \begin{Bmatrix} u_3 \\ u_1 \end{Bmatrix} = \frac{1}{2}\{d\}^{\mathrm{T}}[K]\{d\} \tag{2.6.20}$$

We observe from Eq. (2.6.20) that the strain energy U is a quadratic function of the nodal displacements.

Similar strain energy expressions for springs 2 and 3 are given by

$$U^{(2)} = k_2(u_4 - u_3)^2/2 \quad \text{and} \quad U^{(3)} = k_3(u_2 - u_4)^2/2 \tag{2.6.21}$$

with similar matrix expressions as given by Eq. (2.6.20) for spring 1.

Since the strain energy is a scalar quantity, we can add the energy in each spring to obtain the total strain energy in the system as

$$U = \sum_{i=1}^{3} U^{(e)} \tag{2.6.22}$$

The potential energy of the external nodal forces given in the order of the node numbering for the spring assemblage is

$$\Omega = -(F_{1x}u_1 + F_{3x}u_3 + F_{4x}u_4 + F_{2x}u_2) \tag{2.6.23}$$

Equation (2.6.23) can be expressed in matrix form as

$$\Omega = -[u_1 \ u_2 \ u_3 \ u_4] \begin{Bmatrix} F_{1x} \\ F_{2x} \\ F_{3x} \\ F_{4x} \end{Bmatrix} \tag{2.6.24}$$

The total potential of the assemblage is the sum of the strain energy and the potential energy of the external forces given by adding Eqs. (2.6.19), (2.6.21) and (2.6.23) together as

$$\Pi_p = U + \Omega = \frac{1}{2}k_1(u_3 - u_1)^2 + \frac{1}{2}k_2(u_4 - u_3)^2 + \frac{1}{2}k_3(u_2 - u_4)^2$$
$$- F_{1x}u_1 - F_{2x}u_2 - F_{3x}u_3 - F_{4x}u_4 \tag{2.6.25}$$

Upon minimizing π_p with respect to each nodal displacement, we obtain

$$\frac{\partial \pi_p}{\partial u_1} = -k_1 u_3 + k_1 u_1 - F_{1x} = 0$$

$$\frac{\partial \pi_p}{\partial u_2} = k_3 u_2 - k_3 u_4 - F_{2x} = 0$$

$$\frac{\partial \pi_p}{\partial u_3} = k_1 u_3 - k_1 u_1 - k_2 u_4 + k_2 u_3 - F_{3x} = 0$$

(2.6.26)

$$\frac{\partial \pi_p}{\partial u_4} = k_2 u_4 - k_2 u_3 - k_3 u_2 + k_3 u_4 - F_{4x} = 0$$

In matrix form, Eqs. (2.6.26) become

$$\begin{bmatrix} k_1 & 0 & -k_1 & 0 \\ 0 & k_3 & 0 & -k_3 \\ -k_1 & 0 & k_1 + k_2 & -k_2 \\ 0 & -k_3 & -k_2 & k_2 + k_3 \end{bmatrix} \begin{Bmatrix} u_1 \\ u_2 \\ u_3 \\ u_4 \end{Bmatrix} = \begin{Bmatrix} F_{1x} \\ F_{2x} \\ F_{3x} \\ F_{4x} \end{Bmatrix}$$

(2.6.27)

Substituting numerical values for k_1, k_2, and k_3 into Eq. (2.6.27), we obtain

$$\begin{bmatrix} 200 & 0 & -200 & 0 \\ 0 & 600 & 0 & -600 \\ -200 & 0 & 600 & -400 \\ 0 & -600 & -400 & 1000 \end{bmatrix} \begin{Bmatrix} u_1 \\ u_2 \\ u_3 \\ u_4 \end{Bmatrix} = \begin{Bmatrix} F_{1x} \\ F_{2x} \\ F_{3x} \\ F_{4x} \end{Bmatrix}$$

(2.6.28)

Equation (2.6.28) is identical to Eq. (2.5.18), which was obtained through the direct stiffness method as described in Section 2.4. Hence the assembled equations using the principle of minimum potential energy result in the same equations obtained by the direct stiffness assembly method. ∎

▲ Summary Equations

Definition of an element stiffness matrix:

$$\{f\} = [k]\{d\}$$

(2.1.1)

Definition of global or total stiffness matrix for a structure:

$$\{F\} = [K]\{d\}$$

(2.1.2)

Displacement function assumed for linear spring element:

$$u = a_1 + a_2 x$$

(2.2.2)

Shape functions for linear spring element:

$$N_1 = 1 - x/L \qquad N_2 = x/L \qquad (2.2.10)$$

Basic matrix equation relating nodal forces to nodal displacement for spring element:

$$\begin{Bmatrix} f_{1x} \\ f_{2x} \end{Bmatrix} = \begin{bmatrix} k & -k \\ -k & k \end{bmatrix} \begin{Bmatrix} u_1 \\ u_2 \end{Bmatrix} \qquad (2.2.17)$$

Stiffness matrix for linear spring element:

$$[k] = \begin{bmatrix} k & -k \\ -k & k \end{bmatrix} \qquad (2.2.18)$$

Global equations for a spring assemblage:

$$[F] = [K]\{d\} \qquad (2.2.20)$$

Total potential energy:

$$\pi_p = U + \Omega \qquad (2.6.1)$$

For a system of springs:

$$U = \frac{1}{2}\{d\}^T[K]\{d\} \qquad (2.6.20)$$

▲ References

[1] Turner, M. J., Clough, R. W., Martin, H. C., and Topp, L. J., "Stiffness and Deflection Analysis of Complex Structures," *Journal of the Aeronautical Sciences*, Vol. 23, No. 9, pp. 805–824, Sept. 1956.

[2] Martin, H. C., *Introduction to Matrix Methods of Structural Analysis*, McGraw-Hill, New York, 1966.

[3] Hsieh, Y. Y., *Elementary Theory of Structures*, 2nd ed., Prentice-Hall, Englewood Cliffs, NJ, 1982.

[4] Oden, J. T., and Ripperger, E. A., *Mechanics of Elastic Structures*, 2nd ed., McGraw-Hill, New York, 1981.

[5] Finlayson, B. A., *The Method of Weighted Residuals and Variational Principles*, Academic Press, New York, 1972.

[6] Zienkiewicz, O. C., *The Finite Element Method*, 3rd ed., McGraw-Hill, London, 1977.

[7] Cook, R. D., Malkus, D. S., Plesha, M. E., and Witt, R. J. *Concepts and Applications of Finite Element Analysis*, 4th ed., Wiley, New York, 2002.

[8] Forray, M. J., *Variational Calculus in Science and Engineering*, McGraw-Hill, New York, 1968.

▲ Problems

2.1 **a.** Obtain the global stiffness matrix $[K]$ of the assemblage shown in Figure P2–1 by superimposing the stiffness matrices of the individual springs. Here k_1, k_2, and k_3 are the stiffnesses of the springs as shown.

b. If nodes 1 and 2 are fixed and a force P acts on node 4 in the positive x direction, find an expression for the displacements of nodes 3 and 4.

c. Determine the reaction forces at nodes 1 and 2.

(*Hint:* Do this problem by writing the nodal equilibrium equations and then making use of the force/displacement relationships for each element as done in the first part of Section 2.4. Then solve the problem by the direct stiffness method.)

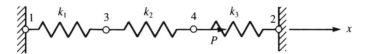

Figure P2–1

2.2 For the spring assemblage shown in Figure P2–2, determine the displacement at node 2 and the forces in each spring element. Also determine the force F_3. Given: Node 3 displaces an amount $\delta = 20$ mm. in the positive x direction because of the force F_3 and $k_1 = k_2 = 100$ N/mm.

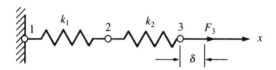

Figure P2–2

2.3 **a.** For the spring assemblage shown in Figure P2–3, obtain the global stiffness matrix by direct superposition.

b. If nodes 1 and 5 are fixed and a force P is applied at node 3, determine the nodal displacements.

c. Determine the reactions at the fixed nodes 1 and 5.

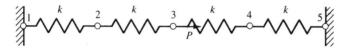

Figure P2–3

2.4 Solve Problem 2.3 with $P = 0$ (no force applied at node 3) and with node 5 given a fixed, known displacement of δ as shown in Figure P2–4.

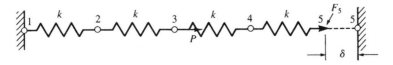

Figure P2–4

2.5 For the spring assemblage shown in Figure P2–5, obtain the global stiffness matrix by the direct stiffness method. Let $k^{(1)} = 200$ N/mm, $k^{(2)} = 400$ N/mm, $k^{(3)} = 600$ N/mm, $k^{(4)} = 800$ N/mm, and $k^{(5)} = 1000$ N/mm.

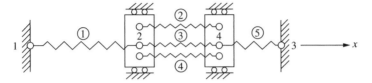

Figure P 2–5

2.6 For the spring assemblage in Figure P2–5, apply a concentrated force of 10,000 N at node 2 in the positive x direction and determine the displacements at nodes 2 and 4.

2.7 Instead of assuming a tension element as in Figure P2–3, now assume a compression element. That is, apply compressive forces to the spring element and derive the stiffness matrix.

2.8–2.16 For the spring assemblages shown in Figures P2–8 through P2–16, determine the nodal displacements, the forces in each element, and the reactions. Use the direct stiffness method for all problems.

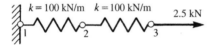

Figure P 2–8

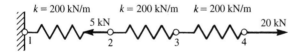

Figure P2–9

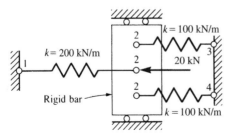

Figure P2–10

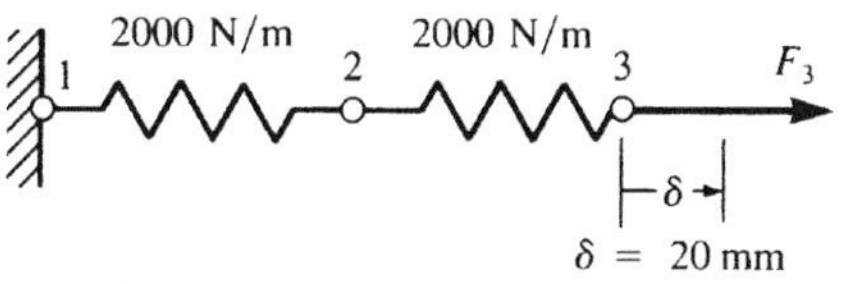

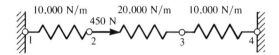

Figure P2–11

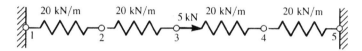

Figure P2–12

2000 N/m 2000 N/m
1 2 3 F_3

$\delta = 20$ mm

10,000 N/m 20,000 N/m 10,000 N/m
450 N
1 2 3 4

20 kN/m 20 kN/m 5 kN 20 kN/m 20 kN/m
1 2 3 4 5

Figure P2–13

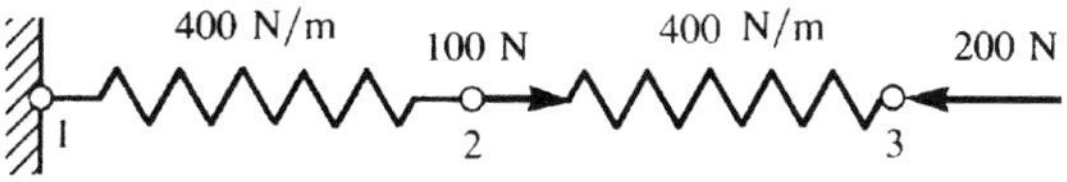

400 N/m 100 N 400 N/m 200 N
1 2 3

Figure P2–14

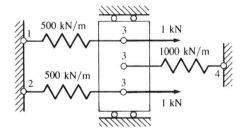

500 kN/m 3 1 kN
1
1000 kN/m
3
500 kN/m 4
2 3
1 kN

Figure P2–15

$k = 20$ kN/m $k = 20$ kN/m $k = 20$ kN/m
1 2 3 4
500 N 500 N

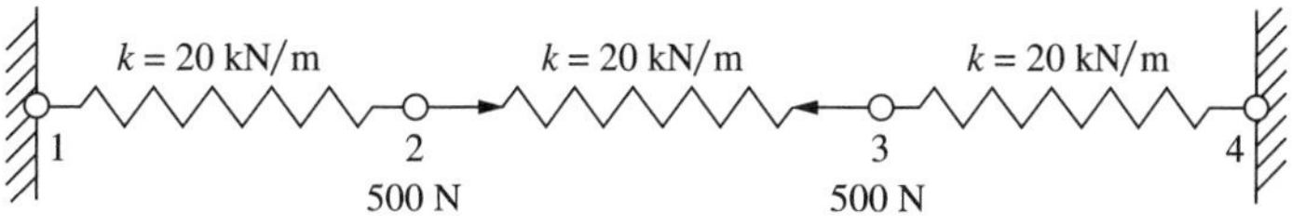

Figure P2–16

2.17 For the five-spring assemblage shown in Figure P2–17, determine the displacements at nodes 2 and 3 and the reactions at nodes 1 and 4. Assume the rigid vertical bars at nodes 2 and 3 connecting the springs remain horizontal at all times but are free to slide or displace left or right. There is an applied force at node 3 of 1000 N to the right.

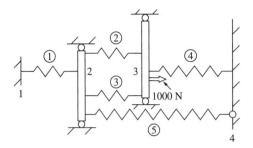

Figure P2–17

Let $k^{(1)} = 500$ N/mm, $k^{(2)} = k^{(3)} = 300$ N/mm, and $k^{(4)} = k^{(5)} = 400$ N/mm.

2.18 Use the principle of minimum potential energy developed in Section 2.6 to solve the spring problems shown in Figure P2–18. That is, plot the total potential energy for variations in the displacement of the free end of the spring to determine the minimum potential energy. Observe that the displacement that yields the minimum potential energy also yields the stable equilibrium position.

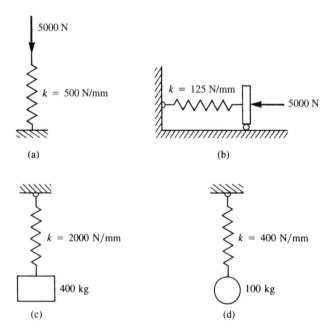

Figure P2–18

2.19 Reverse the direction of the load in Example 2.4 and recalculate the total potential energy. Then use this value to obtain the equilibrium value of displacement.

2.20 The nonlinear spring in Figure P2–20 has the force/deformation relationship $f = k\delta^2$. Express the total potential energy of the spring, and use this potential energy to obtain the equilibrium value of displacement.

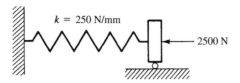

$k = 250$ N/mm

2500 N

Figure P2–20

2.21–2.22 Solve Problems 2.10 and 2.15 by the potential energy approach (see Example 2.5).

2.23 Resistor type elements are often used in electrical circuits. Consider the typical resistor element shown in Figure P2–23 with nodes 1 and 2. One form of Ohm's law says that the potential voltage difference across two points is equal to the current I through the conductor times the resistance R between the two points. In equation form, $V = IR$ where I denotes the current in units of amperes (amps) and V is the potential or voltage drop in units of volts (V) across the conductor of resistance R in units of ohms (Ω). Use the method in Section 2.2 to derive the "stiffness" matrix relating potential drop to current at the nodes shown as

$$\begin{Bmatrix} V_1 \\ V_2 \end{Bmatrix} = R \begin{bmatrix} 1 & -1 \\ -1 & 1 \end{bmatrix} \begin{Bmatrix} I_1 \\ I_2 \end{Bmatrix} \quad \text{or} \quad \{V\} = [K]\{I\}$$

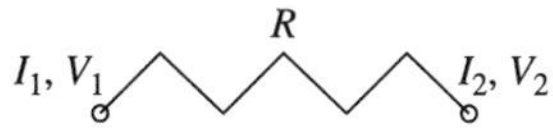

I_1, V_1 R I_2, V_2

Figure P2–23

Development of Truss Equations ▲

CHAPTER OBJECTIVES

- To derive the stiffness matrix for a bar element.
- To illustrate how to solve a bar assemblage by the direct stiffness method.
- To introduce guidelines for selecting displacement functions.
- To describe the concept of transformation of vectors in two different coordinate systems in the plane.
- To derive the stiffness matrix for a bar arbitrarily oriented in the plane.
- To demonstrate how to compute stress for a bar in the plane.
- To show how to solve a plane truss problem.
- To develop the transformation matrix in three-dimensional space and show how to use it to derive the stiffness matrix for a bar arbitrarily oriented in space.
- To demonstrate the solution of space trusses.
- To define symmetry and describe the use of symmetry to solve a problem.
- To introduce and solve problems with inclined supports.
- To derive the bar equations using the theorem of minimum potential energy.
- To compare the finite element solution to an exact solution for a bar.
- To introduce Galerkin's residual method to derive the bar element stiffness matrix and equations.
- To introduce other residual methods and their application to the one-dimensional bar.
- To create a flow chart of a finite element computer program for truss analysis and describe a step-by-step solution from a commercial program.

Introduction

Having set forth the foundation on which the direct stiffness method is based, we will now derive the stiffness matrix for a linear-elastic bar (or truss) element using the general steps outlined in Chapter 1. We will include the introduction of both a local coordinate system, chosen with the element in mind, and a global or reference coordinate system, chosen to be convenient (for numerical purposes) with respect to the overall structure. We will also discuss the transformation of a vector from the local coordinate system to the global coordinate system, using the concept of transformation matrices to express the stiffness matrix of an arbitrarily oriented bar element in terms of the global system. We will solve three example plane truss problems (see Figure 3–1 for a typical railroad trestle plane truss and a lift bridge truss over the Illinois River) to illustrate the procedure of establishing the total stiffness matrix and equations for solution of a structure.

Next we extend the stiffness method to include space trusses. We will develop the transformation matrix in three-dimensional space and analyze two space trusses. Then we describe the concept of symmetry and its use to reduce the size of a problem and facilitate its solution. We will use an example truss problem to illustrate the concept and then describe how to handle inclined, or skewed, supports.

We will then use the principle of minimum potential energy and apply it to rederive the bar element equations. We then compare a finite element solution to an exact solution for a bar subjected to a linear varying distributed load. We will introduce Galerkin's residual method and then apply it to derive the bar element equations. Finally, we will introduce other common residual methods—collocation, subdomain, and least squares—to merely expose you to them. We illustrate these methods by solving a problem of a bar subjected to a linear varying load.

▲ 3.1 Derivation of the Stiffness Matrix for a Bar Element in Local Coordinates ▲

We will now consider the derivation of the stiffness matrix for the linear-elastic, constant cross-sectional area (prismatic) bar element shown in Figure 3–2. The derivation here will be directly applicable to the solution of pin-connected trusses. The bar is subjected to tensile forces T directed along the local axis of the bar and applied at nodes 1 and 2.

The bar element is assumed to have constant cross-sectional area A, modulus of elasticity E, and initial length L. The nodal degrees of freedom are local axial displacements (longitudinal displacements directed along the length of the bar) represented by u_1 and u_2 at the ends of the element as shown in Figure 3–2.

From Hooke's law [Eq. (a)] and the strain/displacement relationship [Eq. (b) or Eq. (1.4.1)], we write

$$\sigma_x = E\varepsilon_x \tag{a}$$

$$\varepsilon_x = \frac{du}{dx} \tag{b}$$

From force equilibrium, we have

$$A\sigma_x = T = \text{constant} \tag{c}$$

for a bar with loads applied only at the ends. (We will consider distributed loading in Section 3.10.) Using Eq. (b) in Eq. (a) and then Eq. (a) in Eq. (c) and differentiating

(a)

(b)

Figure 3–1 (a) A typical railroad trestle plane truss (By Daryl L. Logan); (b) lift bridge truss over the Illinois River (By Daryl L. Logan)

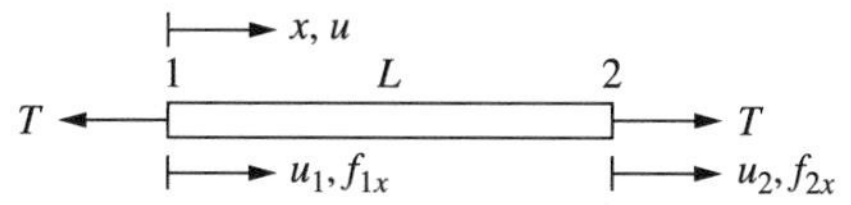

Figure 3–2 Bar subjected to tensile forces T; positive nodal displacements and forces are all in the local x direction

with respect to x, we obtain the differential equation governing the linear-elastic bar behavior as

$$\frac{d}{dx}\left(AE\frac{du}{dx}\right) = 0 \tag{d}$$

where u is the axial displacement function along the element in the x direction and A and E are written as though they were functions of x in the general form of the differential equation, even though A and E will be assumed constant over the whole length of the bar in our derivations to follow.

The following assumptions are used in deriving the bar element stiffness matrix:

1. The bar cannot sustain shear force or bending moment, that is,
   $f_{1y} = 0, f_{2y} = 0, m_1 = 0$ and $m_2 = 0$.
2. Any effect of transverse displacement is ignored.
3. Hooke's law applies; that is, axial stress σ_x is related to axial strain ε_x
   by $\sigma_x = E\varepsilon_x$.
4. No intermediate applied loads.

The steps previously outlined in Chapter 1 are now used to derive the stiffness matrix for the bar element and then to illustrate a complete solution for a bar assemblage.

Step 1 Select the Element Type

Represent the bar by labeling nodes at each end and in general by labeling the element number (Figure 3–2).

Step 2 Select a Displacement Function

Assume a linear displacement variation along the x axis of the bar because a linear function with specified endpoints has a unique path. These specified endpoints are the nodal values u_1 and u_2. (Further discussion regarding the choice of displacement functions is provided in Section 3.2 and References [1–3].) Then

$$u = a_1 + a_2 x \tag{3.1.1}$$

with the total number of coefficients a_i always equal to the total number of degrees of freedom associated with the element. Here the total number of degrees of freedom is two—axial displacements at each of the two nodes of the element. Using the same procedure as in Section 2.2 for the spring element, we express Eq. (3.1.1) as

$$u = \left(\frac{u_2 - u_1}{L}\right)x + u_1 \tag{3.1.2}$$

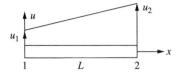

Figure 3–3 Displacement u plotted over the length of the element

The reason we convert the displacement function from the form of Eq. (3.1.1) to Eq. (3.1.2) is that it allows us to express the strain in terms of the nodal displacements using the strain/displacement relationship given by Eq. (3.1.5) and to then relate the nodal forces to the nodal displacements in step 4.

In matrix form, Eq. (3.1.2) becomes

$$u = [N_1 \quad N_2]\begin{Bmatrix} u_1 \\ u_2 \end{Bmatrix} \tag{3.1.3}$$

with shape functions given by

$$N_1 = 1 - \frac{x}{L} \qquad N_2 = \frac{x}{L} \tag{3.1.4}$$

These shape functions are identical to those obtained for the spring element in Section 2.2. The behavior of and some properties of these shape functions were described in Section 2.2. The linear displacement function u (Eq. (3.1.2)), plotted over the length of the bar element, is shown in Figure 3–3.

Step 3 Define the Strain/Displacement and Stress/Strain Relationships

The strain/displacement relationship is

$$\varepsilon_x = \frac{du}{dx} = \frac{u_2 - u_1}{L} \tag{3.1.5}$$

where Eqs. (3.1.3) and (3.1.4) have been used to obtain Eq. (3.1.5), and the stress/strain relationship is

$$\sigma_x = E\varepsilon_x \tag{3.1.6}$$

Step 4 Derive the Element Stiffness Matrix and Equations

The element stiffness matrix is derived as follows. From elementary mechanics, we have

$$T = A\sigma_x \tag{3.1.7}$$

Now, using Eqs. (3.1.5) and (3.1.6) in Eq. (3.1.7), we obtain

$$T = AE\left(\frac{u_2 - u_1}{L}\right) \tag{3.1.8}$$

Also, by the nodal force sign convention of Figure 3–2,

$$f_{1x} = -T \tag{3.1.9}$$

When we substitute Eq. (3.1.8), Eq. (3.1.9) becomes

$$f_{1x} = -\frac{AE}{L}(u_2 - u_1) \tag{3.1.10}$$

Similarly,

$$f_{2x} = T \tag{3.1.11}$$

or, by Eq. (3.1.8), Eq. (3.1.11) becomes

$$f_{2x} = \frac{AE}{L}(u_2 - u_1) \tag{3.1.12}$$

Expressing Eqs. (3.1.10) and (3.1.12) together in matrix form, we have

$$\begin{Bmatrix} f_{1x} \\ f_{2x} \end{Bmatrix} = \frac{AE}{L} \begin{bmatrix} 1 & -1 \\ -1 & 1 \end{bmatrix} \begin{Bmatrix} u_1 \\ u_2 \end{Bmatrix} \tag{3.1.13}$$

Now, because $\{f\} = [k]\{d\}$, we have, from Eq. (3.1.13),

$$[k] = \frac{AE}{L} \begin{bmatrix} 1 & -1 \\ -1 & 1 \end{bmatrix} \tag{3.1.14}$$

Equation (3.1.14) represents the stiffness matrix for a bar element in local coordinates. In Eq. (3.1.14), AE/L for a bar element is analogous to the spring constant k for a spring element.

Step 5 Assemble Element Equations to Obtain Global or Total Equations

Assemble the global stiffness and force matrices and global equations using the direct stiffness method described in Chapter 2 (see Section 3.6 for an example truss). This step applies for structures composed of more than one element such that (again)

$$[K] = \sum_{e=1}^{N} [k^{(e)}] \qquad \text{and} \qquad \{F\} = \sum_{e=1}^{N} \{f^{(e)}\} \tag{3.1.15}$$

where now all local element stiffness matrices $[k^{(e)}]$ must be transformed to global element stiffness matrices $[k]$ (unless the local axes coincide with the global axes) before the direct stiffness method is applied as indicated by Eq. (3.1.15). (This concept of coordinate and stiffness matrix transformations is described in Sections 3.3 and 3.4.)

Step 6 Solve for the Nodal Displacements

Determine the displacements by imposing boundary conditions and simultaneously solving a system of equations, $\{F\} = [K]\{d\}$.

Step 7 Solve for the Element Forces

Finally, determine the strains and stresses in each element by back-substitution of the displacements into equations similar to Eqs. (3.1.5) and (3.1.6).

We will now illustrate a solution for a one-dimensional bar problem.

Example 3.1

For the three-bar assemblage shown in Figure 3–4, determine (a) the global stiffness matrix, (b) the displacements of nodes 2 and 3, and (c) the reactions at nodes 1 and 4. A force of 15,000 N is applied in the x direction at node 2. The length of each element is 0.6 m. Let $E = 2.0 \times 10^{11}$ Pa and $A = 6 \times 10^{-4}$ m^2 for elements 1 and 2, and let $E = 1 \times 10^{11}$ Pa and $A = 12 \times 10^{-4}$ m^2 for element 3. Nodes 1 and 4 are fixed.

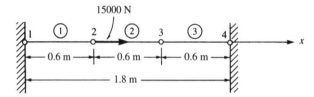

Figure 3–4 Three-bar assemblage

SOLUTION:

(a) Using Eq. (3.1.14), we find that the element stiffness matrices are

$$[k^{(1)}] = [k^{(2)}] = \frac{(6 \times 10^{-4})(2 \times 10^{11})}{0.6} \begin{bmatrix} 1 & -1 \\ -1 & 1 \end{bmatrix} = 2 \times 10^8 \begin{matrix} 1 \\ 2 \end{matrix} \begin{matrix} 2^{(1)} \\ 3^{(2)} \end{matrix} \begin{bmatrix} 1 & -1 \\ -1 & 1 \end{bmatrix} \frac{\text{N}}{\text{m}} \quad (3.1.16)$$

$$[k^{(3)}] = \frac{(12 \times 10^{-4})(1 \times 10^{11})}{0.6} \begin{bmatrix} 1 & -1 \\ -1 & 1 \end{bmatrix} = 2 \times 10^8 \begin{matrix} 3 \\ \end{matrix} \begin{matrix} 4 \\ \end{matrix} \begin{bmatrix} 1 & -1 \\ -1 & 1 \end{bmatrix} \frac{\text{N}}{\text{m}}$$

where, again, the numbers above the matrices in Eqs. (3.1.16) indicate the displacements associated with each matrix. Assembling the element stiffness matrices by the direct stiffness method, we obtain the global stiffness matrix as

$$[K] = 2 \times 10^8 \begin{matrix} u_1 & u_2 & u_3 & u_4 \\ \end{matrix} \begin{bmatrix} 1 & -1 & 0 & 0 \\ -1 & 1+1 & -1 & 0 \\ 0 & -1 & 1+1 & -1 \\ 0 & 0 & -1 & 1 \end{bmatrix} \frac{\text{N}}{\text{m}} \quad (3.1.17)$$

(b) Equation (3.1.17) relates global nodal forces to global nodal displacements as follows:

$$\begin{Bmatrix} F_{1x} \\ F_{2x} \\ F_{3x} \\ F_{4x} \end{Bmatrix} = 2 \times 10^8 \begin{bmatrix} 1 & -1 & 0 & 0 \\ -1 & 2 & -1 & 0 \\ 0 & -1 & 2 & -1 \\ 0 & 0 & -1 & 1 \end{bmatrix} \begin{Bmatrix} u_1 \\ u_2 \\ u_3 \\ u_4 \end{Bmatrix} \qquad (3.1.18)$$

Invoking the boundary conditions, we have

$$u_1 = 0 \qquad u_4 = 0 \qquad\qquad (3.1.19)$$

Using the boundary conditions, substituting known applied global forces into Eq. (3.1.18), and partitioning equations 1 and 4 of Eq. (3.1.18), we solve equations 2 and 3 of Eq. (3.1.18) to obtain

$$\begin{Bmatrix} 15000 \\ 0 \end{Bmatrix} = 2 \times 10^8 \begin{bmatrix} 2 & -1 \\ -1 & 2 \end{bmatrix} \begin{Bmatrix} u_2 \\ u_3 \end{Bmatrix} \qquad (3.1.20)$$

Solving Eq. (3.1.20) simultaneously for the displacements yields

$$u_2 = 5 \times 10^{-5} \text{ m} = 0.05 \text{ mm} \qquad u_3 = 2.5 \times 10^{-5} \text{ m} = 0.025 \text{ mm} \quad (3.1.21)$$

(c) Back-substituting Eqs. (3.1.19) and (3.1.21) into Eq. (3.1.18), we obtain the global nodal forces, which include the reactions at nodes 1 and 4, as follows:

$$F_{1x} = 2 \times 10^8 (u_1 - u_2) = 2 \times 10^8 (0 - 5 \times 10^{-5}) = -10{,}000 \text{ N}$$

$$F_{2x} = 2 \times 10^8 (-u_1 + 2u_2 - u_3) = 2 \times 10^8 [0 + 2(5 \times 10^{-5}) - 2.5 \times 10^{-5}] = 15{,}000 \text{ N}$$

$$F_{3x} = 2 \times 10^8 (-u_2 + 2u_3 - u_4) = 2 \times 10^8 [-5 \times 10^{-5} + 2(2.5 \times 10^{-5}) - 0] = 0$$

$$F_{4x} = 2 \times 10^8 (-u_3 + u_4) = 2 \times 10^8 (-2.5 \times 10^{-5} + 0) = -5000 \text{ N} \qquad (3.1.22)$$

The results of Eqs. (3.1.22) show that the sum of the reactions F_{1x} and F_{4x} is equal in magnitude but opposite in direction to the applied nodal force of 15,000 N at node 2. Equilibrium of the bar assemblage is thus verified. Furthermore, Eqs. (3.1.22) show that $F_{2x} = 15{,}000$ N and $F_{3x} = 0$ are merely the applied nodal forces at nodes 2 and 3, respectively, which further enhances the validity of our solution. ■

▲ 3.2 Selecting Approximation Functions for Displacements ▲

Consider the following guidelines, as they relate to the one-dimensional bar element, when selecting a displacement function. (Further discussion regarding selection of displacement functions and other kinds of approximation functions (such as

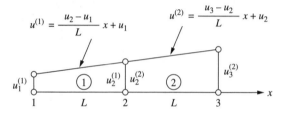

Figure 3–5 Interelement continuity of a two-bar structure

temperature functions) will be provided in Chapter 4 for the beam element, in Chapter 6 for the constant-strain triangular element, in Chapter 8 for the linear-strain triangular element, in Chapter 9 for the axisymmetric element, in Chapter 10 for the three-noded bar element and the rectangular plane element, in Chapter 11 for the three-dimensional stress element, in Chapter 12 for the plate bending element, and in Chapter 13 for the heat transfer problem. More information is also provided in References [1–3].

1. Common approximation functions are usually polynomials such as the simplest one that gives the linear variation of displacement given by Eq. (3.1.1) or equivalently by Eq. (3.1.3), where the function is expressed in terms of the shape functions.

2. The approximation function should be continuous within the bar element. The simple linear function for u of Eq. (3.1.1) certainly is continuous within the element. Therefore, the linear function yields continuous values of u within the element and prevents openings, overlaps, and jumps because of the continuous and smooth variation in u (Figure 3–5).

3. The approximating function should provide interelement continuity for all degrees of freedom at each node for discrete line elements and along common boundary lines and surfaces for two- and three-dimensional elements. For the bar element, we must ensure that nodes common to two or more elements remain common to these elements upon deformation and thus prevent overlaps or voids between elements. For example, consider the two-bar structure shown in Figure 3–5. For the two-bar structure, the linear function for u [Eq. (3.1.2)] within each element will ensure that elements 1 and 2 remain connected; the displacement at node 2 for element 1 will equal the displacement at the same node 2 for element 2; that is, $u_2^{(1)} = u_2^{(2)}$. This rule was also illustrated by Eq. (2.3.3). The linear function is then called a *conforming*, or *compatible, function* for the bar element because it ensures the satisfaction both of continuity between adjacent elements and of continuity within the element.

   In general, the symbol C^m is used to describe the continuity of a piecewise field (such as axial displacement), where the superscript m indicates the degree of derivative that is interelement continuous. A field is then C^0 continuous if the function itself is interelement continuous. For instance, for the field variable being the axial

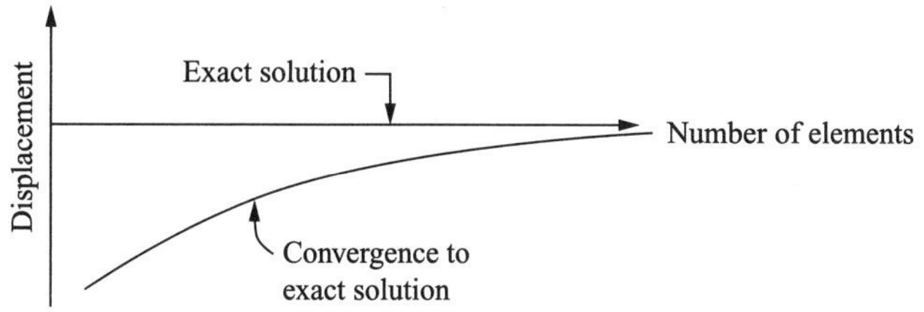

Figure 3–6 Convergence to the exact solution for displacement as the number of elements of a finite element solution is increased

displacement illustrated in Figure 3–5, the displacement is continuous across the common node 2. Hence the displacement field is said to be C^0 continuous. Bar elements, plane elements (see Chapter 7), and solid elements (Chapter 11) are C^0 elements in that they enforce displacement continuity across the common boundaries.

If the function has both its field variable and its first derivative continuous across the common boundary, then the field variable is said to be C^1 continuous. We will later see that the beam (see Chapter 4) and plate (see Chapter 12) elements are C^1 continuous. That is, they enforce both displacement and slope continuity across common boundaries.

4. The approximation function should allow for rigid-body displacement and for a state of constant strain within the element. The one-dimensional displacement function [Eq. (3.1.1)] satisfies these criteria because the a_1 term allows for rigid-body motion (constant motion of the body without straining) and the $a_2 x$ term allows for constant strain because $\varepsilon_x = du/dx = a_2$ is a constant. (This state of constant strain in the element can, in fact, occur if elements are chosen small enough.) The simple polynomial Eq. (3.1.1) satisfying this fourth guideline is then said to be *complete* for the bar element.

This idea of completeness also means in general that the lower-order term cannot be omitted in favor of the higher-order term. For the simple linear function, this means a_1 cannot be omitted while keeping $a_2 x$. Completeness of a function is a necessary condition for convergence to the exact answer, for instance, for displacements and stresses (Figure 3–6) (see Reference [3]). Figure 3–6 illustrates monotonic convergence toward an exact solution for displacement as the number of elements in a finite element solution is increased. Monotonic convergence is then the process in which successive approximation solutions (finite element solutions) approach the exact solution consistently without changing sign or direction.

The idea that the interpolation (approximation) function must allow for a rigid-body displacement means that the function must be capable of yielding a constant value (say, a_1), because such a value can, in fact, occur. Therefore, we must consider the case

$$u = a_1 \tag{3.2.1}$$

For $u = a_1$ requires nodal displacements $u_1 = u_2$ to obtain a rigid-body displacement. Therefore

$$a_1 = u_1 = u_2 \tag{3.2.2}$$

Using Eq. (3.2.2) in Eq. (3.1.3), we have

$$u = N_1 u_1 + N_2 u_2 = (N_1 + N_2)a_1 \tag{3.2.3}$$

From Eqs. (3.2.1) and (3.2.3), we then have

$$u = a_1 = (N_1 + N_2)a_1 \tag{3.2.4}$$

Therefore, by Eq. (3.2.4), we obtain

$$N_1 + N_2 = 1 \tag{3.2.5}$$

Thus Eq. (3.2.5) shows that the displacement interpolation functions must add to unity at every point within the element so that u will yield a constant value when a rigid-body displacement occurs.

▲ 3.3 Transformation of Vectors in Two Dimensions ▲

In many problems it is convenient to introduce both local $(x' - y')$ and global (or reference) $(x - y)$ coordinates. Local coordinates are always chosen to represent the individual element conveniently. Global coordinates are chosen to be convenient for the whole structure.

Given the nodal displacement of an element, represented by the vector **d** in Figure 3–7, we want to relate the components of this vector in one coordinate system to components in another. For general purposes, we will assume in this section that **d** is not coincident with either the local or the global axis. In this case, we want to relate global displacement components to local ones. In so doing, we will develop a transformation matrix that will subsequently be used to develop the global stiffness matrix for a bar element. We define the angle θ to be positive when measured counterclockwise from x to x'. We can express vector displacement **d** in both global and

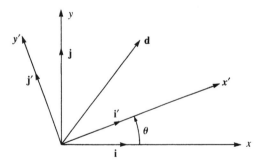

Figure 3–7 General displacement vector **d**

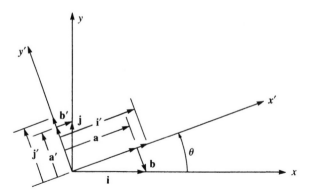

Figure 3–8 Relationship between local and global unit vectors

local coordinates by

$$\mathbf{d} = u_1\mathbf{i} + v_1\mathbf{j} = u_1'\mathbf{i}' + v_1'\mathbf{j}' \tag{3.3.1}$$

where $\mathbf{i}$ and $\mathbf{j}$ are unit vectors in the x and y global directions and $\mathbf{i}'$ and $\mathbf{j}'$ are unit vectors in the x' and y' local directions. We will now relate $\mathbf{i}$ and $\mathbf{j}$ to $\mathbf{i}'$ and $\mathbf{j}'$ through use of Figure 3–8.

Using Figure 3–8 and vector addition, we obtain

$$\mathbf{a} + \mathbf{b} = \mathbf{i} \tag{3.3.2}$$

Also, from the law of cosines,

$$|\mathbf{a}| = |\mathbf{i}| \cos \theta \tag{3.3.3}$$

and because $\mathbf{i}$ is, by definition, a unit vector, its magnitude is given by

$$|\mathbf{i}| = 1 \tag{3.3.4}$$

Therefore, we obtain

$$|\mathbf{a}| = 1 \cos \theta \tag{3.3.5}$$

Similarly,

$$|\mathbf{b}| = 1 \sin \theta \tag{3.3.6}$$

Now $\mathbf{a}$ is in the $\mathbf{i}'$ direction and $\mathbf{b}$ is in the $-\mathbf{j}'$ direction. Therefore,

$$\mathbf{a} = |\mathbf{a}|\mathbf{i}' = (\cos \theta)\mathbf{i}' \tag{3.3.7}$$

and

$$\mathbf{b} = |\mathbf{b}|(-\mathbf{j}') = (\sin \theta)(-\mathbf{j}') \tag{3.3.8}$$

Using Eqs. (3.3.7) and (3.3.8) in Eq. (3.3.2) yields

$$\mathbf{i} = \cos \theta \mathbf{i}' - \sin \theta \mathbf{j}' \tag{3.3.9}$$

Similarly, from Figure 3–8, we obtain the vector equation

$$\mathbf{a}' + \mathbf{b}' = \mathbf{j} \tag{3.3.10}$$

where

$$\mathbf{a}' = \cos \theta \mathbf{j}' \tag{3.3.11}$$

$$\mathbf{b}' = \sin \theta \mathbf{i}' \tag{3.3.12}$$

Using Eqs. (3.3.11) and (3.3.12) in Eq. (3.3.10), we have

$$\mathbf{j} = \sin\theta\mathbf{i}' + \cos\theta\mathbf{j}' \tag{3.3.13}$$

Now, using Eqs. (3.3.9) and (3.3.13) in Eq. (3.3.1), we have

$$u(\cos\theta\mathbf{i}' - \sin\theta\mathbf{j}') + v(\sin\theta\mathbf{i}' + \cos\theta\mathbf{j}') = u'\mathbf{i}' + v'\mathbf{j}' \tag{3.3.14}$$

Combining like coefficients of $\mathbf{i}'$ and $\mathbf{j}'$ in Eq. (3.3.14), we obtain

$$u\cos\theta + v\sin\theta = u'$$

and

$$-u\sin\theta + v\cos\theta = v' \tag{3.3.15}$$

In matrix form, Eqs. (3.3.15) are written as

$$\left\{\begin{matrix} u' \\ v' \end{matrix}\right\} = \begin{bmatrix} C & S \\ -S & C \end{bmatrix} \left\{\begin{matrix} u \\ v \end{matrix}\right\} \tag{3.3.16}$$

where $C = \cos\theta$ and $S = \sin\theta$.

Equation (3.3.16) relates the global displacement matrix $\{d\}$ to the local displacement $\{d'\}$ as

$$\{d'\} = [T]\{d\} \tag{3.3.17}$$

where

$$\{d\} = \left\{\begin{matrix} u \\ v \end{matrix}\right\}, \quad \{d'\} = \left\{\begin{matrix} u' \\ v' \end{matrix}\right\}, \quad [T] = \begin{bmatrix} C & S \\ -S & C \end{bmatrix} \tag{3.3.18}$$

The matrix $[T]$ is called the *transformation* (or *rotation*) *matrix*. For an additional description of this matrix, see Appendix A. It will be used in Section 3.4 to develop the global stiffness matrix for an arbitrarily oriented bar element and to transform global nodal displacements and forces to local ones.

Now, for the case of $v' = 0$, we have, from Eq. (3.3.1),

$$u\mathbf{i} + v\mathbf{j} = u'\mathbf{i}' \tag{3.3.19}$$

Figure 3–9 shows u' expressed in terms of global x and y components. Using trigonometry and Figure 3–9, we then obtain the magnitude of u' as

$$u' = Cu + Sv \tag{3.3.20}$$

Equation (3.3.20) is equivalent to equation 1 of Eq. (3.3.16).

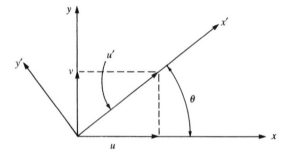

Figure 3–9 Relationship between local and global displacements

Example 3.2

The global nodal displacements at node 2 have been determined to be $u_2 = 2.5$ mm and $v_2 = 5$ mm for the bar element shown in Figure 3–10. Determine the local x displacement at node 2.

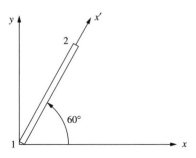

Figure 3–10   Bar element with local axis x' acting along the element

SOLUTION:

Using Eq. (3.3.20), we obtain

$$u_2' = (\cos 60^\circ)(2.5) + (\sin 60^\circ)(5) = 5.58 \text{ mm}$$ ∎

▲ **3.4 Global Stiffness Matrix for Bar Arbitrarily Oriented in the Plane** ▲

We now consider a bar inclined at an angle θ from the global x axis identified by the local axis x' directed from node 1 to node 2 along the direction of the bar, as shown in Figure 3–11. Here positive angle θ is taken counterclockwise from x to x'.

We now use Eq. (3.1.13) where a prime notation is used to denote the local element stiffness matrix $\{k'\}$ which relates the local coordinate nodal forces $\{f'\}$ to local nodal displacements $\{d'\}$ as shown by Eq. (3.4.1).

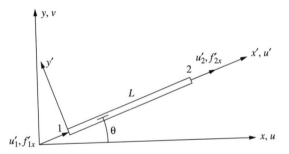

Figure 3–11   Bar element arbitrarily oriented in the global x–y plane

$$\begin{Bmatrix} f'_{1x} \\ f'_{2x} \end{Bmatrix} = \frac{AE}{L} \begin{bmatrix} 1 & -1 \\ -1 & 1 \end{bmatrix} \begin{Bmatrix} u'_1 \\ u'_2 \end{Bmatrix} \qquad (3.4.1)$$

or
$$\{f'\} = [k']\{d'\} \qquad (3.4.2)$$

We now want to relate the global element nodal forces $\{f\}$ to the global nodal displacements $\{d\}$ for a bar element arbitrarily oriented with respect to the global axes as shown in Figure 3–11. This relationship will yield the global stiffness matrix $[k]$ of the element. That is, we want to find a matrix $[k]$ such that

$$\begin{Bmatrix} f_{1x} \\ f_{1y} \\ f_{2x} \\ f_{2y} \end{Bmatrix} = [k] \begin{Bmatrix} u_1 \\ v_1 \\ u_2 \\ v_2 \end{Bmatrix} \qquad (3.4.3)$$

or, in simplified matrix form, Eq. (3.4.3) becomes

$$\{f\} = [k]\{d\} \qquad (3.4.4)$$

We observe from Eq. (3.4.3) that a total of four components of force and four of displacement arise when global coordinates are used. However, a total of two components of force and two of displacement appear for the local-coordinate representation of a spring or a bar, as shown by Eq. (3.4.1). By using relationships between local and global force components and between local and global displacement components, we will be able to obtain the global stiffness matrix. We know from transformation relationship Eq. (3.3.15) that

$$u'_1 = u_1 \cos \theta + v_1 \sin \theta$$
$$u'_2 = u_2 \cos \theta + v_2 \sin \theta \qquad (3.4.5)$$

In matrix form, Eqs. (3.4.5) can be written as

$$\begin{Bmatrix} u'_1 \\ u'_2 \end{Bmatrix} = \begin{bmatrix} C & S & 0 & 0 \\ 0 & 0 & C & S \end{bmatrix} \begin{Bmatrix} u_1 \\ v_1 \\ u_2 \\ v_2 \end{Bmatrix} \qquad (3.4.6)$$

or as
$$\{d'\} = [T^*]\{d\} \qquad (3.4.7)$$

where
$$[T^*] = \begin{bmatrix} C & S & 0 & 0 \\ 0 & 0 & C & S \end{bmatrix} \qquad (3.4.8)$$

Similarly, because forces transform in the same manner as displacements, we replace local and global displacements in Eq. (3.4.6) with local and global forces and obtain

$$\begin{Bmatrix} f'_{1x} \\ f'_{2x} \end{Bmatrix} = \begin{bmatrix} C & S & 0 & 0 \\ 0 & 0 & C & S \end{bmatrix} \begin{Bmatrix} f_{1x} \\ f_{1y} \\ f_{2x} \\ f_{2y} \end{Bmatrix} \qquad (3.4.9)$$

Similar to Eq. (3.4.7), we can write Eq. (3.4.9) as

$$\{f'\} = [T^*]\{f\} \tag{3.4.10}$$

Now, substituting Eq. (3.4.7) into Eq. (3.4.2), we obtain

$$\{f'\} = [k'][T^*]\{d\} \tag{3.4.11}$$

and using Eq. (3.4.10) in Eq. (3.4.11) yields

$$[T^*]\{f\} = [k'][T^*]\{d\} \tag{3.4.12}$$

However, to write the final expression relating global nodal forces to global nodal displacements for an element, we must invert $[T^*]$ in Eq. (3.4.12). This is not immediately possible because $[T^*]$ is not a square matrix. Therefore, we must expand $\{d'\}, \{f'\}$, and $[k']$ to the order that is consistent with the use of global coordinates even though f'_{1y} and v'_{2y} are zero. Using Eq. (3.3.16) for each nodal displacement, we thus obtain

$$\begin{Bmatrix} u'_1 \\ v'_1 \\ u'_2 \\ v'_2 \end{Bmatrix} = \begin{bmatrix} C & S & 0 & 0 \\ -S & C & 0 & 0 \\ 0 & 0 & C & S \\ 0 & 0 & -S & C \end{bmatrix} \begin{Bmatrix} u_1 \\ v_1 \\ u_2 \\ v_2 \end{Bmatrix} \tag{3.4.13}$$

or

$$\{d'\} = [T]\{d\} \tag{3.4.14}$$

where

$$[T] = \begin{bmatrix} C & S & 0 & 0 \\ -S & C & 0 & 0 \\ 0 & 0 & C & S \\ 0 & 0 & -S & C \end{bmatrix} \tag{3.4.15}$$

Similarly, we can write

$$\{f'\} = [T]\{f\} \tag{3.4.16}$$

because forces are like displacements—both are vectors. Also, $[k']$ must be expanded to a 4×4 matrix. Therefore, Eq. (3.4.1) in expanded form becomes

$$\begin{Bmatrix} f'_{1x} \\ f'_{1y} \\ f'_{2x} \\ f'_{2y} \end{Bmatrix} = \frac{AE}{L} \begin{bmatrix} 1 & 0 & -1 & 0 \\ 0 & 0 & 0 & 0 \\ -1 & 0 & 1 & 0 \\ 0 & 0 & 0 & 0 \end{bmatrix} \begin{Bmatrix} u'_1 \\ v'_1 \\ u'_2 \\ v'_2 \end{Bmatrix} \tag{3.4.17}$$

In Eq. (3.4.17), because f'_{1y} and f'_{2y} are zero, rows of zeros corresponding to the row numbers f'_{1y} and f'_{2y} appear in $[k']$. Now, using Eqs. (3.4.14) and (3.4.16) in Eq. (3.4.2), we obtain

$$[T]\{f\} = [k'][T]\{d\} \tag{3.4.18}$$

Equation (3.4.18) is Eq. (3.4.12) expanded. Premultiplying both sides of Eq. (3.4.18) by $[T]^{-1}$, we have

$$\{f\} = [T]^{-1}[k'][T]\{d\} \tag{3.4.19}$$

where $[T]^{-1}$ is the *inverse* of $[T]$. However, it can be shown (see Problem 3.28) that

$$[T]^{-1} = [T]^T \tag{3.4.20}$$

where $[T]^T$ is the *transpose* of $[T]$. The property of square matrices such as $[T]$ given by Eq. (3.4.20) defines $[T]$ to be an orthogonal matrix. For more about orthogonal matrices, see Appendix A. The transformation matrix $[T]$ between rectangular coordinate frames is orthogonal. This property of $[T]$ is used throughout this text. Substituting Eq. (3.4.20) into Eq. (3.4.19), we obtain

$$\{f\} = [T]^T[k'][T]\{d\} \tag{3.4.21}$$

Equating Eqs. (3.4.4) and (3.4.21), we obtain the global stiffness matrix for an element as

$$[k] = [T]^T[k'][T] \tag{3.4.22}$$

Substituting Eq. (3.4.15) for $[T]$ and the expanded form of $[k']$ given in Eq. (3.4.17) into Eq. (3.4.22), we obtain $[k]$ given in explicit form by

$$[k] = \frac{AE}{L} \begin{bmatrix} C^2 & CS & -C^2 & -CS \\ & S^2 & -CS & -S^2 \\ & & C^2 & CS \\ \text{Symmetry} & & & S^2 \end{bmatrix} \tag{3.4.23}$$

Equation (3.4.23) is the explicit stiffness matrix for a bar arbitrarily oriented in the x–y plane.

Now, because the trial displacement function Eq. (3.1.1) was assumed piecewise-continuous element by element, the stiffness matrix for each element can be summed by using the direct stiffness method to obtain

$$\sum_{e=1}^{N} [k^{(e)}] = [K] \tag{3.4.24}$$

where $[K]$ is the total stiffness matrix and N is the total number of elements. Similarly, each element global nodal force matrix can be summed such that

$$\sum_{e=1}^{N} \{f^{(e)}\} = \{F\} \tag{3.4.25}$$

$[K]$ now relates the global nodal forces $\{F\}$ to the global nodal displacements $\{d\}$ for the whole structure by

$$\{F\} = [K]\{d\} \tag{3.4.26}$$

Example 3.3

For the bar element shown in Figure 3–12, evaluate the global stiffness matrix with respect to the x-y coordinate system. Let the bar's cross-sectional area equal 6×10^{-4} m^2, length equal 1.2 m, and modulus of elasticity equal 2×10^{11} Pa. The angle the bar makes with the x axis is 30°.

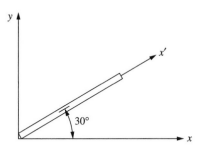

Figure 3–12 Bar element for stiffness matrix evaluation

SOLUTION:

To evaluate the global stiffness matrix $[k]$ for a bar, we use Eq. (3.4.23) with angle θ defined to be positive when measured counterclockwise from x to x'. Therefore,

$$\theta = 30° \qquad C = \cos 30° = \frac{\sqrt{3}}{2} \qquad S = \sin 30° = \frac{1}{2}$$

$$[k] = \frac{(6 \times 10^{-4})(2 \times 10^{11})}{1.2} \begin{bmatrix} \dfrac{3}{4} & \dfrac{\sqrt{3}}{4} & \dfrac{-3}{4} & \dfrac{-\sqrt{3}}{4} \\ & \dfrac{1}{4} & \dfrac{-\sqrt{3}}{4} & \dfrac{-1}{4} \\ & & \dfrac{3}{4} & \dfrac{\sqrt{3}}{4} \\ \text{Symmetry} & & & \dfrac{1}{4} \end{bmatrix} \dfrac{\text{N}}{\text{m}} \qquad (3.4.27)$$

Simplifying Eq. (3.4.27), we have

$$[k] = 10^8 \begin{bmatrix} 0.75 & 0.433 & -0.75 & -0.433 \\ & 0.25 & -0.433 & -0.25 \\ & & 0.75 & 0.433 \\ \text{Symmetry} & & & 0.25 \end{bmatrix} \dfrac{\text{N}}{\text{m}} \qquad (3.4.28)$$

■

▲ 3.5 Computation of Stress for a Bar in the *x-y* Plane ▲

We will now consider the determination of the stress in a bar element. For a bar, the local forces are related to the local displacements by Eq. (3.4.1) or Eq. (3.4.17). This equation is repeated here for convenience.

$$\begin{Bmatrix} f'_{1x} \\ f'_{2x} \end{Bmatrix} = \frac{AE}{L} \begin{bmatrix} 1 & -1 \\ -1 & 1 \end{bmatrix} \begin{Bmatrix} u'_1 \\ u_{2'} \end{Bmatrix} \tag{3.5.1}$$

The usual definition of axial tensile stress is axial force divided by cross-sectional area. Therefore, axial stress is

$$\sigma = \frac{f'_{2x}}{A} \tag{3.5.2}$$

where f'_{2x} is used because it is the axial force that pulls on the bar as shown in Figure 3–13. By Eq. (3.5.1),

$$f'_{2x} = \frac{AE}{L}[-1 \quad 1]\begin{Bmatrix} u'_1 \\ u'_2 \end{Bmatrix} \tag{3.5.3}$$

Therefore, combining Eqs. (3.5.2) and (3.5.3) yields

$$\{\sigma\} = \frac{E}{L}[-1 \quad 1]\{d'\} \tag{3.5.4}$$

Now, using Eq. (3.4.7), we obtain

$$\{\sigma\} = \frac{E}{L}[-1 \quad 1][T^*]\{d\} \tag{3.5.5}$$

Equation (3.5.5) can be expressed in simpler form as

$$\{\sigma\} = [C']\{d\} \tag{3.5.6}$$

where, when we use Eq. (3.4.8) for $[T^*]$,

$$[C'] = \frac{E}{L}[-1 \quad 1]\begin{bmatrix} C & S & 0 & 0 \\ 0 & 0 & C & S \end{bmatrix} \tag{3.5.7}$$

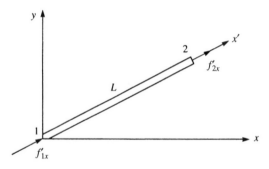

Figure 3–13 Basic bar element with positive nodal forces

After multiplying the matrices in Eq. (3.5.7), we have

$$[C'] = \frac{E}{L}[-C \quad -S \quad C \quad S] \tag{3.5.8}$$

Example 3.4

For the bar shown in Figure 3–14, determine the axial stress. Let $A = 4 \times 10^{-4}$ m^2, $E = 210$ GPa, and $L = 2$ m, and let the angle between x and x' be 60°. Assume the global displacements have been previously determined to be $u_1 = 0.25$ mm, $v_1 = 0.0$, $u_2 = 0.50$ mm, and $v_2 = 0.75$ mm.

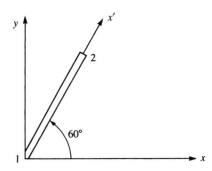

Figure 3–14   Bar element for stress evaluation

SOLUTION:

We can use Eq. (3.5.6) to evaluate the axial stress. Therefore, we first calculate $[C']$ from Eq. (3.5.8) as

$$[C'] = \frac{210 \times 10^6 \text{ kN/m}^2}{2 \text{ m}} \begin{bmatrix} -\frac{1}{2} & -\frac{\sqrt{3}}{2} & \frac{1}{2} & \frac{\sqrt{3}}{2} \end{bmatrix} \tag{3.5.9}$$

where we have used $C = \cos 60° = \frac{1}{2}$ and $S = \sin 60° = \sqrt{3}/2$ in Eq. (3.5.9). Now $\{d\}$ is given by

$$\{d\} = \begin{Bmatrix} u_1 \\ v_1 \\ u_2 \\ v_2 \end{Bmatrix} = \begin{Bmatrix} 0.25 \times 10^{-3} \text{ m} \\ 0.0 \\ 0.50 \times 10^{-3} \text{ m} \\ 0.75 \times 10^{-3} \text{ m} \end{Bmatrix} \tag{3.5.10}$$

Using Eqs. (3.5.9) and (3.5.10) in Eq. (3.5.6), we obtain the bar axial stress as

$$\sigma_x = \frac{210 \times 10^6}{2} \begin{bmatrix} -\frac{1}{2} & -\frac{\sqrt{3}}{2} & \frac{1}{2} & \frac{\sqrt{3}}{2} \end{bmatrix} \begin{Bmatrix} 0.25 \\ 0.0 \\ 0.50 \\ 0.75 \end{Bmatrix} \times 10^{-3}$$

$$= 81.32 \times 10^3 \text{ kN/m}^2 = 81.32 \text{ MPa} \qquad\blacksquare$$

▲ **3.6 Solution of a Plane Truss** ▲

We will now illustrate the use of equations developed in Sections 3.4 and 3.5, along with the direct stiffness method of assembling the total stiffness matrix and equations, to solve the following plane truss example problems. *A* **plane truss** *is a structure composed of bar elements that all lie in a common plane and are connected by frictionless pins.* The plane truss also must have loads acting only in the common plane and all loads must be applied at the nodes or joints.

Example 3.5

For the plane truss composed of the three elements shown in Figure 3–15 subjected to a downward force of 50 kN applied at node 1, determine the x and y displacements at node 1 and the stresses in each element. Let $E = 2 \times 10^{11}$ Pa and $A = 6 \times 10^{-4}$ m^2 for all elements. The lengths of the elements are shown in the figure.

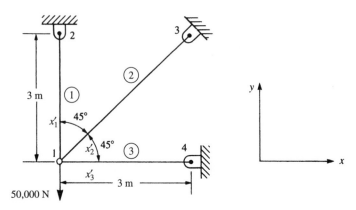

Figure 3–15 Plane truss

SOLUTION:

First, we determine the global stiffness matrices for each element by using Eq. (3.4.23). This requires determination of the angle θ between the global x axis and the local x' axis for each element. In this example, the direction of the x' axis for each element is taken in the direction *from* node 1 *to* the other node as shown in Figure 3–15. The node numbering is arbitrary for each element. However, once the direction is chosen, the angle θ is then established as positive when measured counterclockwise from positive x to x'. For element 1, the local x_1' axis is directed from node 1 to node 2; therefore, $\theta^{(1)} = 90°$. For element 2, the local x_2' axis is directed from node 1 to node 3 and $\theta^{(2)} = 45°$. For element 3, the local x_3' axis is directed from node 1 to node 4 and $\theta^{(3)} = 0°$. It is convenient to construct Table 3–1 to aid in determining each element stiffness matrix.

Table 3–1   Data for the truss of Figure 3–15

Element	$\theta°$	C	S	C^2	S^2	CS
1	90°	0	1	0	1	0
2	45°	$\sqrt{2}/2$	$\sqrt{2}/2$	$\frac{1}{2}$	$\frac{1}{2}$	$\frac{1}{2}$
3	0°	1	0	1	0	0

There are a total of eight nodal components of displacement, or degrees of freedom, for the truss before boundary constraints are imposed. Thus the order of the total stiffness matrix must be 8×8. We could then expand the $[k]$ matrix for each element to the order 8×8 by adding rows and columns of zeros as explained in the first part of Section 2.4. Alternatively, we could label the rows and columns of each element stiffness matrix according to the displacement components associated with it as explained in the latter part of Section 2.4. Using this latter approach, we construct the total stiffness matrix $[K]$ simply by adding terms from the individual element stiffness matrices into their corresponding locations in $[K]$. This approach will be used here and throughout this text.

For element 1, using Eq. (3.4.23), along with Table 3–1 for the direction cosines, we obtain

$$[k^{(1)}] = \frac{(2 \times 10^{11})(6 \times 10^{-4})}{3} \begin{matrix} u_1 & v_1 & u_2 & v_2 \\ \begin{bmatrix} 0 & 0 & 0 & 0 \\ 0 & 1 & 0 & -1 \\ 0 & 0 & 0 & 0 \\ 0 & -1 & 0 & 1 \end{bmatrix} \end{matrix} \qquad (3.6.1)$$

Similarly, for element 2, we have

$$[k^{(2)}] = \frac{(2 \times 10^{11})(6 \times 10^{-4})}{3 \times \sqrt{2}} \begin{matrix} u_1 & v_1 & u_3 & v_3 \\ \begin{bmatrix} 0.5 & 0.5 & -0.5 & -0.5 \\ 0.5 & 0.5 & -0.5 & -0.5 \\ -0.5 & -0.5 & 0.5 & 0.5 \\ -0.5 & -0.5 & 0.5 & 0.5 \end{bmatrix} \end{matrix} \qquad (3.6.2)$$

and for element 3, we have

$$[k^{(3)}] = \frac{(2 \times 10^{11})(6 \times 10^{-4})}{3} \begin{matrix} u_1 & v_1 & u_4 & v_4 \\ \begin{bmatrix} 1 & 0 & -1 & 0 \\ 0 & 0 & 0 & 0 \\ -1 & 0 & 1 & 0 \\ 0 & 0 & 0 & 0 \end{bmatrix} \end{matrix} \qquad (3.6.3)$$

The common factor of $2 \times 10^{11} \times 6 \times 10^{-4}/3$ $(= 4 \times 10^7)$ can be taken from each of Eqs. (3.6.1) through (3.6.3), where each term in the square bracket of Eq. (3.6.2) is now multiplied by $1/\sqrt{2}$. After adding terms from the individual element stiffness matrices into their corresponding locations in $[K]$, we obtain the total stiffness matrix as

$$[K] = (4 \times 10^7) \begin{array}{c} \begin{array}{cccccccc} u_1 & v_1 & u_2 & v_2 & u_3 & v_3 & u_4 & v_4 \end{array} \\ \begin{bmatrix} 1.354 & 0.354 & 0 & 0 & -0.354 & -0.354 & -1 & 0 \\ 0.354 & 1.354 & 0 & -1 & -0.354 & -0.354 & 0 & 0 \\ 0 & 0 & 0 & 0 & 0 & 0 & 0 & 0 \\ 0 & -1 & 0 & 1 & 0 & 0 & 0 & 0 \\ -0.354 & -0.354 & 0 & 0 & 0.354 & 0.354 & 0 & 0 \\ -0.354 & -0.354 & 0 & 0 & 0.354 & 0.354 & 0 & 0 \\ -1 & 0 & 0 & 0 & 0 & 0 & 1 & 0 \\ 0 & 0 & 0 & 0 & 0 & 0 & 0 & 0 \end{bmatrix} \end{array} \quad (3.6.4)$$

The global $[K]$ matrix, Eq. (3.6.4), relates the global forces to the global displacements. We thus write the total structure stiffness equations, accounting for the applied force at node 1 and the boundary constraints at nodes 2–4 as follows:

$$\begin{Bmatrix} 0 \\ -50,000 \\ F_{2x} \\ F_{2y} \\ F_{3x} \\ F_{3y} \\ F_{4x} \\ F_{4y} \end{Bmatrix} = (4 \times 10^7) \begin{bmatrix} 1.354 & 0.354 & 0 & 0 & -0.354 & -0.354 & -1 & 0 \\ 0.354 & 1.354 & 0 & -1 & -0.354 & -0.354 & 0 & 0 \\ 0 & 0 & 0 & 0 & 0 & 0 & 0 & 0 \\ 0 & -1 & 0 & 1 & 0 & 0 & 0 & 0 \\ -0.354 & -0.354 & 0 & 0 & 0.354 & 0.354 & 0 & 0 \\ -0.354 & -0.354 & 0 & 0 & 0.354 & 0.354 & 0 & 0 \\ -1 & 0 & 0 & 0 & 0 & 0 & 1 & 0 \\ 0 & 0 & 0 & 0 & 0 & 0 & 0 & 0 \end{bmatrix}$$

$$\times \begin{Bmatrix} u_1 \\ v_1 \\ u_2 = 0 \\ v_2 = 0 \\ u_3 = 0 \\ v_3 = 0 \\ u_4 = 0 \\ v_4 = 0 \end{Bmatrix} \quad (3.6.5)$$

We could now use the partitioning scheme described in the first part of Section 2.5 to obtain the equations used to determine unknown displacements u_1 and v_1—that is, partition the first two equations from the third through the eighth in Eq. (3.6.5). Alternatively, we could eliminate rows and columns in the total stiffness matrix corresponding to zero displacements as previously described in the latter part of Section 2.5. Here we will use the latter approach; that is, we eliminate rows and column 3–8 in Eq. (3.6.5) because those rows and columns correspond to zero displacements. (Remember, this direct approach must be modified for nonhomogeneous boundary conditions as was indicated in Section 2.5.) We then obtain

$$\begin{Bmatrix} 0 \\ -10{,}000 \end{Bmatrix} = (500{,}000)\begin{bmatrix} 1.354 & 0.354 \\ 0.354 & 1.354 \end{bmatrix}\begin{Bmatrix} u_1 \\ v_1 \end{Bmatrix} \tag{3.6.6}$$

Equation (3.6.6) can now be solved for the displacements by multiplying both sides of the matrix equation by the inverse of the 2 × 2 stiffness matrix or by solving the two equations simultaneously. Using either procedure for solution yields the displacements

$$u_1 = 0.414 \times 10^{-2} \text{ in.} \qquad v_1 = -1.59 \times 10^{-2} \text{ in.}$$

The minus sign in the v_1 result indicates that the displacement component in the y direction at node 1 is in the direction opposite that of the positive y direction based on the assumed global coordinates, that is, a downward displacement occurs at node 1.

Using Eq. (3.5.6) and Table 3–1, we determine the stresses in each element as follows:

$$\sigma^{(1)} = \frac{30 \times 10^6}{120}[0 \quad -1 \quad 0 \quad 1]\begin{Bmatrix} u_1 = 0.414 \times 10^{-2} \\ v_1 = -1.59 \times 10^{-2} \\ u_2 = 0 \\ v_2 = 0 \end{Bmatrix} = 3965 \text{ psi}$$

$$\sigma^{(2)} = \frac{30 \times 10^6}{120\sqrt{2}}\begin{bmatrix} \frac{-\sqrt{2}}{2} & \frac{-\sqrt{2}}{2} & \frac{\sqrt{2}}{2} & \frac{\sqrt{2}}{2} \end{bmatrix}\begin{Bmatrix} u_1 = 0.414 \times 10^{-2} \\ v_1 = -1.59 \times 10^{-2} \\ u_3 = 0 \\ v_3 = 0 \end{Bmatrix}$$

$$= 1471 \text{ psi}$$

$$\sigma^{(3)} = \frac{30 \times 10^6}{120}[-1 \quad 0 \quad 1 \quad 0]\begin{Bmatrix} u_1 = 0.414 \times 10^{-2} \\ v_1 = -1.59 \times 10^{-2} \\ u_4 = 0 \\ v_4 = 0 \end{Bmatrix} = -1035 \text{ psi}$$

We now verify our results by examining force equilibrium at node 1; that is, summing forces in the global x and y directions, we obtain

$$\sum F_x = 0 \qquad (1471 \text{ psi})(2 \text{ in}^2)\frac{\sqrt{2}}{2} - (1035 \text{ psi})(2 \text{ in}^2) = 0$$

$$\sum F_y = 0 \qquad (3965 \text{ psi})(2 \text{ in}^2) + (1471 \text{ psi})(2 \text{ in}^2)\frac{\sqrt{2}}{2} - 10{,}000 = 0 \qquad ∎$$

Example 3.6

For the two-bar truss shown in Figure 3–16, determine the displacement in the y direction of node 1 and the axial force in each element. A force of $P = 1000$ kN is applied at node 1 in the positive y direction while node 1 settles an amount $\delta = 50$ mm in the negative x direction. Let $E = 210$ GPa and $A = 6.00 \times 10^{-4}$ m^2 for each element. The lengths of the elements are shown in the figure.

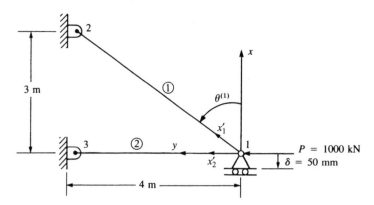

Figure 3–16 Two-bar truss

SOLUTION:

We begin by using Eq. (3.4.23) to determine each element stiffness matrix.

Element 1

$$\cos \theta^{(1)} = \frac{3}{5} = 0.60 \qquad \sin \theta^{(1)} = \frac{4}{5} = 0.80$$

$$[k^{(1)}] = \frac{(6.0 \times 10^{-4} \text{ m}^2)(210 \times 10^6 \text{ kN/m}^2)}{5 \text{ m}} \begin{bmatrix} 0.36 & 0.48 & -0.36 & -0.48 \\ & 0.64 & -0.48 & -0.64 \\ & & 0.36 & 0.48 \\ \text{Symmetry} & & & 0.64 \end{bmatrix} \quad (3.6.7)$$

Simplifying Eq. (3.6.7), we obtain

$$[k^{(1)}] = (25{,}200) \begin{array}{cccc} u_1 & v_1 & u_2 & v_2 \end{array} \begin{bmatrix} 0.36 & 0.48 & -0.36 & -0.48 \\ & 0.64 & -0.48 & -0.64 \\ & & 0.36 & 0.48 \\ \text{Symmetry} & & & 0.64 \end{bmatrix} \quad (3.6.8)$$

Element 2

$$\cos \theta^{(2)} = 0.0 \qquad \sin \theta^{(2)} = 1.0$$

$$[k^{(2)}] = \frac{(6.0 \times 10^{-4})(210 \times 10^6)}{4}
\begin{bmatrix}
0 & 0 & 0 & 0 \\
 & 1 & 0 & -1 \\
 & & 0 & 0 \\
\text{Symmetry} & & & 1
\end{bmatrix} \qquad (3.6.9)$$

$$[k^{(2)}] = (25{,}200)
\begin{array}{cccc}
u_1 & v_1 & u_3 & v_3
\end{array}
\begin{bmatrix}
0 & 0 & 0 & 0 \\
 & 1.25 & 0 & -1.25 \\
 & & 0 & 0 \\
\text{Symmetry} & & & 1.25
\end{bmatrix} \qquad (3.6.10)$$

where, for computational simplicity, Eq. (3.6.10) is written with the same factor (25,200) in front of the matrix as Eq. (3.6.8). Superimposing the element stiffness matrices, Eqs. (3.6.8) and (3.6.10), we obtain the global $[K]$ matrix and relate the global forces to global displacements by

$$\begin{Bmatrix}
F_{1x} \\
F_{1y} \\
F_{2x} \\
F_{2y} \\
F_{3x} \\
F_{3y}
\end{Bmatrix}
= (25{,}200)
\begin{bmatrix}
0.36 & 0.48 & -0.36 & -0.48 & 0 & 0 \\
 & 1.89 & -0.48 & -0.64 & 0 & -1.25 \\
 & & 0.36 & 0.48 & 0 & 0 \\
 & & & 0.64 & 0 & 0 \\
 & & & & 0 & 0 \\
\text{Symmetry} & & & & & 1.25
\end{bmatrix}
\begin{Bmatrix}
u_1 \\
v_1 \\
u_2 \\
v_2 \\
u_3 \\
v_3
\end{Bmatrix} \qquad (3.6.11)$$

We can again partition equations with known displacements and then simultaneously solve those associated with unknown displacements. To do this partitioning, we consider the boundary conditions given by

$$u_1 = \delta \qquad u_2 = 0 \qquad v_2 = 0 \qquad u_3 = 0 \qquad v_3 = 0 \qquad (3.6.12)$$

Therefore, using Eqs. (3.6.12), we partition equation 2 from equations 1, 3, 4, 5, and 6 of Eq. (3.6.11) and are left with

$$P = 25{,}200(0.48\delta + 1.89v_1) \qquad (3.6.13)$$

where $F_{1y} = P$ and $u_1 = \delta$ were substituted into Eq. (3.6.13). Expressing Eq. (3.6.13) in terms of P and δ allows these two influences on v_1 to be clearly separated. Solving Eq. (3.6.13) for v_1, we have

$$v_1 = 0.000021P - 0.254\delta \qquad (3.6.14)$$

Now, substituting the numerical values $P = 1000$ kN and $\delta = -0.05$ m into Eq. (3.6.14), we obtain

$$v_1 = 0.0337 \text{ m} \qquad (3.6.15)$$

where the positive value indicates horizontal displacement to the left.

The local element forces are obtained by using Eq. (3.4.11). We then have the following.

Element 1

$$\left\{ \begin{matrix} f'_{1x} \\ f'_{2x} \end{matrix} \right\} = (25,200) \begin{bmatrix} 1 & -1 \\ -1 & 1 \end{bmatrix} \begin{bmatrix} 0.60 & 0.80 & 0 & 0 \\ 0 & 0 & 0.60 & 0.80 \end{bmatrix} \left\{ \begin{matrix} u_1 = -0.05 \\ v_1 = 0.0337 \\ u_2 = 0 \\ v_2 = 0 \end{matrix} \right\} \qquad (3.6.16)$$

Performing the matrix triple product in Eq. (3.6.16) yields

$$f'_{1x} = -76.6 \text{ kN} \qquad f'_{2x} = 76.6 \text{ kN} \qquad (3.6.17)$$

Element 2

$$\left\{ \begin{matrix} f'_{1x} \\ f'_{3x} \end{matrix} \right\} = (31,500) \begin{bmatrix} 1 & -1 \\ -1 & 1 \end{bmatrix} \begin{bmatrix} 0 & 1 & 0 & 0 \\ 0 & 0 & 0 & 1 \end{bmatrix} \left\{ \begin{matrix} u_1 = -0.05 \\ v_1 = 0.0337 \\ u_3 = 0 \\ v_3 = 0 \end{matrix} \right\} \qquad (3.6.18)$$

Performing the matrix triple product in Eq. (3.6.18), we obtain

$$f'_{1x} = 1061 \text{ kN} \qquad f'_{3x} = -1061 \text{ kN} \qquad (3.6.19)$$

Verification of the computations by checking that equilibrium is satisfied at node 1 is left to your discretion. ∎

Example 3.7

To illustrate how we can combine spring and bar elements in one structure, we now solve the two-bar truss supported by a spring shown in Figure 3–17. Both bars have $E = 210$ GPa and $A = 5.0 \times 10^{-4} \text{ m}^2$. Bar one has a length of 5 m and bar two a length of 10 m. The spring stiffness is $k = 2000$ kN/m.

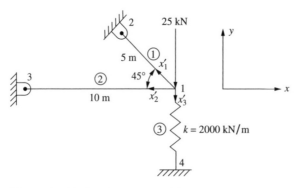

Figure 3–17 Two-bar truss with spring support

SOLUTION:

We begin by using Eq. (3.4.23) to determine each element stiffness matrix.

Element 1

$$\theta^{(1)} = 135°, \quad \cos\theta^{(1)} = -\sqrt{2}/2, \quad \sin\theta^{(1)} = \sqrt{2}/2$$

$$[k^{(1)}] = \frac{(5.0 \times 10^{-4}\,\text{m}^2)(210 \times 10^6\,\text{kN/m}^2)}{5\,\text{m}} \begin{bmatrix} 0.5 & -0.5 & -0.5 & 0.5 \\ -0.5 & 0.5 & 0.5 & -0.5 \\ -0.5 & 0.5 & 0.5 & -0.5 \\ 0.5 & -0.5 & -0.5 & 0.5 \end{bmatrix} \quad (3.6.20)$$

Simplifying Eq. (3.6.20), we obtain

$$[k^{(1)}] = 105 \times 10^2 \begin{matrix} u_1 & v_1 & u_2 & v_2 \\ \begin{bmatrix} 1 & -1 & -1 & 1 \\ -1 & 1 & 1 & -1 \\ -1 & 1 & 1 & -1 \\ 1 & -1 & -1 & 1 \end{bmatrix} \end{matrix} \quad (3.6.21)$$

Element 2

$$\theta^{(2)} = 180°, \quad \cos\theta^{(2)} = -1.0, \quad \sin\theta^{(2)} = 0$$

$$[k^{(2)}] = \frac{(5 \times 10^{-4}\,\text{m}^2)(210 \times 10^6\,\text{kN/m}^2)}{10\,\text{m}} \begin{bmatrix} 1 & 0 & -1 & 0 \\ 0 & 0 & 0 & 0 \\ -1 & 0 & 1 & 0 \\ 0 & 0 & 0 & 0 \end{bmatrix} \quad (3.6.22)$$

Simplifying Eq. (3.6.22), we obtain

$$[k^{(2)}] = 105 \times 10^2 \begin{matrix} u_1 & v_1 & u_3 & v_3 \\ \begin{bmatrix} 1 & 0 & -1 & 0 \\ 0 & 0 & 0 & 0 \\ -1 & 0 & 1 & 0 \\ 0 & 0 & 0 & 0 \end{bmatrix} \end{matrix} \quad (3.6.23)$$

Element 3

$$\theta^{(3)} = 270°, \quad \cos\theta^{(3)} = 0, \quad \sin\theta^{(3)} = -1.0$$

Using Eq. (3.4.23) but replacing AE/L with the spring constant k, we obtain the stiffness matrix of the spring as

$$[k^{(3)}] = 20 \times 10^2 \begin{matrix} u_1 & v_1 & u_2 & v_2 \\ \begin{bmatrix} 0 & 0 & 0 & 0 \\ 0 & 1 & 0 & -1 \\ 0 & 0 & 0 & 0 \\ 0 & -1 & 0 & 1 \end{bmatrix} \end{matrix} \quad (3.6.24)$$

Applying the boundary conditions, we have

$$u_2 = v_2 = u_3 = v_3 = u_4 = v_4 = 0 \tag{3.6.25}$$

Using the boundary conditions in Eq. (3.6.25), the reduced assembled global equations are given by:

$$\left\{ \begin{matrix} F_{1x} = 0 \\ F_{1y} = -25\,\text{kN} \end{matrix} \right\} = 10^2 \begin{bmatrix} 210 & -105 \\ -105 & 125 \end{bmatrix} \left\{ \begin{matrix} u_1 \\ v_1 \end{matrix} \right\} \tag{3.6.26}$$

Solving Eq. (3.6.26) for the global displacements, we obtain

$$u_1 = -1.724 \times 10^{-3}\,\text{m} \qquad v_1 = -3.448 \times 10^{-3}\,\text{m} \tag{3.6.27}$$

We can obtain the stresses in the bar elements by using Eq. (3.5.6) as

$$\sigma^{(1)} = \frac{210 \times 10^3\,\text{MN/m}^2}{5\,\text{m}} [0.707 \quad -0.707 \quad -0.707 \quad 0.707] \left\{ \begin{matrix} -1.724 \times 10^{-3} \\ -3.448 \times 10^{-3} \\ 0 \\ 0 \end{matrix} \right\}$$

Simplifying, we obtain

$$\sigma^{(1)} = 51.2\,\text{MPa}\,(T)$$

Similarly, we obtain the stress in element two as

$$\sigma^{(2)} = \frac{210 \times 10^3\,\text{MN/m}^2}{10\,\text{m}} [1.0 \quad 0 \quad -1.0 \quad 0] \left\{ \begin{matrix} -1.724 \times 10^{-3} \\ -3.448 \times 10^{-3} \\ 0 \\ 0 \end{matrix} \right\}$$

Simplifying, we obtain

$$\sigma^{(2)} = -36.2\,\text{MPa}\,(C) \qquad\blacksquare$$

▲ 3.7 Transformation Matrix and Stiffness Matrix for a Bar in Three-Dimensional Space ▲

We will now derive the transformation matrix necessary to obtain the general stiffness matrix of a bar element arbitrarily oriented in three-dimensional space as shown in Figure 3–18. Let the coordinates of node 1 be taken as x_1, y_1, and z_1, and let those of node 2 be taken as x_2, y_2, and z_2. Also, let θ_x, θ_y, and θ_z be the angles measured from the global x, y, and z axes, respectively, to the local x' axis. Here x' is directed along the element from node 1 to node 2. We must now determine $[T^*]$ such that

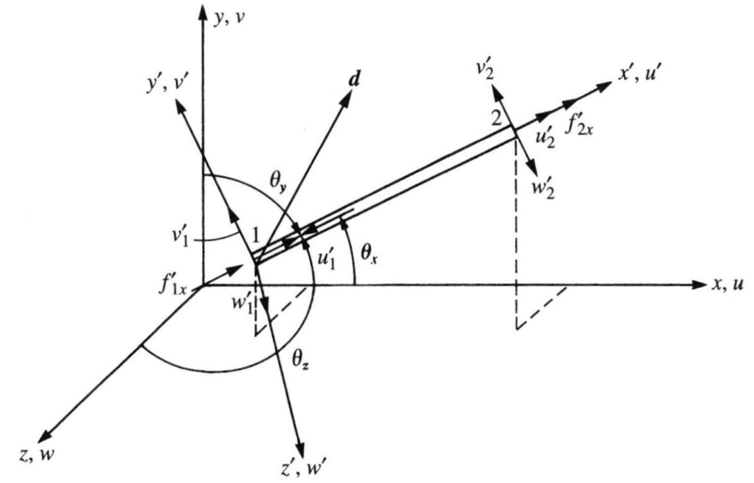

Figure 3–18 Bar in three-dimensional space along with local nodal displacements

$\{d'\} = [T^*]\{d\}$. We begin the derivation of $[T^*]$ by considering the vector $\hat{\mathbf{d}} = \mathbf{d}$ expressed in three dimensions as

$$u'\mathbf{i}' + v'\mathbf{j}' + w'\mathbf{k}' = u\mathbf{i} + v\mathbf{j} + w\mathbf{k} \qquad (3.7.1)$$

where $\mathbf{i}'$, $\mathbf{j}'$, and $\mathbf{k}'$ are unit vectors associated with the local x', y', and z' axes, respectively, and $\mathbf{i}$, $\mathbf{j}$, and $\mathbf{k}$ are unit vectors associated with the global x, y, and z axes. Also w and w' now denote the displacements in the z and z' directions, respectively. Taking the dot product of Eq. (3.7.1) with $\mathbf{i}'$, we have

$$u' + 0 + 0 = u(\mathbf{i}' \cdot \mathbf{i}) + v(\mathbf{i}' \cdot \mathbf{j}) + w(\mathbf{i}' \cdot \mathbf{k}) \qquad (3.7.2)$$

and, by definition of the dot product,

$$\mathbf{i}' \cdot \mathbf{i} = \frac{x_2 - x_1}{L} = C_x$$

$$\mathbf{i}' \cdot \mathbf{j} = \frac{y_2 - y_1}{L} = C_y \qquad (3.7.3)$$

$$\mathbf{i}' \cdot \mathbf{k} = \frac{z_2 - z_1}{L} = C_z$$

where $\qquad L = [(x_2 - x_1)^2 + (y_2 - y_1)^2 + (z_2 - z_1)^2]^{1/2}$

and $\qquad\qquad C_x = \cos\theta_x \qquad C_y = \cos\theta_y \qquad C_z = \cos\theta_z \qquad (3.7.4)$

Here C_x, C_y, and C_z are the projections of $\mathbf{i}'$ on $\mathbf{i}$, $\mathbf{j}$, and $\mathbf{k}$, respectively. Therefore, using Eqs. (3.7.3) in Eq. (3.7.2), we have

$$u' = C_x u + C_y v + C_z w \qquad (3.7.5)$$

For a vector in space directed along the x' axis, Eq. (3.7.5) gives the components of that vector in the global x, y, and z directions. Now, using Eq. (3.7.5), we can write the local axial displacement at node 1 and 2 in explicit form as

$$\left\{ \begin{array}{c} u_1' \\ u_2' \end{array} \right\} = \begin{bmatrix} C_x & C_y & C_z & 0 & 0 & 0 \\ 0 & 0 & 0 & C_x & C_y & C_z \end{bmatrix} \left\{ \begin{array}{c} u_1 \\ v_1 \\ w_1 \\ u_2 \\ v_2 \\ w_2 \end{array} \right\} \tag{3.7.6}$$

Now $\{d'\} = \left\{ \begin{array}{c} u_1' \\ u_2' \end{array} \right\}, \{d\} = \left\{ \begin{array}{c} u_1 \\ v_1 \\ w_1 \\ u_2 \\ v_2 \\ w_2 \end{array} \right\}$, and define $[T^*] = \begin{bmatrix} C_x & C_y & C_z & 0 & 0 & 0 \\ 0 & 0 & 0 & C_x & C_y & C_z \end{bmatrix}$.

$$\tag{3.7.7}$$

Using Eq. (3.7.7), we write Eq. (3.7.6) in matrix form as

$$\{d'\} = [T^*]\{d\} \tag{a}$$

Here $[T^*]$ is the transformation matrix, which enables the local displacement matrix $\{d'\}$ to be expressed in terms of the displacement matrix $\{d\}$ components in the global coordinate system.

Based on Eq. (a), it will be convenient to express the global force matrix in terms of the local force matrix using $[T^*]$ as

$$\{f\} = [T^*]^T\{f'\} \tag{b}$$

Now in local coordinates, the local forces are related to the local displacements by

$$\{f'\} = [k']\{d'\} \tag{c}$$

Upon substituting for $\{d'\}$ from Eq. (a) into Eq. (c) and premultiplying both sides by $\{T^*\}^T$, we have

$$[T^*]^T\{f'\} = [T^*]^T[k'][T^*]\{d\} \tag{d}$$

Now using Eq. (b) in the left side of Eq. (c), we obtain

$$\{f\} = [T^*]^T[k'][T^*]\{d\} \tag{e}$$

The global forces are related to the global displacements by

$$\{f\} = [k]\{d\} \tag{f}$$

Comparing the right sides of Eqs. (e) and (f), we then observe that the global stiffness matrix for a bar arbitrarily oriented in space is

$$[k] = [T^*]^T [k'][T^*] \tag{g}$$

Using Eq. (3.7.7) for $[T^*]$ and Eq. (3.4.2) in Eq. (3.4.1) for $[k']$ we obtain $[k]$ as follows:

$$[k] = \begin{bmatrix} C_x & 0 \\ C_y & 0 \\ C_z & 0 \\ 0 & C_x \\ 0 & C_y \\ 0 & C_z \end{bmatrix} \frac{AE}{L} \begin{bmatrix} 1 & -1 \\ -1 & 1 \end{bmatrix} \begin{bmatrix} C_x & C_y & C_z & 0 & 0 & 0 \\ 0 & 0 & 0 & C_x & C_y & C_z \end{bmatrix} \tag{3.7.8}$$

Simplifying Eq. (3.7.8), we obtain the explicit form of $[k]$ as

$$[k] = \frac{AE}{L} \begin{bmatrix} C_x^2 & C_x C_y & C_x C_z & -C_x^2 & -C_x C_y & -C_x C_z \\ & C_y^2 & C_y C_z & -C_x C_y & -C_y^2 & -C_y C_z \\ & & C_z^2 & -C_x C_z & -C_y C_z & -C_z^2 \\ & & & C_x^2 & C_x C_y & C_x C_z \\ & & & & C_y^2 & C_y C_z \\ \text{Symmetry} & & & & & C_z^2 \end{bmatrix} \tag{3.7.9}$$

Equation (3.7.9) is the basic form of the stiffness matrix for a bar element arbitrarily oriented in three-dimensional space. We will now analyze a simple space truss to illustrate the concepts developed in this section. We will show that the direct stiffness method provides a simple procedure for solving space truss problems.

Example 3.8

Analyze the space truss shown in Figure 3–19. The truss is composed of four nodes, whose coordinates (in millimeters) are shown in the figure, and three elements, whose cross-sectional areas are given in the figure. The modulus of elasticity $E = 8$ GPa for all elements. A load of 5000 N is applied at node 1 in the negative z direction. Nodes 2–4 are supported by ball-and-socket joints and thus constrained from movement in the x, y, and z directions. Node 1 is constrained from movement in the y direction by the roller shown in Figure 3–19.

SOLUTION:

Using Eq. (3.7.9), we will now determine the stiffness matrices of the three elements in Figure 3–19. To simplify the numerical calculations, we first express $[k]$ for each element, given by Eq. (3.7.9), in the form

$$[k] = \frac{AE}{L} \begin{bmatrix} [\lambda] & -[\lambda] \\ -[\lambda] & [\lambda] \end{bmatrix} \tag{3.7.10}$$

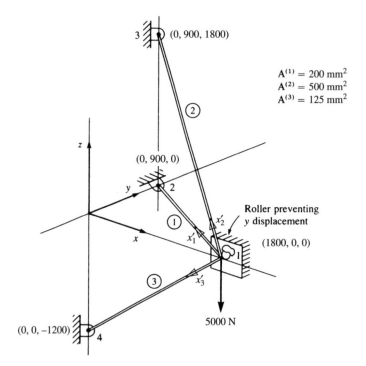

Figure 3–19 Space truss

where $[\lambda]$ is a 3×3 submatrix defined by

$$[\lambda] = \begin{bmatrix} C_x^2 & C_x C_y & C_x C_z \\ C_y C_x & C_y^2 & C_y C_z \\ C_z C_x & C_z C_y & C_z^2 \end{bmatrix} \quad (3.7.11)$$

Therefore, determining $[\lambda]$ will sufficiently describe $[k]$.

Element 3

The direction cosines of element 3 are given, in general, by

$$C_x = \frac{x_4 - x_1}{L^{(3)}} \qquad C_y = \frac{y_4 - y_1}{L^{(3)}} \qquad C_z = \frac{z_4 - z_1}{L^{(3)}} \quad (3.7.12)$$

where the notation x_i, y_i, and z_i is used to denote the coordinates of each node, and $L^{(e)}$ denotes the element length. From the coordinate information given in Figure 3–19, we obtain the length and the direction cosines as

$$L^{(3)} = [(-1.8 \text{ m})^2 + (-1.2 \text{ m})^2]^{1/2} = 2.16 \text{ m}$$

$$C_x = \frac{-1.8}{2.16} = -0.833 \qquad C_y = 0 \qquad C_z = \frac{-1.2}{2.16} = -0.550 \quad (3.7.13)$$

Using the results of Eqs. (3.7.13) in Eq. (3.7.11) yields

$$[\lambda] = \begin{bmatrix} 0.69 & 0 & 0.46 \\ 0 & 0 & 0 \\ 0.46 & 0 & 0.30 \end{bmatrix} \tag{3.7.14}$$

and, from Eq. (3.7.10),

$$[k^{(3)}] = \frac{(12.5 \times 10^{-6})(8 \times 10^9)}{2.16} \begin{array}{cc} u_1 v_1 w_1 & u_4 v_4 w_4 \\ \left[\begin{array}{c|c} [\lambda] & -[\lambda] \\ \hline -[\lambda] & [\lambda] \end{array} \right] \end{array} \tag{3.7.15}$$

Element 1

Similarly, for element 1, we obtain

$$L^{(1)} = 2.01 \text{ m}$$

$$C_x = -0.89 \qquad C_y = 0.45 \qquad C_z = 0$$

$$[\lambda] = \begin{bmatrix} 0.79 & -0.40 & 0 \\ -0.40 & 0.20 & 0 \\ 0 & 0 & 0 \end{bmatrix}$$

and $\qquad [k^{(1)}] = \dfrac{(200 \times 10^{-6})(8 \times 10^9)}{2.01} \begin{array}{cc} u_1 v_1 w_1 & u_2 v_2 w_2 \\ \left[\begin{array}{c|c} [\lambda] & -[\lambda] \\ \hline -[\lambda] & [\lambda] \end{array} \right] \end{array} \qquad$ (3.7.16)

Element 2

Finally, for element 2, we obtain

$$L^{(2)} = 2.7 \text{ m}$$

$$C_x = -0.667 \qquad C_y = 0.33 \qquad C_z = 0.667$$

$$[\lambda] = \begin{bmatrix} 0.45 & -0.22 & -0.45 \\ -0.22 & 0.11 & 0.22 \\ -0.45 & 0.22 & 0.45 \end{bmatrix}$$

and $\qquad [k^{(2)}] = \dfrac{(500 \times 10^{-6})(8 \times 10^9)}{2.7} \begin{array}{cc} u_1 v_1 w_1 & u_3 v_3 w_3 \\ \left[\begin{array}{c|c} [\lambda] & -[\lambda] \\ \hline -[\lambda] & [\lambda] \end{array} \right] \end{array} \qquad$ (3.7.17)

Using the zero-displacement boundary conditions $v_1 = 0, u_2 = v_2 = w_2 = 0, u_3 = v_3 = w_3 = 0$, and $u_4 = v_4 = w_4 = 0$, we can cancel the corresponding rows and columns of each element stiffness matrix. After canceling appropriate rows and columns in Eqs. (3.7.15) through (3.7.17) and then superimposing the resulting element stiffness matrices, we have the total stiffness matrix for the truss as

$$[K] = 10^3 \times \begin{array}{cc} \quad u_1 \qquad w_1 \\ \begin{bmatrix} 1615 & -453 \\ -453 & 805 \end{bmatrix} \end{array} \qquad (3.7.18)$$

The global stiffness equations are then expressed by

$$\begin{Bmatrix} 0 \\ -5000 \end{Bmatrix} = 10^3 \times \begin{bmatrix} 1615 & -453 \\ -453 & 805 \end{bmatrix} \begin{Bmatrix} u_1 \\ w_1 \end{Bmatrix} \qquad (3.7.19)$$

Solving Eq. (3.7.19) for the displacements, we obtain

$$\begin{aligned} u_1 &= -2.07 \times 10^{-3} \text{ m} = -2.07 \text{ mm} \\ w_1 &= -7.38 \times 10^{-3} \text{ m} = -7.38 \text{ mm} \end{aligned} \qquad (3.7.20)$$

where the minus signs in the displacements indicate these displacements to be in the negative x and z directions.

We will now determine the stress in each element. The stresses are determined by using Eq. (3.5.6) expanded to three dimensions. Thus, for an element with first node i and second node j, Eq. (3.5.6) expanded to three dimensions becomes

$$\{\sigma\} = \frac{E}{L}[-C_x \quad -C_y \quad -C_z \quad C_x \quad C_y \quad C_z] \begin{Bmatrix} u_i \\ v_i \\ w_i \\ u_j \\ v_j \\ w_j \end{Bmatrix} \qquad (3.7.21)$$

Derive Eq. (3.7.21) in a manner similar to that used to derive Eq. (3.5.6) (see Problem 3.44, for instance). For element 3, using Eqs. (3.7.13) for the direction cosines, along with the proper length and modulus of elasticity, we obtain the stress as

$$\{\sigma^{(3)}\} = \frac{8 \times 10^9}{2.16}[0.83 \quad 0 \quad 0.55 \quad -0.83 \quad 0 \quad -0.55] \begin{Bmatrix} -0.00207 \\ 0 \\ -0.00738 \\ 0 \\ 0 \\ 0 \end{Bmatrix} \qquad (3.7.22)$$

Simplifying Eq. (3.7.22), we find that the result is

$$\sigma^{(3)} = -21.4 \text{ MPa}$$

where the negative sign in the answer indicates a compressive stress. The stresses in the other elements can be determined in a manner similar to that used for element 3.

For brevity's sake, we will not show the calculations but will merely list these stresses:

$$\sigma^{(1)} = -7.1 \text{ MPa} \qquad \sigma^{(2)} = 10.8 \text{ MPa}$$ ∎

Example 3.9

Analyze the space truss shown in Figure 3–20. The truss is composed of four nodes, whose coordinates (in meters) are shown in the figure, and three elements, whose cross-sectional areas are all $10 \times 10^{-4}\,\text{m}^2$. The modulus of elasticity $E = 210$ GPa for all the elements. A load of 20 kN is applied at node 1 in the global x-direction. Nodes 2–4 are pin supported and thus constrained from movement in the x, y, and z directions.

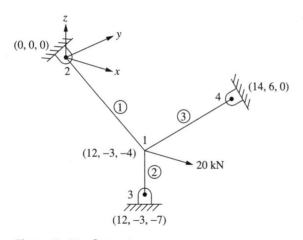

Figure 3–20 Space truss

SOLUTION:

First calculate the element lengths using the distance formula and coordinates given in Figure 3–20 as

$$L^{(1)} = [(0 - 12)^2 + (0 - (-3))^2 + (0 - (-4))^2]^{1/2} = 13\,\text{m}$$

$$L^{(2)} = [(12 - 12)^2 + (-3 + 3)^2 + (-7 + 4)^2]^{1/2} = 3\,\text{m}$$

$$L^{(3)} = [(14 - 12)^2 + (6 + 3)^2 + (0 + 4)^2]^{1/2} = 10.05\,\text{m}$$

For convenience, set up a table of direction cosines, where the local x' axis is taken from node 1 to 2, from 1 to 3 and from 1 to 4 for elements 1, 2, and 3, respectively.

Element Number	$C_x = \frac{x_j - x_i}{L^{(1)}}$	$C_y = \frac{y_j - y_i}{L^{(2)}}$	$C_z = \frac{z_j - z_i}{L^{(3)}}$
1	$-12/13$	$3/13$	$4/13$
2	0	0	-1
3	$2/10.05$	$9/10.05$	$4/10.05$

Now set up a table of products of direction cosines as indicated by the definition of $[\lambda]$ defined by Eq. (3.7.11) as

Element Number	C_x^2	$C_x C_y$	$C_x C_z$	C_y^2	$C_y C_z$	C_z^2
1	0.852	-0.213	-0.284	0.053	-0.071	0.095
2	0	0	0	0	0	1
3	0.040	0.178	0.079	0.802	0.356	0.158

Using Eq. (3.7.11), we express $[\lambda]$ for each element as

$$[\lambda^{(1)}] = \begin{bmatrix} 0.852 & -0.213 & -0.284 \\ -0.213 & 0.053 & 0.071 \\ -0.284 & 0.071 & 0.095 \end{bmatrix} \quad [\lambda^{(2)}] = \begin{bmatrix} 0 & 0 & 0 \\ 0 & 0 & 0 \\ 0 & 0 & 1 \end{bmatrix} \quad [\lambda^{(3)}] = \begin{bmatrix} 0.040 & 0.178 & 0.079 \\ 0.128 & 0.802 & 0.356 \\ 0.079 & 0.356 & 0.158 \end{bmatrix}$$

$$(3.7.23)$$

The boundary conditions are given by

$$u_2 = v_2 = w_2 = 0, \quad u_3 = v_3 = w_3 = 0, \quad u_4 = v_4 = w_4 = 0 \tag{3.7.24}$$

Using the stiffness matrix expressed in terms of $[\lambda]$ in the form of Eq. (3.7.10), we obtain each stiffness matrix as

$$[k^{(1)}] = \frac{AE}{13} \left[\begin{array}{c:c} [\lambda^{(1)}] & -[\lambda^{(1)}] \\ \hdashline -[\lambda^{(1)}] & [\lambda^{(1)}] \end{array} \right] \quad [k^{(2)}] = \frac{AE}{3} \left[\begin{array}{c:c} [\lambda^{(2)}] & -[\lambda^{(2)}] \\ \hdashline -[\lambda^{(2)}] & [\lambda^{(2)}] \end{array} \right] \quad [k^{(3)}] = \frac{AE}{10.05} \left[\begin{array}{c:c} [\lambda^{(3)}] & -[\lambda^{(3)}] \\ \hdashline -[\lambda^{(3)}] & [\lambda^{(3)}] \end{array} \right]$$

$$(3.7.25)$$

Applying the boundary conditions and canceling appropriate rows and columns associated with each zero displacement boundary condition in Eqs. (3.7.25) and then superimposing the resulting element stiffness matrices, we have the total stiffness matrix for the truss as

$$[K] = 210 \begin{bmatrix} 69.519 & 1.327 & -13.985 \\ 1.327 & 83.879 & 40.885 \\ -13.985 & 40.885 & 356.363 \end{bmatrix} \text{kN/m} \tag{3.7.26}$$

The global stiffness equations are then expressed by

$$\left\{ \begin{array}{c} 20\,\text{kN} \\ 0 \\ 0 \end{array} \right\} = 210 \begin{bmatrix} 69.519 & 1.327 & -13.985 \\ 1.327 & 83.879 & 40.885 \\ -13.985 & 40.885 & 356.363 \end{bmatrix} \left\{ \begin{array}{c} u_1 \\ v_1 \\ w_1 \end{array} \right\} \tag{3.7.27}$$

Solving for the displacements, we obtain

$$u_1 = 1.383 \times 10^{-3} \, \text{m}$$

$$v_1 = -5.119 \times 10^{-5} \, \text{m} \tag{3.7.28}$$

$$w_1 = 6.015 \times 10^{-5} \, \text{m}$$

We now determine the element stresses using Eq. (3.7.21) as

$$\sigma^{(1)} = \frac{210 \times 10^6}{13} [12/13 \quad -3/13 \quad -4/13 \quad -12/13 \quad 3/13 \quad 4/13] \begin{Bmatrix} 1.383 \times 10^{-3} \\ -5.119 \times 10^{-5} \\ 6.015 \times 10^{-5} \\ 0 \\ 0 \\ 0 \end{Bmatrix}$$

$$\tag{3.7.29}$$

Simplifying Eq. (3.7.29), we obtain upon converting to MPa units

$$\sigma^{(1)} = 20.51 \, \text{MPa} \tag{3.7.30}$$

The stress in the other elements can be found in a similar manner as

$$\sigma^{(2)} = 4.21 \, \text{MPa} \qquad \sigma^{(3)} = -5.29 \, \text{MPa} \tag{3.7.31}$$

The negative sign in Eq. (3.7.31) indicates a compressive stress in element 3. ■

▲ 3.8 Use of Symmetry in Structures ▲

Different types of symmetry may exist in a structure. These include reflective or mirror, skew, axial, and cyclic. Here we introduce the most common type of symmetry, reflective symmetry. Axial symmetry occurs when a solid of revolution is generated by rotating a plane shape about an axis in the plane. These axisymmetric bodies are common, and hence their analysis is considered in Chapter 9.

In many instances, we can use reflective symmetry to facilitate the solution of a problem. **Reflective symmetry** *means correspondence in size, shape, and position of loads; material properties; and boundary conditions that are on opposite sides of a dividing line or plane.* The use of symmetry allows us to consider a reduced problem instead of the actual problem. Thus, the order of the total stiffness matrix and total set of stiffness equations can be reduced. Longhand solution time is then reduced, and computer solution time for large-scale problems is substantially decreased. Example 3.10 will be used to illustrate reflective symmetry. Additional examples of the use of symmetry are presented in Chapter 4 for beams and in Chapter 7 for plane problems.

Example 3.10

Solve the plane truss problem shown in Figure 3–21. The truss is composed of eight elements and five nodes as shown. A vertical load of $2P$ is applied at node 4. Nodes 1 and 5 are pin supports. Bar elements 1, 2, 7, and 8 have axial stiffnesses of $\sqrt{2}AE$, and bars 3–6 have axial stiffness of AE. Here again, A and E represent the cross-sectional area and modulus of elasticity of a bar.

In this problem, we will use a plane of symmetry. The vertical plane perpendicular to the plane truss passing through nodes 2, 4, and 3 is the plane of reflective symmetry because identical geometry, material, loading, and boundary conditions occur at the corresponding locations on opposite sides of this plane. For loads such as $2P$, occurring in the plane of symmetry, half of the total load must be applied to the reduced structure. For elements occurring in the plane of symmetry, half of the cross-sectional area must be used in the reduced structure. Furthermore, for nodes in the plane of symmetry, the displacement components normal to the plane of symmetry must be set to zero in the reduced structure; that is, we set $u_2 = 0, u_3 = 0$, and $u_4 = 0$. Figure 3–22 shows the reduced structure to be used to analyze the plane truss of Figure 3–21.

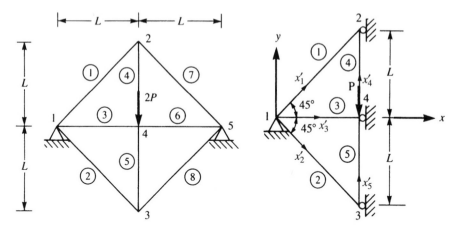

Figure 3–21   Plane truss

Figure 3–22   Truss of Figure 3–21 reduced by symmetry

SOLUTION:

We begin the solution of the problem by determining the angles θ for each bar element. For instance, for element 1, assuming x' to be directed from node 1 to node 2, we obtain $\theta^{(1)} = 45°$ as measured from the global x to the local x' axis. Table 3–2 is used in determining each element stiffness matrix based on the x' axes shown in Figure 3–22 for each element.

There are a total of eight nodal components of displacement for the truss before boundary constraints are imposed. Therefore, $[K]$ must be of order 8×8. For element 1,

Table 3–2 Data for the truss of Figure 3–22

Element	$\theta°$	C	S	C^2	S^2	CS
1	45°	$\sqrt{2}/2$	$\sqrt{2}/2$	1/2	1/2	1/2
2	315°	$\sqrt{2}/2$	$-\sqrt{2}/2$	1/2	1/2	−1/2
3	0°	1	0	1	0	0
4	90°	0	1	0	1	0
5	90°	0	1	0	1	0

using Eq. (3.4.23) along with Table 3–2 for the direction cosines, we obtain

$$[k^{(1)}] = \frac{\sqrt{2}AE}{\sqrt{2}L}
\begin{matrix}
 & u_1 & v_1 & u_2 & v_2 \\
\end{matrix}
\begin{bmatrix}
\frac{1}{2} & \frac{1}{2} & -\frac{1}{2} & -\frac{1}{2} \\
\frac{1}{2} & \frac{1}{2} & -\frac{1}{2} & -\frac{1}{2} \\
-\frac{1}{2} & -\frac{1}{2} & \frac{1}{2} & \frac{1}{2} \\
-\frac{1}{2} & -\frac{1}{2} & \frac{1}{2} & \frac{1}{2}
\end{bmatrix} \tag{3.8.1}$$

Similarly, for elements 2–5, we obtain

$$[k^{(2)}] = \frac{\sqrt{2}AE}{\sqrt{2}L}
\begin{matrix}
 & u_1 & v_1 & u_3 & v_3 \\
\end{matrix}
\begin{bmatrix}
\frac{1}{2} & -\frac{1}{2} & -\frac{1}{2} & \frac{1}{2} \\
-\frac{1}{2} & \frac{1}{2} & \frac{1}{2} & -\frac{1}{2} \\
-\frac{1}{2} & \frac{1}{2} & \frac{1}{2} & -\frac{1}{2} \\
\frac{1}{2} & -\frac{1}{2} & -\frac{1}{2} & \frac{1}{2}
\end{bmatrix} \tag{3.8.2}$$

$$[k^{(3)}] = \frac{AE}{L}
\begin{matrix}
u_1 & v_1 & u_4 & v_4 \\
\end{matrix}
\begin{bmatrix}
1 & 0 & -1 & 0 \\
0 & 0 & 0 & 0 \\
-1 & 0 & 1 & 0 \\
0 & 0 & 0 & 0
\end{bmatrix} \tag{3.8.3}$$

$$[k^{(4)}] = \frac{AE}{L}
\begin{matrix}
u_4 & v_4 & u_2 & v_2 \\
\end{matrix}
\begin{bmatrix}
0 & 0 & 0 & 0 \\
0 & \frac{1}{2} & 0 & -\frac{1}{2} \\
0 & 0 & 0 & 0 \\
0 & -\frac{1}{2} & 0 & \frac{1}{2}
\end{bmatrix} \tag{3.8.4}$$

$$[k^{(5)}] = \frac{AE}{L}
\begin{matrix}
u_3 & v_3 & u_4 & v_4 \\
\end{matrix}
\begin{bmatrix}
0 & 0 & 0 & 0 \\
0 & \frac{1}{2} & 0 & -\frac{1}{2} \\
0 & 0 & 0 & 0 \\
0 & -\frac{1}{2} & 0 & \frac{1}{2}
\end{bmatrix} \tag{3.8.5}$$

where, in Eqs. (3.8.1) through (3.8.5), the column labels indicate the degrees of freedom associated with each element. Also, because elements 4 and 5 lie in the plane of symmetry, half of their original areas have been used in Eqs. (3.8.4) and (3.8.5).

We will limit the solution to determining the displacement components. Therefore, considering the boundary constraints that result in zero-displacement components, we can immediately obtain the reduced set of equations by eliminating rows and columns in each element stiffness matrix corresponding to a zero-displacement component. That is, because $u_1 = 0$ and $v_1 = 0$ (owing to the pin support at node 1 in Figure 3–22) and $u_2 = 0, u_3 = 0$, and $u_4 = 0$ (owing to the symmetry condition), we can cancel rows and columns corresponding to these displacement components in each element stiffness matrix before assembling the total stiffness matrix. The resulting set of stiffness equations is

$$\frac{AE}{L}\begin{bmatrix} 1 & 0 & -\frac{1}{2} \\ 0 & 1 & -\frac{1}{2} \\ -\frac{1}{2} & -\frac{1}{2} & 1 \end{bmatrix}\begin{Bmatrix} v_2 \\ v_3 \\ v_4 \end{Bmatrix} = \begin{Bmatrix} 0 \\ 0 \\ -P \end{Bmatrix} \qquad (3.8.6)$$

On solving Eq. (3.8.6) for the displacements, we obtain

$$v_2 = \frac{-PL}{AE} \qquad v_3 = \frac{-PL}{AE} \qquad v_4 = \frac{-2PL}{AE} \qquad (3.8.7)$$

∎

The ideas presented regarding the use of symmetry should be used sparingly and cautiously in problems of vibration and buckling. For instance, a structure such as a simply supported beam has symmetry about its center but has antisymmetric vibration modes as well as symmetric vibration modes. This will be shown in Chapter 16. If only half the beam were modeled using reflective symmetry conditions, the support conditions would permit only the symmetric vibration modes.

▲ 3.9 Inclined, or Skewed, Supports ▲

In the preceding sections, the supports were oriented such that the resulting boundary conditions on the displacements were in the global directions, x and y.

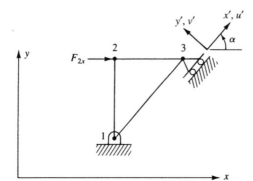

Figure 3–23 Plane truss with inclined boundary conditions at node 3

However, if a support is inclined, or skewed, at an angle α from the global x axis, as shown at node 3 in the plane truss of Figure 3–23, the resulting boundary conditions on the displacements are not in the global x-y directions but are in the local x'-y' directions. We will now describe two methods used to handle inclined supports.

In the first method, to account for inclined boundary conditions, we must perform a transformation of the global displacements at node 3 only into the local nodal coordinate system x'-y', while keeping all other displacements in the x-y global system. We can then enforce the zero-displacement boundary condition v_3' in the force/displacement equations and, finally, solve the equations in the usual manner.

The transformation used is analogous to that for transforming a vector from local to global coordinates. For the plane truss, we use Eq. (3.3.16) applied to node 3 as follows:

$$\left\{ \begin{array}{c} u_3' \\ v_3' \end{array} \right\} = \left[\begin{array}{cc} \cos \alpha & \sin \alpha \\ -\sin \alpha & \cos \alpha \end{array} \right] \left\{ \begin{array}{c} u_3 \\ v_3 \end{array} \right\} \tag{3.9.1}$$

Rewriting Eq. (3.9.1), we have

$$\{d_3'\} = [t_3]\{d_3\} \tag{3.9.2}$$

where

$$[t_3] = \left[\begin{array}{cc} \cos \alpha & \sin \alpha \\ -\sin \alpha & \cos \alpha \end{array} \right] \tag{3.9.3}$$

We now write the transformation for the entire nodal displacement vector as

$$\{d'\} = [T_1]\{d\} \tag{3.9.4}$$

or

$$\{d\} = [T_1]^T \{d'\} \tag{3.9.5}$$

where the transformation matrix for the entire truss is the 6×6 matrix

$$[T_1] = \left[\begin{array}{ccc} [I] & [0] & [0] \\ [0] & [I] & [0] \\ [0] & [0] & [t_3] \end{array} \right] \tag{3.9.6}$$

Each submatrix in Eq. (3.9.6) (the identity matrix $[I]$, the null matrix $[0]$, and matrix $[t_3]$ has the same 2×2 order, that order in general being equal to the number of degrees of freedom at each node.

To obtain the desired displacement vector with global displacement components at nodes 1 and 2 and local displacement components at node 3, we use Eq. (3.9.5) to obtain

$$\left\{ \begin{array}{c} u_1 \\ v_1 \\ u_2 \\ v_2 \\ u_3 \\ v_3 \end{array} \right\} = \left[\begin{array}{ccc} [I] & [0] & [0] \\ [0] & [I] & [0] \\ [0] & [0] & [t_3] \end{array} \right]^T \left\{ \begin{array}{c} u_1' \\ v_1' \\ u_2' \\ v_2' \\ u_3' \\ v_3' \end{array} \right\} \tag{3.9.7}$$

In Eq. (3.9.7), we observe that only the node 3 global components are transformed, as indicated by the placement of the $[t_3]^T$ matrix. We denote the square matrix in Eq. (3.9.7) by $[T_1]^T$. In general, we place a 2×2 $[t]$ matrix in $[T_1]$ wherever the transformation from global to local displacements is needed (where skewed supports exist).

Upon considering Eqs. (3.9.5) and (3.9.6), we observe that only node 3 components of $\{d\}$ are really transformed to local (skewed) axes components. This transformation is indeed necessary whenever the local axes x'-y' fixity directions are known.

Furthermore, the global force vector can also be transformed by using the same transformation as for $\{d'\}$:

$$\{f'\} = [T_1]\{f\} \qquad (3.9.8)$$

In global coordinates, we then have

$$\{f\} = [K]\{d\} \qquad (3.9.9)$$

Premultiplying Eq. (3.9.9) by $[T_1]$, we have

$$[T_1]\{f\} = [T_1][K]\{d\} \qquad (3.9.10)$$

For the truss in Figure 3–23, the left side of Eq. (3.9.10) is

$$\begin{bmatrix} [I] & [0] & [0] \\ [0] & [I] & [0] \\ [0] & [0] & [t_3] \end{bmatrix} \begin{Bmatrix} f_{1x} \\ f_{1y} \\ f_{2x} \\ f_{2y} \\ f_{3x} \\ f_{3y} \end{Bmatrix} = \begin{Bmatrix} f_{1x} \\ f_{1y} \\ f_{2x} \\ f_{2y} \\ f'_{3x} \\ f'_{3y} \end{Bmatrix} \qquad (3.9.11)$$

where the fact that local forces transform similarly to Eq. (3.9.2) as

$$\{f'_3\} = [t_3]\{f_3\} \qquad (3.9.12)$$

has been used in Eq. (3.9.11). From Eq. (3.9.11), we see that only the node 3 components of $\{f\}$ have been transformed to the local axes components, as desired.

Using Eq. (3.9.5) in Eq. (3.9.10), we have

$$[T_1]\{f\} = [T_1][K][T_1]^T\{d'\} \qquad (3.9.13)$$

Using Eq. (3.9.11), we find that the form of Eq. (3.9.13) becomes

$$\begin{Bmatrix} F_{1x} \\ F_{1y} \\ F_{2x} \\ F_{2y} \\ F'_{3x} \\ F'_{3y} \end{Bmatrix} = [T_1][K][T_1]^T \begin{Bmatrix} u_1 \\ v_1 \\ u_2 \\ v_2 \\ u'_3 \\ v'_3 \end{Bmatrix} \qquad (3.9.14)$$

as $u_1 = u'_1, v_1 = v'_1, u_2 = u'_2$, and $v_2 = v'_2$ from Eq. (3.9.7). Equation (3.9.14) is the desired form that allows all known global and inclined boundary conditions to be enforced. The global forces now result in the left side of Eq. (3.9.14). To solve Eq. (3.9.14), first perform the matrix triple product $[T_1][K][T_1]^T$. Then invoke the following boundary conditions (for the truss in Figure 3–23):

$$u_1 = 0 \qquad v_1 = 0 \qquad v_3' = 0 \qquad (3.9.15)$$

Then substitute the known value of the applied force F_{2x} along with $F_{2y} = 0$ and $F_{3x}' = 0$ into Eq. (3.9.14). Finally, partition the equations with known displacements— here equations 1, 2, and 6 of Eq. (3.9.14)—and then simultaneously solve those associated with the unknown displacements u_2, v_2, and u_3'.

After solving for the displacements, return to Eq. (3.9.14) to obtain the global reactions F_{1x} and F_{1y} and the inclined roller reaction F_{3y}'.

Example 3.11

For the plane truss shown in Figure 3–24, determine the displacements and reactions. Let $E = 210$ GPa, $A = 6.00 \times 10^{-4}$ m^2 for elements 1 and 2, and $A = 6\sqrt{2} \times 10^{-4}$ m^2 for element 3.

SOLUTION:

We begin by using Eq. (3.4.23) to determine each element stiffness matrix.

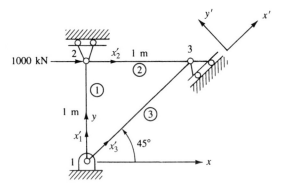

Figure 3–24 Plane truss with inclined support

Element 1

$$\theta^{(1)} = 90°, \quad \cos\theta = 0 \quad \sin\theta = 1$$

$$[k^{(1)}] = \frac{(6.0 \times 10^{-4} \text{ m}^2)(210 \times 10^9 \text{ N/m}^2)}{1 \text{ m}} \begin{array}{cccc} u_1 & v_1 & u_2 & v_2 \\ \begin{bmatrix} 0 & 0 & 0 & 0 \\ & 1 & 0 & -1 \\ & & 0 & 0 \\ \text{Symmetry} & & & 1 \end{bmatrix} \end{array} \qquad (3.9.16)$$

Element 2

$$\theta^{(2)} = 0°, \quad \cos\theta = 1 \quad \sin\theta = 0$$

$$[k^{(2)}] = \frac{(6.0 \times 10^{-4} \text{ m}^2)(210 \times 10^9 \text{ N/m}^2)}{1 \text{ m}} \begin{array}{cccc} u_2 & v_2 & u_3 & v_3 \end{array} \begin{bmatrix} 1 & 0 & -1 & 0 \\ & 0 & 0 & 0 \\ & & 1 & 0 \\ \text{Symmetry} & & & 0 \end{bmatrix} \tag{3.9.17}$$

Element 3

$$\theta^{(3)} = 45°, \quad \cos\theta = \frac{\sqrt{2}}{2} \quad \sin\theta = \frac{\sqrt{2}}{2}$$

$$[k^{(3)}] = \frac{(6\sqrt{2} \times 10^{-4} \text{ m}^2)(210 \times 10^9 \text{ N/m}^2)}{\sqrt{2} \text{ m}} \begin{array}{cccc} u_1 & v_1 & u_3 & v_3 \end{array} \begin{bmatrix} 0.5 & 0.5 & -0.5 & -0.5 \\ & 0.5 & -0.5 & -0.5 \\ & & 0.5 & 0.5 \\ \text{Symmetry} & & & 0.5 \end{bmatrix} \tag{3.9.18}$$

Using the direct stiffness method on Eqs. (3.9.16) through (3.9.18), we obtain the global $[K]$ matrix as

$$[K] = 1260 \times 10^5 \text{ N/m} \begin{bmatrix} 0.5 & 0.5 & 0 & 0 & -0.5 & -0.5 \\ & 1.5 & 0 & -1 & -0.5 & -0.5 \\ & & 1 & 0 & -1 & 0 \\ & & & 1 & 0 & 0 \\ & & & & 1.5 & 0.5 \\ \text{Symmetry} & & & & & 0.5 \end{bmatrix} \tag{3.9.19}$$

Next we obtain the transformation matrix $[T_1]$ using Eq. (3.9.6) to transform the global displacements at node 3 into local nodal coordinates x'-y'. In using Eq. (3.9.6), the angle α is 45°.

$$[T_1] = \begin{bmatrix} 1 & 0 & 0 & 0 & 0 & 0 \\ 0 & 1 & 0 & 0 & 0 & 0 \\ 0 & 0 & 1 & 0 & 0 & 0 \\ 0 & 0 & 0 & 1 & 0 & 0 \\ 0 & 0 & 0 & 0 & \sqrt{2}/2 & \sqrt{2}/2 \\ 0 & 0 & 0 & 0 & -\sqrt{2}/2 & \sqrt{2}/2 \end{bmatrix} \tag{3.9.20}$$

Next we use Eq. (3.9.14) (in general, we would use Eq. (3.9.13)) to express the assembled equations. First define $[K]^* = [T_1][K][T_1]^T$ and evaluate in steps as follows:

$$[T_1][K] = 1260 \times 10^5 \begin{bmatrix} 0.5 & 0.5 & 0 & 0 & -0.5 & -0.5 \\ 0.5 & 1.5 & 0 & -1 & -0.5 & -0.5 \\ 0 & 0 & 1 & 0 & -1 & 0 \\ 0 & -1 & 0 & 1 & 0 & 0 \\ -0.707 & -0.707 & -0.707 & 0 & 1.414 & 0.707 \\ 0 & 0 & 0.707 & 0 & -0.707 & 0 \end{bmatrix} \quad (3.9.21)$$

and

$$[T_1][K][T_1]^T = 1260 \times 10^5 \text{ N/m} \begin{array}{cccccc} u_1 & v_1 & u_2 & v_2 & u_3' & v_3' \end{array}$$

$$\begin{bmatrix} 0.5 & 0.5 & 0 & 0 & -0.707 & 0 \\ 0.5 & 1.5 & 0 & -1 & -0.707 & 0 \\ 0 & 0 & 1 & 0 & -0.707 & 0.707 \\ 0 & -1 & 0 & 1 & 0 & 0 \\ -0.707 & -0.707 & -0.707 & 0 & 1.500 & -0.500 \\ 0 & 0 & 0.707 & 0 & -0.500 & 0.500 \end{bmatrix}$$

$$(3.9.22)$$

Notice in comparing $[K^*]$ in Eq. (3.9.22) to $[K]$ from Eq. (3.9.19) that only the stiffness terms associated with skewed node 3 degrees of freedom have changed as expected.

Applying the boundary conditions, $u_1 = v_1 = v_2 = v_3' = 0$, to Eq. (3.9.22), we obtain

$$\left\{ \begin{array}{l} F_{2x} = 1000 \text{ kN} \\ F_{3x}' = 0 \end{array} \right\} = (126 \times 10^3 \text{ kN/m}) \begin{bmatrix} 1 & -0.707 \\ -0.707 & 1.50 \end{bmatrix} \left\{ \begin{array}{l} u_2 \\ u_3' \end{array} \right\} \quad (3.9.23)$$

Solving Eq. (3.9.23) for the displacements yields

$$u_2 = 11.91 \times 10^{-3} \text{ m} \quad (3.9.24)$$

$$u_3' = 5.613 \times 10^{-3} \text{ m}$$

Postmultiplying the known displacement vector times Eq. (3.9.22) (see Eq. (3.9.14), we obtain the reactions as

$$F_{1x} = -500 \text{ kN}$$

$$F_{1y} = -500 \text{ kN}$$

$$F_{2y} = 0 \quad (3.9.25)$$

$$F_{3y}' = 707 \text{ kN}$$

The free-body diagram of the truss with the reactions is shown in Figure 3–25. You can easily verify that the truss is in equilibrium. ■

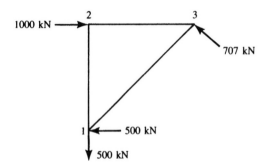

Figure 3–25 Free-body diagram of the truss of Figure 3–24

In the second method used to handle skewed boundary conditions, we use a boundary element of large stiffness to constrain the desired displacement. This is the method used in some computer programs [9].

Boundary elements are used to specify nonzero displacements and rotations to nodes. They are also used to evaluate reactions at rigid and flexible supports. Boundary elements are two-node elements. The line defined by the two nodes specifies the direction along which the force reaction is evaluated or the displacement is specified. In the case of moment reaction, the line specifies the axis about which the moment is evaluated and the rotation is specified.

We consider boundary elements that are used to obtain reaction forces (rigid boundary elements) or specify translational displacements (displacement boundary elements) as truss elements with only one nonzero translational stiffness. Boundary elements used to either evaluate reaction moments or specify rotations behave like beam elements with only one nonzero stiffness corresponding to the rotational stiffness about the specified axis.

The elastic boundary elements are used to model flexible supports and to calculate reactions at skewed or inclined boundaries. Consult Reference [9] for more details about using boundary elements.

▲ 3.10 Potential Energy Approach to Derive Bar Element Equations ▲

We now present the principle of minimum potential energy to derive the bar element equations. Recall from Section 2.6 that the total potential energy π_p was defined as the sum of the internal strain energy U and the potential energy of the external forces Ω:

$$\pi_p = U + \Omega \qquad (3.10.1)$$

To evaluate the strain energy for a bar, we consider only the work done by the internal forces during deformation. Because we are dealing with a one-dimensional bar, the internal force doing work on a differential element of sides Δx, Δy, Δz, is given in Figure 3–26 as $\sigma_x(\Delta y)(\Delta z)$, due only to normal stress σ_x. The displacement of the x face of the element is $\Delta x(\varepsilon_x)$; the displacement of the $x + \Delta x$ face is $\Delta x(\varepsilon_x + d\varepsilon_x)$. The change in

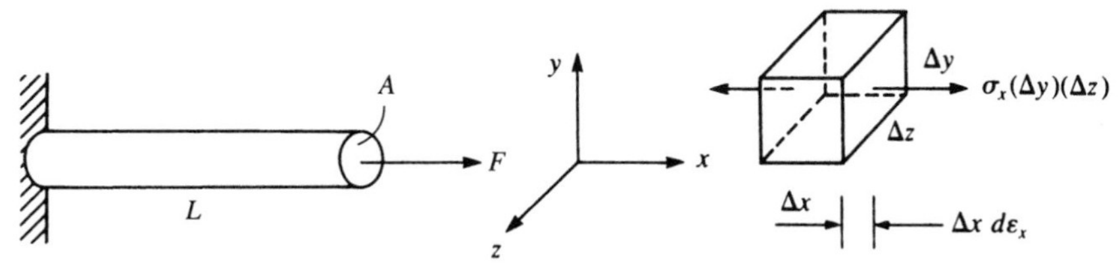

Figure 3–26 Internal force in a one-dimensional bar due to applied external force F

displacement is then $\Delta x \, d\varepsilon_x$, where $d\varepsilon_x$ is the differential change in strain occurring over length Δx. The differential internal work (or strain energy) dU is the internal force multiplied by the displacement through which the force moves, given by

$$dU = \sigma_x(\Delta y)(\Delta z)(\Delta x) \, d\varepsilon_x \qquad (3.10.2)$$

Rearranging and letting the volume of the element approach zero, we obtain, from Eq. (3.10.2),

$$dU = \sigma_x \, d\varepsilon_x \, dV \qquad (3.10.3)$$

For the whole bar, we then have

$$U = \iiint_V \left\{ \int_0^{\varepsilon_x} \sigma_x \, d\varepsilon_x \right\} dV \qquad (3.10.4)$$

Now, for a linear-elastic (Hooke's law) material as shown in Figure 3–27, we see that $\sigma_x = E\varepsilon_x$. Hence substituting this relationship into Eq. (3.10.4), integrating with respect to ε_x, and then resubstituting σ_x for $E\varepsilon_x$, we have

$$U = \frac{1}{2} \iiint_V \sigma_x \varepsilon_x \, dV \qquad (3.10.5a)$$

as the expression for the strain energy for one-dimensional stress.

For a uniform cross-sectional area A of a bar with stress and strain dependent only on the x coordinate, Eq. (3.10.5a) can be simplified to

$$U = \frac{A}{2} \int_x \sigma_x \varepsilon_x \, dx \qquad (3.10.5b)$$

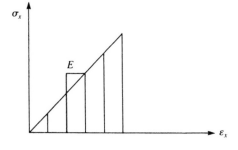

Figure 3–27 Stress-strain curve for Linear-elastic (Hooke's law) material

We observe from the integral in Eq. (3.10.5b) that the strain energy is described as the area under the stress/strain curve.

The potential energy of the external forces, being opposite in sign from the external work expression because the potential energy of external forces is lost when the work is done by the external forces, is given by

$$\Omega = -\iiint_V X_b u \, dV - \iint_{S_1} T_x u_s \, dS - \sum_{i=1}^{M} f_{ix} u_i \qquad (3.10.6)$$

where the first, second, and third terms on the right side of Eq. (3.10.6) represent the potential energy of (1) body forces X_b, typically from the self-weight of the bar (in units of force per unit volume) moving through displacement function u, (2) surface loading or traction T_x, typically from distributed loading acting along the surface of the element (in units of force per unit surface area) moving through displacements u_s, where u_s are the displacements occurring over surface S_1, and (3) nodal concentrated forces f_{ix} moving through nodal displacements u_i. The forces X_b, T_x, and f_{ix} are considered to act in the local x direction of the bar as shown in Figure 3–28. In Eqs. (3.10.5) and (3.10.6), V is the volume of the body and S_1 is the part of the surface S on which surface loading acts. For a bar element with two nodes and one degree of freedom per node, $M = 2$.

We are now ready to describe the finite element formulation of the bar element equations by using the principle of minimum potential energy.

The finite element process seeks a minimum in the potential energy within the constraint of an assumed displacement pattern within each element. The greater the number of degrees of freedom associated with the element (usually meaning increasing the number of nodes), the more closely will the solution approximate the true one and ensure complete equilibrium (provided the true displacement can, in the limit, be approximated). An approximate finite element solution found by using the stiffness method will always provide an approximate value of potential energy greater than or equal to the correct one. This method also results in a structure behavior that is predicted to be physically stiffer than, or at best to have the same stiffness as, the actual one. This is explained by the fact that the structure model is allowed to displace only into shapes defined by the terms of the assumed displacement field within each element of the structure. The correct shape is usually only approximated by the assumed field, although the correct shape can be the same as the assumed field. The assumed field effectively constrains the structure from deforming in its natural manner. This constraint effect stiffens the predicted behavior of the structure.

Apply the following steps when using the principle of minimum potential energy to derive the finite element equations.

1. Formulate an expression for the total potential energy.
2. Assume the displacement pattern to vary with a finite set of undetermined parameters (here these are the nodal displacements u_i), which are substituted into the expression for total potential energy.
3. Obtain a set of simultaneous equations minimizing the total potential energy with respect to these nodal parameters. These resulting equations represent the element equations.

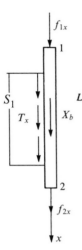

Figure 3–28 General forces acting on a one-dimensional bar

The resulting equations are the approximate (or possibly exact) equilibrium equations whose solution for the nodal parameters seeks to minimize the potential energy when back-substituted into the potential energy expression. The preceding three steps will now be followed to derive the bar element equations and stiffness matrix.

Consider the bar element of length L, with constant cross-sectional area A, shown in Figure 3–28. Using Eqs. (3.10.5) and (3.10.6), we find that the total potential energy, Eq. (3.10.1), becomes

$$\pi_p = \frac{A}{2}\int_0^L \sigma_x \varepsilon_x \, dx - f_{1x}u_1 - f_{2x}u_2 - \iint_{S_1} u_s T_x \, dS - \iiint_V u X_b \, dV \qquad (3.10.7)$$

because A is a constant and variables σ_x and ε_x at most vary with x.

From Eqs. (3.1.3) and (3.1.4), we have the axial displacement function expressed in terms of the shape functions and nodal displacements by

$$u = [N]\{d\} \qquad u_s = [N_S]\{d\} \qquad (3.10.8)$$

where

$$[N] = \left[1 - \frac{x}{L} \quad \frac{x}{L}\right] \qquad (3.10.9)$$

$[N_S]$ is the shape function matrix evaluated over the surface that the distributed surface traction acts and

$$\{d\} = \left\{\begin{array}{c} u_1 \\ u_2 \end{array}\right\} \qquad (3.10.10)$$

Then, using the strain/displacement relationship $\varepsilon_x = du/dx$, we can write the axial strain in matrix form as

$$\{\varepsilon_x\} = \left[-\frac{1}{L} \quad \frac{1}{L}\right]\{d\} \qquad (3.10.11)$$

or

$$\{\varepsilon_x\} = [B]\{d\} \tag{3.10.12}$$

where we define $[B]$ as the gradient matrix

$$[B] = \left[-\frac{1}{L} \quad \frac{1}{L} \right] \tag{3.10.13}$$

The axial stress-strain relationship in matrix form is given by

$$\{\sigma_x\} = [D]\{\varepsilon_x\} \tag{3.10.14}$$

where

$$[D] = [E] \tag{3.10.15}$$

for the one-dimensional stress-strain relationship matrix and E is the modulus of elasticity. Now, by Eq. (3.10.12), we can express Eq. (3.10.14) as

$$\{\sigma_x\} = [D][B]\{d\} \tag{3.10.16}$$

Using Eq. (3.10.7) expressed in matrix notation form, we have the total potential energy given by

$$\pi_p = \frac{A}{2} \int_0^L \{\sigma_x\}^T \{\varepsilon_x\} \, dx - \{d\}^T \{P\} - \iint_{S_1} \{u_s\}^T \{T_x\} \, dS - \iiint_V \{u\}^T \{X_b\} \, dV \tag{3.10.17}$$

where $\{P\}$ now represents the concentrated nodal loads and where in general both $\{\sigma_x\}$ and $\{\varepsilon_x\}$ are column matrices. For proper matrix multiplication, we must place the transpose on $\{\sigma_x\}$. Similarly, $\{u\}$ and $\{T_x\}$ in general are column matrices, so for proper matrix multiplication, $\{u\}$ is transposed in Eq. (3.10.17).

Using Eqs. (3.10.8), (3.10.12), and (3.10.16) in Eq. (3.10.17), we obtain

$$\pi_p = \frac{A}{2} \int_0^L \{d\}^T [B]^T [D]^T [B]\{d\} \, dx - \{d\}^T \{P\}$$
$$- \iint_{S_1} \{d\}^T [N_S]^T \{T_x\} \, dS - \iiint_V \{d\}^T [N]^T \{X_b\} \, dV \tag{3.10.18}$$

In Eq. (3.10.18), π_p is seen to be a function of $\{d\}$; that is, $\pi_p = \pi_p(u_1, u_2)$. However, $[B]$ and $[D]$, Eqs. (3.10.13) and (3.10.15), and the nodal degrees of freedom u_1 and u_2 are not functions of x. Therefore, integrating the first integral in Eq. (3.10.18) with respect to x yields

$$\pi_p = \frac{AL}{2} \{d\}^T [B]^T [D]^T [B]\{d\} - \{d\}^T \{f\} \tag{3.10.19}$$

where

$$\{f\} = \{P\} + \iint_{S_1} [N_S]^T \{T_x\} \, dS + \iiint_V [N]^T \{X_b\} \, dV \tag{3.10.20}$$

From Eq. (3.10.20), we observe three separate types of load contributions from concentrated nodal forces, surface tractions, and body forces, respectively. We define these surface tractions and body-force matrices as

$$\{f_s\} = \iint_{S_1} [N_S]^T \{T_x\} \, dS \tag{3.10.20a}$$

$$\{f_b\} = \iiint_V [N]^T \{X_b\} \, dV \tag{3.10.20b}$$

The expression for $\{f\}$ given by Eq. (3.10.20) then describes how certain loads can be considered to best advantage.

Loads calculated by Eqs. (3.10.20a) and (3.10.20b) are called consistent because they are based on the same shape functions $[N]$ used to calculate the element stiffness matrix. The loads calculated by Eq. (3.10.20a) and (3.10.20b) are also statically equivalent to the original loading; that is, both $\{f_s\}$ and $\{f_b\}$ and the original loads yield the same resultant force and same moment about an arbitrarily chosen point.

The minimization of π_p with respect to each nodal displacement requires that

$$\frac{\partial \pi_p}{\partial u_1} = 0 \quad \text{and} \quad \frac{\partial \pi_p}{\partial u_2} = 0 \tag{3.10.21}$$

Now we explicitly evaluate π_p given by Eq. (3.10.19) to apply Eq. (3.10.21). We define the following for convenience:

$$\{U^*\} = \{d\}^T [B]^T [D]^T [B] \{d\} \tag{3.10.22}$$

Using Eqs. (3.10.10), (3.10.13), and (3.10.15) in Eq. (3.10.22) yields

$$\{U^*\} = [u_1 \quad u_2] \left\{ \begin{array}{c} -\frac{1}{L} \\ \frac{1}{L} \end{array} \right\} [E] \left[-\frac{1}{L} \quad \frac{1}{L} \right] \left\{ \begin{array}{c} u_1 \\ u_2 \end{array} \right\} \tag{3.10.23}$$

Simplifying Eq. (3.10.23), we obtain

$$U^* = \frac{E}{L^2} (u_1^2 - 2u_1 u_2 + u_2^2) \tag{3.10.24}$$

Also, the explicit expression for $\{d\}^T \{f\}$ is

$$\{d\}^T \{f\} = u_1 f_{1x} + u_2 f_{2x} \tag{3.10.25}$$

Therefore, using Eqs. (3.10.24) and (3.10.25) in Eq. (3.10.19) and then applying Eqs. (3.10.21), we obtain

$$\frac{\partial \pi_p}{\partial u_1} = \frac{AL}{2} \left[\frac{E}{L^2} (2u_1 - 2u_2) \right] - f_{1x} = 0 \tag{3.10.26}$$

and

$$\frac{\partial \pi_p}{\partial u_2} = \frac{AL}{2}\left[\frac{E}{L^2}(-2u_1 + 2u_2)\right] - f_{2x} = 0$$

In matrix form, we express Eqs. (3.10.26) as

$$\frac{\partial \pi_p}{\partial\{d\}} = \frac{AE}{L}\begin{bmatrix} 1 & -1 \\ -1 & 1 \end{bmatrix}\begin{Bmatrix} u_1 \\ u_2 \end{Bmatrix} - \begin{Bmatrix} f_{1x} \\ f_{2x} \end{Bmatrix} = \begin{Bmatrix} 0 \\ 0 \end{Bmatrix} \qquad (3.10.27)$$

or, because $\{f\} = [k]\{d\}$, we have the stiffness matrix for the bar element obtained from Eq. (3.10.27) as

$$[k] = \frac{AE}{L}\begin{bmatrix} 1 & -1 \\ -1 & 1 \end{bmatrix} \qquad (3.10.28a)$$

As expected, Eq. (3.10.28a) is identical to the stiffness matrix Eq. (3.1.14) obtained in Section 3.1.

Now that we have derived the bar stiffness matrix by using the theorem of minimum potential energy, we can observe that the strain energy U (the first term on the right side of Eq. (3.10.18)) can also be expressed in the quadratic form $U = 1/2\{d\}^T[k]\{d\}$ as follows:

$$U = \frac{1}{2}\{d\}^T[k]\{d\} = \frac{1}{2}[u_1 u_2]\frac{AE}{L}\begin{bmatrix} 1 & -1 \\ -1 & 1 \end{bmatrix}\begin{Bmatrix} u_1 \\ u_2 \end{Bmatrix} = \frac{AE}{2L}[u_1^2 - 2u_1u_2 + u_2^2] \quad (3.10.28b)$$

Finally, instead of the cumbersome process of explicitly evaluating π_p, we can use the matrix differentiation as given by Eq. (2.6.12) and apply it directly to Eq. (3.10.19) to obtain

$$\frac{\partial \pi_p}{\partial\{d\}} = AL[B]^T[D][B]\{d\} - \{f\} = 0 \qquad (3.10.29)$$

where $[D]^T = [D]$ has been used in writing Eq. (3.10.29). The result of the evaluation of $AL[B]^T[D][B]$ is then equal to $[k]$ given by Eq. (3.10.28a). Throughout this text, we will use this matrix differentiation concept (also see Appendix A), which greatly simplifies the task of evaluating $[k]$.

To illustrate the use of Eq. (3.10.20a) to evaluate the equivalent nodal loads for a bar subjected to axial loading traction T_x, we now solve Example 3.12.

Example 3.12

A bar of length L is subjected to a linearly distributed axial line loading that varies from zero at node 1 to a maximum of CL at node 2 (Figure 3–29). Determine the energy equivalent nodal loads.

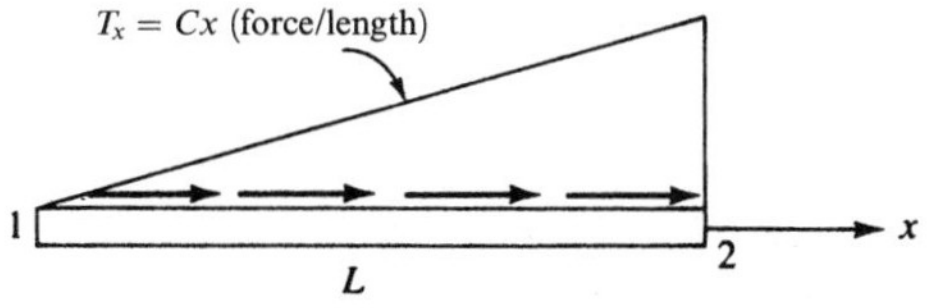

Figure 3–29 Element subjected to linearly varying axial line load

SOLUTION:

Using Eq. (3.10.20a) and shape functions from Eq. (3.10.9), we solve for the energy equivalent nodal forces of the distributed loading as follows:

$$\{f_0\} = \begin{Bmatrix} f_{1x} \\ f_{2x} \end{Bmatrix} = \iint_{S_1} [N]^T \{T_x\} \, dS = \int_0^L \begin{Bmatrix} 1 - \dfrac{x}{L} \\ \dfrac{x}{L} \end{Bmatrix} \{Cx\} \, dx \qquad (3.10.30)$$

$$= \begin{Bmatrix} \dfrac{Cx^2}{2} - \dfrac{Cx^3}{3L} \\ \dfrac{Cx^3}{3L} \end{Bmatrix} \Bigg|_0^L$$

$$= \begin{Bmatrix} \dfrac{CL^2}{6} \\ \dfrac{CL^2}{3} \end{Bmatrix} \qquad (3.10.31)$$

where the integration was carried out over the length of the bar, because T_x is in units of force/length.

Note that the total load is the area under the load distribution given by

$$F = \frac{1}{2}(L)(CL) = \frac{CL^2}{2} \qquad (3.10.32)$$

Therefore, comparing Eq. (3.10.31) with (3.10.32), we find that the equivalent nodal loads for a linearly varying load are

$$f_{1x} = \frac{1}{3} F = \text{one-third of the total load}$$

$$\qquad (3.10.33)$$

$$f_{2x} = \frac{2}{3} F = \text{two-thirds of the total load}$$

In summary, for the simple two-noded bar element subjected to a linearly varying load (triangular loading), place one-third of the total load at the node where the distributed loading begins (zero end of the load) and two-thirds of the total load at the node where the peak value of the distributed load ends. ∎

We now illustrate (Example 3.13) a complete solution for a bar subjected to a surface traction loading.

Example 3.13

For the rod loaded axially as shown in Figure 3–30, determine the axial displacement and axial stress. Let $E = 210$ GPa, $A = 12.5 \times 10^{-4}$ m^2, and $L = 1.5$ m. Use (a) one and (b) two elements in the finite element solutions. (In Section 3.11 one-, two-, four-, and eight-element solutions will be presented from the computer program Algor [9].

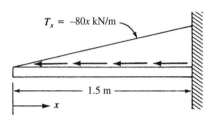

$T_x = -80x$ kN/m

1.5 m

x

Figure 3–30 Rod subjected to triangular load distribution

(a) One-element solution (Figure 3–31).

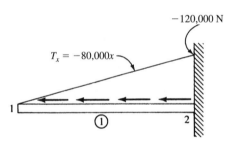

$-120,000$ N

$T_x = -80,000x$

①   2

Figure 3–31 One-element model

SOLUTION:

From Eq. (3.10.20a), the distributed load matrix is evaluated as follows:

$$\{F_0\} = \int_0^L [N]^T \{T_x\}\, dx \tag{3.10.34}$$

where T_x is a line load in units of pounds per inch and $\{f_0\} = \{F_0\}$. Therefore, using Eq. (3.1.4) for $[N]$ in Eq. (3.10.34), we obtain

$$\{F_0\} = \int_0^L \left\{ \begin{array}{c} 1 - \dfrac{x}{L} \\[2mm] \dfrac{x}{L} \end{array} \right\} \{-80,000x\}\, dx \tag{3.10.35}$$

or
$$\left\{ \begin{array}{c} F_{1x} \\ F_{2x} \end{array} \right\} = \left\{ \begin{array}{c} \dfrac{-80,000L^2}{2} + \dfrac{80,000L^2}{3} \\[3mm] \dfrac{-80,000L^2}{3} \end{array} \right\} = \left\{ \begin{array}{c} \dfrac{-80,000L^2}{6} \\[3mm] \dfrac{-80,000L^2}{3} \end{array} \right\} = \left\{ \begin{array}{c} \dfrac{-80,000(1.5)^2}{6} \\[3mm] \dfrac{-80,000(1.5)^2}{3} \end{array} \right\}$$

or
$$F_{1x} = -30,000 \text{ N} \qquad F_{2x} = -60,000 \text{ N} \qquad (3.10.36)$$

Using Eq. (3.10.33), we could have determined the same forces at nodes 1 and 2—that is, one-third of the total load is at node 1 and two-thirds of the total load is at node 2. Using Eq. (3.10.28), we find that the stiffness matrix is given by

$$[k^{(1)}] = 16.67 \times 10^7 \begin{bmatrix} 1 & -1 \\ -1 & 1 \end{bmatrix}$$

The element equations are then

$$16.67 \times 10^7 \begin{bmatrix} 1 & -1 \\ -1 & 1 \end{bmatrix} \begin{Bmatrix} u_1 \\ 0 \end{Bmatrix} = \begin{Bmatrix} -30,000 \\ R_{2x} - 60,000 \end{Bmatrix} \qquad (3.10.37)$$

Solving Eq. 1 of Eq. (3.10.37), we obtain

$$u_1 = -0.00018 \text{ m} = 0.18 \text{ mm} \qquad (3.10.38)$$

The stress is obtained from Eq. (3.10.14) as

$$\{\sigma_x\} = [D]\{\varepsilon_x\}$$
$$= E[B]\{d\}$$
$$= E\begin{bmatrix} -\dfrac{1}{L} & \dfrac{1}{L} \end{bmatrix} \begin{Bmatrix} u_1 \\ u_2 \end{Bmatrix}$$
$$= E\left(\dfrac{u_2 - u_1}{L}\right)$$
$$= (2 \times 10^{11})\left(\dfrac{0 + 0.00018}{1.5}\right)$$
$$= 24 \text{ MPa } (T) \qquad (3.10.39)$$

(b) Two-element solution (Figure 3–32).

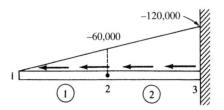

Figure 3–32 Two-element model

We first obtain the element forces. For element 2, we divide the load into a uniform part and a triangular part as shown in Figure 3–32. For the uniform part, half the total uniform load is placed at each node associated with the element. Therefore, the total uniform part is

$$(0.75 \text{ m})(-60,000 \text{ N/m}) = -45,000 \text{ N}$$

and using Eq. (3.10.33) for the triangular part of the load, we have, for element 2,

$$\left\{ \begin{array}{c} f_{2x}^{(2)} \\ f_{3x}^{(2)} \end{array} \right\} = \left\{ \begin{array}{c} -[\frac{1}{2}(45,000) + \frac{1}{3}(22,500)] \\ -[\frac{1}{2}(45,000) + \frac{2}{3}(22,500)] \end{array} \right\} = \left\{ \begin{array}{c} -30,000 \text{ N} \\ -37,500 \text{ N} \end{array} \right\} \tag{3.10.40}$$

For element 1, the total force is from the triangle-shaped distributed load only and is given by

$$\frac{1}{2}(0.75 \text{ m})(-60,000 \text{ N/m}) = -22,500 \text{ N}$$

On the basis of Eq. (3.10.33), this load is separated into nodal forces as shown:

$$\left\{ \begin{array}{c} f_{1x}^{(1)} \\ f_{2x}^{(1)} \end{array} \right\} = \left\{ \begin{array}{c} \frac{1}{3}(-22,500) \\ \frac{2}{3}(-22,500) \end{array} \right\} = \left\{ \begin{array}{c} -7500 \text{ N} \\ -15,000 \text{ N} \end{array} \right\} \tag{3.10.41}$$

The final nodal force matrix is then

$$\left\{ \begin{array}{c} F_{1x} \\ F_{2x} \\ F_{3x} \end{array} \right\} = \left\{ \begin{array}{c} -7500 \\ -30,000 - 15,000 \\ R_{3x} - 37,500 \end{array} \right\} \tag{3.10.42}$$

The element stiffness matrices are now

$$[k^{(1)}] = [k^{(2)}] = \frac{AE}{L/2} \begin{array}{cc} 1 & 2 \\ 2 & 3 \\ \begin{bmatrix} 1 & -1 \\ -1 & 1 \end{bmatrix} \end{array} = 33.34 \times 10^7 \begin{array}{cc} 1 & 2 \\ 2 & 3 \\ \begin{bmatrix} 1 & -1 \\ -1 & 1 \end{bmatrix} \end{array} \tag{3.10.43}$$

The assembled global stiffness matrix is

$$[K] = 33.34 \times 10^7 \begin{bmatrix} 1 & -1 & 0 \\ -1 & 2 & -1 \\ 0 & -1 & 1 \end{bmatrix} \frac{\text{N}}{\text{m}} \tag{3.10.44}$$

The assembled global equations are then

$$33.34 \times 10^7 \begin{bmatrix} 1 & -1 & 0 \\ -1 & 2 & -1 \\ 0 & -1 & 1 \end{bmatrix} \left\{ \begin{array}{c} u_1 \\ u_2 \\ u_3 = 0 \end{array} \right\} = \left\{ \begin{array}{c} -7,500 \\ -45,000 \\ R_{3x} - 37,500 \end{array} \right\} \tag{3.10.45}$$

where the boundary condition $u_3 = 0$ has been substituted into Eq. (3.10.45). Now, solving equations 1 and 2 of Eq. (3.10.45), we obtain

$$u_1 = -0.00018 \text{ m} = -0.18 \text{ mm}$$
$$u_2 = -0.0001575 \text{ m} = -0.1575 \text{ mm} \tag{3.10.46}$$

The element stresses are as follows:

Element 1

$$\sigma_x = E\left[-\frac{1}{(0.75)} \quad \frac{1}{(0.75)}\right]\left\{\begin{array}{l} u_1 = -0.00018 \\ u_2 = -0.0001575 \end{array}\right\}$$

$$= 6 \text{ MPa } (T) \tag{3.10.47}$$

Element 2

$$\sigma_x = E\left[-\frac{1}{(0.75)} \quad \frac{1}{(0.75)}\right]\left\{\begin{array}{l} u_2 = -0.0001575 \\ u_3 = 0 \end{array}\right\}$$

$$= 42 \text{ MPa } (T) \tag{3.10.48}$$

■

▲ 3.11 Comparison of Finite Element Solution to Exact Solution for Bar ▲

We will now compare the finite element solutions for Example 3.13 using one, two, four, and eight elements to model the bar element and the exact solution. The exact solution for displacement is obtained by solving the equation

$$u = \frac{1}{AE}\int_0^x P(x)\,dx \tag{3.11.1}$$

where, using the following free-body diagram,

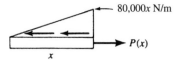

80,000x N/m

P(x)

x

we have

$$P(x) = \tfrac{1}{2}x(80{,}000x) = 40{,}000x^2 \text{ N} \tag{3.11.2}$$

Therefore, substituting Eq. (3.11.2) into Eq. (3.11.1), we have

$$u = \frac{1}{AE}\int_0^x 40{,}000x^2\,dx$$

$$= \frac{40{,}000x^3}{3AE} + C_1 \tag{3.11.3}$$

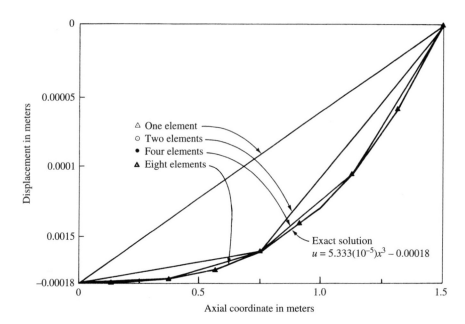

Figure 3–33 Comparison of exact and finite element solutions for axial displacement (along length of bar)

Now, applying the boundary condition at $x = L$, we obtain

$$u(L) = 0 = \frac{40,000L^3}{3AE} + C_1$$

or

$$C_1 = -\frac{40,000L^3}{3AE} \tag{3.11.4}$$

Substituting Eq. (3.11.4) into Eq. (3.11.3) makes the final expression for displacement

$$u = \frac{40,000}{3AE}(x^3 - L^3) \tag{3.11.5}$$

Substituting $A = 12.5 \times 10^{-4}$ m^2, $E = 2 \times 10^{11}$ N/m^2, and $L = 1.5$ m into Eq. (3.11.5), we obtain

$$u = 5.333 \times 10^{-5}x^3 - 0.00018 \tag{3.11.6}$$

The exact solution for axial stress is obtained by solving the equation

$$\sigma(x) = \frac{P(x)}{A} = \frac{40,000x^2}{12.5 \times 10^{-4}} = 32x^2 \text{ N/m}^2 \tag{3.11.7}$$

Figure 3–33 shows a plot of Eq. (3.11.6) along with the finite element solutions (part of which were obtained in Example 3.13). Some conclusions from these results follow.

1. The finite element solutions match the exact solution at the node points. The reason why these nodal values are correct is that the

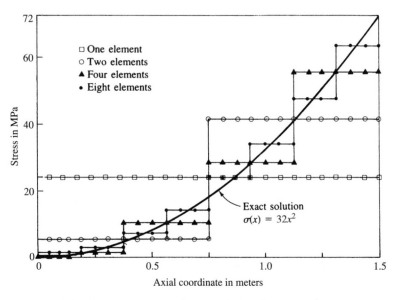

Figure 3–34 Comparison of exact and finite element solutions for axial stress (along length of bar)

element nodal forces were calculated on the basis of being energy-equivalent to the distributed load based on the assumed linear displacement field within each element. (For uniform cross-sectional bars and beams, the nodal degrees of freedom are exact. In general, computed nodal degrees of freedom are not exact.)

2. Although the node values for displacement match the exact solution, the values at locations between the nodes are poor using few elements (see one- and two-element solutions) because we used a linear displacement function within each element, whereas the exact solution, Eq. (3.11.6), is a cubic function. However, because we use increasing numbers of elements, the finite element solution converges to the exact solution (see the four- and eight-element solutions in Figure 3–33).

3. The stress is derived from the slope of the displacement curve as $\sigma = E\varepsilon = E(du/dx)$. Therefore, by the finite element solution, because u is a linear function in each element, axial stress is constant in each element. It then takes even more elements to model the first derivative of the displacement function or, equivalently, the axial stress. This is shown in Figure 3–34, where the best results occur for the eight-element solution.

4. The best approximation of the stress occurs at the midpoint of the element, not at the nodes (Figure 3–34). This is because the derivative of displacement is better predicted between the nodes than at the nodes.

5. The stress is not continuous across element boundaries. Therefore, equilibrium is not satisfied across element boundaries. Also, equilibrium

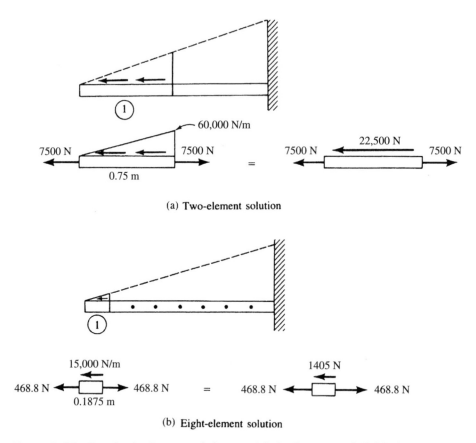

(a) Two-element solution

(b) Eight-element solution

Figure 3–35 Free-body diagram of element 1 in both two- and eight-element models, showing that equilibrium is not satisfied

within each element is, in general, not satisfied. This is shown in Figure 3–35 for element 1 in the two-element solution and element 1 in the eight-element solution [in the eight-element solution the forces are obtained from the Algor computer code [9]]. As the number of elements used increases, the discontinuity in the stress decreases across element boundaries, and the approximation of equilibrium improves.

Finally, in Figure 3–36, we show the convergence of axial stress at the fixed end $(x = L)$ as the number of elements increases.

However, if we formulate the problem in a customary general way, as described in detail in Chapter 4 for beams subjected to distributed loading, we can obtain the exact stress distribution with any of the models used. That is, letting $\{f\} = [k]\{d\} - \{f_0\}$, where $\{f_0\}$ is the initial nodal replacement force system of the distributed load on each element, we subtract the initial replacement force system from the $[k]\{d\}$ result. This yields the nodal forces in each element. For example, considering element 1 of the two-element model, we have [see also

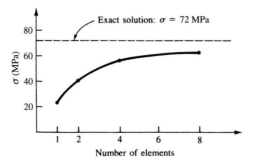

Figure 3–36 Axial stress at fixed end as number of elements increases

Eqs. (3.10.33) and (3.10.41)]

$$\{f_0\} = \left\{ \begin{array}{c} -7,500 \text{ N} \\ -15,000 \text{ N} \end{array} \right\}$$

Using $\{f\} = [k]\{d\} - \{f_0\}$, we obtain

$$\{f\} = \frac{(12.5 \times 10^{-4})(2 \times 10^{11})}{(0.75 \text{ m})} \begin{bmatrix} 1 & -1 \\ -1 & 1 \end{bmatrix} \left\{ \begin{array}{c} -0.00018 \text{ m} \\ -0.0001575 \text{ m} \end{array} \right\} - \left\{ \begin{array}{c} -7,500 \text{ N} \\ -15,000 \text{ N} \end{array} \right\}$$

$$= \left\{ \begin{array}{c} -7,500 + 7,500 \\ 7,500 + 15,000 \end{array} \right\} = \left\{ \begin{array}{c} 0 \\ 22,500 \end{array} \right\}$$

as the actual nodal forces. Drawing a free-body diagram of element 1, we have

$$\sum F_x = 0: \quad -\frac{1}{2}(60,000 \text{ N/m})(0.75 \text{ m}) + 22,500 \text{ N} = 0$$

For other kinds of elements (other than beams), this adjustment is ignored in practice. The adjustment is less important for plane and solid elements than for beams. Also, these adjustments are more difficult to formulate for an element of general shape.

▲ 3.12 Galerkin's Residual Method and Its Use to Derive the One-Dimensional Bar Element Equations ▲

General Formulation

We developed the bar finite element equations by the direct method in Section 3.1 and by the potential energy method (one of a number of variational methods) in Section 3.10. In fields other than structural/solid mechanics, it is quite probable that a variational principle, analogous to the principle of minimum potential energy, for instance, may not be known or even exist. In some flow problems in fluid mechanics and in mass transport problems (Chapter 13), we often have only the differential equation and boundary conditions available. However, the finite element method can still be applied.

The methods of weighted residuals applied directly to the differential equation can be used to develop the finite element equations. In this section, we describe Galerkin's residual method in general and then apply it to the bar element. This development provides the basis for later applications of Galerkin's method to the beam element in Chapter 4 and to the nonstructural heat-transfer element (specifically, the one-dimensional combined conduction, convection, and mass transport element described in Chapter 13). Because of the mass transport phenomena, the variational formulation is not known (or certainly is difficult to obtain), so Galerkin's method is necessarily applied to develop the finite element equations.

There are a number of other residual methods. Among them are collocation, least squares, and subdomain as described in Section 3.13. (For more on these methods, see Reference [5].)

In weighted residual methods, a trial or approximate function is chosen to approximate the independent variable, such as a displacement or a temperature, in a problem defined by a differential equation. This trial function will not, in general, satisfy the governing differential equation. Thus substituting the trial function into the differential equation results in a residual over the whole region of the problem as follows:

$$\iiint_V R \, dV = \text{minimum} \tag{3.12.1}$$

In the residual method, we require that a weighted value of the residual be a minimum over the whole region. The weighting functions allow the weighted integral of residuals to go to zero. If we denote the weighting function by W, the general form of the weighted residual integral is

$$\iiint_V RW \, dV = 0 \tag{3.12.2}$$

Using Galerkin's method, we choose the interpolation function, such as Eq. (3.1.3), in terms of N_i shape functions for the independent variable in the differential equation. In general, this substitution yields the residual $R \neq 0$. By the Galerkin criterion, the shape functions N_i are chosen to play the role of the weighting functions W. Thus for each i, we have

$$\iiint_V RN_i \, dV = 0 \qquad (i = 1, 2, \ldots, n) \tag{3.12.3}$$

Equation (3.12.3) results in a total of n equations. Equation (3.12.3) applies to points within the region of a body without reference to boundary conditions such as specified applied loads or displacements. To obtain boundary conditions, we apply integration by parts to Eq. (3.12.3), which yields integrals applicable for the region and its boundary.

Bar Element Formulation

We now illustrate Galerkin's method to formulate the bar element stiffness equations. We begin with the basic differential equation, without distributed load, derived in

Section 3.1 as

$$\frac{d}{dx}\left(AE\frac{du}{dx}\right) = 0 \tag{3.12.4}$$

where constants A and E are now assumed. The residual R is now defined to be Eq. (3.12.4). Applying Galerkin's criterion [Eq. (3.12.3)] to Eq. (3.12.4), we have

$$\int_0^L \frac{d}{dx}\left(AE\frac{du}{dx}\right)N_i\,dx = 0 \qquad (i = 1,2) \tag{3.12.5}$$

We now apply integration by parts to Eq. (3.12.5). Integration by parts is given in general by

$$\int u\,dv = uv - \int v\,du \tag{3.12.6}$$

where u and v are simply variables in the general equation. Letting

$$u = N_i \qquad du = \frac{dN_i}{dx}\,dx$$
$$dv = \frac{d}{dx}\left(AE\frac{du}{dx}\right)dx \qquad v = AE\frac{du}{dx} \tag{3.12.7}$$

in Eq. (3.12.5) and integrating by parts according to Eq. (3.12.6), we find that Eq. (3.12.5) becomes

$$\left(N_i AE\frac{du}{dx}\right)\Bigg|_0^L - \int_0^L AE\frac{du}{dx}\frac{dN_i}{dx}\,dx = 0 \tag{3.12.8}$$

where the integration by parts introduces the boundary conditions.
    Recall that, because $u = [N]\{d\}$, we have

$$\frac{du}{dx} = \frac{dN_1}{dx}u_1 + \frac{dN_2}{dx}u_2 \tag{3.12.9}$$

or, when Eqs. (3.1.4) are used for $N_1 = 1 - x/L$ and $N_2 = x/L$,

$$\frac{du}{dx} = \begin{bmatrix} -\dfrac{1}{L} & \dfrac{1}{L} \end{bmatrix}\begin{Bmatrix} u_1 \\ u_2 \end{Bmatrix} \tag{3.12.10}$$

Using Eq. (3.12.10) in Eq. (3.12.8), we then express Eq. (3.12.8) as

$$AE\int_0^L \frac{dN_i}{dx}\begin{bmatrix} -\dfrac{1}{L} & \dfrac{1}{L} \end{bmatrix}dx\begin{Bmatrix} u_1 \\ u_2 \end{Bmatrix} = \left(N_i AE\frac{du}{dx}\right)\Bigg|_0^L \qquad (i = 1,2) \tag{3.12.11}$$

Equation (3.12.11) is really two equations (one for $N_i = N_1$ and one for $N_i = N_2$). First, using the weighting function $N_i = N_1$, we have

$$AE\int_0^L \frac{dN_1}{dx}\begin{bmatrix} -\dfrac{1}{L} & \dfrac{1}{L} \end{bmatrix}dx\begin{Bmatrix} u_1 \\ u_2 \end{Bmatrix} = \left(N_1 AE\frac{du}{dx}\right)\Bigg|_0^L \tag{3.12.12}$$

Substituting for dN_1/dx, we obtain

$$AE \int_0^L \left[-\frac{1}{L} \right] \left[-\frac{1}{L} \quad \frac{1}{L} \right] dx \left\{ \begin{matrix} u_1 \\ u_2 \end{matrix} \right\} = f_{1x} \qquad (3.12.13)$$

where $f_{1x} = AE(du/dx)$ because $N_1 = 1$ at $x = 0$ and $N_1 = 0$ at $x = L$. Evaluating Eq. (3.12.13) yields

$$\frac{AE}{L}(u_1 - u_2) = f_{1x} \qquad (3.12.14)$$

Similarly, using $N_i = N_2$, we obtain

$$AE \int_0^L \left[\frac{1}{L} \right] \left[-\frac{1}{L} \quad \frac{1}{L} \right] dx \left\{ \begin{matrix} u_1 \\ u_2 \end{matrix} \right\} = \left(N_2 AE \frac{du}{dx} \right) \Big|_0^L \qquad (3.12.15)$$

Simplifying Eq. (3.12.15) yields

$$\frac{AE}{L}(u_2 - u_1) = f_{2x} \qquad (3.12.16)$$

where $f_{2x} = AE(du/dx)$ because $N_2 = 1$ at $x = L$ and $N_2 = 0$ at $x = 0$. Equations (3.12.14) and (3.12.16) are then seen to be the same as Eqs. (3.1.13) and (3.10.27) derived, respectively, by the direct and the variational method.

▲ 3.13 Other Residual Methods and Their Application to a One-Dimensional Bar Problem

As indicated in Section 3.12 when describing Galerkin's residual method, weighted residual methods are based on assuming an approximate solution to the governing differential equation for the given problem. The assumed or trial solution is typically a displacement or a temperature function that must be made to satisfy the initial and boundary conditions of the problem. This trial solution will not, in general, satisfy the governing differential equation. Thus, substituting the trial function into the differential equation will result in some residuals or errors. Each residual method requires the error to vanish over some chosen intervals or at some chosen points. To demonstrate this concept, we will solve the problem of a rod subjected to a triangular load distribution as shown in Figure 3–30 (see Section 3.10) for which we also have an exact solution for the axial displacement given by Eq. (3.11.5) in Section 3.11. We will illustrate four common weighted residual methods: *collocation, subdomain, least squares,* and *Galerkin's method.*

It is important to note that the primary intent in this section is to introduce you to the general concepts of these other weighted residual methods through a simple

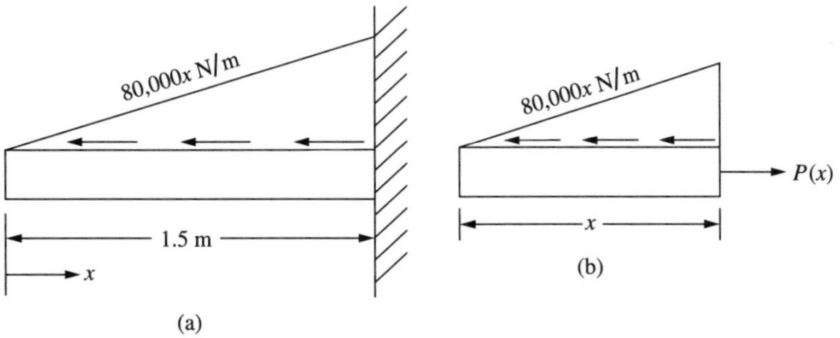

Figure 3–37 (a) Rod subjected to triangular load distribution and (b) free-body diagram of section of rod

example. You should note that we will assume a displacement solution that will in general yield an approximate solution (in our example the assumed displacement function yields an exact solution) over the whole domain of the problem (the rod previously solved in Section 13.10). As you have seen already for the spring and bar elements, we have assumed a linear function over each spring or bar element, and then combined the element solutions as was illustrated in Section 3.10 for the same rod solved in this section. It is common practice to use the simple linear function in each element of a finite element model, with an increasing number of elements used to model the rod yielding a closer and closer approximation to the actual displacement as seen in Figure 3–33.

For clarity's sake, Figure 3–37(a) shows the problem we are solving, along with a free-body diagram of a section of the rod with the internal axial force $P(x)$ shown in Figure 3–37(b).

The governing differential equation for the axial displacement, u, is given by

$$\left(AE\frac{du}{dx}\right) - P(x) = 0 \qquad (3.13.1)$$

where the internal axial force is $P(x) = 40{,}000x^2$. The boundary condition is $u(x = L) = 0$.

The method of weighted residuals requires us to assume an approximation function for the displacement. This approximate solution must satisfy the boundary condition of the problem. Here we assume the following function:

$$u(x) = c_1(x - L) + c_2(x - L)^2 + c_3(x - L)^3 \qquad (3.13.2)$$

where c_1, c_2 and c_3 are unknown coefficients. Equation (3.13.2) also satisfies the boundary condition given by $u(x = L) = 0$.

Substituting Eq. (3.13.2) for u into the governing differential equation, Eq. (3.13.1), results in the following error function, R:

$$AE[c_1 + 2c_2(x - L) + 3c_3(x - L)^2] - 40{,}000x^2 = R \qquad (3.13.3)$$

We now illustrate how to solve the governing differential equation by the four weighted residual methods.

Collocation Method

The *collocation method* requires that the error or residual function, R, be forced to zero at as many points as there are unknown coefficients. Equation (3.13.2) has three unknown coefficients. Therefore, we will make the error function equal zero at three points along the rod. We choose the error function to go to zero at $x = 0$, $x = L/3$, and $x = 2L/3$ as follows:

$$R(c, x = 0) = 0 = AE[c_1 + 2c_2(-L) + 3c_3(-L)^2] = 0$$
$$R(c, x = L/3) = 0 = AE[c_1 + 2c_2(-2L/3) + 3c_3(-2L/3)^2] - 40{,}000(L/3)^2 = 0$$
$$R(c, x = 2L/3) = 0 = AE[c_1 + 2c_2(-L/3) + 3c_3(-L/3)^2] - 40{,}000(2L/3)^2 = 0$$
$$(3.13.4)$$

The three linear equations, Eq. (3.13.4), can now be solved for the unknown coefficients, c_1, c_2 and c_3. The result is

$$c_1 = 40{,}000L^2/(AE) \quad c_2 = 40{,}000L/(AE) \quad c_3 = 40{,}000/(3AE) \qquad (3.13.5)$$

Substituting the numerical values, $A = 12.5 \times 10^{-4}$ m^2, $E = 2 \times 10^{11}$ N/m^2, and $L = 1.5$ m into Eq. (3.13.5), we obtain the c's as:

$$c_1 = 3.43 \times 10^{-4}, \quad c_2 = 2.29 \times 10^{-4}, \quad c_3 = 5.08 \times 10^{-5} \qquad (3.13.6)$$

Substituting the numerical values for the coefficients given in Eq. (3.13.6) into Eq. (3.13.2), we obtain the final expression for the axial displacement as

$$u(x) = 3.43 \times 10^{-4}(x - L) + 2.29 \times 10^{-4}(x - L)^2 + 5.08 \times 10^{-5}(x - L)^3 \quad (3.13.7)$$

Because we have chosen a cubic displacement function, Eq. (3.13.2), and the exact solution, Eq. (3.11.6), is also cubic, the collocation method yields the identical solution as the exact solution. The plot of the solution is shown in Figure 3–33 on page 130.

Subdomain Method

The subdomain method requires that the integral of the error or residual function over some selected subintervals be set to zero. The number of subintervals selected must equal the number of unknown coefficients. Because we have three unknown coefficients in the rod example, we must make the number of subintervals equal to three. We choose the subintervals from 0 to $L/3$, from $L/3$ to $2L/3$, and from $2L/3$ to L as follows:

$$\int_0^{L/3} R\,dx = 0 = \int_0^{L/3} \{AE[c_1 + 2c_2(x - L) + 3c_3(x - L)^2] - 40{,}000x^2\}dx$$

$$\int_{L/3}^{2L/3} R\,dx = 0 = \int_{L/3}^{2L/3} \{AE[c_1 + 2c_2(x - L) + 3c_3(x - L)^2] - 40{,}000x^2\}dx \quad (3.13.8)$$

$$\int_{2L/3}^{L} R\,dx = 0 = \int_{2L/3}^{L} \{AE[c_1 + 2c_2(x - L) + 3c_3(x - L)^2] - 40{,}000x^2\}dx$$

where we have used Eq. (3.13.3) for R in Eqs. (3.13.8).

Integration of Eqs. (3.13.8) results in three simultaneous linear equations that can be solved for the coefficients c_1, c_2 and c_3. Using the numerical values for A, E, and L as previously done, the three coefficients are numerically identical to those given by Eq. (3.13.6). The resulting axial displacement is then identical to Eq. (3.13.7).

Least Squares Method

The least squares method requires the integral over the length of the rod of the error function squared to be minimized with respect to each of the unknown coefficients in the assumed solution, based on the following:

$$\frac{\partial}{\partial c_i} \left(\int_0^L R^2 \, dx \right) = 0 \quad i = 1, 2, \ldots, N \text{ (for } N \text{ unknown coefficients)} \qquad (3.13.9)$$

or equivalently to

$$\int_0^L R \frac{\partial R}{\partial c_i} \, dx = 0 \qquad (3.13.10)$$

Because we have three unknown coefficients in the approximate solution, we will perform the integration three times according to Eq. (3.13.10) with three resulting equations as follows:

$$\int_0^L \{AE[c_1 + 2c_2(x - L) + 3c_3(x - L)^2] - 40{,}000x^2\}AE \, dx = 0$$

$$\int_0^L \{AE[c_1 + 2c_2(x - L) + 3c_3(x - L)^2] - 40{,}000x^2\}AE2(x - L) \, dx = 0 \qquad (3.13.11)$$

$$\int_0^L \{AE[c_1 + 2c_2(x - L) + 3c_3(x - L)^2] - 40{,}000x^2\}AE3(x - L)^2 \, dx = 0$$

In the first, second, and third of Eqs. (3.13.11), respectively, we have used the following partial derivatives:

$$\frac{\partial R}{\partial c_1} = AE, \qquad \frac{\partial R}{\partial c_2} = AE2(x - L), \qquad \frac{\partial R}{\partial c_3} = AE3(x - L)^2 \qquad (3.13.12)$$

where R is again the error function defined by Eq. (3.13.3).

Integration of Eqs. (3.13.11) yields three linear equations that are solved for the three coefficients. The numerical values of the coefficients again are identical to those of Eq. (3.13.6). Hence, the solution is identical to the exact solution.

Galerkin's Method

Galerkin's method requires the error to be orthogonal[1] to some weighting functions W_i as given previously by Eq. (3.12.2). For the rod example, this integral becomes

$$\int_0^L R W_i \, dx = 0 \qquad I = 1, 2, \ldots, N \tag{3.13.13}$$

The weighting functions are chosen to be a part of the approximate solution. Because we have three unknown constants in the approximate solution, we need to generate three equations. Recall that the assumed solution is the cubic given by Eq. (3.13.2); therefore, we select the weighting functions to be

$$W_1 = x - L \qquad W_2 = (x - L)^2 \qquad W_3 = (x - L)^3 \tag{3.13.14}$$

Using the weighting functions from Eq. (3.13.14) successively in Eq. (3.13.13), along with Eq. (3.13.3) for R, we generate the following three equations:

$$\int_0^L \{AE[c_1 + 2c_2(x - L) + 3c_3(x - L)^2] - 40{,}000x^2\}(x - L) \, dx = 0$$

$$\int_0^L \{AE[c_1 + 2c_2(x - L) + 3c_3(x - L)^2] - 40{,}000x^2\}(x - L)^2 \, dx = 0 \tag{3.13.15}$$

$$\int_0^L \{AE[c_1 + 2c_2(x - L) + 3c_3(x - L)^2] - 40{,}000x^2\}(x - L)^3 \, dx = 0$$

Integration of Eqs. (3.13.15) results in three linear equations that can be solved for the unknown coefficients. The numerical values are the same as those given by Eq. (3.13.6). Hence, the solution is identical to the exact solution.

In conclusion, because we assumed the approximate solution in the form of a cubic in x and the exact solution is also a cubic in x, all residual methods have yielded the exact solution. The purpose of this section has still been met to illustrate the four common residual methods to obtain an approximate (or exact in this example) solution to a known differential equation. The exact solution is shown by Eq. (3.11.6) and in Figure 3–33 in Section 3.11.

---

[1] The use of the word orthogonal in this context is a generalization of its use with respect to vectors. Here the ordinary scalar product is replaced by an integral in Eq. (3.13.13). In Eq. (3.13.13), the functions $u(x) = R$ and $v(x) = W_i$ are said to be orthogonal on the interval $0 \le x \le L$ if $\int_0^L u(x)v(x) \, dx$ equals 0.

▲ **3.14 Flowchart for Solution of** ▲
Three-Dimensional Truss Problems

In Figure 3-38, we present a flowchart of a typical finite-element process used for the analysis of three-dimensional truss problems on the basis of the theory presented in Chapter 3.

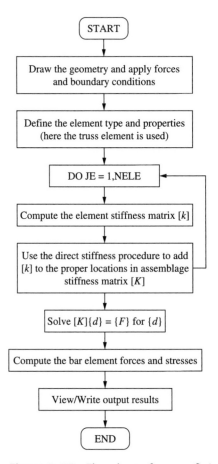

Figure 3–38 Flowchart of a truss finite-element program (NELE represents the number of elements)

▲ **3.15 Computer Program Assisted** ▲
Step-by-Step Solution for Truss Problem

In this section, we present a computer-assisted step-by-step solution of a three-dimensional truss (space truss) problem solved using a computer program; see Reference [9].

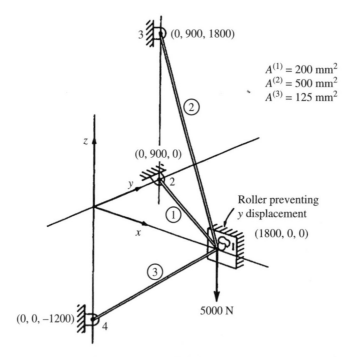

Figure 3–39 Space truss modeled in computer program Algor [9]

The computer-assisted step-by-step problem uses the truss in Example 3.8 and is shown in Figure 3–39.

The following steps have been used to determine the x, y, and z displacement components at node 1 and the stresses in each truss element.

1. The first step is to draw the three truss elements using the standard drawing program in the finite-element program, Algor [9], ANSYS [10], etc. This drawing also could be done using other drawing programs, such as KeyCreator [11] or AutoCAD [12] and then imported into the finite-element program. This drawing requires defining a convenient x, y, z coordinate system and then inputting the x, y, and z coordinates of the two nodes making up each truss element. When we input the nodal coordinates, we are actually defining the description of the overall dimensions of the model truss and the individual elements making up the truss model. When the individual elements, with their associated nodes, are created, we will have defined the topology or connectivity (which nodes are connected to which elements). The element numbering and node numbering are done internally within the computer program. This drawing process is normally the most time-consuming part of finite-element analysis. We often use automatic mesh-generating capabilities for two-and three-dimensional bodies to reduce the time and error involved with modeling.

2. The second step is to select the element type for the kind of analysis to be performed. Here the *truss element* is selected.
3. The third step is to input the geometric properties for the element. Here the *cross-sectional area, A*, is input.
4. The fourth step is to choose the material properties (*modulus of elasticity, E*, for a truss element). Here ASTM A 36 steel is selected, which then means the modulus of elasticity has been input.
5. The fifth step is to apply the boundary conditions to the proper nodes using the proper boundary condition command. Here *pinned boundary condition* is appropriate and applied to the nodes labeled 2, 3, and 4 and roller condition preventing *y* displacement to node 1 in Figure 3–39.
6. The sixth step is to apply the nodal load. Here the *load* of 5000 N is applied in the negative-*z* direction.
7. The seventh step is an optional check of the model. If you choose to perform this step, you will see the boundary conditions represented by a triangle at nodes 2, 3, and 4, a circle at node 1, and the load represented by an arrow pointing in the negative-z direction at node 1.
8. In step eight, we perform the analysis. This means the solution of simultaneous equations of the form $\{F\} = [K]\{d\}$ for displacement components *x*, *y*, and *z* at node 1 are determined. The stresses in each truss element are also determined.
9. In step nine, we select the results relevant for the specific analysis. Here the displacement plot and axial stress plot are the relevant

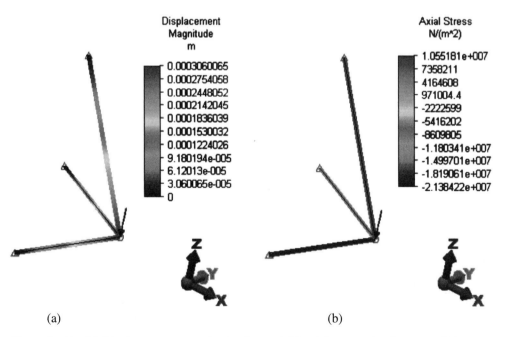

(a)   (b)

Figure 3–40 (a) Displacement magnitude plot and (b) axial stress plot for truss of Figure 3–39

quantities for design. Figures 3–40(a) and (b) show the maximum displacement plot and axial stress plot for the truss. The largest stress of −21.38 MPa (negative sign indicates compressive stress) is in the lower element (element three). The stresses in elements one and two are −7.22 MPa and 10.55 MPa, respectively.

▲ Summary Equations

Displacement function assumed for two-noded bar element:

$$u = a_1 + a_2 x \tag{3.1.1}$$

Shape functions for bar:

$$N_1 = 1 - \frac{x}{L} \qquad N_2 = \frac{x}{L} \tag{3.1.4}$$

Stiffness matrix for bar:

$$[k] = \frac{AE}{L} \begin{bmatrix} 1 & -1 \\ -1 & 1 \end{bmatrix} \tag{3.1.14}$$

Transformation matrix relating vectors in the plane in two different coordinate systems:

$$[T] = \begin{bmatrix} C & S \\ -S & C \end{bmatrix} \tag{3.3.18}$$

Global stiffness matrix for bar arbitrarily oriented in the plane:

$$[k] = \frac{AE}{L} \begin{bmatrix} C^2 & CS & -C^2 & -CS \\ & S^2 & -CS & -S^2 \\ & & C^2 & CS \\ \text{Symmetry} & & & S^2 \end{bmatrix} \tag{3.4.23}$$

Axial stress in a bar:

$$\{\sigma\} = [C']\{d\} \tag{3.5.6}$$

where

$$[C'] = \frac{E}{L}[-C \quad -S \quad C \quad S] \tag{3.5.8}$$

Transformation matrix relating vectors in three-dimensional space:

$$[T^*] = \begin{bmatrix} C_x & C_y & C_z & 0 & 0 & 0 \\ 0 & 0 & 0 & C_x & C_y & C_z \end{bmatrix} \tag{3.7.7}$$

Stiffness matrix for bar element in space:

$$[k] = \frac{AE}{L}\begin{bmatrix} C_x^2 & C_xC_y & C_xC_z & -C_x^2 & -C_xC_y & -C_xC_z \\ & C_y^2 & C_yC_z & -C_xC_y & -C_y^2 & -C_yC_z \\ & & C_z^2 & -C_xC_z & -C_yC_z & -C_z^2 \\ & & & C_x^2 & C_xC_y & C_xC_z \\ & & & & C_y^2 & C_yC_z \\ \text{Symmetry} & & & & & C_z^2 \end{bmatrix} \quad (3.7.9)$$

Total potential energy for bar:

$$\pi_p = \frac{AL}{2}\{d\}^T\{B\}^T[D]^T[B]\{d\} - \{d\}^T\{f\} \quad (3.10.19)$$

where

$$\{f\} = \{P\} + \iint_{S_1} [N_S]^T\{T_x\}ds + \iiint_V [N]^T\{X_b\}dV$$

Quadratic form of bar strain energy:

$$U = \frac{1}{2}\{d\}^T[k]\{d\} = \frac{1}{2}[u_1\,u_2]\frac{AE}{L}\begin{bmatrix} 1 & -1 \\ -1 & 1 \end{bmatrix}\begin{Bmatrix} u_1 \\ u_2 \end{Bmatrix} = \frac{AE}{2L}[u_1^2 - 2u_1u_2 + u_2^2]$$

$$(3.10.28b)$$

▲ References

[1] Turner, M. J., Clough, R. W., Martin, H. C., and Topp, L. J., "Stiffness and Deflection Analysis of Complex Structures," *Journal of the Aeronautical Sciences*, Vol. 23, No. 9, Sept. 1956, pp. 805–824.

[2] Martin, H. C., "Plane Elasticity Problems and the Direct Stiffness Method," *The Trend in Engineering*, Vol. 13, Jan. 1961, pp. 5–19.

[3] Melosh, R. J., "Basis for Derivation of Matrices for the Direct Stiffness Method," *Journal of the American Institute of Aeronautics and Astronautics*, Vol. 1, No. 7, July 1963, pp. 1631–1637.

[4] Oden, J. T., and Ripperger, E. A., *Mechanics of Elastic Structures*, 2nd ed., McGraw-Hill, New York, 1981.

[5] Finlayson, B. A., *The Method of Weighted Residuals and Variational Principles*, Academic Press, New York, 1972.

[6] Zienkiewicz, O. C., *The Finite Element Method*, 3rd ed., McGraw-Hill, London, 1977.

[7] Cook, R. D., Malkus, D. S., Plesha, M. E., and Witt, R. J., *Concepts and Applications of Finite Element Analysis*, 4th ed., Wiley, New York, 2002.

[8] Forray, M. J., *Variational Calculus in Science and Engineering*, McGraw-Hill, New York, 1968.

[9] Linear Stress and Dynamics Reference Division, Docutech On-Line Documentation, Algor Interactive Systems, Pittsburgh, PA.

[10] ANSYS—Engineering Analysis Systems User's Manual, Swanson Analysis Systems, Inc., Johnson Rd., P.O. Box 65 Houston, PA 15342.

[11] KeyCreator, Kubotek, USA, Inc.

[12] Autodesk Inc., McInnis Parkway, San Rafael, CA 94903.

▲ Problems

3.1 **a.** Compute the total stiffness matrix $[K]$ of the assemblage shown in Figure P3–1 by superimposing the stiffness matrices of the individual bars. Note that $[K]$ should be in terms of $A_1, A_2, A_3, E_1, E_2, E_3, L_1, L_2$, and L_3. Here A, E, and L are generic symbols used for cross-sectional area, modulus of elasticity, and length, respectively.

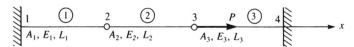

Figure P3–1

b. Now let $A_1 = A_2 = A_3 = A, E_1 = E_2 = E_3 = E$, and $L_1 = L_2 = L_3 = L$. If nodes 1 and 4 are fixed and a force P acts at node 3 in the positive x direction, find expressions for the displacement of nodes 2 and 3 in terms of A, E, L, and P.

c. Now let $A = 6 \times 10^{-4}$ m^2, $E = 70$ GPa, $L = 0.25$ m, and $P = 5000$ N.

   i. Determine the numerical values of the displacements of nodes 2 and 3.

   ii. Determine the numerical values of the reactions at nodes 1 and 4.

   iii. Determine the stresses in elements 1–3.

3.2–3.11 For the bar assemblages shown in Figures P3–2 through P3–11, determine the nodal displacements, the forces in each element, and the reactions. Use the direct stiffness method for these problems.

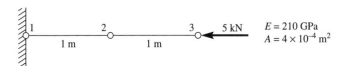

Figure P3–2

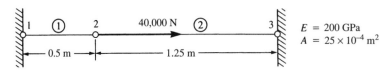

Figure P3–3

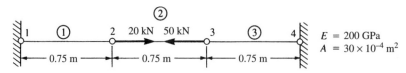

Figure P3–4

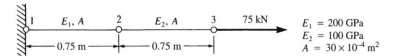

Figure P3–5

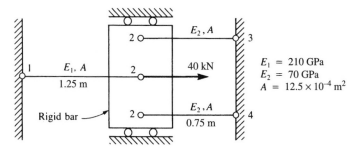

Figure P3–6

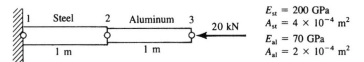

Figure P3–7

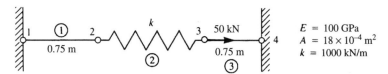

Figure P3–8

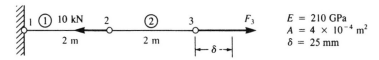

Figure P3–9

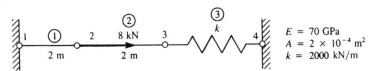

Figure P3–10

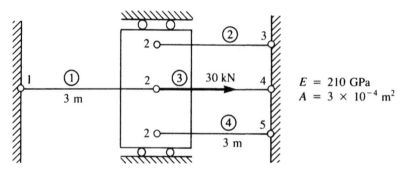

Figure P3–11

3.12 Solve for the axial displacement and stress in the tapered bar shown in Figure P3–12 using one and then two constant-area elements. Evaluate the area at the center of each element length. Use that area for each element. Let $A_0 = 12.5 \times 10^{-4}$ m^2, $L = 0.5$ m, $E = 70$ GPa, and $P = 5000$ N. Compare your finite element solutions with the exact solution.

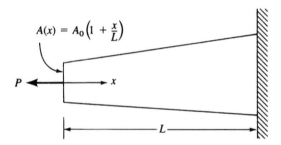

Figure P3–12

3.13 Determine the stiffness matrix for the bar element with end nodes and midlength node shown in Figure P3–13. Let axial displacement $u = a_1 + a_2 x + a_3 x^2$. (This is a higher-order element in that strain now varies linearly through the element.)

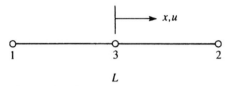

Figure P3–13

3.14 Consider the following displacement function for the two-noded bar element:

$$u = a + bx^2$$

Is this a valid displacement function? Discuss why or why not.

3.15 For each of the bar elements shown in Figure P3–15, evaluate the global x-y stiffness matrix.

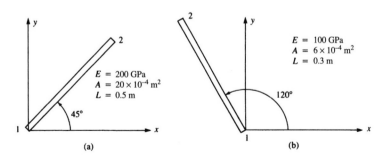

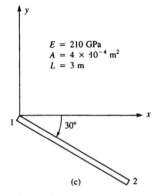

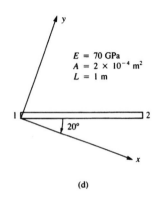

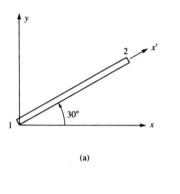

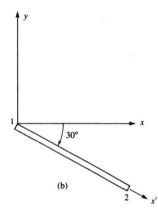

Figure P3–15

3.16 For the bar elements shown in Figure P3–16, the global displacements have been determined to be $u_1 = 10$ mm, $v_1 = 0.0$, $u_2 = 5$ mm, and $v_2 = 15$ mm. Determine the local x displacements at each end of the bars. Let $E = 90$ GPa, $A = 3 \times 10^{-4}$ m^2, and $L = 1.5$ m for each element.

Figure P3–16

3.17 For the bar elements shown in Figure P3–17, the global displacements have been determined to be $u_1 = 0.0$, $v_1 = 2.5$ mm, $u_2 = 5.0$ mm, and $v_2 = 3.0$ mm. Determine the local x' displacements at the ends of each bar. Let $E = 210$ GPa, $A = 10 \times 10^{-4}$ m^2, and $L = 3$ m for each element.

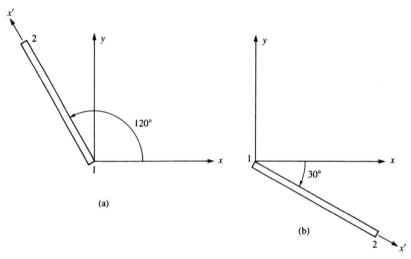

(a)

(b)

Figure P3–17

3.18 Using the method of Section 3.5, determine the axial stress in each of the bar elements shown in Figure P3–18.

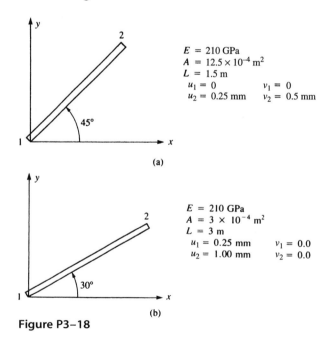

$E = 210$ GPa
$A = 12.5 \times 10^{-4}$ m^2
$L = 1.5$ m
$u_1 = 0$ $v_1 = 0$
$u_2 = 0.25$ mm $v_2 = 0.5$ mm

(a)

$E = 210$ GPa
$A = 3 \times 10^{-4}$ m^2
$L = 3$ m
$u_1 = 0.25$ mm $v_1 = 0.0$
$u_2 = 1.00$ mm $v_2 = 0.0$

(b)

Figure P3–18

3.19 **a.** Assemble the stiffness matrix for the assemblage shown in Figure P3–19 by super-imposing the stiffness matrices of the springs. Here k is the stiffness of each spring.
b. Find the x and y components of deflection of node 1.

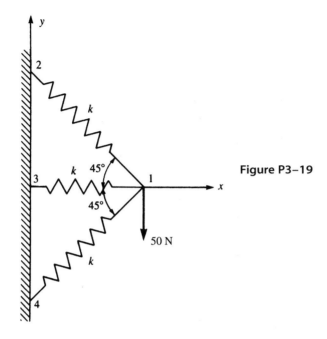

Figure P3–19

3.20 For the plane truss structure shown in Figure P3–20, determine the displacement of node 2 using the stiffness method. Also determine the stress in element 1. Let $A = 30 \times 10^{-4}$ m², $E = 7.5$ GPa, and $L = 2.5$ m

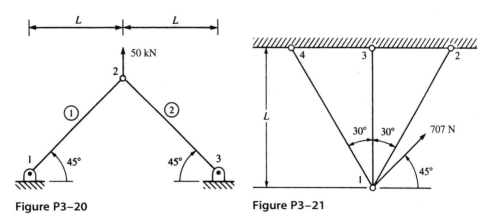

Figure P3–20

Figure P3–21

3.21 Find the horizontal and vertical displacements of node 1 for the truss shown in Figure P3–21. Assume AE is the same for each element.

3.22 For the truss shown in Figure P3–22 solve for the horizontal and vertical components of displacement at node 1 and determine the stress in each element. Also verify force equilibrium at node 1. All elements have $A_1 = 6 \times 10^{-4}$ m^2 and $E = 70$ GPa. Let $L = 2.5$ m.

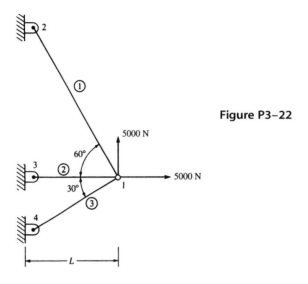

Figure P3–22

3.23 For the truss shown in Figure P3–23, solve for the horizontal and vertical components of displacement at node 1. Also determine the stress in element 1. Let $A = 6 \times 10^{-4}$ m^2, $E = 70 \times 10^9$ N/m^2, and $L = 2.5$ m.

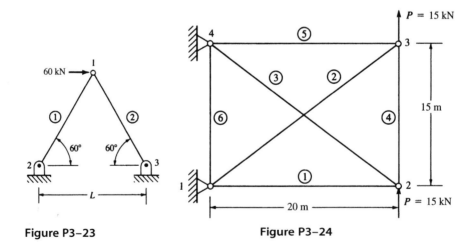

Figure P3–23

Figure P3–24

3.24 Determine the nodal displacements and the element forces for the truss shown in Figure P3–24. Assume all elements have the same AE.

3.25 Now remove the element connecting nodes 2 and 4 in Figure P3–24. Then determine the nodal displacements and element forces.

3.26 Now remove *both* cross elements in Figure P3–24. Can you determine the nodal displacements? If not, why?

3.27 Determine the displacement components at node 3 and the element forces for the plane truss shown in Figure P3–27. Let $A = 50 \times 10^{-4}$ m^2 and $E = 2 \times 10^{11}$ N/m^2 for all elements. Verify force equilibrium at node 3.

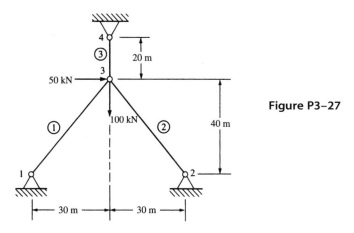

Figure P3–27

3.28 Show that for the transformation matrix $[T]$ of Eq. (3.4.15), $[T]^T = [T]^{-1}$ and hence Eq. (3.4.21) is indeed correct, thus also illustrating that $[k] = [T]^T[k'][T]$ is the expression for the global stiffness matrix for an element.

3.29–3.30 For the plane trusses shown in Figures P3–29 and P3–30, determine the horizontal and vertical displacements of node 1 and the stresses in each element. All elements have $E = 210$ GPa and $A = 4.0 \times 10^{-4}$ m^2.

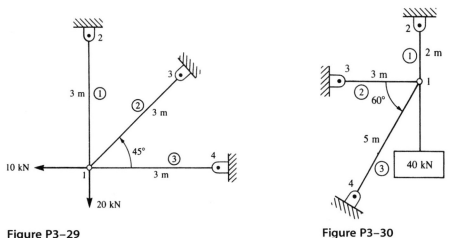

Figure P3–29

Figure P3–30

3.31 Remove element 1 from Figure P3–30 and solve the problem. Compare the displacements and stresses to the results for Problem 3.30.

3.32 For the plane truss shown in Figure P3–32, determine the nodal displacements, the element forces and stresses, and the support reactions. All elements have $E = 70$ GPa and $A = 3.0 \times 10^{-4}$ m². Verify force equilibrium at nodes 2 and 4. Use symmetry in your model.

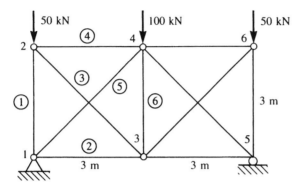

Figure P3–32

3.33 For the plane trusses supported by the spring at node 1 in Figure P3–33 (a) and (b), determine the nodal displacements and the stresses in each element. Let $E = 210$ GPa and $A = 5.0 \times 10^{-4}$ m² for both truss elements.

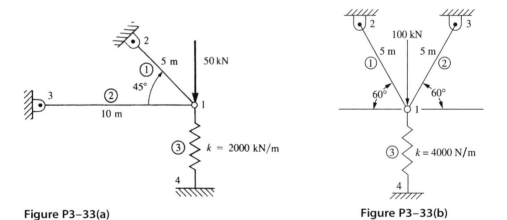

Figure P3–33(a)   **Figure P3–33(b)**

3.34 For the plane truss shown in Figure P3–34, node 2 settles an amount $\delta = 2$ mm. Determine the forces and stresses in each element due to this settlement. Let $E = 210$ GPa and $A = 10 \times 10^{-4}$ m² for each element.

3.35 For the symmetric plane truss shown in Figure P3–35, determine (a) the deflection of node 1 and (b) the stress in element 1. AE/L for element 3 is twice AE/L for the other elements. Let $AE/L = 1.5 \times 10^{10}$ N/m. Then let $A = 5 \times 10^{-4}$ m², $L = 240$ mm, and $E = 70$ GPa to obtain numerical results.

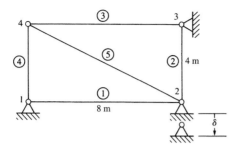

Figure P3–34

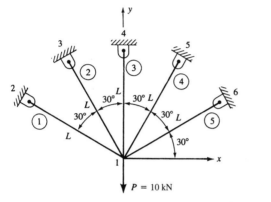

Figure P3–35

3.36–3.37 For the space truss elements shown in Figures P3–36 and P3–37, the global displacements at node 1 have been determined to be $u_1 = 2$ mm, $v_1 = 4$ mm, and $w_1 = 3$ mm. Determine the displacement along the local x' axis at node 1 of the elements. The coordinates, in mm, are shown in the figures.

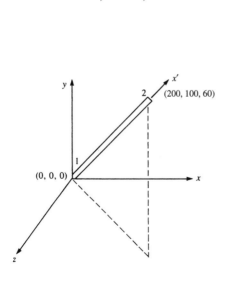

Figure P3–36

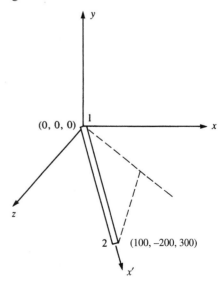

Figure P3–37

3.38–3.39 For the space truss elements shown in Figures P3–38 and P3–39, the global displacements at node 2 have been determined to be $u_2 = 5$ mm, $v_2 = 10$ mm, and $w_2 = 15$ mm. Determine the displacement along the local x' axis at node 2 of the elements. The coordinates, in meters, are shown in the figures.

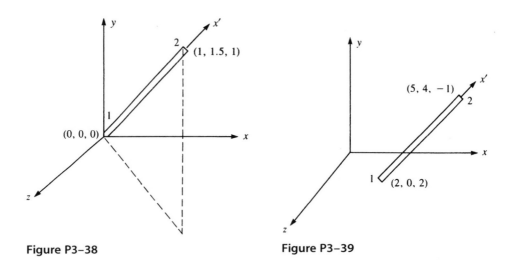

Figure P3–38   **Figure P3–39**

3.40–3.41 For the space trusses shown in Figures P3–40 and P3–41, determine the nodal displacements and the stresses in each element. Let $E = 210$ GPa and $A = 10 \times 10^{-4}$ m^2 for all elements. Verify force equilibrium at node 1. The coordinates of each node, in meters, are shown in the figure. All supports are ball-and-socket joints.

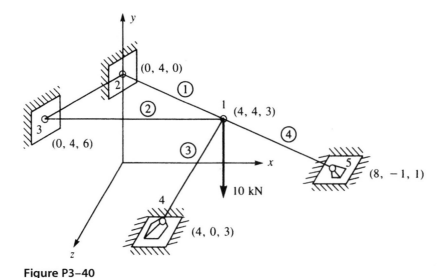

Figure P3–40

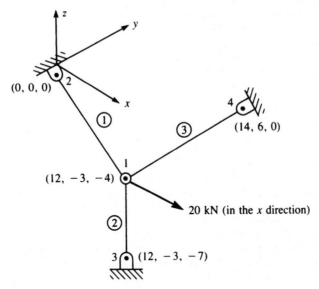

Figure P3–41

3.42 For the space truss subjected to a 10 kN load in the x direction, as shown in Figure P3–42, determine the displacement of node 5. Also determine the stresses in each element. Let $A = 25 \times 10^{-4}$ m^2 and $E = 2 \times 10^{11}$ N/m^2 for all elements. The coordinates of each node, in mm, are shown in the figure. Nodes 1–4 are supported by ball-and-socket joints (fixed supports).

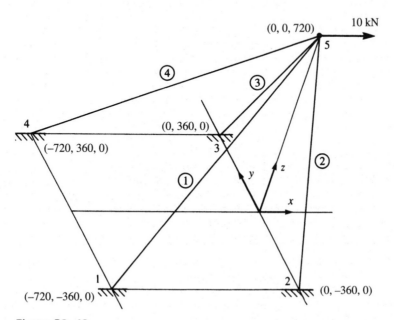

Figure P3–42

3.43 For the space truss subjected to the 40 kN load acting as shown in Figure P3–43, determine the displacement of node 4. Also determine the stresses in each element. Let $A = 40 \times 10^{-4}$ m^2 and $E = 2 \times 10^{11}$ N/m^2 for all elements. The coordinates of each node, in mm, are shown in the figure. Nodes 1–3 are supported by ball-and-socket joints (fixed supports).

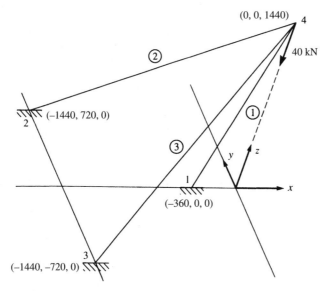

Figure P3–43

3.44 Derive Eq. (3.7.21) for stress in space truss elements by a process similar to that used to derive Eq. (3.5.6) for stress in a plane truss element.

3.45 For the truss shown in Figure P3–45, use symmetry to determine the displacements of the nodes and the stresses in each element. All elements have $E = 200$ GPa. Elements 1, 2, 4, and 5 have $A = 10 \times 10^{-4}$ m^2 and element 3 has $A = 20 \times 10^{-4}$ m^2. Let dimension $a = 2$ m and $P = 40$ kN. The supports at nodes 1 and 4 are defined.

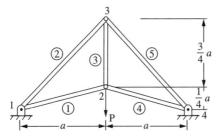

Figure P3–45

3.46 For the truss shown in Figure P3–46, use symmetry to determine the displacements of the nodes and the stresses in each element. All elements have $E = 2 \times 10^{11}$ N/m^2. Elements 1, 2, 4, and 5 have $A = 62.5 \times 10^{-4}$ m^2 and element 3 has $A = 125 \times 10^{-4}$ m^2.

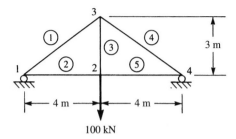

Figure P3–46

3.47 All elements of the structure in Figure P3–47 have the same AE except element 1, which has an axial stiffness of $2AE$. Find the displacements of the nodes and the stresses in elements 2, 3, and 4 by using symmetry. Check equilibrium at node 4. You might want to use the results obtained from the stiffness matrix of Problem 3.24.

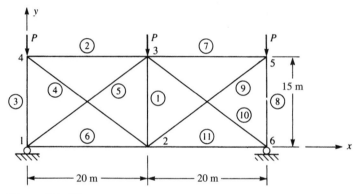

Figure P3–47

3.48 For the roof truss shown in Figure P3–48, use symmetry to determine the displacements of the nodes and the stresses in each element. All elements have $E = 210$ GPa and $A = 10 \times 10^{-4}$ m^2.

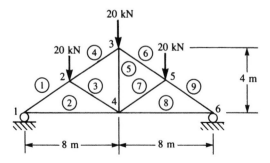

Figure P3–48

3.49–3.51 For the plane trusses with inclined supports shown in Figures P3–49 through P3–51, solve for the nodal displacements and element stresses in the bars. Let $A = 12 \times 10^{-4}$ m^2, $E = 2 \times 10^{11}$ N/m^2, and $L = 0.75$ m for each truss.

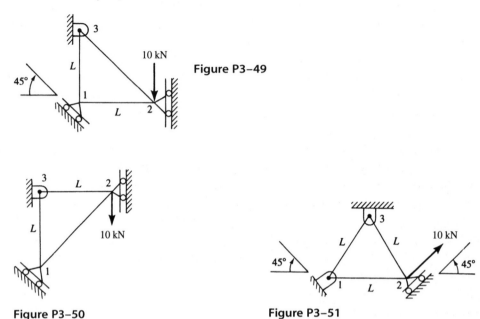

Figure P3–49

Figure P3–50

Figure P3–51

3.52 Use the principle of minimum potential energy developed in Section 3.10 to solve the bar problems shown in Figure P3–52. That is, plot the total potential energy for variations in the displacement of the free end of the bar to determine the minimum potential energy. Observe that the displacement that yields the minimum potential energy also yields the stable equilibrium position. Use displacement increments of 0.05 mm, beginning with $x = -0.1$ mm. Let $E = 2 \times 10^{11}$ N/m^2 and $A = 12 \times 10^{-4}$ m^2 for the bars.

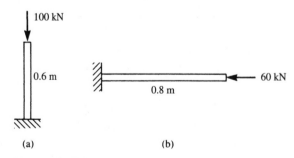

(a)   (b)

Figure P3–52

3.53 Derive the stiffness matrix for the nonprismatic bar shown in Figure P3–53 using the principle of minimum potential energy. Let E be constant.

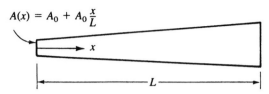

$$A(x) = A_0 + A_0 \frac{x}{L}$$

x

L

Figure P3–53

3.54 For the bar subjected to the linear varying axial load shown in Figure P3–54, determine the nodal displacements and axial stress distribution using (a) two equal-length elements and (b) four equal-length elements. Let $A = 12.5 \times 10^{-4}$ m^2 and $E = 2 \times 10^{11}$ N/m^2. Compare the finite element solution with an exact solution.

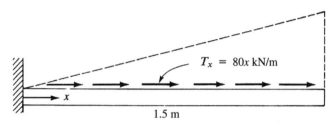

$T_x = 80x$ kN/m

x

1.5 m

Figure P3–54

3.55 For the bar subjected to the uniform line load in the axial direction shown in Figure P3–55, determine the nodal displacements and axial stress distribution using (a) two equal-length elements and (b) four equal-length elements. Compare the finite element results with an exact solution. Let $A = 12.5 \times 10^{-4}$ m^2 and $E = 2 \times 10^{11}$ N/m^2.

3.56 For the bar fixed at both ends and subjected to the uniformly distributed loading shown in Figure P3–56, determine the displacement at the middle of the bar and the stress in the bar. Let $A = 12.5 \times 10^{-4}$ m^2 and $E = 2 \times 10^{11}$ N/m^2.

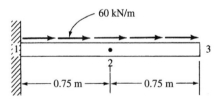

60 kN/m

1

2

3

0.75 m   0.75 m

Figure P3–55

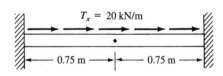

$T_x = 20$ kN/m

0.75 m   0.75 m

Figure P3–56

3.57 For the bar hanging under its own weight shown in Figure P3–57, determine the nodal displacements using (a) two equal-length elements and (b) four equal-length elements. Let $A = 12 \times 10^{-4}$ m^2, $E = 2 \times 10^{11}$ N/m^2, and weight density $\rho_w = 7800$ kg/m^3. (*Hint:* The internal force is a function of x. Use the potential energy approach.)

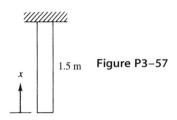

1.5 m   **Figure P3–57**

x

3.58 Determine the energy equivalent nodal forces for the axial distributed loading shown acting on the bar elements in Figure P3–58.

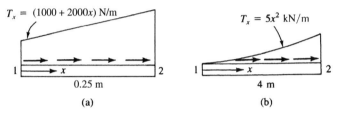

$T_x = (1000 + 2000x)$ N/m

$T_x = 5x^2$ kN/m

1   x 2   1 x 2

0.25 m   4 m

(a)   (b)

Figure P3–58

3.59 Solve Problem 3.55 for the axial displacement in the bar using collocation, subdomain, least squares, and Galerkin's methods. Choose a quadratic polynomial $u(x) = c_1 x + c_2 x^2$ in each method. Compare these weighted residual method solutions to the exact solution.

3.60 For the tapered bar shown in Figure P3–60 with cross sectional areas $A_1 = 12 \times 10^{-4}$ m^2 and $A_2 = 6 \times 10^{-4}$ m^2 at each end, use the collocation, subdomain, least squares, and Galerkin's methods to obtain the displacement in the bar. Compare these weighted residual solutions to the exact solution. Choose a cubic polynomial $u(x) = c_1 x + c_2 x^2 + c_3 x^3$.

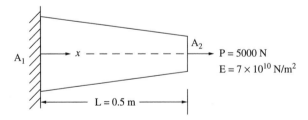

A_1 x - - - - - - - - - A_2 $P = 5000$ N

$E = 7 \times 10^{10}$ N/m^2

$L = 0.5$ m

Figure P3–60

3.61 For the bar shown in Figure P3–61 subjected to the linear varying axial load, determine the displacements and stresses using (a) one and then two finite element models and (b) the collocation, subdomain, least squares, and Galerkin's methods assuming a cubic polynomial of the form $u(x) = c_1 x + c_2 x^2 + c_3 x^3$.

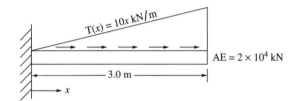

Figure P3–61

3.62–3.70 Use a computer program to solve the truss design problems shown in Figures P3–62 through P3–70. Determine the single most critical cross-sectional area based on maximum allowable yield strength or buckling strength (based on either Euler's or Johnson's formula as relevant) using a factor of safety (FS) listed next to each truss. Recommend a common structural shape and size for each truss. List the largest three nodal displacements and their locations. Also include a plot of the deflected shape of the truss and a principal stress plot.

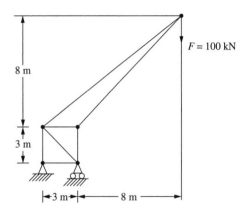

Figure P3–62 Derrick truss (FS = 4.0)

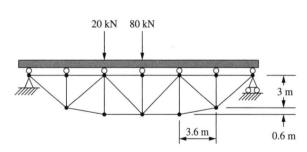

Figure P3–63 Truss bridge (FS = 3.0)

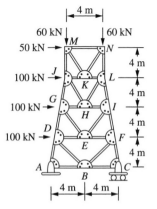

Figure P3–64 Tower (FS = 2.5)

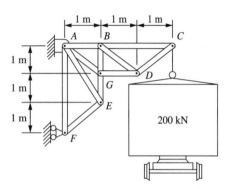

Figure P3–65 Boxcar lift (FS = 3.0)

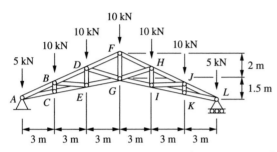

Figure P3-66 Howe scissors roof truss (FS = 2.0)

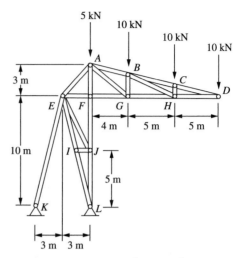

Figure P3-67 Stadium roof truss
(FS = 3.0)

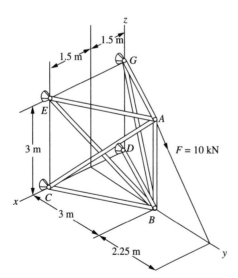

Figure P3-68 Space truss with ball-and-socket joints at C, D, E, and G (FS = 3.0)

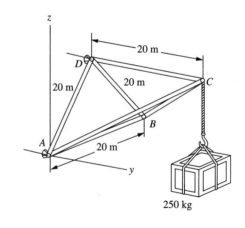

Figure P3-69 Space truss with ball-and-socket joints at A, B, and D
(FS = 2.0)

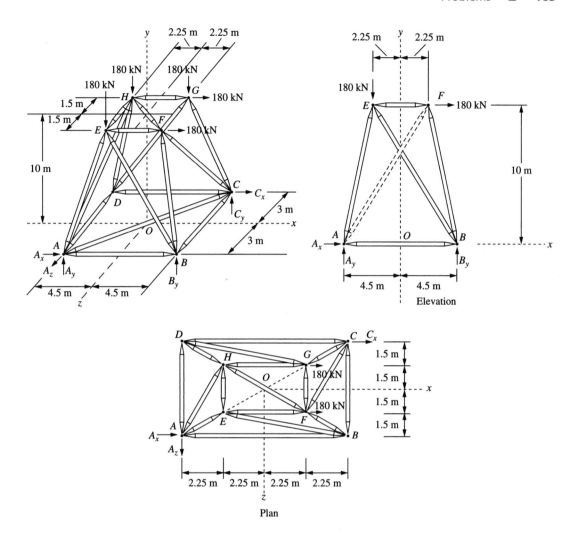

Figure P3–70 Space truss (FS = 2.0)

DEVELOPMENT OF
BEAM EQUATIONS ▲

CHAPTER OBJECTIVES

- To review basic concepts of beam bending.
- To derive the stiffness matrix for a beam element.
- To demonstrate beam analysis using the direct stiffness method.
- To illustrate the effects of shear deformation in shorter beams.
- To introduce the work-equivalence method for replacing distributed loading by a set of discrete loads.
- To introduce the general formulation for solving beam problems with distributed loading acting on them.
- To analyze beams with distributed loading acting on them.
- To compare the finite element solution to an exact solution for a beam.
- To derive the stiffness matrix for the beam element with nodal hinge.
- To show how the potential energy method can be used to derive the beam element equations.
- To apply Galerkin's residual method for deriving the beam element equations.

Introduction

We begin this chapter by developing the stiffness matrix for the bending of a beam element, the most common of all structural elements as evidenced by its prominence in buildings, bridges, towers, and many other structures. The beam element is considered to be straight and to have constant cross-sectional area. We will first derive the beam element stiffness matrix by using the principles developed for simple beam theory.

We will then present simple examples to illustrate the assemblage of beam element stiffness matrices and the solution of beam problems by the direct stiffness method presented in Chapter 2. The solution of a beam problem illustrates that the degrees of freedom associated with a node are a transverse displacement and a rotation. We will include the nodal shear forces and bending moments and the resulting shear force and bending moment diagrams as part of the total solution.

Next, we will discuss procedures for handling distributed loading, because beams and frames are often subjected to distributed loading as well as concentrated nodal loading. We will follow the discussion with solutions of beams subjected to distributed loading and compare a finite element solution to an exact solution for a beam subjected to a distributed loading.

We will then develop the beam element stiffness matrix for a beam element with a nodal hinge and illustrate the solution of a beam with an internal hinge.

To further acquaint you with the potential energy approach for developing stiffness matrices and equations, we will again develop the beam bending element equations using this approach. We hope to increase your confidence in this approach. It will be used throughout much of this text to develop stiffness matrices and equations for more complex elements, such as two-dimensional (plane) stress, axisymmetric stress, and three-dimensional stress.

Finally, the Galerkin residual method is applied to derive the beam element equations.

The concepts presented in this chapter are prerequisite to understanding the concepts for frame analysis presented in Chapter 5.

▲ 4.1 Beam Stiffness ▲

In this section, we will derive the stiffness matrix for a simple beam element. A **beam** *is a long, slender structural member generally subjected to transverse loading that produces significant bending effects as opposed to twisting or axial effects.* This bending deformation is measured as a transverse displacement and a rotation. Hence, the degrees of freedom considered per node are a transverse displacement and a rotation (as opposed to only an axial displacement for the bar element of Chapter 3).

Consider the beam element shown in Figure 4–1. The beam is of length L with axial local coordinate x and transverse local coordinate y. The local transverse nodal displacements are given by v_i's and the rotations by ϕ_i's. The local nodal forces are given by f_{iy}'s and the bending moments by m_i's as shown. We initially neglect all axial effects.

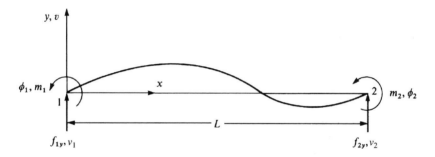

Figure 4–1 Beam element with positive nodal displacements, rotations, forces, and moments

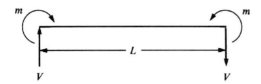

Figure 4–2 Beam theory sign conventions for shear forces and bending moments

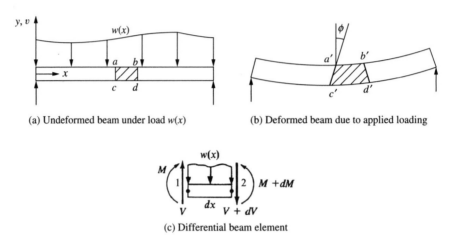

Figure 4–3 Beam under distributed load

At all nodes, the following sign conventions are used:

1. Moments are positive in the counterclockwise direction.
2. Rotations are positive in the counterclockwise direction.
3. Forces are positive in the positive y direction.
4. Displacements are positive in the positive y direction.

Figure 4–2 indicates the sign conventions used in simple beam theory for positive shear forces V and bending moments m.

Beam Stiffness Matrix Based on Euler-Bernoulli Beam Theory (Considering Bending Deformations Only)

The differential equation governing elementary linear-elastic beam behavior [1] (called the Euler-Bernoulli beam as derived by Euler and Bernoulli) is based on plane cross sections perpendicular to the longitudinal centroidal axis of the beam before bending occurs remaining plane and perpendicular to the longitudinal axis after bending occurs. This is illustrated in Figure 4–3, where a plane through vertical line $a-c$ (Figure 4–3(a)) is perpendicular to the longitudinal x axis before bending, and this same plane through $a'-c'$ (rotating through angle ϕ in Figure 4–3(b)) remains perpendicular to the bent x axis after bending. This occurs in practice only when a pure couple or constant moment exists in the beam. However it is a reasonable assumption that yields equations that quite accurately predict beam behavior for most practical beams.

The differential equation is derived as follows. Consider the beam shown in Figure 4–3 subjected to a distributed loading $w(x)$ (force/length). From the force and moment equilibrium of a differential element of the beam, shown in Figure 4–3(c), we have

$$\Sigma F_y = 0: \quad V - (V + dV) - w(x)\, dx = 0 \tag{4.1.1a}$$

Or, simplifying Eq. (4.1.1a), we obtain

$$-w\, dx - dV = 0 \qquad \text{or} \qquad w = -\frac{dV}{dx} . \tag{4.1.1b}$$

$$\Sigma M_2 = 0: \quad -V dx + dM + w(x)\, dx \left(\frac{dx}{2}\right) = 0 \qquad \text{or} \qquad V = \frac{dM}{dx} \tag{4.1.1c}$$

The final form of Eq. (4.1.1c), relating the shear force to the bending moment, is obtained by dividing the left equation by dx and then taking the limit of the equation as dx approaches 0. The $w(x)$ term then disappears.

Also, the curvature κ of the beam is related to the moment by

$$\kappa = \frac{1}{\rho} = \frac{M}{EI} \tag{4.1.1d}$$

where ρ is the radius of the deflected curve shown in Figure 4–4b, v is the transverse displacement function in the y direction (see Figure 4–4a), E is the modulus of elasticity, and I is the principal moment of inertia about the z axis (where the z axis is perpendicular to the x and y axes). ($I = bh^3/12$ for a rectangular cross section of base b and height h shown in Figure 4–4c.)

The curvature for small slopes $\phi = dv/dx$ is given by

$$\kappa = \frac{d^2v}{dx^2} \tag{4.1.1e}$$

Using Eq. (4.1.1e) in (4.1.1d), we obtain

$$\frac{d^2v}{dx^2} = \frac{M}{EI} \tag{4.1.1f}$$

Solving Eq. (4.1.1f) for M and substituting this result into (4.1.1c) and (4.1.1b), we obtain

$$\frac{d^2}{dx^2}\left(EI\frac{d^2v}{dx^2}\right) = -w(x) \tag{4.1.1g}$$

For constant EI and only nodal forces and moments, Eq. (4.1.1g) becomes

$$EI\frac{d^4v}{dx^4} = 0 \tag{4.1.1h}$$

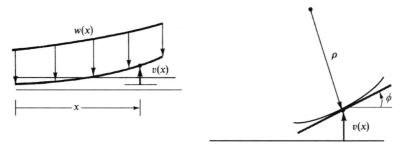

(a) Portion of deflected curve of beam   (b) Radius of deflected curve at $v(x)$

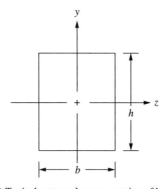

(c) Typical rectangular cross section of beam

Figure 4–4 Deflected curve of beam

We will now follow the steps outlined in Chapter 1 to develop the stiffness matrix and equations for a beam element and then to illustrate complete solutions for beams.

Step 1 Select the Element Type

Represent the beam by labeling nodes at each end and in general by labeling the element number (Figure 4–1).

Step 2 Select a Displacement Function

Assume the transverse displacement variation through the element length to be

$$v(x) = a_1 x^3 + a_2 x^2 + a_3 x + a_4 \tag{4.1.2}$$

The complete cubic displacement function Eq. (4.1.2) is appropriate because there are four total degrees of freedom (a transverse displacement v_i and a small rotation ϕ_i at each node). The cubic function also satisfies the basic beam differential equation—further justifying its selection. In addition, the cubic function also satisfies the conditions of displacement and slope continuity at nodes shared by two elements.

Using the same procedure as described in Section 2.2, we express v as a function of the nodal degrees of freedom v_1, v_2, ϕ_1, and ϕ_2 as follows:

$$v(0) = v_1 = a_4$$

$$\frac{dv(0)}{dx} = \phi_1 = a_3$$

$$v(L) = v_2 = a_1 L^3 + a_2 L^2 + a_3 L + a_4 \qquad (4.1.3)$$

$$\frac{dv(L)}{dx} = \phi_2 = 3a_1 L^2 + 2a_2 L + a_3$$

where $\phi = dv/dx$ for the assumed small rotation ϕ. Solving Eqs. (4.1.3) for a_1 through a_4 in terms of the nodal degrees of freedom and substituting into Eq. (4.1.2), we have

$$v = \left[\frac{2}{L^3}(v_1 - v_2) + \frac{1}{L^2}(\phi_1 + \phi_2) \right] x^3$$

$$+ \left[-\frac{3}{L^2}(v_1 - v_2) - \frac{1}{L}(2\phi_1 + \phi_2) \right] x^2 + \phi_1 x + v_1 \qquad (4.1.4)$$

In matrix form, we express Eq. (4.1.4) as

$$v = [N]\{d\} \qquad (4.1.5)$$

where

$$\{d\} = \begin{Bmatrix} v_1 \\ \phi_1 \\ v_2 \\ \phi_2 \end{Bmatrix} \qquad (4.1.6a)$$

and where

$$[N] = [N_1 \quad N_2 \quad N_3 \quad N_4] \qquad (4.1.6b)$$

and

$$N_1 = \frac{1}{L^3}(2x^3 - 3x^2 L + L^3) \qquad N_2 = \frac{1}{L^3}(x^3 L - 2x^2 L^2 + xL^3)$$

$$\qquad (4.1.7)$$

$$N_3 = \frac{1}{L^3}(-2x^3 + 3x^2 L) \qquad N_4 = \frac{1}{L^3}(x^3 L - x^2 L^2)$$

N_1, N_2, N_3, and N_4 are called the **shape functions** for a beam element. These cubic shape (or interpolation) functions are known as *Hermite cubic interpolation (or cubic spline) functions.* For the beam element, $N_1 = 1$ when evaluated at node 1 and $N_1 = 0$ when evaluated at node 2. Because N_2 is associated with ϕ_1, we have, from the second of Eqs. (4.1.7), $(dN_2/dx) = 1$ when evaluated at node 1. Shape functions N_3 and N_4 have analogous results for node 2.

Step 3 Define the Strain/Displacement
and Stress/Strain Relationships

Assume the following axial strain/displacement relationship to be valid:

$$\varepsilon_x(x, y) = \frac{du}{dx} \qquad (4.1.8)$$

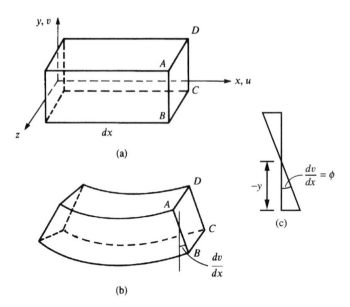

Figure 4–5 Beam segment (a) before deformation and (b) after deformation;
(c) Angle of rotation of cross section *ABCD*

where u is the axial displacement function. From the deformed configuration of the beam shown in Figure 4–5, we relate the axial displacement to the transverse displacement by

$$u = -y\frac{dv}{dx} \tag{4.1.9}$$

where we should recall from elementary beam theory [1] the basic assumption that cross sections of the beam (such as cross section *ABCD*) that are planar before bending deformation remain planar after deformation and, in general, rotate through a small angle (dv/dx). Using Eq. (4.1.9) in Eq. (4.1.8), we obtain

$$\varepsilon_x(x, y) = -y\frac{d^2v}{dx^2} \tag{4.1.10a}$$

Also using Hooke's law $(\sigma_x = E\,\varepsilon_x)$ and substituting Eq. (4.1.1f) for d^2v/dx^2 into Eq. (4.1.10a), we obtain the beam flexure or bending stress formula as

$$\sigma_x = \frac{-My}{I} \tag{4.1.10b}$$

From elementary beam theory, the bending moment and shear force are related to the transverse displacement function. Because we will use these relationships in the derivation of the beam element stiffness matrix, we now present them as

$$m(x) = EI\frac{d^2v}{dx^2} \qquad V = EI\frac{d^3v}{dx^3} \tag{4.1.11}$$

Step 4 Derive the Element Stiffness Matrix and Equations

First, derive the element stiffness matrix and equations using a direct equilibrium approach. We now relate the nodal and beam theory sign conventions for shear forces and bending moments (Figures 4–1 and 4–2), along with Eqs. (4.1.4) and (4.1.11), to obtain

$$f_{1y} = V = EI \frac{d^3v(0)}{dx^3} = \frac{EI}{L^3}(12v_1 + 6L\phi_1 - 12v_2 + 6L\phi_2)$$

$$m_1 = -m = -EI \frac{d^2v(0)}{dx^2} = \frac{EI}{L^3}(6Lv_1 + 4L^2\phi_1 - 6Lv_2 + 2L^2\phi_2)$$

$$f_{2y} = -V = -EI \frac{d^3v(L)}{dx^3} = \frac{EI}{L^3}(-12v_1 - 6L\phi_1 + 12v_2 - 6L\phi_2)$$

$$m_2 = m = EI \frac{d^2v(L)}{dx^2} = \frac{EI}{L^3}(6Lv_1 + 2L^2\phi_1 - 6Lv_2 + 4L^2\phi_2)$$

$$(4.1.12)$$

where the minus signs in the second and third of Eqs. (4.1.12) are the result of opposite nodal and beam theory positive bending moment conventions at node 1 and opposite nodal and beam theory positive shear force conventions at node 2 as seen by comparing Figures 4–1 and 4–2. Equations (4.1.12) relate the nodal forces to the nodal displacements. In matrix form, Eqs. (4.1.12) become

$$\begin{Bmatrix} f_{1y} \\ m_1 \\ f_{2y} \\ m_2 \end{Bmatrix} = \frac{EI}{L^3} \begin{bmatrix} 12 & 6L & -12 & 6L \\ 6L & 4L^2 & -6L & 2L^2 \\ -12 & -6L & 12 & -6L \\ 6L & 2L^2 & -6L & 4L^2 \end{bmatrix} \begin{Bmatrix} v_1 \\ \phi_1 \\ v_2 \\ \phi_2 \end{Bmatrix} \qquad (4.1.13)$$

where the stiffness matrix is then

$$[k] = \frac{EI}{L^3} \begin{bmatrix} 12 & 6L & -12 & 6L \\ 6L & 4L^2 & -6L & 2L^2 \\ -12 & -6L & 12 & -6L \\ 6L & 2L^2 & -6L & 4L^2 \end{bmatrix} \qquad (4.1.14)$$

Equation (4.1.13) indicates that $[k]$ relates transverse forces and bending moments to transverse displacements and rotations, whereas axial effects have been neglected.

In the beam element stiffness matrix (Eq. (4.1.14)) derived in this section, it is assumed that the beam is long and slender; that is, the length, L, to depth, h, dimension ratio of the beam is large. In this case, the deflection due to bending that is predicted by using the stiffness matrix from Eq. (4.1.14) is quite adequate. However, for short, deep beams the transverse shear deformation can be significant and can have the same order of magnitude contribution to the total deformation of the beam. This is seen by the expressions for the bending and shear contributions to the deflection of a beam, where the bending contribution is of order $(L/h)^3$, whereas the shear contribution is only of order (L/h). A general rule for rectangular cross-section beams, is that for a length at least eight times the depth, the transverse shear deflection is less than five percent of the bending deflection [4]. Castigliano's method for finding beam and frame deflections is a convenient way to include the effects of the transverse shear term as shown in Reference [4]. The derivation of the stiffness matrix for a beam including the transverse shear deformation contribution is given in a number of references [5–8].

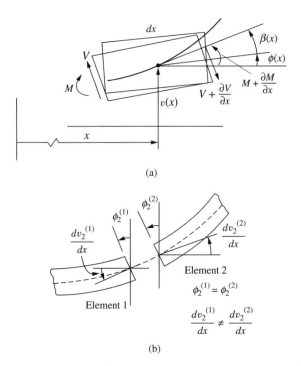

(a)

(b)

Figure 4–6 (a) Element of Timoshenko beam showing shear deformation. Cross sections are no longer perpendicular to the neutral axis line. (b) Two beam elements meeting at node 2

The inclusion of the shear deformation in beam theory with application to vibration problems was developed by Timoshenko and is known as the Timoshenko beam [9–10].

Beam Stiffness Matrix Based on Timoshenko Beam Theory (Including Transverse Shear Deformation)

The shear deformation beam theory is derived as follows. Instead of plane sections remaining plane after bending occurs as shown previously in Figure 4–5, the shear deformation (deformation due to the shear force V) is now included. Referring to Figure 4–6, we observe a section of a beam of differential length dx with the cross section assumed to remain plane but no longer perpendicular to the neutral axis (x axis) due to the inclusion of the shear force resulting in a rotation term indicated by β. The total deflection of the beam at a point x now consists of two parts, one caused by bending and one by shear force, so that the slope of the deflected curve at point x is now given by

$$\frac{dv}{dx} = \phi(x) + \beta(x) \tag{4.1.15a}$$

where rotation due to bending moment and due to transverse shear force are given, respectively, by $\phi(x)$ and $\beta(x)$.

We assume as usual that the linear deflection and angular deflection (slope) are small.

The relation between bending moment and bending deformation (curvature) is now

$$M(x) = EI \frac{d\phi(x)}{dx} \tag{4.1.15b}$$

and the relation between the shear force and shear deformation (rotation due to shear) (shear strain) is given by

$$V(x) = k_s AG\beta(x) \tag{4.1.15c}$$

The difference in dv/dx and ϕ represents the shear strain $\gamma_{yz}(=\beta)$ of the beam as

$$\gamma_{yz} = \frac{dv}{dx} - \phi \tag{4.1.15d}$$

Now consider the differential element in Figure 4–3(c) and Eqs. (4.1.1b) and (4.1.1c) obtained from summing transverse forces and then summing bending moments. We now substitute Eq. (4.1.15c) for V and Eq. (4.1.15b) for M into Eqs. (4.1.1b) and (4.1.1c) along with β from Eq. (4.1.15a) to obtain the two governing differential equations as

$$\frac{d}{dx}\left[k_s AG\left(\frac{dv}{dx} - \phi\right)\right] = -w \tag{4.1.15e}$$

$$\frac{d}{dx}\left(EI\frac{d\phi}{dx}\right) + k_s AG\left(\frac{dv}{dx} - \phi\right) = 0 \tag{4.1.15f}$$

To derive the stiffness matrix for the beam element including transverse shear deformation, we assume the transverse displacement to be given by the cubic function in Eq. (4.1.2). In a manner similar to [8], we choose transverse shear strain γ consistent with the cubic polynomial for $v(x)$, such that γ is a constant given by

$$\gamma = c \tag{4.1.15g}$$

Using the cubic displacement function for v, the slope relation given by Eq. (4.1.15a), and the shear strain Eq. (4.1.15g), along with the bending moment-curvature relation, Eq. (4.1.15b) and the shear force-shear strain relation Eq. (4.1.15c), in the bending moment–shear force relation Eq. (4.1.1c), we obtain

$$c = 6a_1 g \tag{4.1.15h}$$

where $g = EI/k_s AG$ and $k_s A$ is the shear area. Shear areas, A_s, vary with cross-section shapes. For instance, for a rectangular shape A_s is taken as 0.83 times the cross section A, for a solid circular cross section it is taken as 0.9 times the cross section, for a wide-flange cross section it is taken as the thickness of the web times the full depth of the wide-flange, and for thin-walled cross sections it is taken as two times the product of the thickness of the wall times the depth of the cross section.

Using Eqs. (4.1.2), (4.1.15a), (4.1.15g), and (4.1.15h) allow ϕ to be expressed as a polynomial in x as follows:

$$\phi = a_3 + 2a_2 x + (3x^2 + 6g)a_1 \tag{4.1.15i}$$

Using Eqs. (4.1.2) and (4.1.15i), we can now express the coefficients a_1 through a_4 in terms of the nodal displacements v_1 and v_2 and rotations ϕ_1 and ϕ_2 of the beam at the ends $x = 0$ and $x = L$ as previously done to obtain Eq. (4.1.4) when shear deformation was neglected. The expressions for a_1 through a_4 are then given as follows:

$$a_1 = \frac{2v_1 + L\phi_1 - 2v_2 + L\phi_2}{L(L^2 + 12g)}$$

$$a_2 = \frac{-3Lv_1 - (2L^2 + 6g)\phi_1 + 3Lv_2 + (-L^2 + 6g)\phi_2}{L(L^2 + 12g)}$$

$$a_3 = \frac{-12gv_1 + (L^3 + 6gL)\phi_1 + 12gv_2 - 6gL\phi_2}{L(L^2 + 12g)}$$

$$a_4 = v_1$$

(4.1.15j)

Substituting these a's into Eq. (4.1.2), we obtain

$$v = \frac{2v_1 + L\phi_1 - 2v_2 + L\phi_2}{L(L^2 + 12g)} x^3$$

$$\frac{-3Lv_1 - (2L^2 + 6g)\phi_1 + 3Lv_2 + (-L^2 + 6g)\phi_2}{L(L^2 + 12g)} x^2$$

$$\frac{-12gv_1 + (L^3 + 6gL)\phi_1 + 12gv_2 - 6gL\phi_2}{L(L^2 + 12g)} x + v_1$$

(4.1.15k)

In a manner similar to step 4 used to derive the stiffness matrix for the beam element without shear deformation included, we have

$$f_{1y} = V(0) = 6EIa_1 = \frac{EI(12v_1 + 6L\phi_1 - 12v_2 + 6L\phi_2)}{L(L^2 + 12g)}$$

$$m_1 = -m(0) = -2EIa_2 = \frac{EI[6Lv_1 + (4L^2 + 12g)\phi_1 - 6Lv_2 + (2L^2 - 12g)\phi_2]}{L(L^2 + 12g)}$$

$$f_{2y} = -V(L) = \frac{EI(-12v_1 - 6L\phi_1 + 12v_2 - 6L\phi_2)}{L(L^2 + 12g)}$$

$$m_2 = m(L) = \frac{EI[6Lv_1 + (2L^2 - 12g)\phi_1 - 6Lv_2 + (4L^2 + 12g)\phi_2]}{L(L^2 + 12g)}$$

(4.1.15l)

where again the minus signs in the second and third of Eqs. (4.1.15l) are the result of opposite nodal and beam theory positive moment conventions at node 1 and opposite nodal and beam theory positive shear force conventions at node 2, as seen by comparing Figures 4–1 and 4–2. In matrix form Eqs. (4.1.15l) become

$$\begin{Bmatrix} f_{1y} \\ m_1 \\ f_{2y} \\ m_2 \end{Bmatrix} = \frac{EI}{L(L^2 + 12g)} \begin{bmatrix} 12 & 6L & -12 & 6L \\ 6L & (4L^2 + 12g) & -6L & (2L^2 - 12g) \\ -12 & -6L & 12 & -6L \\ 6L & (2L^2 - 12g) & -6L & (4L^2 + 12g) \end{bmatrix} \begin{Bmatrix} v_1 \\ \phi_1 \\ v_2 \\ \phi_2 \end{Bmatrix}$$

(4.1.15m)

where the stiffness matrix, including both bending and shear deformation, is then given by

$$[k] = \frac{EI}{L(L^2 + 12g)} \begin{bmatrix} 12 & 6L & -12 & 6L \\ 6L & (4L^2 + 12g) & -6L & (2L^2 - 12g) \\ 12 & -6L & 12 & -6L \\ 6L & (2L^2 - 12g) & -6L & (4L^2 + 12g) \end{bmatrix} \quad (4.1.15n)$$

In Eq. (4.1.15n) remember that g represents the transverse shear term, and if we set $g = 0$, we obtain Eq. (4.1.14) for the beam stiffness matrix, neglecting transverse shear deformation. To more easily see the effect of the shear correction factor, we define the nondimensional shear correction term as $\varphi = 12EI/(k_s AGL^2) = 12g/L^2$ and rewrite the stiffness matrix as

$$[k] = \frac{EI}{L^3(1 + \varphi)} \begin{bmatrix} 12 & 6L & -12 & 6L \\ 6L & (4 + \varphi)L^2 & -6L & (2 - \varphi)L^2 \\ -12 & -6L & 12 & -6L \\ 6L & (2 - \varphi)L^2 & -6L & (4 + \varphi)L^2 \end{bmatrix} \quad (4.1.15o)$$

Most commercial computer programs, such as [11], will include the shear deformation by having you input the shear area, $A_s = k_s A$.

▲ 4.2 Example of Assemblage of Beam Stiffness Matrices ▲

Step 5 **Assemble the Element Equations to Obtain the Global Equations and Introduce Boundary Conditions**

Consider the beam in Figure 4–7 as an example to illustrate the procedure for assemblage of beam element stiffness matrices. Assume EI to be constant throughout the beam. A force of 5000 N and a moment of 2500 N-m are applied to the beam at midlength. The left end is a fixed support and the right end is a pin support.

First, we discretize the beam into two elements with nodes 1–3 as shown. We include a node at midlength because applied force and moment exist at midlength and,

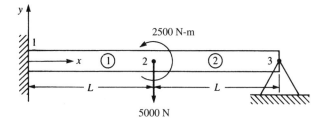

Figure 4–7   Fixed hinged beam subjected to a force and a moment

at this time, loads are assumed to be applied only at nodes. (Another procedure for handling loads applied on elements will be discussed in Section 4.4.)

Using Eq. (4.1.14), we find that the global stiffness matrices for the two elements are now given by

$$[k^{(1)}] = \frac{EI}{L^3} \begin{array}{cccc} v_1 & \phi_1 & v_2 & \phi_2 \end{array} \begin{bmatrix} 12 & 6L & -12 & 6L \\ 6L & 4L^2 & -6L & 2L^2 \\ -12 & -6L & 12 & -6L \\ 6L & 2L^2 & -6L & 4L^2 \end{bmatrix} \tag{4.2.1}$$

and

$$[k^{(2)}] = \frac{EI}{L^3} \begin{array}{cccc} v_2 & \phi_2 & v_3 & \phi_3 \end{array} \begin{bmatrix} 12 & 6L & -12 & 6L \\ 6L & 4L^2 & -6L & 2L^2 \\ -12 & -6L & 12 & -6L \\ 6L & 2L^2 & -6L & 4L^2 \end{bmatrix} \tag{4.2.2}$$

where the degrees of freedom associated with each beam element are indicated by the usual labels above the columns in each element stiffness matrix.

The total stiffness matrix can now be assembled for the beam by using the direct stiffness method. When the total (global) stiffness matrix has been assembled, the external global nodal forces are related to the global nodal displacements. Through direct superposition and Eqs. (4.2.1) and (4.2.2), the governing equations for the beam are thus given by

$$\begin{Bmatrix} F_{1y} \\ M_1 \\ F_{2y} \\ M_2 \\ F_{3y} \\ M_3 \end{Bmatrix} = \frac{EI}{L^3} \begin{bmatrix} 12 & 6L & -12 & 6L & 0 & 0 \\ 6L & 4L^2 & -6L & 2L^2 & 0 & 0 \\ -12 & -6L & 12+12 & -6L+6L & -12 & 6L \\ 6L & 2L^2 & -6L+6L & 4L^2+4L^2 & -6L & 2L^2 \\ 0 & 0 & -12 & -6L & 12 & -6L \\ 0 & 0 & 6L & 2L^2 & -6L & 4L^2 \end{bmatrix} \begin{Bmatrix} v_1 \\ \phi_1 \\ v_2 \\ \phi_2 \\ v_3 \\ \phi_3 \end{Bmatrix} \tag{4.2.3}$$

Now considering the boundary conditions, or constraints, of the fixed support at node 1 and the hinge (pinned) support at node 3, we have

$$\phi_1 = 0 \qquad v_1 = 0 \qquad v_3 = 0 \tag{4.2.4}$$

On considering the third, fourth, and sixth equations of Eqs. (4.2.3) corresponding to the rows with unknown degrees of freedom and using Eqs. (4.2.4), we obtain

$$\begin{Bmatrix} -5000 \\ 5000 \\ 0 \end{Bmatrix} = \frac{EI}{L^3} \begin{bmatrix} 24 & 0 & 6L \\ 0 & 8L^2 & 2L^2 \\ 6L & 2L^2 & 4L^2 \end{bmatrix} \begin{Bmatrix} v_2 \\ \phi_2 \\ \phi_3 \end{Bmatrix} \tag{4.2.5}$$

where $F_{2y} = -5000$ N, $M_2 = 2500$ N-m, and $M_3 = 0$ have been substituted into the reduced set of equations. We could now solve Eq. (4.2.5) simultaneously for the unknown nodal displacement v_2 and the unknown nodal rotations ϕ_2 and ϕ_3. We leave the final solution for you to obtain. Section 4.3 provides complete solutions to beam problems.

▲ 4.3 Examples of Beam Analysis Using the Direct Stiffness Method ▲

We will now perform complete solutions for beams with various boundary supports and loads to illustrate further the use of the equations developed in Section 4.1.

Example 4.1

Using the direct stiffness method, solve the problem of the propped cantilever beam subjected to end load P in Figure 4–8. The beam is assumed to have constant EI and length $2L$. It is supported by a roller at midlength and is built in at the right end.

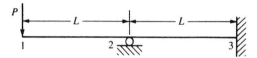

Figure 4–8 Propped cantilever beam

SOLUTION:

We have discretized the beam and established global coordinate axes as shown in Figure 4–8. We will determine the nodal displacements and rotations, the reactions, and the complete shear force and bending moment diagrams.

Using Eq. (4.1.14) for each element, along with superposition, we obtain the structure total stiffness matrix by the same method as described in Section 4.2 for obtaining the stiffness matrix in Eq. (4.2.3). The $[K]$ is

$$
[K] = \frac{EI}{L^3}
\begin{array}{c}
\begin{array}{cccccc}
v_1 & \phi_1 & v_2 & \phi_2 & v_3 & \phi_3
\end{array} \\
\begin{bmatrix}
12 & 6L & -12 & 6L & 0 & 0 \\
 & 4L^2 & -6L & 2L^2 & 0 & 0 \\
 & & 12+12 & -6L+6L & -12 & 6L \\
 & & & 4L^2+4L^2 & -6L & 2L^2 \\
 & & & & 12 & -6L \\
\text{Symmetry} & & & & & 4L^2
\end{bmatrix}
\end{array}
\qquad (4.3.1)
$$

The governing equations for the beam are then given by

$$
\begin{Bmatrix} F_{1y} \\ M_1 \\ F_{2y} \\ M_2 \\ F_{3y} \\ M_3 \end{Bmatrix} = \frac{EI}{L^3} \begin{bmatrix} 12 & 6L & -12 & 6L & 0 & 0 \\ 6L & 4L^2 & -6L & 2L^2 & 0 & 0 \\ -12 & -6L & 24 & 0 & -12 & 6L \\ 6L & 2L^2 & 0 & 8L^2 & -6L & 2L^2 \\ 0 & 0 & -12 & -6L & 12 & -6L \\ 0 & 0 & 6L & 2L^2 & -6L & 4L^2 \end{bmatrix} \begin{Bmatrix} v_1 \\ \phi_1 \\ v_2 \\ \phi_2 \\ v_3 \\ \phi_3 \end{Bmatrix} \tag{4.3.2}
$$

On applying the boundary conditions

$$
v_2 = 0 \qquad v_3 = 0 \qquad \phi_3 = 0 \tag{4.3.3}
$$

and partitioning the equations associated with unknown displacements [the first, second, and fourth equations of Eqs. (4.3.2)] from those equations associated with known displacements in the usual manner, we obtain the final set of equations for a longhand solution as

$$
\begin{Bmatrix} -P \\ 0 \\ 0 \end{Bmatrix} = \frac{EI}{L^3} \begin{bmatrix} 12 & 6L & 6L \\ 6L & 4L^2 & 2L^2 \\ 6L & 2L^2 & 8L^2 \end{bmatrix} \begin{Bmatrix} v_1 \\ \phi_1 \\ \phi_2 \end{Bmatrix} \tag{4.3.4}
$$

where $F_{1y} = -P$, $M_1 = 0$, and $M_2 = 0$ have been used in Eq. (4.3.4). We will now solve Eq. (4.3.4) for the nodal displacement and nodal slopes. We obtain the transverse displacement at node 1 as

$$
v_1 = -\frac{7PL^3}{12EI} \tag{4.3.5}
$$

where the minus sign indicates that the displacement of node 1 is downward.

The slopes are

$$
\phi_1 = \frac{3PL^2}{4EI} \qquad \phi_2 = \frac{PL^2}{4EI} \tag{4.3.6}
$$

where the positive signs indicate counterclockwise rotations at nodes 1 and 2.

We will now determine the global nodal forces. To do this, we substitute the known global nodal displacements and rotations, Eqs. (4.3.5) and (4.3.6), into Eq. (4.3.2). The resulting equations are

$$
\begin{Bmatrix} F_{1y} \\ M_1 \\ F_{2y} \\ M_2 \\ F_{3y} \\ M_3 \end{Bmatrix} = \frac{EI}{L^3} \begin{bmatrix} 12 & 6L & -12 & 6L & 0 & 0 \\ 6L & 4L^2 & -6L & 2L^2 & 0 & 0 \\ -12 & -6L & 24 & 0 & -12 & 6L \\ 6L & 2L^2 & 0 & 8L^2 & -6L & 2L^2 \\ 0 & 0 & -12 & -6L & 12 & -6L \\ 0 & 0 & 6L & 2L^2 & -6L & 4L^2 \end{bmatrix} \begin{Bmatrix} -\dfrac{7PL^3}{12EI} \\ \dfrac{3PL^2}{4EI} \\ 0 \\ \dfrac{PL^2}{4EI} \\ 0 \\ 0 \end{Bmatrix} \tag{4.3.7}
$$

Multiplying the matrices on the right-hand side of Eq. (4.3.7), we obtain the global nodal forces and moments as

$$F_{1y} = -P \qquad M_1 = 0 \qquad F_{2y} = \frac{5}{2}P$$

$$M_2 = 0 \qquad F_{3y} = -\frac{3}{2}P \qquad M_3 = \frac{1}{2}PL \tag{4.3.8}$$

The results of Eqs. (4.3.8) can be interpreted as follows: The value of $F_{1y} = -P$ is the applied force at node 1, as it must be. The values of F_{2y}, F_{3y}, and M_3 are the reactions from the supports as felt by the beam. The moments M_1 and M_2 are zero because no applied or reactive moments are present on the beam at node 1 or node 2.

It is generally necessary to determine the local nodal forces associated with each element of a large structure to perform a stress analysis of the entire structure. We will thus consider the forces in element 1 of this example to illustrate this concept (element 2 can be treated similarly). Using Eqs. (4.3.5) and (4.3.6) in the $\{f\} = [k]\{d\}$ equation for element 1 [also see Eq. (4.1.13)], we have

$$\begin{Bmatrix} f_{1y}^{(1)} \\ m_1^{(1)} \\ f_{2y}^{(1)} \\ m_2^{(1)} \end{Bmatrix} = \frac{EI}{L^3} \begin{bmatrix} 12 & 6L & -12 & 6L \\ 6L & 4L^2 & -6L & 2L^2 \\ -12 & -6L & 12 & -6L \\ 6L & 2L^2 & -6L & 4L^2 \end{bmatrix} \begin{Bmatrix} -\dfrac{7PL^3}{12EI} \\ \dfrac{3PL^2}{4EI} \\ 0 \\ \dfrac{PL^2}{4EI} \end{Bmatrix} \tag{4.3.9}$$

Equation (4.3.9) yields

$$f_{1y} = -P \qquad m_1 = 0 \qquad f_{2y} = P \qquad m_2 = -PL \tag{4.3.10}$$

A free-body diagram of element 1, shown in Figure 4–9(a), should help you to understand the results of Eqs. (4.3.10). The figure shows a nodal transverse force of negative P at node 1 and of positive P and negative moment PL at node 2. These values are consistent with the results given by Eqs. (4.3.10). For completeness, the free-body diagram of element 2 is shown in Figure 4–9(b). We can easily verify the element nodal forces by writing an equation similar to Eq. (4.3.9). From the results of Eqs. (4.3.8), the nodal forces and moments for the whole beam

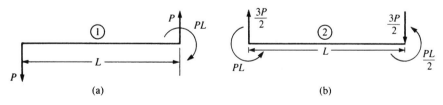

(a)　(b)

Figure 4–9 Free-body diagrams showing forces and moments on (a) element 1 and (b) element 2

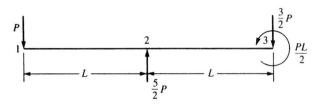

Figure 4–10 Nodal forces and moment on the beam

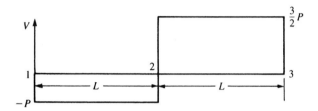

Figure 4–11 Shear force diagram for the beam of Figure 4–10

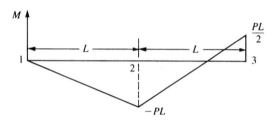

Figure 4–12 Bending moment diagram for the beam of Figure 4–10

are shown on the beam in Figure 4–10. Using the beam sign conventions established in Section 4.1, we obtain the shear force V and bending moment M diagrams shown in Figures 4–11 and 4–12. ∎

In general, for complex beam structures, we will use the element local forces to determine the shear force and bending moment diagrams for each element. We can then use these values for design purposes. Chapter 5 will further discuss this concept as used in computer codes.

Example 4.2

Determine the nodal displacements and rotations, global nodal forces, and element forces for the beam shown in Figure 4–13. We have discretized the beam as indicated by the node numbering. The beam is fixed at nodes 1 and 5 and has a roller support at node 3. Vertical loads of 50 kN each are applied at nodes 2 and 4. Let $E = 210$ GPa and $I = 2 \times 10^{-4}$ m^4 throughout the beam.

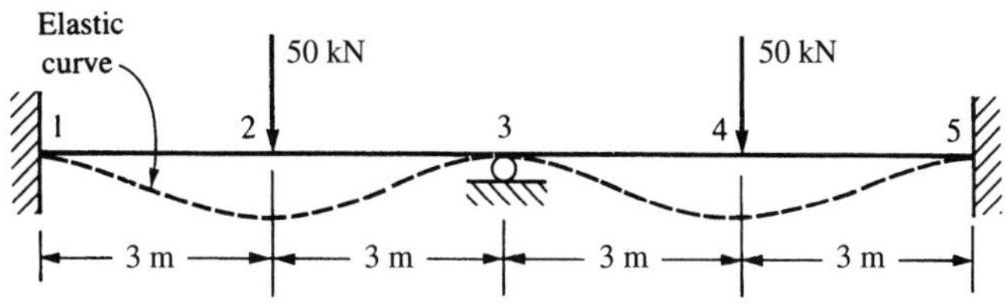

Figure 4–13 Beam example

SOLUTION:

Using Eq. (4.1.10), along with superposition of the four beam element stiffness matrices, we obtain the global stiffness matrix and the global equations as given in Eq. (4.3.11). Here the lengths of each element are the same. Thus, we can factor an L out of the superimposed stiffness matrix.

$$\begin{Bmatrix} F_{1y} \\ M_1 \\ F_{2y} \\ M_2 \\ F_{3y} \\ M_3 \\ F_{4y} \\ M_4 \\ F_{5y} \\ M_5 \end{Bmatrix} = \frac{EI}{L^3} \begin{bmatrix} 12 & 6L & -12 & 6L & 0 & 0 & 0 & 0 & 0 & 0 \\ 6L & 4L^2 & -6L & 2L^2 & 0 & 0 & 0 & 0 & 0 & 0 \\ -12 & -6L & 12+12 & -6L+6L & -12 & 6L & 0 & 0 & 0 & 0 \\ 6L & 2L^2 & -6L+6L & 4L^2+4L^2 & -6L & 2L^2 & 0 & 0 & 0 & 0 \\ 0 & 0 & -12 & -6L & 12+12 & -6L+6L & -12 & 6L & 0 & 0 \\ 0 & 0 & 6L & 2L^2 & -6L+6L & 4L^2+4L^2 & -6L & 2L^2 & 0 & 0 \\ 0 & 0 & 0 & 0 & -12 & -6L & 12+12 & -6L+6L & -12 & 6L \\ 0 & 0 & 0 & 0 & 6L & 2L^2 & -6L+6L & 4L^2+4L^2 & -6L & 2L^2 \\ 0 & 0 & 0 & 0 & 0 & 0 & -12 & -6L & 12 & -6L \\ 0 & 0 & 0 & 0 & 0 & 0 & 6L & 2L^2 & -6L & 4L^2 \end{bmatrix} \begin{Bmatrix} v_1 \\ \phi_1 \\ v_2 \\ \phi_2 \\ v_3 \\ \phi_3 \\ v_4 \\ \phi_4 \\ v_5 \\ \phi_5 \end{Bmatrix}$$

$$(4.3.11)$$

For a longhand solution, we reduce Eq. (4.3.11) in the usual manner by application of the boundary conditions

$$v_1 = \phi_1 = v_3 = v_5 = \phi_5 = 0$$

The resulting equation is

$$\begin{Bmatrix} -50 \times 10^3 \\ 0 \\ 0 \\ -50 \times 10^3 \\ 0 \end{Bmatrix} = \frac{EI}{L^3} \begin{bmatrix} 24 & 0 & 6L & 0 & 0 \\ 0 & 8L^2 & 2L^2 & 0 & 0 \\ 6L & 2L^2 & 8L^2 & -6L & 2L^2 \\ 0 & 0 & -6L^2 & 24 & 0 \\ 0 & 0 & 2L^2 & 0 & 8L^2 \end{bmatrix} \begin{Bmatrix} v_2 \\ \phi_2 \\ \phi_3 \\ v_4 \\ \phi_4 \end{Bmatrix} \qquad (4.3.12)$$

The rotations (slopes) at nodes 2–4 are equal to zero because of symmetry in loading, geometry, and material properties about a plane perpendicular to the beam length and passing through node 3. Therefore, $\phi_2 = \phi_3 = \phi_4 = 0$, and we can further reduce Eq. (4.3.12) to

$$\left\{\begin{array}{c} -50,000 \\ -50,000 \end{array}\right\} = \frac{EI}{L^3}\begin{bmatrix} 24 & 0 \\ 0 & 24 \end{bmatrix}\left\{\begin{array}{c} v_2 \\ v_4 \end{array}\right\} \tag{4.3.13}$$

Solving for the displacements using $L = 3$ m, $E = 210 \times 10^9$ N/m^2, and $I = 2 \times 10^{-4}$ m^4 in Eq. (4.3.13), we obtain

$$v_2 = v_4 = -1.34 \times 10^{-3} \text{ m} = -1.34 \text{ mm} \tag{4.3.14}$$

as expected because of symmetry.

As observed from the solution of this problem, the greater the static redundancy (degrees of static indeterminacy or number of unknown forces and moments that cannot be determined by equations of statics), the smaller the kinematic redundancy (unknown nodal degrees of freedom, such as displacements or slopes)—hence, the fewer the number of unknown degrees of freedom to be solved for. Moreover, the use of symmetry, when applicable, reduces the number of unknown degrees of freedom even further. We can now back-substitute the results from Eq. (4.3.14), along with the numerical values for E, I, and L, into Eq. (4.3.12) to determine the global nodal forces as

$$
\begin{aligned}
F_{1y} &= 25 \text{ kN} & M_1 &= 37,500 \text{ N-m} \\
F_{2y} &= 50 \text{ kN} & M_2 &= 0 \\
F_{3y} &= 50 \text{ kN} & M_3 &= 0 \\
F_{4y} &= 50 \text{ kN} & M_4 &= 0 \\
F_{5y} &= 25 \text{ kN} & M_5 &= -37,500 \text{ N-m}
\end{aligned}
\tag{4.3.15}
$$

Once again, the global nodal forces (and moments) at the support nodes (nodes 1, 3, and 5) can be interpreted as the reaction forces, and the global nodal forces at nodes 2 and 4 are the applied nodal forces.

However, for large structures we must obtain the local element shear force and bending moment at each node end of the element because these values are used in the design/analysis process. We will again illustrate this concept for the element connecting nodes 1 and 2 in Figure 4–13. Using the local equations for this element, for which all nodal displacements have now been determined, we obtain

$$
\left\{\begin{array}{c} f_{1y}^{(1)} \\ m_1^{(1)} \\ f_{2y}^{(1)} \\ m_2^{(1)} \end{array}\right\} = \frac{EI}{L^3}\begin{bmatrix} 12 & 6L & -12 & 6L \\ 6L & 4L^2 & -6L & 2L^2 \\ -12 & -6L & 12 & -6L \\ 6L & 2L^2 & -6L & 4L^2 \end{bmatrix}\left\{\begin{array}{c} v_1 = 0 \\ \phi_1 = 0 \\ v_2 = -1.34 \times 10^{-3} \\ \phi_2 = 0 \end{array}\right\} \tag{4.3.16}
$$

Simplifying Eq. (4.3.16), we have

$$
\begin{Bmatrix} f_{1y}^{(1)} \\ m_1^{(1)} \\ f_{2y}^{(1)} \\ m_2^{(1)} \end{Bmatrix} = \begin{Bmatrix} 25{,}000 \text{ N} \\ 37{,}500 \text{ N-m} \\ -25{,}000 \text{ N} \\ 37{,}500 \text{ N-m} \end{Bmatrix}
\tag{4.3.17}
$$

If you wish, you can draw a free-body diagram to confirm the equilibrium of the element. ■

Finally, you should note that because of reflective symmetry about a vertical plane passing through node 3, we could have initially considered one-half of this beam and used the following model. The fixed support at node 3 is due to the slope being zero at node 3 because of the symmetry in the loading and support conditions.

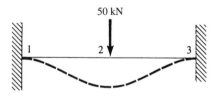

Example 4.3

Determine the nodal displacements and rotations and the global and element forces for the beam shown in Figure 4–14. We have discretized the beam as shown by the node numbering. The beam is fixed at node 1, has a roller support at node 2, and has an elastic spring support at node 3. A downward vertical force of $P = 50$ kN is applied at node 3. Let $E = 210$ GPa and $I = 2 \times 10^{-4}$ m^4 throughout the beam, and let $k = 200$ kN/m.

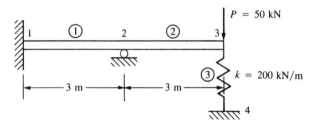

Figure 4–14 Beam example

SOLUTION:

Using Eq. (4.1.14) for each beam element and Eq. (2.2.18) for the spring element as well as the direct stiffness method, we obtain the structure stiffness matrix as

$$
[K] = \frac{EI}{L^3}
\begin{array}{c}
\begin{array}{ccccccc} v_1 & \phi_1 & v_2 & \phi_2 & v_3 & \phi_3 & v_4 \end{array} \\
\begin{bmatrix}
12 & 6L & -12 & 6L & 0 & 0 & 0 \\
 & 4L^2 & -6L & 2L^2 & 0 & 0 & 0 \\
 & & 24 & 0 & -12 & 6L & 0 \\
 & & & 8L^2 & -6L & 2L^2 & 0 \\
 & & & & 12 + \dfrac{kL^3}{EI} & -6L & -\dfrac{kL^3}{EI} \\
 & & & & & 4L^2 & 0 \\
\text{Symmetry} & & & & & & \dfrac{kL^3}{EI}
\end{bmatrix}
\end{array}
\qquad (4.3.18a)
$$

where the spring stiffness matrix $[k_s]$ given below by Eq. (4.3.18b) has been directly added into the global stiffness matrix corresponding to its degrees of freedom at nodes 3 and 4.

$$
[k_s] =
\begin{array}{c}
\begin{array}{cc} v_3 & v_4 \end{array} \\
\begin{bmatrix} k & -k \\ -k & k \end{bmatrix}
\end{array}
\qquad (4.3.18b)
$$

It is easier to solve the problem using the general variables, later making numerical substitutions into the final displacement expressions. The governing equations for the beam are then given by

$$
\begin{Bmatrix}
F_{1y} \\ M_1 \\ F_{2y} \\ M_2 \\ F_{3y} \\ M_3 \\ F_{4y}
\end{Bmatrix}
= \frac{EI}{L^3}
\begin{bmatrix}
12 & 6L & -12 & 6L & 0 & 0 & 0 \\
 & 4L^2 & -6L & 2L^2 & 0 & 0 & 0 \\
 & & 24 & 0 & -12 & 6L & 0 \\
 & & & 8L^2 & -6L & 2L^2 & 0 \\
 & & & & 12 + k' & -6L & -k' \\
 & & & & & 4L^2 & 0 \\
\text{Symmetry} & & & & & & k'
\end{bmatrix}
\begin{Bmatrix}
v_1 \\ \phi_1 \\ v_2 \\ \phi_2 \\ v_3 \\ \phi_3 \\ v_4
\end{Bmatrix}
\qquad (4.3.19)
$$

where $k' = kL^3/(EI)$ is used to simplify the notation. We now apply the boundary conditions

$$
v_1 = 0 \qquad \phi_1 = 0 \qquad v_2 = 0 \qquad v_4 = 0 \qquad (4.3.20)
$$

We delete the first three equations and the seventh equation (corresponding to the boundary conditions given by Eq. (4.3.20)) of Eqs. (4.3.19). The remaining three equations are

$$\left\{\begin{array}{c} 0 \\ -P \\ 0 \end{array}\right\} = \frac{EI}{L^3}\left[\begin{array}{ccc} 8L^2 & -6L & 2L^2 \\ -6L & 12+k' & -6L \\ 2L^2 & -6L & 4L^2 \end{array}\right]\left\{\begin{array}{c} \phi_2 \\ v_3 \\ \phi_3 \end{array}\right\} \qquad (4.3.21)$$

Solving Eqs. (4.3.21) simultaneously for the displacement at node 3 and the rotations at nodes 2 and 3, we obtain

$$v_3 = -\frac{7PL^3}{EI}\left(\frac{1}{12+7k'}\right) \qquad \phi_2 = -\frac{3PL^2}{EI}\left(\frac{1}{12+7k'}\right)$$

$$\phi_3 = -\frac{9PL^2}{EI}\left(\frac{1}{12+7k'}\right) \qquad (4.3.22)$$

The influence of the spring stiffness on the displacements is easily seen in Eq. (4.3.22). Solving for the numerical displacements using $P = 50$ kN, $L = 3$ m, $E = 210$ GPa ($= 210 \times 10^6$ kN/m^2), $I = 2 \times 10^{-4}$ m^4, and $k' = 0.129$ in Eq. (4.3.22), we obtain

$$v_3 = \frac{-7(50 \text{ kN})(3 \text{ m})^3}{(210 \times 10^6 \text{ kN/m}^2)(2 \times 10^{-4} \text{ m}^4)}\left(\frac{1}{12+7(0.129)}\right) = -0.0174 \text{ m} \quad (4.3.23)$$

Similar substitutions into Eq. (4.3.26) yield

$$\phi_2 = -0.00249 \text{ rad} \qquad \phi_3 = -0.00747 \text{ rad} \qquad (4.3.24)$$

We now back-substitute the results from Eqs. (4.3.23) and (4.3.24), along with numerical values for P, E, I, L, and k', into Eq. (4.3.19) to obtain the global nodal forces as

$$F_{1y} = -69.9 \text{ kN} \qquad M_1 = -69.7 \text{ kN}\cdot\text{m}$$

$$F_{2y} = 116.4 \text{ kN} \qquad M_2 = 0.0 \text{ kN}\cdot\text{m} \qquad (4.3.25)$$

$$F_{3y} = -50.0 \text{ kN} \qquad M_3 = 0.0 \text{ kN}\cdot\text{m}$$

For the beam-spring structure, an additional global force F_{4y} is determined at the base of the spring as follows:

$$F_{4y} = -v_3 k = (0.0174)200 = 3.5 \text{ kN} \qquad (4.3.26)$$

This force provides the additional global y force for equilibrium of the structure.

A free-body diagram, including the forces and moments from Eqs. (4.3.25) and (4.3.26) acting on the beam, is shown in Figure 4–15.

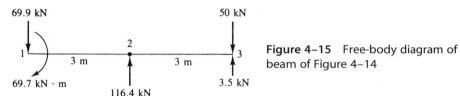

Figure 4–15 Free-body diagram of beam of Figure 4–14

Example 4.4

Determine the displacement and rotation under the force and moment located at the center of the beam shown in Figure 4–16. The beam has been discretized into the two elements shown in Figure 4–16. The beam is fixed at each end. A downward force of 10 kN and an applied moment of 20 kN-m act at the center of the beam. Let $E = 210$ GPa and $I = 4 \times 10^{-4}$ m^4 throughout the beam length.

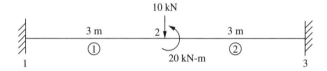

Figure 4–16 Fixed-fixed beam subjected to applied force and moment

SOLUTION:

Using Eq. (4.1.14) for each beam element with $L = 3$ m, we obtain the element stiffness matrices as follows:

$$[k^{(1)}] = \frac{EI}{L^3}\begin{bmatrix} \begin{matrix} v_1 & \phi_1 & v_2 & \phi_2 \end{matrix} \\ 12 & 6L & -12 & 6L \\ & 4L^2 & -6L & 2L^2 \\ & & 12 & -6L \\ \text{Symmetry} & & & 4L^2 \end{bmatrix} \qquad [k^{(2)}] = \frac{EI}{L^3}\begin{bmatrix} \begin{matrix} v_2 & \phi_2 & v_3 & \phi_3 \end{matrix} \\ 12 & 6L & -12 & 6L \\ & 4L^2 & -6L & 2L^2 \\ & & 12 & -6L \\ \text{Symmetry} & & & 4L^2 \end{bmatrix}$$

$$(4.3.27)$$

The boundary conditions are given by

$$v_1 = \phi_1 = v_3 = \phi_3 = 0 \qquad (4.3.28)$$

The global forces are $F_{2y} = -10{,}000$ N and $M_2 = 20{,}000$ N-m.

Applying the global forces and boundary conditions, Eq. (4.3.28), and assembling the global stiffness matrix using the direct stiffness method and Eqs. (4.3.27), we obtain the global equations as:

$$\begin{Bmatrix} -10{,}000 \\ 20{,}000 \end{Bmatrix} = \frac{(210 \times 10^9)(4 \times 10^{-4})}{3^3} \begin{bmatrix} 24 & 0 \\ 0 & 8(3^2) \end{bmatrix} \begin{Bmatrix} v_2 \\ \phi_2 \end{Bmatrix} \qquad (4.3.29)$$

Solving Eq. (4.3.29) for the displacement and rotation, we obtain

$$v_2 = -1.339 \times 10^{-4} \text{ m} \quad \text{and} \quad \phi_2 = 8.928 \times 10^{-5} \text{ rad} \qquad (4.3.30)$$

Using the local equations for each element, we obtain the local nodal forces and moments for element one as follows:

$$\begin{Bmatrix} f_{1y}^{(1)} \\ m_1^{(1)} \\ f_{2y}^{(1)} \\ m_2^{(1)} \end{Bmatrix} = \frac{(210 \times 10^9)(4 \times 10^{-4})}{3^3} \begin{bmatrix} 12 & 6(3) & -12 & 6(3) \\ 6(3) & 4(3^2) & -6(3) & 2(3^2) \\ -12 & -6(3) & 12 & -6(3) \\ 6(3) & 2(3^2) & -6(3) & 4(3^2) \end{bmatrix} \begin{Bmatrix} 0 \\ 0 \\ -1.3339 \times 10^{-4} \\ 8.928 \times 10^{-5} \end{Bmatrix}$$

(4.3.31)

Simplifying Eq. (4.3.31), we have

$$f_{1y}^{(1)} = 10{,}000 \text{ N}, \quad m_1^{(1)} = 12{,}500 \text{ N-m}, \quad f_{2y}^{(1)} = -10{,}000 \text{ N}, \quad m_2^{(1)} = 17{,}500 \text{ N-m}$$

(4.3.32)

Similarly, for element two the local nodal forces and moments are

$$f_{2y}^{(2)} = 0, \quad m_2^{(2)} = 2500 \text{ N-m}, \quad f_{3y}^{(2)} = 0, \quad m_3^{(2)} = -2500 \text{ N-m}$$ (4.3.33)

Using the results from Eqs. (4.3.32) and (4.3.33), we show the local forces and moments acting on each element in Figure 4–16 as follows.

Using the results from Eqs. (4.3.32) and (4.3.33), or Figure 4–17, we obtain the shear force and bending moment diagrams for each element as shown in Figure 4–18.

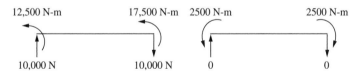

Figure 4–17 Nodal forces and moments acting on each element of Figure 4–16

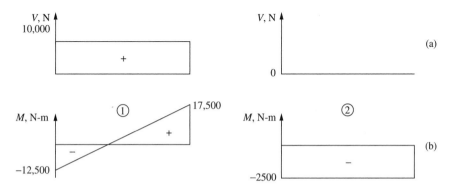

Figure 4–18 Shear force (a) and bending moment (b) diagrams for each element of Figure 4–16

Example 4.5

To illustrate the effects of shear deformation along with the usual bending deformation, we now solve the simple beam shown in Figure 4–19. We will use the beam stiffness matrix given by Eq. (4.1.15o) that includes both the bending and shear deformation contributions for deformation in the x–y plane. The beam is simply supported with a concentrated load of 10,000 N applied at mid-span. We let material properties be $E = 207$ GPa and $G = 80$ GPa. The beam width and height are $b = 25$ mm and $h = 50$ mm, respectively.

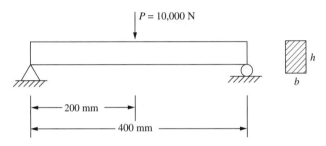

Figure 4–19 Simple beam subjected to concentrated load at center of span

SOLUTION:

We will use symmetry to simplify the solution. Therefore, only one half of the beam will be considered with the slope at the center forced to be zero. Also, one-half of the concentrated load is then used. The model with symmetry enforced is shown in Figure 4–20.

The finite element model will consist of only one beam element. Using Eq. (4.1.15o) for the Timoshenko beam element stiffness matrix, we obtain the global equations as

$$
\frac{EI}{L^3(1+\varphi)}
\begin{bmatrix}
12 & 6L & -12 & 6L \\
6L & (4+\varphi)L^2 & -6L & (2-\varphi)L^2 \\
-12 & -6L & 12 & -6L \\
6L & (2-\varphi)L^2 & -6L & (4+\varphi)L^2
\end{bmatrix}
\begin{Bmatrix}
v_1 = 0 \\
\phi_1 \\
v_2 \\
\phi_2 = 0
\end{Bmatrix}
=
\begin{Bmatrix}
F_{1y} \\
0 \\
-P/2 \\
0
\end{Bmatrix}
\tag{4.3.34}
$$

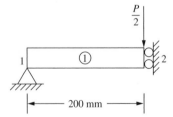

Figure 4–20 Beam with symmetry enforced

Note that the boundary conditions given by $v_1 = 0$ and $\phi_2 = 0$ have been included in Eq. (4.3.34).

Using the second and third equations of Eq. (4.3.34) whose rows are associated with the two unknowns, ϕ_1 and v_2, we obtain

$$v_2 = \frac{-PL^3(4 + \varphi)}{24EI} \quad \text{and} \quad \phi_1 = \frac{-PL^2}{4EI} \tag{4.3.35}$$

As the beam is rectangular in cross section, the moment of inertia is

$$I = bh^3/12$$

Substituting the numerical values for b and h, we obtain I as

$$I = 0.26 \times 10^{-6} \text{ m}^4$$

The shear correction factor is given by

$$\varphi = \frac{12EI}{k_s AGL^2}$$

and k_s for a rectangular cross section is given by $k_s = 5/6$.

Substituting numerical values for E, I, G, L, and k_s, we obtain

$$\varphi = \frac{12 \times 207 \times 10^9 \times 0.26 \times 10^{-6}}{5/6 \times 0.025 \times 0.05 \times 80 \times 10^9 \times 0.2^2} = 0.1938$$

Substituting for $P = 10,000$ N, $L = 0.2$ m, and $\varphi = 0.1938$ into Eq. (4.3.35), we obtain the displacement at the mid-span as

$$v_2 = -2.597 \times 10^{-4} \text{ m} \tag{4.3.36}$$

If we let $l =$ the whole length of the beam, then $l = 2L$ and we can substitute $L = l/2$ into Eq. (4.3.35) to obtain the displacement in terms of the whole length of the beam as

$$v_2 = \frac{-Pl^3(4 + \varphi)}{192EI} \tag{4.3.37}$$

For long slender beams with l about 10 or more times the beam depth, h, the transverse shear correction term φ is small and can be neglected. Therefore, Eq. (4.3.37) becomes

$$v_2 = \frac{-Pl^3}{48EI} \tag{4.3.38}$$

Equation (4.3.38) is the classical beam deflection formula for a simply supported beam subjected to a concentrated load at mid-span.

Using Eq. (4.3.38), the deflection is obtained as

$$v_2 = -2.474 \times 10^{-4} \text{ m} \tag{4.3.39}$$

Comparing the deflections obtained using the shear-correction factor with the deflection predicted using the beam-bending contribution only, we obtain

$$\% \text{ change} = \frac{2.597 - 2.474}{2.474} \times 100 = 4.97\% \text{ difference} \qquad \blacksquare$$

▲ 4.4 Distributed Loading ▲

Beam members can support distributed loading as well as concentrated nodal loading. Therefore, we must be able to account for distributed loading. Consider the fixed-fixed beam subjected to a uniformly distributed loading w shown in Figure 4–21. The reactions, determined from structural analysis theory [2], are shown in Figure 4–22. These reactions are called *fixed-end reactions*. In general, **fixed-end reactions** are those reactions at the ends of an element if the ends of the element are assumed to be fixed—that is, if displacements and rotations are prevented. (Those of you who are unfamiliar with the analysis of indeterminate structures should assume these reactions as given and proceed with the rest of the discussion; we will develop these results in a subsequent presentation of the work-equivalence method.) Therefore, guided by the results from structural analysis for the case of a uniformly distributed load, we replace the load by concentrated nodal forces and moments tending to have the same effect on the beam as the actual distributed load. Figure 4–23 illustrates this idea for a beam. We have replaced the uniformly distributed load by a statically equivalent force system consisting of a concentrated nodal force and moment at each end of the member carrying the distributed load. That is, both the statically equivalent concentrated nodal forces and moments and the original distributed load have the same resultant force and same moment about an arbitrarily chosen point. These statically equivalent forces are always of opposite sign from the fixed-end reactions. If we want to analyze the behavior of loaded member 2–3 in better detail, we can place a node at midspan and use the same procedure just described for each of the two elements representing the horizontal member. That is, to determine the maximum deflection and maximum moment in the beam span, a node 5 is needed at midspan of beam segment 2–3, and work-equivalent forces and moments are applied to each element (from node 2 to node 5 and from node 5 to node 3) shown in Figure 4–23(c).

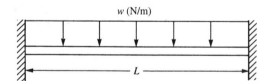

Figure 4–21 Fixed-fixed beam subjected to a uniformly distributed load

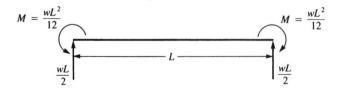

Figure 4–22 Fixed-end reactions for the beam of Figure 4–21

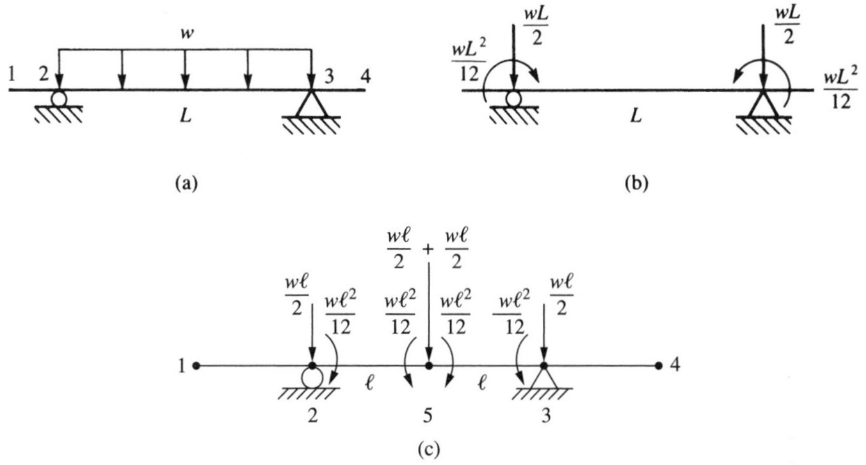

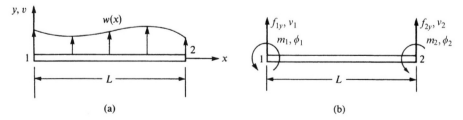

Figure 4–23 (a) Beam with a distributed load, (b) the equivalent nodal force system, and (c) the enlarged beam (for clarity's sake) with equivalent nodal force system when node 5 is added to the midspan

Work-Equivalence Method

We can use the work-equivalence method to replace a distributed load by a set of discrete loads. This method is based on the concept that the work of the distributed load $w(x)$ in going through the displacement field $v(x)$ is equal to the work done by nodal loads f_{iy} and m_i in going through nodal displacements v_i and ϕ_i for arbitrary nodal displacements. To illustrate the method, we consider the example shown in Figure 4–24. The work due to the distributed load is given by

$$W_{\text{distributed}} = \int_0^L w(x)v(x)\,dx \qquad (4.4.1)$$

where $v(x)$ is the transverse displacement given by Eq. (4.1.4). The work due to the discrete nodal forces is given by

$$W_{\text{discrete}} = m_1\phi_1 + m_2\phi_2 + f_{1y}v_1 + f_{2y}v_2 \qquad (4.4.2)$$

Figure 4–24 (a) Beam element subjected to a general load and (b) the statically equivalent nodal force system

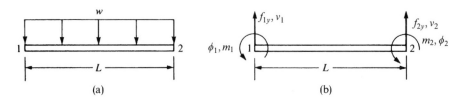

Figure 4–25 (a) Beam subjected to a uniformly distributed loading and (b) the equivalent nodal forces to be determined

We can then determine the nodal moments and forces m_1, m_2, f_{1y}, and f_{2y} used to replace the distributed load by using the concept of work equivalence—that is, by setting $W_{\text{distributed}} = W_{\text{discrete}}$ for arbitrary displacements ϕ_1, ϕ_2, v_1, and v_2.

Example of Load Replacement

To illustrate more clearly the concept of work equivalence, we will now consider a beam subjected to a specified distributed load. Consider the uniformly loaded beam shown in Figure 4–25(a). The support conditions are not shown because they are not relevant to the replacement scheme. By letting $W_{\text{discrete}} = W_{\text{distributed}}$ and by assuming arbitrary ϕ_1, ϕ_2, v_1, and v_2, we will find equivalent nodal forces m_1, m_2, f_{1y}, and f_{2y}. Figure 4–25(b) shows the nodal forces and moments directions as positive based on Figure 4–1.

Using Eqs. (4.4.1) and (4.4.2) for $W_{\text{distributed}} = W_{\text{discrete}}$, we have

$$\int_0^L w(x)v(x)\,dx = m_1\phi_1 + m_2\phi_2 + f_{1y}v_1 + f_{2y}v_2 \tag{4.4.3}$$

where $m_1\phi_1$ and $m_2\phi_2$ are the work due to concentrated nodal moments moving through their respective nodal rotations and $f_{1y}v_1$ and $f_{2y}v_2$ are the work due to the nodal forces moving through nodal displacements. Evaluating the left-hand side of Eq. (4.4.3) by substituting $w(x) = -w$ and $v(x)$ from Eq. (4.1.4), we obtain the work due to the distributed load as

$$\int_0^L w(x)v(x)\,dx = -\frac{Lw}{2}(v_1 - v_2) - \frac{L^2w}{4}(\phi_1 + \phi_2) - Lw(v_2 - v_1)$$

$$+ \frac{L^2w}{3}(2\phi_1 + \phi_2) - \phi_1\left(\frac{L^2w}{2}\right) - v_1(wL) \tag{4.4.4}$$

Now using Eqs. (4.4.3) and (4.4.4) for arbitrary nodal displacements, we let $\phi_1 = 1$, $\phi_2 = 0, v_1 = 0$, and $v_2 = 0$ and then obtain

$$m_1(1) = -\left(\frac{L^2w}{4} - \frac{2}{3}L^2w + \frac{L^2}{2}w\right) = -\frac{wL^2}{12} \tag{4.4.5}$$

Similarly, letting $\phi_1 = 0, \phi_2 = 1, v_1 = 0$, and $v_2 = 0$ yields

$$m_2(1) = -\left(\frac{L^2w}{4} - \frac{L^2w}{3}\right) = \frac{wL^2}{12} \tag{4.4.6}$$

Finally, letting all nodal displacements equal zero except first v_1 and then v_2, we obtain

$$f_{1y}(1) = -\frac{Lw}{2} + Lw - Lw = -\frac{Lw}{2}$$

$$f_{2y}(1) = \frac{Lw}{2} - Lw = -\frac{Lw}{2}$$

(4.4.7)

We can conclude that, in general, for any given load function $w(x)$, we can multiply by $v(x)$ and then integrate according to Eq. (4.4.3) to obtain the concentrated nodal forces (and/or moments) used to replace the distributed load. Moreover, we can obtain the load replacement by using the concept of fixed-end reactions from structural analysis theory. Tables of fixed-end reactions have been generated for numerous load cases and can be found in texts on structural analysis such as Reference [2]. A table of equivalent nodal forces has been generated in Appendix D of this text, guided by the fact that fixed-end reaction forces are of opposite sign from those obtained by the work equivalence method.

Hence, if a concentrated load is applied other than at the natural intersection of two elements, we can use the concept of equivalent nodal forces to replace the concentrated load by nodal concentrated values acting at the beam ends, instead of creating a node on the beam at the location where the load is applied. We provide examples of this procedure for handling concentrated loads on elements in beam Example 4.7 and in plane frame Example 5.3.

General Formulation

In general, we can account for distributed loads or concentrated loads acting on beam elements by starting with the following formulation application for a general structure:

$$\{F\} = [K]\{d\} - \{F_0\}$$

(4.4.8)

where $\{F\}$ are the concentrated nodal forces and $\{F_0\}$ are called the *equivalent nodal forces*, now expressed in terms of global-coordinate components, which are of such magnitude that they yield the same displacements at the nodes as would the distributed load. Using the table in Appendix D of equivalent nodal forces $\{f_0\}$ expressed in terms of local-coordinate components, we can express $\{F_0\}$ in terms of global-coordinate components.

Recall from Section 3.10 the derivation of the element equations by the principle of minimum potential energy. Starting with Eqs. (3.10.19) and (3.10.20), the minimization of the total potential energy resulted in the same form of equation as Eq. (4.4.8) where $\{F_0\}$ now represents the same work-equivalent force replacement system as given by Eq. (3.10.20a) for surface traction replacement. Also, $\{F\} = \{P\}, \{P\}$ [from Eq. (3.10.20)] represents the global nodal concentrated forces. Because we now assume that concentrated nodal forces are not present ($\{F\} = 0$), as we are solving beam problems with distributed loading only in this section, we can write Eq. (4.4.8) as

$$\{F_0\} = [K]\{d\}$$

(4.4.9)

On solving for $\{d\}$ in Eq. (4.4.9) and then substituting the global displacements $\{d\}$ and equivalent nodal forces $\{F_0\}$ into Eq. (4.4.8), we obtain the actual global nodal forces $\{F\}$. For example, using the definition of $\{f_0\}$ and Eqs. (4.4.5) through (4.4.7) (or using load case 4 in Appendix D) for a uniformly distributed load w acting over a one-element beam, we have

$$\{F_0\} = \begin{Bmatrix} \dfrac{-wL}{2} \\ \dfrac{-wL^2}{12} \\ \dfrac{-wL}{2} \\ \dfrac{wL^2}{12} \end{Bmatrix} \tag{4.4.10}$$

This concept can be applied on a local basis to obtain the local nodal forces $\{f\}$ in individual elements of structures by applying Eq. (4.4.8) locally as

$$\{f\} = [k]\{d\} - \{f_0\} \tag{4.4.11}$$

where $\{f_0\}$ are the equivalent local nodal forces.

Examples 4.6 through 4.8 illustrate the method of equivalent nodal forces for solving beams subjected to distributed and concentrated loadings. We will use global-coordinate notation in Examples 4.6 through 4.8—treating the beam as a general structure rather than as an element.

Example 4.6

For the cantilever beam subjected to the uniform load w in Figure 4–26, solve for the right-end vertical displacement and rotation and then for the nodal forces. Assume the beam to have constant EI throughout its length.

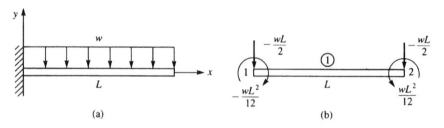

(a)   (b)

Figure 4–26 (a) Cantilever beam subjected to a uniformly distributed load and (b) the work equivalent nodal force system

SOLUTION:

We begin by discretizing the beam. Here only one element will be used to represent the whole beam. Next, the distributed load is replaced by its work-equivalent nodal forces as shown in Figure 4–26(b). The work-equivalent nodal forces are those that result from the uniformly distributed load acting over the whole beam given by Eq. (4.4.10). (Or see appropriate load case 4 in Appendix D.) Using Eq. (4.4.9) and the beam element stiffness matrix, we obtain

$$
\frac{EI}{L^3}
\begin{bmatrix}
12 & 6L & -12 & 6L \\
 & 4L^2 & -6L^2 & 2L^2 \\
 & & 12 & -6L \\
 & & & 4L^2
\end{bmatrix}
\begin{Bmatrix}
v_1 \\ \phi_1 \\ v_2 \\ \phi_2
\end{Bmatrix}
=
\begin{Bmatrix}
F_{1y} - \dfrac{wL}{2} \\[2mm]
M_1 - \dfrac{wL^2}{12} \\[2mm]
\dfrac{-wL}{2} \\[2mm]
\dfrac{wL^2}{12}
\end{Bmatrix}
\tag{4.4.12}
$$

where we have applied the work equivalent nodal forces and moments from Figure 4–26(b).

Applying the boundary conditions $v_1 = 0$ and $\phi_1 = 0$ to Eqs (4.4.12) and then partitioning off the third and fourth equations of Eq. (4.4.12), we obtain

$$
\frac{EI}{L^3}
\begin{bmatrix}
12 & -6L \\
-6L^2 & 4L^2
\end{bmatrix}
\begin{Bmatrix}
v_2 \\ \phi_2
\end{Bmatrix}
=
\begin{Bmatrix}
-\dfrac{wL}{2} \\[2mm]
\dfrac{wL^2}{12}
\end{Bmatrix}
\tag{4.4.13}
$$

Solving Eq. (4.4.13) for the displacements, we obtain

$$
\begin{Bmatrix}
v_2 \\ \phi_2
\end{Bmatrix}
=
\frac{L}{6EI}
\begin{bmatrix}
2L^2 & 3L \\
3L & 6
\end{bmatrix}
\begin{Bmatrix}
\dfrac{-wL}{2} \\[2mm]
\dfrac{wL^2}{12}
\end{Bmatrix}
\tag{4.4.14a}
$$

Simplifying Eq. (4.4.14a), we obtain the displacement and rotation as

$$
\begin{Bmatrix}
v_2 \\ \phi_2
\end{Bmatrix}
=
\begin{Bmatrix}
\dfrac{-wL^4}{8EI} \\[2mm]
\dfrac{-wL^3}{6EI}
\end{Bmatrix}
\tag{4.4.14b}
$$

The negative signs in the answers indicate that v_2 is downward and ϕ_2 is clockwise. In this case, the method of replacing the distributed load by discrete concentrated loads gives exact solutions for the displacement and rotation as could be obtained by classical methods, such as double integration [1]. This is expected, as the work-equivalence method ensures that the nodal displacement and rotation from the finite element method match those from an exact solution.

We will now illustrate the procedure for obtaining the global nodal forces. For convenience, we first define the product $[K]\{d\}$ to be $\{F^{(e)}\}$, where $\{F^{(e)}\}$ are called the *effective global nodal forces*. On using Eq. (4.4.14) for $\{d\}$, we then have

$$\begin{Bmatrix} F_{1y}^{(e)} \\ M_1^{(e)} \\ F_{2y}^{(e)} \\ M_2^{(e)} \end{Bmatrix} = \frac{EI}{L^3} \begin{bmatrix} 12 & 6L & -12 & 6L \\ 6L & 4L^2 & -6L & 2L^2 \\ -12 & -6L & 12 & -6L \\ 6L & 2L^2 & -6L & 4L^2 \end{bmatrix} \begin{Bmatrix} 0 \\ 0 \\ \dfrac{-wL^4}{8EI} \\ \dfrac{-wL^3}{6EI} \end{Bmatrix} \quad (4.4.15)$$

Simplifying Eq. (4.4.15), we obtain

$$\begin{Bmatrix} F_{1y}^{(e)} \\ M_1^{(e)} \\ F_{2y}^{(e)} \\ M_2^{(e)} \end{Bmatrix} = \begin{Bmatrix} \dfrac{wL}{2} \\ \dfrac{5wL^2}{12} \\ \dfrac{-wL}{2} \\ \dfrac{wL^2}{12} \end{Bmatrix} \quad (4.4.16)$$

We then use Eqs. (4.4.10) and (4.4.16) in Eq. (4.4.8) $\{F\} = [K]\{d\} - \{F_0\}$ to obtain the correct global nodal forces as

$$\begin{Bmatrix} F_{1y} \\ M_1 \\ F_{2y} \\ M_2 \end{Bmatrix} = \begin{Bmatrix} \dfrac{wL}{2} \\ \dfrac{5wL^2}{12} \\ \dfrac{-wL}{2} \\ \dfrac{wL^2}{12} \end{Bmatrix} - \begin{Bmatrix} \dfrac{-wL}{2} \\ \dfrac{-wL^2}{12} \\ \dfrac{-wL}{2} \\ \dfrac{wL^2}{12} \end{Bmatrix} = \begin{Bmatrix} wL \\ \dfrac{wL^2}{2} \\ 0 \\ 0 \end{Bmatrix} \quad (4.4.17)$$

In Eq. (4.4.17), F_{1y} is the vertical force reaction and M_1 is the moment reaction as applied by the clamped support at node 1. The results for displacement given by Eq. (4.4.14b) and the global nodal forces given by Eq. (4.4.17) are sufficient to complete the solution of the cantilever beam problem.

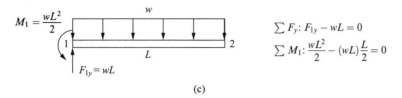

$M_1 = \dfrac{wL^2}{2}$

$\sum F_y: F_{1y} - wL = 0$

$\sum M_1: \dfrac{wL^2}{2} - (wL)\dfrac{L}{2} = 0$

$F_{1y} = wL$

(c)

Figure 4–26 (c) Free-body diagram and equations of equilibrium for beam of Figure 4–26(a).

A free-body diagram of the beam using the reactions from Eq. (4.4.17) verifies both force and moment equilibrium as shown in Figure 4–26(c). ∎

The nodal force and moment reactions obtained by Eq. (4.4.17) illustrate the importance of using Eq. (4.4.8) to obtain the correct global nodal forces and moments. By subtracting the work-equivalent force matrix, $\{F_0\}$, from the product of $[K]$ times $\{d\}$, we obtain the correct reactions at node 1 as can be verified by simple static equilibrium equations. This verification validates the general method as follows:

1. Replace the distributed load by its work-equivalent as shown in Figure 4–26(b) to identify the nodal force and moment used in the solution.
2. Assemble the global force and stiffness matrices and global equations illustrated by Eq. (4.4.12).
3. Apply the boundary conditions to reduce the set of equations as done in previous problems and illustrated by Eq. (4.4.13) where the original four equations have been reduced to two equations to be solved for the unknown displacement and rotation.
4. Solve for the unknown displacement and rotation given by Eq. (4.4.14a) and Eq. (4.4.14b).
5. Use Eq. (4.4.8) as illustrated by Eq. (4.4.17) to obtain the final correct global nodal forces and moments. Those forces and moments at supports, such as the left end of the cantilever in Figure 4–26(a), will be the reactions.

We will solve the following example to illustrate the procedure for handling concentrated loads acting on beam elements at locations other than nodes.

Example 4.7

For the cantilever beam subjected to the concentrated load P in Figure 4–27, solve for the right-end vertical displacement and rotation and the nodal forces, including reactions, by replacing the concentrated load with equivalent nodal forces acting at each end of the beam. Assume EI constant throughout the beam.

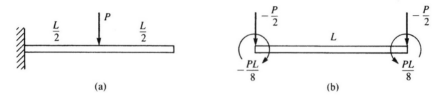

Figure 4–27 (a) Cantilever beam subjected to a concentrated load and (b) the equivalent nodal force replacement system

SOLUTION:

We begin by discretizing the beam. Here only one element is used with nodes at each end of the beam. We then replace the concentrated load as shown in Figure 4–27(b) by using appropriate loading case 1 in Appendix D. Using Eq. (4.4.9) and the beam element stiffness matrix Eq. (4.1.14), we obtain

$$
\frac{EI}{L^3}\begin{bmatrix} 12 & -6L \\ -6L & 4L^2 \end{bmatrix}\begin{Bmatrix} v_2 \\ \phi_2 \end{Bmatrix} = \begin{Bmatrix} \dfrac{-P}{2} \\ \dfrac{PL}{8} \end{Bmatrix} \tag{4.4.18}
$$

where we have applied the nodal forces from Figure 4–27(b) and the boundary conditions $v_1 = 0$ and $\phi_1 = 0$ to reduce the number of matrix equations for the usual longhand solution. Solving Eq. (4.4.18) for the displacements, we obtain

$$
\begin{Bmatrix} v_2 \\ \phi_2 \end{Bmatrix} = \frac{L}{6EI}\begin{bmatrix} 2L^2 & 3L \\ 3L & 6 \end{bmatrix}\begin{Bmatrix} \dfrac{-P}{2} \\ \dfrac{PL}{8} \end{Bmatrix} \tag{4.4.19}
$$

Simplifying Eq. (4.4.19), we obtain the displacement and rotation as

$$
\begin{Bmatrix} v_2 \\ \phi_2 \end{Bmatrix} = \begin{Bmatrix} \dfrac{-5PL^3}{48EI} \\ \dfrac{-PL^2}{8EI} \end{Bmatrix} \begin{matrix} \downarrow \\ \curvearrowright \end{matrix} \tag{4.4.20}
$$

To obtain the unknown nodal forces, we begin by evaluating the effective nodal forces $\{F^{(e)}\} = [K]\{d\}$ as

$$
\begin{Bmatrix} F_{1y}^{(e)} \\ M_1^{(e)} \\ F_{2y}^{(e)} \\ M_2^{(e)} \end{Bmatrix} = \frac{EI}{L^3}\begin{bmatrix} 12 & 6L & -12 & 6L \\ 6L & 4L^2 & -6L & 2L^2 \\ -12 & -6L & 12 & -6L \\ 6L & 2L^2 & -6L & 4L^2 \end{bmatrix}\begin{Bmatrix} 0 \\ 0 \\ \dfrac{-5PL^3}{48EI} \\ \dfrac{-PL^2}{8EI} \end{Bmatrix} \tag{4.4.21}
$$

Simplifying Eq. (4.4.21), we obtain

$$\left\{\begin{array}{c} F_{1y}^{(e)} \\ M_1^{(e)} \\ F_{2y}^{(e)} \\ M_2^{(e)} \end{array}\right\} = \left\{\begin{array}{c} \dfrac{P}{2} \\ \dfrac{3PL}{8} \\ \dfrac{-P}{2} \\ \dfrac{PL}{8} \end{array}\right\} \tag{4.4.22}$$

Then using Eq. (4.4.22) and the equivalent nodal forces from Figure 4–27(b) in Eq. (4.4.8), we obtain the correct nodal forces as

$$\left\{\begin{array}{c} F_{1y} \\ M_1 \\ F_{2y} \\ M_2 \end{array}\right\} = \left\{\begin{array}{c} \dfrac{P}{2} \\ \dfrac{3PL}{8} \\ \dfrac{-P}{2} \\ \dfrac{PL}{8} \end{array}\right\} - \left\{\begin{array}{c} \dfrac{-P}{2} \\ \dfrac{-PL}{8} \\ \dfrac{-P}{2} \\ \dfrac{PL}{8} \end{array}\right\} = \left\{\begin{array}{c} P \\ \dfrac{PL}{2} \\ 0 \\ 0 \end{array}\right\} \tag{4.4.23}$$

We can see from Eq. (4.4.23) that F_{1y} is equivalent to the vertical reaction force and M_1 is the reaction moment as applied by the clamped support at node 1.

Again, the reactions obtained by Eq. (4.4.23) can be verified to be correct by using static equilibrium equations to validate once more the correctness of the general formulation and procedures summarized in the steps given after Example 4.6. ■

To illustrate the procedure for handling concentrated nodal forces and distributed loads acting simultaneously on beam elements, we will solve the following example.

Example 4.8

For the cantilever beam subjected to the concentrated free-end load P and the uniformly distributed load w acting over the whole beam as shown in Figure 4–28, determine the free-end displacements and the nodal forces.

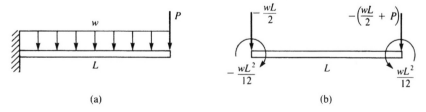

(a) (b)

Figure 4–28 (a) Cantilever beam subjected to a concentrated load and a distributed load and (b) the equivalent nodal force replacement system

SOLUTION:

Once again, the beam is modeled using one element with nodes 1 and 2, and the distributed load is replaced as shown in Figure 4–28(b) using appropriate loading case 4 in Appendix D. Using the beam element stiffness Eq. (4.1.14), we obtain

$$\frac{EI}{L^3}\begin{bmatrix} 12 & -6L \\ -6L & 4L^2 \end{bmatrix}\begin{Bmatrix} v_2 \\ \phi_2 \end{Bmatrix} = \begin{Bmatrix} \dfrac{-wL}{2}-P \\ \dfrac{wL^2}{12} \end{Bmatrix} \tag{4.4.24}$$

where we have applied the nodal forces from Figure 4–28(b) and the boundary conditions $v_1 = 0$ and $\phi_1 = 0$ to reduce the number of matrix equations for the usual long-hand solution. Solving Eq. (4.4.24) for the displacements, we obtain

$$\begin{Bmatrix} v_2 \\ \phi_2 \end{Bmatrix} = \begin{Bmatrix} \dfrac{-wL^4}{8EI}-\dfrac{PL^3}{3EI} \\[2mm] \dfrac{-wL^3}{6EI}-\dfrac{PL^2}{2EI} \end{Bmatrix} \tag{4.4.25}$$

Next, we obtain the effective nodal forces using $\{F^{(e)}\} = [K]\{d\}$ as

$$\begin{Bmatrix} F_{1y}^{(e)} \\ M_1^{(e)} \\ F_{2y}^{(e)} \\ M_2^{(e)} \end{Bmatrix} = \frac{EI}{L^3}\begin{bmatrix} 12 & 6L & -12 & 6L \\ 6L & 4L^2 & -6L & 2L^2 \\ -12 & -6L & 12 & -6L \\ 6L & 2L^2 & -6L & 4L^2 \end{bmatrix}\begin{Bmatrix} 0 \\ 0 \\ \dfrac{-wL^4}{8EI}-\dfrac{PL^3}{3EI} \\[2mm] \dfrac{-wL^3}{6EI}-\dfrac{PL^2}{2EI} \end{Bmatrix} \tag{4.4.26}$$

Simplifying Eq. (4.4.26), we obtain

$$\begin{Bmatrix} F_{1y}^{(e)} \\ M_1^{(e)} \\ F_{2y}^{(e)} \\ M_2^{(e)} \end{Bmatrix} = \begin{Bmatrix} P+\dfrac{wL}{2} \\[2mm] PL+\dfrac{5wL^2}{12} \\[2mm] -P-\dfrac{wL}{2} \\[2mm] \dfrac{wL^2}{12} \end{Bmatrix} \tag{4.4.27}$$

Finally, subtracting the equivalent nodal force matrix [see Figure 4–27(b)] from the effective force matrix of Eq. (4.4.27), we obtain the correct nodal forces as

$$\begin{Bmatrix} F_{1y} \\ M_1 \\ F_{2y} \\ M_2 \end{Bmatrix} = \begin{Bmatrix} P+\dfrac{wL}{2} \\[2mm] PL+\dfrac{5wL^2}{12} \\[2mm] -P-\dfrac{wL}{2} \\[2mm] \dfrac{wL^2}{12} \end{Bmatrix} - \begin{Bmatrix} \dfrac{-wL}{2} \\[2mm] \dfrac{-wL^2}{12} \\[2mm] \dfrac{-wL}{2} \\[2mm] \dfrac{wL^2}{12} \end{Bmatrix} = \begin{Bmatrix} P+wL \\[2mm] PL+\dfrac{wL^2}{2} \\[2mm] -P \\[2mm] 0 \end{Bmatrix} \tag{4.4.28}$$

From Eq. (4.4.28), we see that F_{1y} is equivalent to the vertical reaction force, M_1 is the reaction moment at node 1, and F_{2y} is equal to the applied downward force P at node 2. [Remember that only the equivalent nodal force matrix is subtracted, not the original concentrated load matrix. This is based on the general formulation, Eq. (4.4.8).] ∎

To generalize the work-equivalent method, we apply it to a beam with more than one element as shown in the following Example 4.9.

Example 4.9

For the fixed–fixed beam subjected to the linear varying distributed loading acting over the whole beam shown in Figure 4–29(a) determine the displacement and rotation at the center and the reactions.

SOLUTION:

The beam is now modeled using two elements with nodes 1, 2, and 3 and the distributed load is replaced as shown in Figure 4–29(b) using the appropriate load cases 4 and 5 in Appendix D. Note that load case 5 is used for element one as it has only the linear varying distributed load acting on it with a high end value of $w/2$ as shown in Figure 4–29(a), while both load cases 4 and 5 are used for element two as the distributed load is divided into a uniform part with magnitude $w/2$ and a linear varying part with magnitude at the high end of the load equal to $w/2$ also.

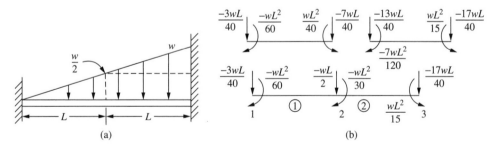

Figure 4–29 (a) Fixed–fixed beam subjected to linear varying line load and (b) the equivalent nodal force replacement system

Using the beam element stiffness Eq. (4.1.14) for each element, we obtain

$$[k^{(1)}] = \frac{EI}{L^3} \begin{bmatrix} 12 & 6L & -12 & 6L \\ 6L & 4L^2 & -6L & 2L^2 \\ -12 & -6L & 12 & -6L \\ 6L & 2L^2 & -6L & 4L^2 \end{bmatrix} \qquad [k^{(2)}] = \frac{EI}{L^3} \begin{bmatrix} 12 & 6L & -12 & 6L \\ 6L & 4L^2 & -6L & 2L^2 \\ -12 & -6L & 12 & -6L \\ 6L & 2L^2 & -6L & 4L^2 \end{bmatrix}$$

$$(4.4.28)$$

The boundary conditions are $v_1 = 0$, $\phi_1 = 0$, $v_3 = 0$, and $\phi_3 = 0$. Using the direct stiffness method and Eqs. (4.4.28) to assemble the global stiffness matrix, and applying the boundary conditions, we obtain

$$
\left\{ \begin{array}{c} F_{2y} \\ M_2 \end{array} \right\} = \left\{ \begin{array}{c} \dfrac{-wL}{2} \\ \dfrac{-wL^2}{20} \end{array} \right\} = \dfrac{EI}{L^3} \begin{bmatrix} 24 & 0 \\ 0 & 8L^2 \end{bmatrix} = \left\{ \begin{array}{c} v_2 \\ \phi_2 \end{array} \right\} \tag{4.4.29}
$$

Solving Eq. (4.4.29) for the displacement and slope, we obtain

$$
v_2 = \dfrac{-wL^4}{48EI} \qquad \phi_2 = \dfrac{-wL^3}{240EI} \tag{4.4.30}
$$

Next, we obtain the effective nodal forces using $\{F^{(e)}\} = [K]\{d\}$ as

$$
\left\{ \begin{array}{c} F_{1y}^{(e)} \\ M_1^{(e)} \\ F_{2y}^{(e)} \\ M_2^{(e)} \\ F_{3y}^{(e)} \\ M_3^{(e)} \end{array} \right\} = \dfrac{EI}{L^3} \begin{bmatrix} 12 & 6L & -12 & 6L & 0 & 0 \\ 6L & 4L^2 & -6L & 2L^2 & 0 & 0 \\ -12 & -6L & 24 & 0 & -12 & 6L \\ 6L & 2L^2 & 0 & 8L^2 & -6L & 2L^2 \\ 0 & 0 & -12 & -6L & 12 & -6L \\ 0 & 0 & 6L & 2L^2 & -6L & 4L^2 \end{bmatrix} \left\{ \begin{array}{c} 0 \\ 0 \\ \dfrac{-wL^4}{48EI} \\ \dfrac{-wL^3}{240EI} \\ 0 \\ 0 \end{array} \right\} \tag{4.4.31}
$$

Solving for the effective forces in Eq. (4.4.31), we obtain

$$
\begin{aligned} F_{1y}^{(e)} &= \dfrac{9wL}{40} & M_1^{(e)} &= \dfrac{7wL^2}{60} \\[2mm] F_{2y}^{(e)} &= \dfrac{-wL}{2} & M_2^{(e)} &= \dfrac{-wL^2}{30} \\[2mm] F_{3y}^{(e)} &= \dfrac{11wL}{40} & M_3^{(e)} &= \dfrac{-2wL^2}{15} \end{aligned} \tag{4.4.32}
$$

Finally, using Eq. (4.4.8) we subtract the equivalent nodal force matrix based on the equivalent load replacement shown in Figure 4–29(b) from the effective force matrix given by the results in Eq. (4.4.32), to obtain the correct nodal forces and moments as

$$
\left\{ \begin{array}{c} F_{1y} \\ M_1 \\ F_{2y} \\ M_2 \\ F_{3y} \\ M_3 \end{array} \right\} = \left\{ \begin{array}{c} \dfrac{9wL}{40} \\ \dfrac{7wL^2}{60} \\ \dfrac{-wL}{2} \\ \dfrac{-wL^2}{30} \\ \dfrac{11wL}{40} \\ \dfrac{-2wL^2}{15} \end{array} \right\} - \left\{ \begin{array}{c} \dfrac{-3wL}{40} \\ \dfrac{-wL^2}{60} \\ \dfrac{-wL}{2} \\ \dfrac{-wL^2}{30} \\ \dfrac{-17wL}{40} \\ \dfrac{wL^2}{15} \end{array} \right\} = \left\{ \begin{array}{c} \dfrac{12wL}{40} \\ \dfrac{8wL^2}{60} \\ 0 \\ 0 \\ \dfrac{28wL}{40} \\ \dfrac{-3wL^2}{15} \end{array} \right\} \tag{4.4.33}
$$

We used symbol L to represent one-half the length of the beam. If we replace L with the actual length $l = 2L$, we obtain the reactions for case 5 in Appendix D, thus verifying the correctness of our result.

In summary, for any structure in which an equivalent nodal force replacement is made, the actual nodal forces acting on the structure are determined by first evaluating the effective nodal forces $\{F^{(e)}\}$ for the structure and then subtracting the equivalent nodal forces $\{F_0\}$ for the structure, as indicated in Eq. (4.4.8). Similarly, for any element of a structure in which equivalent nodal force replacement is made, the actual local nodal forces acting on the element are determined by first evaluating the effective local nodal forces $\{f^{(e)}\}$ for the element and then subtracting the equivalent local nodal forces $\{f_0\}$ associated only with the element, as indicated in Eq. (4.4.11). We provide other examples of this procedure in plane frame Examples 5.2 and 5.3.   ■

▲ 4.5 Comparison of the Finite Element Solution to the Exact Solution for a Beam   ▲

We will now compare the finite element solution to the exact classical beam theory solution for the cantilever beam shown in Figure 4–30 subjected to a uniformly distributed load. Both one- and two-element finite element solutions will be presented and compared to the exact solution obtained by the direct double-integration method. Let $E = 210$ GPa, $I = 4 \times 10^{-5}$ m^4, $L = 2.5$ m, and uniform load $w = 4$ kN/m.

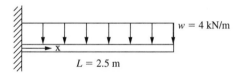

Figure 4–30   Cantilever beam subjected to uniformly distributed load

SOLUTION:

To obtain the solution from classical beam theory, we use the double-integration method [1]. Therefore, we begin with the moment-curvature equation

$$y'' = \frac{M(x)}{EI} \tag{4.5.1}$$

where the double prime superscript indicates differentiation twice with respect to x and M is expressed as a function of x by using a section of the beam as shown:

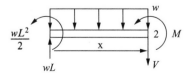

$$\Sigma F_y = 0: \ V(x) = wL - wx$$

$$\Sigma M_2 = 0: \ M(x) = \frac{-wL^2}{2} + wLx - (wx)\left(\frac{x}{2}\right) \qquad (4.5.2)$$

Using Eq. (4.5.2) in Eq. (4.5.1), we have

$$y'' = \frac{1}{EI}\left(\frac{-wL^2}{2} + wLx - \frac{wx^2}{2}\right) \qquad (4.5.3)$$

On integrating Eq. (4.5.3) with respect to x, we obtain an expression for the slope of the beam as

$$y' = \frac{1}{EI}\left(\frac{-wL^2x}{2} + \frac{wLx^2}{2} - \frac{wx^3}{6}\right) + C_1 \qquad (4.5.4)$$

Integrating Eq. (4.5.4) with respect to x, we obtain the deflection expression for the beam as

$$y = \frac{1}{EI}\left(\frac{-wL^2x^2}{4} + \frac{wLx^3}{6} - \frac{wx^4}{24}\right) + C_1x + C_2 \qquad (4.5.5)$$

Applying the boundary conditions $y = 0$ and $y' = 0$ at $x = 0$, we obtain

$$y'(0) = 0 = C_1 \qquad y(0) = 0 = C_2 \qquad (4.5.6)$$

Using Eq. (4.5.6) in Eqs. (4.5.4) and (4.5.5), the final beam theory solution expressions for y' and y are then

$$y' = \frac{1}{EI}\left(\frac{-wx^3}{6} + \frac{wLx^2}{2} - \frac{wL^2x}{2}\right) \qquad (4.5.7)$$

and

$$y = \frac{1}{EI}\left(\frac{-wx^4}{24} + \frac{wLx^3}{6} - \frac{wL^2x^2}{4}\right) \qquad (4.5.8)$$

The one-element finite element solution for slope and displacement is given in variable form by Eqs. (4.4.14b). Using the numerical values of this problem in Eqs. (4.4.14b), we obtain the slope and displacement at the free end (node 2) as

$$\phi_2 = \frac{-wL^3}{6EI} = \frac{-(4000 \ \text{N/m})(2.5 \ \text{m})^3}{6(210 \times 10^9 \ \text{N/m}^2)(4 \times 10^{-5} \ \text{m}^4)} = -0.00124 \ \text{rad}$$

$$\qquad (4.5.9)$$

$$v_2 = \frac{-wL^4}{8EI} = \frac{-(4000 \ \text{N/m})(2.5 \ \text{m})^4}{8(210 \times 10^9 \ \text{N/m}^2)(4 \times 10^{-5} \ \text{m}^4)} = -2.35 \ \text{mm}$$

The slope and displacement given by Eq. (4.5.9) identically match the beam theory values, as Eqs. (4.5.7) and (4.5.8) evaluated at $x = L$ are identical to the variable form of the finite element solution given by Eqs. (4.4.14b). The reason why these nodal values from the finite element solution are correct is that the element nodal forces were calculated on the basis of being energy or work equivalent to the distributed load based on the assumed cubic displacement field within each beam element.

Values of displacement and slope at other locations along the beam for the finite element solution are obtained by using the assumed cubic displacement function [Eq. (4.1.4)] as

$$v(x) = \frac{1}{L^3}(-2x^3 + 3x^2L)v_2 + \frac{1}{L^3}(x^3L - x^2L^2)\phi_2 \qquad (4.5.10)$$

where the boundary conditions $v_1 = \phi_1 = 0$ have been used in Eq. (4.5.10). Using the numerical values in Eq. (4.5.10), we obtain the displacement at the midlength of the beam as

$$v(x = 1.25 \text{ m}) = \frac{1}{(2.5 \text{ m})^3}\left[-2(1.25 \text{ m})^3 + 3(1.25 \text{ m})^2(2.5 \text{ m})\right](-0.00235 \text{ m})$$

$$+ \frac{1}{(2.5 \text{ m})^3}\left[(1.25 \text{ m})^3(2.5 \text{ m}) - (1.25 \text{ m})^2(2.5 \text{ m})^2\right]$$

$$\times (-0.00124 \text{ rad}) = -0.0007875 \text{ m} = -0.788 \text{ mm} \qquad (4.5.11)$$

Using the beam theory [Eq. (4.5.8)], the deflection is

$$y(x = 1.25 \text{ m}) = \frac{4000 \text{ N/m}}{210 \times 10^9 \text{ N/m}^2(4 \times 10^{-5} \text{ m}^4)}$$

$$\times \left[\frac{-(1.25 \text{ m})^4}{24} + \frac{(2.5 \text{ m})(1.25 \text{ m})^3}{6} - \frac{(2.5 \text{ m})^2(1.25 \text{ m})^2}{4}\right]$$

$$= -0.000823 \text{ m} = -0.823 \text{ mm} \qquad (4.5.12)$$

We conclude that the beam theory solution for midlength displacement, $y = -0.823$ mm, is greater than the finite element solution for displacement, $v = -0.788$ mm. In general, the displacements evaluated using the cubic function for v are lower as predicted by the finite element method than by the beam theory except at the nodes. This is always true for beams subjected to some form of distributed load that are modeled using the cubic displacement function. The exception to this result is at the nodes, where the beam theory and finite element results are identical because of the work-equivalence concept used to replace the distributed load by work-equivalent discrete loads at the nodes.

The beam theory solution predicts a quartic (fourth-order) polynomial expression for y [Eq. (4.5.5)] for a beam subjected to uniformly distributed loading, while the finite element solution $v(x)$ assumes a cubic displacement behavior in each beam element under all load conditions. The finite element solution predicts a stiffer structure than the actual one. This is expected, as the finite element model forces the beam into specific modes of displacement and effectively yields a stiffer model than the actual structure. However, as more and more elements are used in the model, the finite element solution converges to the beam theory solution.

For the special case of a beam subjected to only nodal concentrated loads, the beam theory predicts a cubic displacement behavior, as the moment is a linear function and is integrated twice to obtain the resulting cubic displacement function. A simple verification of this cubic displacement behavior would be to solve the cantilevered beam subjected to an end load. In this special case, the finite element solution for displacement matches the beam theory solution for all locations along the beam length, as both functions $y(x)$ and $v(x)$ are then cubic functions.

Monotonic convergence of the solution of a particular problem is discussed in Reference [3], and proof that compatible and complete displacement functions (as described in Section 3.2) used in the displacement formulation of the finite element method yield an upper bound on the true stiffness, hence a lower bound on the displacement of the problem, is discussed in Reference [3].

Under uniformly distributed loading, the beam theory solution predicts a quadratic moment and a linear shear force in the beam. However, the finite element solution using the cubic displacement function predicts a linear bending moment and a constant shear force within each beam element used in the model.

We will now determine the bending moment and shear force in the present problem based on the finite element method. The bending moment is given by

$$M = EIv'' = EI\frac{d^2([N]\{d\})}{dx^2} = EI\frac{(d^2[N])}{dx^2}\{d\} \tag{4.5.13}$$

as $\{d\}$ is not a function of x. Or in terms of the gradient matrix $[B]$ we have

$$M = EI[B]\{d\} \tag{4.5.14}$$

where

$$[B] = \frac{d^2[N]}{dx^2} = \left[\left(-\frac{6}{L^2}+\frac{12x}{L^3}\right)\left(-\frac{4}{L}+\frac{6x}{L^2}\right)\left(\frac{6}{L^2}-\frac{12x}{L^3}\right)\left(-\frac{2}{L}+\frac{6x}{L^2}\right)\right] \tag{4.5.15}$$

The shape functions given by Eq. (4.1.7) are used to obtain Eq. (4.5.15) for the $[B]$ matrix. For the single-element solution, the bending moment is then evaluated by substituting Eq. (4.5.15) for $[B]$ into Eq. (4.5.14) and multiplying $[B]$ by $\{d\}$ to obtain

$$M = EI\left[\left(-\frac{6}{L^2}+\frac{12x}{L^3}\right)v_1 + \left(-\frac{4}{L}+\frac{6x}{L^2}\right)\phi_1 + \left(\frac{6}{L^2}-\frac{12x}{L^3}\right)v_2 + \left(-\frac{2}{L}+\frac{6x}{L^2}\right)\phi_2\right]$$

$$\tag{4.5.16}$$

Evaluating the moment at the wall, $x = 0$, with $v_1 = \phi_1 = 0$, and v_2 and ϕ_2 given by Eq. (4.4.14) in Eq. (4.5.16), we have

$$M(x = 0) = -\frac{10wL^2}{24} = -10,416.7 \text{ N-m} \tag{4.5.17}$$

Using Eq. (4.5.16) to evaluate the moment at $x = 1.25$ m, we have

$$M(x = 1.25 \text{ m}) = -4166.7 \text{ N-m} \tag{4.5.18}$$

Evaluating the moment at $x = 2.5$ m by using Eq. (4.5.16) again, we obtain

$$M(x = 2.5 \text{ m}) = 2083.3 \text{ N-m} \tag{4.5.19}$$

The beam theory solution using Eq. (4.5.2) predicts

$$M(x = 0) = \frac{-wL^2}{2} = -12,500 \text{ N-m} \tag{4.5.20}$$

$$M(x = 1.25 \text{ m}) = -3,125 \text{ N-m}$$

and
$$M(x = 2.5 \text{ m}) = 0$$

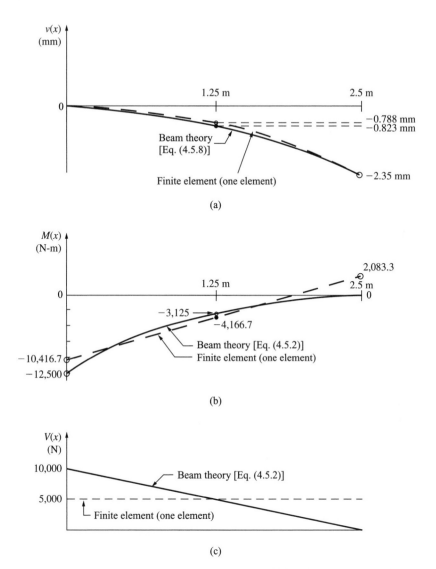

Figure 4–31 Comparison of beam theory and finite element results for a cantilever beam subjected to a uniformly distributed load: (a) displacement diagrams, (b) bending moment diagrams, and (c) shear force diagrams

Figures 4–31(a) through (c) show the plots of the displacement variation, bending moment variation, and shear force variation through the beam length for the beam theory and the one-element finite element solutions. Again, the finite element solution for displacement matches the beam theory solution at the nodes but predicts smaller displacements (less deflection) at other locations along the beam length.

The bending moment is derived by taking two derivatives on the displacement function. It then takes more elements to model the second derivative of the displacement function. Therefore, the finite element solution does not predict the bending moment as well as it does the displacement. For the uniformly loaded beam, the finite element model predicts a linear bending moment variation as shown in Figure 4–31(b). The best approximation for bending moment appears at the midpoint of the element.

The shear force is derived by taking three derivatives on the displacement function. For the uniformly loaded beam, the resulting shear force shown in Figure 4–31(c) is a constant throughout the single-element model. Again, the best approximation for shear force is at the midpoint of the element.

It should be noted that if we use Eq. (4.4.11), that is, $\{f\} = [k]\{d\} - \{f_0\}$, and subtract off the $\{f_0\}$ matrix, we also obtain the correct nodal forces and moments in each element. For instance, from the one-element finite element solution we have for the bending moment at node 1

$$m_1^{(1)} = \frac{EI}{L^3}\left[-6L\left(\frac{-wL^4}{8EI}\right) + 2L^2\left(\frac{-wL^3}{6EI}\right)\right] - \left(\frac{-wL^2}{12}\right) = \frac{wL^2}{2}$$

and at node 2

$$m_2^{(1)} = 0$$

To improve the finite element solution we need to use more elements in the model (refine the mesh) or use a higher-order element, such as a fifth-order approximation for the displacement function, that is, $v(x) = a_1 + a_2x + a_3x^2 + a_4x^3 + a_5x^4 + a_6x^5$, with three nodes (with an extra node at the middle of the element).

We now present the two-element finite element solution for the cantilever beam subjected to a uniformly distributed load. Figure 4–32 shows the beam discretized into two elements of equal length and the work-equivalent load replacement for each element. Using the beam element stiffness matrix [Eq. (4.1.13)], we obtain the element stiffness matrices as follows:

$$[k^{(1)}] = [k^{(2)}] = \frac{EI}{l^3}\begin{array}{cc} & \begin{array}{cccc}1 & & 2 \\ 2 & & 3\end{array} \\ & \begin{bmatrix} 12 & 6l & -12 & 6l \\ 6l & 4l^2 & -6l & 2l^2 \\ -12 & -6l & 12 & -6l \\ 6l & 2l^2 & -6l & 4l^2 \end{bmatrix}\end{array} \qquad (4.5.21)$$

where $l = 1.25$ m is the length of each element and the numbers above the columns indicate the degrees of freedom associated with each element.

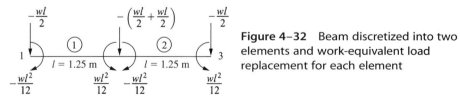

Figure 4–32   Beam discretized into two elements and work-equivalent load replacement for each element

Applying the boundary conditions $v_1 = 0$ and $\phi_1 = 0$ to reduce the number of equations for a normal longhand solution, we obtain the global equations for solution as

$$\frac{EI}{l^3}\begin{bmatrix} 24 & 0 & -12 & 6l \\ 0 & 8l^2 & -6l & 2l^2 \\ -12 & -6l & 12 & -6l \\ 6l & 2l^2 & -6l & 4l^2 \end{bmatrix}\begin{Bmatrix} v_2 \\ \phi_2 \\ v_3 \\ \phi_3 \end{Bmatrix} = \begin{Bmatrix} -wl \\ 0 \\ -wl/2 \\ wl^2/12 \end{Bmatrix} \tag{4.5.22}$$

Solving Eq. (4.5.22) for the displacements and slopes, we obtain

$$v_2 = \frac{-17wl^4}{24EI} \qquad v_3 = \frac{-2wl^4}{EI} \qquad \phi_2 = \frac{-7wl^3}{6EI} \qquad \phi_3 = \frac{-4wl^3}{3EI} \tag{4.5.23}$$

Substituting the numerical values $w = 4000$ N/m, $l = 1.25$ m, $E = 210 \times 10^9$ N/m^2, and $I = 4 \times 10^{-5}$ m^4 into Eq. (4.5.23), we obtain

$$v_2 = -0.823 \text{ mm} \qquad v_3 = -2.35 \text{ mm} \qquad \phi_2 = -0.001085 \text{ rad}$$

$$\phi_3 = -0.00124 \text{ rad}$$

The two-element solution yields nodal displacements that match the beam theory results exactly [see Eqs. (4.5.9) and (4.5.12)]. A plot of the two-element displacement throughout the length of the beam would be a cubic displacement within each element. Within element 1, the plot would start at a displacement of 0 at node 1 and finish at a displacement of -0.823 mm at node 2. A cubic function would connect these values. Similarly, within element 2, the plot would start at a displacement of -0.823 mm and finish at a displacement of -2.35 mm at node 2 [see Figure 4–31(a)]. A cubic function would again connect these values.

▲ 4.6 Beam Element with Nodal Hinge ▲

In some beams an internal hinge may be present. In general, this internal hinge causes a discontinuity in the slope or rotation of the deflection curve at the hinge. Consider the beam shown with two elements and a nodal hinge at node 2 separating the two elements as shown in Figure 4–33(a). In general, $\phi_2^{(1)}$ for element 1 is not equal to $\phi_2^{(2)}$ for element 2, as shown in Figures 4–33(b) and (c). At hinge nodes, rotations are said to be double valued. To model the hinge, we consider the hinge to be placed on either the right end of element 1 or on the left end of element 2 but not on both elements at node 2. Examples 4.10 and 4.11 will illustrate how to solve beam problems with nodal hinges.

Also, the bending moment is zero at the hinge. We could construct other types of connections that release other generalized end forces; that is, connections can be designed to make the shear force or axial force zero at the connection. These special conditions can be treated by starting with the generalized unreleased beam stiffness matrix [Eq. (4.1.14)] and eliminating the known zero force or moment. This yields a modified stiffness matrix with the desired force or moment equal to zero and the corresponding displacement or slope eliminated.

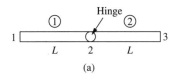

(a)

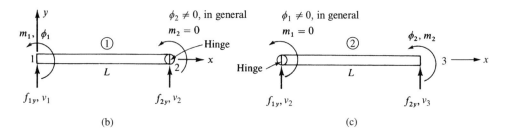

(b)   (c)

Figure 4–33 (a) Beam with 2 elements and nodal hinge, (b) hinge considered to be at right end of element 1, and (c) hinge considered to be at left end of element 2

We now consider the most common cases of a beam element with a nodal hinge at the right end or left end, as shown in Figure 4–33. For the beam element with a hinge at its right end, the moment m_2 is zero and we partition the $[k]$ matrix [Eq. (4.1.14)] to eliminate the degree of freedom ϕ_2 (which is not zero, in general) associated with $m_2 = 0$ as follows:

$$[k] = \frac{EI}{L^3}\begin{bmatrix} 12 & 6L & -12 & \vdots & 6L \\ 6L & 4L^2 & -6L & \vdots & 2L^2 \\ -12 & -6L & 12 & \vdots & -6L \\ \cdots & \cdots & \cdots & & \cdots \\ 6L & 2L^2 & -6L & \vdots & 4L^2 \end{bmatrix} \tag{4.6.1}$$

We condense out the degree of freedom ϕ_2 associated with $m_2 = 0$. Partitioning allows us to condense out the degree of freedom ϕ_2 associated with $m_2 = 0$. That is, Eq. (4.6.1) is partitioned as shown below:

$$[k] = \begin{bmatrix} [K_{11}] & \vdots & [K_{12}] \\ 3 \times 3 & \vdots & 3 \times 1 \\ \cdots & & \cdots \\ [K_{21}] & \vdots & [K_{22}] \\ 1 \times 3 & \vdots & 1 \times 1 \end{bmatrix} \tag{4.6.2}$$

The condensed stiffness matrix is then found by using the equation $\{f\} = [k]\{d\}$ partitioned as follows:

$$\begin{Bmatrix} \{f_1\} \\ 3 \times 1 \\ \cdots \\ \{f_2\} \\ 1 \times 1 \end{Bmatrix} = \begin{bmatrix} [K_{11}] & \vdots & [K_{12}] \\ 3 \times 3 & \vdots & 3 \times 1 \\ \cdots & & \cdots \\ [K_{21}] & \vdots & [K_{22}] \\ 1 \times 3 & \vdots & 1 \times 1 \end{bmatrix} \begin{Bmatrix} \{d_1\} \\ 3 \times 1 \\ \cdots \\ \{d_2\} \\ 1 \times 1 \end{Bmatrix} \tag{4.6.3}$$

where
$$\{d_1\} = \begin{Bmatrix} v_1 \\ \phi_1 \\ v_2 \end{Bmatrix} \qquad \{d_2\} = \{\phi_2\} \tag{4.6.4}$$

Equations (4.6.3) in expanded form are

$$\{f_1\} = [K_{11}]\{d_1\} + [K_{12}]\{d_2\} \tag{4.6.5}$$
$$\{f_2\} = [K_{21}]\{d_1\} + [K_{22}]\{d_2\}$$

Solving for $\{d_2\}$ in the second of Eqs. (4.6.5), we obtain

$$\{d_2\} = [K_{22}]^{-1}(\{f_2\} - [K_{21}]\{d_1\}) \tag{4.6.6}$$

Substituting Eq. (4.6.6) into the first of Eqs. (4.6.5), we obtain

$$\{f_1\} = ([K_{11}] - [K_{12}][K_{22}]^{-1}[K_{21}])\{d_1\} + [K_{12}][K_{22}]^{-1}\{f_2\} \tag{4.6.7}$$

Combining the second term on the right side of Eq. (4.6.7) with $\{f_1\}$, we obtain

$$\{f_c\} = [K_c]\{d_1\} \tag{4.6.8}$$

where the condensed stiffness matrix is

$$[K_c] = [K_{11}] - [K_{12}][K_{22}]^{-1}[K_{21}] \tag{4.6.9}$$

and the condensed force matrix is

$$\{f_c\} = \{f_1\} - [K_{12}][K_{22}]^{-1}\{f_2\} \tag{4.6.10}$$

Substituting the partitioned parts of $[k]$ from Eq. (4.6.1) into Eq. (4.6.9), we obtain the condensed stiffness matrix as

$$[K_c] = [K_{11}] - [K_{12}][K_{22}]^{-1}[K_{21}]$$

$$= \frac{EI}{L^3}\begin{bmatrix} 12 & 6L & -12 \\ 6L & 4L^2 & -6L \\ -12 & -6L & 12 \end{bmatrix} - \frac{EI}{L^3}\begin{Bmatrix} 6L \\ 2L^2 \\ -6L \end{Bmatrix}\frac{1}{4L^2}\begin{bmatrix} 6L & 2L^2 & -6L \end{bmatrix}$$

$$= \frac{3EI}{L^3}\begin{bmatrix} 1 & L & -1 \\ L & L^2 & -L \\ -1 & -L & 1 \end{bmatrix}\begin{matrix} v_1 \\ \phi_1 \\ v_2 \end{matrix} \tag{4.6.11}$$

and the element equations (force/displacement equations) with the hinge at node 2 are

$$\begin{Bmatrix} f_{1y} \\ m_1 \\ f_{2y} \end{Bmatrix} = \frac{3EI}{L^3}\begin{bmatrix} 1 & L & -1 \\ L & L^2 & -L \\ -1 & -L & 1 \end{bmatrix}\begin{Bmatrix} v_1 \\ \phi_1 \\ v_2 \end{Bmatrix} \tag{4.6.12}$$

The generalized rotation ϕ_2 has been eliminated from the equation and will not be calculated using this scheme. However, ϕ_2 is not zero in general. We can expand

Eq. (4.6.12) to include ϕ_2 by adding zeros in the fourth row and column of the $[k]$ matrix to maintain $m_2 = 0$, as follows:

$$\begin{Bmatrix} f_{1y} \\ m_1 \\ f_{2y} \\ m_2 \end{Bmatrix} = \frac{3EI}{L^3} \begin{bmatrix} 1 & L & -1 & 0 \\ L & L^2 & -L & 0 \\ -1 & -L & 1 & 0 \\ 0 & 0 & 0 & 0 \end{bmatrix} \begin{Bmatrix} v_1 \\ \phi_1 \\ v_2 \\ \phi_2 \end{Bmatrix} \qquad (4.6.13)$$

For the beam element with left node 1 and right node 2 with a hinge at its left end, the moment m_1 is zero, and we partition the $[k]$ matrix [Eq. (4.1.14)] to eliminate the zero moment m_1 and its corresponding rotation ϕ_1 to obtain

$$\begin{Bmatrix} f_{1y} \\ f_{2y} \\ m_2 \end{Bmatrix} = \frac{3EI}{L^3} \begin{bmatrix} 1 & -1 & L \\ -1 & 1 & -L \\ L & -L & L^2 \end{bmatrix} \begin{Bmatrix} v_1 \\ v_2 \\ \phi_2 \end{Bmatrix} \qquad (4.6.14)$$

The expanded form of Eq. (4.6.14) including ϕ_1 is

$$\begin{Bmatrix} f_{1y} \\ m_1 \\ f_{2y} \\ m_2 \end{Bmatrix} = \frac{3EI}{L^3} \begin{bmatrix} 1 & 0 & -1 & L \\ 0 & 0 & 0 & 0 \\ -1 & 0 & 1 & -L \\ L & 0 & -L & L^2 \end{bmatrix} \begin{Bmatrix} v_1 \\ \phi_1 \\ v_2 \\ \phi_2 \end{Bmatrix} \qquad (4.6.15)$$

Example 4.10

Determine the displacement and rotation at node 2 and the element forces for the uniform beam with an internal hinge at node 2 shown in Figure 4–34. Let EI be a constant.

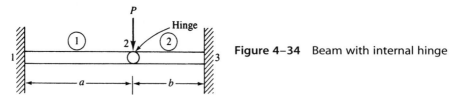

Figure 4–34   Beam with internal hinge

SOLUTION:

We can consider the hinge as part of element 1. Therefore, with the hinge located at the right end of element 1, Eq. (4.6.13) contains the correct stiffness matrix for element 1. The stiffness matrix of element 1 with $L = a$ is then

$$[k^{(1)}] = \frac{3EI}{a^3} \begin{matrix} v_1 & \phi_1 & v_2 & \phi_2 \\ \begin{bmatrix} 1 & a & -1 & 0 \\ a & a^2 & -a & 0 \\ -1 & -a & 1 & 0 \\ 0 & 0 & 0 & 0 \end{bmatrix} \end{matrix} \qquad (4.6.16)$$

As the hinge is considered to be part of element 1, we do not consider it again as part of element 2. So we use the standard beam element stiffness matrix obtained from Eq. (4.1.14) for element 2 as

$$[k^{(2)}] = \frac{EI}{b^3} \begin{matrix} & v_2 & \phi_2 & v_3 & \phi_3 \\ & \begin{bmatrix} 12 & 6b & -12 & 6b \\ 6b & 4b^2 & -6b & 2b^2 \\ -12 & -6b & 12 & -6b \\ 6b & 2b^2 & -6b & 4b^2 \end{bmatrix} \end{matrix} \qquad (4.6.17)$$

Superimposing Eqs. (4.6.16) and (4.6.17) and applying the boundary conditions

$$v_1 = 0, \qquad \phi_1 = 0, \qquad v_3 = 0, \qquad \phi_3 = 0$$

we obtain the total stiffness matrix and total set of equations as

$$EI \begin{bmatrix} \dfrac{3}{a^3} + \dfrac{12}{b^3} & \dfrac{6}{b^2} \\ \dfrac{6}{b^2} & \dfrac{4}{b} \end{bmatrix} \begin{Bmatrix} v_2 \\ \phi_2 \end{Bmatrix} = \begin{Bmatrix} -P \\ 0 \end{Bmatrix} \qquad (4.6.18)$$

Solving Eq. (4.6.18), we obtain

$$v_2 = \frac{-a^3 b^3 P}{3(b^3 + a^3)EI}$$

$$\phi_2 = \frac{a^3 b^2 P}{2(b^3 + a^3)EI} \qquad (4.6.19)$$

The value ϕ_2 is actually that associated with element 2—that is, ϕ_2 in Eq. (4.6.19) is actually $\phi_2^{(2)}$. The value of ϕ_2 at the right end of element 1 ($\phi_2^{(1)}$) is, in general, not equal to $\phi_2^{(2)}$. If we had chosen to assume the hinge to be part of element 2, then we would have used Eq. (4.1.14) for the stiffness matrix of element 1 and Eq. (4.6.15) for the stiffness matrix of element 2. This would have enabled us to obtain $\phi_2^{(1)}$, which is different from $\phi_2^{(2)}$, that is, the slope at node 2 is double valued.

Using Eq. (4.6.12) for element 1, we obtain the element forces as

$$\begin{Bmatrix} f_{1y} \\ m_1 \\ f_{2y} \end{Bmatrix} = \frac{3EI}{a^3} \begin{bmatrix} 1 & a & -1 \\ a & a^2 & -a \\ -1 & -a & 1 \end{bmatrix} \begin{Bmatrix} 0 \\ 0 \\ \dfrac{-a^3 b^3 P}{3(b^3 + a^3)EI} \end{Bmatrix} \qquad (4.6.20)$$

Simplifying Eq. (4.6.20), we obtain the forces as

$$f_{1y} = \frac{b^3 P}{b^3 + a^3}$$

$$m_1 = \frac{ab^3 P}{b^3 + a^3} \qquad (4.6.21)$$

$$f_{2y} = -\frac{b^3 P}{b^3 + a^3}$$

Using Eq. (4.6.17) and the results from Eq. (4.6.19), we obtain the element 2 forces as

$$
\begin{Bmatrix} f_{2y} \\ m_2 \\ f_{3y} \\ m_3 \end{Bmatrix} = \frac{EI}{b^3}
\begin{bmatrix}
12 & 6b & -12 & 6b \\
6b & 4b^2 & -6b & 2b^2 \\
-12 & -6b & 12 & -6b \\
6b & 2b^2 & -6b & 4b^2
\end{bmatrix}
\begin{Bmatrix}
-\dfrac{a^3 b^3 P}{3(b^3 + a^3)EI} \\[2mm]
\dfrac{a^3 b^2 P}{2(b^3 + a^3)EI} \\[2mm]
0 \\[1mm]
0
\end{Bmatrix} \qquad (4.6.22)
$$

Simplifying Eq. (4.6.22), we obtain the element forces as

$$
f_{2y} = -\frac{a^3 P}{b^3 + a^3}
$$

$$
m_2 = 0
$$

$$
f_{3y} = \frac{a^3 P}{b^3 + a^3} \qquad (4.6.23)
$$

$$
m_3 = -\frac{ba^3 P}{b^3 + a^3} \qquad \blacksquare
$$

It should be noted that another way to solve the nodal hinge of Example 4.10 would be to assume a nodal hinge at the right end of element 1 and at the left end of element 2. Hence, we would use the three-equation stiffness matrix of Eq. (4.6.12) for the left element and the three-equation stiffness matrix of Eq. (4.6.14) for the right element. This results in the hinge rotation being condensed out of the global equations. You can verify that we get the same result for the displacement as given by Eq. (4.6.19). However, we must then go back to Eq. (4.6.6) using it separately for each element to obtain the rotation at node 2 for each element. We leave this verification to your discretion.

Example 4.11

Determine the slope at node 2 and the deflection and slope at node 3 for the beam with internal hinge located at node 3, as shown in Figure 4–35. Nodes 1 and 4 are fixed, and there is a knife edge support at node 2. Let $E = 210\,\text{GPa}$ and $I = 2 \times 10^{-4}\,\text{m}^4$.

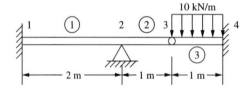

Figure 4–35 Beam with internal hinge and uniformly distributed loading

SOLUTION:

Discretize the beam into three elements, as shown in Figure 4–35. Use Eq. (4.1.14) to determine element one stiffness matrix as

$$[k^{(1)}] = \frac{EI}{8} \begin{array}{c} \begin{array}{cccc} v_1 & \phi_1 & v_2 & \phi_2 \end{array} \\ \begin{bmatrix} 12 & 12 & -12 & 12 \\ 12 & 16 & -12 & 8 \\ -12 & -12 & 12 & -12 \\ 12 & 8 & -12 & 16 \end{bmatrix} \end{array} = EI \begin{bmatrix} 3/2 & 3/2 & -3/2 & 3/2 \\ 3/2 & 2 & -3/2 & 1 \\ -3/2 & -3/2 & 3/2 & -3/2 \\ 3/2 & 1 & -3/2 & 2 \end{bmatrix} \quad (4.6.24)$$

Assume the hinge as part of element 2 and use Eq. (4.6.13) to obtain the element 2 stiffness matrix as

$$[k^{(2)}] = \frac{3EI}{1^3} \begin{array}{c} \begin{array}{cccc} v_2 & \phi_2 & v_3 & \phi_3 \end{array} \\ \begin{bmatrix} 1 & 1 & -1 & 0 \\ 1 & 1 & -1 & 0 \\ -1 & -1 & 1 & 0 \\ 0 & 0 & 0 & 0 \end{bmatrix} \end{array} \quad (4.6.25)$$

As the hinge is considered to be at the right end of element 2, we do not consider it to be part of element 3. So we use Eq. (4.1.14) to obtain the stiffness matrix as

$$[k^{(3)}] = \frac{EI}{1^3} \begin{array}{c} \begin{array}{cccc} v_3 & \phi_3 & v_4 & \phi_4 \end{array} \\ \begin{bmatrix} 12 & 6 & -12 & 6 \\ 6 & 4 & -6 & 2 \\ -12 & -6 & 12 & -6 \\ 6 & 2 & -6 & 4 \end{bmatrix} \end{array} \quad (4.6.26)$$

Using the direct stiffness method and the element stiffness matrices in Eqs. (4.6.24 through 4.6.26), we assemble the global stiffness matrix as

$$[k] = EI \begin{array}{c} \begin{array}{cccccccc} v_1 & \phi_1 & v_2 & \phi_2 & v_3 & \phi_3 & v_4 & \phi_4 \end{array} \\ \begin{bmatrix} 3/2 & 3/2 & -3/2 & 3/2 & 0 & 0 & 0 & 0 \\ 3/2 & 2 & -3/2 & 1 & 0 & 0 & 0 & 0 \\ -3/2 & -3/2 & 9/2 & 3/2 & -3 & 0 & 0 & 0 \\ 3/2 & 1 & 3/2 & 5 & -3 & 0 & 0 & 0 \\ 0 & 0 & -3 & -3 & 15 & 6 & -12 & 6 \\ 0 & 0 & 0 & 0 & 6 & 4 & -6 & 2 \\ 0 & 0 & 0 & 0 & -12 & -6 & 12 & -6 \\ 0 & 0 & 0 & 0 & 6 & 2 & -6 & 4 \end{bmatrix} \end{array} \quad (4.6.27)$$

Applying the boundary conditions $v_1 = \phi_1 = v_2 = v_4 = \phi_4 = 0$, we obtain the reduced stiffness matrix and equations for solution as

$$EI \begin{bmatrix} 5 & -3 & 0 \\ -3 & 15 & 6 \\ 0 & 6 & 4 \end{bmatrix} \begin{Bmatrix} \phi_2 \\ v_3 \\ \phi_3 \end{Bmatrix} = \begin{Bmatrix} 0 \\ -5 \text{ kN} \\ -0.833 \text{ kN} \cdot \text{m} \end{Bmatrix} \tag{4.6.28}$$

where by work equivalence $F_{3y} = \dfrac{-wL}{2} = \dfrac{-10(1)}{2} = -5$ kN and $M_3 = \dfrac{-wL^2}{12} = \dfrac{-5(1)^2}{12} = -0.833$ kN·m.

Substituting numerical values for E and I into Eq. (4.6.28), and solving simultaneous, we obtain

$$v_3 = -2.126 \times 10^{-5} \text{ m}, \quad \phi_2 = -1.276 \times 10^{-5} \text{ rad}, \quad \phi_3 = 2.693 \times 10^{-5} \text{ rad} \tag{4.6.29}$$

Notice that ϕ_3 is actually that associated with element three, that is, ϕ_3 in Eq. (4.6.29) is actually $\phi_3^{(3)}$ as the hinge was assumed to be part of element two and $\phi_3^{(2)}$ was condensed out of the stiffness matrix for element two. ∎

▲ 4.7 Potential Energy Approach to Derive Beam Element Equations ▲

We will now derive the beam element equations using the principle of minimum potential energy. The procedure is similar to that used in Section 3.10 in deriving the bar element equations. Again, our primary purpose in applying the principle of minimum potential energy is to enhance your understanding of the principle. It will be used routinely in subsequent chapters to develop element stiffness equations. We use the same notation here as in Section 3.10.

The total potential energy for a beam is

$$\pi_p = U + \Omega \tag{4.7.1}$$

where the general one-dimensional expression for the strain energy U for a beam is given by

$$U = \iiint_V \frac{1}{2} \sigma_x \varepsilon_x \, dV \tag{4.7.2}$$

and for a single beam element subjected to both distributed and concentrated nodal loads, the potential energy of forces is given by

$$\Omega = -\iint_{S_1} T_y v \, dS - \sum_{i=1}^{2} P_{iy} v_i - \sum_{i=1}^{2} m_i \phi_i \tag{4.7.3}$$

where body forces are now neglected. The terms on the right-hand side of Eq. (4.7.3) represent the potential energy of (1) transverse surface loading T_y (in units of force

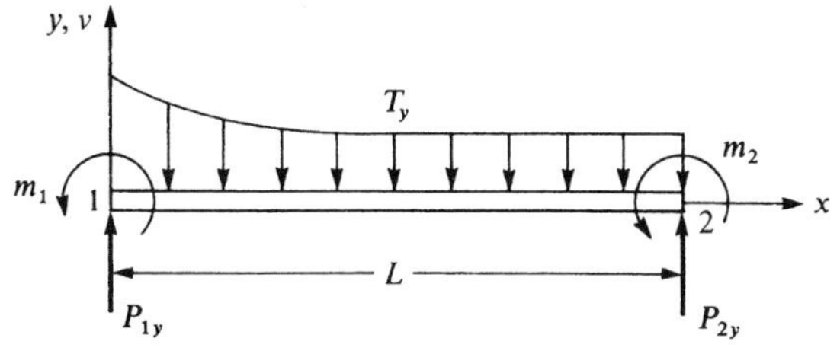

Figure 4–36 Beam element subjected to surface loading and concentrated nodal forces

per unit surface area, acting over surface S_1 and moving through displacements over which T_y act); (2) nodal concentrated force P_{iy} moving through displacements v_i; and (3) moments m_i moving through rotations ϕ_i. Again, v is the transverse displacement function for the beam element of length L shown in Figure 4–36.

Consider the beam element to have constant cross-sectional area A. The differential volume for the beam element can then be expressed as

$$dV = dA\,dx \qquad (4.7.4)$$

and the differential area over which the surface loading acts is

$$dS = b\,dx \qquad (4.7.5)$$

where b is the constant width. Using Eqs. (4.7.4) and (4.7.5) in Eqs. (4.7.1) through (4.7.3), the total potential energy becomes

$$\pi_p = \iiint_{x\ A} \frac{1}{2}\sigma_x\varepsilon_x\,dA\,dx - \int_0^L bT_yv\,dx - \sum_{i=1}^{2}(P_{iy}v_i + m_i\phi_i) \qquad (4.7.6)$$

Substituting Eq. (4.1.4) for v into the strain/displacement relationship Eq. (4.1.10), repeated here for convenience as

$$\varepsilon_x = -y\frac{d^2v}{dx^2} \qquad (4.7.7)$$

we express the strain in terms of nodal displacements and rotations as

$$\{\varepsilon_x\} = -y\left[\frac{12x - 6L}{L^3}\ \frac{6xL - 4L^2}{L^3}\ \frac{-12x + 6L}{L^3}\ \frac{6xL - 2L^2}{L^3}\right]\{d\} \qquad (4.7.8)$$

or

$$\{\varepsilon_x\} = -y[B]\{d\} \qquad (4.7.9)$$

where we define

$$[B] = \left[\frac{12x - 6L}{L^3}\ \frac{6xL - 4L^2}{L^3}\ \frac{-12x + 6L}{L^3}\ \frac{6xL - 2L^2}{L^3}\right] \qquad (4.7.10)$$

The stress/strain relationship is given by

$$\{\sigma_x\} = [D]\{\varepsilon_x\} \qquad (4.7.11)$$

where

$$[D] = [E] \qquad (4.7.12)$$

and E is the modulus of elasticity. Using Eq. (4.7.9) in Eq. (4.7.11), we obtain

$$\{\sigma_x\} = -y[D][B]\{d\} \tag{4.7.13}$$

Next, the total potential energy Eq. (4.7.6) is expressed in matrix notation as

$$\pi_p = \iiint_{x \; A} \frac{1}{2}\{\sigma_x\}^T\{\varepsilon_x\}\, dA\, dx - \int_0^L bT_y[v]^T\, dx - \{d\}^T\{P\} \tag{4.7.14}$$

Using Eqs. (4.1.5), (4.7.9), (4.7.12), and (4.7.13), and defining $w = bT_y$ as the line load (load per unit length) in the y direction, we express the total potential energy, Eq. (4.7.14), in matrix form as

$$\pi_p = \int_0^L \frac{EI}{2}\{d\}^T[B]^T[B]\{d\}\, dx - \int_0^L w\{d\}^T[N]^T\, dx - \{d\}^T\{P\} \tag{4.7.15}$$

where we have used the definition of the moment of inertia

$$I = \iint_A y^2\, dA \tag{4.7.16}$$

to obtain the first term on the right-hand side of Eq. (4.7.15). In Eq. (4.7.15), π_p is now expressed as a function of $\{d\}$.

Differentiating π_p in Eq. (4.7.15) with respect to v_1, ϕ_1, v_2, and ϕ_2 and equating each term to zero to minimize π_p, we obtain four element equations, which are written in matrix form as

$$EI\int_0^L [B]^T[B]\, dx\{d\} - \int_0^L [N]^T w\, dx - \{P\} = 0 \tag{4.7.17}$$

The derivation of the four element equations is left as an exercise (see Problem 4.45). Representing the nodal force matrix as the sum of those nodal forces resulting from distributed loading and concentrated loading, we have

$$\{f\} = \int_0^L [N]^T w\, dx + \{P\} \tag{4.7.18}$$

Using Eq. (4.7.18), the four element equations given by explicitly evaluating Eq. (4.7.17) are then identical to Eq. (4.1.13). The integral term on the right side of Eq. (4.7.18) also represents the work-equivalent replacement of a distributed load by nodal concentrated loads. For instance, letting $w(x) = -w$ (constant), substituting shape functions from Eq. (4.1.7) into the integral, and then performing the integration result in the same nodal equivalent loads as given by Eqs. (4.4.5) through (4.4.7).

Because $\{f\} = [k]\{d\}$, we have, from Eq. (4.7.17),

$$[k] = EI\int_0^L [B]^T[B]\, dx \tag{4.7.19}$$

Using Eq. (4.7.10) in Eq. (4.7.19) and integrating, $[k]$ is evaluated in explicit form as

$$[k] = \frac{EI}{L^3} \begin{bmatrix} 12 & 6L & -12 & 6L \\ & 4L^2 & -6L & 2L^2 \\ & & 12 & -6L \\ \text{Symmetry} & & & 4L^2 \end{bmatrix} \tag{4.7.20}$$

Equation (4.7.20) represents the local stiffness matrix for a beam element. As expected, Eq. (4.7.20) is identical to Eq. (4.1.14) developed previously.

It is worth noting that the strain energy U is the first term on the right side of Eq. (4.7.15) and $\{d\}$ is not a function of x. If we also consider E and I to be constant over each element length L, we can express U as

$$U = \{d\}^T \frac{EI}{2} \int_0^L [B]^T [B] dx \{d\} \tag{4.7.21}$$

By using Eq. (4.7.19), we realize the stiffness matrix, $\{k\}$, is EI times the integral in Eq. (4.7.21).

Therefore, we show U to be expressed again in quadratic form as $U = \frac{1}{2}\{d\}^T [k]\{d\}$.

▲ 4.8 Galerkin's Method for Deriving Beam Element Equations ▲

We will now illustrate Galerkin's method to formulate the beam element stiffness equations. We begin with the basic differential Eq. (4.1.1h) with transverse loading w now included; that is,

$$EI \frac{d^4 v}{dx^4} + w = 0 \tag{4.8.1}$$

We now define the residual R to be Eq. (4.8.1). Applying Galerkin's criterion [Eq. (3.12.3)] to Eq. (4.8.1), we have

$$\int_0^L \left(EI \frac{d^4 v}{dx^4} + w \right) N_i \, dx = 0 \qquad (i = 1, 2, 3, 4) \tag{4.8.2}$$

where the shape functions N_i are defined by Eqs. (4.1.7).

We now apply integration by parts twice to the first term in Eq. (4.8.2) to yield

$$\int_0^L EI(v_{,xxxx}) N_i \, dx = \int_0^L EI(v_{,xx})(N_{i,xx}) \, dx + EI[N_i(v_{,xxx}) - (N_{i,x})(v_{,xx})]_0^L \tag{4.8.3}$$

where the notation of the comma followed by the subscript x indicates differentiation with respect to x. Again, integration by parts introduces the boundary conditions.

Because $v = [N]\{d\}$ as given by Eq. (4.1.5), we have

$$v_{,xx} = \left[\frac{12x - 6L}{L^3} \quad \frac{6xL - 4L^2}{L^3} \quad \frac{-12x + 6L}{L^3} \quad \frac{6xL - 2L^2}{L^3} \right] \{d\} \tag{4.8.4}$$

or, using Eq. (4.7.10),

$$v_{,xx} = [B]\{d\} \qquad (4.8.5)$$

Substituting Eq. (4.8.5) into Eq. (4.8.3), and then Eq. (4.8.3) into Eq. (4.8.2), we obtain

$$\int_0^L (N_{i,xx})EI[B]\,dx\{d\} + \int_0^L N_i w\,dx + [N_i V - (N_{i,x})m]\big|_0^L = 0 \qquad (i = 1, 2, 3, 4)$$

$$(4.8.6)$$

where Eqs. (4.1.11) have been used in the boundary terms. Equation (4.8.6) is really four equations (one each for $N_i = N_1, N_2, N_3$, and N_4). Instead of directly evaluating Eq. (4.8.6) for each N_i, as was done in Section 3.12, we can express the four equations of Eq. (4.8.6) in matrix form as

$$\int_0^L [B]^T EI[B]\,dx\{d\} = \int_0^L -[N]^T w\,dx + ([N]^T_{,x}m - [N]^T V)\big|_0^L \qquad (4.8.7)$$

where we have used the relationship $[N]_{,xx} = [B]$ in Eq. (4.8.7).

Observe that the integral term on the left side of Eq. (4.8.7) is identical to the stiffness matrix previously given by Eq. (4.7.19) and that the first term on the right side of Eq. (4.8.7) represents the equivalent nodal forces due to distributed loading [also given in Eq. (4.7.18)]. The two terms in parentheses on the right side of Eq. (4.8.7) are the same as the concentrated force matrix $\{P\}$ of Eq. (4.7.18). We explain this by evaluating $[N]_{,x}$ and $[N]$, where $[N]$ is defined by Eq. (4.1.6), at the ends of the element as follows:

$$
\begin{aligned}
[N]_{,x}\big|_0 &= [0 \quad 1 \quad 0 \quad 0] & [N]_{,x}\big|_L &= [0 \quad 0 \quad 0 \quad 1] \\
[N]\big|_0 &= [1 \quad 0 \quad 0 \quad 0] & [N]\big|_L &= [0 \quad 0 \quad 1 \quad 0]
\end{aligned}
\qquad (4.8.8)
$$

Therefore, when we use Eqs. (4.8.8) in Eq. (4.8.7), the following terms result:

$$\begin{Bmatrix} 0 \\ 0 \\ 0 \\ 1 \end{Bmatrix} m(L) - \begin{Bmatrix} 0 \\ 1 \\ 0 \\ 0 \end{Bmatrix} m(0) - \begin{Bmatrix} 0 \\ 0 \\ 1 \\ 0 \end{Bmatrix} V(L) + \begin{Bmatrix} 1 \\ 0 \\ 0 \\ 0 \end{Bmatrix} V(0) \qquad (4.8.9)$$

These nodal shear forces and moments are illustrated in Figure 4–37.

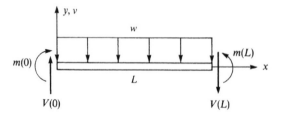

Figure 4–37 Beam element with shear forces, moments, and a distributed load

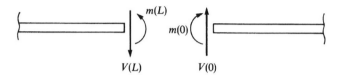

Figure 4–38 Shear forces and moments acting on adjacent elements meeting at a node

Note that when element matrices are assembled, two shear forces and two moments from adjacent elements contribute to the concentrated force and concentrated moment at the node common to the adjacent elements as shown in Figure 4–38. These concentrated shear forces $V(0) - V(L)$ and moments $m(L) - m(0)$ are often zero; that is, $V(0) = V(L)$ and $m(L) = m(0)$ occur except when a concentrated nodal force or moment exists at the node. In the actual computations, we handle the expressions given by Eq. (4.8.9) by including them as concentrated nodal values making up the matrix $\{P\}$.

▲ Summary Equations

Displacement function assumed for beam transverse displacement:

$$v(x) = a_1 x^3 + a_2 x^2 + a_3 x + a_4 \tag{4.1.2}$$

Shape functions for beam element:

$$N_1 = \frac{1}{L^3}(2x^3 - 3x^2 L + L^3) \qquad N_2 = \frac{1}{L^3}(x^3 L - 2x^2 L^2 + xL^3)$$

$$N_3 = \frac{1}{L^3}(-2x^3 + 3x^2 L) \qquad N_4 = \frac{1}{L^3}(x^3 L - x^2 L^2) \tag{4.1.7}$$

Beam bending stress or flexure formula:

$$\sigma_x = \frac{-My}{I} \tag{4.1.10b}$$

Stiffness matrix for beam element:

$$[k] = \frac{EI}{L^3} \begin{bmatrix} 12 & 6L & -12 & 6L \\ 6L & 4L^2 & -6L & 2L^2 \\ -12 & -6L & 12 & -6L \\ 6L & 2L^2 & -6L & 4L^2 \end{bmatrix} \tag{4.1.14}$$

Stiffness matrix including transverse shear deformation (Timoshenko beam theory):

$$[k] = \frac{EI}{L^3(1+\varphi)} \begin{bmatrix} 12 & 6L & -12 & 6L \\ 6L & (4+\varphi)L^2 & -6L & (2-\varphi)L^2 \\ -12 & -6L & 12 & -6L \\ 6L & (2-\varphi)L^2 & -6L & (4+\varphi)L^2 \end{bmatrix} \tag{4.1.15o}$$

Work due to distributed loading:

$$W_{\text{distributed}} = \int_0^L w(x)v(x)\,dx \tag{4.4.1}$$

Work due to discrete nodal forces:

$$W_{\text{discrete}} = m_1\phi_1 + m_2\phi_2 + f_{1y}v_1 + f_{2y}v_2 \tag{4.4.2}$$

General formulation for beam with distributed loading:

$$\{F\} = [K]\{d\} - \{F_0\} \tag{4.4.8}$$

Work-equivalent replacement matrix for beam with uniform load:

$$\{F_0\} = \left\{ \begin{array}{c} \dfrac{-wL}{2} \\[6pt] \dfrac{-wL^2}{12} \\[6pt] \dfrac{-wL}{2} \\[6pt] \dfrac{wL^2}{12} \end{array} \right\} \tag{4.4.10}$$

Beam stiffness matrix with right end nodal hinge:

$$[k] = \frac{3EI}{L^3} \begin{bmatrix} 1 & L & -1 & 0 \\ L & L^2 & -L & 0 \\ -1 & -L & 1 & 0 \\ 0 & 0 & 0 & 0 \end{bmatrix} \tag{4.6.13}$$

Total potential energy for beam element:

$$\pi_p = \int_0^L \frac{EI}{2}\{d\}^T[B]^T[B]\{d\}dx - \int_0^L w\{d\}^T[N]^T dx - \{d\}^T\{P\} \tag{4.7.15}$$

Strain energy expression for beam element:

$$U = \{d\}^T \frac{EI}{2} \int_0^L [B]^T[B]dx\{d\} \tag{4.7.21}$$

▲ References

[1] Gere, J. M., and Goodno, B. J., *Mechanics of Materials*, 7th ed., Cengage Learning, Mason, OH, 2009.

[2] Hsieh, Y. Y., *Elementary Theory of Structures*, 2nd ed., Prentice-Hall, Englewood Cliffs, NJ, 1982.

[3] Fraeijes de Veubeke, B., "Upper and Lower Bounds in Matrix Structural Analysis," *Matrix Methods of Structural Analysis*, AGAR Dograph 72, B. Fraeijes de Veubeke, ed., Macmillan, New York, 1964.

[4] Juvinall, R. C., and Marshek, K. M., *Fundamentals of Machine Component Design*, 4th ed., John Wiley & Sons, New York, 2006.

[5] Przemieneicki, J. S., *Theory of Matrix Structural Analysis*, McGraw-Hill, New York, 1968.

[6] McGuire, W., and Gallagher, R. H., *Matrix Structural Analysis*, John Wiley & Sons, New York, 1979.

[7] Severn, R. T., "Inclusion of Shear Deflection in the Stiffness Matrix for a Beam Element," *Journal of Strain Analysis*, Vol. 5, No. 4, 1970, pp. 239–241.

[8] Narayanaswami, R., and Adelman, H. M., "Inclusion of Transverse Shear Deformation in Finite Element Displacement Formulations," *AIAA Journal*, Vol. 12, No. 11, 1974, pp. 1613–1614.

[9] Timoshenko, S., *Vibration Problems in Engineering*, 3rd. ed., Van Nostrand Reinhold Company, New York, NY, 1955.

[10] Clark, S. K., *Dynamics of Continous Elements*, Prentice Hall, Englewood Cliffs, NJ, 1972.

[11] Algor Interactive Systems, 260 Alpha Dr., Pittsburgh, PA 15238.

[12] Martin, H. C., *Introduction to Matrix Methods of Structural Analysis*, McGraw-Hill, Boston, MA, 1966.

▲ Problems

4.1 Use Eqs. (4.1.7) to plot the shape functions N_1 and N_3 and the derivatives (dN_2/dx) and (dN_4/dx), which represent the shapes (variations) of the slopes ϕ_1 and ϕ_2 over the length of the beam element.

4.2 Derive the element stiffness matrix for the beam element in Figure 4–1 if the rotational degrees of freedom are assumed positive clockwise instead of counterclockwise. Compare the two different nodal sign conventions and discuss. Compare the resulting stiffness matrix to Eq. (4.1.14).

Solve all problems using the finite element stiffness method.

4.3 For the beam shown in Figure P4–3, determine the rotation at pin support A and the rotation and displacement under the load P. Determine the reactions. Draw the shear force and bending moment diagrams. Let EI be constant throughout the beam.

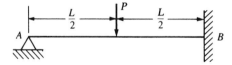
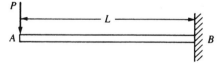

Figure P4–3   **Figure P4–4**

4.4 For the cantilever beam subjected to the free-end load P shown in Figure P4–4, determine the maximum deflection and the reactions. Let EI be constant throughout the beam.

4.5–4.11 For the beams shown in Figures P4–5 through P4–11, determine the displacements and the slopes at the nodes, the forces in each element, and the reactions. Also, draw the shear force and bending moment diagrams.

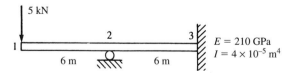

Figure P4–5

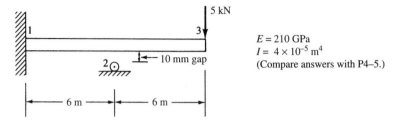

$E = 210$ GPa
$I = 4 \times 10^{-5}$ m^4
(Compare answers with P4–5.)

Figure P4–6

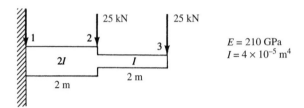

$E = 210$ GPa
$I = 4 \times 10^{-5}$ m^4

Figure P4–7

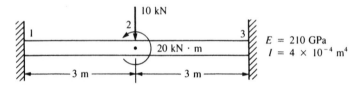

$E = 210$ GPa
$I = 4 \times 10^{-4}$ m^4

Figure P4–8

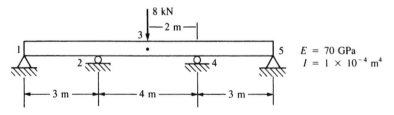

$E = 70$ GPa
$I = 1 \times 10^{-4}$ m^4

Figure P4–9

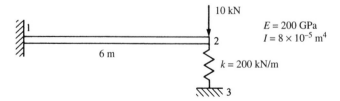

$E = 200$ GPa
$I = 8 \times 10^{-5}$ m^4

$k = 200$ kN/m

Figure P4–10

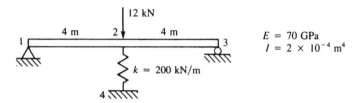

Figure P4–11

4.12 For the fixed-fixed beam subjected to the uniform load w shown in Figure P4–12, determine the midspan deflection and the reactions. Draw the shear force and bending moment diagrams. The middle section of the beam has a bending stiffness of $2EI$; the other sections have bending stiffnesses of EI.

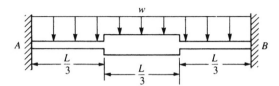

Figure P4–12

4.13 Determine the midspan deflection and the reactions and draw the shear force and bending moment diagrams for the fixed-fixed beam subjected to uniformly distributed load w shown in Figure P4–13. Assume EI constant throughout the beam. Compare your answers with the classical solution (that is, with the appropriate equivalent joint forces given in Appendix D).

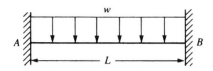

Figure P4–13

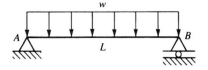

Figure P4–14

4.14 Determine the midspan deflection and the reactions and draw the shear force and bending moment diagrams for the simply supported beam subjected to the uniformly distributed load w shown in Figure P4–14. Assume EI constant throughout the beam.

4.15 For the beam loaded as shown in Figure P4–15, determine the free-end deflection and the reactions and draw the shear force and bending moment diagrams. Assume EI constant throughout the beam.

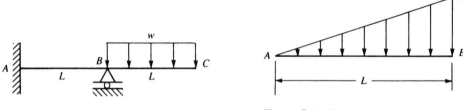

Figure P4–15   **Figure P4–16**

4.16 Using the concept of work equivalence, determine the nodal forces and moments (called *equivalent nodal forces*) used to replace the linearly varying distributed load shown in Figure P4–16.

4.17 For the beam shown in Figure 4–17, determine the displacement and slope at the center and the reactions. The load is symmetrical with respect to the center of the beam. Assume EI constant throughout the beam.

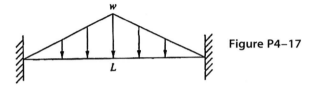

Figure P4–17

4.18 For the beam subjected to the linearly varying line load w shown in Figure P4–18, determine the right-end rotation and the reactions. Assume EI constant throughout the beam.

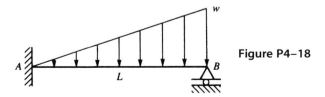

Figure P4–18

4.19–4.24 For the beams shown in Figures P4–19 through P4–24, determine the nodal displacements and slopes, the forces in each element, and the reactions.

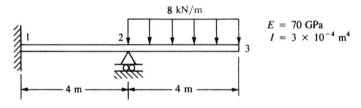

Figure P4–19

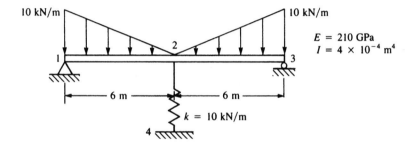

Figure P4–20

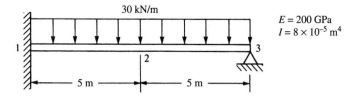

Figure P4–21

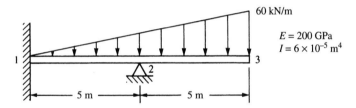

Figure P4–22

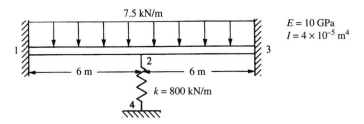

Figure P4–23

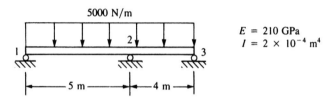

Figure P4–24

4.25–4.30 For the beams shown in Figures P4–25 through P4–30, determine the maximum deflection and maximum bending stress. Let $E = 200$ GPa for all beams as appropriate for the rest of the units in the problem. Let c be the half-depth of each beam.

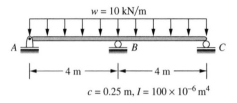

$c = 0.25$ m, $I = 100 \times 10^{-6}$ m^4

Figure P4–25

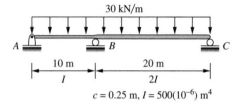

$c = 0.25$ m, $I = 500(10^{-6})$ m^4

Figure P4–26

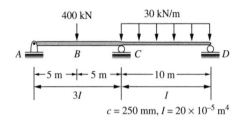

$c = 250$ mm, $I = 20 \times 10^{-5}$ m^4

Figure P4–27

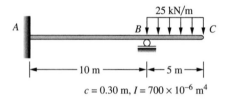

$c = 0.30$ m, $I = 700 \times 10^{-6}$ m^4

Figure P4–28

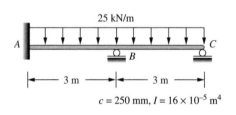

$c = 250$ mm, $I = 16 \times 10^{-5}$ m^4

Figure P4–29

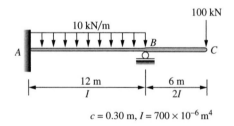

$c = 0.30$ m, $I = 700 \times 10^{-6}$ m^4

Figure P4–30

For the beam design problems shown in Figures P4–31 through P4–36, determine the size of beam to support the loads shown, based on requirements listed next to each beam.

 4.31 Design a beam of ASTM A36 steel with allowable bending stress of 160 MPa to support the load shown in Figure P4–31. Assume a standard wide flange beam from Appendix F, or some other source can be used.

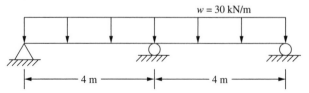

Figure P4–31

4.32 Select a standard steel pipe from Appendix F to support the load shown. The allowable bending stress must not exceed 170 MPa, and the allowable deflection must not exceed $L/360$ of any span.

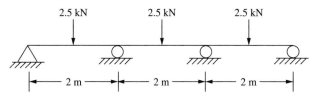

Figure P4–32

4.33 Select a rectangular structural tube from Appendix F to support the loads shown for the beam in Figure P4–33. The allowable bending stress should not exceed 170 MPa.

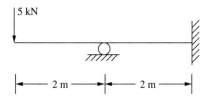

Figure P4–33

4.34 Select a standard W section from Appendix F or some other source to support the loads shown for the beam in Figure P4–34. The bending stress must not exceed 160 MPa.

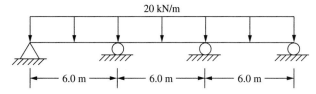

Figure P4–34

4.35 For the beam shown in Figure P4–35, determine a suitable sized W section from Appendix F or from another suitable source such that the bending stress does not exceed 150 MPa and the maximum deflection does not exceed $L/360$ of any span.

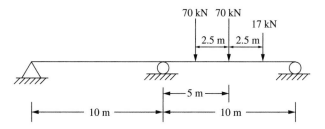

Figure P4–35

4.36 For the stepped shaft shown in Figure P4–36, determine a solid circular cross section for each section shown such that the bending stress does not exceed 160 MPa and the maximum deflection does not exceed $L/360$ of the span.

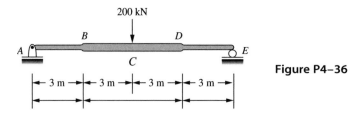

Figure P4–36

4.37 For the beam shown in Figure P4–37 subjected to the concentrated load P and distributed load w, determine the midspan displacement and the reactions. Let EI be constant throughout the beam.

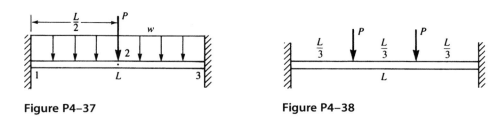

Figure P4–37   Figure P4–38

4.38 For the beam shown in Figure P4–38 subjected to the two concentrated loads P, determine the deflection at the midspan. Use the equivalent load replacement method. Let EI be constant throughout the beam.

4.39 For the beam shown in Figure P4–39 subjected to the concentrated load P and the linearly varying line load w, determine the free-end deflection and rotation and the reactions. Use the equivalent load replacement method. Let EI be constant throughout the beam.

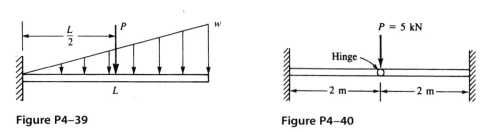

Figure P4–39   Figure P4–40

4.40–4.42 For the beams shown in Figures P4–40 through P4–42, with internal hinge, determine the deflection at the hinge. Let $E = 210$ GPa and $I = 2 \times 10^{-4}$ m⁴.

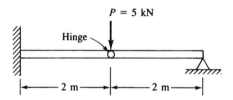

Figure P4–41

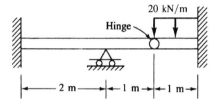

Figure P4–42

4.43 Derive the stiffness matrix for a beam element with a nodal linkage—that is, the shear is 0 at node i, but the usual shear and moment resistance are present at node j (see Figure P4–43).

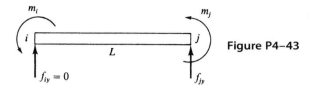

Figure P4–43

4.44 Develop the stiffness matrix for a fictitious pure shear panel element (Figure P4–44) in terms of the shear modulus, G, the shear web area, A_W, and the length, L. Notice that Y and v are the shear force and transverse displacement at each node, respectively.

Given   1) $\tau = G\gamma$,   2) $Y = \tau A_w$,   3) $Y_1 + Y_2 = 0$,   4) $\gamma = \dfrac{v_2 - v_1}{L}$

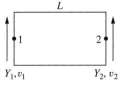

Positive node force
sign convention

Element in equilibrium
(neglect moments)

Figure P4–44

4.45 Explicitly evaluate π_p of Eq. (4.7.15); then differentiate π_p with respect to v_1, ϕ_1, v_2, and ϕ_2 and set each of these equations to zero (that is, minimize π_p) to obtain the four element equations for the beam element. Then express these equations in matrix form.

4.46 Determine the free-end deflection for the tapered beam shown in Figure P4–46. Here $I(x) = I_0(1 + nx/L)$ where I_0 is the moment of inertia at $x = 0$. Compare the exact beam theory solution with a two-element finite element solution for $n = 7$[12].

where $v_1 = \dfrac{1}{17.55}\dfrac{PL^3}{EI_0}, \quad \theta_1 = -\dfrac{1}{9.95}\dfrac{PL^2}{EI_0}$

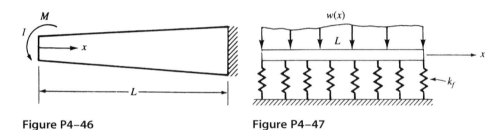

Figure P4–46   **Figure P4–47**

4.47 Derive the equations for the beam element on an elastic foundation (Figure P4–47) using the principle of minimum potential energy. Here k_f is the subgrade spring constant per unit length. The potential energy of the beam is

$$\pi_p = \int_0^L \frac{1}{2} EI(v'')^2 \, dx + \int_0^L \frac{k_f v^2}{2} \, dx - \int_0^L wv \, dx$$

4.48 Derive the equations for the beam element on an elastic foundation (see Figure P4–47) using Galerkin's method. The basic differential equation for the beam on an elastic foundation is

$$(EIv'')'' = -w + k_f v$$

4.49–4.76 Solve Problems 4.5 through 4.11, 4.19 through 4.30, and 4.40 through 4.42 using a suitable computer program.

4.77 For the beam shown in Figure P4–77, use a computer program to determine the deflection at the mid-span using four beam elements, making the shear area zero and then making the shear area equal 5/6 times the cross-sectional area (b times h). Then make the beam have decreasing spans of 200 mm, 100 mm, and 50 mm with zero shear area and then 5/6 times the cross-sectional area. Compare the answers. Based on your program answers, can you conclude whether your program includes the effects of transverse shear deformation?

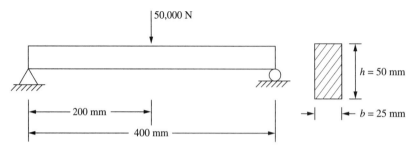

Figure P4–77

4.78 For the beam shown in Figure P4–77, use a longhand solution to solve the problem. Compare answers using the beam stiffness matrix, Eq. (4.1.14), without transverse shear deformation effects and then Eq. (4.1.15o), which includes the transverse shear effects.

CHAPTER 5

FRAME AND GRID EQUATIONS ▲

CHAPTER OBJECTIVES

- To derive the two-dimensional arbitrarily oriented beam element stiffness matrix.
- To demonstrate solutions of rigid plane frames by the direct stiffness method.
- To describe how to handle inclined or skewed supports.
- To derive the stiffness matrix and equations for grid analysis.
- To provide equations to determine torsional constants for various cross sections.
- To illustrate the solution of grid structures.
- To develop the stiffness matrix for a beam element arbitrarily oriented in space.
- To present the solution of a space frame.
- To introduce the concept of substructuring.

Introduction

Many structures, such as buildings (Figure 5–1) and bridges, are composed of frames and/or grids. This chapter develops the equations and methods for solution of plane and space frames and grids.

First, we will develop the stiffness matrix for a beam element arbitrarily oriented in a plane. We will then include the axial nodal displacement degree of freedom in the local beam element stiffness matrix. Then we will combine these results to develop the stiffness matrix, including axial deformation effects, for an arbitrarily oriented beam element, thus making it possible to analyze plane frames. Specific examples of plane frame analysis follow. We will then consider frames with inclined or skewed supports.

Next, we will develop the grid element stiffness matrix. We will present the solution of a grid deck system to illustrate the application of the grid equations. We will then develop the stiffness matrix for a beam element arbitrarily oriented in space. We will also consider the concept of substructure analysis.

▲ 5.1 Two-Dimensional Arbitrarily Oriented Beam Element ▲

We can derive the stiffness matrix for an arbitrarily oriented beam element, as shown in Figure 5–2, in a manner similar to that used for the bar element in Chapter 3. The local axes x' and y' are located along the beam element and transverse to the beam

(a)   (b)

Figure 5–1 (a) The Arizona Cardinals' football stadium under construction—a rigid building frame (Courtesy Ed Yack) and (b) Mini Baja space frame constructed of tubular steel members welded together (By Daryl L. Logan)

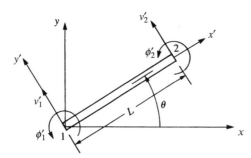

Figure 5–2 Arbitrarily oriented beam element

element, respectively, and the global axes x and y are located to be convenient for the total structure.

Recall that we can relate local displacements to global displacements by using Eq. (3.3.16), repeated here for convenience as

$$\begin{Bmatrix} u' \\ v' \end{Bmatrix} = \begin{bmatrix} C & S \\ -S & C \end{bmatrix} \begin{Bmatrix} u \\ v \end{Bmatrix} \tag{5.1.1}$$

Using the second equation of Eqs. (5.1.1) for the beam element, we relate local nodal degrees of freedom to global degrees of freedom by

$$\begin{Bmatrix} v_1' \\ \phi_1' \\ v_2' \\ \phi_2' \end{Bmatrix} = \begin{bmatrix} -S & C & 0 & 0 & 0 & 0 \\ 0 & 0 & 1 & 0 & 0 & 0 \\ 0 & 0 & 0 & -S & C & 0 \\ 0 & 0 & 0 & 0 & 0 & 1 \end{bmatrix} \begin{Bmatrix} u_1 \\ v_1 \\ \phi_1 \\ u_2 \\ v_2 \\ \phi_2 \end{Bmatrix} \tag{5.1.2}$$

where, for a beam element, we define

$$[T] = \begin{bmatrix} -S & C & 0 & 0 & 0 & 0 \\ 0 & 0 & 1 & 0 & 0 & 0 \\ 0 & 0 & 0 & -S & C & 0 \\ 0 & 0 & 0 & 0 & 0 & 1 \end{bmatrix} \tag{5.1.3}$$

as the *transformation matrix*. The axial effects are not yet included. Equation (5.1.2) indicates that rotation is invariant with respect to either coordinate system. For example, $\phi_1' = \phi_1$, and moment $m_1' = m_1$ can be considered to be a vector pointing normal to the x'-y' plane or to the x-y plane by the usual right-hand rule. From either viewpoint, the moment is in the $z' = z$ direction. Therefore, moment is unaffected as the element changes orientation in the x-y plane.

Substituting Eq. (5.1.3) for $[T]$ and Eq. (4.1.14) for local $[k']$ into Eq. (3.4.22), $[k] = [T]^T[k'][T]$, we obtain the global element stiffness matrix as

$$[k] = \frac{EI}{L^3}
\begin{array}{cccccc}
\;\;u_1 & \;\;v_1 & \;\;\phi_1 & \;\;u_2 & \;\;v_2 & \;\;\phi_2 \\
\end{array}
\begin{bmatrix}
12S^2 & -12SC & -6LS & -12S^2 & 12SC & -6LS \\
 & 12C^2 & 6LC & 12SC & -12C^2 & 6LC \\
 & & 4L^2 & 6LS & -6LC & 2L^2 \\
 & & & 12S^2 & -12SC & 6LS \\
 & & & & 12C^2 & -6LC \\
\text{Symmetry} & & & & & 4L^2
\end{bmatrix}
\tag{5.1.4}$$

where, again, $C = \cos\theta$ and $S = \sin\theta$. It is not necessary here to expand $[T]$ given by Eq. (5.1.3) to make it a square matrix to be able to use Eq. (3.4.22). Because Eq. (3.4.22) is a generally applicable equation, the matrices used must merely be of the correct order for matrix multiplication (see Appendix A for more on matrix multiplication). The stiffness matrix Eq. (5.1.4) is the global element stiffness matrix for a beam element that includes shear and bending resistance. Local axial effects are not yet included. The transformation from local to global stiffness by multiplying matrices $[T]^T[k'][T]$, as done in Eq. (5.1.4), is usually done on the computer.

We will now include the axial effects in the element, as shown in Figure 5–3. The element now has three degrees of freedom per node (u_i', v_i', ϕ_i'). For axial effects, we recall from Eq. (3.1.13),

$$\left\{ \begin{array}{c} f_{1x}' \\ f_{2x}' \end{array} \right\} = \frac{AE}{L} \begin{bmatrix} 1 & -1 \\ -1 & 1 \end{bmatrix} \left\{ \begin{array}{c} u_1' \\ u_2' \end{array} \right\}
\tag{5.1.5}$$

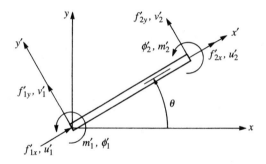

Figure 5–3 Local forces and displacements acting on a beam element

Combining the axial effects of Eq. (5.1.5) with the shear and principal bending moment effects of Eq. (4.1.13), we have, in local coordinates,

$$
\begin{Bmatrix} f'_{1x} \\ f'_{1y} \\ m'_1 \\ f'_{2x} \\ f'_{2y} \\ m'_2 \end{Bmatrix} = \begin{bmatrix} C_1 & 0 & 0 & -C_1 & 0 & 0 \\ 0 & 12C_2 & 6C_2L & 0 & -12C_2 & 6C_2L \\ 0 & 6C_2L & 4C_2L^2 & 0 & -6C_2L & 2C_2L^2 \\ -C_1 & 0 & 0 & C_1 & 0 & 0 \\ 0 & -12C_2 & -6C_2L & 0 & 12C_2 & -6C_2L \\ 0 & 6C_2L & 2C_2L^2 & 0 & -6C_2L & 4C_2L^2 \end{bmatrix} \begin{Bmatrix} u'_1 \\ v'_1 \\ \phi'_1 \\ u'_2 \\ v'_2 \\ \phi'_2 \end{Bmatrix} \quad (5.1.6)
$$

where
$$
C_1 = \frac{AE}{L} \quad \text{and} \quad C_2 = \frac{EI}{L^3} \quad (5.1.7)
$$

and, therefore,

$$
[k'] = \begin{bmatrix} C_1 & 0 & 0 & -C_1 & 0 & 0 \\ 0 & 12C_2 & 6C_2L & 0 & -12C_2 & 6C_2L \\ 0 & 6C_2L & 4C_2L^2 & 0 & -6C_2L & 2C_2L^2 \\ -C_1 & 0 & 0 & C_1 & 0 & 0 \\ 0 & -12C_2 & -6C_2L & 0 & 12C_2 & -6C_2L \\ 0 & 6C_2L & 2C_2L^2 & 0 & -6C_2L & 4C_2L^2 \end{bmatrix} \quad (5.1.8)
$$

The $[k']$ matrix in Eq. (5.1.8) now has three degrees of freedom per node and now includes axial effects (in the x' direction), as well as shear force effects (in the y' direction) and principal bending moment effects (about the $z' = z$ axis). Using Eqs. (5.1.1) and (5.1.2), we now relate the local to the global displacements by

$$
\begin{Bmatrix} u'_1 \\ v'_1 \\ \phi'_1 \\ u'_2 \\ v'_2 \\ \phi'_2 \end{Bmatrix} = \begin{bmatrix} C & S & 0 & 0 & 0 & 0 \\ -S & C & 0 & 0 & 0 & 0 \\ 0 & 0 & 1 & 0 & 0 & 0 \\ 0 & 0 & 0 & C & S & 0 \\ 0 & 0 & 0 & -S & C & 0 \\ 0 & 0 & 0 & 0 & 0 & 1 \end{bmatrix} \begin{Bmatrix} u_1 \\ v_1 \\ \phi_1 \\ u_2 \\ v_2 \\ \phi_2 \end{Bmatrix} \quad (5.1.9)
$$

where $[T]$ has now been expanded to include local axial deformation effects as

$$
[T] = \begin{bmatrix} C & S & 0 & 0 & 0 & 0 \\ -S & C & 0 & 0 & 0 & 0 \\ 0 & 0 & 1 & 0 & 0 & 0 \\ 0 & 0 & 0 & C & S & 0 \\ 0 & 0 & 0 & -S & C & 0 \\ 0 & 0 & 0 & 0 & 0 & 1 \end{bmatrix} \quad (5.1.10)
$$

Substituting $[T]$ from Eq. (5.1.10) and $[k']$ from Eq. (5.1.8) into Eq. (3.4.22), $([k] = [T]^T[k'][T])$ we obtain the general transformed global stiffness matrix for a beam element that includes axial force, shear force, and bending moment effects as follows:

$$[k] = \frac{E}{L} \times$$

$$
\begin{bmatrix}
AC^2+\dfrac{12I}{L^2}S^2 & \left(A-\dfrac{12I}{L^2}\right)CS & -\dfrac{6I}{L}S & -\left(AC^2+\dfrac{12I}{L^2}S^2\right) & -\left(A-\dfrac{12I}{L^2}\right)CS & -\dfrac{6I}{L}S \\
 & AS^2+\dfrac{12I}{L^2}C^2 & \dfrac{6I}{L}C & -\left(A-\dfrac{12I}{L^2}\right)CS & -\left(AS^2+\dfrac{12I}{L^2}C^2\right) & \dfrac{6I}{L}C \\
 & & 4I & \dfrac{6I}{L}S & -\dfrac{6I}{L}C & 2I \\
 & & & AC^2+\dfrac{12I}{L^2}S^2 & \left(A-\dfrac{12I}{L^2}\right)CS & \dfrac{6I}{L}S \\
 & & & & AS^2+\dfrac{12I}{L^2}C^2 & -\dfrac{6I}{L}C \\
\text{Symmetry} & & & & & 4I
\end{bmatrix}
$$

(5.1.11)

The analysis of a rigid plane frame can be undertaken by applying stiffness matrix Eq. (5.1.11). *A* **rigid plane frame** *is defined here as a series of beam elements rigidly connected to each other; that is, the original angles made between elements at their joints remain unchanged after the deformation due to applied loads or applied displacements.*

Furthermore, moments are transmitted from one element to another at the joints. Hence, moment continuity exists at the rigid joints. In addition, the element centroids, as well as the applied loads, lie in a common plane (*x-y* plane). From Eq. (5.1.11), we observe that the element stiffnesses of a frame are functions of *E, A, L, I*, and the angle of orientation θ of the element with respect to the global-coordinate axes. It should be noted that computer programs often refer to the frame element as a beam element, with the understanding that the program is using the stiffness matrix in Eq. (5.1.11) for plane frame analysis.

▲ 5.2 Rigid Plane Frame Examples ▲

To illustrate the use of the equations developed in Section 5.1, we will now perform complete solutions for the following rigid plane frames.

Example 5.1

As the first example of rigid plane frame analysis, solve the simple "bent" shown in Figure 5–4.

SOLUTION:

The frame is fixed at nodes 1 and 4 and subjected to a positive horizontal force of 40 kN applied at node 2 and to a positive moment of 500 N-m applied

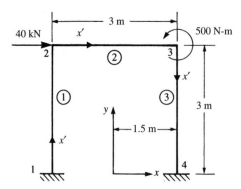

Figure 5–4 Plane frame for analysis, also showing local x' axis for each element

at node 3. The global-coordinate axes and the element lengths are shown in Figure 5–4.

Let $E = 200$ GPa and $A = 6500$ mm² for all elements, and let $I = 80 \times 10^6$ mm⁴ for elements 1 and 3, and $I = 40 \times 10^6$ mm⁴ for element 2.

Using Eq. (5.1.11), we obtain the global stiffness matrices for each element.

Element 1

For element 1, the angle between the global x and the local x' axes is 90° (counterclockwise) because x' is assumed to be directed from node 1 to node 2. Therefore,

$$C = \cos 90° = \frac{x_2 - x_1}{L^{(1)}} = \frac{-1.5 - (-1.5)}{3} = 0$$

$$S = \sin 90° = \frac{y_2 - y_1}{L^{(1)}} = \frac{3 - 0}{3} = 1$$

Also,

$$\frac{12I}{L^2} = \frac{12(80 \times 10^6)}{(3000)^2} = 106.67 \text{ mm}^2 \qquad (5.2.1)$$

$$\frac{6I}{L} = \frac{6(80 \times 10^6)}{3000} = 160{,}000 \text{ mm}^3$$

$$\frac{E}{L} = \frac{210 \times 10^3}{3000} = 66.67 \text{ N/mm}^3$$

Then, using Eqs. (5.2.1) to help in evaluating Eq. (5.1.11) for element 1, we obtain the element global stiffness matrix as

$$[k^{(1)}] = 66.67 \times 10^3 \begin{array}{c} \begin{matrix} u_1 & \quad v_1 & \quad \phi_1 & \quad u_2 & \quad v_2 & \quad \phi_2 \end{matrix} \\ \begin{bmatrix} 0.106 & 0 & -160 & -0.106 & 0 & -160 \\ 0 & 6.5 & 0 & 0 & -6.5 & 0 \\ -160 & 0 & 320{,}000 & 160 & 0 & 160{,}000 \\ -0.106 & 0 & 160 & 0.106 & 0 & 160 \\ 0 & -6.5 & 0 & 0 & 6.5 & 0 \\ -160 & 0 & 160{,}000 & 160 & 0 & 320{,}000 \end{bmatrix} \begin{array}{c} \\ \\ \frac{\text{N}}{\text{mm}} \\ \\ \\ \end{array} \end{array} \qquad (5.2.2)$$

where all diagonal terms are positive.

Element 2

For element 2, the angle between x and x' is zero because x' is directed from node 2 to node 3. Therefore,

$$C = 1 \qquad S = 0$$

Also,

$$\frac{12I}{L^2} = \frac{12(40 \times 10^6)}{(3000)^2} = 0.0835 \text{ in.}^2$$

$$\frac{6I}{L} = \frac{6(40 \times 10^6)}{3000} = 80{,}000 \text{ mm}^3 \qquad (5.2.3)$$

$$\frac{E}{L} = 66.67 \times 10^6 \ \frac{\text{N}}{\text{mm}^3}$$

Using the quantities obtained in Eqs. (5.2.3) in evaluating Eq. (5.1.11) for element 2, we obtain

$$[k^{(2)}] = 66.67 \times 10^3 \begin{array}{c} \begin{array}{cccccc} u_2 & v_2 & \phi_2 & u_3 & v_3 & \phi_3 \end{array} \\ \begin{bmatrix} 6.5 & 0 & 0 & -6.5 & 0 & 0 \\ 0 & 0.0533 & 80 & 0 & -0.0533 & 80 \\ 0 & 80 & 160{,}000 & 0 & -80 & 80{,}000 \\ -6.5 & 0 & 0 & 6.5 & 0 & 0 \\ 0 & -0.0533 & -80 & 0 & 0.0533 & -80 \\ 0 & 80 & 80{,}000 & 0 & -80 & 160{,}000 \end{bmatrix} \end{array} \frac{\text{N}}{\text{mm}} \qquad (5.2.4)$$

Element 3

For element 3, the angle between x and x' is $270°$ (or $-90°$) because x' is directed from node 3 to node 4. Therefore,

$$C = 0 \qquad S = -1$$

Therefore, evaluating Eq. (5.1.11) for element 3, we obtain

$$[k^{(3)}] = 66.67 \times 10^3 \begin{array}{c} \begin{array}{cccccc} u_3 & v_3 & \phi_3 & u_4 & v_4 & \phi_4 \end{array} \\ \begin{bmatrix} 0.106 & 0 & 160 & -0.106 & 0 & 160 \\ 0 & 6.5 & 0 & 0 & -6.5 & 0 \\ 160 & 0 & 320{,}000 & -160 & 0 & 160{,}000 \\ -0.106 & 0 & -160 & 0.106 & 0 & -160 \\ 0 & -6.5 & 0 & 0 & 6.5 & 0 \\ 160 & 0 & 160{,}000 & -160 & 0 & 320{,}000 \end{bmatrix} \end{array} \frac{\text{N}}{\text{mm}} \qquad (5.2.5)$$

Superposition of Eqs. (5.2.2), (5.2.4), and (5.2.5) and application of the boundary conditions $u_1 = v_1 = \phi_1 = 0$ and $u_4 = v_4 = \phi_4 = 0$ at nodes 1 and 4 yield the reduced set of equations for a longhand solution as

$$
\begin{Bmatrix} 4 \times 10^4 \\ 0 \\ 0 \\ 0 \\ 0 \\ 5 \times 10^5 \end{Bmatrix} = 66.67 \times 10^3 \begin{bmatrix} 6.606 & 0 & 160 & -6.5 & 0 & 0 \\ 0 & 6.5553 & 80 & 0 & -0.0533 & 80 \\ 160 & 80 & 480,000 & 0 & -80 & 80,000 \\ -6.5 & 0 & 0 & 6.606 & 0 & 160 \\ 0 & -0.0533 & -80 & 0 & 6.5553 & -80 \\ 0 & 80 & 80,000 & 160 & -80 & 480,000 \end{bmatrix} \begin{Bmatrix} u_2 \\ v_2 \\ \phi_2 \\ u_3 \\ v_3 \\ \phi_3 \end{Bmatrix}
$$

(5.2.6)

Solving Eq. (5.2.6) for the displacements and rotations, we have

$$
\begin{Bmatrix} u_2 \\ v_2 \\ \phi_2 \\ u_3 \\ v_3 \\ \phi_3 \end{Bmatrix} = \begin{Bmatrix} 5.007 \text{ mm} \\ 0.0345 \text{ mm} \\ -0.00144 \text{ rad} \\ 4.961 \text{ mm} \\ -0.0345 \text{ mm} \\ -0.00140 \text{ rad} \end{Bmatrix}
$$

(5.2.7)

The results indicate that the top of the frame moves to the right with negligible vertical displacement and small rotations of elements at nodes 2 and 3.

The element forces can now be obtained using $\{f'\} = [k'][T]\{d\}$ for each element, as was previously done in solving truss and beam problems. We will illustrate this procedure only for element 1. For element 1, on using Eq. (5.1.10) for $[T]$ and Eq. (5.2.7) for the displacements at node 2, we have

$$
[T]\{d\} = \begin{bmatrix} 0 & 1 & 0 & 0 & 0 & 0 \\ -1 & 0 & 0 & 0 & 0 & 0 \\ 0 & 0 & 1 & 0 & 0 & 0 \\ 0 & 0 & 0 & 0 & 1 & 0 \\ 0 & 0 & 0 & -1 & 0 & 0 \\ 0 & 0 & 0 & 0 & 0 & 1 \end{bmatrix} \begin{Bmatrix} u_1 = 0 \\ v_1 = 0 \\ \phi_1 = 0 \\ u_2 = 5.007 \text{ mm} \\ v_2 = 0.0345 \text{ mm} \\ \phi_2 = -0.00144 \text{ rad} \end{Bmatrix}
$$

(5.2.8)

On multiplying the matrices in Eq. (5.2.8), we obtain

$$
[T]\{d\} = \begin{Bmatrix} 0 \\ 0 \\ 0 \\ 0.0345 \text{ mm} \\ -5.007 \text{ mm} \\ -0.00140 \text{ rad} \end{Bmatrix}
$$

(5.2.9)

Then using $[k']$ from Eq. (5.1.8), we obtain element 1 local forces as

$$\{f'\} = [k'][T]\{d\} = 66.67 \times 10^3 \begin{bmatrix} 6.5 & 0 & 0 & -6.5 & 0 & 0 \\ 0 & 0.106 & 160 & 0 & -0.106 & 160 \\ 0 & 160 & 320{,}000 & 0 & -160 & 160{,}000 \\ -6.5 & 0 & 0 & 6.5 & 0 & 0 \\ 0 & -0.106 & -160 & 0 & 0.106 & -160 \\ 0 & 160 & 160{,}000 & 0 & -160 & 320{,}000 \end{bmatrix} \begin{Bmatrix} 0 \\ 0 \\ 0 \\ 0.0345 \\ -5.007 \\ -0.00144 \end{Bmatrix}$$

(5.2.10)

Simplifying Eq. (5.2.10), we obtain the local forces acting on element 1 as

$$\begin{Bmatrix} f'_{1x} \\ f'_{1y} \\ m'_1 \\ f'_{2x} \\ f'_{2y} \\ m'_2 \end{Bmatrix} = \begin{Bmatrix} -14950 \text{ N} \\ 20023 \text{ N} \\ 38049902 \text{ N-mm} \\ 14950 \text{ N} \\ -20023 \text{ N} \\ 22689134 \text{ N-mm} \end{Bmatrix}$$

(5.2.11)

A free-body diagram of each element is shown in Figure 5–5 along with equilibrium verification. In Figure 5–5, the x' axis is directed from node 1 to node 2—consistent with the order of the nodal degrees of freedom used in developing the stiffness matrix for the element. Since the x-y plane was initially established as shown in Figure 5–4, the z axis is directed outward—consequently, so is the z' axis (recall $z' = z$). The y' axis is then established such that x' cross y' yields the direction of z'. The signs on the resulting element forces in Eq. (5.2.11) are thus consistently shown in Figure 5–5. The forces in elements 2 and 3 can be obtained in a manner similar to that used to obtain Eq. (5.2.11) for the nodal forces in element 1. Here we report only the final results for the forces in elements 2 and 3 and leave it to your discretion to perform the detailed calculations. The element forces (shown in Figures 5–5(b) and (c)) are as follows:

Element 2

$$f'_{2x} = 20243 \text{ N} \qquad f'_{2y} = -14950 \text{ N} \qquad m'_2 = -22689134 \text{ N-mm}$$
$$f'_{3x} = -20243 \text{ N} \qquad f'_{3y} = 14950 \text{ N} \qquad m'_3 = -2264437 \text{ N-mm}$$

(5.2.12a)

Element 3

$$f'_{3x} = 14950 \text{ N} \qquad f'_{3y} = 20243 \text{ N} \qquad m'_3 = 22870420 \text{ N-mm}$$
$$f'_{4x} = -14950 \text{ N} \qquad f'_{4y} = -20243 \text{ N} \qquad m'_4 = 37948705 \text{ N-mm}$$

(5.2.12b)

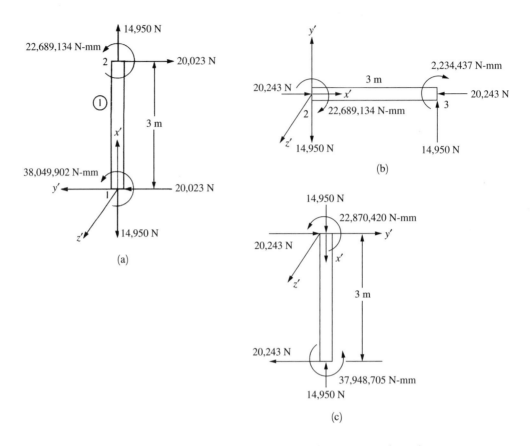

Figure 5–5 Free-body diagrams of (a) element 1, (b) element 2, and (c) element 3

Considering the free body of element 1, the equilibrium equations are

$$\sum F_{x'}: \ -20023 + 20023 = 0$$

$$\sum F_{y'}: \ -14950 + 14950 = 0$$

$$\sum M_2: 38,049,902 + 22,689,134 - 20,023(3000) \cong 0$$

Considering moment equilibrium at node 2, we see from Eqs. (5.2.12a) and (5.2.12b) that on element 1, $m_2' = 22,689,134$ N-mm, and the opposite value, $-22,689,134$ N-mm, occurs on element 2. Similarly, moment equilibrium is satisfied at node 3, as m_3' from elements 2 and 3 add to the 505,983 N-m applied moment. That is, from Eqs. (5.2.12a) and (5.2.12b) we have

$$-22,364,437 + 22,870,420 = 505,983 \text{ N-mm}$$

$$\cong 500 \text{ N-m} \qquad \blacksquare$$

Example 5.2

To illustrate the procedure for solving frames subjected to distributed loads, solve the rigid plane frame shown in Figure 5–6. The frame is fixed at nodes 1 and 3 and subjected to a uniformly distributed load of 13 kN/m applied downward over element 2. The global-coordinate axes have been established at node 1. The element lengths are shown in the figure. Let $E = 200$ GPa, $A = 0.06$ m², and $I = 3.6 \times 10^{-4}$ m⁴ for both elements of the frame.

SOLUTION:

We begin by replacing the distributed load acting on element 2 by nodal forces and moments acting at nodes 2 and 3. Using Eqs. (4.4.5)–(4.4.7) (or Appendix D), the equivalent nodal forces and moments are calculated as

$$f_{2y} = -\frac{wL}{2} = -\frac{(13 \times 10^3)12}{2} = -78,000 \text{ N} = -78 \text{ kN}$$

$$m_2 = -\frac{wL^2}{12} = -\frac{(13 \times 10^3)12^2}{12} = -156,000 \text{ N-m} = -156 \text{ kN-m}$$

(5.2.13)

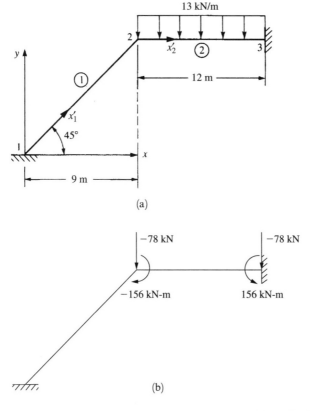

(a)

(b)

Figure 5–6 (a) Plane frame for analysis and (b) equivalent nodal forces on frame

$$f_{3y} = -\frac{wL}{2} = -\frac{(13 \times 10^3)12}{2} = -78,000 \text{ N} = -78 \text{ kN}$$

$$m_3 = \frac{wL^2}{12} = \frac{(13 \times 10^3)12^2}{12} = 156,000 \text{ N-m} = 156 \text{ Kn-m}$$

We then use Eq. (5.1.11) to determine each element stiffness matrix:

Element 1

$$\theta^{(1)} = 45° \qquad C = 0.707 \qquad S = 0.707 \qquad L^{(1)} = 12.72 \text{ m} = 12720.0 \text{ mm}$$

$$\frac{E}{L} = \frac{200 \times 10^3}{12720} = 15.72 \frac{\text{N}}{\text{mm}^3}$$

$$[k^{(1)}] = 15.72 \begin{bmatrix} 30,010 & 29,980 & 12,005 \\ 29,980 & 30,010 & -12,005 \\ 12,005 & -12,005 & 1.44 \times 10^9 \end{bmatrix} \frac{\text{N}}{\text{mm}^3} \qquad (5.2.14)$$

Simplifying Eq. (5.2.14), we obtain

$$[k^{(1)}] = 15.72 \times 10^5 \begin{array}{ccc} u_2 & v_2 & \phi_2 \end{array} \\ \begin{bmatrix} 0.3001 & 0.2998 & 0.12005 \\ 0.2998 & 0.3001 & -0.12005 \\ 0.12005 & -0.12005 & 14,400 \end{bmatrix} \frac{\text{N}}{\text{mm}^3} \qquad (5.2.15)$$

where only the parts of the stiffness matrix associated with degrees of freedom at node 2 are included because node 1 is fixed.

Element 2

$$\theta^{(2)} = 0° \qquad C = 1 \qquad S = 0 \qquad L^{(2)} = 12 \text{ m} = 12,000 \text{ mm}$$

$$\frac{E}{L} = \frac{200 \times 10^3}{12,000} = 16.67 \frac{\text{N}}{\text{mm}^3}$$

$$[k^{(2)}] = 16.67 \begin{bmatrix} 60,000 & 0 & 0 \\ 0 & 30 & 180,000 \\ 0 & 180,000 & 1.44 \times 10^9 \end{bmatrix} \frac{\text{N}}{\text{mm}^3} \qquad (5.2.16)$$

Simplifying Eq. (5.2.16), we obtain

$$[k^{(2)}] = 16.67 \times 10^5 \begin{array}{ccc} u_2 & v_2 & \phi_2 \end{array} \\ \begin{bmatrix} 0.6 & 0 & 0 \\ 0 & 0.0003 & 1.8 \\ 0 & 1.8 & 14,400 \end{bmatrix} \frac{\text{N}}{\text{mm}^3} \qquad (5.2.17)$$

where, again, only the parts of the stiffness matrix associated with degrees of freedom at node 2 are included because node 3 is fixed. On superimposing the stiffness matrices of the elements, using Eqs. (5.2.15) and (5.2.17), and using Eq. (5.2.13) for the nodal forces and moments only at node 2 (because the structure is fixed at node 3), we have

$$
\begin{Bmatrix} F_{2x} = 0 \\ F_{2y} = -78 \times 10^3 \\ M_2 = -156 \times 10^3 \end{Bmatrix} = (10^3) \begin{bmatrix} 1471.95 & 471.28 & 1887.18 \\ 471.28 & 472.25 & 113.41 \\ 1887.18 & 113.41 & 46{,}641{,}600 \end{bmatrix} \begin{Bmatrix} u_2 \\ v_2 \\ \phi_2 \end{Bmatrix} \quad (5.2.18)
$$

Solving Eq. (5.2.18) for the displacements and the rotation at node 2, we obtain

$$
\begin{Bmatrix} u_2 \\ v_2 \\ \phi_2 \end{Bmatrix} = \begin{Bmatrix} 0.0803 \text{ mm} \\ -0.2374 \text{ mm} \\ -0.0033 \text{ rad} \end{Bmatrix} \quad (5.2.19)
$$

The results indicate that node 2 moves to the right ($u_2 = 0.0803$ mm) and down ($v_2 = -0.2374$ mm) and the rotation of the joint is clockwise ($\phi_2 = -0.0033$ rad).

The local forces in each element can now be determined. The procedure for elements that are subjected to a distributed load must be applied to element 2. Recall that the local forces are given by $\{f'\} = [k'][T]\{d\}$. For element 1, we then have

$$
[T]\{d\} = \begin{bmatrix} 0.707 & 0.707 & 0 & 0 & 0 & 0 \\ -0.707 & 0.707 & 0 & 0 & 0 & 0 \\ 0 & 0 & 1 & 0 & 0 & 0 \\ 0 & 0 & 0 & 0.707 & 0.707 & 0 \\ 0 & 0 & 0 & -0.707 & 0.707 & 0 \\ 0 & 0 & 0 & 0 & 0 & 1 \end{bmatrix} \begin{Bmatrix} 0 \\ 0 \\ 0 \\ 0.0803 \\ -0.2374 \\ -0.0033 \end{Bmatrix} \quad (5.2.20)
$$

Simplifying Eq. (5.2.20) yields

$$
[T]\{d\} = \begin{Bmatrix} 0 \\ 0 \\ 0 \\ -0.11108 \\ -0.224632 \\ -0.003342 \end{Bmatrix} \quad (5.2.21)
$$

Using Eq. (5.2.21) and Eq. (5.1.8) for $[k']$, we obtain

$$
\begin{Bmatrix} f'_{1x} \\ f'_{1y} \\ m'_1 \\ f'_{2x} \\ f'_{2y} \\ m'_2 \end{Bmatrix} = \begin{bmatrix} 5893 & 0 & 0 & -5893 & 0 & 0 \\ & 2.730 & 694.8 & 0 & -2.730 & 694.8 \\ & & 117{,}900 & 0 & -694.8 & 117{,}900 \\ & & & 5893 & 0 & 0 \\ & & & & 2.730 & -694.8 \\ \text{Symmetry} & & & & & 235{,}800 \end{bmatrix} \begin{Bmatrix} 0 \\ 0 \\ 0 \\ -0.11108 \\ -0.224632 \\ -0.003342 \end{Bmatrix}
$$

$$(5.2.22)$$

Simplifying Eq. (5.2.22) yields the local forces in element 1 as

$$f'_{1x} = 104.77 \text{ kN} \qquad f'_{1y} = -8.827 \text{ kN} \qquad m'_{1x} = -37.22 \text{ kN-m}$$

$$f'_{2x} = -104.77 \text{ kN} \qquad f'_{2y} = 8.827 \text{ kN} \qquad m'_{2x} = -75.05 \text{ kN-m}$$

(5.2.23)

For element 2, the local forces are given by Eq. (4.4.11) because a distributed load is acting on the element. From Eqs. (5.1.10) and (5.2.19), we then have

$$[T]\{d\} = \begin{bmatrix} 1 & 0 & 0 & 0 & 0 & 0 \\ 0 & 1 & 0 & 0 & 0 & 0 \\ 0 & 0 & 1 & 0 & 0 & 0 \\ 0 & 0 & 0 & 1 & 0 & 0 \\ 0 & 0 & 0 & 0 & 1 & 0 \\ 0 & 0 & 0 & 0 & 0 & 1 \end{bmatrix} \begin{Bmatrix} 0.0803 \\ -0.2374 \\ -0.0033 \\ 0 \\ 0 \\ 0 \end{Bmatrix}$$

(5.2.24)

Simplifying Eq. (5.2.24), we obtain

$$\begin{Bmatrix} 0.0803 \\ -0.2374 \\ -0.0033 \\ 0 \\ 0 \\ 0 \end{Bmatrix}$$

(5.2.25)

Using Eq. (5.2.25) and Eq. (5.1.8) for $[k']$, we have

$$[k']\{d'\} = [k'][T]\{d\} = \begin{bmatrix} 6250 & 0 & 0 & -6250 & 0 & 0 \\ & 3.25 & 781.1 & 0 & -3.25 & 781.1 \\ & & 250{,}000 & 0 & -781.1 & 125{,}000 \\ & & & 6250 & 0 & 0 \\ & & & & 3.25 & -781.1 \\ \text{Symmetry} & & & & & 250{,}000 \end{bmatrix} \begin{Bmatrix} 0.0803 \\ -0.2374 \\ -0.0033 \\ 0 \\ 0 \\ 0 \end{Bmatrix}$$

(5.2.26)

Simplifying Eq. (5.2.26) yields

$$[k']\{d'\} = \begin{Bmatrix} 80.316 \text{ kN} \\ -10.018 \text{ kN} \\ -79.928 \text{ kN-m} \\ -80.316 \text{ kN} \\ -10.018 \text{ kN} \\ -40.320 \text{ kN-m} \end{Bmatrix}$$

(5.2.27)

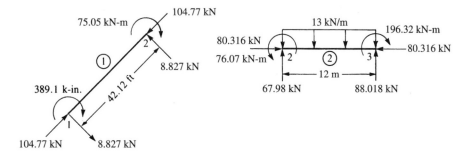

Figure 5–7 Free-body diagrams of elements 1 and 2

To obtain the actual element local nodal forces, we apply Eq. (4.4.11); that is, we must subtract the equivalent nodal forces [Eqs. (5.2.13)] from Eq. (5.2.27) to yield

$$
\begin{Bmatrix} f'_{2x} \\ f'_{2y} \\ m'_2 \\ f'_{3x} \\ f'_{3y} \\ m'_3 \end{Bmatrix} = \begin{Bmatrix} 80.316 \\ -10.018 \\ -79.928 \\ -80.316 \\ 10.018 \\ -40.32 \end{Bmatrix} - \begin{Bmatrix} 0 \\ -78 \\ -156 \\ 0 \\ -78 \\ 156 \end{Bmatrix} \tag{5.2.28}
$$

Simplifying Eq. (5.2.28), we obtain

$$
f'_{2x} = 80.316 \text{ kN} \qquad f'_{2y} = 67.98 \text{ kN} \qquad m'_2 = 76.07 \text{ kN-m}
$$
$$
f'_{3x} = -80.316 \text{ kN} \qquad f'_{3y} = 88.018 \text{ kN} \qquad m'_3 = -196.32 \text{ kN-m}
$$

(5.2.29)

Using Eqs. (5.2.23) and (5.2.29) for the local forces in each element, we can construct the free-body diagram for each element, as shown in Figure 5–7. From the free-body diagrams, one can confirm the equilibrium of each element, the total frame, and joint 2 as desired. ∎

In Example 5.3, we will illustrate the equivalent joint force replacement method for a frame subjected to a load acting on an element instead of at one of the joints of the structure. Since no distributed loads are present, the point of application of the concentrated load could be treated as an extra joint in the analysis, and we could solve the problem in the same manner as Example 5.1.

This approach has the disadvantage of increasing the total number of joints, as well as the size of the total structure stiffness matrix $[K]$. For small structures solved by computer, this does not pose a problem. However, for very large structures, this might reduce the maximum size of the structure that could be analyzed. Certainly, this additional node greatly increases the longhand solution time for the structure. Hence, we will illustrate a standard procedure based on the concept of equivalent joint forces applied to the case of concentrated loads. We will again use Appendix D.

Example 5.3

Solve the frame shown in Figure 5–8(a). The frame consists of the three elements shown and is subjected to a 65-kN horizontal load applied at midlength of element 1. Nodes 1, 2, and 3 are fixed, and the dimensions are shown in the figure. Let $E = 200$ GPa, $I = 3.0 \times 10^{-4}$ m^4, and $A = 5.0 \times 10^{-3}$ m^2 for all elements.

SOLUTION:

1. We first express the applied load in the element 1 local coordinate system (here x' is directed from node 1 to node 4). This is shown in Figure 5–8(b).

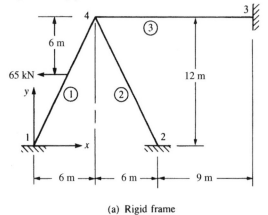

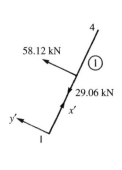

(a) Rigid frame

(b) Applied load expressed in element 1 local-coordinate system

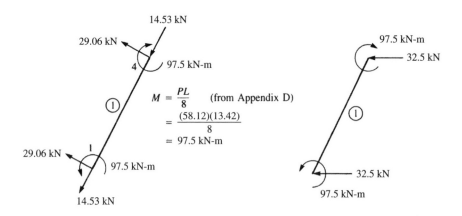

$$M = \frac{PL}{8} \quad \text{(from Appendix D)}$$
$$= \frac{(58.12)(13.42)}{8}$$
$$= 97.5 \text{ kN-m}$$

(c) Equivalent joint forces expressed in local-coordinate system

(d) Final equivalent joint forces expressed in global-coordinate system

Figure 5–8 Rigid frame with a load applied on an element

2. Next, we determine the equivalent joint forces $\{f_0\}$ at each end of element 1, using the table in Appendix D. (These forces are of opposite sign from what are traditionally known as *fixed-end forces* in classical structural analysis theory [1].) These equivalent forces (and moments) are shown in Figure 5–8(c).
3. We then transform the equivalent joint forces from the present local-coordinate-system forces into the global-coordinate-system forces, using the equation $\{f\} = [T]^T\{f'\}$, where $[T]$ is defined by Eq. (5.1.10). These global joint forces are shown in Figure 5–8(d).
4. Then we analyze the structure in Figure 5–8(d), using the equivalent joint forces (plus actual joint forces, if any) in the usual manner.
5. We obtain the final internal forces developed at the ends of each element that has an applied load (here element 1 only) by subtracting step 2 joint forces from step 4 joint forces; that is, Eq. (4.4.11) $(\{f\} = \{f^{(e)}\} - \{f_0\})$ is applied locally to all elements that originally had loads acting on them.

The solution of the structure as shown in Figure 5–8(d) now follows. Using Eq. (5.1.11), we obtain the global stiffness matrix for each element.

Element 1

For element 1, the angle between the global x and the local x' axes is 63.43° because x' is assumed to be directed from node 1 to node 4. Therefore,

$$C = \cos 63.43° = \frac{x_4 - x_1}{L^{(1)}} = \frac{6 - 0}{13.42} = 0.447$$

$$S = \sin 63.43° = \frac{y_4 - y_1}{L^{(1)}} = \frac{12 - 0}{13.42} = 0.895$$

$$\frac{12I}{L^2} = \frac{12(3 \times 10^{-4})}{(13.42)^2} = 1.998 \times 10^{-5} \qquad \frac{6I}{L} = \frac{6(3 \times 10^{-4})}{13.42} = 1.341 \times 10^{-4}$$

$$\frac{E}{L} = \frac{200 \times 10^9}{13.42} = 1.49 \times 10^{10}$$

Using the preceding results in Eq. (5.1.11) for $[k]$, we obtain

$$[k^{(1)}] = (10^6) \begin{array}{ccc} u_4 & v_4 & \phi_4 \end{array} \\ \begin{bmatrix} 15.12 & 29.78 & 1.78 \\ 29.78 & 59.73 & -0.88 \\ 1.78 & -0.88 & 17.88 \end{bmatrix} \qquad (5.2.30)$$

where only the parts of the stiffness matrix associated with degrees of freedom at node 4 are included because node 1 is fixed and, hence, not needed in the solution for the nodal displacements.

Element 3

For element 3, the angle between x and x' is zero because x' is directed from node 4 to node 3. Therefore,

$$C = 1 \quad S = 0 \quad \frac{12I}{L^2} = \frac{12(3.0 \times 10^{-4})}{(15)^2} = 1.6 \times 10^{-5}$$

$$\frac{6I}{L} = \frac{6(3.0 \times 10^{-4})}{15} = 1.2 \times 10^{-4} \quad \frac{E}{L} = \frac{200 \times 10^9}{15} = 1.33 \times 10^{10}$$

Substituting these results into $[k]$, we obtain

$$[k^{(3)}] = (10^6) \begin{array}{c} \begin{array}{ccc} u_4 & v_4 & \phi_4 \end{array} \\ \begin{bmatrix} 66.67 & 0 & 0 \\ 0 & 0.21 & 1.59 \\ 0 & 1.59 & 15.99 \end{bmatrix} \end{array} \qquad (5.2.31)$$

since node 3 is fixed.

Element 2

For element 2, the angle between x and x' is 116.57° because x' is directed from node 2 to node 4. Therefore,

$$C = \frac{6 - 12}{13.42} = -0.447 \quad S = \frac{12 - 0}{13.42} = 0.895$$

$$\frac{12I}{L^2} = 1.998 \times 10^{-5} \quad \frac{6I}{L} = 1.341 \times 10^{-4} \quad \frac{E}{L} = 1.49 \times 10^{10}$$

since element 2 has the same properties as element 1. Substituting these results into $[k]$, we obtain

$$[k^{(2)}] = (10^6) \begin{array}{c} \begin{array}{ccc} u_4 & v_4 & \phi_4 \end{array} \\ \begin{bmatrix} 15.12 & -29.78 & 1.78 \\ -29.78 & 59.73 & 0.88 \\ 1.78 & 0.88 & 17.88 \end{bmatrix} \end{array} \qquad (5.2.32)$$

since node 2 is fixed. On superimposing the stiffness matrices given by Eqs. (5.2.30), (5.2.31), and (5.2.32), and using the nodal forces given in Figure 5–8(d) at node 4 only, we have

$$\begin{Bmatrix} -32.5 \text{ kN} \\ 0 \\ -97.5 \text{ kN-m} \end{Bmatrix} = \begin{bmatrix} 96.91 & 0 & 3.56 \\ 0 & 119.67 & 1.59 \\ 3.56 & 1.59 & 51.75 \end{bmatrix} \begin{Bmatrix} u_4 \\ v_4 \\ \phi_4 \end{Bmatrix} \qquad (5.2.33)$$

Simultaneously solving the three equations in Eq. (5.2.33), we obtain

$$u_4 = -0.267 \text{ mm}$$

$$v_4 = 0.025 \text{ mm} \qquad (5.2.34)$$

$$\phi_4 = -0.001866 \text{ rad}$$

Next, we determine the element forces by again using $\{f'\} = [k'][T]\{d\}$. In general, we have

$$[T]\{d\} = \begin{bmatrix} C & S & 0 & 0 & 0 & 0 \\ -S & C & 0 & 0 & 0 & 0 \\ 0 & 0 & 1 & 0 & 0 & 0 \\ 0 & 0 & 0 & C & S & 0 \\ 0 & 0 & 0 & -S & C & 0 \\ 0 & 0 & 0 & 0 & 0 & 1 \end{bmatrix} \begin{Bmatrix} u_i \\ v_i \\ \phi_i \\ u_j \\ v_j \\ \phi_j \end{Bmatrix}$$

Thus, the preceding matrix multiplication yields

$$[T]\{d\} = \begin{Bmatrix} Cu_i + Sv_i \\ -Su_i + Cv_i \\ \phi_i \\ Cu_j + Sv_j \\ -Su_j + Cv_j \\ \phi_j \end{Bmatrix} \tag{5.2.35}$$

Element 1

$$[T]\{d\} = \begin{Bmatrix} 0 \\ 0 \\ 0 \\ (0.447)(-2.67 \times 10^{-4}) + (0.895)(2.5 \times 10^{-5}) \\ (-0.895)(-2.67 \times 10^{-4}) + (0.447)(2.5 \times 10^{-5}) \\ -0.001866 \end{Bmatrix} = \begin{Bmatrix} 0 \\ 0 \\ 0 \\ -9.697 \times 10^{-5} \\ 2.5 \times 10^{-4} \\ -0.001186 \end{Bmatrix} \tag{5.2.36}$$

Using Eq. (5.1.8) for $[k']$ and Eq. (5.2.36), we obtain

$$[k'][T]\{d\} = (10^6) \begin{bmatrix} 74.51 & 0 & 0 & -74.51 & 0 & 0 \\ 0 & 0.297 & 1.996 & 0 & -0.297 & 1.996 \\ 0 & 1.996 & 17.86 & 0 & -1.996 & 8.93 \\ -74.51 & 0 & 0 & 74.51 & 0 & 0 \\ 0 & -0.297 & -1.996 & 0 & 0.297 & -1.996 \\ 0 & 1.996 & 8.93 & 0 & -1.996 & 17.86 \end{bmatrix} \times \begin{Bmatrix} 0 \\ 0 \\ 0 \\ -0.0969 \\ 0.025 \text{ mm} \\ -0.001186 \text{ rad} \end{Bmatrix} \tag{5.2.37}$$

These values are now called *effective nodal forces* $\{f^{(e)}\}$. Multiplying the matrices of Eq. (5.2.37) and using Eq. (4.4.11) to subtract the equivalent nodal forces in local coordinates for the element shown in Figure 5–8(c), we obtain the final nodal forces in

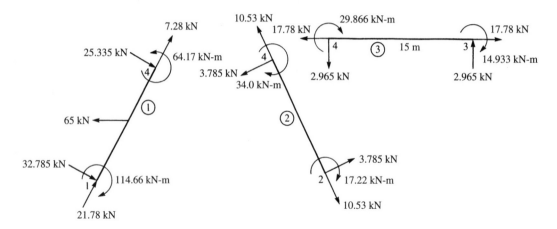

Figure 5–9 Free-body diagrams of all elements of the frame in Figure 5–8(a)

element 1 as

$$\{f'^{(1)}\} = \begin{Bmatrix} 7.25 \text{ kN} \\ -3.725 \text{ kN} \\ -16.66 \text{ kN-m} \\ -7.25 \text{ kN} \\ 3.725 \text{ kN} \\ -33.33 \text{ kN} \end{Bmatrix} - \begin{Bmatrix} -14.53 \text{ kN} \\ 29.06 \text{ kN} \\ 97.5 \text{ kN-m} \\ -14.53 \text{ kN} \\ 29.06 \text{ kN} \\ -97.5 \text{ kN-m} \end{Bmatrix} = \begin{Bmatrix} 21.78 \text{ kN} \\ -32.785 \text{ kN} \\ -114.66 \text{ kN-m} \\ 7.28 \text{ kN} \\ -25.335 \text{ kN} \\ 64.17 \text{ kN-m} \end{Bmatrix} \quad (5.2.38)$$

Similarly, we can use Eqs. (5.2.35) and (5.1.8) for elements 3 and 2 to obtain the local nodal forces in these elements. Since these elements do not have any applied loads on them, the final nodal forces in local coordinates associated with each element are given by $\{f'\} = [k'][T]\{d\}$. These forces have been determined as follows:

Element 3

$$f'_{4x} = -17.78 \text{ kN} \qquad f'_{4y} = -2.695 \text{ kN} \qquad m'_4 = -29.866 \text{ kN-m}$$
$$f'_{3x} = 17.78 \text{ kN} \qquad f'_{3y} = 2.965 \text{ kN} \qquad m'_3 = -14.933 \text{ kN-m} \qquad (5.2.39)$$

Element 2

$$f'_{2x} = -10.53 \text{ kN} \qquad f'_{2y} = -3.785 \text{ kN} \qquad m'_2 = -17.22 \text{ kN-m}$$
$$f'_{4x} = 10.53 \text{ kN} \qquad f'_{4y} = 3.785 \text{ kN} \qquad m'_4 = -34.0 \text{ kN-m} \qquad (5.2.40)$$

Free-body diagrams of all elements are shown in Figure 5–9. Each element has been determined to be in equilibrium, as often occurs even if errors are made in the long-hand calculations. However, equilibrium at node 4 and equilibrium of the whole frame are also satisfied. For instance, using the results of Eqs. (5.2.38) through (5.2.40) to check equilibrium at node 4, which is implicit in the formulation of the global equations, we have

$$\sum M_4 = 64.17 - 29.866 - 34.0 = 0.304 \text{ kN-m} \quad \text{(close to zero)}$$

$$\sum F_x = 7.28(0.447) + 25.335(0.895) - 10.53(0.447)$$

$$-3.785(0.895) - 17.78 = -0.05 \text{ kN} \quad \text{(close to zero)}$$

$$\sum F_y = 7.28(0.895) - 25.335(0.447) + 10.53(0.895)$$

$$-3.785(0.447) - 2.965 = 0.039 \text{ kN} \quad \text{(close to zero)}$$

Thus, the solution has been verified to be correct within the accuracy associated with a longhand solution.   ■

To illustrate the solution of a problem involving both bar and frame elements, we will solve the following example.

Example 5.4

The bar element 2 is used to stiffen the cantilever beam element 1, as shown in Figure 5–10. Determine the displacements at node 1 and the element forces. For the bar, let $A = 1.0 \times 10^{-3}$ m². For the beam, let $A = 2 \times 10^{-3}$ m², $I = 5 \times 10^{-5}$ m⁴, and $L = 3$ m. For both the bar and the beam elements, let $E = 210$ GPa. Let the angle between the beam and the bar be 45°. A downward force of 500 kN is applied at node 1.

SOLUTION:

For brevity's sake, since nodes 2 and 3 are fixed, we keep only the parts of $[k]$ for each element that are needed to obtain the global $[K]$ matrix necessary for solution of the nodal degrees of freedom. Using Eq. (3.4.23), we obtain $[k]$ for the bar as

$$[k^{(2)}] = \frac{(1 \times 10^{-3})(210 \times 10^6)}{(3/\cos 45°)} \begin{bmatrix} 0.5 & 0.5 \\ 0.5 & 0.5 \end{bmatrix}$$

or, simplifying this equation, we obtain

$$[k^{(2)}] = 70 \times 10^3 \begin{matrix} u_1 & v_1 \\ \begin{bmatrix} 0.354 & 0.354 \\ 0.354 & 0.354 \end{bmatrix} \end{matrix} \frac{\text{kN}}{\text{m}} \qquad (5.2.41)$$

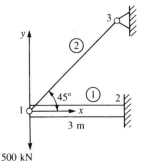

Figure 5–10 Cantilever beam with a bar element support

Using Eq. (5.1.11), we obtain $[k]$ for the beam (including axial effects) as

$$[k^{(1)}] = 70 \times 10^3 \begin{array}{ccc} u_1 & v_1 & \phi_1 \end{array} \\ \begin{bmatrix} 2 & 0 & 0 \\ 0 & 0.067 & 0.10 \\ 0 & 0.10 & 0.20 \end{bmatrix} \frac{\text{kN}}{\text{m}} \tag{5.2.42}$$

where $(E/L) \times 10^{-3}$ has been factored out in evaluating Eq. (5.2.42).

We assemble Eqs. (5.2.41) and (5.2.42) in the usual manner to obtain the global stiffness matrix as

$$[K] = 70 \times 10^3 \begin{bmatrix} 2.354 & 0.354 & 0 \\ 0.354 & 0.421 & 0.10 \\ 0 & 0.10 & 0.20 \end{bmatrix} \frac{\text{kN}}{\text{m}} \tag{5.2.43}$$

The global equations are then written for node 1 as

$$\begin{Bmatrix} F_{1x} \\ F_{1y} \\ M_1 \end{Bmatrix} = \begin{Bmatrix} 0 \\ -500 \\ 0 \end{Bmatrix} = 70 \times 10^3 \begin{bmatrix} 2.354 & 0.354 & 0 \\ 0.354 & 0.421 & 0.10 \\ 0 & 0.10 & 0.20 \end{bmatrix} \begin{Bmatrix} u_1 \\ v_1 \\ \phi_1 \end{Bmatrix} \tag{5.2.44}$$

Solving Eq. (5.2.44), we obtain

$$u_1 = 0.00338 \text{ m} \qquad v_1 = -0.0225 \text{ m} \qquad \phi_1 = 0.0113 \text{ rad} \tag{5.2.45}$$

In general, the local element forces are obtained using $\{f'\} = [k'][T]\{d\}$. For the bar element, we then have

$$\begin{Bmatrix} f'_{1x} \\ f'_{3x} \end{Bmatrix} = \frac{AE}{L} \begin{bmatrix} 1 & -1 \\ -1 & 1 \end{bmatrix} \begin{bmatrix} C & S & 0 & 0 \\ 0 & 0 & C & S \end{bmatrix} \begin{Bmatrix} u_1 \\ v_1 \\ u_3 \\ v_3 \end{Bmatrix} \tag{5.2.46}$$

The matrix triple product of Eq. (5.2.46) yields (as one equation)

$$f'_{1x} = \frac{AE}{L}(Cu_1 + Sv_1) \tag{5.2.47}$$

Substituting the numerical values into Eq. (5.2.47), we obtain

$$f'_{1x} = \frac{(1 \times 10^{-3} \text{ m}^2)(210 \times 10^6 \text{ kN/m}^2)}{4.24 \text{ m}} \left[\frac{\sqrt{2}}{2}(0.00338 - 0.0225) \right] \tag{5.2.48}$$

Simplifying Eq. (5.2.48), we obtain the axial force in the bar (element 2) as

$$f'_{1x} = -670 \text{ kN} \tag{5.2.49}$$

where the negative sign means f'_{1x} is in the direction opposite x' for element 2. Similarly, we obtain

$$f'_{3x} = 670 \text{ kN} \tag{5.2.50}$$

which means the bar is in tension as shown in Figure 5–11. Since the local and global axes are coincident for the beam element, we have $\{f'\} = \{f\}$ and $\{d'\} = \{d\}$. Therefore, from Eq. (5.1.6), we have at node 1

$$\begin{Bmatrix} f'_{1x} \\ f'_{1y} \\ m'_1 \end{Bmatrix} = \begin{bmatrix} C_1 & 0 & 0 \\ 0 & 12C_2 & 6C_2L \\ 0 & 6C_2L & 4C_2L^2 \end{bmatrix} \begin{Bmatrix} u_1 \\ v_1 \\ \phi_1 \end{Bmatrix} \qquad (5.2.51)$$

where only the upper left 3×3 part of the stiffness matrix is needed because the displacements at node 2 are equal to zero. Substituting numerical values into Eq. (5.2.51), we obtain

$$\begin{Bmatrix} f'_{1x} \\ f'_{1y} \\ m'_1 \end{Bmatrix} = 70 \times 10^3 \begin{bmatrix} 2 & 0 & 0 \\ 0 & 0.067 & 0.10 \\ 0 & 0.10 & 0.20 \end{bmatrix} \begin{Bmatrix} 0.00338 \\ -0.0225 \\ 0.0113 \end{Bmatrix}$$

The matrix product then yields

$$f'_{1x} = 473 \text{ kN} \qquad f'_{1y} = -26.5 \text{ kN} \qquad m'_1 = 0.0 \text{ kN} \cdot \text{m} \qquad (5.2.52)$$

Similarly, using the lower left 3×3 part of Eq. (5.1.6), we have at node 2,

$$\begin{Bmatrix} f'_{2x} \\ f'_{2y} \\ m'_2 \end{Bmatrix} = 70 \times 10^3 \begin{bmatrix} -2 & 0 & 0 \\ 0 & -0.067 & -0.10 \\ 0 & 0.10 & 0.10 \end{bmatrix} \begin{Bmatrix} 0.00338 \\ -0.0225 \\ 0.0113 \end{Bmatrix}$$

The matrix product then yields

$$f'_{2x} = -473 \text{ kN} \qquad f'_{2y} = 26.5 \text{ kN} \qquad m'_2 = -78.3 \text{ kN} \cdot \text{m} \qquad (5.2.53)$$

To help interpret the results of Eqs. (5.2.49), (5.2.50), (5.2.52), and (5.2.53), free-body diagrams of the bar and beam elements are shown in Figure 5–11. To further verify the results, we can show a check on equilibrium of node 1 to be satisfied. You should also verify that moment equilibrium is satisfied in the beam.

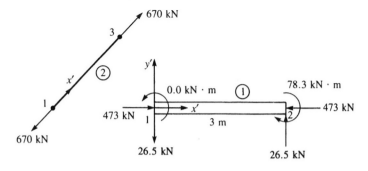

Figure 5–11 Free-body diagrams of the bar (element 2) and beam (element 1) elements of Figure 5–10 ■

▲ **5.3 Inclined or Skewed Supports—**
Frame Element

For the frame element with inclined support at node 3 in Figure 5–12, the transformation matrix $[T]$ used to transform global to local nodal displacements is given by Eq. (5.1.10).

In the example shown in Figure 5–12, we use $[T]$ applied to node 3 as follows:

$$\begin{Bmatrix} u_3' \\ v_3' \\ \phi_3' \end{Bmatrix} = \begin{bmatrix} \cos\alpha & \sin\alpha & 0 \\ -\sin\alpha & \cos\alpha & 0 \\ 0 & 0 & 1 \end{bmatrix} \begin{Bmatrix} u_3 \\ v_3 \\ \phi_3 \end{Bmatrix} \tag{5.3.1}$$

The same steps as given in Section 3.9 then follow for the plane frame. The resulting equations for the plane frame in Figure 5–12 are (see also Eq. (3.9.13))

$$[T_i]\{f\} = [T_i][K][T_i]^T\{d\} \tag{5.3.2}$$

or

$$\begin{Bmatrix} F_{1x} \\ F_{1y} \\ M_1 \\ F_{2x} \\ F_{2y} \\ M_2 \\ F_{3x}' \\ F_{3y}' \\ M_3 \end{Bmatrix} = [T_i][K][T_i]^T \begin{Bmatrix} u_1 = 0 \\ v_1 = 0 \\ \phi_1 = 0 \\ u_2 \\ v_2 \\ \phi_2 \\ u_3' \\ v_3' = 0 \\ \phi_3' = \phi_3 \end{Bmatrix} \tag{5.3.3}$$

where

$$[T_i] = \begin{bmatrix} [I] & [0] & [0] \\ [0] & [I] & [0] \\ [0] & [0] & [t_3] \end{bmatrix} \tag{5.3.4}$$

and

$$[t_3] = \begin{bmatrix} \cos\alpha & \sin\alpha & 0 \\ -\sin\alpha & \cos\alpha & 0 \\ 0 & 0 & 1 \end{bmatrix} \tag{5.3.5}$$

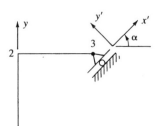

Figure 5–12 Frame with inclined support

▲ 5.4 Grid Equations ▲

A **grid** *is a structure on which loads are applied perpendicular to the plane of the structure, as opposed to a plane frame, where loads are applied in the plane of the structure.* We will now develop the grid element stiffness matrix. The elements of a grid are assumed to be rigidly connected, so that the original angles between elements connected together at a node remain unchanged. Both torsional and bending moment continuity then exist at the node point of a grid. Examples of grids include floor and bridge deck systems. A typical grid structure subjected to loads F_1, F_2, F_3, and F_4 is shown in Figure 5–13.

We will now consider the development of the grid element stiffness matrix and element equations. A representative grid element with the nodal degrees of freedom and nodal forces is shown in Figure 5–14. The degrees of freedom at each node for a grid are a vertical deflection v_i' (normal to the grid), a torsional rotation ϕ_{ix}' about the x' axis, and a bending rotation ϕ_{iz}' about the z' axis. Any effect of axial displacement is ignored; that is, $u_i' = 0$. The nodal forces consist of a transverse force f_{iy}', a torsional moment m_{ix}' about the x' axis, and a bending moment m_{iz}' about the z' axis. Grid elements do not resist axial loading; that is $f_{ix}' = 0$.

To develop the local stiffness matrix for a grid element, we need to include the torsional effects in the basic beam element stiffness matrix Eq. (4.1.14). Recall that Eq. (4.1.14) already accounts for the bending and shear effects.

We can derive the torsional bar element stiffness matrix in a manner analogous to that used for the axial bar element stiffness matrix in Chapter 3. In the derivation, we simply replace f_{ix}' with m_{ix}', u_i' with ϕ_{ix}', E with G (the shear modulus), A with J (the torsional constant, or stiffness factor), σ with τ (shear stress), and ε with γ (shear strain).

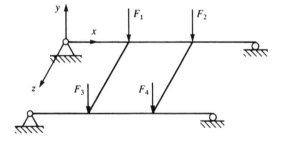

Figure 5–13 Typical grid structure

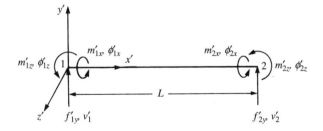

Figure 5–14 Grid element with nodal degrees of freedom and nodal forces

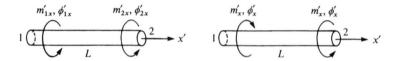

Figure 5–15   Nodal and element torque sign conventions

The actual derivation is briefly presented as follows. We assume a circular cross section with radius R for simplicity but without loss of generalization.

Step 1

Figure 5–15 shows the sign conventions for nodal torque and angle of twist and for element torque.

Step 2

We assume a linear angle-of-twist variation along the x' axis of the bar such that

$$\phi' = a_1 + a_2 x' \qquad (5.4.1)$$

Using the usual procedure of expressing a_1 and a_2 in terms of unknown nodal angles of twist ϕ'_{1x} and ϕ'_{2x}, we obtain

$$\phi' = \left(\frac{\phi'_{2x} - \phi'_{1x}}{L}\right) x' + \phi'_{1x} \qquad (5.4.2)$$

or, in matrix form, Eq. (5.4.2) becomes

$$\phi' = [N_1 \quad N_2] \begin{Bmatrix} \phi'_{1x} \\ \phi'_{2x} \end{Bmatrix} \qquad (5.4.3)$$

with the shape functions given by

$$N_1 = 1 - \frac{x'}{L} \qquad N_2 = \frac{x'}{L} \qquad (5.4.4)$$

Step 3

We obtain the shear strain γ/angle of twist ϕ' relationship by considering the torsional deformation of the bar segment shown in Figure 5–16. Assuming that all radial lines, such as OA, remain straight during twisting or torsional deformation, we observe that the arc length $\widehat{AB}$ is given by

$$\widehat{AB} = \gamma_{\max} \, dx' = R \, d\phi'$$

Solving for the maximum shear strain $\gamma_{\max}$, we obtain

$$\gamma_{\max} = \frac{R \, d\phi'}{dx'}$$

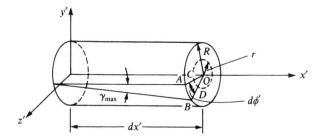

Figure 5–16 Torsional deformation of a bar segment

Similarly, at any radial position r, we then have, from similar triangles OAB and OCD,

$$\gamma = r\frac{d\phi'}{dx'} = \frac{r}{L}(\phi'_{2x} - \phi'_{1x})$$ (5.4.5)

where we have used Eq. (5.4.2) to derive the final expression in Eq. (5.4.5).

The shear stress τ/shear strain γ relationship for linear-elastic isotropic materials is given by

$$\tau = G\gamma$$ (5.4.6)

where G is the shear modulus of the material.

Step 4

We derive the element stiffness matrix in the following manner. From elementary mechanics, we have the shear stress related to the applied torque by

$$m'_x = \frac{\tau J}{R}$$ (5.4.7)

where J is called the *polar moment of inertia* for the circular cross section or, generally, the *torsional constant* for noncircular cross sections. Using Eqs. (5.4.5) and (5.4.6) in Eq. (5.4.7), we obtain

$$m'_x = \frac{GJ}{L}(\phi'_{2x} - \phi'_{1x})$$ (5.4.8)

By the nodal torque sign convention of Figure 5–15,

$$m'_{1x} = -m'_x$$ (5.4.9)

or, by using Eq. (5.4.8) in Eq. (5.4.9), we obtain

$$m'_{1x} = \frac{GJ}{L}(\phi'_{1x} - \phi'_{2x})$$ (5.4.10)

Similarly,
$$m'_{2x} = m'_x$$ (5.4.11)

or
$$m'_{2x} = \frac{GJ}{L}(\phi'_{2x} - \phi'_{1x})$$ (5.4.12)

Expressing Eqs. (5.4.10) and (5.4.12) together in matrix form, we have the resulting torsion bar stiffness matrix equation:

$$\begin{Bmatrix} m'_{1x} \\ m'_{2x} \end{Bmatrix} = \frac{GJ}{L} \begin{bmatrix} 1 & -1 \\ -1 & 1 \end{bmatrix} \begin{Bmatrix} \phi'_{1x} \\ \phi'_{2x} \end{Bmatrix} \tag{5.4.13}$$

Hence, the stiffness matrix for the torsion bar is

$$[k'] = \frac{GJ}{L} \begin{bmatrix} 1 & -1 \\ -1 & 1 \end{bmatrix} \tag{5.4.14}$$

The cross sections of various structures, such as bridge decks, are often not circular. However, Eqs. (5.4.13) and (5.4.14) are still general; to apply them to other cross sections, we simply evaluate the torsional constant J for the particular cross section. For instance, for cross sections made up of thin rectangular shapes such as channels, angles, or I shapes, we approximate J by

$$J = \sum \frac{1}{3} b_i t_i^3 \tag{5.4.15}$$

where b_i is the length of any element of the cross section and t_i is the thickness of any element of the cross section. In Table 5–1, we list values of J for various common cross sections. The first four cross sections are called *open sections*. Equation (5.4.15) applies only to these open cross sections. (For more information on the J concept, consult References [2] and [3], and for an extensive table of torsional constants for various cross-sectional shapes, consult Reference [4].) We assume the loading to go through the shear center of these open cross sections in order to prevent twisting of the cross section. For more on the shear center consult References [2] and [5].

On combining the torsional effects of Eq. (5.4.13) with the shear and bending effects of Eq. (4.1.13), we obtain the local stiffness matrix equation for a grid element as

$$\begin{Bmatrix} f'_{1y} \\ m'_{1x} \\ m'_{1z} \\ f'_{2y} \\ m'_{2x} \\ m'_{2z} \end{Bmatrix} = \begin{bmatrix} \dfrac{12EI}{L^3} & 0 & \dfrac{6EI}{L^2} & \dfrac{-12EI}{L^3} & 0 & \dfrac{6EI}{L^2} \\ & \dfrac{GJ}{L} & 0 & 0 & \dfrac{-GJ}{L} & 0 \\ & & \dfrac{4EI}{L} & \dfrac{-6EI}{L^2} & 0 & \dfrac{2EI}{L} \\ & & & \dfrac{12EI}{L^3} & 0 & \dfrac{-6EI}{L^2} \\ & & & & \dfrac{GJ}{L} & 0 \\ \text{Symmetry} & & & & & \dfrac{4EI}{L} \end{bmatrix} \begin{Bmatrix} v'_1 \\ \phi'_{1x} \\ \phi'_{1z} \\ v'_2 \\ \phi'_{2x} \\ \phi'_{2z} \end{Bmatrix} \tag{5.4.16}$$

Table 5–1   Torsional constants J and shear centers SC for various cross sections

Cross Section	Torsional Constant
1. Channel 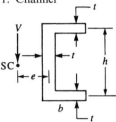	$J = \dfrac{t^3}{3}(h + 2b)$ $e = \dfrac{h^2 b^2 t}{4I}$
2. Angle 	$J = \frac{1}{3}(b_1 t_1^3 + b_2 t_2^3)$
3. Z section 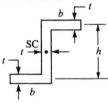	$J = \dfrac{t^3}{3}(2b + h)$
4. Wide-flanged beam with unequal flanges 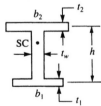	$J = \frac{1}{3}(b_1 t_1^3 + b_2 t_2^3 + h t_w^3)$
5. Solid circular 	$J = \dfrac{\pi}{2} r^4$
6. Closed hollow rectangular 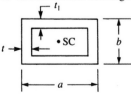	$J = \dfrac{2 t t_1 (a - t)^2 (b - t_1)^2}{at + b t_1 - t^2 - t_1^2}$

where, from Eq. (5.4.16), the local stiffness matrix for a grid element is

$$
[k'_G] =
\begin{array}{cccccc}
v'_1 & \phi'_{1x} & \phi'_{1z} & v'_2 & \phi'_{2x} & \phi'_{2z}
\end{array}
\begin{bmatrix}
\dfrac{12EI}{L^3} & 0 & \dfrac{6EI}{L^2} & \dfrac{-12EI}{L^3} & 0 & \dfrac{6EI}{L^2} \\[2mm]
0 & \dfrac{GJ}{L} & 0 & 0 & \dfrac{-GJ}{L} & 0 \\[2mm]
\dfrac{6EI}{L^2} & 0 & \dfrac{4EI}{L} & \dfrac{-6EI}{L^2} & 0 & \dfrac{2EI}{L} \\[2mm]
\dfrac{-12EI}{L^3} & 0 & \dfrac{-6EI}{L^2} & \dfrac{12EI}{L^3} & 0 & \dfrac{-6EI}{L^2} \\[2mm]
0 & \dfrac{-GJ}{L} & 0 & 0 & \dfrac{GJ}{L} & 0 \\[2mm]
\dfrac{6EI}{L^2} & 0 & \dfrac{2EI}{L} & \dfrac{-6EI}{L^2} & 0 & \dfrac{4EI}{L}
\end{bmatrix}
\qquad (5.4.17)
$$

and the degrees of freedom are in the order (1) vertical deflection, (2) torsional rotation, and (3) bending rotation, as indicated by the notation used above the columns of Eq. (5.4.17).

The transformation matrix relating local to global degrees of freedom for a grid is given by

$$
[T_G] =
\begin{bmatrix}
1 & 0 & 0 & 0 & 0 & 0 \\
0 & C & S & 0 & 0 & 0 \\
0 & -S & C & 0 & 0 & 0 \\
0 & 0 & 0 & 1 & 0 & 0 \\
0 & 0 & 0 & 0 & C & S \\
0 & 0 & 0 & 0 & -S & C
\end{bmatrix}
\qquad (5.4.18)
$$

where θ is now positive, taken counterclockwise from x to x' in the x-z plane (Figure 5–17) and

$$
C = \cos\theta = \frac{x_j - x_i}{L} \qquad S = \sin\theta = \frac{z_j - z_i}{L}
$$

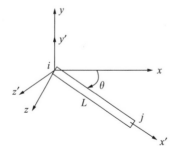

Figure 5–17 Grid element arbitrarily oriented in the x-z plane

where L is the length of the element from node i to node j. As indicated by Eq. (5.4.18) for a grid, the vertical deflection v' is invariant with respect to a coordinate transformation (that is, $y = y'$) (Figure 5–17).

The global stiffness matrix for a grid element arbitrarily oriented in the x-z plane is then given by using Eqs. (5.4.17) and (5.4.18) in

$$[k_G] = [T_G]^T [k'_G][T_G] \qquad (5.4.19)$$

Now that we have formulated the global stiffness matrix for the grid element, the procedure for solution then follows in the same manner as that for the plane frame.

To illustrate the use of the equations developed in Section 5.4, we will now solve the following grid structures.

Example 5.5

Analyze the grid shown in Figure 5–18. The grid consists of three elements, is fixed at nodes 2, 3, and 4, and is subjected to a downward vertical force (perpendicular to the x-z plane passing through the grid elements) of 400 kN. The global-coordinate axes have been established at node 3, and the element lengths are shown in the figure. Let $E = 200$ GPa, $G = 80$ GPa, $I = 150 \times 10^6$ mm^4, and $J = 40 \times 10^6$ mm^4 for all elements of the grid.

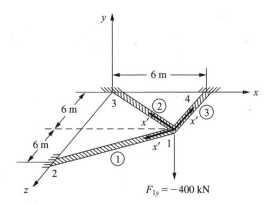

Figure 5–18 Grid for analysis showing local x' axis for each element

SOLUTION:

Substituting Eq. (5.4.17) for the local stiffness matrix and Eq. (5.4.18) for the transformation matrix into Eq. (5.4.19), we can obtain each element global stiffness matrix. To expedite the longhand solution, the boundary conditions at nodes 2, 3, and 4

$$v_2 = \phi_{2x} = \phi_{2z} = 0 \qquad v_3 = \phi_{3x} = \phi_{3z} = 0 \qquad v_4 = \phi_{4x} = \phi_{4z} = 0 \qquad (5.4.20)$$

make it possible to use only the upper left-hand 3×3 partitioned part of the local stiffness and transformation matrices associated with the degrees of freedom at node 1. Therefore, the global stiffness matrices for each element are as follows:

Element 1

For element 1, we assume the local x' axis to be directed from node 1 to node 2 for the formulation of the element stiffness matrix. We need the following expressions to evaluate the element stiffness matrix:

$$C = \cos \theta = \frac{x_2 - x_1}{L^{(1)}} = \frac{-6 - 0}{6.708} = -0.894$$

$$S = \sin \theta = \frac{z_2 - z_1}{L^{(1)}} = \frac{3 - 0}{6.708} = 0.447$$

$$\frac{12EI}{L^3} = \frac{12(200 \times 10^3)(150 \times 10^6)}{(6.708 \times 10^3)^3} = 0.00119 \times 10^6$$

$$\frac{6EI}{L^2} = \frac{6(200 \times 10^3)(150 \times 10^6)}{(6.708 \times 10^3)^2} = 4.0 \times 10^6$$

$$\frac{GJ}{L} = \frac{(80 \times 10^3)(40 \times 10^6)}{(6.708 \times 10^3)} = 477.04 \times 10^6$$

$$\frac{4EI}{L} = \frac{4(200 \times 10^3)(150 \times 10^6)}{(6.708 \times 10^3)} = 17889 \times 10^6$$

(5.4.21)

Considering the boundary condition Eqs. (5.4.20), using the results of Eqs. (5.4.21) in Eq. (5.4.17) for $[k_G']$ and Eq. (5.4.18) for $[T_G]$, and then applying Eq. (5.4.19), we obtain the upper left-hand 3×3 partitioned part of the global stiffness matrix for element 1 as

$$[k_G^{(1)}] = \begin{bmatrix} 1 & 0 & 0 \\ 0 & -0.894 & -0.447 \\ 0 & 0.447 & -0.894 \end{bmatrix} (10^6) \begin{bmatrix} 0.0012 & 0 & 4.0 \\ 0 & 477.04 & 0 \\ 4.0 & 0 & 17889 \end{bmatrix} \begin{bmatrix} 1 & 0 & 0 \\ 0 & -0.894 & 0.447 \\ 0 & -0.447 & -0.894 \end{bmatrix}$$

Performing the matrix multiplications, we obtain the global element grid stiffness matrix

$$[k_G^{(1)}] = (10^3) \begin{matrix} v_1 & \phi_1 & \phi_2 \\ \begin{bmatrix} 0.0012 & -1.788 & -3.576 \\ -1.788 & 3950 & 6958.13 \\ -3.576 & 6958.13 & 14392.84 \end{bmatrix} \end{matrix} \frac{\text{kN}}{\text{mm}}$$

(5.4.22)

where the labels next to the columns indicate the degrees of freedom.

Element 2

For element 2, we assume the local x' axis to be directed from node 1 to node 3 for the formulation of the element stiffness matrix. We need the following expressions to evaluate the element stiffness matrix:

$$C = \frac{x_3 - x_1}{L^{(2)}} = \frac{-6 - 0}{6.708} = -0.894$$

$$S = \frac{z_3 - z_1}{L^{(2)}} = \frac{-6 - 0}{6.708} = -0.447$$

(5.4.23)

Other expressions used in Eq. (5.4.17) are identical to those in Eqs. (5.4.21) for element 1 because $E, G, I, J,$ and L are identical. Evaluating Eq. (5.4.19) for the global stiffness matrix for element 2, we obtain

$$[k_G^{(2)}] = \begin{bmatrix} 1 & 0 & 0 \\ 0 & -0.894 & 0.447 \\ 0 & -0.447 & -0.894 \end{bmatrix}(10^6)\begin{bmatrix} 0.0012 & 0 & 4.0 \\ 0 & 477.04 & 0 \\ 4.0 & 0 & 17889 \end{bmatrix}\begin{bmatrix} 1 & 0 & 0 \\ 0 & -0.894 & -0.447 \\ 0 & 0.447 & -0.894 \end{bmatrix}$$

Simplifying, we obtain

$$[k_G^{(2)}] = (10^3)\begin{bmatrix} & v_1 & \phi_{1x} & \phi_{1z} & \\ & 0.0012 & 1.788 & -3.576 & \\ & 1.788 & 3950 & -6958.13 & \\ & -3.576 & -6958.13 & 14392.84 & \end{bmatrix}\frac{kN}{mm}$$

(5.4.24)

Element 3

For element 3, we assume the local x' axis to be directed from node 1 to node 4. We need the following expressions to evaluate the element stiffness matrix:

$$C = \frac{x_4 - x_1}{L^{(3)}} = \frac{6 - 6}{3} = 0$$

$$S = \frac{z_4 - z_1}{L^{(3)}} = \frac{0 - 3}{3} = -1$$

(5.4.25)

$$\frac{12EI}{L^3} = \frac{12(200 \times 10^3)(150 \times 10^6)}{(3 \times 10^3)^3} = 0.0133 \times 10^6$$

$$\frac{6EI}{L^2} = \frac{6(200 \times 10^3)(150 \times 10^6)}{(3 \times 10^3)^2} = 20 \times 10^6$$

$$\frac{GJ}{L} = \frac{(80 \times 10^3)(40 \times 10^6)}{(3 \times 10^3)} = 1066 \times 10^6$$

$$\frac{4EI}{L} = \frac{4(200 \times 10^3)(150 \times 10^6)}{(3 \times 10^6)} = 40,000 \times 10^6$$

Using Eqs. (5.4.25), we can obtain the upper part of the global stiffness matrix for element 3 as

$$[k_G^{(3)}] = (10^6) \begin{array}{c} \begin{array}{ccc} v_1 \qquad\;\; & \phi_{1x} \quad & \phi_{1z} \end{array} \\ \begin{bmatrix} 0.0133 & 20 & 0 \\ 20 & 40,000 & 0 \\ 0 & 0 & 1066 \end{bmatrix} \end{array} \frac{kN}{mm} \qquad (5.4.26)$$

Superimposing the global stiffness matrices from Eqs. (5.4.22), (5.4.24), and (5.4.26), we obtain the total stiffness matrix of the grid (with boundary conditions applied) as

$$[K_G] = (10^3) \begin{array}{c} \begin{array}{ccc} v_1 \qquad\;\; & \phi_{1x} \quad & \phi_{1z} \end{array} \\ \begin{bmatrix} 0.0157 & 20 & -7.152 \\ 20 & 47900 & 0 \\ -7.152 & 0 & 29851.68 \end{bmatrix} \end{array} \frac{kN}{mm} \qquad (5.4.27)$$

The grid matrix equation then becomes

$$\begin{Bmatrix} F_{1y} = -400 \\ M_{1x} = 0 \\ M_{1z} = 0 \end{Bmatrix} = (10^3) \begin{bmatrix} 0.0157 & 20 & -7.152 \\ 20 & 47900 & 0 \\ -7.152 & 0 & 29851.68 \end{bmatrix} \begin{Bmatrix} v_1 \\ \phi_{1x} \\ \phi_{1z} \end{Bmatrix} \qquad (5.4.28)$$

The force F_{1y} is negative because the load is applied in the negative y direction. Solving for the displacement and the rotations in Eq. (5.4.28), we obtain

$$v_1 = -70.96 \text{ mm}$$

$$\phi_{1x} = 0.0296 \text{ rad} \qquad (5.4.29)$$

$$\phi_{1z} = -0.0169 \text{ rad}$$

The results indicate that the y displacement at node 1 is downward as indicated by the minus sign, the rotation about the x axis is positive, and the rotation about the z axis is negative. Based on the downward loading location with respect to the supports, these results are expected.

Having solved for the unknown displacement and the rotations, we can obtain the local element forces on formulating the element equations in a manner similar to that for the beam and the plane frame. The local forces (which are needed in the design/analysis stage) are found by applying the equation $\{f'\} = [k'_G][T_G]\{d\}$ for each element as follows:

Element 1

Using Eqs. (5.4.17) and (5.4.18) for $[k'_G]$ and $[T_G]$ and Eq. (5.4.29), we obtain

$$[T_G]\{d\} = \begin{bmatrix} 1 & 0 & 0 & 0 & 0 & 0 \\ 0 & -0.894 & 0.447 & 0 & 0 & 0 \\ 0 & -0.447 & -0.894 & 0 & 0 & 0 \\ 0 & 0 & 0 & 1 & 0 & 0 \\ 0 & 0 & 0 & 0 & -0.894 & 0.447 \\ 0 & 0 & 0 & 0 & -0.447 & -0.894 \end{bmatrix} \begin{Bmatrix} -70.96 \\ 0.0296 \\ -0.0169 \\ 0 \\ 0 \\ 0 \end{Bmatrix}$$

Multiplying the matrices, we obtain

$$[T_G]\{d\} = \begin{Bmatrix} -70.96 \\ -0.03401 \\ 0.001877 \\ 0 \\ 0 \\ 0 \end{Bmatrix} \tag{5.4.30}$$

Then $\{f'\} = [k'_G][T_G]\{d\}$ becomes

$$\begin{Bmatrix} f'_{1y} \\ m'_{1x} \\ m'_{1z} \\ f'_{2y} \\ m'_{2x} \\ m'_{2z} \end{Bmatrix} = \begin{bmatrix} 0.0012 & 0 & 4.0 & -0.0012 & 0 & 4.0 \\ 0 & 477.04 & 0 & 0 & -477.04 & 0 \\ 4.0 & 0 & 17889 & -4.0 & 0 & 8944.5 \\ -0.0012 & 0 & -4.0 & 0.0012 & 0 & -4.0 \\ 0 & -477.04 & 0 & 0 & 477.04 & 0 \\ 4.0 & 0 & 8944.5 & -4.0 & 0 & 17889 \end{bmatrix} \begin{Bmatrix} -70.96 \\ -0.03401 \\ 0.001877 \\ 0 \\ 0 \\ 0 \end{Bmatrix} \tag{5.4.31}$$

Multiplying the matrices in Eq. (5.4.31), we obtain the local element forces as

$$\begin{Bmatrix} f'_{1y} \\ m'_{1x} \\ m'_{1z} \\ f'_{2y} \\ m'_{2x} \\ m'_{2z} \end{Bmatrix} = \begin{Bmatrix} -77.644 \text{ kN} \\ -16.244 \text{ kN-m} \\ -250.262 \text{ kN-m} \\ 77.644 \text{ kN} \\ 16.244 \text{ kN-m} \\ -267.05 \text{ kN-m} \end{Bmatrix} \tag{5.4.32}$$

The directions of the forces acting on element 1 are shown in the free-body diagram of element 1 in Figure 5–19.

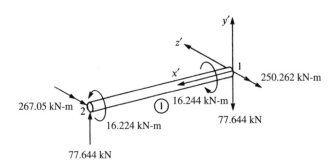

Figure 5–19 Free-body diagrams of the elements of Figure 5–18 showing local-coordinate systems for each

Element 2

Similarly, using $\{f'\} = [k'_G][T_G]\{d\}$ for element 2, with the direction cosines in Eqs. (5.4.23), we obtain

$$
\begin{Bmatrix} f'_{1y} \\ m'_{1x} \\ m'_{1z} \\ f'_{3y} \\ m'_{3x} \\ m'_{3z} \end{Bmatrix} =
\begin{bmatrix}
0.0012 & 0 & 4.0 & -0.0012 & 0 & 4.0 \\
0 & 477.04 & 0 & 0 & -477.04 & 0 \\
4.0 & 0 & 17889 & -4.0 & 0 & 8944.5 \\
-0.0012 & 0 & -4.0 & 0.0012 & 0 & -4.0 \\
0 & -477.04 & 0 & 0 & 477.04 & 0 \\
4.0 & 0 & 8944.5 & -4.0 & 0 & 17889
\end{bmatrix}
$$

$$
\times
\begin{bmatrix}
1 & 0 & 0 & 0 & 0 & 0 \\
0 & -0.894 & -0.447 & 0 & 0 & 0 \\
0 & 0.447 & -0.894 & 0 & 0 & 0 \\
0 & 0 & 0 & 1 & 0 & 0 \\
0 & 0 & 0 & 0 & -0.894 & -0.447 \\
0 & 0 & 0 & 0 & 0.447 & -0.894
\end{bmatrix}
\begin{Bmatrix}
-70.96 \\
0.0296 \\
-0.0169 \\
0 \\
0 \\
0
\end{Bmatrix}
$$

$$(5.4.33)$$

Multiplying the matrices in Eq. (5.4.33), we obtain the local element forces as

$$f'_{1y} = 28.207 \text{ kN}$$

$$m'_{1x} = -9.019 \text{ kN-m}$$

$$m'_{1z} = 223.13 \text{ kN-m}$$

$$f'_{3y} = -28.207 \text{ kN} \qquad (5.4.34)$$

$$m'_{3x} = 9.019 \text{ kN-m}$$

$$m'_{3z} = -30.354 \text{ kN-m}$$

Element 3

Finally, using the direction cosines in Eqs. (5.4.25), we obtain the local element forces as

$$
\begin{Bmatrix} f'_{1y} \\ m'_{1x} \\ m'_{1z} \\ f'_{3y} \\ m'_{3x} \\ m'_{3z} \end{Bmatrix}
=
\begin{bmatrix}
0.0133 & 0 & 20 & -0.0133 & 0 & 20 \\
0 & 1066 & 0 & 0 & -1066 & 0 \\
20 & 0 & 40{,}000 & -20 & 0 & 20{,}000 \\
-0.0133 & 0 & -20 & 0.01333 & 0 & -20 \\
0 & -1066 & 0 & 0 & 1066 & 0 \\
20 & 0 & 20{,}000 & -20 & 0 & 40{,}000
\end{bmatrix}
$$

$$
\times
\begin{bmatrix}
1 & 0 & 0 & 0 & 0 & 0 \\
0 & 0 & -1 & 0 & 0 & 0 \\
0 & 1 & 0 & 0 & 0 & 0 \\
0 & 0 & 0 & 1 & 0 & 0 \\
0 & 0 & 0 & 0 & 0 & -1 \\
0 & 0 & 0 & 0 & 1 & 0
\end{bmatrix}
\begin{Bmatrix}
-70.96 \\
0.0296 \\
-0.0169 \\
0 \\
0 \\
0
\end{Bmatrix}
\qquad (5.4.35)
$$

Multiplying the matrices in Eq. (5.4.35), we obtain the local element forces as

$$f'_{1y} = 351.76 \text{ kN}$$

$$m'_{1x} = 18.015 \text{ kN-m}$$

$$m'_{1z} = -235.2 \text{ kN-m}$$

$$f'_{4y} = 351.76 \text{ kN} \qquad (5.4.36)$$

$$m'_{4x} = -18.015 \text{ kN-m}$$

$$m'_{4z} = -827.2 \text{ kN-m}$$

Free-body diagrams for all elements are shown in Figure 5–19. Each element is in equilibrium. For each element, the x' axis is shown directed from the first node to the

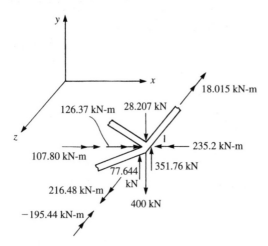

Figure 5–20 Free-body diagram of node 1 of Figure 5–18

second node, the y' axis coincides with the global y axis, and the z' axis is perpendicular to the x'-y' plane with its direction given by the right-hand rule.

To verify equilibrium of node 1, we draw a free-body diagram of the node showing all forces and moments transferred from node 1 of each element, as in Figure 5–20. In Figure 5–20, the local forces and moments from each element have been transformed to global components, and any applied nodal forces have been included. To perform this transformation, recall that, in general, $\{f'\} = [T]\{f\}$, and therefore $\{f\} = [T]^T\{f'\}$ because $[T]^T = [T]^{-1}$. Since we are transforming forces at node 1 of each element, only the upper 3×3 part of Eq. (5.4.18) for $[T_G]$ need be applied. Therefore, by premultiplying the local element forces and moments at node 1 by the transpose of the transformation matrix for each element, we obtain the global nodal forces and moments as follows:

Element 1

$$\begin{Bmatrix} f_{1y} \\ m_{1x} \\ m_{1z} \end{Bmatrix} = \begin{bmatrix} 1 & 0 & 0 \\ 0 & -0.894 & -0.447 \\ 0 & 0.447 & -0.894 \end{bmatrix} \begin{Bmatrix} 77.644 \\ -16.244 \\ -250.262 \end{Bmatrix}$$

Simplifying, we obtain the global-coordinate force and moments as

$$f_{1y} = -77.664 \text{ kN} \qquad m_{1x} = 126.37 \text{ kN-m} \qquad m_{1z} = 216.48 \text{ kN-m} \qquad (5.4.37)$$

where $f_{1y} = f'_{1y}$ because $y = y'$.

Element 2

$$\begin{Bmatrix} f_{1y} \\ m_{1x} \\ m_{1z} \end{Bmatrix} = \begin{bmatrix} 1 & 0 & 0 \\ 0 & -0.894 & 0.447 \\ 0 & -0.447 & -0.894 \end{bmatrix} \begin{Bmatrix} 28.702 \\ -9.019 \\ 228.13 \end{Bmatrix}$$

Simplifying, we obtain the global-coordinate force and moments as

$$f_{1y} = 28.207 \text{ kN} \qquad m_{1x} = 107.80 \text{ kN-m} \qquad m_{1z} = -195.44 \text{ kN-m} \qquad (5.4.38)$$

Element 3

$$\left\{ \begin{array}{c} f_{1y} \\ m_{1x} \\ m_{1z} \end{array} \right\} = \begin{bmatrix} 1 & 0 & 0 \\ 0 & 0 & 1 \\ 0 & -1 & 0 \end{bmatrix} \left\{ \begin{array}{c} -351.76 \text{ kN-m} \\ 18.015 \text{ kN-m} \\ -235.2 \text{ kN-m} \end{array} \right\}$$

Simplifying, we obtain the global-coordinate force and moments as

$$f_{1y} = -351.76 \text{ kN} \qquad m_{1x} = -235.2 \text{ kN-m} \qquad m_{1z} = 18.015 \text{ kN-m} \qquad (5.4.39)$$

Then forces and moments from each element that are equal in magnitude but opposite in sign will be applied to node 1. Hence, the free-body diagram of node 1 is shown in Figure 5–20. Force and moment equilibrium are verified as follows:

$$\sum F_{1y} = -400 - 28.207 + 77.644 + 351.76 = 1.197 \text{ kN} \qquad \text{(close to zero)}$$

$$\sum M_{1x} = -126.37 - 107.80 + 235 = 0.83 \text{ kN} \qquad \text{(close to zero)}$$

$$\sum M_{1z} = -216.48 + 195.44 + 18.015 = -3.025 \text{ kN} \qquad \text{(close to zero)}$$

Thus, we have verified the solution to be correct within the accuracy associated with a longhand solution. ∎

Example 5.6

Analyze the grid shown in Figure 5–21. The grid consists of two elements, is fixed at nodes 1 and 3, and is subjected to a downward vertical load of 22 kN. The global-coordinate axes and element lengths are shown in the figure. Let $E = 210$ GPa, $G = 84$ GPa, $I = 16.6 \times 10^{-5} \text{ m}^4$, and $J = 4.6 \times 10^{-5} \text{ m}^4$.

SOLUTION:

As in Example 5.5, we use the boundary conditions and express only the part of the stiffness matrix associated with the degrees of freedom at node 2. The boundary conditions at nodes 1 and 3 are

$$v_1 = \phi_{1x} = \phi_{1z} = 0 \qquad v_3 = \phi_{3x} = \phi_{3z} = 0 \qquad (5.4.40)$$

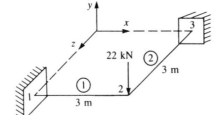

Figure 5–21   Grid example

The global stiffness matrices for each element are obtained as follows:

Element 1

For element 1, we have the local x' axis coincident with the global x axis. Therefore, we obtain

$$C = \frac{x_2 - x_1}{L^{(1)}} = \frac{3}{3} = 1 \qquad S = \frac{z_2 - z_1}{L^{(1)}} = \frac{3 - 3}{3} = 0$$

Other expressions needed to evaluate the stiffness matrix are

$$\frac{12EI}{L^3} = \frac{12(210 \times 10^6 \text{ kN/m}^2)(16.6 \times 10^{-5} \text{ m}^4)}{(3 \text{ m})^3} = 1.55 \times 10^4$$

$$\frac{6EI}{L^2} = \frac{6(210 \times 10^6)(16.6 \times 10^{-5})}{(3)^2} = 2.32 \times 10^4$$

$$\frac{GJ}{L} = \frac{(84 \times 10^6)(4.6 \times 10^{-5})}{3} = 1.28 \times 10^3 \tag{5.4.41}$$

$$\frac{4EI}{L} = \frac{4(210 \times 10^6)(16.6 \times 10^{-5})}{3} = 4.65 \times 10^4$$

Considering the boundary condition Eqs. (5.4.40), using the results of Eqs. (5.4.41) in Eq. (5.4.17) for $[k'_G]$ and Eq. (5.4.18) for $[T_G]$, and then applying Eq. (5.4.19), we obtain the reduced part of the global stiffness matrix associated only with the degrees of freedom at node 2 as

$$[k_G^{(1)}] = \begin{bmatrix} 1 & 0 & 0 \\ 0 & 1 & 0 \\ 0 & 0 & 1 \end{bmatrix} \begin{bmatrix} 1.55 & 0 & -2.32 \\ 0 & 0.128 & 0 \\ -2.32 & 0 & 4.65 \end{bmatrix} (10^4) \begin{bmatrix} 1 & 0 & 0 \\ 0 & 1 & 0 \\ 0 & 0 & 1 \end{bmatrix}$$

Since the local axes associated with element 1 are parallel to the global axes, we observe that $[T_G]$ is merely the identity matrix; therefore, $[k_G] = [k'_G]$. Performing the matrix multiplications, we obtain

$$[k_G^{(1)}] = \begin{bmatrix} 1.55 & 0 & -2.32 \\ 0 & 0.128 & 0 \\ -2.32 & 0 & 4.65 \end{bmatrix} (10^4) \frac{\text{kN}}{\text{m}} \tag{5.4.42}$$

Element 2

For element 2, we assume the local x' axis to be directed from node 2 to node 3 for the formulation of $[k_G]$. Therefore,

$$C = \frac{x_3 - x_2}{L^{(2)}} = \frac{0 - 0}{3} = 0 \qquad S = \frac{z_3 - z_2}{L^{(2)}} = \frac{0 - 3}{3} = -1 \tag{5.4.43}$$

Other expressions used in Eq. (5.4.17) are identical to those obtained in Eqs. (5.4.41) for element 1. Evaluating Eq. (5.4.19) for the global stiffness matrix, we obtain

$$[k_G^{(2)}] = \begin{bmatrix} 1 & 0 & 0 \\ 0 & 0 & 1 \\ 0 & -1 & 0 \end{bmatrix} \begin{bmatrix} 1.55 & 0 & 2.32 \\ 0 & 0.128 & 0 \\ 2.32 & 0 & 4.65 \end{bmatrix} (10^4) \begin{bmatrix} 1 & 0 & 0 \\ 0 & 0 & -1 \\ 0 & 1 & 0 \end{bmatrix}$$

where the reduced part of $[k_G]$ is now associated with node 2 for element 2. Again performing the matrix multiplications, we have

$$[k_G^{(2)}] = \begin{bmatrix} 1.55 & 2.32 & 0 \\ 2.32 & 4.65 & 0 \\ 0 & 0 & 0.128 \end{bmatrix} (10^4) \frac{kN}{m} \tag{5.4.44}$$

Superimposing the global stiffness matrices from Eqs. (5.4.42) and (5.4.44), we obtain the total global stiffness matrix (with boundary conditions applied) as

$$[K_G] = \begin{bmatrix} 3.10 & 2.32 & -2.32 \\ 2.32 & 4.78 & 0 \\ -2.32 & 0 & 4.78 \end{bmatrix} (10^4) \frac{kN}{m} \tag{5.4.45}$$

The grid matrix equation becomes

$$\begin{Bmatrix} F_{2y} = -22 \\ M_{2x} = 0 \\ M_{2z} = 0 \end{Bmatrix} = \begin{bmatrix} 3.10 & 2.32 & -2.32 \\ 2.32 & 4.78 & 0 \\ -2.32 & 0 & 4.78 \end{bmatrix} \begin{Bmatrix} v_2 \\ \phi_{2x} \\ \phi_{2z} \end{Bmatrix} (10^4) \tag{5.4.46}$$

Solving for the displacement and the rotations in Eq. (5.4.46), we obtain

$$v_2 = -0.259 \times 10^{-2} \text{ m}$$

$$\phi_{2x} = 0.126 \times 10^{-2} \text{ rad} \tag{5.4.47}$$

$$\phi_{2z} = -0.126 \times 10^{-2} \text{ rad}$$

We determine the local element forces by applying the local equation $\{f'\} = [k_G'][T_G]\{d\}$ for each element as follows:

Element 1

Using Eq. (5.4.17) for $[k_G']$, Eq. (5.4.18) for $[T_G]$, and Eqs. (5.4.47), we obtain

$$[T_G]\{d\} = \begin{bmatrix} 1 & 0 & 0 & 0 & 0 & 0 \\ 0 & 1 & 0 & 0 & 0 & 0 \\ 0 & 0 & 1 & 0 & 0 & 0 \\ 0 & 0 & 0 & 1 & 0 & 0 \\ 0 & 0 & 0 & 0 & 1 & 0 \\ 0 & 0 & 0 & 0 & 0 & 1 \end{bmatrix} \begin{Bmatrix} 0 \\ 0 \\ 0 \\ -0.259 \times 10^{-2} \\ 0.126 \times 10^{-2} \\ -0.126 \times 10^{-2} \end{Bmatrix}$$

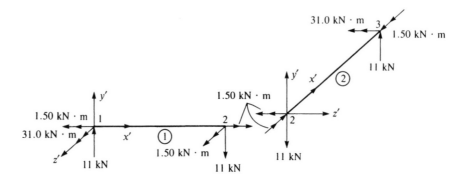

Figure 5–22 Free-body diagram of each element of Figure 5–21

Multiplying the matrices, we have

$$[T_G]\{d\} = \begin{Bmatrix} 0 \\ 0 \\ 0 \\ -0.259 \times 10^{-2} \\ 0.126 \times 10^{-2} \\ -0.126 \times 10^{-2} \end{Bmatrix} \tag{5.4.48}$$

Using Eqs. (5.4.17), (5.4.41), and (5.4.48), we obtain the local element forces as

$$\begin{Bmatrix} f'_{1y} \\ m'_{1x} \\ m'_{1z} \\ f'_{2y} \\ m'_{2x} \\ m'_{2z} \end{Bmatrix} = (10^4) \begin{bmatrix} 1.55 & 0 & 2.32 & -1.55 & 0 & 2.32 \\ & 0.128 & 0 & 0 & -0.128 & 0 \\ & & 4.65 & -2.32 & 0 & 2.33 \\ & & & 1.55 & 0 & -2.32 \\ & & & & 0.128 & 0 \\ \text{Symmetry} & & & & & 4.65 \end{bmatrix} \begin{Bmatrix} 0 \\ 0 \\ 0 \\ -0.259 \times 10^{-2} \\ 0.126 \times 10^{-2} \\ -0.126 \times 10^{-2} \end{Bmatrix} \tag{5.4.49}$$

Multiplying the matrices in Eq. (5.4.49), we obtain

$$f'_{1y} = 11.0 \text{ kN} \qquad m'_{1x} = -1.50 \text{ kN} \cdot \text{m} \qquad m'_{1z} = 31.0 \text{ kN} \cdot \text{m}$$
$$f'_{2y} = -11.0 \text{ kN} \qquad m'_{2x} = 1.50 \text{ kN} \cdot \text{m} \qquad m'_{2z} = 1.50 \text{ kN} \cdot \text{m} \tag{5.4.50}$$

Element 2

We can obtain the local element forces for element 2 in a similar manner. Because the procedure is the same as that used to obtain the element 1 local forces, we will not show the details but will only list the final results:

$$f'_{2y} = -11.0 \text{ kN} \qquad m'_{2x} = 1.50 \text{ kN} \cdot \text{m} \qquad m'_{2z} = -1.50 \text{ kN} \cdot \text{m}$$
$$f'_{3y} = 11.0 \text{ kN} \qquad m'_{3x} = -1.50 \text{ kN} \cdot \text{m} \qquad m'_{3z} = -31.0 \text{ kN} \cdot \text{m} \tag{5.4.51}$$

Free-body diagrams showing the local element forces are shown in Figure 5–22. ∎

▲ 5.5 Beam Element Arbitrarily Oriented in Space ▲

In this section, we develop the stiffness matrix for the beam element arbitrarily oriented in space, or three dimensions. This element can then be used to analyze frames in three-dimensional space.

First we consider bending about two axes, as shown in Figure 5–23.

We establish the following sign convention for the axes. Now we choose positive x' from node 1 to 2. Then y' is the principal axis for which the moment of inertia is minimum, I_y. By the right-hand rule we establish z', and the maximum moment of inertia is I_z.

Bending in x'-z' Plane

First consider bending in the x'-z' plane due to m'_y. Then clockwise rotation ϕ'_y is in the same sense as before for single bending. The stiffness matrix due to bending in the x'-z' plane is then

$$[k'_y] = \frac{EI_y}{L^4} \begin{bmatrix} 12L & -6L^2 & -12L & -6L^2 \\ & 4L^3 & 6L^2 & 2L^3 \\ & & 12L & 6L^2 \\ \text{Symmetry} & & & 4L^3 \end{bmatrix} \quad (5.5.1)$$

where I_y is the moment of inertia of the cross section about the principal axis y', the weak axis; that is, $I_y < I_z$.

Bending in the x'-y' Plane

Now we consider bending in the x'-y' plane due to m'_z. Now positive rotation ϕ'_z is counterclockwise instead of clockwise. Therefore, some signs change in the stiffness

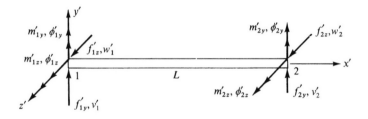

Figure 5–23 Bending about two axes y' and z'

matrix for bending in the x'-z' plane. The resulting stiffness matrix is

$$[k'_z] = \frac{EI_z}{L^4} \begin{bmatrix} 12L & 6L^2 & -12L & 6L^2 \\ & 4L^3 & -6L^2 & 2L^3 \\ & & 12L & -6L^2 \\ \text{Symmetry} & & & 4L^3 \end{bmatrix} \tag{5.5.2}$$

Direct superposition of Eqs. (5.5.1) and (5.5.2) with the axial stiffness matrix Eq. (3.1.14) and the torsional stiffness matrix Eq. (5.4.14) yields the element stiffness matrix for the beam or frame element in three-dimensional space as

$$[k'] = $$

	u'_1	v'_1	w'_1	ϕ'_{1x}	ϕ'_{1y}	ϕ'_{1z}	u'_2	v'_2	w'_2	ϕ'_{2x}	ϕ'_{2y}	ϕ'_{2z}
$\dfrac{AE}{L}$	0	0	0	0	0	$-\dfrac{AE}{L}$	0	0	0	0	0
0	$\dfrac{12EI_z}{L^3}$	0	0	0	$\dfrac{6EI_z}{L^2}$	0	$-\dfrac{12EI_z}{L^3}$	0	0	0	$\dfrac{6EI_z}{L^2}$
0	0	$\dfrac{12EI_y}{L^3}$	0	$-\dfrac{6EI_y}{L^2}$	0	0	0	$-\dfrac{12EI_y}{L^3}$	0	$-\dfrac{6EI_y}{L^2}$	0
0	0	0	$\dfrac{GJ}{L}$	0	0	0	0	0	$-\dfrac{GJ}{L}$	0	0
0	0	$-\dfrac{6EI_y}{L^2}$	0	$\dfrac{4EI_y}{L}$	0	0	0	$\dfrac{6EI_y}{L^2}$	0	$\dfrac{2EI_y}{L}$	0
0	$\dfrac{6EI_z}{L^2}$	0	0	0	$\dfrac{4EI_z}{L}$	0	$-\dfrac{6EI_z}{L^2}$	0	0	0	$\dfrac{2EI_z}{L}$
$-\dfrac{AE}{L}$	0	0	0	0	0	$\dfrac{AE}{L}$	0	0	0	0	0
0	$-\dfrac{12EI_z}{L^3}$	0	0	0	$-\dfrac{6EI_z}{L^2}$	0	$\dfrac{12EI_z}{L^3}$	0	0	0	$-\dfrac{6EI_z}{L^2}$
0	0	$-\dfrac{12EI_y}{L^3}$	0	$\dfrac{6EI_y}{L^2}$	0	0	0	$\dfrac{12EI_y}{L^3}$	0	$\dfrac{6EI_y}{L^2}$	0
0	0	0	$-\dfrac{GJ}{L}$	0	0	0	0	0	$\dfrac{GJ}{L}$	0	0
0	0	$-\dfrac{6EI_y}{L^2}$	0	$\dfrac{2EI_y}{L}$	0	0	0	$\dfrac{6EI_y}{L^2}$	0	$\dfrac{4EI_y}{L}$	0
0	$\dfrac{6EI_z}{L^2}$	0	0	0	$\dfrac{2EI_z}{L}$	0	$-\dfrac{6EI_z}{L^2}$	0	0	0	$\dfrac{4EI_z}{L}$

$$(5.5.3)$$

The transformation from local to global axis system is accomplished as follows:

$$[k] = [T]^T[k'][T] \tag{5.5.4}$$

where $[k']$ is given by Eq. (5.5.3) and $[T]$ is given by

$$[T] = \begin{bmatrix} [\lambda]_{3\times3} & & & \\ & [\lambda]_{3\times3} & & \\ & & [\lambda]_{3\times3} & \\ & & & [\lambda]_{3\times3} \end{bmatrix} \tag{5.5.5}$$

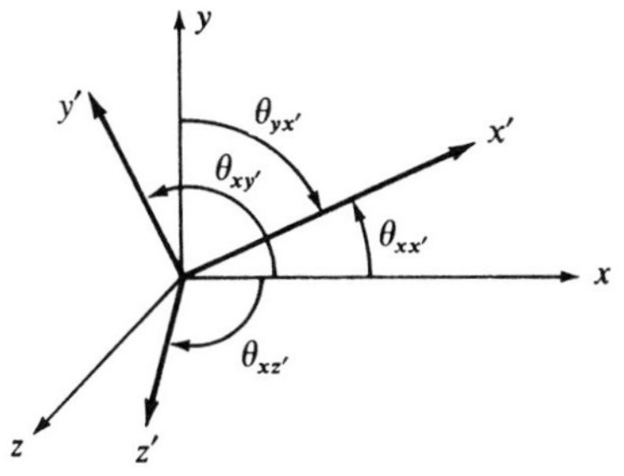

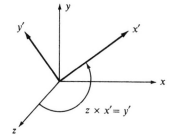

Figure 5–24 Direction cosines associated with the x axis

Figure 5–25 Illustration showing how local y' axis is determined

where

$$[\lambda] = \begin{bmatrix} C_{xx'} & C_{yx'} & C_{zx'} \\ C_{xy'} & C_{yy'} & C_{zy'} \\ C_{xz'} & C_{yz'} & C_{zz'} \end{bmatrix}$$

(5.5.6)

Here $C_{yx'}$ and $C_{xy'}$ are not necessarily equal. The direction cosines are shown in part in Figure 5–24.

Remember that direction cosines of the x' axis member are

$$x' = \cos\theta_{xx'}\mathbf{i} + \cos\theta_{yx'}\mathbf{j} + \cos\theta_{zx'}\mathbf{k}$$

(5.5.7)

where

$$\cos\theta_{xx'} = \frac{x_2 - x_1}{L} = l$$

$$\cos\theta_{yx'} = \frac{y_2 - y_1}{L} = m$$

(5.5.8)

$$\cos\theta_{zx'} = \frac{z_2 - z_1}{L} = n$$

The y' axis is selected to be perpendicular to the x' and z axes in such a way that the cross product of global z with x' results in the y' axis, as shown in Figure 5–25. Therefore,

$$z \times x' = y' = \frac{1}{D}\begin{vmatrix} \mathbf{i} & \mathbf{j} & \mathbf{k} \\ 0 & 0 & 1 \\ l & m & n \end{vmatrix}$$

(5.5.9)

$$y' = -\frac{m}{D}\mathbf{i} + \frac{l}{D}\mathbf{j}$$

(5.5.10)

and

$$D = (l^2 + m^2)^{1/2}$$

The z' axis will be determined by the orthogonality condition $z' = x' \times y'$ as follows:

$$z' = x' \times y' = \frac{1}{D}\begin{vmatrix} \mathbf{i} & \mathbf{j} & \mathbf{k} \\ l & m & n \\ -m & l & 0 \end{vmatrix}$$

(5.5.11)

or

$$z' = -\frac{ln}{D}\mathbf{i} - \frac{mn}{D}\mathbf{j} + D\mathbf{k} \tag{5.5.12}$$

Combining Eqs. (5.5.7), (5.5.10), and (5.5.12), the 3×3 transformation matrix becomes

$$[\lambda]_{3 \times 3} = \begin{bmatrix} l & m & n \\ -\dfrac{m}{D} & \dfrac{l}{D} & 0 \\ -\dfrac{ln}{D} & -\dfrac{mn}{D} & D \end{bmatrix} \tag{5.5.13}$$

This vector $[\lambda]$ rotates a vector from the local coordinate system into the global one. This is the $[\lambda]$ used in the $[T]$ matrix. In summary, we have

$$\cos\theta_{xy'} = -\frac{m}{D}$$

$$\cos\theta_{yy'} = \frac{l}{D}$$

$$\cos\theta_{zy'} = 0$$

$$\cos\theta_{xz'} = -\frac{ln}{D} \tag{5.5.14}$$

$$\cos\theta_{yz'} = -\frac{mn}{D}$$

$$\cos\theta_{zz'} = D$$

Two exceptions arise when local and global axes have special orientations with respect to each other. If the local x' axis coincides with the global z axis, then the member is parallel to the global z axis and the y' axis becomes uncertain, as shown in Figure 5–26(a). In this case the local y' axis is selected as the global y axis. Then, for

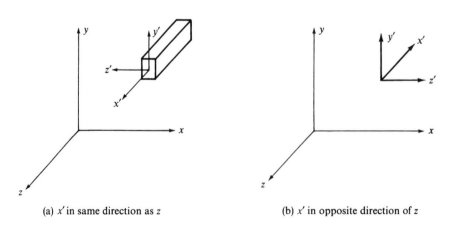

(a) x' in same direction as z

(b) x' in opposite direction of z

Figure 5–26 Special cases of transformation matrices

the positive x' axis in the same direction as the global z, $[\lambda]$ becomes

$$[\lambda] = \begin{bmatrix} 0 & 0 & 1 \\ 0 & 1 & 0 \\ -1 & 0 & 0 \end{bmatrix} \tag{5.5.15}$$

For the positive x' axis opposite the global z [Figure 5–26(b)], $[\lambda]$ becomes

$$[\lambda] = \begin{bmatrix} 0 & 0 & -1 \\ 0 & 1 & 0 \\ 1 & 0 & 0 \end{bmatrix} \tag{5.5.16}$$

Example 5.7

Determine the direction cosines and the rotation matrix of the local x', y', z' axes in reference to the global x, y, z axes for the beam element oriented in space with end nodal coordinates of 1 (0, 0, 0) and 2 (3, 4, 12), as shown in Figure 5–27.

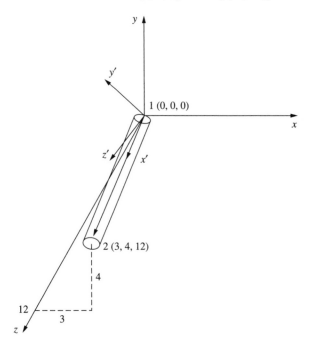

Figure 5–27 Beam element oriented in space

SOLUTION:

First we determine the length of the element as

$$L = \sqrt{3^2 + 4^2 + 12^2} = 13$$

Now using Eq. (5.5.8), we obtain the direction cosines of the x' axis as follows:

$$l_x = \frac{x_2 - x_1}{L} = \frac{3 - 0}{13} = \frac{3}{13}$$

$$m_x = \frac{y_2 - y_1}{L} = \frac{4 - 0}{13} = \frac{4}{13} \tag{5.5.17}$$

$$n_x = \frac{z_2 - z_1}{L} = \frac{12 - 0}{13} = \frac{12}{13}$$

By Eq. (5.5.10) or (5.5.14), we obtain the direction cosines of the y' axis as follows:

$$D = (l^2 + m^2)^{1/2} = \left[\left(\frac{3}{13}\right)^2 + \left(\frac{4}{13}\right)^2\right]^{1/2} = \frac{5}{13} \tag{5.5.18}$$

Define the direction cosines of the y' axis as l_y, m_y, and n_y, where

$$l_y = -\frac{m}{D} = -\frac{4}{5}$$

$$m_y = \frac{l}{D} = \frac{3}{5} \tag{5.5.19}$$

$$n_y = 0$$

For the z' axis, define the direction cosines as l_z, m_z, n_z and again use Eq. (5.5.12) or (5.5.14) as follows:

$$l_z = -\frac{ln}{D} = \frac{\left(-\frac{3}{13}\right)\left(\frac{12}{13}\right)}{\frac{5}{13}} = -\frac{36}{65}$$

$$m_z = -\frac{mn}{D} = \frac{\left(-\frac{4}{13}\right)\left(\frac{12}{13}\right)}{\frac{5}{13}} = -\frac{48}{65} \tag{5.5.20}$$

$$n_z = D = \frac{5}{13}$$

Now check that $l^2 + m^2 + n^2 = 1$.

$$\text{For } x' : \frac{3^2 + 4^2 + 12^2}{13^2} = 1$$

$$\text{For } y' : \frac{(-4)^2 + 3^2}{5^2} = 1 \tag{5.5.21}$$

$$\text{For } z' : \left(-\frac{36}{65}\right)^2 + \left(-\frac{48}{65}\right)^2 + \left(\frac{25}{65}\right)^2 = 1$$

By Eq. (5.5.13), the rotation matrix is

$$[\lambda]_{3 \times 3} = \begin{bmatrix} \frac{3}{13} & \frac{4}{13} & \frac{12}{13} \\ -\frac{4}{5} & \frac{3}{5} & 0 \\ -\frac{36}{65} & -\frac{48}{65} & \frac{5}{13} \end{bmatrix} \tag{5.5.22}$$

Based on the resulting direction cosines from Eqs. (5.5.17), (5.5.19), and (5.5.20), the local axes are also shown in Figure 5–27. ∎

Example 5.8

Determine the displacements and rotations at the free node (node 1) and the element local forces and moments for the space frame shown in Figure 5–28. Also verify equilibrium at node 1. Let $E = 200$ GPa, $G = 60$ GPa, $J = 20 \times 10^{-6}$ m⁴, $I_y = 40 \times 10^{-6}$ m⁴, $I_z = 40 \times 10^{-6}$ m⁴, $A = 6.25 \times 10^{-3}$ m², and $L = 2.5$ m for all three beam elements.

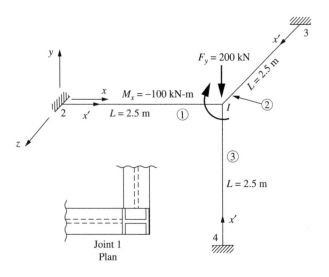

Figure 5–28 Space frame for analysis

SOLUTION:

Use Eq. (5.5.4) to obtain the global stiffness matrix for each element. This requires us to first use Eq. (5.5.3) to obtain each local stiffness matrix, Eq. (5.5.5) to obtain the transformation matrix for each element, and Eqs. (5.5.6) and (5.5.14) to obtain the direction cosine matrix for each element.

Element 1

We establish the local x' axis to go from node 2 to node 1 as shown in Figure 5–28. Therefore, using Eq. (5.5.8), we obtain the direction cosines of the x' axis as follows:

$$l = 1 \quad m = 0 \quad n = 0 \tag{5.5.23}$$

Also,

$$D = (l^2 + m^2)^{1/2} = 1$$

Using Eqs. (5.5.10) and (5.5.14), we obtain the direction cosines of the y' axis as follows:

$$l_y = -\frac{m}{D} = 0 \qquad m_y = \frac{l}{D} = 1 \qquad n_y = 0 \tag{5.5.24}$$

Using Eqs. (5.5.12) and (5.5.14), we obtain the direction cosines of the z' axis as follows:

$$l_z = -\frac{ln}{D} = 0 \qquad m_z = -\frac{mn}{D} = 0 \qquad n_z = D = 1 \qquad (5.5.25)$$

Using Eqs. (5.5.23) through (5.5.25) in Eq. (5.5.13), we have

$$[\lambda] = \begin{bmatrix} 1 & 0 & 0 \\ 0 & 1 & 0 \\ 0 & 0 & 1 \end{bmatrix} \qquad (5.5.26)$$

Using Eq. (5.5.3), we obtain the local stiffness matrix for element 1 as

	u_2'	v_2'	w_2'	ϕ_{2x}'	ϕ_{2y}'	ϕ_{2z}'	u_1'	v_1'	w_1'	ϕ_{1x}'	ϕ_{1y}'	ϕ_{1z}'
	500.00	0	0	0	0	0	-500.00	0	0	0	0	0
	0	6.144	0	0	0	7.68	0	-6.144	0	0	0	7.68
	0	0	6.144	0	-7.68	0	0	0	-6.144	0	-7.68	0
	0	0	0	0.486	0	0	0	0	0	-0.486	0	0
	0	0	-7.68	0	12.8	0	0	0	7.68	0	6.40	0
$[k'^{(1)}] =$	0	7.68	0	0	0	12.8	0	-7.68	0	0	0	6.40
	-500.00	0	0	0	0	0	500.00	0	0	0	0	0
	0	-6.144	0	0	0	-7.68	0	6.144	0	0	0	-7.68
	0	0	-6.144	0	7.68	0	0	0	6.144	0	7.68	0
	0	0	0	-0.486	0	0	0	0	0	0.486	0	0
	0	0	-7.68	0	6.4	0	0	0	7.68	0	12.8	0
	0	7.68	0	0	0	6.4	0	-7.68	0	0	0	12.8

$$(5.5.27)$$

Using Eq. (5.5.26) in Eq. (5.5.5), we obtain the transformation matrix from local to global axis system as

$$[T] = \begin{bmatrix}
1 & 0 & 0 & 0 & 0 & 0 & 0 & 0 & 0 & 0 & 0 & 0 \\
0 & 1 & 0 & 0 & 0 & 0 & 0 & 0 & 0 & 0 & 0 & 0 \\
0 & 0 & 1 & 0 & 0 & 0 & 0 & 0 & 0 & 0 & 0 & 0 \\
0 & 0 & 0 & 1 & 0 & 0 & 0 & 0 & 0 & 0 & 0 & 0 \\
0 & 0 & 0 & 0 & 1 & 0 & 0 & 0 & 0 & 0 & 0 & 0 \\
0 & 0 & 0 & 0 & 0 & 1 & 0 & 0 & 0 & 0 & 0 & 0 \\
0 & 0 & 0 & 0 & 0 & 0 & 1 & 0 & 0 & 0 & 0 & 0 \\
0 & 0 & 0 & 0 & 0 & 0 & 0 & 1 & 0 & 0 & 0 & 0 \\
0 & 0 & 0 & 0 & 0 & 0 & 0 & 0 & 1 & 0 & 0 & 0 \\
0 & 0 & 0 & 0 & 0 & 0 & 0 & 0 & 0 & 1 & 0 & 0 \\
0 & 0 & 0 & 0 & 0 & 0 & 0 & 0 & 0 & 0 & 1 & 0 \\
0 & 0 & 0 & 0 & 0 & 0 & 0 & 0 & 0 & 0 & 0 & 1
\end{bmatrix} \qquad (5.5.28)$$

Finally, using Eq. (5.5.4), we obtain the global stiffness matrix for element 1 as

$$[k^{(1)}] = [T]^T [k'^{(1)}][T] =$$

	u_2	v_2	w_2	ϕ_{2x}	ϕ_{2y}	ϕ_{2z}	u_1	v_1	w_1	ϕ_{1x}	ϕ_{1y}	ϕ_{1z}
	500	0	0	0	0	0	-500	0	0	0	0	0
	0	6.144	0	0	0	7.68	0	-6.144	0	0	0	7.68
	0	0	6.144	0	-7.68	0	0	0	-6.144	0	-7.68	0
	0	0	0	0.486	0	0	0	0	0	-0.486	0	0
	0	0	-7.68	0	12.8	0	0	0	7.68	0	6.40	0
	0	7.68	0	0	0	12.8	0	-7.68	0	0	0	6.40
	-500	0	0	0	0	0	500	0	0	0	0	0
	0	-6.144	0	0	0	-7.68	0	6.144	0	0	0	-7.68
	0	0	-6.144	0	7.68	0	0	0	6.144	0	7.68	0
	0	0	0	-0.486	0	0	0	0	0	0.486	0	0
	0	0	-7.68	0	6.4	0	0	0	7.68	0	12.8	0
	0	7.68	0	0	0	6.4	0	-7.68	0	0	0	12.8

$$(5.5.29)$$

Element 2

We establish the local x' axis from node 3 to node 1 as shown in Figure 5–28. We note that the local x' axis coincides with the global z axis. Therefore, by Eq. (5.5.15), we obtain

$$[\lambda] = \begin{bmatrix} 0 & 0 & 1 \\ 0 & 1 & 0 \\ -1 & 0 & 0 \end{bmatrix} \qquad (5.5.30)$$

The local stiffness matrix is the same as the one in Eq. (5.5.27) as all properties are the same as for element 1. However, we must remember that the degrees of freedom are for node 3 and then node 1.

Using Eq. (5.5.30) in Eq. (5.5.5), we obtain the transformation matrix as follows:

$$[T] = \begin{bmatrix} 0 & 0 & 1 & 0 & 0 & 0 & 0 & 0 & 0 & 0 & 0 & 0 \\ 0 & 1 & 0 & 0 & 0 & 0 & 0 & 0 & 0 & 0 & 0 & 0 \\ -1 & 0 & 0 & 0 & 0 & 0 & 0 & 0 & 0 & 0 & 0 & 0 \\ 0 & 0 & 0 & 0 & 0 & 1 & 0 & 0 & 0 & 0 & 0 & 0 \\ 0 & 0 & 0 & 0 & 1 & 0 & 0 & 0 & 0 & 0 & 0 & 0 \\ 0 & 0 & 0 & -1 & 0 & 0 & 0 & 0 & 0 & 0 & 0 & 0 \\ 0 & 0 & 0 & 0 & 0 & 0 & 0 & 0 & 1 & 0 & 0 & 0 \\ 0 & 0 & 0 & 0 & 0 & 0 & 0 & 1 & 0 & 0 & 0 & 0 \\ 0 & 0 & 0 & 0 & 0 & 0 & -1 & 0 & 0 & 0 & 0 & 0 \\ 0 & 0 & 0 & 0 & 0 & 0 & 0 & 0 & 0 & 0 & 0 & 1 \\ 0 & 0 & 0 & 0 & 0 & 0 & 0 & 0 & 0 & 0 & 1 & 0 \\ 0 & 0 & 0 & 0 & 0 & 0 & 0 & 0 & 0 & -1 & 0 & 0 \end{bmatrix} \qquad (5.5.31)$$

Finally, using Eq. (5.5.31) in Eq. (5.5.4), we obtain the global stiffness matrix for element 2 as

$$[k^{(2)}] =$$

	u_3	v_3	w_3	ϕ_{3x}	ϕ_{3y}	ϕ_{3z}	u_1	v_1	w_1	ϕ_{1x}	ϕ_{1y}	ϕ_{1z}
	6.144	0	0	0	7.68	0	-6.144	0	0	0	7.68	0
	0	6.144	0	-7.68	0	0	0	-6.144	0	-7.68	0	0
	0	0	500	0	0	0	0	0	-500	0	0	0
	0	-7.68	0	12.8	0	0	0	7.68	0	6.4	0	0
	7.68	0	0	0	12.8	0	-7.68	0	0	0	6.4	0
	0	0	0	0	0	0.486	0	0	0	0	0	-0.486
	-6.144	0	0	0	-7.68	0	6.144	0	0	0	-7.68	0
	0	-6.144	0	7.68	0	0	0	6.144	0	7.68	0	0
	0	0	-500	0	0	0	0	0	500	0	0	0
	0	-7.68	0	6.4	0	0	0	7.68	0	12.8	0	0
	7.68	0	0	0	6.4	0	-7.68	0	0	0	12.8	0
	0	0	0	0	0	-0.486	0	0	0	0	0	0.486

$$(5.5.32)$$

Element 3

We establish the local x' axis from node 4 to node 1 for element 3 as shown in Figure 5–28. The direction cosines are now

$$l = \frac{0-0}{2.5} = 0 \qquad m = \frac{0-(-2.5)}{2.5} = 1 \qquad n = \frac{0-0}{2.5} = 0 \qquad (5.5.33)$$

Also, $D = 1$.
Using Eq. (5.5.14), we obtain the rest of the direction cosines as

$$l_y = -\frac{m}{D} = -1 \qquad m_y = \frac{L}{D} = 0 \qquad n_y = 0 \qquad (5.5.34)$$

and

$$l_z = -\frac{ln}{D} = 0 \qquad m_z = -\frac{mn}{D} = 0 \qquad n_z = D = 1 \qquad (5.5.35)$$

Using Eqs. (5.5.33) through (5.5.35), we obtain

$$[\lambda] = \begin{bmatrix} 0 & 1 & 0 \\ -1 & 0 & 0 \\ 0 & 0 & 1 \end{bmatrix} \qquad (5.5.36)$$

The transformation matrix for element 3 is then obtained by using Eq. (5.5.5) as:

$$[T] = \begin{bmatrix}
0 & 1 & 0 & 0 & 0 & 0 & 0 & 0 & 0 & 0 & 0 & 0 \\
-1 & 0 & 0 & 0 & 0 & 0 & 0 & 0 & 0 & 0 & 0 & 0 \\
0 & 0 & 1 & 0 & 0 & 0 & 0 & 0 & 0 & 0 & 0 & 0 \\
0 & 0 & 0 & 0 & 1 & 0 & 0 & 0 & 0 & 0 & 0 & 0 \\
0 & 0 & 0 & -1 & 0 & 0 & 0 & 0 & 0 & 0 & 0 & 0 \\
0 & 0 & 0 & 0 & 0 & 1 & 0 & 0 & 0 & 0 & 0 & 0 \\
0 & 0 & 0 & 0 & 0 & 0 & 0 & 1 & 0 & 0 & 0 & 0 \\
0 & 0 & 0 & 0 & 0 & 0 & -1 & 0 & 0 & 0 & 0 & 0 \\
0 & 0 & 0 & 0 & 0 & 0 & 0 & 0 & 1 & 0 & 0 & 0 \\
0 & 0 & 0 & 0 & 0 & 0 & 0 & 0 & 0 & 0 & 1 & 0 \\
0 & 0 & 0 & 0 & 0 & 0 & 0 & 0 & 0 & -1 & 0 & 0 \\
0 & 0 & 0 & 0 & 0 & 0 & 0 & 0 & 0 & 0 & 0 & 1
\end{bmatrix} \tag{5.5.37}$$

The element 3 properties are identical to the element 1 properties; therefore, the local stiffness matrix is identical to the one in Eq. (5.5.27). We must remember that the degrees of freedom are now in the order node 4 and then node 1.

Using Eq. (5.5.37) in Eq. (5.5.4), we obtain the global stiffness matrix for element 3 as

$$[k^{(3)}] = \begin{array}{cccccccccccc}
u_4 & v_4 & w_4 & \phi_{4x} & \phi_{4y} & \phi_{4z} & u_1 & v_1 & w_1 & \phi_{1x} & \phi_{1y} & \phi_{1z} \\
\end{array}$$

u_4	v_4	w_4	ϕ_{4x}	ϕ_{4y}	ϕ_{4z}	u_1	v_1	w_1	ϕ_{1x}	ϕ_{1y}	ϕ_{1z}
6.144	0	0	0	0	−7.68	−6.144	0	0	0	0	−7.68
0	500	0	0	0	0	0	−500	0	0	0	0
0	0	6.144	7.68	0	0	0	0	−6.144	7.68	0	0
0	0	7.68	12.8	0	0	0	0	−7.68	6.4	0	0
0	0	0	0	12.8	0	0	0	0	0	−0.486	0
−7.68	0	0	0	0	12.8	7.68	0	0	0	0	6.4
−6.144	0	0	0	0	7.68	6.144	0	0	0	0	7.68
0	−500	0	0	0	0	0	500	0	0	0	0
0	0	−6.144	−7.68	0	0	0	0	6.144	−7.68	0	0
0	0	7.68	6.4	0	0	0	0	−7.68	12.8	0	0
0	0	0	0	−12.8	0	0	0	0	0	0.486	0
−7.68	0	0	0	0	6.4	7.68	0	0	0	0	12.8

(5.5.38)

Applying the boundary conditions that displacements in the x, y, and z directions are all zero at nodes 2, 3, and 4, and rotations about the x, y, and z axes are all zero at nodes 2, 3, and 4, we obtain the reduced global stiffness matrix. Also, the applied global force is directed in the negative y direction at node 1 and so expressed as $F_{1y} = -200$ kN, and the global moment about the x axis at node 1 is $M_{1x} = -100$ kN-m.

With these considerations, the final global equations are

$$(10^3)\begin{Bmatrix} 0 \\ -200 \\ 0 \\ -100 \\ 0 \\ 0 \end{Bmatrix} = (10^6)\begin{bmatrix} 512.288 & 0 & 0 & 0 & -7.68 & 7.68 \\ 0 & 512.288 & 0 & 7.68 & 0 & -7.68 \\ 0 & 0 & 512.288 & -7.68 & 7.68 & 0 \\ 0 & 7.68 & -7.68 & 26.086 & 0 & 0 \\ -7.68 & 0 & 7.68 & 0 & 26.086 & 0 \\ 7.68 & -7.68 & 0 & 0 & 0 & 26.086 \end{bmatrix}\begin{Bmatrix} u_1 \\ v_1 \\ w_1 \\ \phi_{1x} \\ \phi_{1y} \\ \phi_{1z} \end{Bmatrix} \quad (5.5.39)$$

Finally, solving simultaneously for the displacements and rotations at node 1, we obtain

$$\{d\} = \begin{bmatrix} 1.747 \times 10^{-6}\,\text{m} \\ -3.357 \times 10^{-4}\,\text{m} \\ -5.65 \times 10^{-5}\,\text{m} \\ -3.751 \times 10^{-3}\,\text{rad} \\ 1.714 \times 10^{-5}\,\text{rad} \\ -9.935 \times 10^{-5}\,\text{rad} \end{bmatrix} \quad (5.5.40)$$

We now determine the element local forces and moments using the equation $\{f'\} = [k'][T]\{d\}$ for each element as previously done for plane frames and trusses. As we are dealing with space frame elements, these element local forces and moments are now the normal force, two shear forces, torsional moment, and two bending moments at each end of each element.

Element 1

Using Eq. (5.5.27) for the local stiffness matrix, Eq. (5.5.28) for the transformation matrix, $[T]$, and Eq. (5.5.40) for the displacements, we obtain the local element forces and moments as

$$\{f'^{(1)}\} = \begin{Bmatrix} -0.873\,\text{kN} \\ 1.299\,\text{kN} \\ 0.215\,\text{kN} \\ 1.822\,\text{kN-m} \\ -0.324\,\text{kN-m} \\ 1.942\,\text{kN-m} \\ 0.873\,\text{kN} \\ -1.299\,\text{kN} \\ -0.215\,\text{kN} \\ -1.822\,\text{kN-m} \\ -0.214\,\text{kN-m} \\ 1.306\,\text{kN-m} \end{Bmatrix} \quad (5.5.41)$$

Element 2

Using Eq. (5.5.27) for the local stiffness matrix, Eq. (5.5.28) for the transformation matrix, and Eq. (5.5.40) for the displacements, we obtain the local forces and moments as

$$\{f'^{(2)}\} = \begin{Bmatrix} 28.250\,\text{kN} \\ 30.87\,\text{kN} \\ -0.120\,\text{kN} \\ 0.048\,\text{kN-m} \\ 0.096\,\text{kN-m} \\ 26.584\,\text{kN-m} \\ -28.25\,\text{kN} \\ -30.87\,\text{kN} \\ 0.120\,\text{kN} \\ -0.048\,\text{kN-m} \\ 0.205\,\text{kN-m} \\ 50.59\,\text{kN-m} \end{Bmatrix} \qquad (5.5.42)$$

Element 3

Similarly, using Eqs. (5.5.27), (5.5.37), and (5.5.40), we obtain the local forces and moments as

$$\{f'^{(3)}\} = \begin{Bmatrix} 167.85\,\text{kN} \\ -0.752\,\text{kN} \\ -28.46\,\text{kN} \\ -0.0083\,\text{kN-m} \\ 23.57\,\text{kN-m} \\ -0.622\,\text{kN-m} \\ -167.85\,\text{kN} \\ 0.752\,\text{kN} \\ 28.46\,\text{kN} \\ 0.0083\,\text{kN-m} \\ 47.57\,\text{kN-m} \\ -1.258\,\text{kN-m} \end{Bmatrix} \qquad (5.5.43)$$

We can verify equilibrium of node 1 by considering the node 1 forces and moments from each element that transfer to the node. We use the results from Eqs. (5.5.41), (5.5.42), and (5.5.43) to establish the proper forces and moments transferred to node 1. (Note that based on Newton's third law, the opposite forces and moments from each element are sent to node 1.) For instance, we observe from summing forces in the global y direction (shown in the diagram that follows)

$$1.299\ \text{kN} + 30.87\ \text{kN} + 167.85\ \text{kN} - 200\ \text{kN} = 0.019 \quad (\text{close to zero}) \quad (5.5.44)$$

In Eq. (5.5.44), 1.299 kN is from the element 1 local y' force that is coincident with the global y direction; 30.87 kN is from the element 2 local y' force that is co-incident with the global y direction, while 167.85 kN is from the element 3 local x' direction that is coincident with the global y direction. We observe these axes from

Figure 5–28. Verification of the other equilibrium equations is left to your discretion.

Global y force equilibrium ■

An example using the frame element in three-dimensional space is shown in Figure 5–29. Figure 5–29 shows a bus frame subjected to a static roof-crush analysis. In this model, 599 frame elements and 357 nodes were used. A total downward load of 100 kN was uniformly spread over the 56 nodes of the roof portion of the frame. Figure 5–30 shows the rear of the frame and the displaced view of the rear frame. Other frame models with additional loads simulating rollover and front-end collisions were studied in Reference [6].

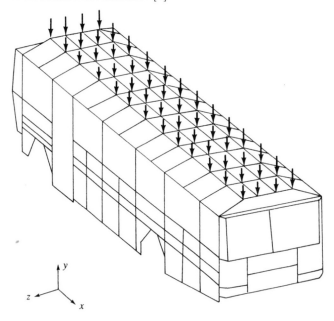

Figure 5–29 Finite element model of bus frame subjected to roof load [6]

▲ **5.6 Concept of Substructure Analysis** ▲

The problem of exceeding memory capacity on today's personal computers has decreased significantly for most applications. However, for those structures that are too large to be analyzed as a single system or treated as a whole; that is, the final

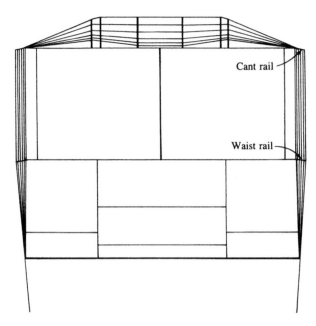

Cant rail

Waist rail

Figure 5–30 Displaced view of the frame of Figure 5–29 made of square section members

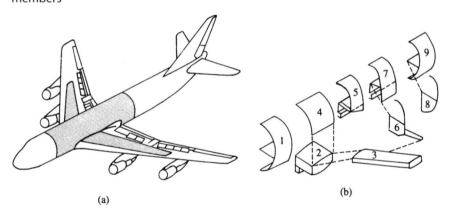

(a)

(b)

Figure 5–31 Airplane frame showing substructuring. (a) Boeing 747 aircraft (shaded area indicates portion of the airframe analyzed by finite element method). (b) Substructures for finite element analysis of shaded region

stiffness matrix and equations for solution exceed the memory capacity of the computer, the concept of **substructure** analysis can be used. The procedure to overcome this problem is to separate the whole structure into smaller units called **substructures**. For example, the space frame of an airplane, as shown in Figure 5–31(a), may require thousands of nodes and elements to model and describe completely the response of the whole structure. If we separate the aircraft into substructures, such as parts of the fuselage or body, wing sections, and so on, as shown in Figure 5–31(b), then we can solve the problem more readily and on computers with limited memory.

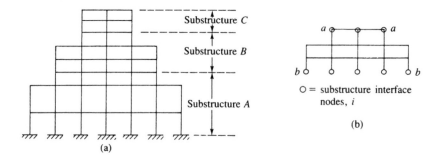

Figure 5–32 (a) Rigid frame for substructure analysis and (b) substructure B

The analysis of the airplane frame is performed by treating each substructure separately while ensuring force and displacement compatibility at the intersections where partitioning occurs.

To describe the procedure of substructuring, consider the rigid frame shown in Figure 5–32 (even though this frame could be analyzed as a whole). First we define individual separate substructures. Normally, we make these substructures of similar size, and to reduce computations, we make as few cuts as possible. We then separate the frame into three parts, A, B, and C.

We now analyze a typical substructure B shown in Figure 5–32(b). This substructure includes the beams at the top (a-a), but the beams at the bottom (b-b) are included in substructure A, although the beams at top could be included in substructure C and the beams at the bottom could be included in substructure B.

The force/displacement equations for substructure B are partitioned with the interface displacements and forces separated from the interior ones as follows:

$$\left\{ \begin{matrix} \{F_i^B\} \\ \{F_e^B\} \end{matrix} \right\} = \left[\begin{matrix} [K_{ii}^B] & [K_{ie}^B] \\ [K_{ei}^B] & [K_{ee}^B] \end{matrix} \right] \left\{ \begin{matrix} \{d_i^B\} \\ \{d_e^B\} \end{matrix} \right\} \qquad (5.6.1)$$

where the superscript B denotes the substructure B, subscript i denotes the interface nodal forces and displacements, and subscript e denotes the interior nodal forces and displacements to be eliminated by static condensation. Using static condensation, Eq. (5.6.1) becomes

$$\{F_i^B\} = [K_{ii}^B]\{d_i^B\} + [K_{ie}^B]\{d_e^B\} \qquad (5.6.2)$$

$$\{F_e^B\} = [K_{ei}^B]\{d_i^B\} + [K_{ee}^B]\{d_e^B\} \qquad (5.6.3)$$

We eliminate the interior displacements $\{d_e\}$ by solving Eq. (5.6.3) for $\{d_e^B\}$, as follows:

$$\{d_e^B\} = [K_{ee}^B]^{-1}[\{F_e^B\} - [K_{ei}^B]\{d_i^B\}] \qquad (5.6.4)$$

Then we substitute Eq. (5.6.4) for $\{d_e^B\}$ into Eq. (5.6.2) to obtain

$$\{F_i^B\} - [K_{ie}^B][K_{ee}^B]^{-1}\{F_e^B\} = \left([K_{ii}^B] - [K_{ie}^B][K_{ee}^B]^{-1}[K_{ei}^B]\right)\{d_i^B\} \qquad (5.6.5)$$

We define

$$\{\bar{F}_i^B\} = [K_{ie}^B][K_{ee}^B]^{-1}\{F_e^B\} \quad \text{and} \quad [\bar{K}_{ii}^B] = [K_{ii}^B] - [K_{ie}^B][K_{ee}^B]^{-1}[K_{ei}^B] \quad (5.6.6)$$

Substituting Eq. (5.6.6) into (5.6.5), we obtain

$$\{F_i^B\} - \{\bar{F}_i^B\} = [\bar{K}_{ii}^B]\{d_i^B\} \tag{5.6.7}$$

Similarly, we can write force/displacement equations for substructures A and C. These equations can be partitioned in a manner similar to Eq. (5.6.1) to obtain

$$\left\{ \begin{array}{c} \{F_i^A\} \\ \hline \{F_e^A\} \end{array} \right\} = \left[\begin{array}{c|c} \{K_{ii}^A\} & \{K_{ie}^A\} \\ \hline \{K_{ei}^A\} & \{K_{ee}^A\} \end{array} \right] \left\{ \begin{array}{c} \{d_i^A\} \\ \hline \{d_e^A\} \end{array} \right\} \tag{5.6.8}$$

Eliminating $\{d_e^A\}$, we obtain

$$\{F_i^A\} - \{\bar{F}_i^A\} = [\bar{K}_{ii}^A]\{d_i^A\} \tag{5.6.9}$$

Similarly, for substructure C, we have

$$\{F_i^C\} - \{\bar{F}_i^C\} = [\bar{K}_{ii}^C]\{d_i^C\} \tag{5.6.10}$$

The whole frame is now considered to be made of superelements A, B, and C connected at interface nodal points (each superelement being made up of a collection of individual smaller elements). Using compatibility, we have

$$\left\{d_{i\,\text{top}}^A\right\} = \left\{d_{i\,\text{bottom}}^B\right\} \quad \text{and} \quad \left\{d_{i\,\text{top}}^B\right\} = \left\{d_{i\,\text{bottom}}^C\right\} \tag{5.6.11}$$

That is, the interface displacements at the common locations where cuts were made must be the same.

The response of the whole structure can now be obtained by direct superposition of Eqs. (5.6.7), (5.6.9), and (5.6.10), where now the final equations are expressed in terms of the interface displacements at the eight interface nodes only [Figure 5–32(b)] as

$$\{F_i\} - \{\bar{F}_i\} = [\bar{K}_{ii}]\{d_i\} \tag{5.6.12}$$

The solution of Eq. (5.6.12) gives the displacements at the interface nodes. To obtain the displacements within each substructure, we use the force-displacement Eqs. (5.6.4) for $\{d_e^B\}$ with similar equations for substructures A and C. Example 5.9 illustrates the concept of substructure analysis. In order to solve by hand, a relatively simple structure is used.

Example 5.9

Solve for the displacement and rotation at node 3 for the beam in Figure 5–33 by using substructuring. Let $E = 200$ GPa and $I = 4.0 \times 10^{-4}$ m^4.

SOLUTION:

To illustrate the substructuring concept, we divide the beam into two substructures, labeled 1 and 2 in Figure 5–34. The 45 kN force has been assigned to node 3 of substructure 2, although it could have been assigned to either substructure or a fraction of it assigned to each substructure.

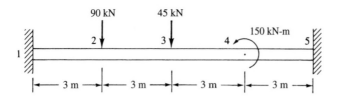

Figure 5–33 Beam analyzed by substructuring

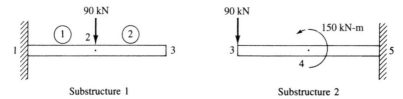

Substructure 1   Substructure 2

Figure 5–34 Beam of Figure 5–33 separated into substructures

The stiffness matrix for each beam element is given by Eq. (4.1.14) as

$$
[k^{(1)}] = [k^{(2)}] = [k^{(3)}] = [k^{(4)}] = \frac{(200 \times 10^9)(4.0 \times 10^{-4})}{(3)^3}
\begin{matrix} 1 \\ 2 \\ 3 \\ 4 \end{matrix}
\begin{bmatrix}
12 & 6(3) & -12 & 6(3) \\
6(3) & 4(3)^2 & -6(3) & 2(3)^2 \\
-12 & -6(3) & 12 & -6(3) \\
6(3) & 2(3)^2 & -6(3) & 4(3)^2
\end{bmatrix}
\begin{matrix} 2 \\ 3 \\ 4 \\ 5 \end{matrix}
$$

(5.6.13)

$$
= (2.96 \times 10^6)
\begin{bmatrix}
12 & 18 & -12 & 18 \\
18 & 36 & -18 & 18 \\
-12 & -18 & 12 & -18 \\
18 & 18 & -18 & 36
\end{bmatrix}
$$

(5.6.14)

For substructure 1, we add the stiffness matrices of elements 1 and 2 together. The equations are

$$
(2.96 \times 10^6)
\begin{bmatrix}
12+12 & -18+18 & -12 & 18 \\
-18+18 & 18+18 & -18 & 18 \\
-12 & -18 & 12 & -18 \\
18 & 18 & -18 & 36
\end{bmatrix}
\begin{Bmatrix}
v_2 \\ \phi_2 \\ v_3 \\ \phi_3
\end{Bmatrix}
=
\begin{Bmatrix}
-90 \times 10^3 \\ 0 \\ 0 \\ 0
\end{Bmatrix}
$$

(5.6.15)

where the boundary conditions $v_1 = \phi_1 = 0$ were used to reduce the equations.

Rewriting Eq. (5.6.15) with the interface displacements first allows us to use Eq. (5.6.6) to condense out, or eliminate, the interior degrees of freedom, v_2 and ϕ_2. These reordered equations are

$$(2.96 \times 10^6)(12v_3 - 18\phi_3 - 12v_2 - 18\phi_2) = 0$$

$$(2.96 \times 10^6)(-18v_3 + 36\phi_3 + 18v_2 + 18\phi_2) = 0$$

$$(2.96 \times 10^6)(-12v_3 + 18\phi_3 + 24v_2 + \phi_2) = -90 \times 10^3 \qquad (5.6.16)$$

$$(2.96 \times 10^6)(-18v_3 + 18\phi_3 + 0v_2 + 72\phi_2) = 0$$

Using Eq. (5.6.6), we obtain equations for the interface degrees of freedom as

$$(2.96 \times 10^6)\left\{ \begin{bmatrix} 12 & -18 \\ -18 & 36 \end{bmatrix} - \begin{bmatrix} -12 & -18 \\ 18 & 18 \end{bmatrix} \begin{bmatrix} 24 & 0 \\ 0 & 72 \end{bmatrix}^{-1} \begin{bmatrix} -12 & 18 \\ -18 & 18 \end{bmatrix} \right\} \begin{Bmatrix} v_3 \\ \phi_3 \end{Bmatrix}$$

$$= \begin{Bmatrix} 0 \\ 0 \end{Bmatrix} - \begin{bmatrix} -12 & -18 \\ 18 & 18 \end{bmatrix} \begin{bmatrix} 24 & 0 \\ 0 & 72 \end{bmatrix}^{-1} \begin{Bmatrix} -90 \times 10^3 \\ 0 \end{Bmatrix} \qquad (5.6.17)$$

Simplifying Eq. (5.6.17), we obtain

$$\begin{bmatrix} 1.416 & -4.392 \\ -4.392 & 17.856 \end{bmatrix} \begin{Bmatrix} v_3 \\ \phi_3 \end{Bmatrix} = \begin{Bmatrix} -15.324 \times 10^{-3} \\ 22.98 \times 10^{-3} \end{Bmatrix} \qquad (5.6.18)$$

For substructure 2, we add the stiffness matrices of elements 3 and 4 together. The equations are

$$(2.96 \times 10^6) \begin{bmatrix} 12 & 18 & -12 & 18 \\ 18 & 36 & -18 & 18 \\ -12 & -18 & 12+12 & -18+18 \\ 18 & 18 & -18+18 & 36+36 \end{bmatrix} \begin{Bmatrix} v_3 \\ \phi_3 \\ v_4 \\ \phi_4 \end{Bmatrix} = \begin{Bmatrix} -45 \times 10^3 \\ 0 \\ 0 \\ 1.5 \times 10^5 \end{Bmatrix}$$

$$(5.6.19)$$

where boundary conditions $v_5 = \phi_5 = 0$ were used to reduce the equations.

Using static condensation, Eq. (5.6.6), we obtain equations with only the interface displacements v_3 and ϕ_3. These equations are

$$(2.96 \times 10^6)\left\{ \begin{bmatrix} 12 & 18 \\ 18 & 36 \end{bmatrix} - \begin{bmatrix} -12 & 18 \\ -18 & 18 \end{bmatrix} \begin{bmatrix} 24 & 0 \\ 0 & 72 \end{bmatrix}^{-1} \begin{bmatrix} -12 & -18 \\ 18 & 18 \end{bmatrix} \right\} \begin{Bmatrix} v_3 \\ \phi_3 \end{Bmatrix}$$

$$= \begin{Bmatrix} -10 \\ 0 \end{Bmatrix} - \begin{bmatrix} -12 & 18 \\ -18 & 18 \end{bmatrix} \begin{bmatrix} 24 & 0 \\ 0 & 72 \end{bmatrix}^{-1} \begin{Bmatrix} 0 \\ 1.5 \times 10^5 \end{Bmatrix} \qquad (5.6.20)$$

Simplifying Eq. (5.6.20), we obtain

$$\begin{bmatrix} 1.416 & 4.392 \\ 4.392 & 17.856 \end{bmatrix} \begin{Bmatrix} v_3 \\ \phi_3 \end{Bmatrix} = \begin{Bmatrix} -27.97 \times 10^{-3} \\ -12.77 \times 10^{-3} \end{Bmatrix} \qquad (5.6.21)$$

Adding Eqs. (5.6.18) and (5.6.21), we obtain the final nodal equilibrium equations at the interface degrees of freedom as

$$\begin{bmatrix} 2.832 & 0 \\ 0 & 35.712 \end{bmatrix} \begin{Bmatrix} v_3 \\ \phi_3 \end{Bmatrix} = \begin{Bmatrix} -43.294 \times 10^{-3} \\ -10.21 \times 10^{-3} \end{Bmatrix} \qquad (5.6.22)$$

Solving Eq. (5.6.22) for the displacement and rotation at node 3, we obtain

$$v_3 = -0.0152\ 3 = 15.2\ \text{mm}$$

$$\phi_3 = 0.000286\ \text{rad}$$

(5.6.23)

We could now return to Eq. (5.6.15) or Eq. (5.6.16) to obtain v_2 and ϕ_2 and to Eq. (5.6.19) to obtain v_4 and ϕ_4. ∎

We emphasize that this example is used as a simple illustration of substructuring and is not typical of the size of problems where substructuring is normally performed. Generally, substructuring is used when the number of degrees of freedom is very large, as might occur, for instance, for very large structures such as the airframe in Figure 5–31.

▲ Summary Equations

Stiffness matrix for rigid plane frame beam element:

$$[k] = \frac{E}{L} \times$$

$$\begin{bmatrix}
AC^2 + \frac{12I}{L^2}S^2 & \left(A - \frac{12I}{L^2}\right)CS & -\frac{6I}{L}S & -\left(AC^2 + \frac{12I}{L^2}S^2\right) & -\left(A - \frac{12I}{L^2}\right)CS & -\frac{6I}{L}S \\
 & AS^2 + \frac{12I}{L^2}C^2 & \frac{6I}{L}C & -\left(A - \frac{12I}{L^2}\right)CS & -\left(AS^2 + \frac{12I}{L^2}C^2\right) & \frac{6I}{L}C \\
 & & 4I & \frac{6I}{L}S & -\frac{6I}{L}C & 2I \\
 & & & AC^2 + \frac{12I}{L^2}S^2 & \left(A - \frac{12I}{L^2}\right)CS & \frac{6I}{L}S \\
 & & & & AS^2 + \frac{12I}{L^2}C^2 & -\frac{6I}{L}C \\
\text{Symmetry} & & & & & 4I
\end{bmatrix}$$

(5.1.11)

Equations for plane frame with inclined support:

$$\begin{Bmatrix} F_{1x} \\ F_{1y} \\ M_1 \\ F_{2x} \\ F_{2y} \\ M_2 \\ F'_{3x} \\ F'_{3y} \\ M_3 \end{Bmatrix} = [T_i][K][T_i]^T \begin{Bmatrix} u_1 = 0 \\ v_1 = 0 \\ \phi_1 = 0 \\ u_2 \\ v_2 \\ \phi_2 \\ u'_3 \\ v'_3 = 0 \\ \phi'_3 = \phi_3 \end{Bmatrix} \qquad (5.3.3)$$

where

$$[T_i] = \begin{bmatrix} [I] & [0] & [0] \\ [0] & [I] & [0] \\ [0] & [0] & [t_3] \end{bmatrix}$$ (5.3.4)

and

$$[t_3] = \begin{bmatrix} \cos\alpha & \sin\alpha & 0 \\ -\sin\alpha & \cos\alpha & 0 \\ 0 & 0 & 1 \end{bmatrix}$$ (5.3.5)

Stiffness matrix for torsion bar element:

$$[k'] = \frac{GJ}{L} \begin{bmatrix} 1 & -1 \\ -1 & 1 \end{bmatrix}$$ (5.4.14)

See Table 5–1 for torsional constants for various cross-sectional shapes:

$$J = \sum \frac{1}{3} b_i t_i^3$$ (5.4.15)

Stiffness matrix for grid element:

$$[k_G] = \begin{array}{cccccc} v_1' & \phi_{1x}' & \phi_{1z}' & v_2' & \phi_{2x}' & \phi_{2z}' \\ \begin{bmatrix} \dfrac{12EI}{L^3} & 0 & \dfrac{6EI}{L^2} & \dfrac{-12EI}{L^3} & 0 & \dfrac{6EI}{L^2} \\[2mm] 0 & \dfrac{GJ}{L} & 0 & 0 & \dfrac{-GJ}{L} & 0 \\[2mm] \dfrac{6EI}{L^2} & 0 & \dfrac{4EI}{L} & \dfrac{-6EI}{L^2} & 0 & \dfrac{2EI}{L} \\[2mm] \dfrac{-12EI}{L^3} & 0 & \dfrac{-6EI}{L^2} & \dfrac{12EI}{L^3} & 0 & \dfrac{-6EI}{L^2} \\[2mm] 0 & \dfrac{-GJ}{L} & 0 & 0 & \dfrac{GJ}{L} & 0 \\[2mm] \dfrac{6EI}{L^2} & 0 & \dfrac{2EI}{L} & \dfrac{-6EI}{L^2} & 0 & \dfrac{4EI}{L} \end{bmatrix} \end{array}$$ (5.4.17)

Transformation matrix for grid element:

$$[T_G] = \begin{bmatrix} 1 & 0 & 0 & 0 & 0 & 0 \\ 0 & C & S & 0 & 0 & 0 \\ 0 & -S & C & 0 & 0 & 0 \\ 0 & 0 & 0 & 1 & 0 & 0 \\ 0 & 0 & 0 & 0 & C & S \\ 0 & 0 & 0 & 0 & -S & C \end{bmatrix}$$ (5.4.18)

Global stiffness matrix for grid element:

$$[k_G] = [T_G]^T [k_G'][T_G]$$ (5.4.19)

Stiffness matrix for beam or frame element in three-dimensional space:

$$
[k'] =
\begin{bmatrix}
\dfrac{AE}{L} & 0 & 0 & 0 & 0 & 0 & -\dfrac{AE}{L} & 0 & 0 & 0 & 0 & 0 \\[6pt]
0 & \dfrac{12EI_z}{L^3} & 0 & 0 & 0 & \dfrac{6EI_z}{L^2} & 0 & -\dfrac{12EI_z}{L^3} & 0 & 0 & 0 & \dfrac{6EI_z}{L^2} \\[6pt]
0 & 0 & \dfrac{12EI_y}{L^3} & 0 & -\dfrac{6EI_y}{L^2} & 0 & 0 & 0 & -\dfrac{12EI_y}{L^3} & 0 & -\dfrac{6EI_y}{L^2} & 0 \\[6pt]
0 & 0 & 0 & \dfrac{GJ}{L} & 0 & 0 & 0 & 0 & 0 & -\dfrac{GJ}{L} & 0 & 0 \\[6pt]
0 & 0 & -\dfrac{6EI_y}{L^2} & 0 & \dfrac{4EI_y}{L} & 0 & 0 & 0 & \dfrac{6EI_y}{L^2} & 0 & \dfrac{2EI_y}{L} & 0 \\[6pt]
0 & \dfrac{6EI_z}{L^2} & 0 & 0 & 0 & \dfrac{4EI_z}{L} & 0 & -\dfrac{6EI_z}{L^2} & 0 & 0 & 0 & \dfrac{2EI_z}{L} \\[6pt]
-\dfrac{AE}{L} & 0 & 0 & 0 & 0 & 0 & \dfrac{AE}{L} & 0 & 0 & 0 & 0 & 0 \\[6pt]
0 & -\dfrac{12EI_z}{L^3} & 0 & 0 & 0 & -\dfrac{6EI_z}{L^2} & 0 & \dfrac{12EI_z}{L^3} & 0 & 0 & 0 & -\dfrac{6EI_z}{L^2} \\[6pt]
0 & 0 & -\dfrac{12EI_y}{L^3} & 0 & \dfrac{6EI_y}{L^2} & 0 & 0 & 0 & \dfrac{12EI_y}{L^3} & 0 & \dfrac{6EI_y}{L^2} & 0 \\[6pt]
0 & 0 & 0 & -\dfrac{GJ}{L} & 0 & 0 & 0 & 0 & 0 & \dfrac{GJ}{L} & 0 & 0 \\[6pt]
0 & 0 & -\dfrac{6EI_y}{L^2} & 0 & \dfrac{2EI_y}{L} & 0 & 0 & 0 & \dfrac{6EI_y}{L^2} & 0 & \dfrac{4EI_y}{L} & 0 \\[6pt]
0 & \dfrac{6EI_z}{L^2} & 0 & 0 & 0 & \dfrac{2EI_z}{L} & 0 & -\dfrac{6EI_z}{L^2} & 0 & 0 & 0 & \dfrac{4EI_z}{L}
\end{bmatrix}
$$

(with column headings u_1', v_1', w_1', ϕ_{1x}', ϕ_{1y}', ϕ_{1z}', u_2', v_2', w_2', ϕ_{2x}', ϕ_{2y}', ϕ_{2z}')

(5.5.3)

Global stiffness matrix for the beam or frame element in three-dimensional space:

$$[k] = [T]^T [k'][T] \tag{5.5.4}$$

where

$$
[T] =
\begin{bmatrix}
[\lambda]_{3\times3} & & & \\
& [\lambda]_{3\times3} & & \\
& & [\lambda]_{3\times3} & \\
& & & [\lambda]_{3\times3}
\end{bmatrix}
\tag{5.5.5}
$$

and

$$
[\lambda] =
\begin{bmatrix}
C_{xx'} & C_{yx'} & C_{zx'} \\
C_{xy'} & C_{yy'} & C_{zy'} \\
C_{xz'} & C_{yz'} & C_{zx'}
\end{bmatrix}
\tag{5.5.6}
$$

▲ References

[1] Kassimali, A., *Structural Analysis*, 2nd ed., Brooks/Cole Publishers, Pacific Grove, CA, 1999.
[2] Budynas, R. G., *Advanced Strength and Applied Stress Analysis*, 2nd ed., McGraw-Hill, New York, 1999.

[3] Allen, H. G., and Bulson, P. S., *Background to Buckling*, McGraw-Hill, London, 1980.

[4] Young, W. C., and Budynas, R. G., *Roark's Formulas for Stress and Strain*, 7th ed., McGraw-Hill, New York, 2002.

[5] Gere, J. M., and Goodno, B. J., *Mechanics of Materials*, 7th ed., Cengage Learning, Mason, OH, 2009.

[6] Parakh, Z. K., *Finite Element Analysis of Bus Frames under Simulated Crash Loadings*, M.S. Thesis, Rose-Hulman Institute of Technology, Terre Haute, Indiana, May 1989.

[7] Martin, H. C., *Introduction to Matrix Methods of Structural Analysis*, McGraw-Hill, New York, 1966.

[8] Juvinall, R. C., and Marshek, K. M., *Fundamentals of Machine Component Design*, 4th ed., p. 198, Wiley, 2005.

[9] *Machinery's Handbook*, Oberg, E., et. al., 26th ed., Industrial Press, N.Y., 2000.

[10] *Standard Specifications for Highway Bridges*, American Association of State Highway and Transportation Officials (AASHTO), Washington, D.C.

▲ Problems

Solve all problems using the finite element stiffness method.

5.1 For the rigid frame shown in Figure P5–1, determine (1) the displacement components and the rotation at node 2, (2) the support reactions, and (3) the forces in each element. Then check equilibrium at node 2. Let $E = 210 \times 10^9$ N/m^2, $A = 6.25 \times 10^{-3}$ m^2, and $I = 1.95 \times 10^{-4}$ m^4 for both elements.

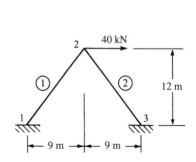

Figure P5–1

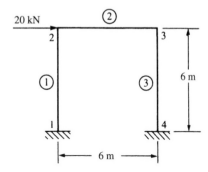

Figure P5–2

5.2 For the rigid frame shown in Figure P5–2, determine (1) the nodal displacement components and rotations, (2) the support reactions, and (3) the forces in each element. Let $E = 210$ GPa, $A = 6.25 \times 10^{-3}$ m^2, and $I = 7.8 \times 10^{-5}$ m^4 for all elements.

5.3 For the rigid stairway frame shown in Figure P5–3, determine (1) the displacements at node 2, (2) the support reactions, and (3) the local nodal forces acting on each element. Draw the bending moment diagram for the whole frame. Remember that the angle between elements 1 and 2 is preserved as deformation takes place; similarly for the angle between elements 2 and 3. Furthermore, owing to symmetry, $u_2 = -u_3$, $v_2 = v_3$, and $\phi_2 = -\phi_3$. What size A36 steel channel section would be needed to keep the allowable bending stress less than two-thirds of the yield stress? (For A36 steel, the yield stress is 240 MPa.)

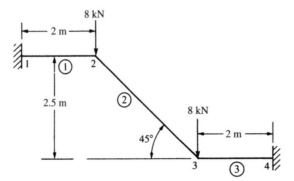

Figure P5–3

5.4 For the rigid frame shown in Figure P5–4, determine (1) the nodal displacements and rotation at node 4, (2) the reactions, and (3) the forces in each element. Then check equilibrium at node 4. Finally, draw the shear force and bending moment diagrams for each element. Let $E = 210$ GPa, $A = 5 \times 10^{-3}$ m^2, and $I = 3 \times 10^{-4}$ m^4 for all elements.

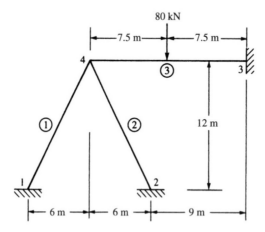

Figure P5–4

5.5–5.15 For the rigid frames shown in Figures P5–5 through P5–15, determine the displacements and rotations of the nodes, the element forces, and the reactions. The values of E, A, and I to be used are listed next to each figure.

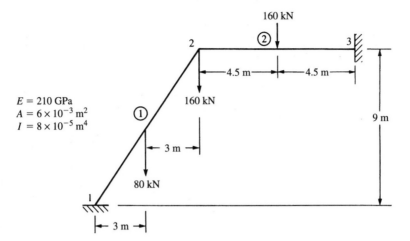

Figure P5–5

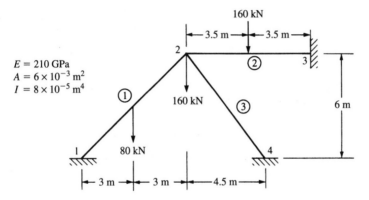

Figure P5–6

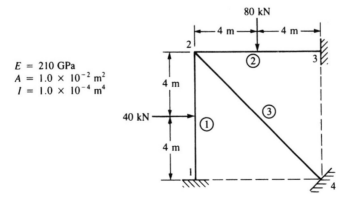

Figure P5–7

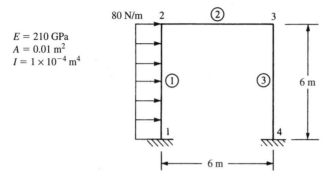

$E = 210$ GPa
$A = 0.01$ m^2
$I = 1 \times 10^{-4}$ m^4

80 N/m

6 m

6 m

Figure P5-8

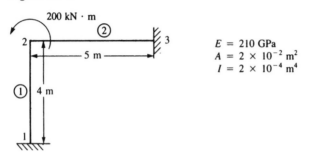

200 kN · m

5 m

4 m

$E = 210$ GPa
$A = 2 \times 10^{-2}$ m^2
$I = 2 \times 10^{-4}$ m^4

Figure P5-9

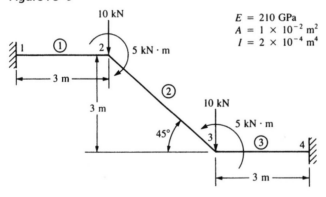

10 kN

5 kN · m

3 m

3 m

45°

10 kN

5 kN · m

3 m

$E = 210$ GPa
$A = 1 \times 10^{-2}$ m^2
$I = 2 \times 10^{-4}$ m^4

Figure P5-10

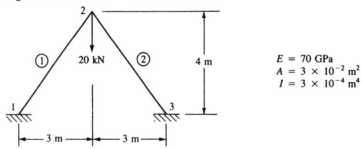

2

20 kN

4 m

3 m

3 m

$E = 70$ GPa
$A = 3 \times 10^{-2}$ m^2
$I = 3 \times 10^{-4}$ m^4

Figure P5-11

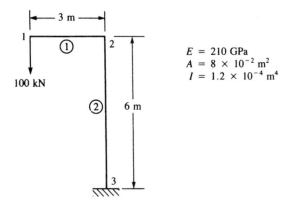

$E = 210$ GPa
$A = 8 \times 10^{-2}$ m^2
$I = 1.2 \times 10^{-4}$ m^4

Figure P5–12

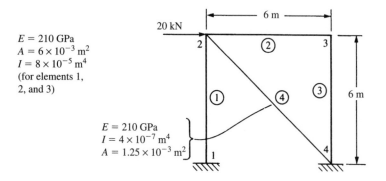

$E = 210$ GPa
$A = 6 \times 10^{-3}$ m^2
$I = 8 \times 10^{-5}$ m^4
(for elements 1, 2, and 3)

$E = 210$ GPa
$I = 4 \times 10^{-7}$ m^4
$A = 1.25 \times 10^{-3}$ m^2

Figure P5–13

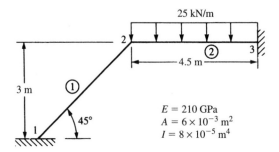

$E = 210$ GPa
$A = 6 \times 10^{-3}$ m^2
$I = 8 \times 10^{-5}$ m^4

Figure P5–14

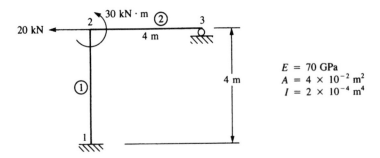

Figure P5–15

5.16–5.18 Solve the structures in Figures P5–16 through P5–18 by using substructuring.

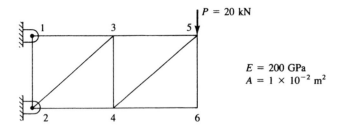

Figure P5–16 (Substructure the truss at nodes 3 and 4)

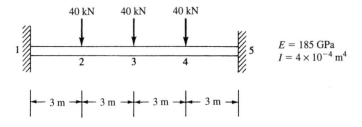

Figure P5–17 (Substructure the beam at node 3)

Figure P5–18 (Substructure the beam at node 2)

Solve Problems 5.19 through 5.39 by using a computer program.

5.19 For the rigid frame shown in Figure P5–19, determine (1) the nodal displacement components and (2) the support reactions. (3) Draw the shear force and bending moment diagrams. For all elements, let $E = 210$ GPa, $I = 8 \times 10^{-5}$ m^4, and $A = 6 \times 10^{-3}$ m^2

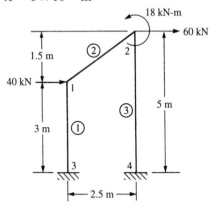

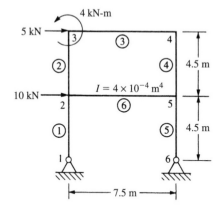

Figure P5–19   **Figure P5–20**

5.20 For the rigid frame shown in Figure P5–20, determine (1) the nodal displacement components and (2) the support reactions. (3) Draw the shear force and bending moment diagrams. Let $E = 240$ GPa, $I = 8 \times 10^{-5}$ m^4, and $A = 6 \times 10^{-3}$ m^2 for all elements, except as noted in the figure.

5.21 For the slant-legged rigid frame shown in Figure P5–21, size the structure for minimum weight based on a maximum bending stress of 140 MPa in the horizontal beam elements and a maximum compressive stress (due to bending and direct axial load) of 105 MPa in the slant-legged elements. Use the same element size for the two slant-legged elements and the same element size for the two 3-m sections of the horizontal element. Assume A36 steel is used.

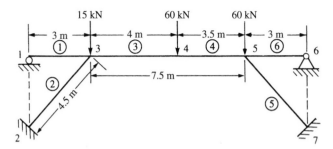

Figure P5–21

5.22 For the rigid building frame shown in Figure P5–22, determine the forces in each element and calculate the bending stresses. Assume all the vertical elements have $A = 6 \times 10^{-3}$ m^2 and $I = 4 \times 10^{-5}$ m^4 and all horizontal elements have $A = 9 \times 10^{-3}$ m^2 and $I = 6 \times 10^{-5}$ m^4 Let $E = 200$ GPa for all elements. Let $c = 125$ mm for the vertical elements and $c = 150$ mm for the horizontal elements, where c denotes the distance from the neutral axis to the top or bottom of the beam cross section, as used in the bending stress formula $\sigma = (Mc/I)$.

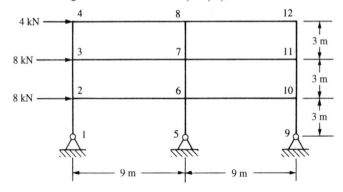

Figure P5–22

5.23–5.38 For the rigid frames or beams shown in Figures P5–23 through P5–38, determine the displacements and rotations at the nodes, the element forces, and the reactions.

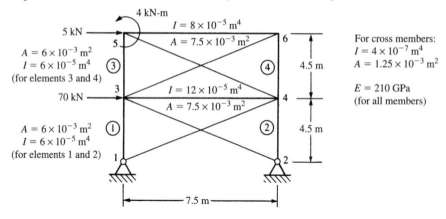

Figure P5–23

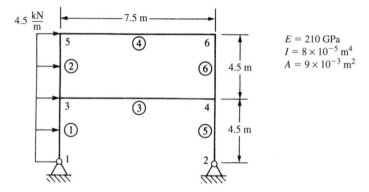

Figure P5–24

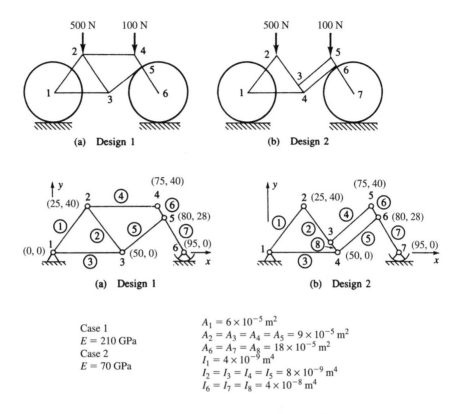

$$A_1 = 6 \times 10^{-5} \, \text{m}^2$$

Case 1
E = 210 GPa
Case 2
E = 70 GPa

$$A_2 = A_3 = A_4 = A_5 = 9 \times 10^{-5} \, \text{m}^2$$
$$A_6 = A_7 = A_8 = 18 \times 10^{-5} \, \text{m}^2$$
$$I_1 = 4 \times 10^{-9} \, \text{m}^4$$
$$I_2 = I_3 = I_4 = I_5 = 8 \times 10^{-9} \, \text{m}^4$$
$$I_6 = I_7 = I_8 = 4 \times 10^{-8} \, \text{m}^4$$

Figure P5–25 Two bicycle frame models (coordinates shown in inches)

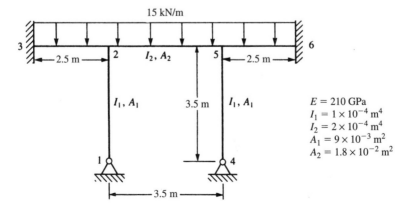

Figure P5–26

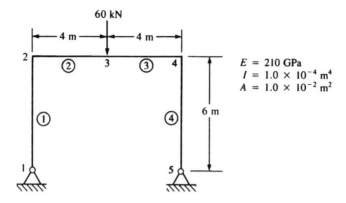

Figure P5-27

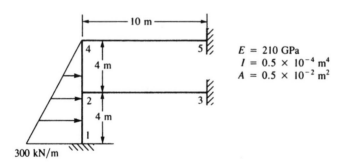

Figure P5-28

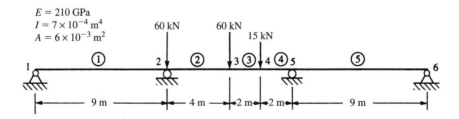

Figure P5-29

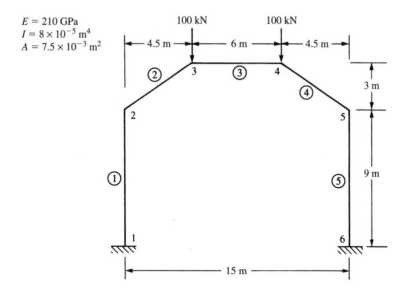

$E = 210$ GPa
$I = 8 \times 10^{-5}$ m⁴
$A = 7.5 \times 10^{-3}$ m²

100 kN   100 kN

4.5 m   6 m   4.5 m

3 m

9 m

15 m

Figure P5–30

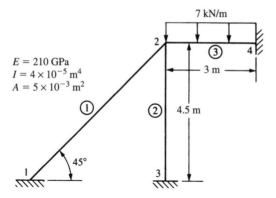

7 kN/m

$E = 210$ GPa
$I = 4 \times 10^{-5}$ m⁴
$A = 5 \times 10^{-3}$ m²

3 m

4.5 m

45°

Figure P5–31

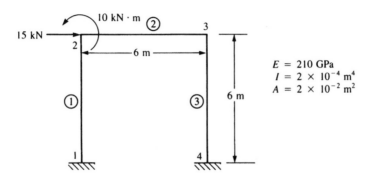

10 kN · m

15 kN

6 m

$E = 210$ GPa
$I = 2 \times 10^{-4}$ m⁴
$A = 2 \times 10^{-2}$ m²

6 m

Figure P5–32

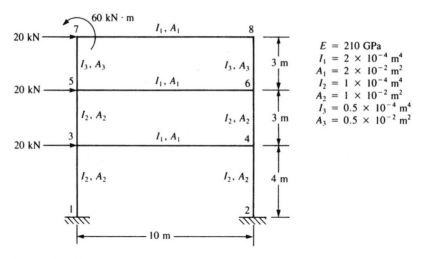

$E = 210$ GPa
$I_1 = 2 \times 10^{-4}$ m^4
$A_1 = 2 \times 10^{-2}$ m^2
$I_2 = 1 \times 10^{-4}$ m^4
$A_2 = 1 \times 10^{-2}$ m^2
$I_3 = 0.5 \times 10^{-4}$ m^4
$A_3 = 0.5 \times 10^{-2}$ m^2

Figure P5–33

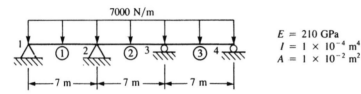

$E = 210$ GPa
$I = 1 \times 10^{-4}$ m^4
$A = 1 \times 10^{-2}$ m^2

Figure P5–34

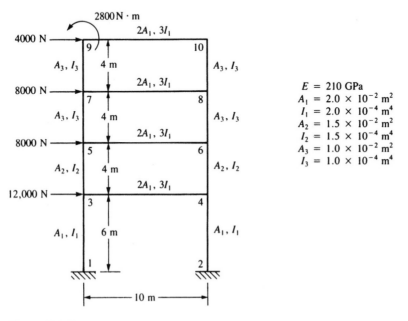

$E = 210$ GPa
$A_1 = 2.0 \times 10^{-2}$ m^2
$I_1 = 2.0 \times 10^{-4}$ m^4
$A_2 = 1.5 \times 10^{-2}$ m^2
$I_2 = 1.5 \times 10^{-4}$ m^4
$A_3 = 1.0 \times 10^{-2}$ m^2
$I_3 = 1.0 \times 10^{-4}$ m^4

Figure P5–35

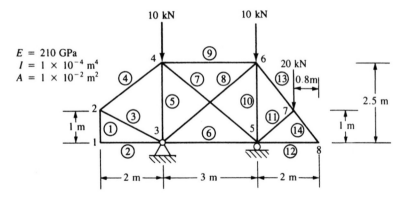

Figure P5-36

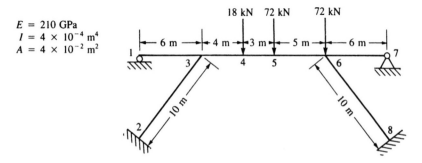

Figure P5-37

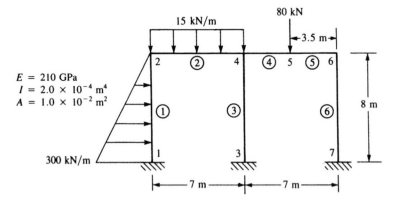

Figure P5-38

5.39 Consider the plane structure shown in Figure P5–39. First assume the structure to be a plane frame with rigid joints, and analyze using a frame element. Then assume the structure to be pin-jointed and analyze as a plane truss, using a truss element. If the structure is actually a truss, is it appropriate to model it as a rigid frame?

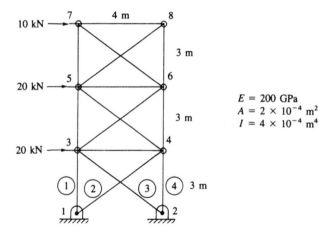

Figure P5-39

How can you model the truss using the frame (or beam) element? In other words, what idealization could you make in your model to use the beam element to approximate a truss?

 5.40 For the two-story, two-bay rigid frame shown in Figure P5-40, determine (1) the nodal displacement components and (2) the shear force and bending moments in each member. Let $E = 200$ GPa, $I = 2 \times 10^{-4}$ m^4 for each horizontal member and $I = 1.5 \times 10^{-4}$ m^4 for each vertical member.

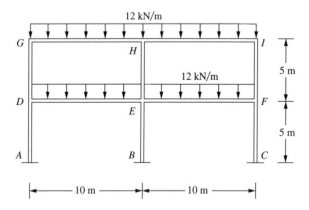

Figure P5-40

 5.41 For the two-story, three-bay rigid frame shown in Figure P5-41, determine (1) the nodal displacements and (2) the member end shear forces and bending moments. (3) Draw the shear force and bending moment diagrams for each member. Let $E = 200$ GPa, $I = 1.29 \times 10^{-4}$ m^4 for the beams and $I = 0.462 \times 10^{-4}$ m^4 for the columns.

The properties for I correspond to a W 610×155 and a W 410×114 wide-flange section, respectively, in metric units.

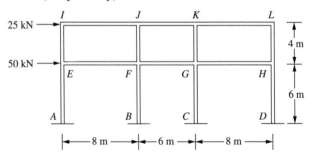

Figure P5–41

5.42 For the rigid frame shown in Figure P5–42, determine (1) the nodal displacements and rotations and (2) the member shear forces and bending moments. Let $E = 200$ GPa, $I = 0.795 \times 10^{-4}\,\mathrm{m}^4$ for the horizontal members, and $I = 0.316 \times 10^{-4}\,\mathrm{m}^4$ for the vertical members. These I values correspond to a W 460×158 and a W 410×85 wide-flange section, respectively.

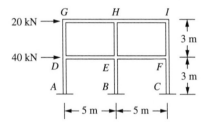

Figure P5–42

5.43 For the rigid frame shown in Figure P5–43, determine (1) the nodal displacements and rotations and (2) the shear force and bending moments in each member. Let $E = 200$ GPa, $I = 1 \times 10^{-3}\,\mathrm{m}^4$ for the horizontal members and $I = 4 \times 10^{-4}\,\mathrm{m}^4$ for the vertical members. The I values correspond to a W 0.6×2.6 (A $= 0.0179\,\mathrm{m}^2$) and a W 16×77 (A $= 0.0114\,\mathrm{m}^2$).

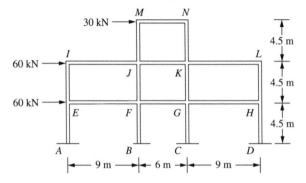

Figure P5–43

5.44 A structure is fabricated by welding together three lengths of I-shaped members as shown in Figure P5–44. The yield strength of the members is 250 MPa, $E = 200$ GPa, and Poisson's ratio is 0.3. The members all have cross-section properties corresponding to a W460 × 52. That is, $A = 22.3$ in^2, depth of section is $d = 18.21$ in., $I_x = 1330$ in.^{4} $S_x = 146$ in.^{3} $I_y = 152$ in.^{4} and $S_y = 27.6$ in.^{3} Determine whether a load of $Q = 40$ kN downward is safe against general yielding of the material. The factor of safety against general yielding is to be 2.0. Also, determine the maximum vertical and horizontal deflections of the structure.

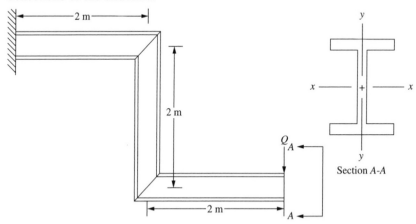

Figure P5–44

5.45 For the tapered beam shown in Figure P5–45, determine the maximum deflection using one, two, four, and eight elements. Calculate the moment of inertia at the mid-length station for each element. Let $E = 210$ GPa, $I_0 = 4 \times 10^{-5}$ m^4, and $L = 2.5$ m. Run cases where $n = 1, 3$, and 7. Use a beam element. The analytical solution for deflection and slope at the free end for $n = 7$ is given by Reference [7] as shown below:

$$v_1 = \frac{PL^3}{49EI_0}(1/7 \ln 8 + 2.5) = \frac{1}{17.55}\frac{PL^3}{EI_0}$$

$$\theta_1 = \frac{PL^2}{49EI_0}(\ln 8 - 7) = -\frac{1}{9.95}\frac{PL^2}{EI_0}$$

$$I(x) = I_0\left(1 + n\frac{x}{L}\right)$$

where n = arbitrary numerical factor and I_0 = moment of inertia of section at $x = 0$.

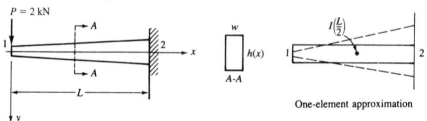

Figure P5–45 Tapered cantilever beam

5.46 Derive the stiffness matrix for the nonprismatic torsion bar shown in Figure P5–46. The radius of the shaft is given by

$$r = r_0 + (x/L)r_0, \text{ where } r_0 \text{ is the radius at } x = 0.$$

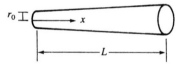

Figure P5–46

5.47 Derive the total potential energy for the prismatic circular cross-section torsion bar shown in Figure P5–47. Also determine the equivalent nodal torques for the bar subjected to uniform torque per unit length (kN-m/m). Let G be the shear modulus and J be the polar moment of inertia of the bar.

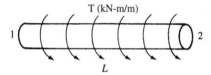

Figure P5–47

5.48 For the grid shown in Figure P5–48, determine the nodal displacements and the local element forces. Let $E = 200$ GPa, $G = 80$ GPa, $I = 8 \times 10^{-5}$ m^4, $A = 1 \times 10^{-2}$ m^2 and $J = 4 \times 10^{-5}$ m^4 for both elements.

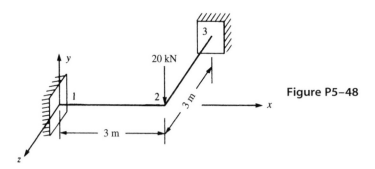

Figure P5–48

5.49 Resolve Problem 5–48 with an additional nodal moment of 100 kN-m applied about the x axis at node 2.

5.50–5.51 For the grids shown in Figures P5–50 and P5–51, determine the nodal displacements and the local element forces. Let $E = 210$ GPa, $G = 84$ GPa, $I = 2 \times 10^{-4}$ m^4, $J = 1 \times 10^{-4}$ m^4, and $A = 1 \times 10^{-2}$ m^2.

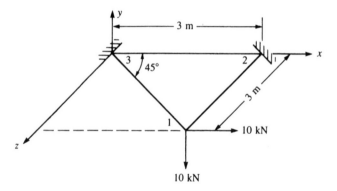

Figure P5–50

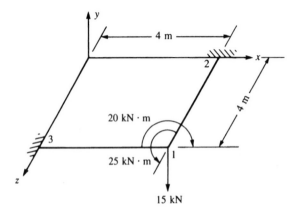

Figure P5–51

5.52–5.57 Solve the grid structures shown in Figures P5–52 through P5–57 using a computer program. For grids P5–52—P5–54, let $E = 210$ GPa, $G = 84$ GPa, $I = 8 \times 10^{-5}$ m^4, $A = 1 \times 10^{-2}$ m^2 and $J = 4 \times 10^{-5}$ m^4, except as noted in the figures. In Figure P5–54,

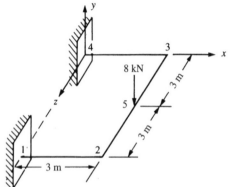

Figure P5–52

let the cross elements have $I = 20 \times 10^{-6}$ m⁴ and $J = 8 \times 10^{-6}$ m⁴, with dimensions and loads as in Figure P5–53. For grids P5–55 through P5–57, let $E = 210$ GPa, $G = 84$ GPa, $I = 2 \times 10^{-4}$ m⁴, $J = 1 \times 10^{-4}$ m⁴, and $A = 1 \times 10^{-2}$ m².

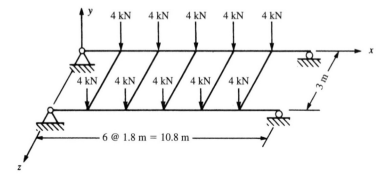

Figure P5–53

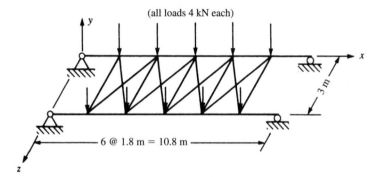

Figure P5–54

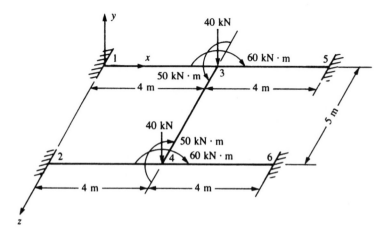

Figure P5–55

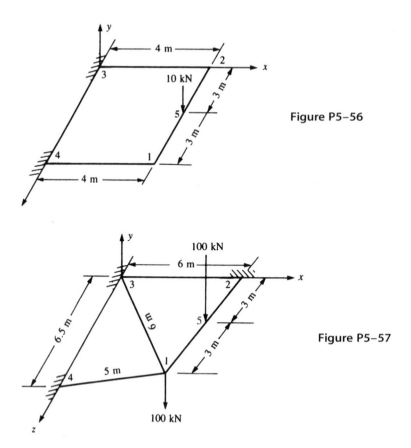

Figure P5-56

Figure P5-57

5.58-5.59 Determine the displacements and reactions for the space frames shown in Figures P5-58 and P5-59. Let $I_x = 4 \times 10^{-5}$ m^4, $I_y = 8 \times 10^{-5}$ m^4, $I_z = 4 \times 10^{-4}$ m^4, $E = 200$ GPa, $G = 77$ GPa, and $A = 0.06$ m^2 for both frames.

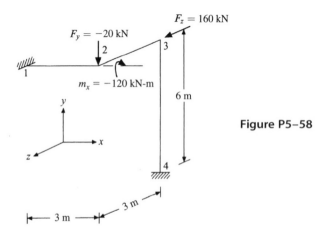

Figure P5-58

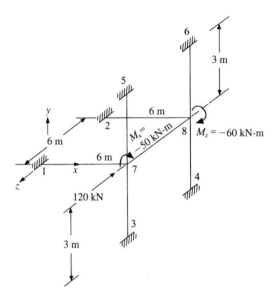

Figure P5–59

Use a computer program to assist in the design problems in Problems 5.60 through 5.72.

5.60 Design a jib crane as shown in Figure P5–60 that will support a downward load of 24 kN. Choose a common structural steel shape for all members. Use allowable stresses of $0.66S_y$ (S_y is the yield strength of the material) in bending, and $0.60S_y$ in

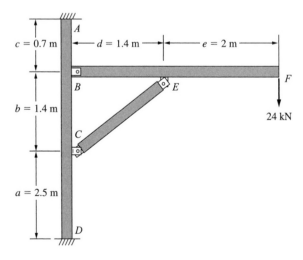

Figure P5–60

tension on gross areas. The maximum deflection should not exceed 1/360 of the length of the horizontal beam. Buckling should be checked using Euler's or Johnson's method as applicable.

5.61 Design the support members AB and CD for the platform lift shown in Figure P5–61. Select a mild steel and choose suitable cross-sectional shapes with no more than a 4:1 ratio of moments of inertia between the two principal directions of the cross section. You may choose two different cross sections to make up each arm to reduce weight. The actual structure has four support arms, but the loads shown are for one side of the platform with the two arms shown. The loads shown are under operating conditions. Use a factor of safety of 2 for human safety. In developing the finite element model, remove the platform and replace it with statically equivalent loads at the joints at B and D. Use truss elements or beam elements with low bending stiffness to model the arms from B to D, the intermediate connection E to F, and the hydraulic actuator. The allowable stresses are $0.66S_y$ in bending and $0.60S_y$ in tension. Check buckling using either Euler's method or Johnson's method as appropriate. Also check maximum deflections. Any deflection greater than 1/360 of the length of member AB is considered too large.

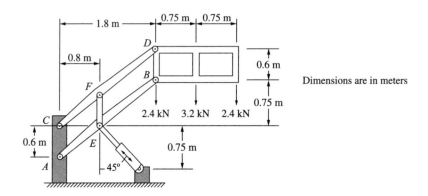

Figure P5–61

5.62 A two-story building frame is to be designed as shown in Figure P5–62. The members are all to be I-beams with rigid connections. We would like the floor joists beams to have a 0.38 m depth and the columns to have a 0.25 m width. The material is to be A36 structural steel. Two horizontal loads and vertical loads are shown. Select members such that the allowable bending in the beams is 170 MPa. Check buckling in the columns using Euler's or Johnson's method as appropriate. The allowable deflection in the beams should not exceed 1/360 of each beam span. The overall sway of the frame should not exceed 12.5 mm.

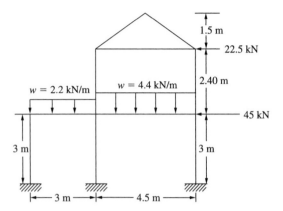

Figure P5–62

5.63 A pulpwood loader as shown in Figure P5–63 is to be designed to lift 10 kN. Select a steel and determine a suitable tubular cross section for the main upright member *BF* that has attachments for the hydraulic cylinder actuators *AE* and *DG*. Select a steel and determine a suitable box section for the horizontal load arm *AC*. The horizontal load arm may have two different cross sections *AB* and *BC* to reduce weight. The finite element model should use beam elements for all members except the hydraulic cylinders, which should be truss elements. The pinned joint at *B* between the upright and the horizontal beam is best modeled with end release of the end node of the top element on the upright member. The allowable bending stress is $0.66S_y$ in members *AB* and *BC*. Member *BF* should be checked for buckling. The allowable deflection at *C* should be less than 1/360 of the length of *BC*. As a bonus, the client would like you to select the size of the hydraulic cylinders *AE* and *DG*.

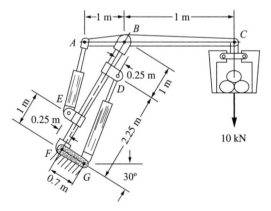

Figure P5–63

5.64 A piston ring (with a split as shown in Figure P5–64) is to be expanded by a tool to facilitate its installation. The ring is sufficiently thin (5 mm depth) to justify using conventional straight-beam bending formulas. The ring requires a displacement of 2.5 mm at its separation for installation. Determine the force required to produce this separation. In addition, determine the largest stress in the ring. Let $E = 125$ GPa, $G = 50$ GPa, cross-sectional area $A = 40$ mm², and principal moment of inertia $I = 20 \times 10^{-9}$ m⁴. The inner radius is 46 mm, and the outer radius is 54 mm. Use models with 4, 6, 8, 10, and 20 elements in a symmetric model until convergence to the same results occurs. Plot the displacement versus the number of elements for a constant force F predicted by the conventional beam theory equation of Reference [8].

$$\delta = \frac{3\pi F R^3}{EI} + \frac{\pi F R}{EA} + \frac{6\pi F R}{5GA} \qquad \text{where } R = 50 \text{ mm and } \delta = 2.5 \text{ mm}$$

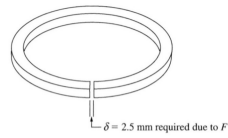

Figure P5–64

$\delta = 2.5$ mm required due to F

5.65 A small hydraulic floor crane as shown in Figure P5–65 carries a 20-kN load. Determine the size of the beam and column needed. Select either a standard box section or a wide-flange section. Assume a rigid connection between the beam and column. The column is rigidly connected to the floor. The allowable bending stress in the beam is $0.60S_y$. The allowable deflection is $1/360$ of the beam length. Check the column for buckling.

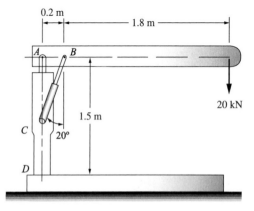

Figure P5–65

5.66 Determine the size of a solid round shaft such that the maximum angle of twist between C and B is 0.26 degrees per meter of length and the deflection of the beam is less than 0.0127 cm under the pulley C for the loads, as shown in Figure P5–66. Assume simple supports at bearings A and B. Assume the shaft is made from cold-rolled AISI 1020 steel. (Recommended angles of twist in driven shafts can be found in *Machinery's Handbook*, Oberg, E., et. al., 26th ed., Industrial Press, N.Y., 2000.)

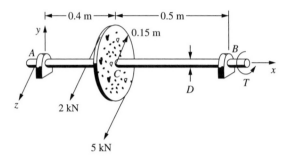

Figure P5–66

5.67 The shaft in Figure P5–67 supports a winch load of 3.5 kN and a torsional moment of 900 N-m at F (650 mm from the center of the bearing at A). In addition, a radial load of 2.25 kN and an axial load of 1.8 kN act at point E from a worm gearset. Assume the maximum stress in the shaft cannot be larger than that obtained from the maximum distortional energy theory with a factor of safety of 2.5. Also make sure the angle of twist is less than 1.5 deg between A and D. In your model, assume the bearing at A to be frozen when calculating the angle of twist. Bearings at B, C, and D can be assumed as simple supports. Determine the required shaft diameter.

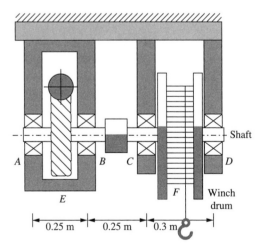

Figure P5–67

5.68 Design the gabled frame subjected to the external wind load shown in Figure P5–68 (comparable to an 125 kmph wind speed) for an industrial building. Assume this is one of a typical frame spaced every 6 m. Select a wide flange section based on allowable bending stress of 140 MPa and an allowable compressive stress of 70 MPa in any member. Neglect the possibility of buckling in any members. Use ASTM A36 steel.

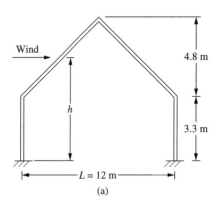

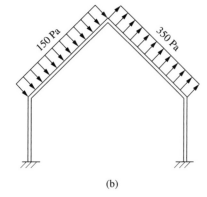

Figure P5–68

5.69 Design the gabled frame shown for a balanced snow load shown in Figure P5–69 (typical of the Midwest) for an apartment building. Select a wide flange section for the frame. Assume the allowable bending stress not to exceed 140 MPa. Use ASTM A36 steel.

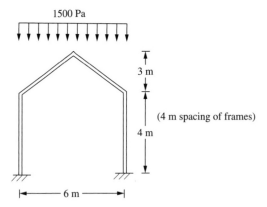

Figure P5–69

5.70 Design a gantry crane that must be able to lift 90 kN as it must lift compressors, motors, heat exchangers, and controls. This load should be placed at the center of one of the main 3.6 m-long beams as shown in Figure P5–70, by the hoisting device location. Note that this beam is on one side of the crane. Assume you are using

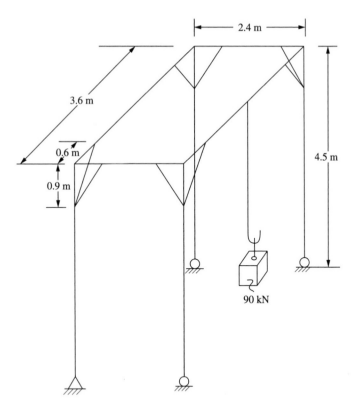

Figure P5–70

ASTM A36 structural steel. The crane must be 3.6 m long, 2.4 m wide, and 4.5 m high. The beams should all be the same size, the columns all the same size, and the bracing all the same size. The corner bracing can be wide flange sections or some other common shape. You must verify that the structure is safe by checking the beam's bending strength and allowable deflection, the column's buckling strength, and the bracing's buckling strength. Use a factor of safety against material yielding of the beams of 5. Verify that the beam deflection is less than L/360, where L is the span of the beam. Check Euler buckling of the long columns and the bracing. Use a factor of safety against buckling of 5. Assume the column-to-beam joints to be rigid while the bracing (a total of eight braces) is pinned to the column and beam at each of the four corners. Also assume the gantry crane is on rollers with one roller locked down to behave as a pin support as shown.

 5.71 Design the rigid highway bridge frame structure shown in Figure P5–71 for a moving truck load (shown below) simulating a truck moving across the bridge. Use the load shown and place it along the top girder at various locations. Use the allowable stresses in bending and compression and allowable deflection given in the *Standard Specifications for Highway Bridges*, American Association of State Highway and Transportation Officials (AASHTO), Washington, D.C. or use some other reasonable values.

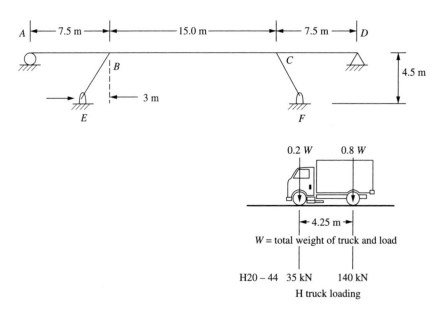

Figure P5–71

 5.72 For the tripod space frame shown in Figure P5–72, determine standard steel pipe sections such that the maximum bending stress must not exceed 150 MPa, the compressive stress to prevent buckling must not exceed that given by the Euler buckling formula with a factor of safety of 2, and the maximum deflection will not exceed $L/360$ in any span, L. Assume the three bottom supports to be fixed. All coordinates shown in units of millimeters.

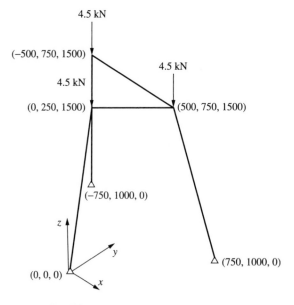

Figure P5–72

5.73 The curved semi-circular frame shown in Figure P5–73 is supported by a pin on the left end and a roller on the right end and is subjected to a load $P = 4.5$ kN at its apex. The frame has a radius to centerline cross section of $R = 3000$ mm. Select a structural steel W shape from Appendix F such that the maximum stress does not exceed 150 MPa. Perform a finite element analysis using 4, 8, and then 16 elements in your finite element model. Also, determine the maximum deflection for each model. It is suggested that the finite element answers for deflection be compared to the solution obtained by classical methods, such as using Castigliano's theorem. The expression for deflection under the load is given by using Castigliano's theorem as

$$\delta_y = \frac{0.178PR^3}{EI} + \frac{0.393PR}{AE} + \frac{0.393PR}{A_vG}$$

where A is the cross sectional area of the W shape, A_v is the shear area of the W shape (use depth of web times thickness of web for the shear area); $E = 210$ FPa, and $G = 80$ GPa.

Now change the radius of the frame to 500 mm and repeat the problem. Run the finite element model with the shear area included in your computer program input and then without. Comment on the difference in results and compare to the predicted analytical deflection by using the equation above for δ_y.

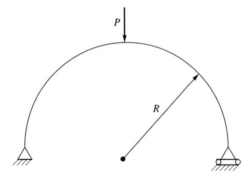

Figure P5–73

DEVELOPMENT OF THE PLANE STRESS AND PLANE STRAIN STIFFNESS EQUATIONS

CHAPTER OBJECTIVES

- To review basic concepts of plane stress and plane strain.
- To derive the constant-strain triangular (CST) element stiffness matrix and equations.
- To demonstrate how to determine the stiffness matrix and stresses for a constant strain element.
- To describe how to treat body and surface forces for two-dimensional elements.
- To evaluate the explicit stiffness matrix for the constant-strain triangle element.
- To perform a detailed finite element solution of a plane stress problem.
- To derive the bilinear four-noded rectangular (Q4) element stiffness matrix.
- To compare the CST and Q4 model results for a beam bending problem and describe some of the CST and Q4 element defects.

Introduction

In Chapters 2 through 5, we considered only line elements. Two or more line elements are connected only at common nodes, forming framed or articulated structures such as trusses, frames, and grids. Line elements have geometric properties such as cross-sectional area and moment of inertia associated with their cross sections. However, only one local coordinate x along the length of the element is required to describe a position along the element (hence, they are called *line elements* or one-dimensional elements). Nodal compatibility is then enforced during the formulation of the nodal equilibrium equations for a line element.

This chapter considers the two-dimensional finite element. Two-dimensional (planar) elements are defined by three or more nodes in a two-dimensional plane (that is, x-y). The elements are connected at common nodes and/or along common edges to form continuous structures such as those shown in Figures 1–3, 1–4, 1–6, 6–2a, and 6–6(b). Nodal displacement compatibility is then enforced during the formulation of the nodal equilibrium equations for two-dimensional elements. If proper displacement functions are chosen, compatibility along common edges is also obtained.

The two-dimensional element is extremely important for (1) plane stress analysis, which includes problems such as plates with holes, fillets, or other changes in geometry that are loaded in their plane resulting in local stress concentrations, as illustrated in Figure 6–1; and (2) plane strain analysis, which includes problems such as a long underground box culvert subjected to a uniform load acting constantly over its length, as illustrated in Figure 1–3, a long, cylindrical control rod subjected to a load that remains constant over the rod length (or depth), as illustrated in Figure 1–4, and dams and pipes subjected to loads that remain constant over their lengths, as shown in Figure 6–2.

We begin this chapter with the development of the stiffness matrix for a basic two-dimensional or plane finite element, called the *constant-strain triangular element*. We consider the constant-strain triangle (CST) stiffness matrix because its derivation is the simplest among the available two-dimensional elements. The element is called a CST because it has a constant strain throughout it.

We will derive the CST stiffness matrix by using the principle of minimum potential energy because the energy formulation is the most feasible for the development of the equations for both two- and three-dimensional finite elements.

We will then present a simple, thin-plate plane stress example problem to illustrate the assemblage of the plane element stiffness matrices using the direct stiffness method as presented in Chapter 2. We will present the total solution, including the stresses within the plate.

Finally, we will develop the stiffness matrix for the simple four-noded rectangular (Q4) element and compare the finite element solution to a beam bending problem modeled using the CST and Q4 elements.

▲ 6.1 Basic Concepts of Plane Stress and Plane Strain   ▲

In this section, we will describe the concepts of plane stress and plane strain. These concepts are important because the developments in this chapter are directly applicable only to systems assumed to behave in a plane stress or plane strain manner. Therefore, we will now describe these concepts in detail.

Plane Stress

Plane stress *is defined to be a state of stress in which the normal stress and the shear stresses directed perpendicular to the plane are assumed to be zero.* For instance, in Figures 6–1(a) and 6–1(b), the plates in the x-y plane shown subjected to surface tractions T (pressure acting on the surface edge or face of a member in units of force/area) in the plane are under a state of plane stress; that is, the normal stress σ_z and the shear stresses τ_{xz} and τ_{yz} are assumed to be zero. Generally, members that are thin (those with a small z dimension compared to the in-plane x and y dimensions) and whose loads act only in the x-y plane can be considered to be under plane stress.

Plane Strain

Plane strain *is defined to be a state of strain in which the strain normal to the x-y plane ε_z and the shear strains γ_{xz} and γ_{yz} are assumed to be zero.* The assumptions of plane

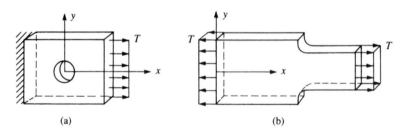

Figure 6–1 Plane stress problems: (a) plate with hole; (b) plate with fillet

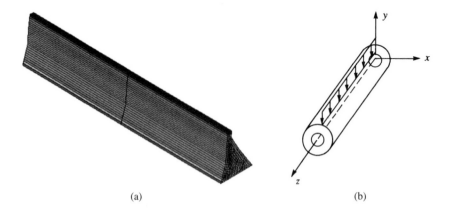

Figure 6–2 Plane strain problems: (a) dam subjected to horizontal loading (See the full-color insert for a color version of this figure.); (b) pipe subjected to a vertical load

strain are realistic for long bodies (say, in the z direction) with constant cross-sectional area subjected to loads that act only in the x and/or y directions and do not vary in the z direction. Some plane strain examples are shown in Figure 6–2 [and in Figures 1–3 (a long underground box culvert) and 1–4 (a hydraulic cylinder rod end)]. In these examples, only a unit thickness (1 cm or 1 m) of the structure is considered because each unit thickness behaves identically (except near the ends). The finite element models of the structures in Figure 6–2 consist of appropriately discretized cross sections in the x-y plane with the loads acting over unit thicknesses in the x and/or y directions only.

Two-Dimensional State of Stress and Strain

The concept of a two-dimensional state of stress and strain and the stress–strain relationships for plane stress and plane strain are necessary to understand fully the development and applicability of the stiffness matrix for the plane stress/plane strain triangular element. Therefore, we briefly outline the essential concepts of two-dimensional stress and strain (see References [1] and [2] and Appendix C for more details on this subject).

First, we illustrate the two-dimensional state of stress using Figure 6–3. The infinitesimal element with sides dx and dy has normal stresses σ_x and σ_y acting in the

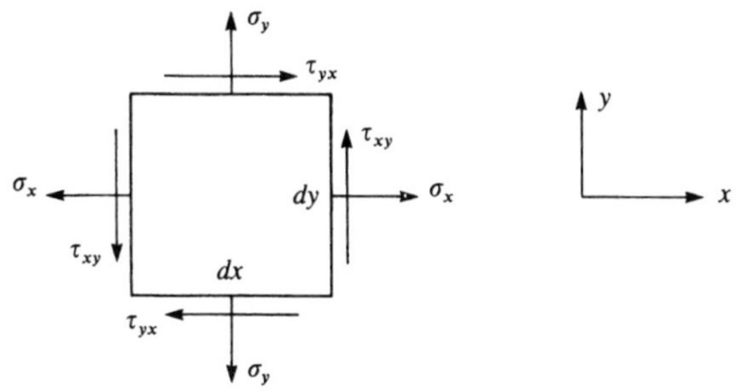

Figure 6–3 Two-dimensional state of stress

x and y directions (here on the vertical and horizontal faces), respectively. The shear stress τ_{xy} acts on the x edge (vertical face) in the y direction. The shear stress τ_{yx} acts on the y edge (horizontal face) in the x direction. Moment equilibrium of the element results in τ_{xy} being equal in magnitude to τ_{yx}. See Appendix C.1 for proof of this equality. Hence, three independent stresses exist and are represented by the vector column matrix

$$\{\sigma\} = \begin{Bmatrix} \sigma_x \\ \sigma_y \\ \tau_{xy} \end{Bmatrix} \tag{6.1.1}$$

The element equilibrium equations are derived in Appendix C.1.

The stresses given by Eq. (6.1.1) will be expressed in terms of the nodal displacement degrees of freedom. Hence, once the nodal displacements are determined, these stresses can be evaluated directly.

Recall from strength of materials [2] that the **principal stresses**, which are the maximum and minimum normal stresses in the two-dimensional plane, can be obtained from the following expressions:

$$\sigma_1 = \frac{\sigma_x + \sigma_y}{2} + \sqrt{\left(\frac{\sigma_x - \sigma_y}{2}\right)^2 + \tau_{xy}^2} = \sigma_{max}$$

$$\sigma_2 = \frac{\sigma_x + \sigma_y}{2} - \sqrt{\left(\frac{\sigma_x - \sigma_y}{2}\right)^2 + \tau_{xy}^2} = \sigma_{min}$$
$$\tag{6.1.2}$$

Also, the **principal angle** θ_p, which defines the normal whose direction is perpendicular to the plane on which the maximum or minimum principal stress acts, is defined by

$$\tan 2\theta_p = \frac{2\tau_{xy}}{\sigma_x - \sigma_y} \tag{6.1.3}$$

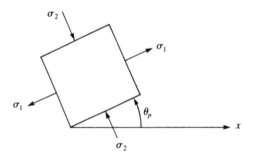

Figure 6-4 Principal stresses and their directions

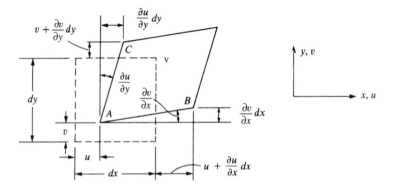

Figure 6-5 Displacements and rotations of lines of an element in the x-y plane

Figure 6–4 shows the principal stresses σ_1 and σ_2 and the angle θ_p. Recall (as Figure 6–4 indicates) that the shear stress is zero on the planes having principal (maximum and minimum) normal stresses.

In Figure 6–5, we show an infinitesimal element used to represent the general two-dimensional state of strain at some point in a structure. The element is shown to be displaced by amounts u and v in the x and y directions at point A, and to displace or extend an additional (incremental) amount $(\partial u/\partial x)\,dx$ along line AB, and $(\partial v/\partial y)\,dy$ along line AC in the x and y directions, respectively. Furthermore, observing lines AB and AC, we see that point B moves upward an amount $(\partial v/\partial x)\,dx$ with respect to A, and point C moves to the right an amount $(\partial u/\partial y)\,dy$ with respect to A.

From the general definitions of normal and shear strains and the use of Figure 6–5, we obtain

$$\varepsilon_x = \frac{\partial u}{\partial x} \qquad \varepsilon_y = \frac{\partial v}{\partial y} \qquad \gamma_{xy} = \frac{\partial u}{\partial y} + \frac{\partial v}{\partial x} \tag{6.1.4}$$

Appendix C.2 shows a detailed derivation of Eqs. (6.1.4). Hence, recall that the strains ε_x and ε_y are the changes in length per unit length of material fibers originally parallel to the x and y axes, respectively, when the element undergoes deformation. These strains are then called *normal* (or *extensional* or *longitudinal*) strains. The strain γ_{xy} is the change in the original right angle made between dx and dy when the element undergoes deformation. The strain γ_{xy} is then called a *shear strain*.

The strains given by Eqs. (6.1.4) are generally represented by the vector column matrix

$$\{\varepsilon\} = \left\{ \begin{array}{c} \varepsilon_x \\ \varepsilon_y \\ \gamma_{xy} \end{array} \right\} \tag{6.1.5}$$

The relationships between strains and displacements referred to the x and y directions given by Eqs. (6.1.4) are sufficient for your understanding of subsequent material in this chapter.

We now present the stress–strain relationships for isotropic materials for both plane stress and plane strain. For plane stress, we assume the following stresses to be zero:

$$\sigma_z = \tau_{xz} = \tau_{yz} = 0 \tag{6.1.6}$$

Applying Eq. (6.1.6) to the three-dimensional stress–strain relationship [see Appendix C, Eq. (C.3.10)], the shear strains $\gamma_{xz} = \gamma_{yz} = 0$, but $\varepsilon_z \neq 0$. For plane stress conditions, we then have

$$\{\sigma\} = [D]\{\varepsilon\} \tag{6.1.7}$$

where

$$[D] = \frac{E}{1 - v^2} \begin{bmatrix} 1 & v & 0 \\ v & 1 & 0 \\ 0 & 0 & \dfrac{1 - v}{2} \end{bmatrix} \tag{6.1.8}$$

is called the *stress–strain matrix* (or *constitutive matrix*), E is the modulus of elasticity, and v is Poisson's ratio. In Eq. (6.1.7), $\{\sigma\}$ and $\{\varepsilon\}$ are defined by Eqs. (6.1.1) and (6.1.5), respectively.

For plane strain, we assume the following strains to be zero:

$$\varepsilon_z = \gamma_{xz} = \gamma_{yz} = 0 \tag{6.1.9}$$

Applying Eq. (6.1.9) to the three-dimensional stress–strain relationship [Eq. (C.3.10)], the shear stresses $\tau_{xz} = \tau_{yz} = 0$, but $\sigma_z \neq 0$. The stress–strain matrix then becomes

$$[D] = \frac{E}{(1 + v)(1 - 2v)} \begin{bmatrix} 1 - v & v & 0 \\ v & 1 - v & 0 \\ 0 & 0 & \dfrac{1 - 2v}{2} \end{bmatrix} \tag{6.1.10}$$

The $\{\sigma\}$ and $\{\varepsilon\}$ matrices remain the same as for the plane stress case. The basic partial differential equations for plane stress, as derived in Reference [1], are

$$\frac{\partial^2 u}{\partial x^2} + \frac{\partial^2 u}{\partial y^2} = \frac{1+v}{2}\left(\frac{\partial^2 u}{\partial y^2} - \frac{\partial^2 v}{\partial x \partial y}\right)$$

$$\frac{\partial^2 v}{\partial x^2} + \frac{\partial^2 v}{\partial y^2} = \frac{1+v}{2}\left(\frac{\partial^2 v}{\partial x^2} - \frac{\partial^2 u}{\partial x \partial y}\right)$$

(6.1.11)

▲ 6.2 Derivation of the Constant-Strain Triangular Element Stiffness Matrix and Equations

To illustrate the steps and introduce the basic equations necessary for the plane triangular element, consider the thin plate subjected to tensile surface traction loads T_S in Figure 6–6(a).

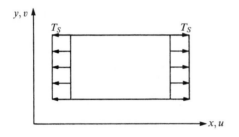

Figure 6–6(a)   Thin plate in tension

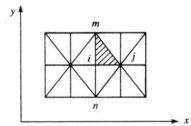

Figure 6–6(b)   Discretized plate of Figure 6–6(a) using triangular elements

Step 1   Select Element Type

To analyze the plate, we consider the basic triangular element in Figure 6–7 taken from the discretized plate, as shown in Figure 6–6(b). The discretized plate has been

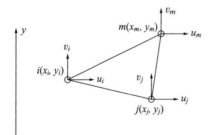

Figure 6–7   Basic triangular element showing degrees of freedom

divided into triangular elements, each with nodes such as i, j, and m. We use triangular elements because boundaries of irregularly shaped bodies can be closely approximated in this way, and because the expressions related to the triangular element are comparatively simple. This discretization is called a *coarse-mesh generation* if a few large elements are used. Each node has two degrees of freedom—an x and a y displacement. We will let u_i and v_i represent the node i displacement components in the x and y directions, respectively.

Here all formulations are based on this counterclockwise system of labeling of nodes, although a formulation based on a clockwise system of labeling could be used. Remember that a consistent labeling procedure for the whole body is necessary to avoid problems in the calculations such as negative element areas. Here (x_i, y_i), (x_j, y_j), and (x_m, y_m) are the known nodal coordinates of nodes i, j, and m, respectively. The nodal displacement matrix is given by

$$\{d\} = \begin{Bmatrix} \{d_i\} \\ \{d_j\} \\ \{d_m\} \end{Bmatrix} = \begin{Bmatrix} u_i \\ v_i \\ u_j \\ v_j \\ u_m \\ v_m \end{Bmatrix} \tag{6.2.1}$$

Step 2   Select Displacement Functions

We select a linear displacement function for each element as

$$u(x, y) = a_1 + a_2 x + a_3 y$$
$$v(x, y) = a_4 + a_5 x + a_6 y \tag{6.2.2}$$

where $u(x, y)$ and $v(x, y)$ describe displacements at any interior point (x_i, y_i) of the element.

The linear function ensures that compatibility will be satisfied. A linear function with specified endpoints has only one path through which to pass—that is, through the two points. Hence, the linear function ensures that the displacements along the edge and at the nodes shared by adjacent elements, such as edge i-j of the two elements shown in Figure 6–6(b), are equal. Using Eqs. (6.2.2), the general displacement function $\{\psi\}$, which stores the functions u and v, can be expressed as

$$\{\psi\} = \begin{Bmatrix} a_1 + a_2 x + a_3 y \\ a_4 + a_5 x + a_6 y \end{Bmatrix} = \begin{bmatrix} 1 & x & y & 0 & 0 & 0 \\ 0 & 0 & 0 & 1 & x & y \end{bmatrix} \begin{Bmatrix} a_1 \\ a_2 \\ a_3 \\ a_4 \\ a_5 \\ a_6 \end{Bmatrix} \tag{6.2.3}$$

To obtain the a's in Eqs. (6.2.2), we begin by substituting the coordinates of the nodal points into Eqs. (6.2.2) to yield

$$
\begin{aligned}
u_i &= u(x_i, y_i) = a_1 + a_2 x_i + a_3 y_i \\
u_j &= u(x_j, y_j) = a_1 + a_2 x_j + a_3 y_j \\
u_m &= u(x_m, y_m) = a_1 + a_2 x_m + a_3 y_m \\
v_i &= v(x_i, y_i) = a_4 + a_5 x_i + a_6 y_i \\
v_j &= v(x_j, y_j) = a_4 + a_5 x_j + a_6 y_j \\
v_m &= v(x_m, y_m) = a_4 + a_5 x_m + a_6 y_m
\end{aligned}
\tag{6.2.4}
$$

We can solve for the a's beginning with the first three of Eqs. (6.2.4) expressed in matrix form as

$$
\left\{ \begin{array}{c} u_i \\ u_j \\ u_m \end{array} \right\} =
\begin{bmatrix} 1 & x_i & y_i \\ 1 & x_j & y_j \\ 1 & x_m & y_m \end{bmatrix}
\left\{ \begin{array}{c} a_1 \\ a_2 \\ a_3 \end{array} \right\}
\tag{6.2.5}
$$

or, solving for the a's, we have

$$
\{a\} = [x]^{-1} \{u\}
\tag{6.2.6}
$$

where $[x]$ is the 3×3 matrix on the right side of Eq. (6.2.5). The method of cofactors (Appendix A) is one possible method for finding the inverse of $[x]$. Thus,

$$
[x]^{-1} = \frac{1}{2A}
\begin{bmatrix} \alpha_i & \alpha_j & \alpha_m \\ \beta_i & \beta_j & \beta_m \\ \gamma_i & \gamma_j & \gamma_m \end{bmatrix}
\tag{6.2.7}
$$

where

$$
2A = \begin{vmatrix} 1 & x_i & y_i \\ 1 & x_j & y_j \\ 1 & x_m & y_m \end{vmatrix}
\tag{6.2.8}
$$

is the determinant of $[x]$, which on evaluation is

$$
2A = x_i(y_j - y_m) + x_j(y_m - y_i) + x_m(y_i - y_j)
\tag{6.2.9}
$$

Here A is the area of the triangle, and

$$
\begin{array}{lll}
\alpha_i = x_j y_m - y_j x_m & \alpha_j = y_i x_m - x_i y_m & \alpha_m = x_i y_j - y_i x_j \\
\beta_i = y_j - y_m & \beta_j = y_m - y_i & \beta_m = y_i - y_j \\
\gamma_i = x_m - x_j & \gamma_j = x_i - x_m & \gamma_m = x_j - x_i
\end{array}
\tag{6.2.10}
$$

Having determined $[x]^{-1}$, we can now express Eq. (6.2.6) in expanded matrix form as

$$\left\{ \begin{array}{c} a_1 \\ a_2 \\ a_3 \end{array} \right\} = \frac{1}{2A} \begin{bmatrix} \alpha_i & \alpha_j & \alpha_m \\ \beta_i & \beta_j & \beta_m \\ \gamma_i & \gamma_j & \gamma_m \end{bmatrix} \left\{ \begin{array}{c} u_i \\ u_j \\ u_m \end{array} \right\} \tag{6.2.11}$$

Similarly, using the last three of Eqs. (6.2.4), we can obtain

$$\left\{ \begin{array}{c} a_4 \\ a_5 \\ a_6 \end{array} \right\} = \frac{1}{2A} \begin{bmatrix} \alpha_i & \alpha_j & \alpha_m \\ \beta_i & \beta_j & \beta_m \\ \gamma_i & \gamma_j & \gamma_m \end{bmatrix} \left\{ \begin{array}{c} v_i \\ v_j \\ v_m \end{array} \right\} \tag{6.2.12}$$

We will derive the general x displacement function $u(x, y)$ of $\{\psi\}$ (v will follow analogously) in terms of the coordinate variables x and y, known coordinate variables $\alpha_i, \alpha_j, \ldots, \gamma_m$, and unknown nodal displacements u_i, u_j, and u_m. Beginning with Eqs. (6.2.2) expressed in matrix form, we have

$$\{u\} = \begin{bmatrix} 1 & x & y \end{bmatrix} \left\{ \begin{array}{c} a_1 \\ a_2 \\ a_3 \end{array} \right\} \tag{6.2.13}$$

Substituting Eq. (6.2.11) into Eq. (6.2.13), we obtain

$$\{u\} = \frac{1}{2A} \begin{bmatrix} 1 & x & y \end{bmatrix} \begin{bmatrix} \alpha_i & \alpha_j & \alpha_m \\ \beta_i & \beta_j & \beta_m \\ \gamma_i & \gamma_j & \gamma_m \end{bmatrix} \left\{ \begin{array}{c} u_i \\ u_j \\ u_m \end{array} \right\} \tag{6.2.14}$$

Expanding Eq. (6.2.14), we have

$$\{u\} = \frac{1}{2A} \begin{bmatrix} 1 & x & y \end{bmatrix} \left\{ \begin{array}{c} \alpha_i u_i + \alpha_j u_j + \alpha_m u_m \\ \beta_i u_i + \beta_j u_j + \beta_m u_m \\ \gamma_i u_i + \gamma_j u_j + \gamma_m u_m \end{array} \right\} \tag{6.2.15}$$

Multiplying the two matrices in Eq. (6.2.15) and rearranging, we obtain

$$u(x, y) = \frac{1}{2A} \{ (\alpha_i + \beta_i x + \gamma_i y) u_i + (\alpha_j + \beta_j x + \gamma_j y) u_j + (\alpha_m + \beta_m x + \gamma_m y) u_m \} \tag{6.2.16}$$

Similarly, replacing u_i by v_i, u_j by v_j, and u_m by v_m in Eq. (6.2.16), we have the y displacement given by

$$v(x, y) = \frac{1}{2A} \{ (\alpha_i + \beta_i x + \gamma_i y) v_i + (\alpha_j + \beta_j x + \gamma_j y) v_j + (\alpha_m + \beta_m x + \gamma_m y) v_m \} \tag{6.2.17}$$

To express Eqs. (6.2.16) and (6.2.17) for u and v in simpler form, we define

$$N_i = \frac{1}{2A}(\alpha_i + \beta_i x + \gamma_i y)$$

$$N_j = \frac{1}{2A}(\alpha_j + \beta_j x + \gamma_j y)$$

$$(6.2.18)$$

$$N_m = \frac{1}{2A}(\alpha_m + \beta_m x + \gamma_m y)$$

Thus, using Eqs. (6.2.18), we can rewrite Eqs. (6.2.16) and (6.2.17) as

$$u(x, y) = N_i u_i + N_j u_j + N_m u_m$$

$$v(x, y) = N_i v_i + N_j v_j + N_m v_m$$

$$(6.2.19)$$

Expressing Eqs. (6.2.19) in matrix form, we obtain

$$\{\psi\} = \left\{ \begin{array}{c} u(x, y) \\ v(x, y) \end{array} \right\} = \left\{ \begin{array}{c} N_i u_i + N_j u_j + N_m u_m \\ N_i v_i + N_j v_j + N_m v_m \end{array} \right\}$$

or

$$\{\psi\} = \begin{bmatrix} N_i & 0 & N_j & 0 & N_m & 0 \\ 0 & N_i & 0 & N_j & 0 & N_m \end{bmatrix} \begin{Bmatrix} u_i \\ v_i \\ u_j \\ v_j \\ u_m \\ v_m \end{Bmatrix}$$

$$(6.2.20)$$

Finally, expressing Eq. (6.2.20) in abbreviated matrix form, we have

$$\{\psi\} = [N]\{d\}$$

$$(6.2.21)$$

where $[N]$ is given by

$$[N] = \begin{bmatrix} N_i & 0 & N_j & 0 & N_m & 0 \\ 0 & N_i & 0 & N_j & 0 & N_m \end{bmatrix}$$

$$(6.2.22)$$

We have now expressed the general displacements as functions of $\{d\}$, in terms of the shape functions $N_i, N_j,$ and N_m. The shape functions represent the shape of $\{\psi\}$ when plotted over the surface of a typical element. For instance, N_i represents the shape of the variable u when plotted over the surface of the element for $u_i = 1$ and all other degrees of freedom equal to zero; that is, $u_j = u_m = v_i = v_j = v_m = 0$. In addition, $u(x_i, y_i)$ must be equal to u_i. Therefore, we must have $N_i = 1, N_j = 0,$ and $N_m = 0$ at (x_i, y_i). Similarly, $u(x_j, y_j) = u_j$. Therefore, $N_i = 0, N_j = 1,$ and $N_m = 0$ at (x_j, y_j). Figure 6–8 shows the shape variation of N_i plotted over the surface of a typical element. Note that N_i does not equal zero except along a line connecting and including nodes j and m.

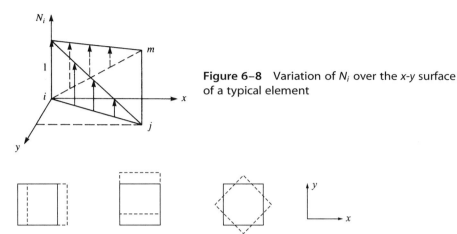

Figure 6–8 Variation of N_i over the x-y surface of a typical element

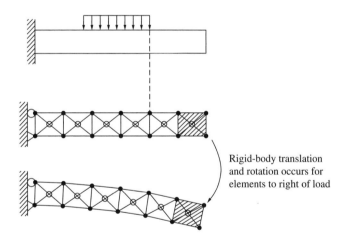

(a) Rigid-body modes of a plane stress element (from left to right, pure translation in x and y directions and pure rotation)

Rigid-body translation and rotation occurs for elements to right of load

(b) Cantilever beam modeled using constant-strain triangle elements; elements to the right of the loading are stress-free

Figure 6–9 Unstressed elements in a cantilever beam modeled with CST

Finally, $N_i + N_j + N_m = 1$ for all x and y locations on the surface of the element so that u and v will yield a constant value when rigid-body displacement occurs. The proof of this relationship follows that given for the bar element in Section 3.2 and is left as an exercise (Problem 6.1). The shape functions are also used to determine the body and surface forces at element nodes, as described in Section 6.3.

The requirement of completeness for the constant-strain triangle element used in a two-dimensional plane stress element is illustrated in Figure 6–9. The element must be able to translate uniformly in either the x or y direction in the plane and to rotate without straining as shown in Figure 6–9(a). The reason that the element must be able to translate as a rigid body and to rotate stress-free is illustrated in the example of a cantilever beam modeled with plane stress elements as shown in Figure 6–9(b). By simple statics, the beam elements beyond the loading are stress free. Hence these elements must be free to translate and rotate without stretching or changing shape.

Step 3 Define the Strain/Displacement and Stress/Strain Relationships

We express the element strains and stresses in terms of the unknown nodal displacements.

Element Strains

The strains associated with the two-dimensional element are given by

$$\{\varepsilon\} = \left\{ \begin{array}{c} \varepsilon_x \\ \varepsilon_y \\ \gamma_{xy} \end{array} \right\} = \left\{ \begin{array}{c} \dfrac{\partial u}{\partial x} \\[2mm] \dfrac{\partial v}{\partial y} \\[2mm] \dfrac{\partial u}{\partial y} + \dfrac{\partial v}{\partial x} \end{array} \right\} \tag{6.2.23a}$$

Substituting displacement functions for u and v from Eqs. (6.2.2) into Eq. (6.2.23a), we have

$$\varepsilon_x = a_2 \qquad \varepsilon_y = a_6 \qquad \gamma_{xy} = a_3 + a_5 \tag{6.2.23b}$$

We observe from Eq. (6.2.23b) that the strains in the element are constant. The element is then called a constant-strain triangle (CST). It should be also noted that based on the assumption of choosing displacement functions that are linear in x and y, all lines in the triangle element remain straight as the element deforms.

Using Eqs. (6.2.19) for the displacements, we have

$$\frac{\partial u}{\partial x} = u_{,x} = \frac{\partial}{\partial x}(N_i u_i + N_j u_j + N_m u_m) \tag{6.2.24}$$

or $\qquad\qquad u_{,x} = N_{i,x} u_i + N_{j,x} u_j + N_{m,x} u_m \tag{6.2.25}$

where the comma followed by a variable indicates differentiation with respect to that variable. We have used $u_{i,x} = 0$ because $u_i = u(x_i, y_i)$ is a constant value; similarly, $u_{j,x} = 0$ and $u_{m,x} = 0$.

Using Eqs. (6.2.18), we can evaluate the expressions for the derivatives of the shape functions in Eq. (6.2.25) as follows:

$$N_{i,x} = \frac{1}{2A}\frac{\partial}{\partial x}(\alpha_i + \beta_i x + \gamma_i y) = \frac{\beta_i}{2A} \tag{6.2.26}$$

Similarly,
$$N_{j,x} = \frac{\beta_j}{2A} \quad \text{and} \quad N_{m,x} = \frac{\beta_m}{2A} \tag{6.2.27}$$

Therefore, using Eqs. (6.2.26) and (6.2.27) in Eq. (6.2.25), we have

$$\frac{\partial u}{\partial x} = \frac{1}{2A}(\beta_i u_i + \beta_j u_j + \beta_m u_m) \tag{6.2.28}$$

Similarly, we can obtain

$$\frac{\partial v}{\partial y} = \frac{1}{2A}(\gamma_i v_i + \gamma_j v_j + \gamma_m v_m)$$

$$\frac{\partial u}{\partial y} + \frac{\partial v}{\partial x} = \frac{1}{2A}(\gamma_i u_i + \beta_i v_i + \gamma_j u_j + \beta_j v_j + \gamma_m u_m + \beta_m v_m) \tag{6.2.29}$$

Using Eqs. (6.2.28) and (6.2.29) in Eq. (6.2.23a), we obtain

$$\{\varepsilon\} = \frac{1}{2A} \begin{bmatrix} \beta_i & 0 & \beta_j & 0 & \beta_m & 0 \\ 0 & \gamma_i & 0 & \gamma_j & 0 & \gamma_m \\ \gamma_i & \beta_i & \gamma_j & \beta_j & \gamma_m & \beta_m \end{bmatrix} \begin{Bmatrix} u_i \\ v_i \\ u_j \\ v_j \\ u_m \\ v_m \end{Bmatrix} \tag{6.2.30}$$

or
$$\{\varepsilon\} = [[B_i]\ [B_j]\ [B_m]] \begin{Bmatrix} \{d_i\} \\ \{d_j\} \\ \{d_m\} \end{Bmatrix} \tag{6.2.31}$$

where

$$[B_i] = \frac{1}{2A}\begin{bmatrix} \beta_i & 0 \\ 0 & \gamma_i \\ \gamma_i & \beta_i \end{bmatrix} \qquad [B_j] = \frac{1}{2A}\begin{bmatrix} \beta_j & 0 \\ 0 & \gamma_j \\ \gamma_j & \beta_j \end{bmatrix} \qquad [B_m] = \frac{1}{2A}\begin{bmatrix} \beta_m & 0 \\ 0 & \gamma_m \\ \gamma_m & \beta_m \end{bmatrix} \tag{6.2.32}$$

Finally, in simplified matrix form, Eq. (6.2.31) can be written as

$$\{\varepsilon\} = [B]\{d\} \tag{6.2.33}$$

where
$$[B] = [[B_i]\ [B_j]\ [B_m]] \tag{6.2.34}$$

The $[B]$ matrix (sometimes called a gradient matrix) is independent of the x and y coordinates. It depends solely on the element nodal coordinates, as seen from Eqs. (6.2.32) and (6.2.10). The strains in Eq. (6.2.33) will be constant (consistent with the simple expressions previously given by Eq. (6.2.23b)).

Stress–Strain Relationship

In general, the in-plane stress–strain relationship is given by

$$\left\{ \begin{array}{c} \sigma_x \\ \sigma_y \\ \tau_{xy} \end{array} \right\} = [D] \left\{ \begin{array}{c} \varepsilon_x \\ \varepsilon_y \\ \gamma_{xy} \end{array} \right\} \tag{6.2.35}$$

where $[D]$ is given by Eq. (6.1.8) for plane stress problems and by Eq. (6.1.10) for plane strain problems. Using Eq. (6.2.33) in Eq. (6.2.35), we obtain the in-plane stresses in terms of the unknown nodal degrees of freedom as

$$\{\sigma\} = [D][B]\{d\} \tag{6.2.36}$$

where the stresses $\{\sigma\}$ are also constant everywhere within the element.

Step 4 Derive the Element Stiffness Matrix and Equations

Using the principle of minimum potential energy, we can generate the equations for a typical constant-strain triangular element. Keep in mind that for the basic plane stress element, the total potential energy is now a function of the nodal displacements $u_i, v_i, u_j, \ldots, v_m$ (that is, $\{d\}$) such that

$$\pi_p = \pi_p(u_i, v_i, u_j, \ldots, v_m) \tag{6.2.37}$$

Here the total potential energy is given by

$$\pi_p = U + \Omega_b + \Omega_p + \Omega_s \tag{6.2.38}$$

where the strain energy is given by

$$U = \frac{1}{2} \iiint_V \{\varepsilon\}^T \{\sigma\} \, dV \tag{6.2.39}$$

or, using Eq. (6.2.35), we have

$$U = \frac{1}{2} \iiint_V \{\varepsilon\}^T [D] \{\varepsilon\} \, dV \tag{6.2.40}$$

where we have used $[D]^T = [D]$ in Eq. (6.2.40).

The potential energy of the body forces is given by

$$\Omega_b = - \iiint_V \{\psi\}^T \{X\} \, dV \tag{6.2.41}$$

where $\{\psi\}$ is again the general displacement function, and $\{X\}$ is the body weight/unit volume or weight density matrix (typically, in units of pounds per cubic inch or kilonewtons per cubic meter).

The potential energy of concentrated loads is given by

$$\Omega_p = -\{d\}^T\{P\} \tag{6.2.42}$$

where $\{d\}$ represents the usual nodal displacements, and $\{P\}$ now represents the concentrated external loads.

The potential energy of distributed loads (or surface tractions) moving through respective surface displacements is given by

$$\Omega_s = -\iint_S \{\psi_S\}^T\{T_S\}\,dS \tag{6.2.43}$$

where $\{T_S\}$ represents the surface tractions (typically in units of pounds per square inch or kilonewtons per square meter), $\{\psi_S\}$ represents the field of surface displacements through which the surface tractions act, and S represents the surfaces over which the tractions $\{T_S\}$ act. Similar to Eq. (6.2.21), we express $\{\psi_S\}$ as $\{\psi_S\} = [N_S]\{d\}$, where $[N_S]$ represents the shape function matrix evaluated along the surface where the surface traction acts.

Using Eq. (6.2.21) for $\{\psi\}$ and Eq. (6.2.33) for the strains in Eqs. (6.2.40) through (6.2.43), we have

$$\pi_p = \frac{1}{2}\iiint_V \{d\}^T[B]^T[D][B]\{d\}\,dV - \iiint_V \{d\}^T[N]^T\{X\}\,dV$$

$$- \{d\}^T\{P\} - \iint_S \{d\}^T[N_S]^T\{T_S\}\,dS \tag{6.2.44}$$

The nodal displacements $\{d\}$ are independent of the general x-y coordinates, so $\{d\}$ can be taken out of the integrals of Eq. (6.2.44). Therefore,

$$\pi_p = \frac{1}{2}\{d\}^T\iiint_V [B]^T[D][B]\,dV\{d\} - \{d\}^T\iiint_V [N]^T\{X\}\,dV$$

$$- \{d\}^T\{P\} - \{d\}^T\iint_S [N_S]^T\{T_S\}\,dS \tag{6.2.45}$$

From Eqs. (6.2.41) through (6.2.43), we can see that the last three terms of Eq. (6.2.45) represent the total load system $\{f\}$ on an element; that is,

$$\{f\} = \iiint_V [N]^T\{X\}\,dV + \{P\} + \iint_S [N_S]^T\{T_S\}\,dS \tag{6.2.46}$$

where the first, second, and third terms on the right side of Eq. (6.2.46) represent the body forces, the concentrated nodal forces, and the surface tractions, respectively. Using Eq. (6.2.46) in Eq. (6.2.45), we obtain

$$\pi_p = \frac{1}{2}\{d\}^T\iiint_V [B]^T[D][B]\,dV\{d\} - \{d\}^T\{f\} \tag{6.2.47}$$

Taking the first variation, or equivalently, as shown in Chapters 2 and 3, the partial derivative of π_p with respect to the nodal displacements since $\pi_p = \pi_p(\{d\})$ (as was previously done for the bar and beam elements in Chapters 3 and 4, respectively), we obtain

$$\frac{\partial \pi_p}{\partial \{d\}} = \left[\iiint_V [B]^T [D][B] \, dV \right] \{d\} - \{f\} = 0 \tag{6.2.48}$$

Rewriting Eq. (6.2.48), we have

$$\iiint_V [B]^T [D][B] \, dV \{d\} = \{f\} \tag{6.2.49}$$

where the partial derivative with respect to matrix $\{d\}$ was previously defined by Eq. (2.6.12). From Eq. (6.2.49) we can see that

$$[k] = \iiint_V [B]^T [D][B] \, dV \tag{6.2.50}$$

For an element with constant thickness, t, Eq. (6.2.50) becomes

$$[k] = t \iint_A [B]^T [D][B] \, dx \, dy \tag{6.2.51}$$

where the integrand is not a function of x or y for the constant-strain triangular element and thus can be taken out of the integral to yield

$$[k] = tA[B]^T [D][B] \tag{6.2.52}$$

where A is given by Eq. (6.2.9), $[B]$ is given by Eq. (6.2.34), and $[D]$ is given by Eq. (6.1.8) or Eq. (6.1.10). We will assume elements of constant thickness. (This assumption is convergent to the actual situation as the element size is decreased.)

From Eq. (6.2.52) we see that $[k]$ is a function of the nodal coordinates (because $[B]$ and A are defined in terms of them) and of the mechanical properties E and v (of which $[D]$ is a function). The expansion of Eq. (6.2.52) for an element is

$$[k] = \begin{bmatrix} [k_{ii}] & [k_{ij}] & [k_{im}] \\ [k_{ji}] & [k_{jj}] & [k_{jm}] \\ [k_{mi}] & [k_{mj}] & [k_{mm}] \end{bmatrix} \tag{6.2.53}$$

where the 2×2 submatrices are given by

$$[k_{ii}] = [B_i]^T [D][B_i] tA$$
$$[k_{ij}] = [B_i]^T [D][B_j] tA \tag{6.2.54}$$
$$[k_{im}] = [B_i]^T [D][B_m] tA$$

and so forth. In Eqs. (6.2.54), $[B_i]$, $[B_j]$, and $[B_m]$ are defined by Eqs. (6.2.32). The $[k]$ matrix is seen to be a 6×6 matrix (equal in order to the number of degrees of freedom per node, two, times the total number of nodes per element, three).

In general, Eq. (6.2.46) must be used to evaluate the surface and body forces. When Eq. (6.2.46) is used to evaluate the surface and body forces, these forces are called *consistent loads* because they are derived from the consistent (energy) approach. For higher-order elements, typically with quadratic or cubic displacement functions, Eq. (6.2.46) should be used. However, for the CST element, the body and surface forces can be lumped at the nodes with equivalent results (this is illustrated in Section 6.3) and added to any concentrated nodal forces to obtain the element force matrix. The element equations are then given by

$$
\begin{Bmatrix} f_{1x} \\ f_{1y} \\ f_{2x} \\ f_{2y} \\ f_{3x} \\ f_{3y} \end{Bmatrix} = \begin{bmatrix} k_{11} & k_{12} & \cdots & k_{16} \\ k_{21} & k_{22} & \cdots & k_{26} \\ \vdots & \vdots & & \vdots \\ k_{61} & k_{62} & \cdots & k_{66} \end{bmatrix} \begin{Bmatrix} u_1 \\ v_1 \\ u_2 \\ v_2 \\ u_3 \\ v_3 \end{Bmatrix}
\tag{6.2.55}
$$

Finally, realizing that the strain energy U is the first term on the right side of Eq. (6.2.47) and using the expression for the stiffness matrix given by Eq. (6.2.50), we can again express the strain energy in the quadratic form $U = \frac{1}{2} \{d\}^T [k]\{d\}$.

Step 5   Assemble the Element Equations to Obtain the Global Equations and Introduce Boundary Conditions

We obtain the global structure stiffness matrix and equations by using the direct stiffness method as

$$
[K] = \sum_{e=1}^{N} [k^{(e)}]
\tag{6.2.56}
$$

and

$$
\{F\} = [K]\{d\}
\tag{6.2.57}
$$

where, in Eq. (6.2.56), all element stiffness matrices are defined in terms of the global x-y coordinate system, $\{d\}$ is now the total structure displacement matrix, and

$$
\{F\} = \sum_{e=1}^{N} \{f^{(e)}\}
\tag{6.2.58}
$$

is the column of equivalent global nodal loads obtained by lumping body forces and distributed loads at the proper nodes (as well as including concentrated nodal loads) or by consistently using Eq. (6.2.46). (Further details regarding the treatment of body forces and surface tractions will be given in Section 6.3.)

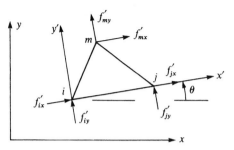

Figure 6–10 Triangular element with local axes not parallel to global axes

In the formulation of the element stiffness matrix Eq. (6.2.52), the matrix has been derived for a general orientation in global coordinates. Equation (6.2.52) then applies for all elements. All element matrices are expressed in the global-coordinate orientation. Therefore, no transformation from local to global equations is necessary. However, for completeness, we will now describe the method to use if the local axes for the constant-strain triangular element are not parallel to the global axes for the whole structure.

If the local axes for the constant-strain triangular element are not parallel to the global axes for the whole structure, we must apply rotation-of-axes transformations similar to those introduced in Chapter 3 by Eq. (3.3.16) to the element stiffness matrix, as well as to the element nodal force and displacement matrices. We illustrate the transformation of axes for the triangular element shown in Figure 6–10, considering the element to have local axes x'-y' not parallel to global axes x-y. Local nodal forces are shown in the figure. The transformation from local to global equations follows the procedure outlined in Section 3.4. We have the same general expressions, Eqs. (3.4.14), (3.4.16), and (3.4.22), to relate local to global displacements, forces, and stiffness matrices, respectively; that is,

$$\{d'\} = [T]\{d\} \qquad \{f'\} = [T]\{f\} \qquad [k] = [T]^T[k'][T] \qquad (6.2.59)$$

where Eq. (3.4.15) for the transformation matrix $[T]$ used in Eqs. (6.2.59) must be expanded because two additional degrees of freedom are present in the constant-strain triangular element. Thus, Eq. (3.4.15) is expanded to

$$[T] = \begin{bmatrix} C & S & 0 & 0 & 0 & 0 \\ -S & C & 0 & 0 & 0 & 0 \\ 0 & 0 & C & S & 0 & 0 \\ 0 & 0 & -S & C & 0 & 0 \\ 0 & 0 & 0 & 0 & C & S \\ 0 & 0 & 0 & 0 & -S & C \end{bmatrix} \begin{matrix} u_i \\ v_i \\ u_j \\ v_j \\ u_m \\ v_m \end{matrix} \qquad (6.2.60)$$

where $C = \cos\theta$, $S = \sin\theta$, and θ is shown in Figure 6–10.

Step 6 Solve for the Nodal Displacements

We determine the unknown global structure nodal displacements by solving the system of algebraic equations given by Eq. (6.2.57).

Step 7 Solve for the Element Forces (Stresses)

Having solved for the nodal displacements, we obtain the strains and stresses in the global x and y directions in the elements by using Eqs. (6.2.33) and (6.2.36). Finally, we determine the maximum and minimum in-plane principal stresses σ_1 and σ_2 by using the transformation Eqs. (6.1.2), where these stresses are usually assumed to act at the centroid of the element. The angle that one of the principal stresses makes with the x axis is given by Eq. (6.1.3).

Example 6.1

Evaluate the stiffness matrix for the element shown in Figure 6–11. The coordinates are shown in units of mm. Assume plane stress conditions. Let $E = 210$ GPa, $v = 0.25$, and thickness $t = 20$ mm. Assume the element nodal displacements have been determined to be $u_1 = 0.0$, $v_1 = 0.05$ mm, $u_2 = 0.025$ mm, $v_2 = 0.0$, $u_3 = 0.0$, and $v_3 = 0.05$ mm. Determine the element stresses.

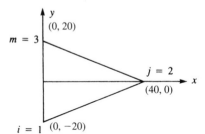

Figure 6–11 Plane stress element for stiffness matrix evaluation

SOLUTION:

We use Eq. (6.2.52) to obtain the element stiffness matrix. To evaluate $[k]$, we first use Eqs. (6.2.10) to obtain the β's and γ's as follows:

$$\beta_i = y_j - y_m = 0 - 20 = -20 \qquad \gamma_i = x_m - x_j = 0 - 40 = -40$$

$$\beta_j = y_m - y_i = 20 - (-20) = 40 \qquad \gamma_j = x_i - x_m = 0 - 0 = 0 \qquad (6.2.61)$$

$$\beta_m = y_i - y_j = -20 - 0 = -20 \qquad \gamma_m = x_j - x_i = 40 - 0 = 40$$

Using Eqs. (6.2.32) and (6.2.34), we obtain matrix $[B]$ as

$$[B] = \frac{10^2}{2(8)} \begin{bmatrix} -2 & 0 & 4 & 0 & -2 & 0 \\ 0 & -4 & 0 & 0 & 0 & 4 \\ -4 & -2 & 0 & 4 & 4 & -1 \end{bmatrix} \frac{1}{m} \qquad (6.2.62)$$

where we have used $A = 8 \times 10^{-4}$ m^2 in Eq. (6.2.62).

Using Eq. (6.1.8) for plane stress conditions,

$$[D] = \frac{210 \times 10^9}{1 - (0.25)^2} \begin{bmatrix} 1 & 0.25 & 0 \\ 0.25 & 1 & 0 \\ 0 & 0 & \dfrac{1 - 0.25}{2} \end{bmatrix} \frac{N}{m^2} \qquad (6.2.63)$$

Substituting Eqs. (6.2.62) and (6.2.63) into Eq. (6.2.52), we obtain

$$[k] = \frac{(20 \times 10^{-3})(8 \times 10^{-4})(210 \times 10^9)(10^2)}{16(0.9375)} \begin{bmatrix} -2 & 0 & -4 \\ 0 & -4 & -2 \\ 4 & 0 & 0 \\ 0 & 0 & 4 \\ -2 & 0 & 4 \\ 0 & 4 & -2 \end{bmatrix}$$

$$\times \begin{bmatrix} 1 & 0.25 & 0 \\ 0.25 & 1 & 0 \\ 0 & 0 & 0.375 \end{bmatrix} \frac{10^2}{2(8)} \begin{bmatrix} -2 & 0 & 4 & 0 & -2 & 0 \\ 0 & -4 & 0 & 0 & 0 & 4 \\ -4 & -2 & 0 & 4 & 4 & -2 \end{bmatrix}$$

Performing the matrix triple product, we have

$$[k] = (56 \times 10^7) \begin{bmatrix} 2.5 & 1.25 & -2 & -1.5 & -0.5 & 0.25 \\ 1.25 & 4.375 & -1 & -0.75 & -0.25 & -3.625 \\ -2 & -1 & 4 & 0 & -2 & 1 \\ -1.5 & -0.75 & 0 & 1.5 & 1.5 & -0.75 \\ -0.5 & -0.25 & -2 & 1.5 & 2.5 & -1.25 \\ 0.25 & -3.625 & 1 & -0.75 & -1.25 & 4.375 \end{bmatrix} \frac{N}{m} \qquad (6.2.64)$$

To evaluate the stresses, we use Eq. (6.2.36). Substituting Eqs. (6.2.62) and (6.2.63), along with the given nodal displacements, into Eq. (6.2.36), we obtain

$$
\left\{
\begin{array}{c}
\sigma_x \\
\sigma_y \\
\tau_{xy}
\end{array}
\right\}
=
\frac{210 \times 10^9}{1 - (0.25)^2}
\begin{bmatrix}
1 & 0.25 & 0 \\
0.25 & 1 & 0 \\
0 & 0 & 0.375
\end{bmatrix}
$$

$$
\times
\frac{10^2}{2(8)}
\begin{bmatrix}
-2 & 0 & 4 & 0 & -2 & 0 \\
0 & -4 & 0 & 0 & 0 & 4 \\
-4 & -2 & 0 & 4 & 4 & -2
\end{bmatrix}
\left\{
\begin{array}{c}
0.0 \\
0.05 \\
0.025 \\
0.0 \\
0.0 \\
0.05
\end{array}
\right\}
\times 10^{-3}
\qquad (6.2.65)
$$

Performing the matrix triple product in Eq. (6.2.65), we have

$$
\sigma_x = 140 \text{ MPa} \qquad \sigma_y = 35 \text{ MPa} \qquad \tau_{xy} = -105 \text{ MPa} \qquad (6.2.66)
$$

Finally, the principal stresses and principal angle are obtained by substituting the results from Eqs. (6.2.66) into Eqs. (6.1.2) and (6.1.3) as follows:

$$
\sigma_1 = \frac{140 + 35}{2} + \left[\left(\frac{140 - 35}{2} \right)^2 + (-105)^2 \right]^{1/2}
$$

$$
= 204.89 \text{ MPa}
$$

$$
\sigma_2 = \frac{140 + 35}{2} - \left[\left(\frac{140 - 35}{2} \right)^2 + (-105)^2 \right]^{1/2} \qquad (6.2.67)
$$

$$
= -29.89 \text{ MPa}
$$

$$
\theta_p = \frac{1}{2} \tan^{-1} \left[\frac{2(-105)}{140 - 35} \right] = -31.7° \qquad ■
$$

▲ 6.3 Treatment of Body and Surface Forces ▲

Body Forces

Using the first term on the right side of Eq. (6.2.46), we can evaluate the body forces at the nodes as

$$
\{f_b\} = \iiint_V [N]^T \{X\} \, dV \qquad (6.3.1)
$$

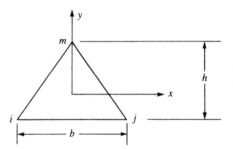

Figure 6–12 Element with centroidal coordinate axes

where
$$\{X\} = \left\{ \begin{array}{c} X_b \\ Y_b \end{array} \right\}$$
(6.3.2)

and X_b and Y_b are the weight densities in the x and y directions in units of force/unit volume, respectively. These forces may arise, for instance, because of actual body weight (gravitational forces), angular velocity (called *centrifugal body forces*, as described in Chapter 9), or inertial forces in dynamics.

In Eq. (6.3.1), $[N]$ is a linear function of x and y; therefore, the integration must be carried out. Without lack of generality, the integration is simplified if the origin of the coordinates is chosen at the centroid of the element. For example, consider the element with coordinates shown in Figure 6–12. With the origin of the coordinate placed at the centroid of the element, we have, from the definition of the centroid, $\iint x\,dA = \iint y\,dA = 0$ and therefore,

$$\iint \beta_i x\,dA = \iint \gamma_i y\,dA = 0$$
(6.3.3)

and
$$\alpha_i = \alpha_j = \alpha_m = \frac{2A}{3}$$
(6.3.4)

Using Eqs. (6.3.2) through (6.3.4) in Eq. (6.3.1), the body force at node i is then represented by

$$\{f_{bi}\} = \left\{ \begin{array}{c} X_b \\ Y_b \end{array} \right\} \frac{tA}{3}$$
(6.3.5)

Similarly, considering the j and m node body forces, we obtain the same results as in Eq. (6.3.5). In matrix form, the element body forces are

$$\{f_b\} = \left\{ \begin{array}{c} f_{bix} \\ f_{biy} \\ f_{bjx} \\ f_{bjy} \\ f_{bmx} \\ f_{bmy} \end{array} \right\} = \left\{ \begin{array}{c} X_b \\ Y_b \\ X_b \\ Y_b \\ X_b \\ Y_b \end{array} \right\} \frac{At}{3}$$
(6.3.6)

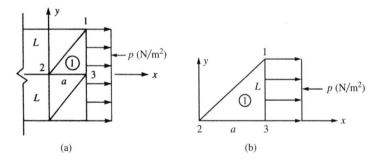

Figure 6–13 (a) Elements with uniform surface traction acting on one edge and (b) element 1 with uniform surface traction along edge 1–3

From the results of Eq. (6.3.6), we can conclude that the body forces are distributed to the nodes in three equal parts. The signs depend on the directions of X_b and Y_b with respect to the positive x and y global coordinates. For the case of body weight only, because of the gravitational force associated with the y direction, we have only Y_b $(X_b = 0)$.

Surface Forces

Using the third term on the right side of Eq. (6.2.46), we can evaluate the surface forces at the nodes as

$$\{f_s\} = \iint_S [N_S]^T \{T_S\} \, dS \tag{6.3.7}$$

We emphasize that the subscript S in $[N_S]$ in Eq. (6.3.7) means the shape functions evaluated along the surface where the surface traction is applied.

We will now illustrate the use of Eq. (6.3.7) by considering the example of a uniform stress p (say, in pounds per square inch) acting between nodes 1 and 3 on the edge of element 1 in Figure 6–13(b). In Eq. (6.3.7), the surface traction now becomes

$$\{T_S\} = \begin{Bmatrix} p_x \\ p_y \end{Bmatrix} = \begin{Bmatrix} p \\ 0 \end{Bmatrix} \tag{6.3.8}$$

and

$$[N_S]^T = \begin{bmatrix} N_1 & 0 \\ 0 & N_1 \\ N_2 & 0 \\ 0 & N_2 \\ N_3 & 0 \\ 0 & N_3 \end{bmatrix} \quad \text{evaluated at } x = a, \; y = y \tag{6.3.9}$$

As the surface traction p acts along the edge at $x = a$ and $y = y$ from $y = 0$ to $y = L$, we evaluate the shape functions at $x = a$ and $y = y$ and integrate over the surface from 0 to L in the y direction and from 0 to t in the z direction, as shown by Eq. (6.3.10).

Using Eqs. (6.3.8) and (6.3.9), we express Eq. (6.3.7) as

$$\{f_s\} = \int_0^t \int_0^L \begin{bmatrix} N_1 & 0 \\ 0 & N_1 \\ N_2 & 0 \\ 0 & N_2 \\ N_3 & 0 \\ 0 & N_3 \end{bmatrix} \begin{Bmatrix} p \\ 0 \end{Bmatrix} dz\, dy \qquad (6.3.10)$$

evaluated at $x = a,\ y = y$

Simplifying Eq. (6.3.10), we obtain

$$\{f_s\} = t \int_0^L \begin{bmatrix} N_1 p \\ 0 \\ N_2 p \\ 0 \\ N_3 p \\ 0 \end{bmatrix} dy \qquad (6.3.11)$$

evaluated at $x = a,\ y = y$

Now, by Eqs. (6.2.18) (with $i = 1$), we have

$$N_1 = \frac{1}{2A}(\alpha_1 + \beta_1 x + \gamma_1 y) \qquad (6.3.12)$$

For convenience, we choose the coordinate system for the element as shown in Figure 6–14. Using the definition Eqs. (6.2.10), we obtain

$$\alpha_i = x_j y_m - y_j x_m$$

or, with $i = 1, j = 2$, and $m = 3$,

$$\alpha_1 = x_2 y_3 - y_2 x_3 \qquad (6.3.13)$$

Substituting the coordinates into Eq. (6.3.13), we obtain

$$\alpha_1 = 0 \qquad (6.3.14)$$

Similarly, again using Eqs. (6.2.10), we obtain

$$\beta_1 = 0 \qquad \gamma_1 = a \qquad (6.3.15)$$

Therefore, substituting Eqs. (6.3.14) and (6.3.15) into Eq. (6.3.12), we obtain

$$N_1 = \frac{ay}{2A} \qquad (6.3.16)$$

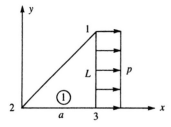

Figure 6–14 Representative element subjected to edge surface traction p

Similarly, using Eqs. (6.2.18), we can show that

$$N_2 = \frac{L(a - x)}{2A} \quad \text{and} \quad N_3 = \frac{Lx - ay}{2A} \tag{6.3.17}$$

On substituting Eqs. (6.3.16) and (6.3.17) for N_1, N_2, and N_3 into Eq. (6.3.11), evaluating N_1, N_2, and N_3 at $x = a$ and $y = y$ (the coordinates corresponding to the location of the surface load p), and then integrating with respect to y, we obtain

$$\{f_s\} = \frac{t}{2(aL/2)} \begin{Bmatrix} a\left(\dfrac{L^2}{2}\right)p \\ 0 \\ 0 \\ 0 \\ \left(L^2 - \dfrac{L^2}{2}\right)ap \\ 0 \end{Bmatrix} \tag{6.3.18}$$

where the shape function $N_2 = 0$ between nodes 1 and 3, as should be the case according to the definitions of the shape functions. Simplifying Eq. (6.3.18), we finally obtain

$$\{f_s\} = \begin{Bmatrix} f_{s1x} \\ f_{s1y} \\ f_{s2x} \\ f_{s2y} \\ f_{s3x} \\ f_{s3y} \end{Bmatrix} = \begin{Bmatrix} pLt/2 \\ 0 \\ 0 \\ 0 \\ pLt/2 \\ 0 \end{Bmatrix} \tag{6.3.19}$$

Figure 6–15 illustrates the results for the surface load equivalent nodal forces for both elements 1 and 2.

We can conclude that for a constant-strain triangle, a distributed load on an element edge can be treated as concentrated loads acting at the nodes associated with the loaded edge by making the two kinds of load statically equivalent [which is equivalent to applying Eq. (6.3.7)]. However, for higher-order elements such as the

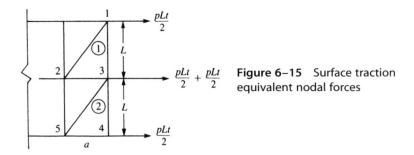

Figure 6–15 Surface traction equivalent nodal forces

linear-strain triangle (discussed in Chapter 8), the load replacement should be made by using Eq. (6.3.7), which was derived by the principle of minimum potential energy. For higher-order elements, this load replacement by use of Eq. (6.3.7) is generally not equal to the apparent statically equivalent one; however, it is consistent in that this replacement results directly from the energy approach.

We now recognize the force matrix $\{f_s\}$ defined by Eq. (6.3.7), and based on the principle of minimum potential energy, to be equivalent to that based on work equivalence, which we previously used in Chapter 4 when discussing distributed loads acting on beams.

▲ 6.4 Explicit Expression for the Constant-Strain Triangle Stiffness Matrix ▲

Although the stiffness matrix is generally formulated internally in most computer programs by performing the matrix triple product indicated by Eq. (6.4.1), it is still a valuable learning experience to evaluate the stiffness matrix explicitly for the constant-strain triangular element. Hence, we will consider the plane strain case specifically in this development.

First, recall that the stiffness matrix is given by

$$[k] = tA[B]^T[D][B] \tag{6.4.1}$$

where, for the plane strain case, $[D]$ is given by Eq. (6.1.10) and $[B]$ is given by Eq. (6.2.34). On substituting the matrices $[D]$ and $[B]$ into Eq. (6.4.1), we obtain

$$[k] = \frac{tE}{4A(1+v)(1-2v)} \begin{bmatrix} \beta_i & 0 & \gamma_i \\ 0 & \gamma_i & \beta_i \\ \beta_j & 0 & \gamma_j \\ 0 & \gamma_j & \beta_j \\ \beta_m & 0 & \gamma_m \\ 0 & \gamma_m & \beta_m \end{bmatrix}$$

$$\times \begin{bmatrix} 1-v & v & 0 \\ v & 1-v & 0 \\ 0 & 0 & \dfrac{1-2v}{2} \end{bmatrix} \begin{bmatrix} \beta_i & 0 & \beta_j & 0 & \beta_m & 0 \\ 0 & \gamma_i & 0 & \gamma_j & 0 & \gamma_m \\ \gamma_i & \beta_i & \gamma_j & \beta_j & \gamma_m & \beta_m \end{bmatrix} \tag{6.4.2}$$

On multiplying the matrices in Eq. (6.4.2), we obtain Eq. (6.4.3), the explicit constant-strain triangle stiffness matrix for the plane strain case. Note that $[k]$ is a function of the difference in the x and y nodal coordinates, as indicated by the γ's and β's, of the material properties E and v, and of the thickness t and surface area A of the element.

$$[k] = \frac{tE}{4A(1+v)(1-2v)}$$

$$\times \begin{bmatrix} \beta_i^2(1-v) + \gamma_i^2\left(\frac{1-2v}{2}\right) & \beta_i\gamma_i v + \beta_i\gamma_i\left(\frac{1-2v}{2}\right) & \beta_i\beta_j(1-v) + \gamma_i\gamma_j\left(\frac{1-2v}{2}\right) \\ & \gamma_i^2(1-v) + \beta_i^2\left(\frac{1-2v}{2}\right) & \beta_j\gamma_i v + \beta_i\gamma_j\left(\frac{1-2v}{2}\right) \\ & & \beta_j^2(1-v) + \gamma_j^2\left(\frac{1-2v}{2}\right) \\ \\ \text{Symmetry} \end{bmatrix}$$

$$\begin{matrix} \beta_i\gamma_j v + \beta_j\gamma_i\left(\frac{1-2v}{2}\right) & \beta_i\beta_m(1-v) + \gamma_i\gamma_m\left(\frac{1-2v}{2}\right) & \beta_i\gamma_m v + \beta_m\gamma_i\left(\frac{1-2v}{2}\right) \\ \gamma_i\gamma_j(1-v) + \beta_i\beta_j\left(\frac{1-2v}{2}\right) & \beta_m\gamma_i v + \beta_i\gamma_m\left(\frac{1-2v}{2}\right) & \gamma_i\gamma_m(1-v) + \beta_i\beta_m\left(\frac{1-2v}{2}\right) \\ \beta_j\gamma_j v + \beta_j\gamma_j\left(\frac{1-2v}{2}\right) & \beta_j\beta_m(1-v) + \gamma_j\gamma_m\left(\frac{1-2v}{2}\right) & \beta_j\gamma_m v + \gamma_j\beta_m\left(\frac{1-2v}{2}\right) \\ \gamma_j^2(1-v) + \beta_j^2\left(\frac{1-2v}{2}\right) & \beta_m\gamma_j v + \beta_j\gamma_m\left(\frac{1-2v}{2}\right) & \gamma_j\gamma_m(1-v) + \beta_j\beta_m\left(\frac{1-2v}{2}\right) \\ & \beta_m^2(1-v) + \gamma_m^2\left(\frac{1-2v}{2}\right) & \gamma_m\beta_m v + \beta_m\gamma_m\left(\frac{1-2v}{2}\right) \\ & & \gamma_m^2(1-v) + \beta_m^2\left(\frac{1-2v}{2}\right) \end{matrix}$$

$$(6.4.3)$$

For the plane stress case, we need only replace $1-v$ by 1, $(1-2v)/2$ by $(1-v)/2$, and $(1+v)(1-2v)$ outside the brackets by $1-v^2$ in Eq. (6.4.3).

Finally, it should be noted that for Poisson's ratio v approaching 0.5, as in rubber-like materials and plastic solids, for instance, a material becomes incompressible [2]. For plane strain, as v approaches 0.5, the denominator becomes zero in the material property matrix [see Eq. (6.1.10)] and hence in the stiffness matrix, Eq. (6.4.3). A value of v

near 0.5 can cause ill-conditioned structural equations. A special formulation (called a *penalty formulation* [3]) has been used in this case.

▲ 6.5 Finite Element Solution of a Plane Stress Problem ▲

To illustrate the finite element method for a plane stress problem, we now present a detailed solution.

Example 6.2

For a thin plate subjected to the surface traction shown in Figure 6–16, determine the nodal displacements and the element stresses. The plate thickness $t = 20$ mm, $E = 210$ GPa, and $v = 0.30$.

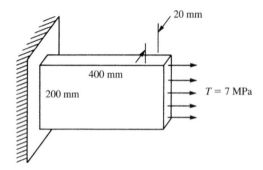

Figure 6–16 Thin plate subjected to tensile stress

SOLUTION:

Discretization

To illustrate the finite element method solution for the plate, we first discretize the plate into two elements, as shown in Figure 6–17. It should be understood that the coarseness of the mesh will not yield as true a predicted behavior of the plate as would a finer mesh, particularly near the fixed edge. However, since we are performing a longhand solution, we will use a coarse discretization for simplicity (but without loss of generality of the method).

In Figure 6–17, the original tensile surface traction in Figure 6–16 has been converted to nodal forces as follows:

$$F = \frac{1}{2}TA$$

$$F = \frac{1}{2}(7 \times 10^6)(20 \times 10^{-3})(200 \times 10^{-3})$$

$$F = 14,000 \text{ N}$$

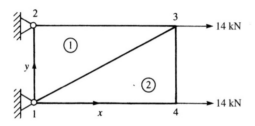

Figure 6–17 Discretized plate

In general, for higher-order elements, Eq. (6.3.7) should be used to convert distributed surface tractions to nodal forces. However, for the CST element, we have shown in Section 6.3 that a statically equivalent force replacement can be used directly, as has been done here.

The governing global matrix equation is

$$\{F\} = [K]\{d\} \tag{6.5.1}$$

Expanding matrices in Eq. (6.5.1), we obtain

$$
\begin{Bmatrix}
F_{1x} \\
F_{1y} \\
F_{2x} \\
F_{2y} \\
F_{3x} \\
F_{3y} \\
F_{4x} \\
F_{4y}
\end{Bmatrix}
=
\begin{Bmatrix}
R_{1x} \\
R_{1y} \\
R_{2x} \\
R_{2y} \\
14{,}000 \\
0 \\
14{,}000 \\
0
\end{Bmatrix}
= [K]
\begin{Bmatrix}
u_1 \\
v_1 \\
u_2 \\
v_2 \\
u_3 \\
v_3 \\
u_4 \\
v_4
\end{Bmatrix}
= [K]
\begin{Bmatrix}
0 \\
0 \\
0 \\
0 \\
u_3 \\
v_3 \\
u_4 \\
v_4
\end{Bmatrix}
\tag{6.5.2}
$$

where $[K]$ is an 8×8 matrix (two degrees of freedom per node with four nodes) before deleting rows and columns to account for the fixed boundary support conditions at nodes 1 and 2.

Assemblage of the Stiffness Matrix

We assemble the global stiffness matrix by superposition of the individual element stiffness matrices. By Eq. (6.2.52), the stiffness matrix for an element is

$$[k] = tA[B]^T[D][B] \tag{6.5.3}$$

In Figure 6–18 for element 1, we have coordinates $x_i = 0$, $y_i = 0$, $x_j = 400$ mm, $y_j = 200$ mm, $x_m = 0$, and $y_m = 200$ mm, since the global coordinate axes are set up at node 1, and

$$A = \frac{1}{2}bh$$

$$A = \left(\frac{1}{2}\right)(400)(200) = 40{,}000 \text{ mm}^2 = 0.04 \text{ m}^2$$

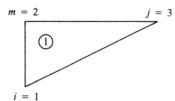

$m = 2$ $j = 3$

①

$i = 1$

Figure 6–18 Element 1 of the discretized plate

or, in general, A can be obtained equivalently by the nodal coordinate formula of Eq. (6.2.9).

We will now evaluate $[B]$, where $[B]$ is given by Eq. (6.2.34), expanded here as

$$[B] = \frac{1}{2A} \begin{bmatrix} \beta_i & 0 & \beta_j & 0 & \beta_m & 0 \\ 0 & \gamma_i & 0 & \gamma_j & 0 & \gamma_m \\ \gamma_i & \beta_i & \gamma_j & \beta_j & \gamma_m & \beta_m \end{bmatrix} \qquad (6.5.4)$$

and, from Eqs. (6.2.10),

$$\beta_i = y_j - y_m = 200 - 200 = 0$$
$$\beta_j = y_m - y_i = 200 - 0 = 200$$
$$\beta_m = y_i - y_j = 0 - 200 = -200$$
$$\gamma_i = x_m - x_j = 0 - 400 = -400 \qquad (6.5.5)$$
$$\gamma_j = x_i - x_m = 0 - 0 = 0$$
$$\gamma_m = x_j - x_i = 400 - 0 = 400$$

Therefore, substituting Eqs. (6.5.5) into Eq. (6.5.4), we obtain

$$[B] = \frac{(20 \times 10^{-3})}{(2 \times 4 \times 10^{-2})} \begin{bmatrix} 0 & 0 & 10 & 0 & -10 & 0 \\ 0 & -20 & 0 & 0 & 0 & 20 \\ -20 & 0 & 0 & 10 & 20 & -10 \end{bmatrix} \frac{1}{m} \qquad (6.5.6)$$

For plane stress, the $[D]$ matrix is conveniently expressed here as

$$[D] = \frac{E}{(1 - v^2)} \begin{bmatrix} 1 & v & 0 \\ v & 1 & 0 \\ 0 & 0 & \dfrac{1-v}{2} \end{bmatrix} \qquad (6.5.7)$$

With $v = 0.3$ and $E = 210 \times 10^9$ N/m^2, we obtain

$$[D] = \frac{210 \times 10^9}{0.91} \begin{bmatrix} 1 & 0.3 & 0 \\ 0.3 & 1 & 0 \\ 0 & 0 & 0.35 \end{bmatrix} \frac{N}{m^2} \qquad (6.5.8)$$

Then
$$[B]^T[D] = \frac{(20 \times 10^{-3})(210 \times 10^9)}{(2 \times 4 \times 10^{-2})(0.91)} \begin{bmatrix} 0 & 0 & -20 \\ 0 & -20 & 0 \\ 10 & 0 & 0 \\ 0 & 0 & 10 \\ -10 & 0 & 20 \\ 0 & 20 & -10 \end{bmatrix} \begin{bmatrix} 1 & 0.3 & 0 \\ 0.3 & 1 & 0 \\ 0 & 0 & 0.35 \end{bmatrix}$$
(6.5.9)

Simplifying Eq. (6.5.9) yields

$$[B]^T[D] = \frac{(52.5)(10^9)}{0.91} \begin{bmatrix} 0 & 0 & -7 \\ -6 & -20 & 0 \\ 10 & 3 & 0 \\ 0 & 0 & 3.5 \\ -10 & -3 & 7 \\ 6 & 20 & -3.5 \end{bmatrix}$$
(6.5.10)

Using Eqs. (6.5.10) and (6.5.6) in Eq. (6.5.3), we have the stiffness matrix for element 1 as

$$[k^{(1)}] = \frac{(52.5)(10^9)}{0.91} \times (20 \times 10^{-3})(0.04) \begin{bmatrix} 0 & 0 & -7 \\ -6 & -20 & 0 \\ 10 & 3 & 0 \\ 0 & 0 & 3.5 \\ -10 & -3 & 7 \\ 6 & 20 & -3.5 \end{bmatrix}$$

$$\times \frac{20 \times 10^{-3}}{(2 \times 4 \times 10^{-2})} \begin{bmatrix} 0 & 0 & 10 & 0 & -10 & 0 \\ 0 & -20 & 0 & 0 & 0 & 20 \\ -20 & 0 & 0 & 10 & 20 & -10 \end{bmatrix}$$
(6.5.11)

Finally, simplifying Eq. (6.5.11) yields

$$[k^{(1)}] = \frac{(10.5 \times 10^6)}{0.91} \begin{array}{cccccc} u_1 & v_1 & u_3 & v_3 & u_2 & v_2 \end{array} \begin{bmatrix} 140 & 0 & 0 & -70 & -140 & 70 \\ 0 & 400 & -60 & 0 & 60 & -400 \\ 0 & -60 & 100 & 0 & -100 & 60 \\ -70 & 0 & 0 & 35 & 70 & -35 \\ -140 & 60 & -100 & 70 & 240 & -130 \\ 70 & -400 & 60 & -35 & -130 & 435 \end{bmatrix} \frac{N}{m}$$
(6.5.12)

where the labels above the columns indicate the counterclockwise nodal order of the degrees of freedom in the element 1 stiffness matrix.

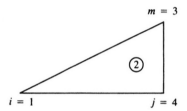

Figure 6–19 Element 2 of the discretized plate

In Figure 6–19 for element 2, we have $x_i = 0$, $y_i = 0$, $x_j = 400$ mm, $y_j = 0$, $x_m = 400$ mm, and $y_m = 200$ mm. Then, from Eqs. (6.2.10), we have

$$\beta_i = y_j - y_m = 0 - 200 = -200$$

$$\beta_j = y_m - y_i = 200 - 0 = 200$$

$$\beta_m = y_i - y_j = 0 - 0 = 0$$

$$\gamma_i = x_m - x_j = 400 - 400 = 0 \qquad (6.5.13)$$

$$\gamma_j = x_i - x_m = 0 - 400 = -400$$

$$\gamma_m = x_j - x_i = 400 - 0 = 400$$

Therefore, using Eqs. (6.5.13) in Eq. (6.5.4) yields

$$[B] = \frac{20 \times 10^{-3}}{(2)(0.04)} \begin{bmatrix} -10 & 0 & 10 & 0 & 0 & 0 \\ 0 & 0 & 0 & -20 & 0 & 20 \\ 0 & -10 & -20 & 10 & 20 & 0 \end{bmatrix} \frac{1}{m} \qquad (6.5.14)$$

The $[D]$ matrix is again given by

$$[D] = \frac{(210 \times 10^9)}{0.91} \begin{bmatrix} 1 & 0.3 & 0 \\ 0.3 & 1 & 0 \\ 0 & 0 & 0.35 \end{bmatrix} \frac{N}{m^2} \qquad (6.5.15)$$

Then, using Eqs. (6.5.14) and (6.5.15), we obtain

$$[B]^T[D] = \frac{(20 \times 10^{-3})(210 \times 10^9)}{(2)(0.04)(0.91)} \begin{bmatrix} -10 & 0 & 0 \\ 0 & 0 & -10 \\ 10 & 0 & -20 \\ 0 & -20 & 10 \\ 0 & 0 & 20 \\ 0 & 20 & 0 \end{bmatrix} \begin{bmatrix} 1 & 0.3 & 0 \\ 0.3 & 1 & 0 \\ 0 & 0 & 0.35 \end{bmatrix} \qquad (6.5.16)$$

Simplifying Eq. (6.5.16) yields

$$[B]^T[D] = \frac{(52.5 \times 10^9)}{0.91} \begin{bmatrix} -10 & -3 & 0 \\ 0 & 0 & -3.5 \\ 10 & 3 & -7 \\ -6 & -20 & 3.5 \\ 0 & 0 & 7 \\ 6 & 20 & 0 \end{bmatrix} \qquad (6.5.17)$$

Finally, substituting Eqs. (6.5.17) and (6.5.14) into Eq. (6.5.3), we obtain the stiffness matrix for element 2 as

$$[k^{(2)}] = (20 \times 10^{-3})(0.04) \frac{(52.5 \times 10^9)}{0.91} \begin{bmatrix} -10 & -3 & 0 \\ 0 & 0 & -3.5 \\ 10 & 3 & -7 \\ -6 & -20 & 3.5 \\ 0 & 0 & 7 \\ 6 & 20 & 0 \end{bmatrix}$$

$$\times \frac{(20 \times 10^{-3})}{(2)(0.04)} \begin{bmatrix} -10 & 0 & 10 & 0 & 0 & 0 \\ 0 & 0 & 0 & -20 & 0 & 20 \\ 0 & -10 & -20 & 10 & 20 & 0 \end{bmatrix} \qquad (6.5.18)$$

Equation (6.5.18) simplifies to

$$[k^{(2)}] = \frac{(10.5 \times 10^6)}{0.91} \begin{array}{c} \begin{matrix} u_1 \quad\;\; v_1 \quad\;\; u_4 \quad\;\; v_4 \quad\;\; u_3 \quad\;\; v_3 \end{matrix} \\ \begin{bmatrix} 100 & 0 & -100 & 60 & 0 & -60 \\ 0 & 35 & 70 & -35 & -70 & 0 \\ -100 & 70 & 240 & -130 & -140 & 60 \\ 60 & -35 & -130 & 435 & 70 & -400 \\ 0 & -70 & -140 & 70 & 140 & 0 \\ -60 & 0 & 60 & -400 & 0 & 400 \end{bmatrix} \dfrac{N}{m} \end{array} \qquad (6.5.19)$$

where the degrees of freedom in the element 2 stiffness matrix are shown above the columns in Eq. (6.5.19). Rewriting the element stiffness matrices, Eqs. (6.5.12) and (6.5.19), expanded to the order of, and rearranged according to, increasing nodal degrees of freedom of the total $[K]$ matrix (where we have factored out a constant 5), we obtain

Element 1

$$[k^{(1)}] = \frac{52.5 \times 10^6}{0.91}
\begin{array}{c}
\begin{array}{cccccccc} u_1 & \;v_1 & \;\;u_2 & \;\;v_2 & \;\;u_3 & \;\;v_3 & \;u_4 & v_4 \end{array} \\
\begin{bmatrix}
28 & 0 & -28 & 14 & 0 & -14 & 0 & 0 \\
0 & 80 & 12 & -80 & -12 & 0 & 0 & 0 \\
-28 & 12 & 48 & -26 & -20 & 14 & 0 & 0 \\
14 & -80 & -26 & 87 & 12 & -7 & 0 & 0 \\
0 & -12 & -20 & 12 & 20 & 0 & 0 & 0 \\
-14 & 0 & 14 & -7 & 0 & 7 & 0 & 0 \\
0 & 0 & 0 & 0 & 0 & 0 & 0 & 0 \\
0 & 0 & 0 & 0 & 0 & 0 & 0 & 0
\end{bmatrix}
\end{array}
\frac{N}{m} \quad (6.5.20)$$

Element 2

$$[k^{(2)}] = \frac{52.5 \times 10^6}{0.91}
\begin{array}{c}
\begin{array}{cccccccc} u_1 & \;v_1 & u_2 & v_2 & \;u_3 & \;\;v_3 & \;u_4 & v_4 \end{array} \\
\begin{bmatrix}
20 & 0 & 0 & 0 & 0 & -12 & -20 & 12 \\
0 & 7 & 0 & 0 & -14 & 0 & 14 & -7 \\
0 & 0 & 0 & 0 & 0 & 0 & 0 & 0 \\
0 & 0 & 0 & 0 & 0 & 0 & 0 & 0 \\
0 & -14 & 0 & 0 & 28 & 0 & -28 & 14 \\
-12 & 0 & 0 & 0 & 0 & 80 & 12 & -80 \\
-20 & 14 & 0 & 0 & -28 & 12 & 48 & -26 \\
12 & -7 & 0 & 0 & 14 & -80 & -26 & 87
\end{bmatrix}
\end{array}
\frac{N}{m} \quad (6.5.21)$$

Using superposition of the element stiffness matrices, Eqs. (6.5.20) and (6.5.21), now that the orders of the degrees of freedom are the same, we obtain the total global stiffness matrix as

$$[K] = \frac{52.5 \times 10^6}{0.91}
\begin{array}{c}
\begin{array}{cccccccc} u_1 & \;\;v_1 & \;\;u_2 & \;\;v_2 & \;\;u_3 & \;\;v_3 & \;\;u_4 & v_4 \end{array} \\
\begin{bmatrix}
48 & 0 & -28 & 14 & 0 & -26 & -20 & 12 \\
0 & 87 & 12 & -80 & -26 & 0 & 14 & -7 \\
-28 & 12 & 48 & -26 & -20 & 14 & 0 & 0 \\
14 & -80 & -26 & 87 & 12 & -7 & 0 & 0 \\
0 & -26 & -20 & 12 & 48 & 0 & -28 & 14 \\
-26 & 0 & 14 & -7 & 0 & 87 & 12 & -80 \\
-20 & 14 & 0 & 0 & -28 & 12 & 48 & -26 \\
12 & -7 & 0 & 0 & 14 & -80 & -26 & 87
\end{bmatrix}
\end{array}
\frac{N}{m} \quad (6.5.22)$$

[Alternatively, we could have applied the direct stiffness method to Eqs. (6.5.12) and (6.5.19) to obtain Eq. (6.5.22).] Substituting $[K]$ into $\{F\} = [K]\{d\}$ of Eq. (6.5.2), we have

$$
\begin{Bmatrix} R_{1x} \\ R_{1y} \\ R_{2x} \\ R_{2y} \\ 14 \times 10^3 \\ 0 \\ 14 \times 10^3 \\ 0 \end{Bmatrix} = \frac{52.5 \times 10^9}{0.91}
\begin{bmatrix}
48 & 0 & -28 & 14 & 0 & -26 & -20 & 12 \\
0 & 87 & 12 & -80 & -26 & 0 & 14 & -7 \\
-28 & 12 & 48 & -26 & -20 & 14 & 0 & 0 \\
14 & -80 & -26 & 87 & 12 & -7 & 0 & 0 \\
0 & -26 & -20 & 12 & 48 & 0 & -28 & 14 \\
-26 & 0 & 14 & -7 & 0 & 87 & 12 & -80 \\
-20 & 14 & 0 & 0 & -28 & 12 & 48 & -26 \\
12 & -7 & 0 & 0 & 14 & -80 & -26 & 87
\end{bmatrix}
\begin{Bmatrix} 0 \\ 0 \\ 0 \\ 0 \\ u_3 \\ v_3 \\ u_4 \\ v_4 \end{Bmatrix}
$$

$$(6.5.23)$$

Applying the support or boundary conditions by eliminating rows and columns corresponding to displacement matrix rows and columns equal to zero [namely, rows and columns 1–4 in Eq. (6.5.23)], we obtain

$$
\begin{Bmatrix} 14 \times 10^3 \\ 0 \\ 14 \times 10^3 \\ 0 \end{Bmatrix} = \frac{52.5 \times 10^6}{0.91}
\begin{bmatrix}
48 & 0 & -28 & 14 \\
0 & 87 & 12 & -80 \\
-28 & 12 & 48 & -26 \\
14 & -80 & -26 & 87
\end{bmatrix}
\begin{Bmatrix} u_3 \\ v_3 \\ u_4 \\ v_4 \end{Bmatrix}
$$

$$(6.5.24)$$

Premultiplying both sides of Eq. (6.5.24) by $[K]^{-1}$, we have

$$
\begin{Bmatrix} u_3 \\ v_3 \\ u_4 \\ v_4 \end{Bmatrix} = \frac{0.91}{52.5 \times 10^6}
\begin{bmatrix}
48 & 0 & -28 & 14 \\
0 & 87 & 12 & -80 \\
-28 & 12 & 48 & -26 \\
14 & -80 & -26 & 87
\end{bmatrix}^{-1}
\begin{Bmatrix} 14 \times 10^3 \\ 0 \\ 14 \times 10^3 \\ 0 \end{Bmatrix}
$$

$$(6.5.25)$$

Solving for the displacements in Eq. (6.5.25), we obtain

$$
\begin{Bmatrix} u_3 \\ v_3 \\ u_4 \\ v_4 \end{Bmatrix} = \frac{0.91}{3750}
\begin{Bmatrix} 0.05024 \\ 0.00034 \\ 0.05470 \\ 0.00878 \end{Bmatrix}
$$

$$(6.5.26)$$

Simplifying Eq. (6.5.26), the final displacements are given by

$$
\begin{Bmatrix} u_3 \\ v_3 \\ u_4 \\ v_4 \end{Bmatrix} = \begin{Bmatrix} 12.19 \\ 0.083 \\ 13.27 \\ 2.08 \end{Bmatrix} \times 10^{-6} \text{ m}
$$

$$(6.5.27)$$

Comparing the finite element solution to an analytical solution, as a first approximation, we have the axial displacement given by

$$
\delta = \frac{PL}{AE} = \frac{(28{,}000)(400 \times 10^{-3})}{(20 \times 10^{-3})(200 \times 10^{-3})(210 \times 10^9)} = 13.33 \times 10^{-6} \text{ m}
$$

for a one-dimensional bar subjected to tensile force. Hence, the nodal x displacement components of Eq. (6.5.27) for the two-dimensional plate appear to be reasonably correct, considering the coarseness of the mesh and the directional stiffness bias of the model. (For more on this subject see Section 7.5.) The y displacement would be expected to be downward at the top (node 3) and upward at the bottom (node 4) as a result of the Poisson effect. However, the directional stiffness bias due to the coarse mesh accounts for this unexpected poor result.

We now determine the stresses in each element by using Eq. (6.2.36):

$$\{\sigma\} = [D][B]\{d\} \tag{6.5.28}$$

In general, for element 1, we then have

$$\{\sigma\} = \frac{E}{(1-v^2)} \begin{bmatrix} 1 & v & 0 \\ v & 1 & 0 \\ 0 & 0 & \dfrac{1-v}{2} \end{bmatrix} \times \left(\frac{1}{2A}\right) \begin{bmatrix} \beta_1 & 0 & \beta_3 & 0 & \beta_2 & 0 \\ 0 & \gamma_1 & 0 & \gamma_3 & 0 & \gamma_2 \\ \gamma_1 & \beta_1 & \gamma_3 & \beta_3 & \gamma_2 & \beta_2 \end{bmatrix} \begin{Bmatrix} u_1 \\ v_1 \\ u_3 \\ v_3 \\ u_2 \\ v_2 \end{Bmatrix}$$

$$\tag{6.5.29}$$

Substituting numerical values for $[B]$, given by Eq. (6.5.6); for $[D]$, given by Eq. (6.5.8); and the appropriate part of $\{d\}$, given by Eq. (6.5.27), we obtain

$$\{\sigma\} = \frac{(210 \times 10^9)(10^{-6})}{0.91(4)} \begin{bmatrix} 1 & 0.3 & 0 \\ 0.3 & 1 & 0 \\ 0 & 0 & 0.35 \end{bmatrix}$$

$$\times \begin{bmatrix} 0 & 0 & 10 & 0 & -10 & 0 \\ 0 & -20 & 0 & 0 & 0 & 20 \\ -20 & 0 & 0 & 10 & 20 & -10 \end{bmatrix} \begin{Bmatrix} 0 \\ 0 \\ 12.19 \\ 0.083 \\ 0 \\ 0 \end{Bmatrix} \tag{6.5.30}$$

Simplifying Eq. (6.5.30), we obtain

$$\begin{Bmatrix} \sigma_x \\ \sigma_y \\ \tau_{xy} \end{Bmatrix} = \begin{Bmatrix} 7032.41 \\ 2109.72 \\ 16.75 \end{Bmatrix} \text{kPa} \tag{6.5.31}$$

In general, for element 2, we have

$$\{\sigma\} = \frac{E}{(1-v^2)}\left(\frac{1}{2A}\right)\begin{bmatrix} 1 & v & 0 \\ v & 1 & 0 \\ 0 & 0 & \frac{1-v}{2} \end{bmatrix} \times \begin{bmatrix} \beta_1 & 0 & \beta_4 & 0 & \beta_3 & 0 \\ 0 & \gamma_1 & 0 & \gamma_4 & 0 & \gamma_3 \\ \gamma_1 & \beta_1 & \gamma_4 & \beta_4 & \gamma_3 & \beta_3 \end{bmatrix}\begin{Bmatrix} u_1 \\ v_1 \\ u_4 \\ v_4 \\ u_3 \\ v_3 \end{Bmatrix}$$

(6.5.32)

Substituting numerical values into Eq. (6.5.32), we obtain

$$\{\sigma\} = \frac{(210 \times 10^9)(10^{-6})}{0.91(4)}\begin{bmatrix} 1 & 0.3 & 0 \\ 0.3 & 1 & 0 \\ 0 & 0 & 0.35 \end{bmatrix}$$

$$\times \begin{bmatrix} -10 & 0 & 10 & 0 & 0 & 0 \\ 0 & 0 & 0 & -20 & 0 & 20 \\ 0 & -10 & -20 & 10 & 20 & 0 \end{bmatrix}\begin{Bmatrix} 0 \\ 0 \\ 13.27 \\ 2.08 \\ 12.19 \\ 0.083 \end{Bmatrix}$$

(6.5.33)

Simplifying Eq. (6.5.33), we obtain

$$\begin{Bmatrix} \sigma_x \\ \sigma_y \\ \tau_{xy} \end{Bmatrix} = \begin{Bmatrix} 6964.5 \\ -7.5 \\ -16.1 \end{Bmatrix}\text{kPa}$$

(6.5.34)

The principal stresses can now be determined from Eq. (6.1.2), and the principal angle made by one of the principal stresses can be determined from Eq. (6.1.3). (The other principal stress will be directed 90° from the first.) We determine these principal stresses for element 2 (those for element 1 will be similar) as

$$\sigma_1 = \frac{\sigma_x + \sigma_y}{2} + \left[\left(\frac{\sigma_x - \sigma_y}{2}\right)^2 + \tau_{xy}^2\right]^{1/2}$$

$$\sigma_1 = \frac{6964.5 + (-7.5)}{2} + \left[\left(\frac{6964.5 - (-7.5)}{2}\right)^2 + (-16.1)^2\right]^{1/2}$$

(6.5.35)

$$\sigma_1 = 3478.5 + 3486.03 = 6964.53 \text{ kPa}$$

$$\sigma_2 = \frac{6964.5 + (-7.5)}{2} - 3486.03 = -7.53 \text{ kPa}$$

The principal angle is then

$$\theta_p = \frac{1}{2}\tan^{-1}\left[\frac{2\tau_{xy}}{\sigma_x - \sigma_y}\right]$$

(6.5.36)

or $\qquad \theta_p = \frac{1}{2}\tan^{-1}\left[\frac{2(-16.1)}{6964.5 - (-7.5)}\right] = 0.13° \simeq 0°$

Owing to the uniform stress of 7 MPa acting only in the x direction on the edge of the plate, we would expect the stress $\sigma_x(= \sigma_1)$ to be near 7 MPa in each element. Thus, the results from Eqs. (6.5.31) and (6.5.34) for σ_x are quite good. We would expect the stress σ_y to be very small (at least near the free edge). The restraint of element 1 at nodes 1 and 2 causes a relatively large element stress σ_y, whereas the restraint of element 2 at only one node causes a very small stress σ_y. The shear stresses τ_{xy} remain close to zero, as expected. Had the number of elements been increased, with smaller ones used near the support edge, even more realistic results would have been obtained. However, a finer discretization would result in a cumbersome longhand solution and thus was not used here. Use of a computer program is recommended for a detailed solution to this plate problem and certainly for solving more complex stress–strain problems. ∎

The maximum distortion energy theory [4] (also called the *von Mises* or *von Mises-Hencky theory*) for ductile materials subjected to static loading predicts that a material will fail if the von Mises stress (also called *equivalent* or *effective stress*) reaches the yield strength, S_y, of the material. The von Mises stress as derived in [4], for instance, is given in terms of the three principal stresses by

$$\sigma_{vm} = \frac{1}{\sqrt{2}}\left[(\sigma_1 - \sigma_2)^2 + (\sigma_2 - \sigma_3)^2 + (\sigma_3 - \sigma_1)^2\right]^{1/2}$$

(6.5.37a)

or equivalently in terms of the x-y-z components as

$$\sigma_{vm} = \frac{1}{\sqrt{2}}\left[(\sigma_x - \sigma_y)^2 + (\sigma_y - \sigma_z)^2 + (\sigma_z - \sigma_x)^2 + 6(\tau_{xy}^2 + \tau_{yz}^2 + \tau_{zx}^2)\right]^{1/2}$$

(6.5.37b)

Thus for yielding to occur, the von Mises stress must become equal to or greater than the yield strength of the material as given by

$$\sigma_{vm} \geq S_y$$

(6.5.38)

We can see from Eqs. (6.5.37a or 6.5.37b) that the von Mises stress is a scalar that measures the intensity of the entire stress state as it includes the three principal stresses or the three normal stresses in the x, y, and z directions, along with the shear stresses on the x, y, and z planes. Other stresses, such as the maximum principal one, do not provide the most accurate way of predicting failure.

Most computer programs incorporate this failure theory and, as an optional result, the user can request a plot of the von Mises stress throughout the material model being analyzed. If the von Mises stress value is equal to or greater than the yield strength of the material being considered, then another material with greater yield strength can be selected or other design changes can be made.

For brittle materials, such as glass and cast iron, with different tension and compression properties, it is recommended to use the Coulomb-Mohr theory to predict failure. For more on this theory consult [4].

▲ 6.6 Rectangular Plane Element (Bilinear Rectangle, Q4) ▲

We will now develop the four-noded rectangular plane element stiffness matrix. We will later refer to this element in the isoparametric formulation of a general quadrilateral element in Section 10.2. This element is also called the bilinear rectangle because of the linear terms in x and y for the x and y displacement functions shown in Eq. (6.6.2). The "Q4" symbol represents the element as a quadrilateral with four corner nodes.

Two advantages of the rectangular element over the triangular element are ease of data input and simpler interpretation of output stresses. A disadvantage of the rectangular element is that the simple linear-displacement rectangle with its associated straight sides poorly approximates the real boundary condition edges.

The usual steps outlined in Chapter 1 will be followed to obtain the element stiffness matrix and related equations.

Step 1   Select Element Type

Consider the rectangular element shown in Figure 6–20 (all interior angles are 90°) with corner nodes 1–4 (again labeled counterclockwise) and base and height dimensions $2b$ and $2h$, respectively.

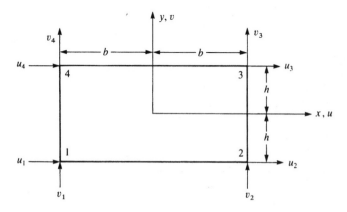

Figure 6–20   Basic four-node rectangular element with nodal degrees of freedom

The unknown nodal displacements are now given by

$$\{d\} = \begin{Bmatrix} u_1 \\ v_1 \\ u_2 \\ v_2 \\ u_3 \\ v_3 \\ u_4 \\ v_4 \end{Bmatrix} \tag{6.6.1}$$

Step 2 Select Displacement Functions

For a compatible displacement field, the element displacement functions u and v must be linear along each edge because only two points (the corner nodes) exist along each edge. We then select the linear displacement functions as

$$u(x, y) = a_1 + a_2 x + a_3 y + a_4 xy$$
$$v(x, y) = a_5 + a_6 x + a_7 y + a_8 xy \tag{6.6.2}$$

There are a total of eight generalized degrees of freedom (a's) in Eq. (6.6.2) and a total of eight specific degrees of freedom (u_1, v_1 at node 1 through u_4, v_4 at node 4) for the element.

We can proceed in the usual manner to eliminate the a_i's from Eqs. (6.6.2) to obtain

$$u(x, y) = \frac{1}{4bh}[(b - x)(h - y)u_1 + (b + x)(h - y)u_2$$

$$+ (b + x)(h + y)u_3 + (b - x)(h + y)u_4]$$

$$v(x, y) = \frac{1}{4bh}[(b - x)(h - y)v_1 + (b + x)(h - y)v_2 \tag{6.6.3}$$

$$+ (b + x)(h + y)v_3 + (b - x)(h + y)v_4]$$

These displacement expressions, Eqs. (6.6.3), can be expressed equivalently in terms of the shape functions and unknown nodal displacements as

$$\{\psi\} = [N]\{d\} \tag{6.6.4}$$

where the shape functions are given by

$$N_1 = \frac{(b - x)(h - y)}{4bh} \qquad N_2 = \frac{(b + x)(h - y)}{4bh}$$

$$N_3 = \frac{(b + x)(h + y)}{4bh} \qquad N_4 = \frac{(b - x)(h + y)}{4bh} \tag{6.6.5}$$

and the N_i's are again such that $N_1 = 1$ at node 1 and $N_1 = 0$ at all the other nodes, with similar requirements for the other shape functions. In expanded form, Eq. (6.6.4) becomes

$$\left\{ \begin{array}{c} u \\ v \end{array} \right\} = \begin{bmatrix} N_1 & 0 & N_2 & 0 & N_3 & 0 & N_4 & 0 \\ 0 & N_1 & 0 & N_2 & 0 & N_3 & 0 & N_4 \end{bmatrix} \left\{ \begin{array}{c} u_1 \\ v_1 \\ u_2 \\ v_2 \\ u_3 \\ v_3 \\ u_4 \\ v_4 \end{array} \right\} \tag{6.6.6}$$

Step 3   Define the Strain–Displacement and Stress–Strain Relationships

Again the element strains for the two-dimensional stress state are given by

$$\left\{ \begin{array}{c} \varepsilon_x \\ \varepsilon_y \\ \gamma_{xy} \end{array} \right\} = \left\{ \begin{array}{c} \dfrac{\partial u}{\partial x} \\[2ex] \dfrac{\partial v}{\partial y} \\[2ex] \dfrac{\partial u}{\partial y} + \dfrac{\partial v}{\partial x} \end{array} \right\} \tag{6.6.7a}$$

Using Eq. (6.6.2) in Eq. (6.6.7a), we express the strains in terms of the a's as

$$\begin{aligned} \varepsilon_x &= a_2 + a_4 y \\ \varepsilon_y &= a_7 + a_8 x \\ \gamma_{xy} &= (a_3 + a_6) + a_4 x + a_8 y \end{aligned} \tag{6.6.7b}$$

Using Eq. (6.6.6) in Eq. (6.6.7a) and taking the derivatives of u and v as indicated, we can express the strains in terms of the unknown nodal displacements as

$$\{\varepsilon\} = [B]\{d\} \tag{6.6.8}$$

where

$$[B] = \frac{1}{4bh} \begin{bmatrix} -(h-y) & 0 & (h-y) & 0 \\ 0 & -(b-x) & 0 & -(b+x) \\ -(b-x) & -(h-y) & -(b+x) & (h-y) \end{bmatrix}$$

$$\begin{matrix} (h+y) & 0 & -(h+y) & 0 \\ 0 & (b+x) & 0 & (b-x) \\ (b+x) & (h+y) & (b-x) & -(h+y) \end{matrix} \tag{6.6.9}$$

From Eqs. (6.6.7b), (6.6.8), and (6.6.9), we observe that ε_x is a function of y, ε_y is a function of x, and γ_{xy} is a function of both x and y. The stresses are again given by the formulas in Eq. (6.2.36), where $[B]$ is now that of Eq. (6.6.9) and $\{d\}$ is that of Eq. (6.6.1).

Step 4 Derive the Element Stiffness Matrix and Equations

The stiffness matrix is determined by

$$[k] = \int_{-h}^{h} \int_{-b}^{b} [B]^T [D][B] t \, dx \, dy \tag{6.6.10}$$

with $[D]$ again given by the usual plane stress or plane strain conditions, Eq. (6.1.8) or (6.1.10). Because the $[B]$ matrix is a function of x and y, integration of Eq. (6.6.10) must be performed. The $[k]$ matrix for the rectangular element is now of order 8×8. A numerical evaluation of Eq. (6.6.10) for $[k]$ is shown in Eq. (6.6.11) using $b = 100$ mm, $h = 50$ mm, $t = 25$ mm, $E = 210$ GPa, and $v = 0.3$. This double integral was solved using Mathcad [5].

$$[k] = \begin{bmatrix} 5.077\text{e}6 & 1.5\text{e}6 & 1.385\text{e}6 & -115{,}384.0 & -2.538\text{e}6 & -1.5\text{e}6 & -3.923\text{e}6 & 115{,}384.0 \\ 1.5\text{e}6 & 1.258\text{e}7 & 115{,}384.0 & 5.885\text{e}6 & -1.5\text{e}6 & -6.288\text{e}6 & -115{,}384.0 & -1.217\text{e}7 \\ 1.385\text{e}6 & 115{,}384.0 & 5.077\text{e}6 & -1.5\text{e}6 & -3.923\text{e}6 & -115{,}384.0 & -2.538\text{e}6 & 1.5\text{e}6 \\ -115{,}384.0 & 5.885\text{e}6 & -1.5\text{e}6 & 1.258\text{e}7 & 115{,}384.0 & -1.217\text{e}7 & 1.5\text{e}6 & -6.288\text{e}6 \\ -2.538\text{e}6 & -1.5\text{e}6 & -3.923\text{e}6 & 115{,}384.0 & 5.077\text{e}6 & 1.5\text{e}6 & 1.385\text{e}6 & -115{,}384.0 \\ -1.5\text{e}6 & -6.288\text{e}6 & -115{,}384.0 & -1.217\text{e}7 & 1.5\text{e}6 & 1.258\text{e}7 & 115{,}384.0 & 5.885\text{e}6 \\ -3.923\text{e}6 & -115{,}384.0 & -2.538\text{e}6 & 1.5\text{e}6 & 1.385\text{e}6 & 115{,}384.0 & 5.077\text{e}6 & -1.5\text{e}6 \\ 115{,}384.0 & -1.217\text{e}7 & 1.5\text{e}6 & -6.288\text{e}6 & -115{,}384.0 & 5.885\text{e}6 & -1.5\text{e}6 & 1.258\text{e}7 \end{bmatrix} \frac{\text{N}}{\text{m}^2}$$

$$\tag{6.6.11}$$

The element force matrix is determined by Eq. (6.2.46) as

$$\{f\} = \iiint_V [N]^T \{X\} \, dV + \{P\} + \iint_S [N_s]^T \{T\} \, dS \tag{6.6.12}$$

where $[N]$ is the rectangular matrix in Eq. (6.6.6), and N_1 through N_4 are given by Eqs. (6.6.5). The element equations are then given by

$$\{f\} = [k]\{d\} \tag{6.6.13}$$

Steps 5 through 7

Steps 5 through 7, which involve assembling the global stiffness matrix and equations, determining the unknown nodal displacements, and calculating the stress, are identical to those in Section 6.2 for the CST. However, the stresses within each element now vary in both the x and y directions.

Numerical Comparison of CST to Q4 Element Models and Element Defects

Table 6–1 compares the free end deflection and maximum principal stress for a cantilevered beam modeled with 2, 4, and 8 rows of either all triangular CST elements or all rectangular Q4 elements.

Table 6–1 Table comparing free-end deflections and largest principal stresses for CST and Q4 element models (end force = 4000 N, length = 1 m, $I = 1 \times 10^{-5}$ m^4, thickness = 0.12 m, E = 200 GPa)

Plane Element Used/Rows	Number of Nodes	Number of Degrees of Freedom	Free End Displ., m	Principal Stress, MPa
Q4/2	60	120	6.708×10^{-4}	19.35
Q4/4	200	400	6.729×10^{-4}	20.30
Q4/8	720	1440	6.729×10^{-4}	21.72
CST/2	60	120	3.630×10^{-4}	7.80
CST/4	200	400	5.537×10^{-4}	13.76
CST/8	720	1440	6.385×10^{-4}	17.61
Classical beam theory			6.667×10^{-4}	20.00

Typical Q4 and CST models:

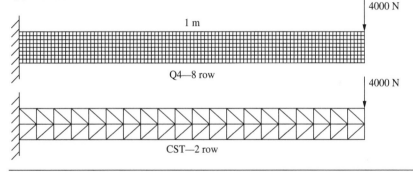

Numbers in Table by Wes Campbell.

Displacement Results   We observe from the displacement results that the CST element models produce stiffer models than the actual beam behavior, as the deflections are predicted to be smaller than classical beam theory predicts. We also observe that the CST model converges very slowly to the classical beam theory solution. This is partly due to the element predicting only constant stress within each element when for a bending problem; the stress actually varies linearly through the depth of the beam. This problem is rectified by using the linear-strain triangle (LST) element as described in Chapter 8.

The results indicate that the Q4 element model predicts more accurate deflection behavior than the CST element model. The two-row model of Q4 elements yields deflections very close to that predicted by the classical beam deflection equation, whereas the two-row model of CST elements is quite inaccurate in predicting the deflection. As the number of rows is increased to four and then eight, the deflections are predicted increasingly more accurately for the CST and Q4 element models. The two-noded beam element model gives the identical deflection as the classical equation ($\delta = PL^3/3EI$) as expected (see discussion in Section 4.5) and is the most appropriate model for this problem when you are not concerned, for instance, with stress concentrations.

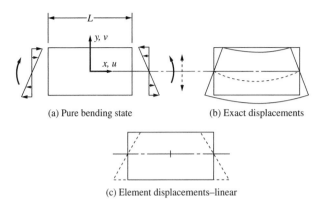

(a) Pure bending state　　(b) Exact displacements

(c) Element displacements–linear

Figure 6–21　(a) Pure bending state, (b) exact bending displacement, and (c) Q4 element displacements—linear edge displacements

As further shown in [3] for a beam subjected to pure bending, the CST has a spurious or false shear stress and hence a spurious shear strain in parts of the model that should not have any shear stress or shear strain. This spurious shear strain absorbs energy; therefore, some of the energy that should go into bending is lost. The CST is then too stiff in bending, and the resulting deformation is smaller than it actually should be. This phenomenon of excessive stiffness developing in one or more modes of deformation is sometimes described as *shear locking* or *parasitic shear*.

Furthermore, in problems where plane strain conditions exist (recall this means when $\varepsilon_z = 0$) and the Poisson's ratio approaches 0.5, a mesh can actually lock, which means the mesh then cannot deform at all.

It should be noted that using a single row of Q4 elements with their linear edge displacement is not recommended to accurately predict the stress gradient through the depth of the beam. This is illustrated in Figure 6–21, where for the pure bending state (approximated by this example), the exact displacement is shown in Figure 6–21(b), while the Q4 element displacement is shown in Figure 6–21(c), which is not capable of pure bending deformation.

Stress Results　As mentioned previously, the CST element has constant strain and stress within it, while the Q4 element normal strain, ε_x, and hence the normal stress, σ_x, is linear in the y direction. (Also see Eq. (6.6.7b).) Therefore, the CST is not able to simulate the bending behavior nearly as well as the Q4 element. The classical beam theory/bending stress equation predicts a linear stress variation through the depth of the beam given by $\sigma_x = -My/I$—shown in Section 4.1 as well. As shown when comparing the principal stresses for each model, as more rows are used, the stresses approach the classical bending stress of 20 MPa with the Q4 approaching the classical solution much faster as indicated by comparing the two-row solutions for Q4 and CST models.

Finally, the eight-noded quadratic edge displacement element (Q8) predicts bending behavior better than both the CST and Q4 elements. Thus, fewer Q8 elements can be used and faster convergence to the proper solution are obtained using this element. In fact, using even a single row of Q8 elements yields reasonable results in bending, as shown in [3]. Again, Section 10.5 describes the Q8 element.

This brief description of some of the limitations in using the CST and Q4 elements does not prevent us from using them to model plane stress and plane strain problems. It just requires us to use a fine mesh as opposed to a coarse one, particularly where bending occurs and where in general large stress gradients will results. Also, we must make sure our computer program can handle Poisson's ratios that approach 0.5 (if that is desired, such as in rubber-like materials). For common materials, such as metals, Poisson's ratio is around 0.3, so locking should not be of concern.

▲ Summary Equations

Stress vector for two-dimensional stress state:

$$\{\sigma\} = \begin{Bmatrix} \sigma_x \\ \sigma_y \\ \tau_{xy} \end{Bmatrix} \tag{6.1.1}$$

Principal stresses for two-dimensional stress state:

$$\sigma_1 = \frac{\sigma_x + \sigma_y}{2} + \sqrt{\left(\frac{\sigma_x - \sigma_y}{2}\right)^2 + \tau_{xy}^2} = \sigma_{max}$$

$$\sigma_2 = \frac{\sigma_x + \sigma_y}{2} - \sqrt{\left(\frac{\sigma_x - \sigma_y}{2}\right)^2 + \tau_{xy}^2} = \sigma_{min} \tag{6.1.2}$$

Principal angle:

$$\tan 2\theta_p = \frac{2\tau_{xy}}{\sigma_x - \sigma_y} \tag{6.1.3}$$

Strain-displacement equations for two-dimensional stress state:

$$\varepsilon_x = \frac{\partial u}{\partial x} \qquad \varepsilon_y = \frac{\partial v}{\partial y} \qquad \gamma_{xy} = \frac{\partial u}{\partial y} + \frac{\partial v}{\partial x} \tag{6.1.4}$$

Strain vector for two-dimensional stress state:

$$\{\varepsilon\} = \begin{Bmatrix} \varepsilon_x \\ \varepsilon_y \\ \gamma_{xy} \end{Bmatrix} \tag{6.1.5}$$

Stress–strain relationship for two-dimensional stress state:

$$\{\sigma\} = [D]\{\varepsilon\} \tag{6.1.7}$$

Stress–strain or constitutive matrix for plane stress condition:

$$[D] = \frac{E}{1 - v^2} \begin{bmatrix} 1 & v & 0 \\ v & 1 & 0 \\ 0 & 0 & \frac{1 - v}{2} \end{bmatrix} \tag{6.1.8}$$

Stress–strain matrix for plane strain condition:

$$[D] = \frac{E}{(1+v)(1-2v)} \begin{bmatrix} 1-v & v & 0 \\ v & 1-v & 0 \\ 0 & 0 & \frac{1-2v}{2} \end{bmatrix} \tag{6.1.10}$$

Displacement functions for three-noded triangular element:

$$\begin{aligned} u(x,y) &= a_1 + a_2 x + a_3 y \\ v(x,y) &= a_4 + a_5 x + a_6 y \end{aligned} \tag{6.2.2}$$

Shape functions for three-noded triangular element:

$$\begin{aligned} N_i &= \frac{1}{2A}(\alpha_i + \beta_i x + \gamma_i y) \\ N_j &= \frac{1}{2A}(\alpha_j + \beta_j x + \gamma_j y) \\ N_m &= \frac{1}{2A}(\alpha_m + \beta_m x + \gamma_m y) \end{aligned} \tag{6.2.18}$$

where

$$\begin{aligned} \alpha_i &= x_j y_m - y_j x_m & \alpha_j &= y_i x_m - x_i y_m & \alpha_m &= x_i y_j - y_i x_j \\ \beta_i &= y_j - y_m & \beta_j &= y_m - y_i & \beta_m &= y_i - y_j \\ \gamma_i &= x_m - x_j & \gamma_j &= x_i - x_m & \gamma_m &= x_j - x_i \end{aligned} \tag{6.2.10}$$

Shape function matrix for three-noded triangular element:

$$[N] = \begin{bmatrix} N_i & 0 & N_j & 0 & N_m & 0 \\ 0 & N_i & 0 & N_j & 0 & N_m \end{bmatrix} \tag{6.2.22}$$

Strain-displacement equations in matrix form:

$$\{\varepsilon\} = [[B_i]\ [B_j]\ [B_m]] \begin{Bmatrix} \{d_i\} \\ \{d_j\} \\ \{d_m\} \end{Bmatrix} \tag{6.2.31}$$

where the gradient matrix is

$$[B_i] = \frac{1}{2A} \begin{bmatrix} \beta_i & 0 \\ 0 & \gamma_i \\ \gamma_i & \beta_i \end{bmatrix} \qquad [B_j] = \frac{1}{2A} \begin{bmatrix} \beta_j & 0 \\ 0 & \gamma_j \\ \gamma_j & \beta_j \end{bmatrix} \qquad [B_m] = \frac{1}{2A} \begin{bmatrix} \beta_m & 0 \\ 0 & \gamma_m \\ \gamma_m & \beta_m \end{bmatrix} \tag{6.2.32}$$

$$[B] = [[B_i]\ [B_j]\ [B_m]] \tag{6.2.34}$$

Stress–strain relationship as function of displacement matrix:

$$\{\sigma\} = [D][B]\{d\} \tag{6.2.36}$$

Total potential energy for two-dimensional stress state:

$$\pi_p = U + \Omega_b + \Omega_p + \Omega_s \tag{6.2.38}$$

where
Strain energy is

$$U = \frac{1}{2} \iiint_V \{\varepsilon\}^T \{\sigma\} dV \qquad (6.2.39)$$

Potential energy of body forces is

$$\Omega_b = -\iiint_V \{\psi\}^T \{X\} dV \qquad (6.2.41)$$

Potential energy of concentrated loads is

$$\Omega_p = -\{d\}^T \{P\} \qquad (6.2.42)$$

Potential energy of surface tractions is

$$\Omega_s = -\iint_S \{\psi_S\}^T \{T_S\} dS \qquad (6.2.43)$$

Stiffness matrix for CST element:

$$[k] = tA[B]^T[D][B] \qquad (6.2.52)$$

Explicit body forces:

$$\{f_b\} = \begin{Bmatrix} f_{bix} \\ f_{biy} \\ f_{bjx} \\ f_{bjy} \\ f_{bmx} \\ f_{bmy} \end{Bmatrix} = \begin{Bmatrix} X_b \\ Y_b \\ X_b \\ Y_b \\ X_b \\ Y_b \end{Bmatrix} \frac{At}{3} \qquad (6.3.6)$$

Explicit surface forces for uniform surface traction in x-direction along side 1–3:

$$\{f_s\} = \begin{Bmatrix} f_{s1x} \\ f_{s1y} \\ f_{s2x} \\ f_{s2y} \\ f_{s3x} \\ f_{s3y} \end{Bmatrix} = \begin{Bmatrix} pLt/2 \\ 0 \\ 0 \\ 0 \\ pLt/2 \\ 0 \end{Bmatrix} \qquad (6.3.19)$$

Explicit expression for constant-strain triangle (CST) stiffness matrix (See Eq. (6.4.3)):
von Mises stress:

$$\sigma_{vm} = \frac{1}{\sqrt{2}} \left[(\sigma_1 - \sigma_2)^2 + (\sigma_2 - \sigma_3)^2 + (\sigma_3 - \sigma_1)^2 \right]^{1/2} \qquad (6.5.37a)$$

Failure based on maximum distortion energy theory:

$$\sigma_{vm} \geq S_y \qquad (6.5.38)$$

Displacement functions for bilinear four-noded rectangle element:

$$u(x, y) = a_1 + a_2 x + a_3 y + a_4 xy$$

$$v(x, y) = a_5 + a_6 x + a_7 y + a_8 xy \tag{6.6.2}$$

Shape functions for four-noded rectangle element:

$$N_1 = \frac{(b - x)(h - y)}{4bh} \qquad N_2 = \frac{(b + x)(h - y)}{4bh}$$

$$N_3 = \frac{(b + x)(h + y)}{4bh} \qquad N_4 = \frac{(b - x)(h + y)}{4bh} \tag{6.6.5}$$

Strain-displacement equations for four-noded rectangle element in terms of a's:

$$\varepsilon_x = a_2 + a_4 y$$

$$\varepsilon_y = a_7 + a_8 x \tag{6.6.7b}$$

$$\gamma_{xy} = (a_3 + a_6) + a_4 x + a_8 y$$

Strain-displacement equations in matrix form:

$$\{\varepsilon\} = [B]\{d\} \tag{6.6.8}$$

where the gradient matrix is

$$[B] = \frac{1}{4bh} \begin{bmatrix} -(h - y) & 0 & (h - y) & 0 \\ 0 & -(b - x) & 0 & -(b + x) \\ -(b - x) & -(h - y) & -(b + x) & (h - y) \end{bmatrix}$$

$$\begin{matrix} (h + y) & 0 & -(h + y) & 0 \\ 0 & (b + x) & 0 & (b - x) \\ (b + x) & (h + y) & (b - x) & -(h + y) \end{matrix} \tag{6.6.9}$$

Stiffness matrix for four-noded rectangular element:

$$[k] = \int_{-h}^{h} \int_{-b}^{b} [B]^T [D][B] t \, dx \, dy \tag{6.6.10}$$

Element force matrix for four-noded rectangular element:

$$\{f\} = \iiint_V [N]^T \{X\} \, dV + \{P\} + \iint_S [N_s]^T \{T\} \, dS \tag{6.6.12}$$

▲ References

[1] Timoshenko, S., and Goodier, J., *Theory of Elasticity*, 3rd ed., McGraw-Hill, New York 1970.

[2] Gere, J. M., and Goodno, B. J., *Mechanics of Materials*, 7th ed., Cengage Learning, Mason, OH, 2009.

[3] Cook, R. D., Malkus, D. S., Plesha, M. E., and Witt, R. J., *Concepts and Applications of Finite Element Analysis*, 4th ed., Wiley, New York, 2002.

[4] Shigley, J. E., Mischke, C. R., and Budynas, R. G., *Mechanical Engineering Design*, 7th ed., McGraw-Hill, New York, 2004.

[5] Mathcad 14.0, Parametric Technology Corp.

▲ Problems

6.1 Sketch the variations of the shape functions N_j and N_m, given by Eqs. (6.2.18), over the surface of the triangular element with nodes i, j, and m. Check that $N_i + N_j + N_m = 1$ anywhere on the element.

6.2 For a simple three-noded triangular element, show explicitly that differentiation of Eq. (6.2.47) indeed results in Eq. (6.2.48); that is, substitute the expression for $[B]$ and the plane stress condition for $[D]$ into Eq. (6.2.47), and then differentiate π_p with respect to each nodal degree of freedom in Eq. (6.2.47) to obtain Eq. (6.2.48).

6.3 Evaluate the stiffness matrix for the elements shown in Figure P6–3. The coordinates are in units of cm. Assume plane stress conditions. Let $E = 210$ GPa, $v = 0.25$, and thickness $t = 1$ cm.

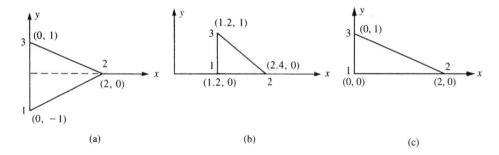

Figure P6–3

6.4 For the elements given in Problem 6.3, the nodal displacements are given as

$$u_1 = 0.0 \text{ mm} \qquad v_1 = 0.0625 \text{ mm} \qquad u_2 = 0.03 \text{ mm}$$

$$v_2 = 0.0 \text{ mm} \qquad u_3 = 0.0 \text{ mm} \qquad v_3 = 0.0625 \text{ mm}$$

Determine the element stresses $\sigma_x, \sigma_y, \tau_{xy}, \sigma_1$, and σ_2 and the principal angle θ_p. Use the values of E, v, and t given in Problem 6.3.

6.5 Determine the von Mises stress for Problem 6.4.

6.6 Evaluate the stiffness matrix for the elements shown in Figure P6–6. The coordinates are given in units of millimeters. Assume plane stress conditions. Let $E = 210$ GPa, $v = 0.25$, and $t = 10$ mm.

6.7 For the elements given in Problem 6.6, the nodal displacements are given as

$$u_1 = 2.0 \text{ mm} \qquad v_1 = 1.0 \text{ mm} \qquad u_2 = 0.5 \text{ mm}$$

$$v_2 = 0.0 \text{ mm} \qquad u_3 = 3.0 \text{ mm} \qquad v_3 = 1.0 \text{ mm}$$

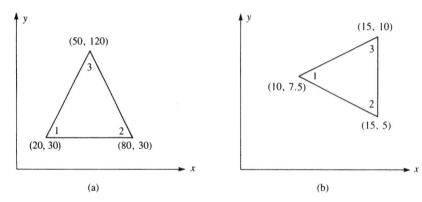

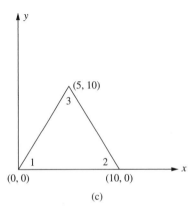

Figure P6–6

Determine the element stresses $\sigma_x, \sigma_y, \tau_{xy}, \sigma_1$, and σ_2 and the principal angle θ_p. Use the values of E, v, and t given in Problem 6.6.

6.8 Determine the von Mises stress for Problem 6.7.

6.9 For the plane strain elements shown in Figure P6–9, the nodal displacements are given as

$$u_1 = 0.001 \text{ cm} \qquad v_1 = 0.005 \text{ cm} \qquad u_2 = 0.001 \text{ cm}$$

$$v_2 = 0.0025 \text{ cm} \qquad u_3 = 0.0 \text{ cm} \qquad v_3 = 0.0 \text{ cm}$$

Determine the element stresses $\sigma_x, \sigma_y, \tau_{xy}, \sigma_1$, and σ_2 and the principal angle θ_p. Let $E = 210$ GPa and $v = 0.25$, and use unit thickness for plane strain. All coordinates are in cm.

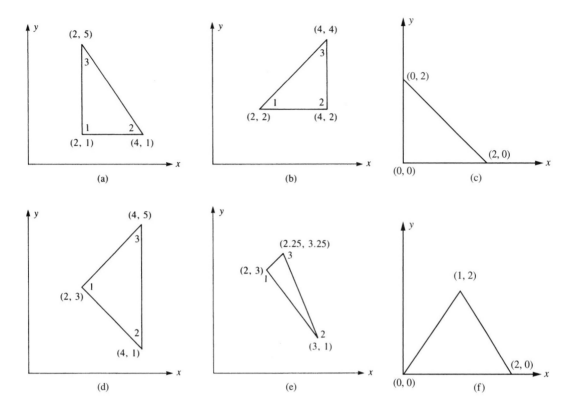

Figure P6–9

6.10 For the plane strain elements shown in Figure P6–10, the nodal displacements are given as

$$u_1 = 0.005 \text{ mm} \qquad v_1 = 0.002 \text{ mm} \qquad u_2 = 0.0 \text{ mm}$$

$$v_2 = 0.0 \text{ mm} \qquad u_3 = 0.005 \text{ mm} \qquad v_3 = 0.0 \text{ mm}$$

Determine the element stresses $\sigma_x, \sigma_y, \tau_{xy}, \sigma_1$, and σ_2 and the principal angle θ_p. Let $E = 70$ GPa and $v = 0.3$, and use unit thickness for plane strain. All coordinates are in millimeters.

6.11 Determine the nodal forces for (1) a linearly varying pressure p_x on the edge of the triangular element shown in Figure P6–11(a); and (2) the quadratic varying pressure shown in Figure P6–11(b) by evaluating the surface integral given by Eq. (6.3.7). Assume the element thickness is equal to t.

6.12 Determine the nodal forces for (1) the quadratic varying pressure loading shown in Figure P6–12(a) and (2) the sinusoidal varying pressure loading shown in Figure P6–12(b)

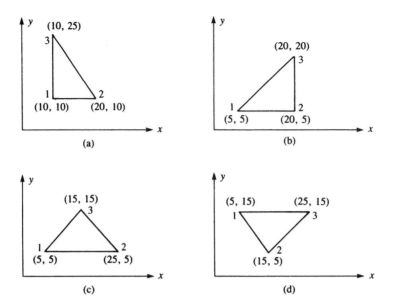

Figure P6–10

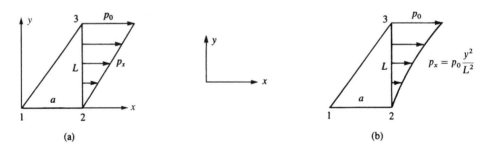

Figure P6–11

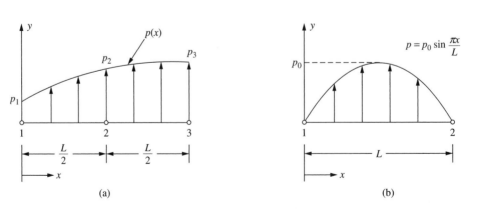

Figure P6–12

by the work equivalence method (use the surface integral expression given by Eq. (6.3.7)). Assume the element thickness to be t.

6.13 Determine the nodal displacements and the element stresses, including principal stresses, for the thin plate of Section 6.5 with a uniform shear load (instead of a tensile load) acting on the right edge, as shown in Figure P6–13. Use $E = 210$ GPa, $v = 0.30$, and $t = 25$ mm. (*Hint:* The $[K]$ matrix derived in Section 6.5 and given by Eq. (6.5.22) can be used to solve the problem.)

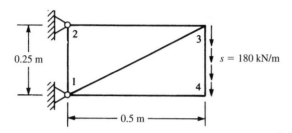

Figure P6–13

6.14 Determine the nodal displacements and the element stresses, including principal stresses, due to the loads shown for the thin plates in Figure P6–14 on the next page. Use $E = 210$ GPa, $v = 0.30$, and $t = 5$ mm. Assume plane stress conditions apply. The recommended discretized plates are shown in the figures.

6.15 Evaluate the body force matrix for the plates shown in Figures P6–14(a) and (c). Assume the weight density to be 77.1 kN/m^3.

6.16 Why is the triangular stiffness matrix derived in Section 6.2 called a constant-strain triangle?

6.17 How do the stresses vary within the constant-strain triangle element?

6.18 Can you use the plane stress or plane strain element to model the following:
 a. a flat slab floor of a building with vertical loading perpendicular to the slab
 b. a wall subjected to wind loading (the wall acts as a shear wall with loads in the plane of the wall)
 c. a tensile plate with a hole drilled transversely through it
 d. an eyebar with loads in the plane of the eyebar
 e. a soil mass subjected to a strip footing loading
 f. a wrench subjected to a force in the plane of the wrench
 g. a wrench subjected to twisting forces (the twisting forces act out of the plane of the wrench)
 h. a triangular plate connection with loads in the plane of the triangle
 i. a triangular plate connection with out-of-plane loads

6.19 The plane stress element only allows for in-plane displacements, while the frame or beam element resists displacements and rotations. How can we combine the plane stress and beam elements and still ensure compatibility?

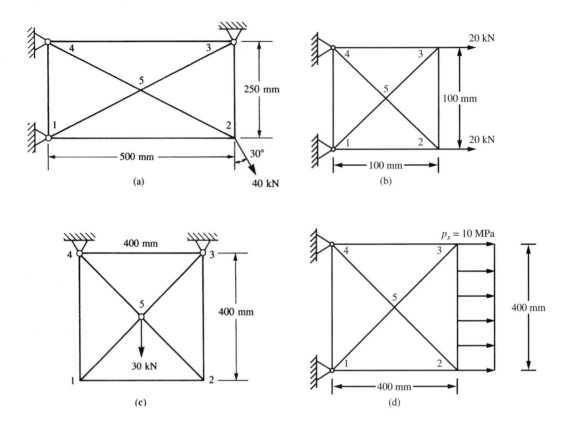

Figure P6–14

6.20 For the plane structures modeled by triangular elements shown in Figure P6–20, show that numbering in the direction that has fewer nodes, as in Figure P6–20(a) (as opposed to numbering in the direction that has more nodes), results in a reduced bandwidth. Illustrate this fact by filling in, with X's, the occupied elements in $[K]$ for each mesh, as was done in Appendix B.4. Compare the bandwidths for each case.

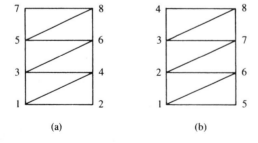

Figure P6–20

6.21 Go through the detailed steps to evaluate Eq. (6.3.6).

6.22 How would you treat a linearly varying thickness for a three-noded triangle?

6.23 Compute the stiffness matrix of element 1 of the two-triangle element model of the rectangular plate in plane stress shown in Figure P6–23. Then use it to compute the stiffness matrix of element 2.

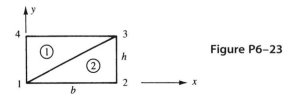

Figure P6–23

6.24 Show that the sum $N_1 + N_2 + N_3 + N_4$ is equal to 1 anywhere on a rectangular element, where N_1 through N_4 are defined by Eqs. (6.6.5).

6.25 For the rectangular element of Figure 6–20 on page 367 the nodal displacements are given by

$$u_1 = 0 \text{ mm} \qquad v_1 = 0 \text{ mm} \qquad u_2 = 0.1 \text{ mm}$$

$$v_2 = 0.05 \text{ mm} \qquad u_3 = 0.05 \text{ mm} \qquad v_3 = -0.05 \text{ mm}$$

$$u_4 = 0 \text{ mm} \qquad v_4 = 0 \text{ mm}$$

For $b = 40$ mm, $h = 1$ in., $E = 210$ GPa, and $v = 0.3$, determine the element strains and stresses at the centroid of the element and at the corner nodes.

Practical Considerations in Modeling; Interpreting Results; and Examples of Plane Stress–Strain Analysis ▲

CHAPTER OBJECTIVES

- To present concepts that should be considered when modeling for a solution by the finite element method, such as aspect ratio, symmetry, natural subdivisions, sizing of elements and the h, p, and r methods of refinement, concentrated loads and infinite stress, infinite medium, and connecting different kinds of elements.
- To describe some of the approximations inherent in finite element solutions.
- To illustrate convergence of solution and introduce the patch test for convergence of solution.
- To discuss the interpretation of stresses in an element, including a common method of averaging the nodal values (also called smoothing).
- To consider the concept of static condensation as a method used in some computer programs to develop the stiffness matrix of a quadrilateral element.
- To present a flowchart of a typical finite element process used for the analysis of plane stress and plane strain.
- To demonstrate various real-world applications where plane stress–strain element models are applicable. Such examples include a bicycle wrench, a connecting rod with notch and hole stress concentrations, an irregularly shaped overload protection device, an irregularly shaped dam, and a beam welded to a column with surface contact elements to allow the separation of surface between beam and column during the beam flexing.

Introduction

In this chapter, we will describe some modeling guidelines, including generally recommended mesh size, natural subdivisions modeling around concentrated loads, and more on use of symmetry and associated boundary conditions. This is followed by discussion of equilibrium, compatibility, and convergence of solution. We will then consider interpretation of stress results.

Next, we introduce the concept of static condensation, which enables us to apply the concept of the basic constant-strain triangle stiffness matrix to a quadrilateral element. Thus, both three-sided and four-sided two-dimensional elements can be used in the finite element models of actual bodies.

We then show some computer program results. A computer program facilitates the solution of complex, large-number-of-degrees-of-freedom plane stress/plane strain problems that generally cannot be solved longhand because of the larger number of equations involved. Also, problems for which longhand solutions do not exist (such as those involving complex geometries and complex loads or where unrealistic, often gross, assumptions were previously made to simplify the problem to allow it to be described via a classical differential equation approach) can now be solved with a higher degree of confidence in the results by using the finite element approach (with its resulting system of algebraic equations).

▲ 7.1 Finite Element Modeling ▲

We will now discuss various concepts that should be considered when modeling any problem for solution by the finite element method.

General Considerations

Finite element modeling is partly an art guided by visualizing physical interactions taking place within the body. One appears to acquire good modeling techniques through experience and by working with experienced people. General-purpose programs provide some guidelines for specific types of problems [12, 15]. In subsequent parts of this section, some significant concepts that should be considered are described.

In modeling, the user is first confronted with the sometimes difficult task of understanding the physical behavior taking place and understanding the physical behavior of the various elements available for use. Choosing the proper type of element or elements to match as closely as possible the physical behavior of the problem is one of the numerous decisions that must be made by the user. Understanding the boundary conditions imposed on the problem can, at times, be a difficult task. Also, it is often difficult to determine the kinds of loads that must be applied to a body and their magnitudes and locations. Again, working with more experienced users and searching the literature can help overcome these difficulties.

Aspect Ratio and Element Shapes

*The **aspect ratio** is defined as the ratio of the longest dimension to the shortest dimension of a quadrilateral element.* In many cases, as the aspect ratio increases, the inaccuracy of the solution increases. To illustrate this point, Figure 7–1(a) shows five different finite element models used to analyze a beam subjected to bending. The element used

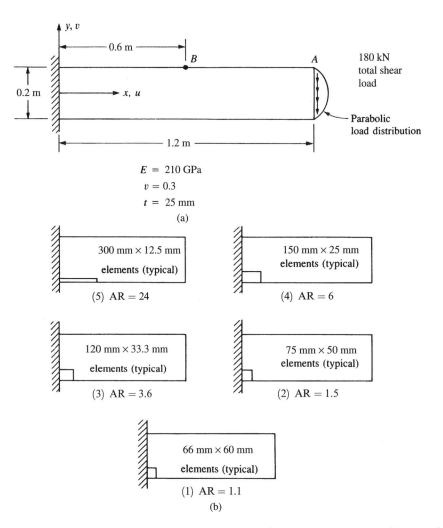

Figure 7–1 (a) Beam with loading (b) effects of the aspect ratio (AR) illustrated by five cases with different aspect ratios

here is the rectangular one described in Section 6.6. Figure 7–1(b) is a plot of the resulting error in the displacement at point A of the beam versus the aspect ratio. Table 7–1 reports a comparison of results for the displacements at points A and B for the five models, and the exact solution [2].

There are exceptions for which aspect ratios approaching 50 still produce satis-factory results; for example, if the stress gradient is close to zero at some location of the actual problem, then large aspect ratios at that location still produce reasonable results.

In general, an element yields best results if its shape is compact and regular. Although different elements have different sensitivities to shape distortions, try to maintain (1) aspect ratios low as in Figure 7–1, cases 1 and 2, and (2) corner angles

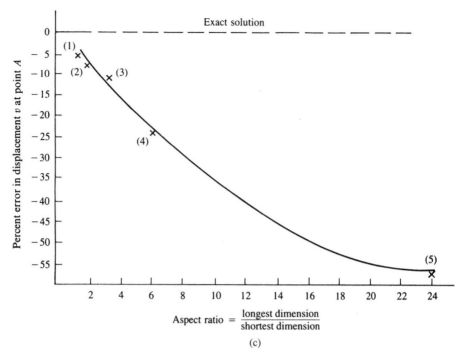

Figure 7–1 (c) Inaccuracy of solution as a function of the aspect ratio (numbers in parentheses correspond to the cases listed in Table 7–1)

of quadrilaterals near 90°. Figure 7–2 on the next page shows elements with poor shapes that tend to promote poor results. If few of these poor element shapes exist in a model, then usually only results near these elements are poor. In the Algor program [12], when $\alpha \geqslant 170°$ in Figure 7–2(c), the program automatically divides the quadrilateral into two triangles.

Table 7–1 Comparison of results for various aspect ratios

Case	Aspect Ratio	Number of Nodes	Number of Elements	Vertical Displacement, v (mm) Point A	Point B	Percent Error in Displacement at A
1	1.1	84	60	−27.76	−8.78	5.2
2	1.5	85	64	−27.38	−8.61	6.4
3	3.6	77	60	−25.75	−8.33	11.9
4	6.0	81	64	−22.50	−7.11	23.0
5	24.0	85	64	−12.7	−4.01	56.0
Exact solution [2]				−29.26	−9.14	

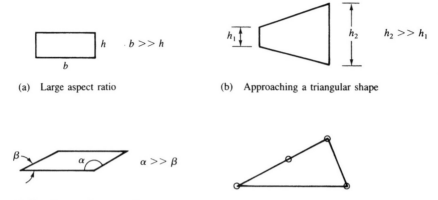

(a) Large aspect ratio

(b) Approaching a triangular shape

(c) Very large and very small
corner angles

(d) Triangular quadrilateral

Figure 7–2 Elements with poor shapes

Use of Symmetry

The appropriate use of symmetry* will often expedite the modeling of a problem. Use of symmetry allows us to consider a reduced problem instead of the actual problem. Thus, we can use a finer subdivision of elements with less labor and computer costs. For another discussion on the use of symmetry, see Reference [3].

Figures 7–3 through 7–5 illustrate the use of symmetry in modeling a soil mass subjected to foundation loading, a uniaxially loaded member with a fillet, and a plate with a hole subjected to internal pressure, respectively. Note that at the plane of symmetry the displacement in the direction perpendicular to the plane must be equal to zero. This is modeled by the rollers at nodes 2–6 in Figure 7–3, where the plane of symmetry is the vertical plane passing through nodes 1–6, perpendicular to the plane of the model. In Figures 7–4(a) and 7–5(a), there are two planes of symmetry. Thus, we need model only one-fourth of the actual members, as shown in Figures 7–4(b) and 7–5(b). Therefore, rollers are used at nodes along both the vertical and horizontal planes of symmetry.

As previously indicated in Chapter 3, in vibration and buckling problems, symmetry must be used with caution since symmetry in geometry does not imply symmetry in all vibration or buckling modes.

Natural Subdivisions at Discontinuities

Figure 7–6 illustrates various natural subdivisions for finite element discretization. For instance, nodes are required at locations of concentrated loads or discontinuity in loads, as shown in Figure 7–6(a) and (b). Nodal lines are defined by abrupt changes of plate thickness, as in Figure 7–6(c), and by abrupt changes of material properties,

---

* Again, reflective *symmetry* means correspondence in size, shape, and position of loads; material properties; and boundary conditions that are on opposite sides of a dividing line or plane.

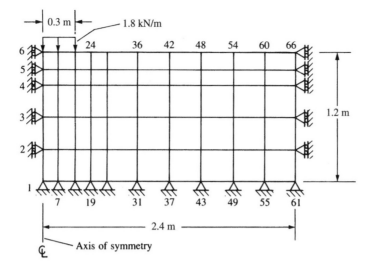

Figure 7–3 Use of symmetry applied to a soil mass subjected to foundation loading (number of nodes = 66, number of elements = 50)

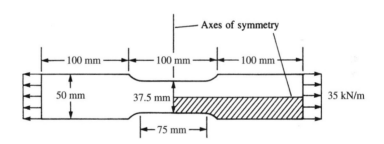

(a) Plane stress uniaxially loaded member with fillet

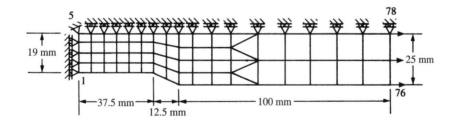

(b) Enlarged finite element model of the cross-hatched quarter of the member (number of nodes = 78, number of elements = 60)

Figure 7–4 Use of symmetry applied to a uniaxially loaded member with a fillet

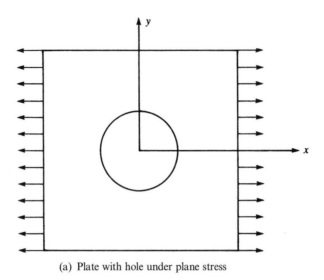

(a) Plate with hole under plane stress

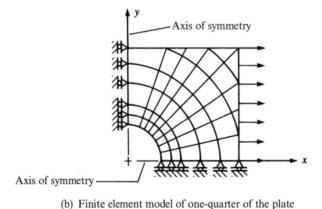

(b) Finite element model of one-quarter of the plate

Figure 7–5 Problem reduction using axes of symmetry applied to a plate with a hole subjected to tensile force

as in Figure 7–6(d) and (e). Other natural subdivisions occur at re-entrant corners, as in Figure 7–6(f), and along holes in members, as in Figure 7–5.

Sizing of Elements and the *h, p,* and *r* Methods of Refinement

For structural problems, to obtain displacements, rotations, stresses, and strains, many computer programs include two basic solution methods and some a third. (These same methods apply to nonstructural problems as well.) These are called the *h method*, the *p method,* and the *r method*. These methods are then used to revise or refine a finite

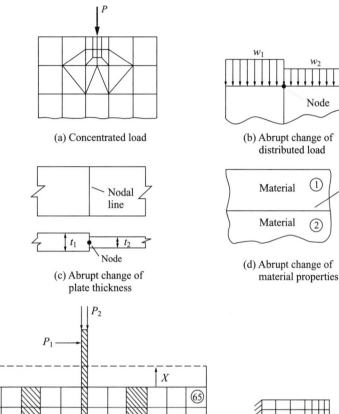

(a) Concentrated load

(b) Abrupt change of distributed load

(c) Abrupt change of plate thickness

(d) Abrupt change of material properties

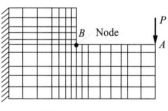

P_2

P_1

X

65

1

13

(e) Basic model of an implant (cross-hatched) in bone, located at various depths X beneath the bony surface, using rectangular elements

(f) Re-entrant corner, B

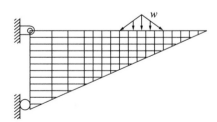

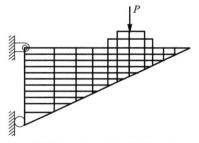

(g) Structure with a distributed load

(h) Using elements to distribute the loading and spread the concentrated load

Figure 7–6 Natural subdivisions at discontinuities

element mesh to improve the results in the next refined analysis. The goal of the analyst is to refine the mesh to obtain the necessary accuracy by using only as many degrees of freedom as necessary. The final objective of this so called *adaptive refinement* is to obtain equal distribution of an error indicator over all elements.

The discretization depends on the geometry of the structure, the loading pattern, and the boundary conditions. For instance, regions of stress concentration or high stress gradient due to fillets, holes, or re-entrant corners require a finer mesh near those regions, as indicated in Figures 7–4, 7–5, and 7–6(f).

We will briefly describe the *h*, *p*, and *r* methods of refinement and provide references for those interested in more in-depth understanding of these methods.

h **Method of Refinement**   In the *h* method of refinement, we use the particular element based on the shape functions for that element (for example, linear functions for the bar, quadratic for the beam, bilinear for the CST). We then start with a baseline mesh to provide a baseline solution for error estimation and to provide guidance for mesh revision. We then add elements of the same kind to refine or make smaller elements in the model. Sometimes a uniform refinement is done where the original element size (Figure 7–7(a)) is perhaps divided in two in both directions as shown in Figure 7–7(b). More often, the refinement is a nonuniform *h* refinement as shown in Figure 7–7(c) (perhaps even a local refinement used to capture some physical phenomenon, such as a shock wave or a thin boundary layer in fluids) [19]. The mesh refinement is continued until the results from one mesh compare closely to those of the previously refined mesh. It is also possible that part of the mesh can be enlarged instead of refined. For instance, in regions where the stresses do not change or change slowly, larger elements may be quite acceptable. The *h*-type mesh refinement strategy had its beginnings in [20–23]. Many commercial computer codes, such as [12], are based on the *h* refinement.

p **Method of Refinement**   In the *p* method of refinement [24–28], the polynomial *p* is increased from perhaps quadratic to a higher-order polynomial based on the degree of accuracy specified by the user. In the *p* method of refinement, the *p* method adjusts the order of the polynomial or the *p* level in the element field quantity, such as displacement, to better fit the conditions of the problem, such as the boundary conditions, the loading, and the geometry changes. A problem is solved at a given *p* level, and then the order of the polynomial is normally increased while the element geometry remains the same and the problem is solved again. The results of the iterations are compared to some set of convergence criteria specified by the user. Higher-order polynomials normally yield better solutions. This iteration process is done automatically within the computer program. Therefore, the user does not need to manually change the size of elements by creating a finer mesh, as must be done in the *h* method. (The *h* refinement can be automated using a remeshing algorithm within the finite element software.) Depending on the problem, a coarse mesh will often yield acceptable results. An extensive discussion of error indicators and estimates is given in the literature [19].

The *p* refinement may consist of adding degrees of freedom to existing nodes, adding nodes on existing boundaries between elements, and/or adding internal degrees of freedom. A uniform *p* refinement (same refinement performed on all

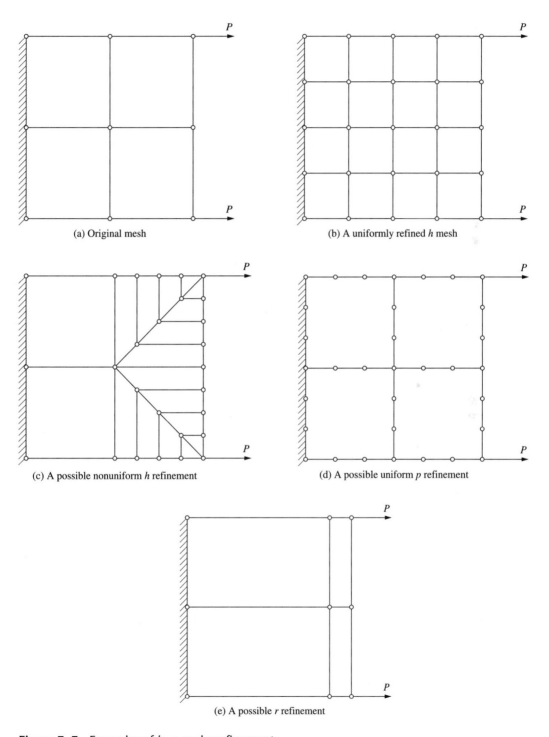

(a) Original mesh

(b) A uniformly refined h mesh

(c) A possible nonuniform h refinement

(d) A possible uniform p refinement

(e) A possible r refinement

Figure 7–7 Examples of h, p, and r refinement

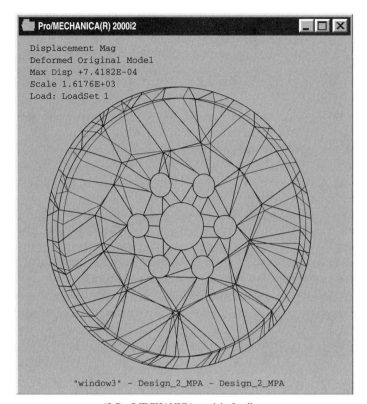

(f) Pro/MECHANICA model of pulley.

Figure 7–7 (*Continued*)

elements) is shown in Figure 7–7(d). One of the more common commercial computer programs, Pro/MECHANICA [29], uses the *p* method exclusively. A typical discretized finite element model of a pulley using Pro/MECHANICA is shown in Figure 7–7f.

r **Method of Refinement** In the *r* method of refinement, the nodes are rearranged or relocated without changing the number of elements or the polynomial degree of their field quantities, i.e., displacements. Figure 7–7(e) illustrates a possible *r* refinement of the original coarse mesh shown in Figure 7–7(a). Notice in this *r* refinement that we have refined the mesh closer to the loads with resulting coarseness in the mesh away from the end loads.

Transition Triangles

Figure 7–4 illustrates the use of triangular elements for transitions from smaller quadrilaterals to larger quadrilaterals. This transition is necessary because for simple CST elements, intermediate nodes along element edges are inconsistent with the energy formulation of the CST equations. If intermediate nodes were used, no assurance of compatibility would be possible, and resulting holes could occur in the deformed model. Using higher-order elements, such as the linear-strain triangle described in Chapter 8, allows us to use intermediate nodes along element edges and maintain compatibility.

Concentrated or Point Loads and Infinite Stress

Concentrated or point loads can be applied to nodes of an element provided the element supports the degree of freedom associated with the load. For instance, truss elements and two- and three-dimensional elements support only translational degrees of freedom, and therefore concentrated nodal moments cannot be applied to these elements; only concentrated forces can be applied. However, we should realize that physically concentrated forces are usually an idealization and mathematical convenience that represent a distributed load of high intensity acting over a small area.

According to classical linear theories of elasticity for beams, plates, and solid bodies [2, 16, 17], at a point loaded by a concentrated normal force there is finite displacement and stress in a beam, finite displacement but infinite stress in a plate, and both infinite displacement and stress in a two- or three-dimensional solid body. These results are the consequences of the differing assumptions about the stress fields in standard linear theories of beams, plates, and solid elastic bodies. A truly concentrated force would cause material under the load to yield, and linear elastic theories do not predict yielding.

In a finite element analysis, when a concentrated force is applied to a node of a finite element model, infinite displacement and stress are never computed. A concentrated force on a plane stress or strain model has a number of equivalent distributed loadings, which would not be expected to produce infinite displacements or infinite stresses. Infinite displacements and stresses can be approached only as the mesh around the load is highly refined. The best we can hope for is that we can highly refine the mesh in the vicinity of the concentrated load as shown in Figure 7–6(a), with the understanding that the deformations and stresses will be approximate around the load, or that these stresses near the concentrated force are not the object of study, while stresses near another point away from the force, such as B in Figure 7–6(f), are of concern. The preceding remarks about concentrated forces apply to concentrated reactions as well.

Finally, another way to model with a concentrated force is to use additional elements and a single concentrated load as shown in Figure 7–6(h). The shape of the distribution used to simulate a distributed load can be controlled by the relative stiffness of the elements above the loading plane to the actual structure by changing the modulus of elasticity of these elements. This method spreads the concentrated load over a number of elements of the actual structure.

Infinite stress based on elasticity solutions may also exist for special geometries and loadings, such as the re-entrant corner shown in Figure 7–6(f). The stress is predicted to be infinite at the re-entrant corner. Hence, the finite element method based on linear elastic material models will never yield convergence (no matter how many times you refine the mesh) to a correct stress level at the re-entrant corner [18].

We must either change the sharp re-entrant corner to one with a radius or use a theory that accounts for plastic or yielding behavior in the material.

Infinite Medium

Figure 7–3 shows a typical model used to represent an infinite medium (a soil mass subjected to a foundation load). The guideline for the finite element model is that enough material must be included such that the displacements at nodes and stresses within the elements become negligibly small at locations far from the foundation load. Just how much of the medium should be modeled can be determined by a trial-and-error procedure in which the horizontal and vertical distances from the load are varied and the resulting effects on the displacements and stresses are observed. Alternatively, the experiences of other investigators working on similar problems may prove helpful. For a homogeneous soil mass, experience has shown that the influence of the footing becomes insignificant if the horizontal distance of the model is taken as approximately four to six times the width of the footing and the vertical distance is taken as approximately four to ten times the width of the footing [4–6]. Also, the use of infinite elements is described in Reference [13].

   After choosing the horizontal and vertical dimensions of the model, we must idealize the boundary conditions. Usually, the horizontal displacement becomes negligible far from the load, and we restrain the horizontal movement of all the nodal points on that boundary (the right-side boundary in Figure 7–3). Hence, rollers are used to restrain the horizontal motion along the right side. The bottom boundary can be completely fixed, as is modeled in Figure 7–3 by using pin supports at each nodal point along the bottom edge. Alternatively, the bottom can be constrained only against vertical movement. The choice depends on the soil conditions at the bottom of the model. Usually, complete fixity is assumed if the lower boundary is taken as bedrock.

   In Figure 7–3, the left-side vertical boundary is taken to be directly under the center of the load because symmetry has been assumed. As we said before when discussing symmetry, all nodal points along the line of symmetry are restrained against horizontal displacement.

   Finally, Reference [11] is recommended for additional discussion regarding guidelines in modeling with different element types, such as beams, plane stress/plane strain, and three-dimensional solids.

Connecting (Mixing) Different Kinds of Elements

Sometimes it becomes necessary in a model to mix different kinds of elements, such as beams and plane elements, such as CSTs. The problem with mixing these elements is that they have different degrees of freedom at each node. The beam allows for transverse displacement and rotation at each node, while the plane element only has in-plane displacements at each node. The beam can resist a concentrated moment at a node, whereas a plane element (CST) cannot. Therefore, if a beam element is connected to a plane element at a single node as shown in Figure 7–8(a), the result will be a hinge connection at A. This means only a force can be transmitted through the

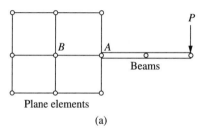
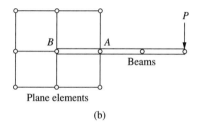

Figure 7–8 Connecting beam element to plane elements: (a) no moment is transferred, (b) moment is transferred

node between the two kinds of elements. This also creates a mechanism, as shown by the stiffness matrix being singular. This problem can be corrected by extending the beam into the plane element by adding one or more beam elements, shown as *AB*, for one beam element in Figure 7–8(b). Moment can now be transferred through the beam to the plane element. This extension assures that translational degrees of freedom of beam and plane element are connected at nodes *A* and *B*. Nodal rotations are associated with only the beam element, *AB*. The calculated stresses in the plane element will not normally be accurate near node *A*.

For more examples of connecting different kinds of elements see Figures 1–5, 11–10, 12–10 and 16–31. These figures show examples of beam and plate elements connected together (Figures 1–5, 12–10, and 16–31) and solid (brick) elements connected to plates (Figure 11–10).

Checking the Model

The discretized finite element model should be checked carefully before results are computed. Ideally, a model should be checked by an analyst not involved in the preparation of the model, who is then more likely to be objective.

Preprocessors with their detailed graphical display capabilities (Figure 7–9) now make it comparatively easy to find errors, particularly the more obvious ones involved with a misplaced node or missing element or a misplaced load or boundary support. Preprocessors include such niceties as color, shrink plots, rotated views, sectioning, exploded views, and removal of hidden lines to aid in error detection.

Most commercial codes also include warnings regarding overly distorted element shapes and checking for sufficient supports. However, the user must still select the proper element types, place supports and forces in proper locations, use consistent units, etc., to obtain a successful analysis.

Checking the Results and Typical Postprocessor Results

The results should be checked for consistency by making sure that intended support nodes have zero displacement, as required. If symmetry exists, then stresses and displacements should exhibit this symmetry. Computed results from the finite element program should be compared with results from other available techniques, even if

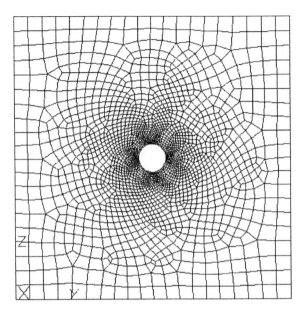

Figure 7–9 Plate of ASTM-A36 steel (2.5 m long, 2.5 m wide, 0.1 m thick, and with a hole radius 0.05 m) discretized using a preprocessor program [15] with automatic mesh generation

these techniques may be cruder than the finite element results. For instance, approximate mechanics of material formulas, experimental data, and numerical analysis of simpler but similar problems may be used for comparison, particularly if you have no real idea of the magnitude of the answers. Remember to use all results with some degree of caution, as errors can crop up in such sources as textbook or handbook comparison solutions and experimental results.

In the end, the analyst should probably spend as much time processing, checking, and analyzing results as is spent in data preparation.

Finally, we present some typical postprocessor results for the plane stress problem of Figure 7–9 (Figures 7–10 and 7–11, on page 400). Other examples with results are shown in Section 7.7.

▲ 7.2 Equilibrium and Compatibility of Finite Element Results ▲

An approximate solution for a stress analysis problem using the finite element method based on assumed displacement fields does not generally satisfy all the requirements for equilibrium and compatibility that an exact theory-of-elasticity solution satisfies. However, remember that relatively few exact solutions exist. Hence, the finite element method is a very practical one for obtaining reasonable, but approximate, numerical solutions. Recall the advantages of the finite element method as described in Chapter 1 and as illustrated numerous times throughout this text.

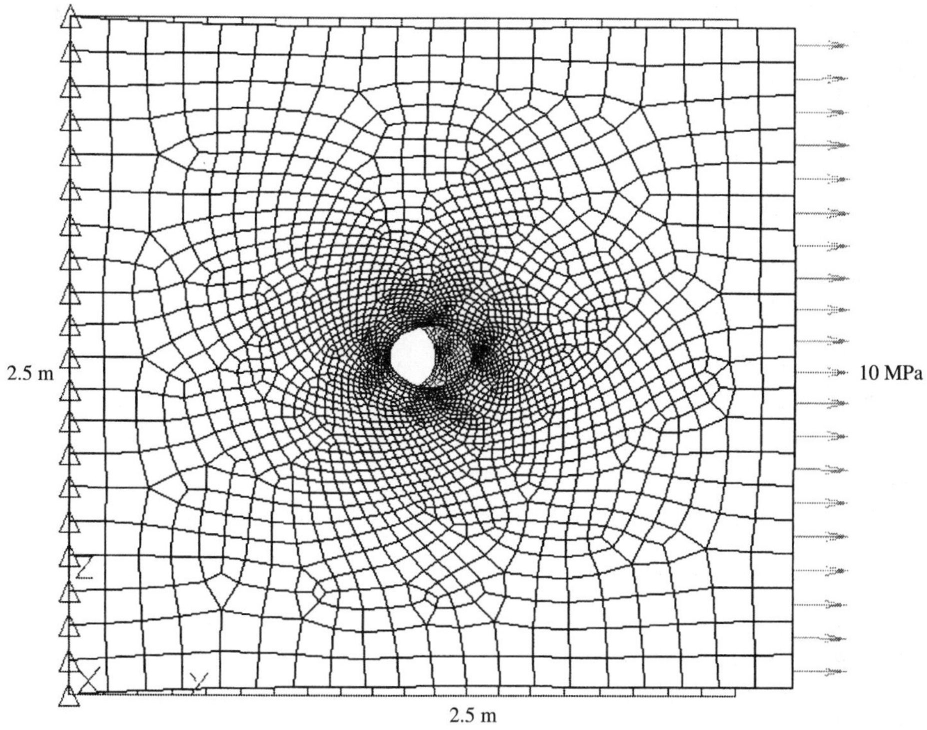

2.5 m

10 MPa

2.5 m

Figure 7–10 Plate with a hole showing the deformed shape of a plate superimposed over an undeformed shape. Plate is fixed on the left edge and subjected to 10 MPa tensile stress along the right edge. Maximum horizontal displacement is 2.41 mm at the center of the right edge. (Plate is steel with 0.1 m thickness)

We now describe some of the approximations generally inherent in finite element solutions.

1. Equilibrium of nodal forces and moments is satisfied. This is true because the global equation $\{F\} = [K]\{d\}$ is a nodal equilibrium equation whose solution for $\{d\}$ is such that the sums of all forces and moments applied to each node are zero. Equilibrium of the whole structure is also satisfied because the structure reactions are included in the global forces and hence in the nodal equilibrium equations. Numerous example problems, particularly involving truss and frame analysis in Chapter 3 and 5, respectively, have illustrated the equilibrium of nodes and of total structures.

2. Equilibrium within an element is not always satisfied. However, for the constant-strain bar of Chapter 3 and the constant-strain triangle of Chapter 6, element equilibrium is satisfied. Also the cubic displacement function is shown to satisfy the basic beam equilibrium differential equation in Chapter 4 and hence to satisfy element force and

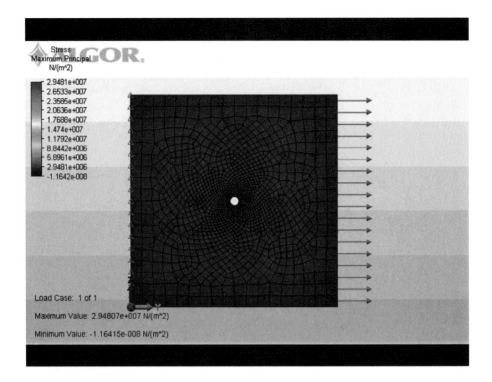

Figure 7–11 Maximum principal stress contour (shrink fit plot) for a plate with hole. Largest principal stresses of 29.48 MPa occur at the top and bottom of the hole, which indicates a stress concentration of 2.948. Stresses were obtained by using an average of the nodal values (called smoothing). (Same plate properties as in Figure 7–10.) (See the full-color insert for a color version of this figure.)

moment equilibrium. However, elements such as the linear-strain triangle of Chapter 8, the axisymmetric element of Chapter 9, and the rectangular element of Chapter 10 usually only approximately satisfy the element equilibrium equations.

3. Equilibrium is not usually satisfied between elements. A differential element including parts of two adjacent finite elements is usually not in equilibrium (Figure 7–12). For line elements, such as used for truss and frame analysis, interelement equilibrium is satisfied, as shown in example problems in Chapters 3 through 5. However, for two- and three-dimensional elements, interelement equilibrium is not usually satisfied. For instance, the results of Example 6.2 indicate that the normal stress along the diagonal edge between the two elements is different in the two elements. Also, the coarseness of the mesh causes this lack of interelement equilibrium to be even more pronounced. The normal and shear stresses at a free edge usually are not zero even though theory predicts them to be. Again, Example 6.2 illustrates this, with free-edge stresses σ_y and τ_{xy} not equal to zero. However, as more

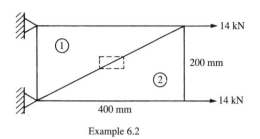

Example 6.2

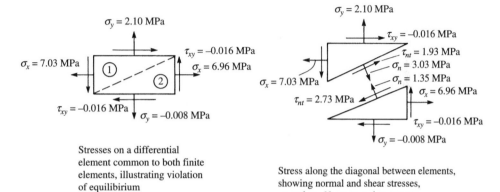

Stresses on a differential element common to both finite elements, illustrating violation of equilibirium

Stress along the diagonal between elements, showing normal and shear stresses, σ_n and τ_{nt}. *Note: σ_n and τ_{nt} are not equal in magnitude but are opposite in sign for the two elements, and so interelement equilibirium is not satisfied*

Figure 7–12 Example 6.2, illustrating violation of equilibrium of a differential element and along the diagonal edge between two elements (the coarseness of the mesh amplifies the violation of equilibrium)

elements are used (refined mesh) the σ_y and τ_{xy} stresses on the stress-free edges will approach zero.

4. Compatibility is satisfied within an element as long as the element displacement field is continuous. Hence, individual elements do not tear apart.

5. In the formulation of the element equations, compatibility is invoked at the nodes. Hence, elements remain connected at their common nodes. Similarly, the structure remains connected to its support nodes because boundary conditions are invoked at these nodes.

6. Compatibility may or may not be satisfied along interelement boundaries. For line elements such as bars and beams, interelement boundaries are merely nodes. Therefore, the preceding statement 5 applies for these line elements. The constant-strain triangle and the rectangular element of Chapter 6 remain straight-sided when deformed. Therefore, interelement compatibility exists for these elements; that is, these plane elements deform along common lines without openings, overlaps, or discontinuities. Incompatible elements,

those that allow gaps or overlaps between elements, can be acceptable and even desirable. Incompatible element formulations, in some cases, have been shown to converge more rapidly to the exact solution [1]. (For more on this special topic, consult References [7] and [8].)

▲ 7.3 Convergence of Solution ▲

In Section 3.2, we presented guidelines for the selection of so-called compatible and complete displacement functions as they related to the bar element. Those four guidelines are generally applicable, and satisfaction of them has been shown to ensure monotonic convergence of the solution of a particular problem [9]. Furthermore, it has been shown [10] that these compatible and complete displacement functions used in the displacement formulation of the finite element method yield an upper bound on the true stiffness, and hence a lower bound on the displacement of the problem, as shown in Figure 7–13.

Hence, as the mesh size is reduced—that is, as the number of elements is increased—we are ensured of monotonic convergence of the solution when compatible and complete displacement functions are used. Examples of this convergence are given in References [1] and [11], and in Table 7–2 for the beam with loading shown in Figure 7–1(a). All elements in the table are rectangular. The results in Table 7–2 indicate the influence of the number of elements (or the number of degrees of freedom as measured by the number of nodes) on the convergence toward a common solution,

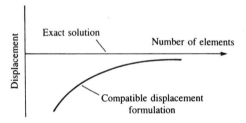

Figure 7–13 Convergence of a finite element solution based on the compatible displacement formulation

Table 7–2 Comparison of results for different numbers of elements

Case	Number of Nodes	Number of Elements	Aspect Ratio	Vertical Displacement, v (mm) Point A
1	21	12	2	−18.80
2	39	24	1	−24.90
3	45	32	3	−22.22
4	85	64	1.5	−27.38
5	105	80	1.2	−29.74
Exact solution [2]				−29.26

in this case the exact one. We again observe the influence of the aspect ratio. The higher the aspect ratio, even with a larger number of degrees of freedom, the worse the answer, as indicated by comparing cases 2 and 3.

Patch Test

To guarantee the convergence of a solution, the element being used in your model must pass a test called the *patch test*. This test was originally developed by Irons [30, 31] and is further discussed by Taylor, et al. [32], MacNeal and Harder [33], and Cook, et al. [7] (among others). The patch test is based in part on the same requirements described in Section 3.2; that the element must be able to accommodate both rigid-body motion and constant states of strain as both are possible within a structure. The patch test then can be used to determine if an element satisfies convergence requirements. It also can be applied to determine if sufficient Gauss points have been used in the numerical integration process to evaluate the stiffness matrix when the concept of isoparametric formulation of stiffness matrices is used as described in Chapter 10.

The patch test is performed by considering a simple finite element model composed of four irregular shaped elements of the same material with at least one node inside of the patch (called the patch node), as shown in Figure 7–14. The elements should be irregular, as some regular elements (such as rectangular) may pass the test whereas irregular ones will not. The elements may be all triangles or quadrilaterals or a mix of both. The boundary can be a rectangle though.

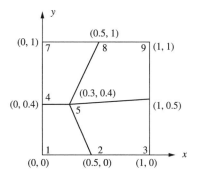

Figure 7–14 A patch of quadrilateral elements used for displacement patch test

The "displacement" patch test can be used to check if the elements can represent rigid-body motion and a constant state of strain. To verify if the elements can represent rigid-body motion, we do the following:

Step 1 Set the x-displacements of all nodes on the boundary to a value (say 1). That is, let $u_1 = u_2 = u_3 = u_4 = u_6 = u_7 = u_8 = u_9 = 1$ (x-translational rigid-body motion check). Set the y-displacement of these nodes to zero.

Step 2 Set the applied forces to zero at all nodes, including the interior node 5.

Step 3 Using the displacement values of 1 from step 1, set up the finite element equations using the stiffness method.

Step 4 Solve for the unknown displacements at node 5.

Step 5 To pass the rigid-body motion test, the computed x- and y-displacements at node 5 should be equal to $u_5 = 1$ and $v_5 = 0$.

Step 6 Repeat the steps by now setting all v_i's to 1 except at node 5 (y-translational rigid-body motion check). The displacement at node 5 should now become $u_5 = 0$ and $v_5 = 1$.

Step 7 Repeat the steps with all u_i's $= 1$ and all v_i's $= 1$ except at node 5 (x-y diagonal rigid-body motion check). The displacements at node 5 should be $u_5 = 1$ and $v_5 = 1$.

Step 8 The strains should be calculated within each element and should be zero.

To verify that the elements can represent a state of constant strain, the following steps are taken:

Step 1 As strains are derivatives of displacements, constant strain conditions can be obtained by assuming linear displacement functions. So set $u(x) = x$ and $v(x) = 0$. This yields $\varepsilon_x = \partial u/\partial x = 1$. The other in-plane strains, $\varepsilon_y = \partial v/\partial y = 0$ and $\gamma_{xy} = \partial u/\partial y + \partial v/\partial x = 0$. The displacement at each node must then be equal to its x-coordinate. In order to pass the patch test, the calculated x-displacement at node 5 must equal its x-coordinate; that is, $u_5 = 0.3$ and $v_5 = 0.0$.

Step 2 Repeat step 1 with $u(y) = 0$ and $v(y) = y$. This yields, $\varepsilon_x = 0$, $\varepsilon_y = 1$, and $\gamma_{xy} = 0$. The displacement at each node must then be equal to its y-coordinate. In order to pass the patch test, the calculated y-displacement at node 5 must equal its y-coordinate; that is, $u_5 = 0.0$ and $v_5 = 0.4$.

Step 3 Repeat the steps again with the shear strain becoming $\gamma_{xy} = 1$ and the normal strains equal to zero.

The "force" patch test validates that errors associated with the applied loads do not occur. The steps are as follows:

Step 1 Assume a uniform stress state of $\sigma_x = 1$ or some convenient constant value is applied along the right side of the patch. Replace this stress with its work-equivalent nodal load.

Step 2 Internal node 5 is not loaded.

Step 3 The patch has just enough supports to prevent rigid-body motion. In Figure 7–15, the left edge has one pin support and two rollers. (One roller would be sufficient to prevent rigid-body motion.) The roller supports allow for strain ε_y due to the Poisson effect. This strain will occur for Poisson's ratio not equal to zero and therefore should be accounted for.

Step 4 The finite element direct stiffness method is again used to obtain the displacements and element stresses. The uniform stress $\sigma_x = 1$ should be

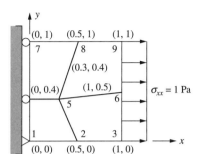

Figure 7–15 A patch of quadrilateral elements for the force patch test

obtained within each element. The other in-plane stresses, σ_y and τ_{xy}, should be zero.

Step 5 Repeat the steps assuming first that $\sigma_y = 1$ and the other stresses are zero. Then assume that $\tau_{xy} = 1$ and the two normal stresses equal zero.

The patch test also can be applied to other element types. For instance, a patch of solid elements described for a three-dimensional stress state in Chapter 11 should be able to properly yield all six constant states of stress. The patch test for the plate bending analysis described in Chapter 12 should yield constant bending moments M_x and M_y and constant twisting moment M_{xy}.

An element that passes the patch test is capable of meeting the following requirements:

(a) Predicting rigid-body motion without strain when this state exists.
(b) Predicting states of constant strain if they occur.
(c) Compatibility with adjacent elements when a state of constant strain exists in adjacent elements.

When these requirements are met, it is sufficient to guarantee that a mesh of these elements will yield convergence to the solution as the mesh is continually refined.

The patch test is then a standard test for developers of new elements to test whether the element has the necessary convergence properties. But the test does not indicate how well an element works in other applications. An element passing the patch test may still yield poor accuracy in a coarse mesh or show slow convergence as the mesh is refined.

▲ 7.4 Interpretation of Stresses ▲

In the stiffness or displacement formulation of the finite element method used throughout this text, the primary quantities determined are the interelement nodal displacements of the assemblage. The secondary quantities, such as strain and stress in an element, are then obtained through use of $\{\varepsilon\} = [B]\{d\}$ and $\{\sigma\} = [D][B]\{d\}$. For elements using linear-displacement models, such as the bar and the constant-strain triangle, $[B]$ is constant, and since we assume $[D]$ to be constant, the stresses are constant

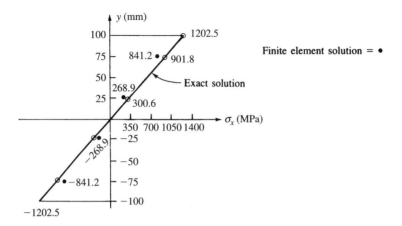

Figure 7–16 Comparison of the finite element solution and the exact solution of bending stress through a beam cross section

over the element. In this case, it is common practice to assign the stress to the centroid of the element with acceptable results.

However, as illustrated in Section 3.11 for the axial member, stresses are not predicted as accurately as the displacements (see Figures 3–32 and 3–33). For example, remember the constant-strain or constant-stress element has been used in modeling the beam in Figure 7–1. Therefore, the stress in each element is assumed constant. Figure 7–16 compares the exact beam theory solution for bending stress through the beam depth at the centroidal location of the elements next to the wall with the finite element solution of case 4 in Table 7–2. This finite element model consists of four elements through the beam depth. Therefore, only four stress values are obtained through the depth. Again, the best approximation of the stress appears to occur at the midpoint of each element, since the derivative of displacement is better predicted between the nodes than at the nodes.

For higher-order elements, such as the linear-strain triangle of Chapter 8, [B], and hence the stresses, are functions of the coordinates. The common practice is then to directly evaluate the stresses at the centroid of the element.

An alternative procedure sometimes is to use an average (possibly weighted) value of the stresses evaluated at each node of the element. This averaging method is often based on evaluating the stresses at the Gauss points located within the element (described in Chapter 10) and then interpolating to the element nodes using the shape functions of the specific element. Then these stresses in all elements at a common node are averaged to represent the stress at the node. This averaging process is called *smoothing*. Figure 7–11 shows a maximum principal stress "fringe carpet" (dithered) contour plot obtained by smoothing.

Smoothing results in a pleasing, continuous plot which may not indicate some serious problems with the model and the results. You should always view the unsmoothed contour plots as well. Highly discontinuous contours between elements in a region of an unsmoothed plot indicate modeling problems and typically require additional refinement of the element mesh in the suspect region.

If the discontinuities in an unsmoothed contour plot are small or are in regions of little consequence, a smoothed contour plot can normally be used with a high degree of confidence in the results. There are, however, exceptions when smoothing leads to erroneous results. For instance, if the thickness or material stiffness changes significantly between adjacent elements, the stresses will normally be different from one element to the next. Smoothing will likely hide the actual results. Also, for shrink-fit problems involving one cylinder being expanded enough by heating to slip over the smaller one, the circumferential stress between the mating cylinders is normally quite different [16].

The computer program examples in Section 7.7 show additional results, such as displaced models, along with smoothed stress plots. The stresses to be plotted can be von Mises (used in the maximum distortion energy theory to predict failure of ductile materials subjected to static loading as described in Section 6.5); Tresca (used in the Tresca or maximum shear stress theory also to predict failure of ductile materials subjected to static loading) [14, 16], and maximum and minimum principal stresses.

▲ 7.5 Static Condensation ▲

We will now consider the concept of static condensation because this concept is used in developing the stiffness matrix of a quadrilateral element in many computer programs.

Consider the basic quadrilateral element with external nodes 1–4 shown in Figure 7–17. An imaginary node 5 is temporarily introduced at the intersection of the diagonals of the quadrilateral to create four triangles. We then superimpose the stiffness matrices of the four triangles to create the stiffness matrix of the quadrilateral element, where the internal imaginary node 5 degrees of freedom are said to be *condensed out* so as never to enter the final equations. Hence, only the degrees of freedom associated with the four *actual* external corner nodes enter the equations.

We begin the static condensation procedure by partitioning the equilibrium equations as

$$\begin{bmatrix} [k_{11}] & | & [k_{12}] \\ \hline [k_{21}] & | & [k_{22}] \end{bmatrix} \begin{Bmatrix} \{d_a\} \\ \{d_i\} \end{Bmatrix} = \begin{Bmatrix} \{F_a\} \\ \{F_i\} \end{Bmatrix} \tag{7.5.1}$$

where $\{d_i\}$ is the vector of internal displacements corresponding to the imaginary internal node (node 5 in Figure 7–17), $\{F_i\}$ is the vector of loads at the internal node, and $\{d_a\}$ and $\{F_a\}$ are the actual nodal degrees of freedom and loads, respectively, at the actual nodes. Rewriting Eq. (7.5.1), we have

$$[k_{11}]\{d_a\} + [k_{12}]\{d_i\} = \{F_a\} \tag{7.5.2}$$

$$[k_{21}]\{d_a\} + [k_{22}]\{d_i\} = \{F_i\} \tag{7.5.3}$$

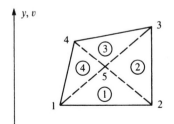

Figure 7–17 Quadrilateral with an internal node

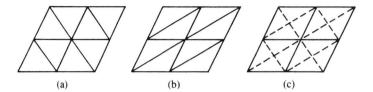

Figure 7–18 Skew effects in finite element modeling

Solving for $\{d_i\}$ in Eq. (7.5.3), we obtain

$$\{d_i\} = -[k_{22}]^{-1}[k_{21}]\{d_a\} + [k_{22}]^{-1}\{F_i\} \tag{7.5.4}$$

Substituting Eq. (7.5.4) into Eq. (7.5.2), we obtain the condensed equilibrium equation

$$[k_c]\{d_a\} = \{F_c\} \tag{7.5.5}$$

where

$$[k_c] = [k_{11}] - [k_{12}][k_{22}]^{-1}[k_{21}] \tag{7.5.6}$$

$$\{F_c\} = \{F_a\} - [k_{12}][k_{22}]^{-1}\{F_i\} \tag{7.5.7}$$

and $[k_c]$ and $\{F_c\}$ are called the *condensed stiffness matrix* and the *condensed load vector*, respectively. Equation (7.5.5) can now be solved for the actual corner node displacements in the usual manner of solving simultaneous linear equations.

Both constant-strain triangular (CST) and constant-strain quadrilateral elements are used to analyze plane stress/plane strain problems. The quadrilateral element has the stiffness of four CST elements. An advantage of the four-CST quadrilateral is that the solution becomes less dependent on the *skew* of the subdivision mesh, as shown in Figure 7–18. Here **skew** means the *directional stiffness bias* that can be built into a model through certain discretization patterns, since the stiffness matrix of an element is a function of its nodal coordinates, as indicted by Eq. (6.2.52). The four-CST mesh of Figure 7–18(c) represents a reduction in the skew effect over the meshes of Figures 7–18(a) and (b). Figure 7–18(b) is generally worse than Figure 7–18(a) because the use of long, narrow triangles results in an element stiffness matrix that is stiffer along the narrow direction of the triangle.

The resulting stiffness matrix of the quadrilateral element will be an 8×8 matrix consisting of the stiffnesses of four triangles, as was shown in Figure 7–17. The stiffness matrix is first assembled according to the usual direct stiffness method. Then we apply static condensation as outlined in Eqs. (7.5.1) through (7.5.7) to remove the internal node 5 degrees of freedom.

The stiffness matrix of a typical triangular element (labeled element 1 in Figure 7–17) with nodes 1, 2, and 5 is given in general form by

$$[k^{(1)}] = \begin{bmatrix} [k_{11}^{(1)}] & [k_{12}^{(1)}] & [k_{15}^{(1)}] \\ [k_{21}^{(1)}] & [k_{22}^{(1)}] & [k_{25}^{(1)}] \\ [k_{51}^{(1)}] & [k_{52}^{(1)}] & [k_{55}^{(1)}] \end{bmatrix} \tag{7.5.8}$$

where the superscript in parentheses again refers to the element number, and each submatrix $[k_{ij}^{(1)}]$ is of order 2×2. The stiffness matrix of the quadrilateral, assembled using Eq. (7.5.8) along with similar stiffness matrices for elements 2–4 of Figure 7–17, is given by the following (before static condensation is used):

$$
[k] =
\begin{bmatrix}
\begin{matrix}[k_{11}^{(1)}] \\ + \\ [k_{11}^{(4)}]\end{matrix} & [k_{12}^{(1)}] & [0] & [k_{14}^{(4)}] & [k_{15}^{(1)}] + [k_{15}^{(4)}] \\[2em]
[k_{21}^{(1)}] & \begin{matrix}[k_{22}^{(1)}] \\ + \\ [k_{22}^{(2)}]\end{matrix} & [k_{23}^{(2)}] & [0] & [k_{25}^{(1)}] + [k_{25}^{(2)}] \\[2em]
[0] & [k_{32}^{(2)}] & \begin{matrix}[k_{33}^{(2)}] \\ + \\ [k_{33}^{(3)}]\end{matrix} & [k_{34}^{(3)}] & [k_{35}^{(2)}] + [k_{35}^{(3)}] \\[2em]
[k_{41}^{(4)}] & [0] & [k_{43}^{(3)}] & \begin{matrix}[k_{44}^{(3)}] \\ + \\ [k_{44}^{(4)}]\end{matrix} & [k_{45}^{(3)}] + [k_{45}^{(4)}] \\[2em]
\begin{matrix}[k_{51}^{(1)}] \\ + \\ [k_{51}^{(4)}]\end{matrix} & \begin{matrix}[k_{52}^{(1)}] \\ + \\ [k_{52}^{(2)}]\end{matrix} & \begin{matrix}[k_{53}^{(2)}] \\ + \\ [k_{53}^{(3)}]\end{matrix} & \begin{matrix}[k_{54}^{(3)}] \\ + \\ [k_{54}^{(4)}]\end{matrix} & \begin{matrix}([k_{55}^{(1)}] + [k_{55}^{(2)}]) \\ + \\ ([k_{55}^{(3)}] + [k_{55}^{(4)}])\end{matrix}
\end{bmatrix}
\tag{7.5.9}
$$

with column headings $(u_1, v_1)\ (u_2, v_2)\ (u_3, v_3)\ (u_4, v_4)\ (u_5, v_5)$

where the orders of the degrees of freedom are shown above the columns of the stiffness matrix and the partitioning scheme used in static condensation is indicated by the dotted lines. Before static condensation is applied, the stiffness matrix is of order 10×10.

Example 7.1

Consider the quadrilateral with internal node 5 and dimensions as shown in Figure 7–19 to illustrate the application of static condensation.

Recall that the original stiffness matrix of the quadrilateral is 10×10, but static condensation will result in an 8×8 stiffness matrix after removal of the degrees of freedom (u_5, v_5) at node 5.

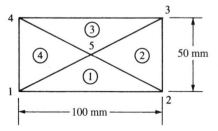

Figure 7–19 Quadrilateral with an internal node

Using the CST stiffness matrix of Eq. (6.4.3) for plane strain, we have

$$[k^{(1)}] = [k^{(3)}] = \frac{E}{0.0416}
\begin{array}{ccccccc}
3 & 4 & 5 \\
1 & 2 & 5 \\
\end{array}
\begin{bmatrix}
1.5 & 1.0 & 0.1 & 0.2 & -1.6 & -1.2 \\
 & 3.0 & -0.2 & 2.6 & -0.8 & -5.6 \\
 & & 1.5 & -1.0 & -1.6 & 1.2 \\
 & & & 3.0 & 0.8 & -5.6 \\
 & & & & 3.2 & 0.0 \\
\text{Symmetry} & & & & & 11.2
\end{bmatrix}$$

(7.5.10)

Similarly, from Figure 7–19, we can show that

$$[k^{(2)}] = [k^{(4)}] = \frac{E}{0.0416}
\begin{array}{ccc}
2 & 3 & 5 \\
4 & 1 & 5 \\
\end{array}
\begin{bmatrix}
1.5 & -1.0 & -0.1 & 0.2 & -1.4 & 0.8 \\
 & 3.0 & -0.2 & -2.6 & 1.2 & -0.4 \\
 & & 1.5 & 1.0 & -1.4 & -0.8 \\
 & & & 3.0 & -1.2 & -0.4 \\
 & & & & 2.8 & 0.0 \\
\text{Symmetry} & & & & & 0.8
\end{bmatrix}$$

(7.5.11)

where the numbers above the columns in Eqs. (7.5.10) and (7.5.11) indicate the orders of the degrees of freedom associated with each stiffness matrix. Here the quantity in the denominator of Eq. (6.4.3), $4A(1 + v)(1 - 2v)$, is equal to 0.0416 m Eqs. (7.5.10) and (7.5.11) because $A = 12.5 \times 10^{-4}$ m^2 and v is taken to be 0.3. Also, the thickness t of the element has been taken as 10 mm. Now we can superimpose the stiffness terms as indicated by Eq. (7.5.9) to obtain the general expression for a four-CST element. The resulting assembled total stiffness matrix before static condensation is applied is given by

$$[k] = \frac{E}{0.0416}
\begin{bmatrix}
3.0 & 2.0 & 0.1 & 0.2 & 0.0 & 0.0 & -0.1 & -0.2 & -3.0 & -2.0 \\
 & 6.0 & -0.2 & 2.6 & 0.0 & 0.0 & 0.2 & -2.6 & -2.0 & -6.0 \\
 & & 3.0 & -2.0 & -0.1 & 0.2 & 0.0 & 0.0 & -3.0 & 2.0 \\
 & & & 6.0 & -0.2 & -2.6 & 0.0 & 0.0 & 2.0 & -6.0 \\
 & & & & 3.0 & 2.0 & 0.1 & 0.2 & -3.0 & -2.0 \\
 & & & & & 6.0 & -0.2 & 2.6 & -2.0 & -6.0 \\
 & & & & & & 3.0 & -2.0 & -3.0 & 2.0 \\
 & & & & & & & 6.0 & 2.0 & -6.0 \\
 & & & & & & & & 12.0 & 0.0 \\
\text{Symmetry} & & & & & & & & & 24.0
\end{bmatrix}$$

(7.5.12)

After we partition Eq. (7.5.12) and use Eq. (7.5.6), the condensed stiffness matrix is given by

$$
[k_c] = \frac{E}{0.0416}
\begin{array}{cccccccc}
u_1 & v_1 & u_2 & v_2 & u_3 & v_3 & u_4 & v_4
\end{array}
\begin{bmatrix}
2.08 & 1.00 & -0.48 & 0.20 & -0.92 & -1.00 & -0.68 & -0.20 \\
 & 4.17 & -0.20 & 1.43 & -1.00 & -1.83 & 0.20 & -3.77 \\
 & & 2.08 & -1.00 & -0.68 & 0.20 & -0.92 & 1.00 \\
 & & & 4.17 & -0.20 & -3.77 & 1.00 & -1.83 \\
 & & & & 2.08 & 1.00 & -0.48 & 0.20 \\
 & & & & & 4.17 & -0.20 & 1.43 \\
 & & & & & & 2.08 & -1.00 \\
\text{Symmetry} & & & & & & & 4.17
\end{bmatrix}
$$

(7.5.13)

■

▲ 7.6 Flowchart for the Solution of Plane Stress–Strain Problems ▲

In Figure 7–20 on the next page, we present a flowchart of a typical finite element process used for the analysis of plane stress and plane strain problems on the basis of the theory presented in Chapter 6.

▲ 7.7 Computer Program–Assisted Step-by-Step Solution, Other Models, and Results for Plane Stress–Strain Problems ▲

In this section, we present a computer-assisted step-by-step solution of a plane stress problem, along with results of some plane stress–strain problems solved using a computer program [12]. These results illustrate the various kinds of difficult problems that can be solved using a general-purpose computer program.

The computer-assisted step-by-step problem is the bicycle wrench shown in Figure 7–21(a) on page 413. The following steps have been used to solve for the stresses in the wrench. (Some of these steps may be interchanged.)

Step 1 The first step is to draw the outline of the wrench using a standard drawing program as shown in Figure 7–21(a). The exact dimensions of the wrench are obtained from Figure P7–38, where the overall depth of the wrench is 2.0 cm, the length is 14 cm, and the sides of the hexagons are 9 mm long for the middle one and 7 mm long for the side ones. The radius of the enclosed ends is 1.50 cm.

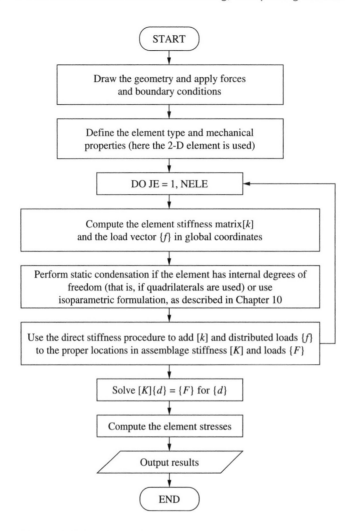

Figure 7–20 Flowchart of plane stress/strain finite element process

Step 2 The second step is to use a two-dimensional mesh generator to create the model mesh as shown in Figure 7–21(b).

Step 3 The third step is to apply the boundary conditions to the proper nodes using the proper boundary condition command. This is shown in Figure 7–21(c) as indicated by the small @ signs at the nodes on the inside of the left hexagonal shaped hole. The @ sign indicates complete fixity for a node. (Algor more recently has changed to the triangle symbol to denote complete fixity for a node.) This means these nodes are constrained from translating in the y and z directions in the plane of the wrench.

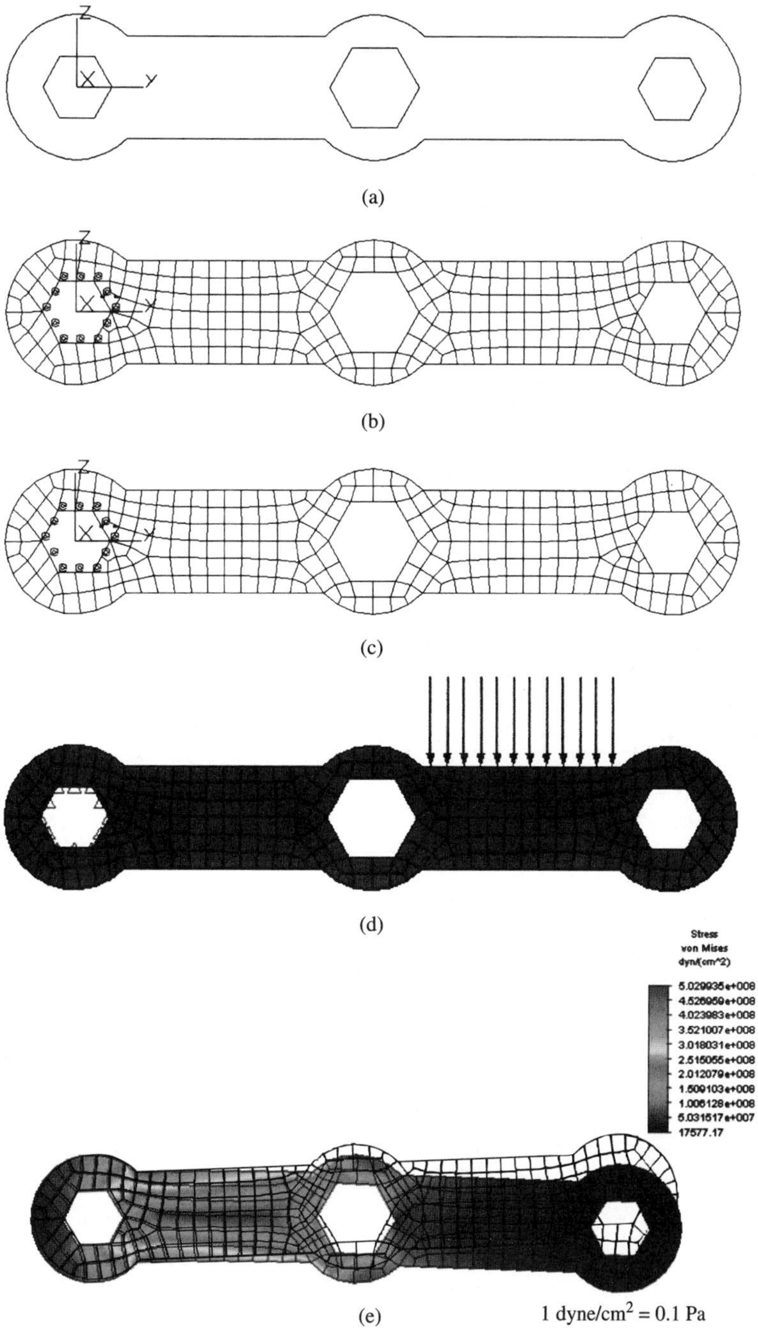

Figure 7–21 Bicycle wrench (a) Outline drawing of wrench, (b) meshed model of wrench, (c) boundary conditions and selecting surface where surface traction will be applied, (d) checked model showing the boundary conditions and surface traction, and (e) von Mises stress plot (Compliments of Angela Moe) (See the full-color insert for a color version of this figure.)

Step 4 The fourth step requires us to select the surface where the distributed loading is to be applied and then the magnitude of the surface traction. This is the upper surface between the middle and right hexagonal holes where the surface traction of 100 N/cm^2 is applied as shown in Figure 7–21(d). In the computer program this surface changes to the color red as selected by the user (Figure 7–21(c)).

Step 5 In step 5 we choose the material properties. Here ASTM A-514 steel has been selected, as this is quenched and tempered steel with high yield strength and will allow for the thickness to be minimized.

Step 6 In step six we select the element type for the kind of analysis to be performed. Here we select the plane stress element, as this is a good approximation to the kind of behavior that is produced in a plane stress analysis. For the plane stress element a thickness is required. An initial guess of 1 cm is made. This thickness appears to be compatible with the other dimensions of the wrench.

Step 7 The seventh step is an optional check of the model. If you choose to perform this step you will see the boundary conditions now appear as triangles at the left nodes corresponding to the @ signs for full fixity and the surface traction arrows, indicating the location and direction of the surface traction shown also in Figure 7–21(d).

Step 8 In step 8 we perform the stress analysis of the model.

Step 9 In step 9 we select the results, such as the displacement plot, the principal stress plot, and the von Mises stress plot. The von Mises stress plot is used to determine the failure of the wrench based on the maximum distortion energy theory as described in Section 6.5. The von Mises stress plot is shown in Figure 7–21(e). The maximum von Mises stress indicated in Figure 7–21(e) is 502 MPa, and the yield strength of the ASTM A-514 steel is 690 MPa. Therefore, the wrench is safe from yielding. Additional trials can be made if the factor of safety is satisfied and if the maximum deflection appears to be satisfactory.

Figure 7–22(a) shows a finite element model of a steel connecting rod that is fixed on its left edge and loading around the right inner edge of the hole with a total force of 15 kN. For more details, including the geometry of this rod, see Figure P7–15 at the end of this chapter. Figure 7–22(b) shows the resulting maximum principal stress. The largest principal stress of 96.9 MPa occurs at the top and bottom inside edges of the hole.

Figure 7–23 shows a finite element model along with the von Mises stress plot of an overload protection device (see Problem 7.33 for details of this problem). The upper member of the device was modeled. Node *S* at the shear pin location was constrained from vertical motion, five nodes along the left side of pin hole *B* were constrained in both the horizontal and vertical directions, and all nodes at the pin hole at *A* were constrained in the vertical direction. A load of 700 N was spread over the three lowest nodes at the inner side of the right section hanging down near point *E*

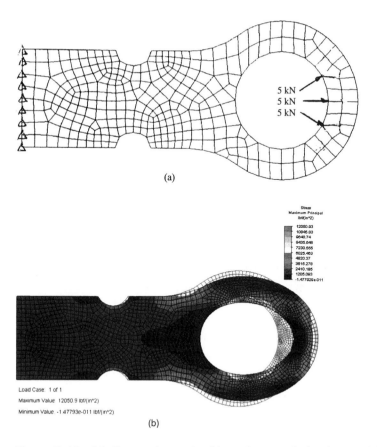

(a)

(b)

Figure 7–22 (a) Connecting rod subjected to tensile loading and (b) resulting principal stress throughout the rod

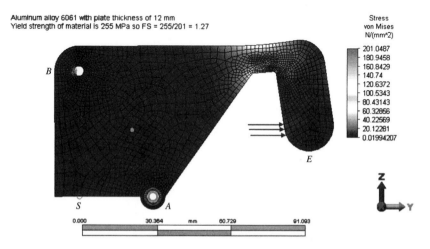

Figure 7–23 von Mises stress plot of overload protection device (See the full-color insert for a color version of this figure.)

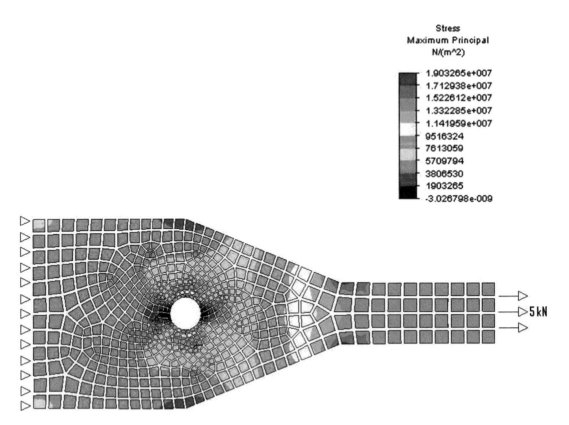

Figure 7–24   Shrink fit plot of principal stresses in a tapered plate with hole

to simulate the load to fail the pin at S in shear at a stress of 40 MPa. The largest von Mises stress of 201 MPa occurs at the inner edge of the cutout section for a device with thickness of 12 mm.

Figure 7–24 shows the shrink plot of a finite element analysis of a tapered plate with a hole in it, subjected to tensile loading along the right edge. The left edge was fixed. For details of this problem see Problem 7.26. The shrink plot separates the elements for a clear look at the model. The largest principal stress of 6.29 MPa occurs at the edge of the hole, whereas the second largest principal stress of 5.67 MPa occurs at the elbow between the smallest cross section and where the taper begins.

Figure 7–25 shows the plot of the minimum principal stresses in a dam subjected to hydrostatic and self weight loading. The minimum principal stress of 1.86 MPa occurs at the inside edge. For more details of this problem see Problem 7.27.

Finally, Figure 7–26(a) shows a finite element model and the von Mises stress plot for a beam welded to a column by top and bottom fillet welds. A surface contact between the beam and column was used that allowed the beam and column to separate wherever tension existed along the surfaces in contact. The beam is 70 mm thick by 120 mm deep by 200 mm long with a load of 10 kN applied vertically 160 mm from the left end of the beam and at 60 mm down from the top edge of the beam.

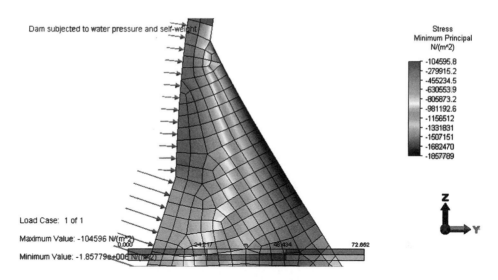

Figure 7–25 Plot of minimum principal stress with largest absolute value of 1.86 MPa located on back side of dam subjected to both hydrostatic and self-weight loading

The material is steel with $E = 205$ GPa and $v = 0.25$ for all material. The weld leg size is 6 mm.

After mesh refinement around the top weld to double the number of elements in the weld, the maximum von Mises stress was determined to be 87.3 MPa at the toe of the top weld as seen best in the zoomed-in Figure 7–26(b). This value compares reasonably well with that obtained by the classical method shown on pages 458–460 of Reference [30] where a value of 94 MPa was obtained.

▲ References

[1] Desai, C. S., and Abel, J. F., *Introduction to the Finite Element Method*, Van Nostrand Reinhold, New York, 1972.

[2] Timoshenko, S., and Goodier, J., *Theory of Elasticity*, 3rd ed., McGraw-Hill, New York, 1970.

[3] Glockner, P. G., "Symmetry in Structural Mechanics," *Journal of the Structural Division*, American Society of Civil Engineers, Vol. 99, No. ST1, pp. 71–89, 1973.

[4] Yamada, Y., "Dynamic Analysis of Civil Engineering Structures," *Recent Advances in Matrix Methods of Structural Analysis and Design*, R. H. Gallagher, Y. Yamada, and J. T. Oden, eds., University of Alabama Press, Tuscaloosa, AL, pp. 487–512, 1970.

[5] Koswara, H., *A Finite Element Analysis of Underground Shelter Subjected to Ground Shock Load*, M.S. Thesis, Rose-Hulman Institute of Technology, Terre Haute, IN, 1983.

[6] Dunlop, P., Duncan, J. M., and Seed, H. B., "Finite Element Analyses of Slopes in Soil," *Journal of the Soil Mechanics and Foundations Division*, Proceedings of the American Society of Civil Engineers, Vol. 96, No. SM2, March 1970.

[7] Cook, R. D., Malkus, D. S., Plesha, M. E., and Witt, R. J., *Concepts and Applications of Finite Element Analysis*, 4th ed., Wiley, New York, 2002.

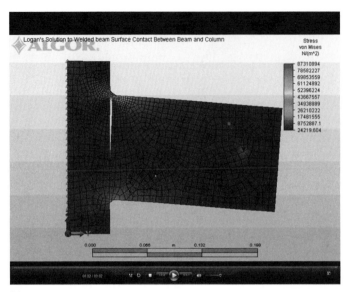

(a)

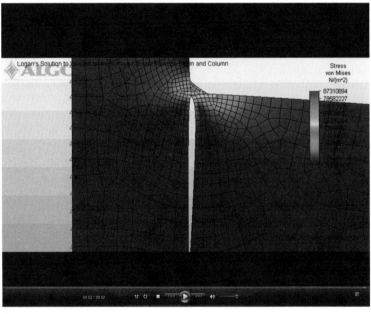

(b)

Figure 7–26 (a) von Mises stress plot of beam welded to column (largest von Mises stress of 87.3 MPa is located at toe of top fillet weld as shown by (b) zoomed-in view of top fillet weld (notice also that a surface contact was used between the beam and column that allowed for the gap to form where the beam separated from the column) (See the full-color insert for a color version of this figure.)

[8] Taylor, R. L., Beresford, P. J., and Wilson, E. L., "A Nonconforming Element for Stress Analysis," *International Journal for Numerical Methods in Engineering*, Vol. 10, No. 6, pp. 1211–1219, 1976.

[9] Melosh, R. J., "Basis for Derivation of Matrices for the Direct Stiffness Method," *Journal of the American Institute of Aeronautics and Astronautics*, Vol. 1, No. 7, pp. 1631–1637, July 1963.

[10] Fraeijes de Veubeke, B., "Upper and Lower Bounds in Matrix Structural Analysis," *Matrix Methods of Structural Analysis*, AGARDograph 72, B. Fraeijes de Veubeke, ed., Macmillan, New York, 1964.

[11] Dunder, V., and Ridlon, S., "Practical Applications of Finite Element Method," *Journal of the Structural Division*, American Society of Civil Engineers, No. ST1, pp. 9–21, 1978.

[12] *Linear Stress and Dynamics Reference Division*, Docutech On-line Documentation, Algor, Inc., Pittsburgh, PA 15238.

[13] Bettess, P., "More on Infinite Elements," *International Journal for Numerical Methods in Engineering*, Vol. 15, pp. 1613–1626, 1980.

[14] Gere, J. M., *Mechanics of Materials*, 5th ed., Brooks/Cole Publishers, Pacific Grove, CA, 2001.

[15] *Superdraw Reference Division*, Docutech On-line Documentation, Algor, Inc., Pittsburgh, PA 15238.

[16] Cook, R. D., and Young, W. C., *Advanced Mechanics of Materials*, Macmillan, New York, 1985.

[17] Cook, R. D., *Finite Element Modeling for Stress Analysis*, Wiley, New York, 1995.

[18] Kurowski, P., "Easily Made Errors Mar FEA Results," Machine Design, September. 13, 2001.

[19] Huebner, K. H., Dewirst, D. L., Smith, D. E., and Byrom, T. G., *The Finite Element Method for Engineers*, Wiley, New York, 2001.

[20] Demkowicz, L., Devloo, P., and Oden, J. T., "On an *h*-Type Mesh-Refinement Strategy Based on Minimization of Interpolation Errors," *Comput. Methods Appl. Mech. Eng.*, Vol. 53, pp. 67–89, 1985.

[21] Löhner, R., Morgan, K., and Zienkiewicz, O. C., "An Adaptive Finite Element Procedure for Compressible High Speed Flows," *Comput. Methods Appl. Mech. Eng.*, Vol. 51, pp. 441–465, 1985.

[22] Löhner, R., "An Adaptive Finite Element Scheme for Transient Problems in CFD," *Comput. Methods Appl. Mech. End.*, Vol. 61, pp. 323–338, 1987.

[23] Ramakrishnan, R., Bey, K. S., and Thornton, E. A., "Adaptive Quadrilateral and Triangular Finite Element Scheme for Compressible Flows," *AIAA J.*, Vol. 28, No. 1, pp. 51–59, 1990.

[24] Peano, A. G., "Hierarchies of Conforming Finite Elements for Plane Elasticity and Plate Bending," *Comput. Match. Appl.*, Vol. 2, pp. 211–224, 1976.

[25] Szabó, B. A., "Some Recent Developments in Finite Element Analysis," *Comput. Match. Appl.*, Vol. 5, pp. 99–115, 1979.

[26] Peano, A. G., Pasini, A., Riccioni., R. and Sardella, L., "Adaptive Approximation in Finite Element Structural Analysis," *Comput. Struct.*, Vol. 10, pp. 332–342, 1979.

[27] Zienkiewicz, O. C., Gago, J. P. de S. R., and Kelly, D. W., "The Hierarchical Concept in Finite Element Analysis," *Comput. Struct.*, Vol. 16, No. 1–4, pp. 53–65, 1983.

[28] Szabó, B. A., "Mesh Design for the *p*-Version of the Finite Element Method," *Comput. Methods Appl. Mech. Eng.*, Vol. 55, pp. 181–197, 1986.

[29] Toogood, Roger, *Pro/MECHANICA, Structural Tutorial*, SDC Publications, 2001.

[30] Bazeley, G. P., Cheung, Y. K., Irons, B. M., and Zienkiewicz, O. C., "Triangular Elements in Plate Bending—Conforming and Nonconforming Solutions," *Proceedings of the Conference on Matrix Methods in Structural Mechanics*, Wright Patterson Air Force Base, Dayton, Ohio, pp. 547–576, 1965.

[31] Irons, B. M., and Razzaque, A., "Experience with the Patch Test for Convergence of Finite Element Methods," *The Mathematical Foundations of Finite Element Method with Applications to Partial Differential Equations*, A. R. Aziz, ed., Academic Press, New York, pp. 557–587, 1972.

[32] MacNeal, R. H., and Harder, R. L., "A Proposed Standard Set of Problems to Test Finite Element Accuracy," *Finite Elements in Analysis and Design*, Vol. 1, No. 1, pp. 3–20, 1985.

[33] Taylor, R. L., Simo, J. C., Zienkiewicz, O. C., and Chan, A. C. H., "The Patch Test—A Condition for Assessing FEM Convergence," *International Journal for Numerical Methods in Engineering*, Vol. 22, No. 1, pp. 39–62, 1986.

▲ Problems

7.1 For the finite element mesh shown in Figure P7–1, comment on the appropriateness of the mesh. Indicate the mistakes in the model. Explain and show how to correct them.

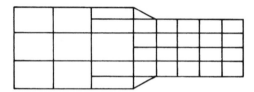

Figure P7–1

7.2 Comment on the mesh sizing in Figure P7–2. Is it reasonable? If not, explain why not.

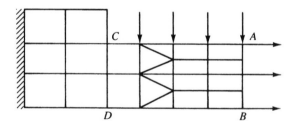

Figure P7–2

7.3 What happens if the material property $v = 0.5$ in the plane strain case? Is this possible? Explain.

7.4 Under what conditions is the structure in Figure P7–4 a plane strain problem? Under what conditions is the structure a plane stress problem?

7.5 When do problems occur using the smoothing (averaging of stress at the nodes from elements connected to the node) method for obtaining stress results?

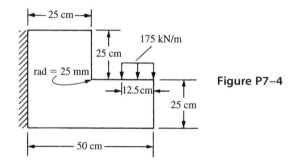

Figure P7–4

7.6 What thickness do you think is used in computer programs for plane strain problems?

7.7 Which one of the CST models shown in Figure P7–7 is expected to give the best results for a cantilever beam subjected to an end shear load? Why?

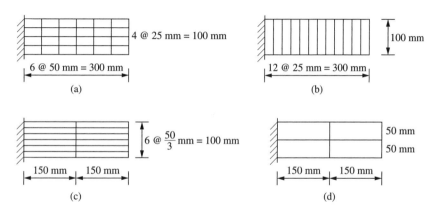

Figure P7–7

7.8 Show that Eq. (7.5.13) is obtained by static condensation of Eq. (7.5.12).

7.9 In considering the patch test, answer the following questions:
  a. Can elements of different mechanical properties be used? Why?
  b. Can the patch be arbitrary in shape? Why?
  c. Can we mix triangular and quadrilateral elements in the patch test?
  d. Can we mix bar elements with triangular or quadrilateral elements? Why?
  e. When should the patch test be applied?

7.10 Consider the bar with two elements shown in Figure P7.10. Perform a patch test using these two elements. Let $E = 200$ GPa, and $A = 1 \times 10^{-4}$ m^2. Use the standard bar element stiffness matrix (Eq. (3.1.14)) derived using the shape functions $N_1(x) = x/L$ and $N_2(x) = 1 - x/L$.

  a. For the rigid body motion test, set $u_1 = 1$ m and $u_3 = 1$ m and verify that $u_2 = 1$ m by using the direct stiffness method.

b. For the constant strain test, assume linear displacement function $u(x) = x$ for the nodes at the boundaries, such that $u_1 = 0$ and $u_3 = 2$ m, and verify that $u_2 = 0.6$ m.

Figure P7–10

Solve the following problems using a computer program. In some of these problems, we suggest that students be assigned separate parts (or models) to facilitate parametric studies.

7.11 Consider the rectangular plate shown (Figure P7–11) in plane stress. Using a computer program, verify that the plane stress element of the code satisfied the patch test. That is, apply constant displacement of $u = 0.005$ m to the right-side nodes, 3, 6, and 9, and determine the displacement at interior node 5. Use $E = 200$ GPa, $v = 0.3$, and plate thickness of 0.1 m. Explain your results.

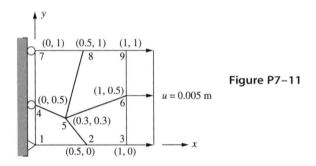

Figure P7–11

7.12 Determine the free-end displacements and the element stresses for the plate discretized into four triangular elements and subjected to the tensile forces shown in Figure P7–12. Compare your results to the solution given in Section 6.5. Why are these results different? Let $E = 210$ GPa, $v = 0.30$, and $t = 20$ mm.

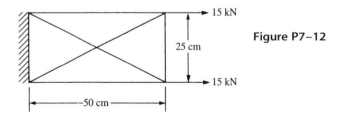

Figure P7–12

7.13 Determine the stresses in the plate with the hole subjected to the tensile stress shown in Figure P7–13. Graph the stress variation σ_x versus the distance y from the hole. Let $E = 200$ GPa, $v = 0.25$, and $t = 25$ mm. (Use approximately 25, 50, 75, 100, and then 120 nodes in your finite element model.) Use symmetry as appropriate.

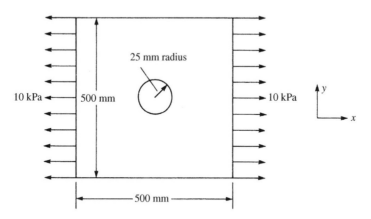

Figure P7–13

7.14 Solve the following problem of a steel tensile plate with a concentrated load applied at the top, as shown in Figure P7–14. Determine at what depth the effect of the load dies out. Plot stress σ_y versus distance from the load. At distances of 25 mm, 50 mm, 100 mm, 150 mm, 250 mm, 375 mm, 500 mm, and 750 mm from the load, list σ_y

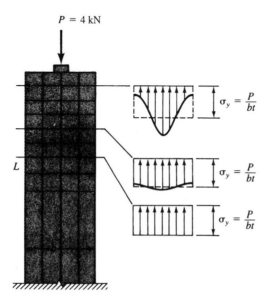

Figure P7–14

versus these distances. Let the width of the plate be $b = 100$ mm, thickness of the plate be $t = 6.25$ mm, and length be $L = 1000$ mm. Look up the concept of St. Venant's principle to see how it explains the stress behavior in this problem.

7.15 For the connecting rod shown in Figure P7–15, determine the maximum principal stresses and their location. Let $E = 210$ GPa, $v = 0.25$, $t = 25$ mm, and $P = 4$ kN.

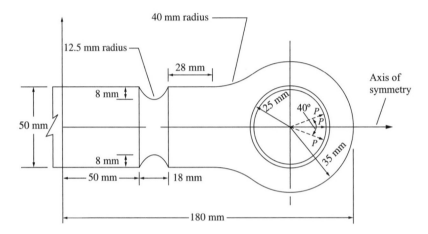

Figure P7–15

7.16 Determine the maximum principal stresses and their locations for the member with fillet subjected to tensile forces shown in Figure P7–16. Let $E = 200$ GPa and $v = 0.25$. Then let $E = 73$ GPa and $v = 0.30$. Let $t = 25$ mm for both cases. Compare your answers for the two cases.

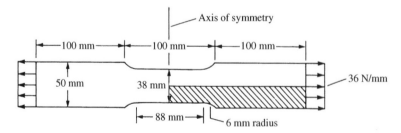

Figure P7–16

7.17 Determine the stresses in the member with a re-entrant corner as shown in Figure P7–17. At what location are the principal stresses largest? Let $E = 210$ GPa and $v = 0.25$. Use plane strain conditions.

7.18 Determine the stresses in the soil mass subjected to the strip footing load shown in Figure P7–18. Use a width of $2D$ and depth of D, where D is 3, 4, 6, 8, and 10 m.

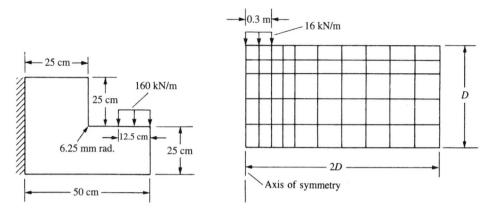

Figure P7–17   **Figure P7–18**

Plot the maximum stress contours on your finite element model for each case. Compare your results. Comment regarding your observations on modeling infinite media. Let $E = 200$ MPa and $v = 0.30$. Use plane strain conditions.

7.19 For the tooth implant subjected to loads shown in Figure P7–19, determine the maximum principal stresses. Let $E = 12$ GPa and $v = 0.3$ for the dental restorative implant material (cross-hatched), and let $E = 7.5$ GPa and $v = 0.35$ for the bony material. Let $X = 1.25$ mm, 2.5 mm, 5.0 mm, 7.5 mm, and 12.5 mm, where X represents the various depths of the implant beneath the bony surface. Rectangular elements are used in the finite element model shown in Figure P7–19. Assume the thickness of each element to be $t = 6.25$ mm.

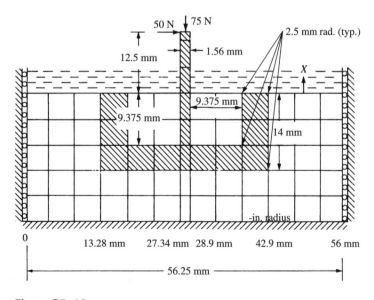

Figure P7–19

7.20 Determine the middepth deflection at the free end and the maximum principal stresses and their location for the beam subjected to the shear load variation shown in Figure P7–20. Do this using 64 rectangular elements all of size 300 mm × 12.5 mm; then all of size 150 mm × 25 mm; then all of size 75 mm × 50 mm. Then use 60 rectangular elements all of size 60 mm × 66.67 mm; then all of size 120 mm × 33.33 mm. Compare the free-end deflections and the maximum principal stresses in each case to the exact solution. Let $E = 210$ GPa, $v = 0.3$, and $t = 25$ mm. Comment on the accuracy of both displacements and stresses.

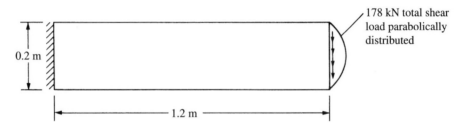

Figure P7–20

7.21 Determine the stresses in the shear wall shown in Figure P7–21. At what location are the principal stresses largest? Let $E = 21$ GPa, $v = 0.25$, $t_{wall} = 0.10$ m, and $t_{beam} = 0.20$ m. Use 0.1 m radii at the re-entrant corners.

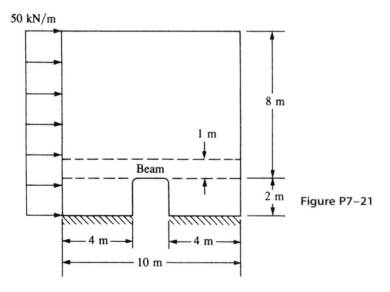

50 kN/m

8 m

1 m

Beam

2 m   **Figure P7–21**

4 m   4 m

10 m

7.22 Determine the stresses in the plates with the round and square holes subjected to the tensile stresses shown in Figure P7–22. Compare the largest principal stresses for each plate. Let $E = 210$ GPa, $v = 0.25$, and $t = 5$ mm.

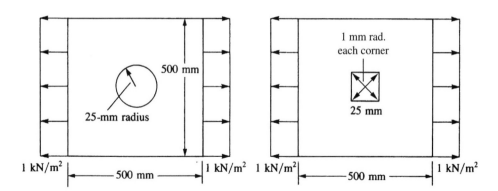

Figure P7–22

 7.23 For the concrete overpass structure shown in Figure P7–23, determine the maximum principal stresses and their locations. Assume plane strain conditions. Let $E = 20$ GPa and $v = 0.30$.

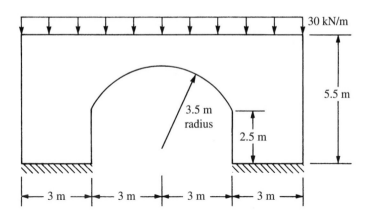

Figure P7–23

 7.24 For the steel culvert shown in Figure P7–24 on the next page, determine the maximum principal stresses and their locations and the largest displacement and its location. Let $E_{steel} = 210$ GPa and let $v = 0.30$.

 7.25 For the tensile member shown in Figure P7–25 on the next page with two holes, determine the maximum principal stresses and their locations. Let $E = 210$ GPa, $v = 0.25$, and $t = 10$ mm. Then let $E = 70$ GPa and $v = 0.30$. Compare your results. Use 20 kN spread uniformly over the right side.

 7.26 For the plate shown in Figure P7–26 on the next page, determine the maximum principal stresses and their locations. Let $E = 210$ GPa and $v = 0.25$.

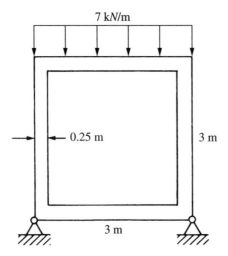

Figure P7–24

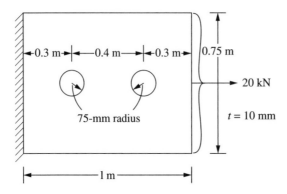

Figure P7–25

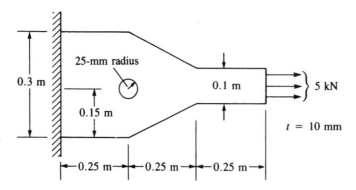

Figure P7–26

7.27 For the concrete dam shown subjected to water pressure in Figure P7–27, determine the principal stresses. Let $E = 25$ GPa and $v = 0.15$. Assume plane strain conditions. Perform the analysis for self-weight and then for hydrostatic (water) pressure against the dam vertical face as shown.

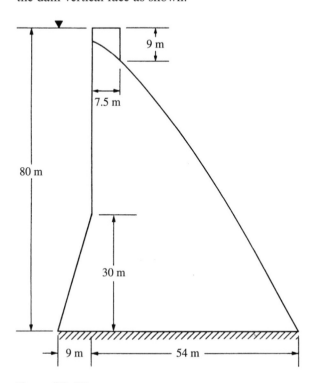

Figure P7–27

7.28 Determine the stresses in the wrench shown in Figure P7–28 on the next page. Let $E = 200$ GPa and $v = 0.25$, and assume uniform thickness $t = 10$ mm.

7.29 Determine the principal stresses in the blade implant and the bony material shown in Figure P7–29 on page 431. Let $E_{blade} = 20$ GPa, $v_{blade} = 0.30$, $E_{bone} = 12$ GPa, and $v_{bone} = 0.35$. Assume plane stress conditions with $t = 5$ mm.

7.30 Determine the stresses in the plate shown in Figure P7–30 on page 431. Let $E = 210$ GPa and $v = 0.25$. The element thickness is 10 mm.

7.31 For the 12.5 mm thick canopy hook shown in Figure P7–31 on page 432, used to hold down an aircraft canopy, determine the maximum von Mises stress and maximum deflection. The hook is subjected to a concentrated upward load of 100 kN as shown. Assume boundary conditions of fixed supports over the lower half of the inside hole diameter. The hook is made from AISI 4130 steel, quenched and tempered at 200°C. (This problem is compliments of Mr. Steven Miller.)

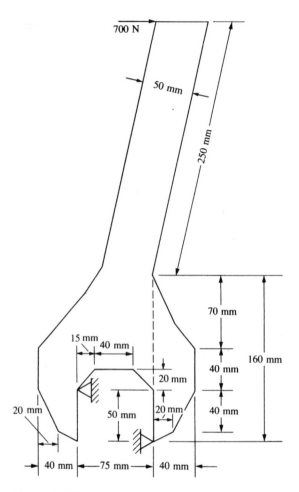

Figure P7–28

7.32 For the 6.25 mm thick L-shaped steel bracket shown in Figure P7–32 on page 432, show that the stress at the 90° re-entrant corner never converges. Try models with increasing numbers of elements to show this while plotting the maximum principal stress in the bracket. That is, start with one model, then refine the mesh around the re-entrant corner and see what happens, say, after two refinements. Why? Then add a fillet, say, of radius 12.5 mm and see what happens as you refine the mesh. Again plot the maximum principal stress for each refinement.

Use a computer program to help solve the design-type problems, 7.33 through 7.39.

7.33 The machine shown in Figure P7–33 on page 433 is an overload protection device that releases the load when the shear pin S fails. Determine the maximum von Mises stress in the upper part ABE if the pin shears when its shear stress is 40 MPa. Assume the upper part to have a uniform thickness of 6 mm. Assume plane stress conditions

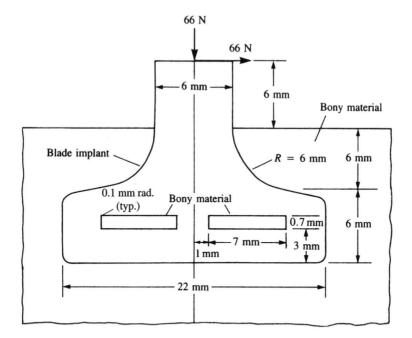

Figure P7–29

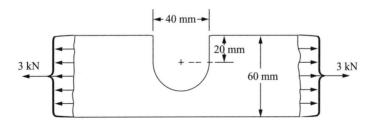

Figure P7–30

for the upper part. The part is made of 6061 aluminum alloy. Is the thickness sufficient to prevent failure based on the maximum distortion energy theory? If not, suggest a better thickness. (Scale all dimensions as needed.)

7.34 The steel triangular plate 6 mm thick shown in Figure P7–34 on page 433 is bolted to a steel column with 18 mm-diameter bolts in the pattern shown. Assuming the column and bolts are very rigid relative to the plate and neglecting friction forces between the column and plate, determine the highest load exerted on any bolt. The bolts should not be included in the model, just fix the nodes around the bolt circles and consider the reactions at these nodes as the bolt loads. If 18 mm-diameter bolts are not sufficient, recommend another standard diameter. Assume a standard material for the bolts. Compare the reactions from the finite element results to those found by classical methods.

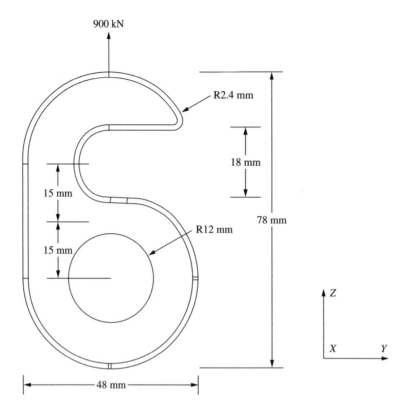

Figure P7–31

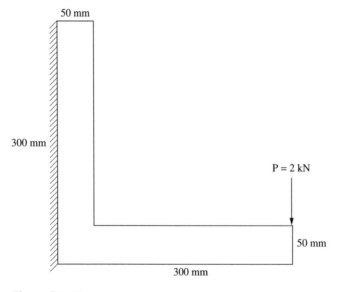

Figure P7–32

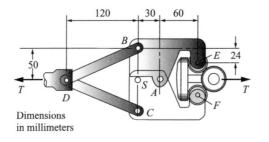
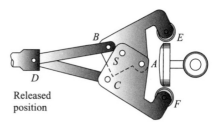

Dimensions
in millimeters

Released
position

Figure P7–33

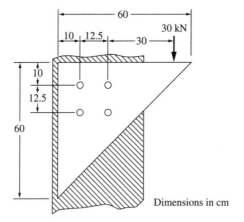

Dimensions in cm

Figure P7–34

7.35 A 6 mm thick machine part supports an end load of 4.5 kN as shown in Figure P7–35. Determine the stress concentration factors for the two changes in geometry located at the radii shown on the lower side of the part. Compare the stresses you get to classical beam theory results with and without the change in geometry, that is, with a uniform depth of 24 mm instead of the additional material depth of 36 mm. Assume standard mild steel is used for the part. Recommend any changes you might make in the geometry.

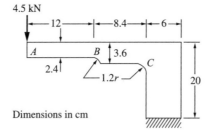

Dimensions in cm

Figure P7–35

7.36 A plate with an off-center hole is shown in Figure P7–36. Determine how close to the top edge the hole can be placed before yielding of the A36 steel occurs (based on the maximum distortion energy theory). The applied tensile stress is 70,000 kPa, and the plate thickness is 6 mm. Now if the plate is made of 6061-T6 aluminum alloy with a yield strength of 255 MPa, does this change your answer? If the plate thickness is changed to 12 mm, how does this change the results? Use same total load as when the plate is 6 mm thick.

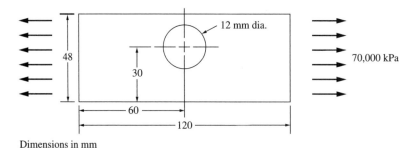

Dimensions in mm

Figure P7–36

7.37 One arm of a crimper tool shown in Figure P7–37 is to be designed of 1080 as-rolled steel. The loads and boundary conditions are shown in the figure. Select a thickness

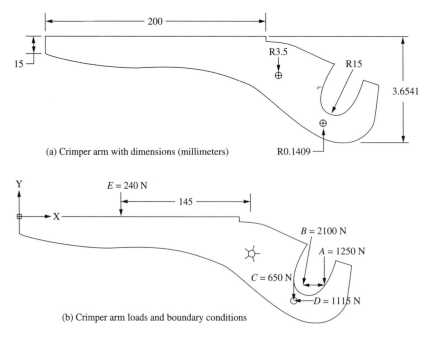

(a) Crimper arm with dimensions (millimeters)

(b) Crimper arm loads and boundary conditions

Figure P7–37

for the arm based on the material not yielding with a factor of safety of 1.5. Recommend any other changes in the design. (Scale any other dimensions that you need.) Remember stresses at concentrated loads are false.

7.38 Design the bicycle wrench with the approximate dimensions shown in Figure P7–38. If you need to change dimensions explain why. The wrench should be made of steel or aluminum alloy. Determine the thickness needed based on the maximum distortion energy theory. Plot the deformed shape of the wrench and the principal stress and von Mises stress. The boundary conditions are shown in the figure, and the loading is shown as a distributed load acting over the right part of the wrench. Use a factor of safety of 1.5 against yielding. Round each corner with a 0.1 cm radius.

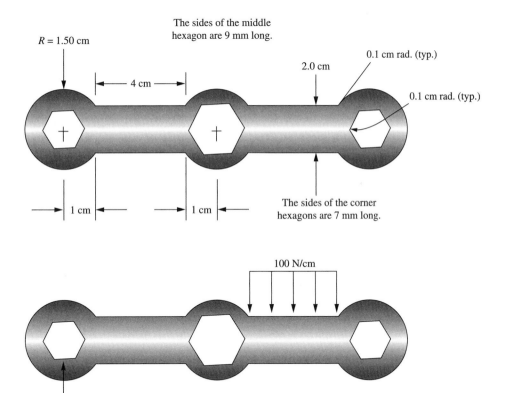

Figure P7–38

7.39 For the various parts shown in Figure P7–39 on the next page determine the best one to relieve stress. Make the original part have a small radius of 2.5 mm at the inside re-entrant corners. Place a uniform pressure load of 6.4×10^6 N/m^2 on the right end of each part and fix the left end. All units shown are taken in millimeters. Let the material be A36 steel.

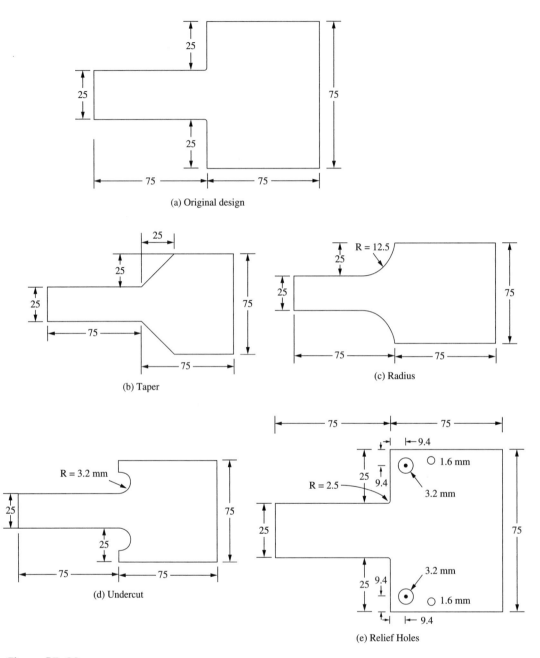

CHAPTER 8

DEVELOPMENT OF THE LINEAR-STRAIN TRIANGLE EQUATIONS ▲

CHAPTER OBJECTIVES

- To develop the linear-strain triangular (LST) element stiffness matrix.
- To describe how the LST stiffness matrix can be determined.
- To compare the difference in results using the CST and LST elements.

Introduction

In this chapter, we consider the development of the stiffness matrix and equations for a higher-order triangular element, called the *linear-strain triangle* (LST). This element is available in many commercial computer programs and has some advantages over the constant-strain triangle described in Chapter 6.

The LST element has six nodes and twelve unknown displacement degrees of freedom. The displacement functions for the element are quadratic instead of linear (as in the CST).

The procedures for development of the equations for the LST element follow the same steps as those used in Chapter 6 for the CST element. However, the number of equations now becomes twelve instead of six, making a longhand solution extremely cumbersome. Hence, we will use a computer to perform many of the mathematical operations.

After deriving the element equations, we will compare results from problems solved using the LST element with those solved using the CST element. The introduction of the higher-order LST element will illustrate the possible advantages of higher-order elements and should enhance your general understanding of the concepts involved with finite element procedures.

▲ 8.1 Derivation of the Linear-Strain Triangular Element Stiffness Matrix and Equations ▲

We will now derive the LST stiffness matrix and element equations. The steps used here are identical to those used for the CST element, and much of the notation is the same.

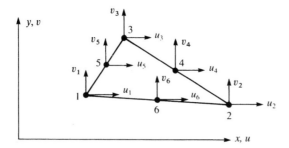

Figure 8–1 Basic six-node triangular element showing degrees of freedom

Step 1 Select Element Type

Consider the triangular element shown in Figure 8–1 with the usual end nodes and three additional nodes conveniently located at the midpoints of the sides. Thus, a computer program can automatically compute the midpoint coordinates once the coordinates of the corner nodes are given as input.

The unknown nodal displacements are now given by

$$\{d\} = \begin{Bmatrix} \{d_1\} \\ \{d_2\} \\ \{d_3\} \\ \{d_4\} \\ \{d_5\} \\ \{d_6\} \end{Bmatrix} = \begin{Bmatrix} u_1 \\ v_1 \\ u_2 \\ v_2 \\ u_3 \\ v_3 \\ u_4 \\ v_4 \\ u_5 \\ v_5 \\ u_6 \\ v_6 \end{Bmatrix} \tag{8.1.1}$$

Step 2 Select a Displacement Function

We now select a quadratic displacement function in each element as

$$u(x, y) = a_1 + a_2 x + a_3 y + a_4 x^2 + a_5 xy + a_6 y^2$$
$$v(x, y) = a_7 + a_8 x + a_9 y + a_{10} x^2 + a_{11} xy + a_{12} y^2 \tag{8.1.2}$$

Again, the number of coefficients $a_i(12)$ equals the total number of degrees of freedom for the element. The displacement compatibility among adjoining elements is satisfied because three nodes are located along each side and a parabola is defined by three points on its path. Since adjacent elements are connected at common nodes, their displacement compatibility across the boundaries will be maintained.

In general, when considering triangular elements, we can use a complete polynomial in Cartesian coordinates to describe the displacement field within an element.

Terms in Pascal Triangle	Polynomial Degree	Number of Terms	Triangle
1	0 (constant)	1	
x y	1 (linear)	3	CST  (Chap. 6)
x^2 xy y^2	2 (quadratic)	6	LST 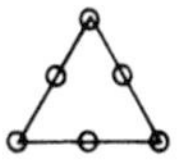 (Chap. 8)
x^3 x^2y xy^2 y^3	3 (cubic)	10	QST 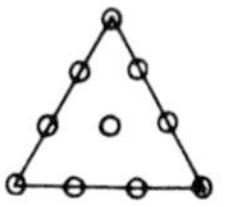

Figure 8–2 Relation between type of plane triangular element and polynomial coefficients based on a Pascal triangle

Using internal nodes as necessary for the higher-order cubic and quartic elements, we use all terms of a truncated Pascal triangle in the displacement field or, equivalently, the shape functions, as shown by Figure 8–2; that is, a complete linear function is used for the CST element considered previously in Chapter 6. The complete quadratic function is used for the LST of this chapter. The complete cubic function is used for the quadratic-strain triangle (QST), with an internal node necessary as the tenth node.

The general displacement functions, Eqs. (8.1.2), expressed in matrix form are now

$$\{\psi\} = \left\{ \begin{array}{c} u \\ v \end{array} \right\} = \begin{bmatrix} 1 & x & y & x^2 & xy & y^2 & 0 & 0 & 0 & 0 & 0 & 0 \\ 0 & 0 & 0 & 0 & 0 & 0 & 1 & x & y & x^2 & xy & y^2 \end{bmatrix} \left\{ \begin{array}{c} a_1 \\ a_2 \\ \vdots \\ a_{12} \end{array} \right\} \tag{8.1.3}$$

Alternatively, we can express Eq. (8.1.3) as

$$\{\psi\} = [M^*]\{a\} \tag{8.1.4}$$

where $[M^*]$ is defined to be the first matrix on the right side of Eq. (8.1.3). The coefficients a_1 through a_{12} can be obtained by substituting the coordinates into u and v as follows:

$$\left\{ \begin{array}{c} u_1 \\ u_2 \\ \vdots \\ u_6 \\ v_1 \\ \vdots \\ v_5 \\ v_6 \end{array} \right\} = \begin{bmatrix} 1 & x_1 & y_1 & x_1^2 & x_1y_1 & y_1^2 & 0 & 0 & 0 & 0 & 0 & 0 \\ 1 & x_2 & y_2 & x_2^2 & x_2y_2 & y_2^2 & 0 & 0 & 0 & 0 & 0 & 0 \\ \vdots & \vdots & \vdots & \vdots & \vdots & \vdots & \vdots & \vdots & \vdots & \vdots & \vdots & \vdots \\ 1 & x_6 & y_6 & x_6^2 & x_6y_6 & y_6^2 & 0 & 0 & 0 & 0 & 0 & 0 \\ 0 & 0 & 0 & 0 & 0 & 0 & 1 & x_1 & y_1 & x_1^2 & x_1y_1 & y_1^2 \\ \vdots & \vdots & \vdots & \vdots & \vdots & \vdots & \vdots & \vdots & \vdots & \vdots & \vdots & \vdots \\ 0 & 0 & 0 & 0 & 0 & 0 & 1 & x_5 & y_5 & x_5^2 & x_5y_5 & y_5^2 \\ 0 & 0 & 0 & 0 & 0 & 0 & 1 & x_6 & y_6 & x_6^2 & x_6y_6 & y_6^2 \end{bmatrix} \left\{ \begin{array}{c} a_1 \\ a_2 \\ \vdots \\ a_6 \\ a_7 \\ \vdots \\ a_{11} \\ a_{12} \end{array} \right\} \tag{8.1.5}$$

Solving for the a_i's, we have

$$
\begin{Bmatrix} a_1 \\ \vdots \\ a_6 \\ a_7 \\ \vdots \\ a_{12} \end{Bmatrix} = \begin{bmatrix} 1 & x_1 & y_1 & x_1^2 & x_1y_1 & y_1^2 & 0 & 0 & 0 & 0 & 0 & 0 \\ \vdots & \vdots & \vdots & \vdots & \vdots & \vdots & \vdots & \vdots & \vdots & \vdots & \vdots & \vdots \\ 1 & x_6 & y_6 & x_6^2 & x_6y_6 & y_6^2 & 0 & 0 & 0 & 0 & 0 & 0 \\ 0 & 0 & 0 & 0 & 0 & 0 & 1 & x_1 & y_1 & x_1^2 & x_1y_1 & y_1^2 \\ \vdots & \vdots & \vdots & \vdots & \vdots & \vdots & \vdots & \vdots & \vdots & \vdots & \vdots & \vdots \\ 0 & 0 & 0 & 0 & 0 & 0 & 1 & x_6 & y_6 & x_6^2 & x_6y_6 & y_6^2 \end{bmatrix}^{-1} \begin{Bmatrix} u_1 \\ \vdots \\ u_6 \\ v_1 \\ \vdots \\ v_6 \end{Bmatrix}
$$

$$(8.1.6)$$

or, alternatively, we can express Eq. (8.1.6) as

$$\{a\} = [X]^{-1}\{d\} \qquad (8.1.7)$$

where $[X]$ is the 12×12 matrix on the right side of Eq. (8.1.6). It is best to invert the $[X]$ matrix by using a digital computer. Then the a_i's, in terms of nodal displacements, are substituted into Eq. (8.1.4). Note that only the 6×6 part of $[X]$ in Eq. (8.1.6) really must be inverted. Finally, using Eq. (8.1.7) in Eq. (8.1.4), we can obtain the general displacement expressions in terms of the shape functions and the nodal degrees of freedom as

$$\{\psi\} = [N]\{d\} \qquad (8.1.8)$$

where

$$[N] = [M^*][X]^{-1} \qquad (8.1.9)$$

Step 3 Define the Strain–Displacement and Stress–Strain Relationships

The element strains are again given by

$$
\{\varepsilon\} = \begin{Bmatrix} \varepsilon_x \\ \varepsilon_y \\ \gamma_{xy} \end{Bmatrix} = \begin{Bmatrix} \dfrac{\partial u}{\partial x} \\[2mm] \dfrac{\partial v}{\partial y} \\[2mm] \dfrac{\partial v}{\partial x} + \dfrac{\partial u}{\partial y} \end{Bmatrix} \qquad (8.1.10)
$$

or, using Eq. (8.1.3) for u and v in Eq. (8.1.10), we obtain the strain-generalized displacement equations as

$$
\{\varepsilon\} = \begin{bmatrix} 0 & 1 & 0 & 2x & y & 0 & 0 & 0 & 0 & 0 & 0 & 0 \\ 0 & 0 & 0 & 0 & 0 & 0 & 0 & 0 & 1 & 0 & x & 2y \\ 0 & 0 & 1 & 0 & x & 2y & 0 & 1 & 0 & 2x & y & 0 \end{bmatrix} \begin{Bmatrix} a_1 \\ a_2 \\ \vdots \\ a_{12} \end{Bmatrix} \qquad (8.1.11)
$$

We observe that Eq. (8.1.11) yields a linear strain variation in the element. Therefore, the element is called a *linear-strain triangle* (LST). Rewriting Eq. (8.1.11), we have

$$\{\varepsilon\} = [M']\{a\} \qquad (8.1.12)$$

where $[M']$ is the first matrix on the right side of Eq. (8.1.11). Substituting Eq. (8.1.6) for the a_i's into Eq. (8.1.12), we have $\{\varepsilon\}$ in terms of the nodal displacements as

$$\{\varepsilon\} = [B]\{d\} \tag{8.1.13}$$

where $[B]$ is a function of the variables x and y and the coordinates (x_1, y_1) through (x_6, y_6) given by

$$[B] = [M'][X]^{-1} \tag{8.1.14}$$

where Eq. (8.1.7) has been used in expressing Eq. (8.1.14). Note that $[B]$ is now a matrix of order 3×12.

The stresses are again given by

$$\begin{Bmatrix} \sigma_x \\ \sigma_y \\ \tau_{xy} \end{Bmatrix} = [D] \begin{Bmatrix} \varepsilon_x \\ \varepsilon_y \\ \gamma_{xy} \end{Bmatrix} = [D][B]\{d\} \tag{8.1.15}$$

where $[D]$ is given by Eq. (6.1.8) for plane stress or by Eq. (6.1.10) for plane strain. These stresses are now linear functions of x and y coordinates.

Step 4   Derive the Element Stiffness Matrix and Equations

We determine the stiffness matrix in a manner similar to that used in Section 6.2 by using Eq. (6.2.50) repeated here as

$$[k] = \iiint_V [B]^T [D][B]\, dV \tag{8.1.16}$$

However, the $[B]$ matrix is now a function of x and y as given by Eq. (8.1.14). Therefore, we must perform the integration in Eq. (8.1.16). Finally, the $[B]$ matrix is of the form

$$[B] = \frac{1}{2A} \begin{bmatrix} \beta_1 & 0 & \beta_2 & 0 & \beta_3 & 0 & \beta_4 & 0 & \beta_5 & 0 & \beta_6 & 0 \\ 0 & \gamma_1 & 0 & \gamma_2 & 0 & \gamma_3 & 0 & \gamma_4 & 0 & \gamma_5 & 0 & \gamma_6 \\ \gamma_1 & \beta_1 & \gamma_2 & \beta_2 & \gamma_3 & \beta_3 & \gamma_4 & \beta_4 & \gamma_5 & \beta_5 & \gamma_6 & \beta_6 \end{bmatrix} \tag{8.1.17}$$

where the β's and γ's are now functions of x and y as well as of the nodal coordinates, as is illustrated for a specific linear-strain triangle in Section 8.2 by Eq. (8.2.8). The stiffness matrix is then seen to be a 12×12 matrix on multiplying the matrices in Eq. (8.1.16). The stiffness matrix, Eq. (8.1.16), is very cumbersome to obtain in explicit form, so it will not be given here. However, if the origin of the coordinates is considered to be at the centroid of the element, the integrations become amenable [9]. Alternatively, area coordinates [3, 8, 9] can be used to obtain an explicit form of the stiffness matrix. However, even the use of area coordinates usually involves tedious calculations. Therefore, the integration is best carried out numerically. (Numerical integration is described in Section 10.3.)

The element body forces and surface forces should not be automatically lumped at the nodes, but for a consistent formulation (one that is formulated from the same shape functions used to formulate the stiffness matrix), Eqs. (6.3.1) and (6.3.7), respectively, should be used. (Problems 8.3 and 8.4 illustrate this concept.) These forces can be added to any concentrated nodal forces to obtain the element force matrix. Here the element force matrix is of order 12×1 because, in general, there could be an x and a y component of force at each of the six nodes associated with the element. The element equations are then given by

$$
\begin{Bmatrix} f_{1x} \\ f_{1y} \\ \vdots \\ f_{6y} \end{Bmatrix} = \begin{bmatrix} k_{11} & \cdots & k_{1,12} \\ k_{21} & & k_{2,12} \\ \vdots & & \vdots \\ k_{12,1} & \cdots & k_{12,12} \end{bmatrix} \begin{Bmatrix} u_1 \\ v_1 \\ \vdots \\ v_6 \end{Bmatrix} \tag{8.1.18}
$$
$$
(12 \times 1) \qquad\qquad (12 \times 12) \qquad\qquad (12 \times 1)
$$

Steps 5 through 7

Steps 5 through 7, which involve assembling the global stiffness matrix and equations, determining the unknown global nodal displacements, and calculating the stresses, are identical to those in Section 6.2 for the CST. However, instead of constant stresses in each element, we now have a linear variation of the stresses in each element. Common practice was to use the centroidal element stresses. Current practice is to use the average of the nodal element stresses.

▲ 8.2 Example LST Stiffness Determination ▲

To illustrate some of the procedures outlined in Section 8.1 for deriving an LST stiffness matrix, consider the following example. Figure 8–3 shows a specific LST and its coordinates. The triangle is of base dimension b and height h, with midside nodes.

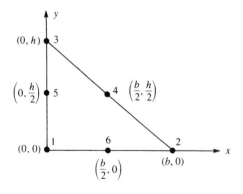

Figure 8–3 LST triangle for evaluation of a stiffness matrix

Using the first six equations of Eq. (8.1.5), we calculate the coefficients a_1 through a_6 by evaluating the displacement u at each of the six known coordinates of each node as follows:

$$u_1 = u(0,0) = a_1$$

$$u_2 = u(b,0) = a_1 + a_2 b + a_4 b^2$$

$$u_3 = u(0,h) = a_1 + a_3 h + a_6 h^2$$

$$u_4 = u\left(\frac{b}{2}, \frac{h}{2}\right) = a_1 + a_2 \frac{b}{2} + a_3 \frac{h}{2} + a_4 \left(\frac{b}{2}\right)^2 + a_5 \frac{bh}{4} + a_6 \left(\frac{h}{2}\right)^2 \quad (8.2.1)$$

$$u_5 = u\left(0, \frac{h}{2}\right) = a_1 + a_3 \frac{h}{2} + a_6 \left(\frac{h}{2}\right)^2$$

$$u_6 = u\left(\frac{b}{2}, 0\right) = a_1 + a_2 \frac{b}{2} + a_4 \left(\frac{b}{2}\right)^2$$

Solving Eqs. (8.2.1) simultaneously for the a_i's, we obtain

$$a_1 = u_1 \qquad a_2 = \frac{4u_6 - 3u_1 - u_2}{b} \qquad a_3 = \frac{4u_5 - 3u_1 - u_3}{h}$$

$$a_4 = \frac{2(u_2 - 2u_6 + u_1)}{b^2} \qquad a_5 = \frac{4(u_1 + u_4 - u_5 - u_6)}{bh} \quad (8.2.2)$$

$$a_6 = \frac{2(u_3 - 2u_5 + u_1)}{h^2}$$

Substituting Eqs. (8.2.2) into the displacement expression for u from Eqs. (8.1.2), we have

$$u = u_1 + \left[\frac{4u_6 - 3u_1 - u_2}{b}\right]x + \left[\frac{4u_5 - 3u_1 - u_3}{h}\right]y + \left[\frac{2(u_2 - 2u_6 + u_1)}{b^2}\right]x^2$$

$$+ \left[\frac{4(u_1 + u_4 - u_5 - u_6)}{bh}\right]xy + \left[\frac{2(u_3 - 2u_5 + u_1)}{h^2}\right]y^2 \quad (8.2.3)$$

Similarly, solving for a_7 through a_{12} by evaluating the displacement v at each of the six nodes and then substituting the results into the expression for v from Eqs. (8.1.2), we obtain

$$v = v_1 + \left[\frac{4v_6 - 3v_1 - v_2}{b}\right]x + \left[\frac{4v_5 - 3v_1 - v_3}{h}\right]y + \left[\frac{2(v_2 - 2v_6 + v_1)}{b^2}\right]x^2$$

$$+ \left[\frac{4(v_1 + v_4 - v_5 - v_6)}{bh}\right]xy + \left[\frac{2(v_3 - 2v_5 + v_1)}{h^2}\right]y^2 \quad (8.2.4)$$

Using Eqs. (8.2.3) and (8.2.4), we can express the general displacement expressions in terms of the shape functions as

$$\begin{Bmatrix} u \\ v \end{Bmatrix} = \begin{bmatrix} N_1 & 0 & N_2 & 0 & N_3 & 0 & N_4 & 0 & N_5 & 0 & N_6 & 0 \\ 0 & N_1 & 0 & N_2 & 0 & N_3 & 0 & N_4 & 0 & N_5 & 0 & N_6 \end{bmatrix} \begin{Bmatrix} u_1 \\ v_1 \\ \vdots \\ v_6 \end{Bmatrix}$$

(8.2.5)

where the shape functions are obtained by collecting coefficients that multiply each u_i term in Eq. (8.2.3). For instance, collecting all terms that multiply by u_1 in Eq. (8.2.3), we obtain N_1. These shape functions are then given by

$$N_1 = 1 - \frac{3x}{b} - \frac{3y}{h} + \frac{2x^2}{b^2} + \frac{4xy}{bh} + \frac{2y^2}{h^2} \qquad N_2 = \frac{-x}{b} + \frac{2x^2}{b^2}$$

$$N_3 = \frac{-y}{h} + \frac{2y^2}{h^2} \qquad N_4 = \frac{4xy}{bh} \qquad N_5 = \frac{4y}{h} - \frac{4xy}{bh} - \frac{4y^2}{h^2} \qquad (8.2.6)$$

$$N_6 = \frac{4x}{b} - \frac{4x^2}{b^2} - \frac{4xy}{bh}$$

Using Eq. (8.2.5) in Eq. (8.1.10), and performing the differentiations indicated on u and v, we obtain

$$\{\varepsilon\} = [B]\{d\} \qquad (8.2.7)$$

where $[B]$ is of the form of Eq. (8.1.17), with the resulting β's and γ's in Eq. (8.1.17) given by

$$\beta_1 = -3h + \frac{4hx}{b} + 4y \qquad \beta_2 = -h + \frac{4hx}{b} \qquad \beta_3 = 0$$

$$\beta_4 = 4y \qquad \beta_5 = -4y \qquad \beta_6 = 4h - \frac{8hx}{b} - 4y$$

$$\gamma_1 = -3b + 4x + \frac{4by}{h} \qquad \gamma_2 = 0 \qquad \gamma_3 = -b + \frac{4by}{h}$$

$$\gamma_4 = 4x \qquad \gamma_5 = 4b - 4x - \frac{8by}{h} \qquad \gamma_6 = -4x$$

(8.2.8)

These β's and γ's are specific to the element in Figure 8–3. Specifically, using Eqs. (8.1.1) and (8.1.17) in Eq. (8.2.7), we obtain

$$\varepsilon_x = \frac{1}{2A}[\beta_1 u_1 + \beta_2 u_2 + \beta_3 u_3 + \beta_4 u_4 + \beta_5 u_5 + \beta_6 u_6]$$

$$\varepsilon_y = \frac{1}{2A}[\gamma_1 v_1 + \gamma_2 v_2 + \gamma_3 v_3 + \gamma_4 v_4 + \gamma_5 v_5 + \gamma_6 v_6] \qquad (8.2.9)$$

$$\gamma_{xy} = \frac{1}{2A}[\gamma_1 u_1 + \beta_1 v_1 + \cdots + \beta_6 v_6]$$

The stiffness matrix for a constant-thickness element can now be obtained on substituting Eqs. (8.2.8) into Eq. (8.1.17) to obtain $[B]$, then substituting $[B]$ into

Eq. (8.1.16) and using calculus to set up the appropriate integration. The explicit expression for the 12×12 stiffness matrix, being extremely cumbersome to obtain, is not given here. Stiffness matrix expressions for higher-order elements are found in References [1] and [2].

▲ 8.3 Comparison of Elements ▲

For a given number of nodes, a better representation of true stress and displacement is generally obtained using the LST element than is obtained with the same number of nodes using a much finer subdivision into simple CST elements. For example, using one LST yields better results than using four CST elements with the same number of nodes (Figure 8–4) and hence the same number of degrees of freedom (except for the case when constant stress exists).

We now present results to compare the CST of Chapter 6 with the LST of this chapter. Consider the cantilever beam subjected to a parabolic load variation acting as shown in Figure 8–5. Let $E = 210$ GPa, $v = 0.25$, and $t = 25.4$ mm.

Table 8–1 lists the series of tests run to compare results using the CST and LST elements. Table 8–2 shows comparisons of free-end (tip) deflection and stress σ_x for each element type used to model the cantilever beam. From Table 8–2, we can observe that the larger the number of degrees of freedom for a given type of triangular element, the closer the solution converges to the exact one (compare run A-1 to run A-2, and B-1 to B-2). For a given number of nodes, the LST analysis yields somewhat better results for displacement than the CST analysis (compare run A-1 to run B-1).

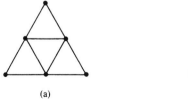

(a)   (b)

Figure 8–4 Basic triangular element: (a) four-CST and (b) one-LST

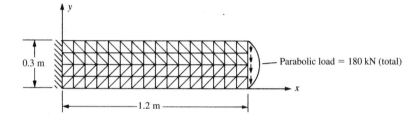

Parabolic load = 180 kN (total)

0.3 m

1.2 m

Figure 8–5 Cantilever beam used to compare the CST and LST elements with a 4×16 mesh

Table 8–1 Models used to compare CST and LST results for the cantilever beam of Figure 8–5 using ANSYS computer program [10]

Series of Tests Run	Number of Nodes	Number of Degrees of Freedom, n_d	Number of Triangular Elements
A-1 4×16 mesh	85	160	128 CST
A-2 8×32	297	576	512 CST
B-1 2×8	85	160	32 LST
B-2 4×16	297	576	128 LST

Table 8–2 Comparison of CST and LST results for the cantilever beam of Figure 8–5

Run	n_d	Bandwidth[1] n_b	Tip Deflection (mm)	σ_x (MPa)	Location (mm), x, y
A-1	160	14	−7.51	463.6	5.71, 28.57
A-2	576	22	−8.60	560.6	2.86, 29.54
B-1	160	18	−8.50	406.0	11.43, 26.67
B-2	576	22	−8.93	482.3	5.71, 28.57
Exact solution			−9.18	551.6	0, 30

[1] Bandwidth is described in Appendix B.4.

However, one of the reasons that the bending stress σ_x predicted by the LST model B-1 compared to CST model A-1 is not as accurate is as follows. Recall that the stress is calculated at the centroid of the element. We observe from the table that the location of the bending stress is closer to the wall and closer to the top for the CST model A-1 compared to the LST model B-1. As the classical bending stress is a linear function with increasing positive linear stress from the neutral axis for the downward applied load in this example, we expect the largest stress to be at the very top of the beam. So the model A-1 with more and smaller elements (with eight elements through the beam depth) has its centroid closer to the top (at 18.75 mm from the top) than model B-1 with few elements (two elements through the beam depth) with centroidal stress located at 37.5 mm from the top. Similarly, comparing A-2 to B-2 we observe the same trend in the results—displacement at the top end being more accurately predicted by the LST model, but stresses being calculated at the centroid making the A-2 model appear more accurate than the LST model due to the location where the stress is reported.

Although the CST element is rather poor in modeling bending, we observe from Table 8–2 that the element can be used to model a beam in bending if a sufficient number of elements are used through the depth of the beam. In general, both LST and CST analyses yield results good enough for most plane stress/strain problems, provided a sufficient number of elements are used. In fact, most commercial programs incorporate the use of CST and/or LST elements for plane stress/strain problems, although these elements are used primarily as transition elements (usually during mesh

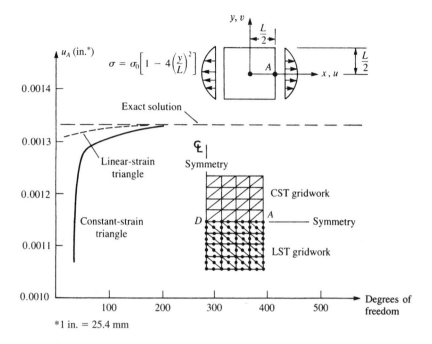

Figure 8–6 Plates subjected to parabolically distributed edge loads; comparison of results for triangular elements. (*Gallagher, Richard H., Finite Element Analysis: Fundamentals, 1st, © 1975*. Printed and Electronically reproduced by permission of Pearson Education, Inc., Upper Saddle River, New Jersey.)

generation). The four-sided isoparametric plane stress/strain element is most frequently used in commercial programs and is described in Chapter 10.

Also, recall that finite element displacements will always be less than (or equal to) the exact ones, because finite element models are normally predicted to be stiffer than the actual structures when the displacement formulation of the finite element method is used. (The reason for the stiffer model was discussed in Sections 3.10 and 7.3. Proof of this assertion can be found in References [4–7].

Finally, Figure 8–6 (from Reference [8]) illustrates a comparison of CST and LST models of a plate subjected to parabolically distributed edge loads. Figure 8–6 shows that the LST model converges to the exact solution for horizontal displacement at point *A* faster than does the CST model. However, the CST model is quite acceptable even for modest numbers of degrees of freedom. For example, a CST model with 100 nodes (200 degrees of freedom) often yields nearly as accurate a solution as does an LST model with the same number of degrees of freedom.

In conclusion, the results of Table 8–2 and Figure 8–6 indicate that the LST model might be preferred over the CST model for plane stress applications when relatively small numbers of nodes are used. However, the use of triangular elements of higher order, such as the LST, is not visibly advantageous when large numbers of nodes are used, particularly when the cost of formation of the element stiffnesses, equation bandwidth, and overall complexities involved in the computer modeling are considered.

▲ Summary Equations

Displacement functions for linear-strain triangle (LST) element:

$$u(x,y) = a_1 + a_2x + a_3y + a_4x^2 + a_5xy + a_6y^2$$
$$v(x,y) = a_1 + a_8x + a_9y + a_{10}x^2 + a_{11}xy + a_{12}y^2$$

$$(8.1.2)$$

Shape function matrix:

$$[N] = [M^*][X]^{-1} \qquad (8.1.9)$$

where

$$[M^*] = \begin{bmatrix} 1 & x & y & x^2 & xy & y^2 & 0 & 0 & 0 & 0 & 0 & 0 \\ 0 & 0 & 0 & 0 & 0 & 0 & 1 & x & y & x^2 & xy & y^2 \end{bmatrix}$$

and

$$[X]^{-1} = \begin{bmatrix} 1 & x_1 & y_1 & x_1^2 & x_1y_1 & y_1^2 & 0 & 0 & 0 & 0 & 0 & 0 \\ \vdots & \vdots & \vdots & \vdots & \vdots & \vdots & \vdots & \vdots & \vdots & \vdots & \vdots & \vdots \\ 1 & x_6 & y_6 & x_6^2 & x_6y_6 & y_6^2 & 0 & 0 & 0 & 0 & 0 & 0 \\ 0 & 0 & 0 & 0 & 0 & 0 & 1 & x_1 & y_1 & x_1^2 & x_1y_1 & y_1^2 \\ \vdots & \vdots & \vdots & \vdots & \vdots & \vdots & \vdots & \vdots & \vdots & \vdots & \vdots & \vdots \\ 0 & 0 & 0 & 0 & 0 & 0 & 1 & x_6 & y_6 & x_6^2 & x_6y_6 & y_6^2 \end{bmatrix}^{-1}$$

Strain-generalized displacement equations:

$$\{\varepsilon\} = \begin{bmatrix} 0 & 1 & 0 & 2x & y & 0 & 0 & 0 & 0 & 0 & 0 & 0 \\ 0 & 0 & 0 & 0 & 0 & 0 & 0 & 0 & 1 & 0 & x & 2y \\ 0 & 0 & 1 & 0 & x & 2y & 0 & 1 & 0 & 2x & y & 0 \end{bmatrix} \begin{Bmatrix} a_1 \\ a_2 \\ \vdots \\ a_{12} \end{Bmatrix} \qquad (8.1.11)$$

▲ References

[1] Pederson, P., "Some Properties of Linear Strain Triangles and Optimal Finite Element Models," *International Journal for Numerical Methods in Engineering*, Vol. 7, pp. 415–430, 1973.

[2] Tocher, J. L., and Hartz, B. J., "Higher-Order Finite Element for Plane Stress," *Journal of the Engineering Mechanics Division*, Proceedings of the American Society of Civil Engineers, Vol. 93, No. EM4, pp. 149–174, August 1967.

[3] Bowes, W. H., and Russell, L. T., *Stress Analysis by the Finite Element Method for Practicing Engineers*, Lexington Books, Toronto, 1975.

[4] Fraeijes de Veubeke, B., "Upper and Lower Bounds in Matrix Structural Analysis," *Matrix Methods of Structural Analysis*, AGAR-Dograph 72, B. Fraeijes de Veubeke, ed., Macmillan, New York, 1964.

[5] McLay, R. W., *Completeness and Convergence Properties of Finite Element Displacement Functions: A General Treatment*, American Institute of Aeronautics and Astronautics Paper No. 67–143, AIAA 5th Aerospace Meeting, New York, 1967.

[6] Tong, P., and Pian, T. H. H., "The Convergence of Finite Element Method in Solving Linear Elastic Problems," *International Journal of Solids and Structures*, Vol. 3, pp. 865–879, 1967.

[7] Cowper, G. R., "Variational Procedures and Convergence of Finite-Element Methods," *Numerical and Computer Methods in Structural Mechanics*, S. J. Fenves, N. Perrone, A. R. Robinson, and W. C. Schnobrich, eds., Academic Press, New York, 1973.

[8] Gallagher, R., *Finite Element Analysis Fundamentals*, Prentice Hall, Englewood Cliffs, NJ, 1975.

[9] Zienkiewicz, O. C., *The Finite Element Method*, 3rd ed., McGraw-Hill, New York, 1977.

[10] ANSYS—Engineering Analysis Systems, Johnson Rd., P.O. Box 65, Houston, PA 15342.

▲ Problems

8.1 Evaluate the shape functions given by Eq. (8.2.6). Sketch the variation of each function over the surface of the triangular element shown in Figure 8–3.

8.2 Express the strains ε_x, ε_y, and γ_{xy} for the element of Figure 8–3 by using the results given in Section 8.2. Evaluate these strains at the centroid of the element; then evaluate the stresses at the centroid in terms of E and v. Assume plane stress conditions apply.

8.3 For the element of Figure 8–3 (shown again as Figure P8–3) subjected to the uniform pressure shown acting over the vertical side, determine the nodal force replacement system using Eq. (6.3.7). Assume an element thickness of t.

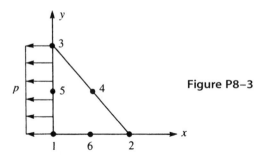

Figure P8–3

8.4 For the element of Figure 8–3 (shown as Figure P8–4) subjected to the linearly varying line load shown acting over the vertical side, determine the nodal force replacement system using Eq. (6.3.7). Compare this result to that of Problem 6.11. Are these results expected? Explain.

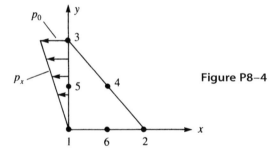

Figure P8–4

8.5 For the linear-strain elements shown in Figure P8–5, determine the strains $\varepsilon_x, \varepsilon_y$, and γ_{xy}. Evaluate the stresses σ_x, σ_y, and τ_{xy} at the centroids. The coordinates of the nodes are shown in units of centimeters. Let $E = 210$ GPa, $v = 0.25$, and $t = 6$ mm for both elements. Assume plane stress conditions apply. The nodal displacements are given as

$$u_1 = 0.0 \qquad\qquad v_1 = 0.0$$

$$u_2 = 2.5 \times 10^{-3} \text{ cm} \qquad v_2 = 5 \times 10^{-3} \text{ cm}$$

$$u_3 = 1.25 \times 10^{-3} \text{ cm} \qquad v_3 = 5 \times 10^{-4} \text{ cm}$$

$$u_4 = 5 \times 10^{-4} \text{ cm} \qquad v_4 = 2.5 \times 10^{-4} \text{ cm}$$

$$u_5 = 0.0 \qquad\qquad v_5 = 2.5 \times 10^{-4} \text{ cm}$$

$$u_6 = 1.25 \times 10^{-3} \text{ cm} \qquad v_6 = 2.5 \times 10^{-3} \text{ cm}$$

(*Hint*: Use the results of Section 8.2.)

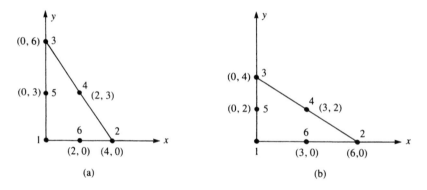

(a)   (b)

Figure P8–5

8.6 For the linear-strain element shown in Figure P8–6, determine the strains $\varepsilon_x, \varepsilon_y$, and γ_{xy}. Evaluate these strains at the centroid of the element; then evaluate the stresses σ_x, σ_y, and τ_{xy} at the centroid. The coordinates of the nodes are shown in units of millimeters. Let $E = 210$ GPa, $v = 0.25$, and $t = 10$ mm. Assume plane stress conditions apply. Use the nodal displacements given in Problem 8.5 (converted to

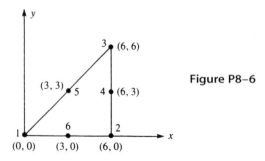

Figure P8–6

millimeters). Note that the β's and γ's from the example in Section 8.2 cannot be used here as the element in Figure P8–6 is oriented differently than the one in Figure 8–3.

8.7 Evaluate the shape functions for the linear-strain triangle shown in Figure P8–7. Then evaluate the $[B]$ matrix. Units are millimeters.

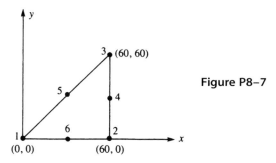

Figure P8–7

8.8 Use the LST element to solve Example 6.2. Compare the results.

8.9 Write a computer program to solve plane stress problems using the LST element.

CHAPTER OBJECTIVES

- To review the basic concepts and theory of elasticity equations for axisymmetric behavior.
- To drive the axisymmetric element stiffness matrix, body force, and surface traction equations.
- To demonstrate the solution of an axisymmetric pressure vessel using the stiffness method.
- To compare the finite element solution to an exact solution for a cylindrical pressure vessel.
- To illustrate some practical applications of axisymmetric elements.

Introduction

In previous chapters, we have been concerned with line or one-dimensional elements (Chapters 2 through 5) and two-dimensional elements (Chapters 6 through 8). In this chapter, we consider a special two-dimensional element called the *axisymmetric element*. This element is quite useful when symmetry with respect to geometry and loading exists about an axis of the body being analyzed. Problems that involve soil masses subjected to circular footing loads or thick-walled pressure vessels can often be analyzed using the element developed in this chapter.

We begin with the development of the stiffness matrix for the simplest axisymmetric element, the triangular torus, whose vertical cross section is a plane triangle.

We then present the longhand solution of a thick-walled pressure vessel to illustrate the use of the axisymmetric element equations. This is followed by a description of some typical large-scale problems that have been modeled using the axisymmetric element.

▲ 9.1 Derivation of the Stiffness Matrix ▲

In this section, we will derive the stiffness matrix and the body and surface force matrices for the axisymmetric element. However, before the development, we will first present some fundamental concepts prerequisite to the understanding of the derivation. Axisymmetric elements are triangular tori such that each element is symmetric

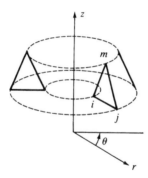

Figure 9–1 Typical axisymmetric element *ijm*

with respect to geometry and loading about an axis such as the z axis in Figure 9–1. Hence, the z axis is called the *axis of symmetry* or the *axis of revolution*. Each vertical cross section of the element is a plane triangle. The nodal points of an axisymmetric triangular element describe circumferential lines, as indicated in Figure 9–1.

In plane stress problems, stresses exist only in the *x-y* plane. In axisymmetric problems, the radial displacements develop circumferential strains that induce stresses σ_r, σ_θ, σ_z, and τ_{rz}, where r, θ, and z indicate the radial, circumferential, and longitudinal directions, respectively. Triangular torus elements are often used to idealize the axisymmetric system because they can be used to simulate complex surfaces and are simple to work with. For instance, the axisymmetric problem of a semi-infinite half-space loaded by a circular area (circular footing) shown in Figure 9–2(a), the domed pressure vessel shown in Figure 9–2(b), and the engine valve stem shown in Figure 9–2(c) can be solved using the axisymmetric element developed in this chapter.

Because of symmetry about the z axis, the stresses are independent of the θ coordinate. Therefore, all derivatives with respect to θ vanish, and the displacement component v (tangent to the θ direction), the shear strains $\gamma_{r\theta}$ and $\gamma_{\theta z}$, and the shear stresses $\tau_{r\theta}$ and $\tau_{\theta z}$ are all zero.

Figure 9–3 shows an axisymmetric ring element and its cross section to represent the general state of strain for an axisymmetric problem. It is most convenient to express the displacements of an element $ABCD$ in the plane of a cross section in cylindrical coordinates. We then let u and w denote the displacements in the radial and longitudinal directions, respectively. The side AB of the element is displaced an amount u, and side CD is then displaced an amount $u + (\partial u/\partial r)\, dr$ in the radial direction. The normal strain in the radial direction is then given by

$$\varepsilon_r = \frac{\partial u}{\partial r} \tag{9.1.1a}$$

In general, the strain in the tangential direction depends on the tangential displacement v and on the radial displacement u. However, for axisymmetric deformation behavior, recall that the tangential displacement v is equal to zero. Hence, the tangential strain is due only to the radial displacement. Having only radial displacement u, the new length of the arc $\widehat{AB}$ is $(r + u)\, d\theta$, and the tangential strain is then given by

$$\varepsilon_\theta = \frac{(r+u)\, d\theta - r\, d\theta}{r\, d\theta} = \frac{u}{r} \tag{9.1.1b}$$

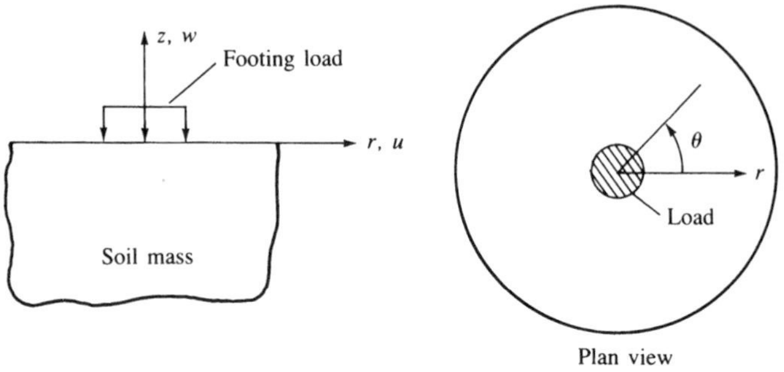

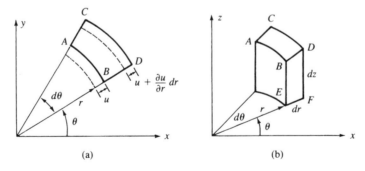

(a) soil mass

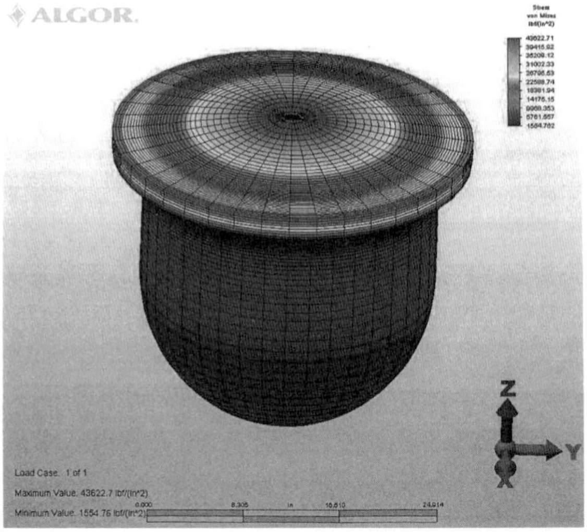

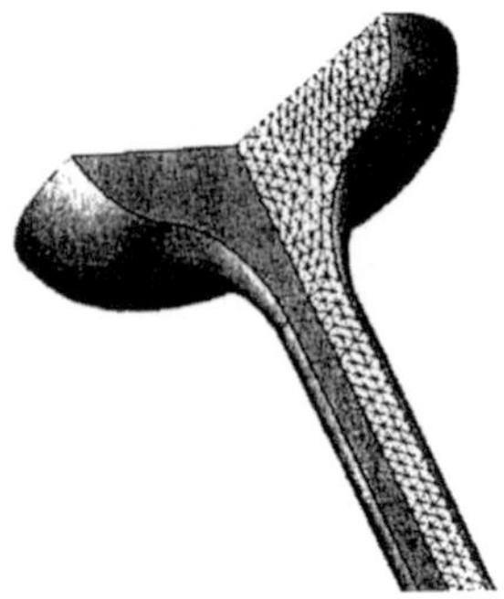

(b) enclosed pressure vessel (c) engine valve stem

Figure 9–2 Examples of axisymmetric problems: (a) semi-infinite half-space (soil mass) modeled by axisymmetric elements, (b) enclosed pressure vessel (Courtesy of Algor, Inc.) (See the full-color insert for a color version of this figure.), and (c) an engine valve stem

Figure 9–3 (a) Plane cross section of (b) axisymmetric element

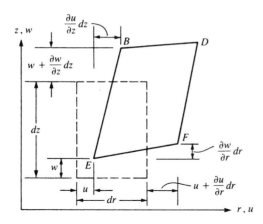

Figure 9–4 Displacement and rotations of lines of element in the *r-z* plane

Next, we consider the longitudinal element *BDEF* to obtain the longitudinal strain and the shear strain. In Figure 9–4, the element is shown to displace by amounts *u* and *w* in the radial and longitudinal directions at point *E*, and to displace additional amounts $(\partial w/\partial z)\,dz$ along line *BE* and $(\partial u/\partial r)\,dr$ along line *EF*. Furthermore, observing lines *EF* and *BE*, we see that point *F* moves upward an amount $(\partial w/\partial r)\,dr$ with respect to point *E* and point *B* moves to the right an amount $(\partial u/\partial z)\,dz$ with respect to point *E*. Again, from the basic definitions of normal and shear strain, we have the longitudinal normal strain given by

$$\varepsilon_z = \frac{\partial w}{\partial z} \tag{9.1.1c}$$

and the shear strain in the *r-z* plane given by

$$\gamma_{rz} = \frac{\partial u}{\partial z} + \frac{\partial w}{\partial r} \tag{9.1.1d}$$

Summarizing the strain–displacement relationships of Eqs. (9.1.1a–d) in one equation for easier reference, we have

$$\varepsilon_r = \frac{\partial u}{\partial r} \qquad \varepsilon_\theta = \frac{u}{r} \qquad \varepsilon_z = \frac{\partial w}{\partial z} \qquad \gamma_{rz} = \frac{\partial u}{\partial z} + \frac{\partial w}{\partial r} \tag{9.1.1e}$$

The isotropic stress–strain relationship, obtained by simplifying the general stress–strain relationships given in Appendix C, is

$$\begin{Bmatrix} \sigma_r \\ \sigma_z \\ \sigma_\theta \\ \tau_{rz} \end{Bmatrix} = \frac{E}{(1+v)(1-2v)} \begin{bmatrix} 1-v & v & v & 0 \\ v & 1-v & v & 0 \\ v & v & 1-v & 0 \\ 0 & 0 & 0 & \dfrac{1-2v}{2} \end{bmatrix} \begin{Bmatrix} \varepsilon_r \\ \varepsilon_z \\ \varepsilon_\theta \\ \gamma_{rz} \end{Bmatrix} \tag{9.1.2}$$

The theoretical development follows that of the plane stress–strain problem given in Chapter 6.

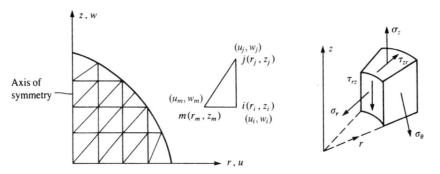

(a) Typical slice through an axisymmetric solid discretized into triangular elements

(b) Stresses in the axisymmetric problem

Figure 9–5 Discretized axisymmetric solid

Step 1   Select Element Type

An axisymmetric solid is shown discretized in Figure 9–5(a), along with a typical triangular element. The element has three nodes with two degrees of freedom per node (that is, u_i, w_i at node i). The stresses in the axisymmetric problem are shown in Figure 9–5(b).

Step 2   Select Displacement Functions

The element displacement functions are taken to be

$$u(r, z) = a_1 + a_2 r + a_3 z$$
$$w(r, z) = a_4 + a_5 r + a_6 z \tag{9.1.3}$$

so that we have the same linear displacement functions as used in the plane stress, constant-strain triangle. Again, the total number of a_i's (six) introduced in the displacement functions is the same as the total number of degrees of freedom for the element. The nodal displacements are

$$\{d\} = \begin{Bmatrix} \{d_i\} \\ \{d_j\} \\ \{d_m\} \end{Bmatrix} = \begin{Bmatrix} u_i \\ w_i \\ u_j \\ w_j \\ u_m \\ w_m \end{Bmatrix} \tag{9.1.4}$$

and u evaluated at node i is

$$u(r_i, z_i) = u_i = a_1 + a_2 r_i + a_3 z_i \tag{9.1.5}$$

Using Eq. (9.1.3), the general displacement function is then expressed in matrix form as

$$\{\psi\} = \left\{\begin{matrix} u \\ w \end{matrix}\right\} = \left\{\begin{matrix} a_1 + a_2 r + a_3 z \\ a_4 + a_5 r + a_6 z \end{matrix}\right\} = \begin{bmatrix} 1 & r & z & 0 & 0 & 0 \\ 0 & 0 & 0 & 1 & r & z \end{bmatrix} \left\{\begin{matrix} a_1 \\ a_2 \\ a_3 \\ a_4 \\ a_5 \\ a_6 \end{matrix}\right\} \tag{9.1.6}$$

Substituting the coordinates of the nodal points shown in Figure 9–5(a) into Eq. (9.1.6), we can solve for the a_i's in a manner similar to that in Section 6.2. The resulting expressions are

$$\left\{\begin{matrix} a_1 \\ a_2 \\ a_3 \end{matrix}\right\} = \begin{bmatrix} 1 & r_i & z_i \\ 1 & r_j & z_j \\ 1 & r_m & z_m \end{bmatrix}^{-1} \left\{\begin{matrix} u_i \\ u_j \\ u_m \end{matrix}\right\} \tag{9.1.7}$$

and

$$\left\{\begin{matrix} a_4 \\ a_5 \\ a_6 \end{matrix}\right\} = \begin{bmatrix} 1 & r_i & z_i \\ 1 & r_j & z_j \\ 1 & r_m & z_m \end{bmatrix}^{-1} \left\{\begin{matrix} w_i \\ w_j \\ w_m \end{matrix}\right\} \tag{9.1.8}$$

Performing the inversion operations in Eqs. (9.1.7) and (9.1.8), we have

$$\left\{\begin{matrix} a_1 \\ a_2 \\ a_3 \end{matrix}\right\} = \frac{1}{2A} \begin{bmatrix} \alpha_i & \alpha_j & \alpha_m \\ \beta_i & \beta_j & \beta_m \\ \gamma_i & \gamma_j & \gamma_m \end{bmatrix} \left\{\begin{matrix} u_i \\ u_j \\ u_m \end{matrix}\right\} \tag{9.1.9}$$

and

$$\left\{\begin{matrix} a_4 \\ a_5 \\ a_6 \end{matrix}\right\} = \frac{1}{2A} \begin{bmatrix} \alpha_i & \alpha_j & \alpha_m \\ \beta_i & \beta_j & \beta_m \\ \gamma_i & \gamma_j & \gamma_m \end{bmatrix} \left\{\begin{matrix} w_i \\ w_j \\ w_m \end{matrix}\right\} \tag{9.1.10}$$

where

$$\alpha_i = r_j z_m - z_j r_m \qquad \alpha_j = r_m z_i - z_m r_i \qquad \alpha_m = r_i z_j - z_i r_j$$
$$\beta_i = z_j - z_m \qquad \beta_j = z_m - z_i \qquad \beta_m = z_i - z_j \tag{9.1.11}$$
$$\gamma_i = r_m - r_j \qquad \gamma_j = r_i - r_m \qquad \gamma_m = r_j - r_i$$

We define the shape functions, similar to Eqs. (6.2.18), as

$$N_i = \frac{1}{2A} (\alpha_i + \beta_i r + \gamma_i z)$$

$$N_j = \frac{1}{2A} (\alpha_j + \beta_j r + \gamma_j z) \tag{9.1.12}$$

$$N_m = \frac{1}{2A} (\alpha_m + \beta_m r + \gamma_m z)$$

Substituting Eqs. (9.1.7) and (9.1.8) into Eq. (9.1.6), along with the shape function Eqs. (9.1.12), we find that the general displacement function is

$$\{\psi\} = \begin{Bmatrix} u(r, z) \\ w(r, z) \end{Bmatrix} = \begin{bmatrix} N_i & 0 & N_j & 0 & N_m & 0 \\ 0 & N_i & 0 & N_j & 0 & N_m \end{bmatrix} \begin{Bmatrix} u_i \\ w_i \\ u_j \\ w_j \\ u_m \\ w_m \end{Bmatrix} \tag{9.1.13}$$

or

$$\{\psi\} = [N]\{d\} \tag{9.1.14}$$

Step 3 Define the Strain/Displacement and Stress/Strain Relationships

When we use Eqs. (9.1.3) in (9.1.1e), the strains become

$$\{\varepsilon\} = \begin{Bmatrix} a_2 \\ a_6 \\ \dfrac{a_1}{r} + a_2 + \dfrac{a_3 z}{r} \\ a_3 + a_5 \end{Bmatrix} \tag{9.1.15}$$

Rewriting Eq. (9.1.15) with the a_i's as a separate column matrix, we have

$$\begin{Bmatrix} \varepsilon_r \\ \varepsilon_z \\ \varepsilon_\theta \\ \gamma_{rz} \end{Bmatrix} = \begin{bmatrix} 0 & 1 & 0 & 0 & 0 & 0 \\ 0 & 0 & 0 & 0 & 0 & 1 \\ \dfrac{1}{r} & 1 & \dfrac{z}{r} & 0 & 0 & 0 \\ 0 & 0 & 1 & 0 & 1 & 0 \end{bmatrix} \begin{Bmatrix} a_1 \\ a_2 \\ a_3 \\ a_4 \\ a_5 \\ a_6 \end{Bmatrix} \tag{9.1.16}$$

Substituting Eqs. (9.1.9) and (9.1.10) into Eq. (9.1.16) and simplifying, we obtain

$$\{\varepsilon\} = \frac{1}{2A} \begin{bmatrix} \beta_i & 0 & \beta_j & 0 & \beta_m & 0 \\ 0 & \gamma_i & 0 & \gamma_j & 0 & \gamma_m \\ \dfrac{\alpha_i}{r} + \beta_i + \dfrac{\gamma_i z}{r} & 0 & \dfrac{\alpha_j}{r} + \beta_j + \dfrac{\gamma_j z}{r} & 0 & \dfrac{\alpha_m}{r} + \beta_m + \dfrac{\gamma_m z}{r} & 0 \\ \gamma_i & \beta_i & \gamma_j & \beta_j & \gamma_m & \beta_m \end{bmatrix} \begin{Bmatrix} u_i \\ w_i \\ u_j \\ w_j \\ u_m \\ w_m \end{Bmatrix}$$

$$\tag{9.1.17}$$

or, rewriting Eq. (9.1.17) in simplified matrix form,

$$\{\varepsilon\} = [[B_i] \ [B_j] \ [B_m]] \begin{Bmatrix} u_i \\ w_i \\ u_j \\ w_j \\ u_m \\ w_m \end{Bmatrix} \qquad (9.1.18)$$

where

$$[B_i] = \frac{1}{2A} \begin{bmatrix} \beta_i & 0 \\ 0 & \gamma_i \\ \dfrac{\alpha_i}{r} + \beta_i + \dfrac{\gamma_i z}{r} & 0 \\ \gamma_i & \beta_i \end{bmatrix} \qquad (9.1.19)$$

Similarly, we obtain submatrices $[B_j]$ and $[B_m]$ by replacing the subscript i with j and then with m in Eq. (9.1.19). Rewriting Eq. (9.1.18) in compact matrix form, we have

$$\{\varepsilon\} = [B]\{d\} \qquad (9.1.20)$$

where

$$[B] = [[B_i] \ [B_j] \ [B_m]] \qquad (9.1.21)$$

is called the gradient matrix.

Note that $[B]$ is a function of the r and z coordinates. Therefore, in general, the strain ε_θ will not be constant.

The stresses are given by

$$\{\sigma\} = [D][B]\{d\} \qquad (9.1.22)$$

where $[D]$ is given by the first matrix on the right side of Eq. (9.1.2). (As mentioned in Chapter 6, for $v = 0.5$, a special formula must be used; see Reference [9].)

Step 4 Derive the Element Stiffness Matrix and Equations

The stiffness matrix is

$$[k] = \iiint_V [B]^T [D][B] \, dV \qquad (9.1.23)$$

or

$$[k] = 2\pi \iint_A [B]^T [D][B] r \, dr \, dz \qquad (9.1.24)$$

after integrating along the circumferential boundary. The $[B]$ matrix, Eq. (9.1.21), is a function of r and z. Therefore, $[k]$ is a function of r and z and is of order 6×6.

We can evaluate Eq. (9.1.24) for $[k]$ by one of three methods:

1. Numerical integration (Gaussian quadrature) as discussed in Chapter 10.
2. Explicit multiplication and term-by-term integration [1].
3. Evaluate $[B]$ for a centroidal point $(\bar{r}, \bar{z})$ of the element

$$r = \bar{r} = \frac{r_i + r_j + r_m}{3} \qquad z = \bar{z} = \frac{z_i + z_j + z_m}{3} \tag{9.1.25}$$

and define $[B(\bar{r}, \bar{z})] = [\bar{B}]$. Therefore, as a first approximation,

$$[k] = 2\pi\bar{r}A[\bar{B}]^T[D][\bar{B}] \tag{9.1.26}$$

If the triangular subdivisions are consistent with the final stress distribution (that is, small elements in regions of high stress gradients), then acceptable results can be obtained by method 3.

Distributed Body Forces

Loads such as gravity (in the direction of the z axis) or centrifugal forces in rotating machine parts (in the direction of the r axis) are considered to be body forces (as shown in Figure 9–6). The body forces can be found by

$$\{f_b\} = 2\pi \iint_A [N]^T \begin{Bmatrix} R_b \\ Z_b \end{Bmatrix} r \, dr \, dz \tag{9.1.27}$$

where $R_b = \omega^2 \rho r$ for a machine part moving with a constant angular velocity ω about the z axis, with material mass density ρ and radial coordinate r, and where Z_b is the body force per unit volume due to the force of gravity.

Considering the body force at node i, we have

$$\{f_{bi}\} = 2\pi \iint_A [N_i]^T \begin{Bmatrix} R_b \\ Z_b \end{Bmatrix} r \, dr \, dz \tag{9.1.28}$$

where

$$[N_i]^T = \begin{bmatrix} N_i & 0 \\ 0 & N_i \end{bmatrix} \tag{9.1.29}$$

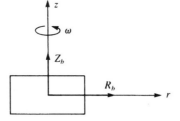

Figure 9–6 Axisymmetric element with body forces per unit volume

Multiplying and integrating in Eq. (9.1.28), we obtain

$$\{f_{bi}\} = \frac{2\pi}{3} \left\{ \begin{array}{c} \bar{R}_b \\ \bar{Z}_b \end{array} \right\} A\bar{r} \tag{9.1.30}$$

where the origin of the coordinates has been taken as the centroid of the element, and $\bar{R}_b$ is the radially directed body force per unit volume evaluated at the centroid of the element. The body forces at nodes j and m are identical to those given by Eq. (9.1.30) for node i. Hence, for an element, we have

$$\{f_b\} = \frac{2\pi \bar{r} A}{3} \left\{ \begin{array}{c} \bar{R}_b \\ \bar{Z}_b \\ \bar{R}_b \\ \bar{Z}_b \\ \bar{R}_b \\ \bar{Z}_b \end{array} \right\} \tag{9.1.31}$$

where
$$\bar{R}_b = \omega^2 \rho \bar{r} \tag{9.1.32}$$

Equation (9.1.31) is a first approximation to the radially directed body force distribution.

Surface Forces

Surface forces can be found by

$$\{f_s\} = \iint_S [N_s]^T \{T\} \, dS \tag{9.1.33}$$

where again $[N_s]$ denotes the shape function matrix evaluated along the surface where the surface traction acts.

For radial and axial pressures p_r and p_z, respectively, we have

$$\{f_s\} = \iint_S [N_s]^T \left\{ \begin{array}{c} p_r \\ p_z \end{array} \right\} \, dS \tag{9.1.34}$$

For example, along the vertical face jm of an element, let uniform loads p_r and p_z be applied, as shown in Figure 9–7 along surface $r = r_j$. We can use Eq. (9.1.34)

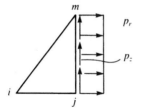

Figure 9–7 Axisymmetric element with surface forces

written for each node separately. For instance, for node j, substituting N_j from Eqs. (9.1.12) into Eq. (9.1.34), we have

$$\{f_{sj}\} = \int_{z_j}^{z_m} \frac{1}{2A} \begin{bmatrix} \alpha_j + \beta_j r + \gamma_j z & 0 \\ 0 & \alpha_j + \beta_j r + \gamma_j z \end{bmatrix} \begin{Bmatrix} p_r \\ p_z \end{Bmatrix} 2\pi r_j \, dz \qquad (9.1.35)$$

evaluated at $r = r_j$, $z = z$

Performing the integration of Eq. (9.1.35) explicitly, along with similar evaluations for f_{si} and f_{sm}, we obtain the total distribution of surface force to nodes i, j, and m as

$$\{f_s\} = \frac{2\pi r_j (z_m - z_j)}{2} \begin{Bmatrix} 0 \\ 0 \\ p_r \\ p_z \\ p_r \\ p_z \end{Bmatrix} \qquad (9.1.36)$$

Steps 5 through 7

Steps 5 through 7, which involve assembling the total stiffness matrix, total force matrix, and total set of equations; solving for the nodal degrees of freedom; and calculating the element stresses, are analogous to those of Chapter 6 for the CST element, except the stresses are not constant in each element. They are usually determined by one of two methods that we use to determine the LST element stresses. Either we determine the centroidal element stresses, or we determine the nodal stresses for the element and then average them. The latter method has been shown to be more accurate in some cases [2].

Example 9.1

For the element of an axisymmetric body rotating with a constant angular velocity $\omega = 100$ rev/min as shown in Figure 9–8, evaluate the approximate body force matrix. Include the weight of the material, where the weight density ρ_w is 7800 kg/m^3. The coordinates of the element (in cm) are shown in the figure.

We need to evaluate Eq. (9.1.31) to obtain the approximate body force matrix. Therefore, the body masses per unit volume evaluated at the centroid of the element are

$$Z_b = 7800 \text{ kg/m}^3$$

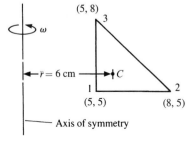

Figure 9–8 Axisymmetric element subjected to angular velocity

and by Eq. (9.1.32), we have

$$\bar{R}_b = \omega^2 \rho \bar{r} = \left[\left(100 \frac{\text{rev}}{\text{min}} \right) \left(2\pi \frac{\text{rad}}{\text{rev}} \right) \left(\frac{1 \text{ min}}{60 \text{ s}} \right) \right]^2 (7800 \text{ kg/m}^3)(0.06 \text{ m})$$

$$\bar{R}_b = 51322 \text{ N/m}^3$$

$$\frac{2\pi \bar{r} A}{3} = \frac{2\pi (0.06)(4.5 \times 10^{-4})}{3} = 5.65 \times 10^{-5} \text{ m}^3$$

$$f_{b1r} = (5.65 \times 10^{-5})(51322) = 2.9 \text{ N}$$

$$f_{b1z} = (-5.65 \times 10^{-5})(7800)(9.81) = -4.32 \text{ N} \qquad \text{(downward)}$$

Because we are using the first approximation Eq. (9.1.31), all r-directed nodal body forces are equal, and all z-directed body forces are equal. Therefore,

$$f_{b2r} = 2.9 \text{ N} \qquad f_{b2z} = -4.32 \text{ N}$$
$$f_{b3r} = 2.9 \text{ N} \qquad f_{b3z} = -4.32 \text{ N} \qquad \blacksquare$$

▲ 9.2 Solution of an Axisymmetric Pressure Vessel

To illustrate the use of the equations developed in Section 9.1, we will now solve an axisymmetric stress problem.

Example 9.2

For the long, thick-walled cylinder under internal pressure p equal to 1 psi shown in Figure 9–9, determine the displacements and stresses.

Discretization

To illustrate the finite element solution for the cylinder, we first discretize the cylinder into four triangular elements, as shown in Figure 9–10. A horizontal slice of the cylinder represents the total cylinder behavior. Because we are performing a longhand

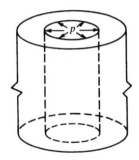

Figure 9–9 Thick-walled cylinder subjected to internal pressure

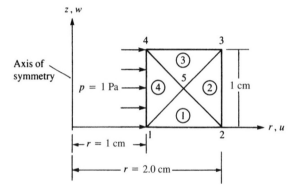

Figure 9–10 Discretized cylinder slice

solution, a coarse mesh of elements is used for simplicity's sake (but without loss of generality of the method). The governing global matrix equation is

$$\begin{Bmatrix} F_{1r} \\ F_{1z} \\ F_{2r} \\ F_{2z} \\ F_{3r} \\ F_{3z} \\ F_{4r} \\ F_{4z} \\ F_{5r} \\ F_{5z} \end{Bmatrix} = [K] \begin{Bmatrix} u_1 \\ w_1 \\ u_2 \\ w_2 \\ u_3 \\ w_3 \\ u_4 \\ w_4 \\ u_5 \\ w_5 \end{Bmatrix} \tag{9.2.1}$$

where the $[K]$ matrix is of order 10×10.

Assemblage of the Stiffness Matrix

We assemble the $[K]$ matrix in the usual manner by superposition of the individual element stiffness matrices. For simplicity's sake, we will use the first approximation method given by Eq. (9.1.26) to evaluate the element matrices. Therefore,

$$[k] = 2\pi\bar{r}A[\bar{B}]^T[D][\bar{B}] \tag{9.2.2}$$

For element 1 (Figure 9–11), the coordinates are $r_i = 1.0$, $z_i = 0$, $r_j = 2.0$, $z_j = 0$, $r_m = 1.5$, and $z_m = 0.5$ ($i = 1, j = 2$, and $m = 5$ for element 1) for the global-coordinate axes as set up in Figure 9–10.

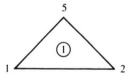

Figure 9–11 Element 1 of the discretized cylinder

We now evaluate $[\bar{B}]$, where $[\bar{B}]$ is given by Eq. (9.1.19) evaluated at the centroid of the element $r = \bar{r}$, $z = \bar{z}$, and expanded here as

$$[\bar{B}] = \frac{1}{2A} \begin{bmatrix} \beta_i & 0 & \beta_j & 0 & \beta_m & 0 \\ 0 & \gamma_i & 0 & \gamma_j & 0 & \gamma_m \\ \frac{\alpha_i}{\bar{r}} + \beta_i + \frac{\gamma_i \bar{z}}{\bar{r}} & 0 & \frac{\alpha_j}{\bar{r}} + \beta_j + \frac{\gamma_j \bar{z}}{\bar{r}} & 0 & \frac{\alpha_m}{\bar{r}} + \beta_m + \frac{\gamma_m \bar{z}}{\bar{r}} & 0 \\ \gamma_i & \beta_i & \gamma_j & \beta_j & \gamma_m & \beta_m \end{bmatrix} \qquad (9.2.3)$$

where, using element coordinates in Eqs. (9.1.11), we have

$$\alpha_i = r_j z_m - z_j r_m = (2.0)(0.5) - (0.0)(1.5) = 1.0 \text{ cm}^2$$

$$\alpha_j = r_m z_i - z_m r_i = (1.5)(0) - (0.5)(1.0) = -0.5 \text{ cm}^2$$

$$\alpha_m = r_i z_j - z_i r_j = (1.0)(0.0) - (0)(2.0) = 0.0 \text{ cm}^2$$

$$\beta_i = z_j - z_m = 0.0 - 0.5 = -0.5 \text{ cm}$$

$$\beta_j = z_m - z_i = 0.5 - 0 = 0.5 \text{ cm} \qquad (9.2.4)$$

$$\beta_m = z_i - z_j = 0.0 - 0.0 = 0.0 \text{ cm}$$

$$\gamma_i = r_m - r_j = 1.5 - 2.0 = -0.5 \text{ cm}$$

$$\gamma_j = r_i - r_m = 1.0 - 1.5 = -0.5 \text{ cm}$$

$$\gamma_m = r_j - r_i = 2.0 - 1.0 = 1.0 \text{ cm}$$

and $$\bar{r} = 1.0 + \frac{1}{2}(1.0) = 1.5 \text{ cm} \qquad \bar{z} = \frac{1}{3}(0.5) = 0.167 \text{ cm}$$

$$A = \frac{1}{2}(1.0)(0.5) = 0.25 \text{ cm}^2$$

Substituting the results from Eqs. (9.2.4) into Eq. (9.2.3), we obtain

$$[\bar{B}] = \frac{10^2}{0.5} \begin{bmatrix} -0.5 & 0 & 0.5 & 0 & 0 & 0 \\ 0 & -0.5 & 0 & -0.5 & 0 & 1.0 \\ 0.11 & 0 & 0.11 & 0 & 0.11 & 0 \\ -0.5 & -0.5 & -0.5 & 0.5 & 1.0 & 0 \end{bmatrix} \frac{1}{m} \qquad (9.2.5)$$

For the axisymmetric stress case, the matrix $[D]$ is given in Eq. (9.1.2) as

$$[D] = \frac{E}{(1+v)(1-2v)} \begin{bmatrix} 1-v & v & v & 0 \\ v & 1-v & v & 0 \\ v & v & 1-v & 0 \\ 0 & 0 & 0 & \dfrac{1-2v}{2} \end{bmatrix} \qquad (9.2.6)$$

With $v = 0.3$ and $E = 200$ GPa, we obtain

$$[D] = \frac{200 \times 10^9}{(1 + 0.3)[1 - 2(0.3)]} \begin{bmatrix} 1 - 0.3 & 0.3 & 0.3 & 0 \\ 0.3 & 1 - 0.3 & 0.3 & 0 \\ 0.3 & 0.3 & 1 - 0.3 & 0 \\ 0 & 0 & 0 & \dfrac{1 - 2(0.3)}{2} \end{bmatrix} \quad (9.2.7)$$

or, simplifying Eq. (9.2.7),

$$[D] = 384.6 \times 10^9 \begin{bmatrix} 0.7 & 0.3 & 0.3 & 0 \\ 0.3 & 0.7 & 0.3 & 0 \\ 0.3 & 0.3 & 0.7 & 0 \\ 0 & 0 & 0 & 0.2 \end{bmatrix} \frac{\text{N}}{\text{m}^2} \quad (9.2.8)$$

Using Eqs. (9.2.5) and (9.2.8), we obtain

$$[\bar{B}]^T[D] = \frac{(10^2)(384.6 \times 10^9)}{0.50} \begin{bmatrix} -0.317 & -0.117 & -0.073 & -0.10 \\ -0.15 & -0.35 & -0.15 & -0.10 \\ 0.383 & 0.183 & 0.227 & -0.10 \\ -0.15 & -0.35 & -0.15 & 0.10 \\ 0.033 & 0.033 & 0.077 & 0.20 \\ 0.30 & 0.70 & 0.30 & 0 \end{bmatrix} \quad (9.2.9)$$

Substituting Eqs. (9.2.5) and (9.2.9) into Eq. (9.2.2), we obtain the stiffness matrix for element 1 as

$$[k^{(1)}] = (10^6) \begin{array}{c} \begin{matrix} i = 1 \quad\quad\quad\quad j = 2 \quad\quad\quad\quad m = 5 \end{matrix} \\ \begin{bmatrix} 36.25 & 19.66 & -21.11 & 1.54 & -19.57 & -21.2 \\ 19.66 & 40.78 & -7.52 & 22.65 & -21.12 & -63.43 \\ -21.11 & -7.52 & 48.20 & -25.64 & -13.59 & 33.16 \\ 1.54 & 22.65 & -25.64 & 40.78 & 15.13 & -63.43 \\ -19.57 & -21.12 & -13.59 & 15.13 & 37.69 & 5.98 \\ -21.2 & -63.43 & 33.16 & -63.43 & 5.98 & 126.86 \end{bmatrix} \end{array} \frac{\text{N}}{\text{m}} \quad (9.2.10)$$

where the numbers above the columns indicate the nodal orders of degrees of freedom in the element 1 stiffness matrix.

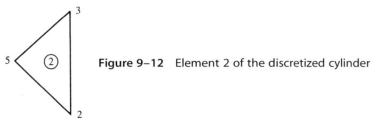

Figure 9–12 Element 2 of the discretized cylinder

For element 2 (Figure 9–12), the coordinates are $r_i = 2.0$, $z_i = 0.0$, $r_j = 2.0$, $z_j = 1.0$, $r_m = 1.5$, and $z_m = 0.5$ ($i = 2$, $j = 3$, and $m = 5$ for element 2). Therefore,

$$\alpha_i = (2.0)(0.5) - (1.0)(1.5) = -0.5 \text{ cm}^2$$

$$\alpha_j = (1.5)(0.0) - (0.5)(2.0) = -1.0 \text{ cm}^2 \qquad (9.2.11)$$

$$\alpha_m = (2.0)(1.0) - (0.0)(2.0) = 2.0 \text{ cm}^2$$

$$\beta_i = 1.0 - 0.5 = 0.5 \text{ cm} \qquad \beta_j = 0.5 - 0.0 = 0.5 \text{ cm}$$

$$\beta_m = 0.0 - 1.0 = -1.0 \text{ cm} \qquad \gamma_i = 1.5 - 2.0 = -0.5 \text{ cm}$$

$$\gamma_j = 2.0 - 1.5 = 0.5 \text{ cm} \qquad \gamma_m = 2.0 - 2.0 = 0.0 \text{ cm}$$

and $\qquad \bar{r} = 1.833 \text{ cm} \qquad \bar{z} = 0.5 \text{ cm} \qquad A = 0.25 \text{ cm}^2$

Using Eqs. (9.2.11) in Eq. (9.2.2) and proceeding as for element 1, we obtain the stiffness matrix for element 2 as

$$
[k^{(2)}] = (10^6)
\begin{array}{ccc}
\quad i = 2 \qquad\qquad\quad j = 3 \qquad\qquad\quad m = 5 \\
\begin{bmatrix}
57.14 & -30.7 & 34.99 & 8.55 & -79.25 & 22.14 \\
-30.7 & 49.75 & -8.55 & -27.68 & 30.2 & -22.14 \\
34.99 & -8.55 & 57.14 & 30.7 & -79.25 & -22.14 \\
8.55 & -27.68 & 30.7 & 49.53 & -30.2 & -22.14 \\
-79.2 & 30.2 & -79.25 & -30.2 & 144.25 & 0 \\
14.54 & -14.54 & -14.54 & -14.54 & 0 & 44.28
\end{bmatrix}
\dfrac{\text{N}}{\text{m}}
\end{array}
$$

$$(9.2.12)$$

We obtain the stiffness matrices for elements 3 and 4 in a manner similar to that used to obtain the stiffness matrices for elements 1 and 2. Thus,

$$
[k^{(3)}] = (10^6)
\begin{array}{ccc}
\quad i = 3 \qquad\qquad\quad j = 4 \qquad\qquad\quad m = 5 \\
\begin{bmatrix}
48.36 & 25.68 & -21.05 & 7.57 & -13.55 & -33.20 \\
25.68 & 40.76 & -1.52 & 22.65 & -15.11 & -63.38 \\
-21.05 & -1.52 & 32.68 & -19.65 & -19.58 & 21.15 \\
7.57 & 22.65 & -19.65 & 40.76 & 21.15 & -63.38 \\
-13.55 & -15.11 & -19.58 & 21.15 & 37.80 & -6.06 \\
-33.20 & -63.38 & 21.15 & -63.38 & -6.06 & 126.67
\end{bmatrix}
\dfrac{\text{N}}{\text{m}}
\end{array}
$$

$$(9.2.13)$$

and

$$i = 4 \qquad\qquad j = 1 \qquad\qquad m = 5$$

$$[k^{(4)}] = (10^6) \begin{bmatrix} 27.68 & -14.61 & 13.62 & 0.51 & -44.27 & 14.10 \\ -14.61 & 31.69 & -0.51 & -17.62 & 24.17 & -14.10 \\ 13.62 & -0.51 & 27.68 & 14.61 & -44.27 & -14.10 \\ 0.51 & -17.62 & 14.61 & 31.69 & -24.16 & -14.10 \\ -44.27 & 24.17 & -44.27 & -24.16 & 112.60 & 0 \\ 14.10 & -14.10 & -14.10 & -14.10 & 0 & 28.20 \end{bmatrix} \frac{\text{N}}{\text{m}}$$

(9.2.14)

Using superposition of the element stiffness matrices [Eqs. (9.2.10) and (9.2.12) through (9.2.14)], where we rearrange the elements of each stiffness matrix in order of increasing nodal degrees of freedom, we obtain the global stiffness matrix as

$$[K] = (10^6) \begin{bmatrix} 63.93 & 34.27 & -21.11 & 1.54 & 0 & 0 & 13.62 & -0.51 & -63.84 & -35.3 \\ 34.27 & 71.47 & -7.52 & 22.65 & 0 & 0 & 0.51 & -17.62 & -45.28 & -77.53 \\ -21.11 & -7.52 & 105.34 & -56.34 & 34.99 & 8.55 & 0 & 0 & -92.84 & 55.30 \\ 1.54 & 22.65 & -56.34 & 90.53 & -8.55 & -27.68 & 0 & 0 & 45.33 & -85.57 \\ 0 & 0 & 34.99 & -8.55 & 105.5 & 56.38 & -21.05 & 7.54 & -92.80 & -55.34 \\ 0 & 0 & 8.55 & -27.68 & 56.38 & 90.59 & -1.52 & 22.65 & -45.31 & -85.52 \\ 13.62 & 0.51 & 0 & 0 & -21.05 & -1.52 & 63.96 & -34.26 & -63.85 & 35.25 \\ -0.51 & -17.62 & 0 & 0 & 7.57 & 22.65 & -34.26 & 72.45 & 45.32 & -77.48 \\ -63.84 & -45.28 & -92.84 & 45.33 & -92.80 & -45.31 & -63.85 & 45.32 & 332.34 & 0 \\ -35.3 & -77.53 & 55.30 & -85.57 & -55.34 & -85.52 & 35.25 & -77.48 & 0 & 326.01 \end{bmatrix} \frac{\text{N}}{\text{m}}$$

(9.2.15)

The applied nodal forces are given by Eq. (9.1.36) as

$$F_{1r} = F_{4r} = \frac{2\pi(0.01)(0.01)}{2}(1) = 0.0003141 \text{ N}$$

(9.2.16)

All other nodal forces are zero. Using Eq. (9.2.15) for $[K]$ and Eq. (9.2.16) for the nodal forces in Eq. (9.2.1), and solving for the nodal displacements, we obtain

$$u_1 = 0.0967 \times 10^{-9} \text{ mm} \qquad w_1 = -0.0071 \times 10^{-9} \text{ mm}$$

$$u_2 = 0.0658 \times 10^{-9} \text{ mm} \qquad w_2 = 0.0104 \times 10^{-9} \text{ mm}$$

$$u_3 = 0.0658 \times 10^{-9} \text{ mm} \qquad w_3 = -0.0104 \times 10^{-9} \text{ mm} \qquad (9.2.17)$$

$$u_4 = 0.0967 \times 10^{-9} \text{ mm} \qquad w_4 = -0.0075 \times 10^{-9} \text{ mm}$$

$$u_5 = 0.0731 \times 10^{-9} \text{ mm} \qquad w_5 = 0.00417 \times 10^{-9} \text{ mm}$$

The results for nodal displacements are as expected because radial displacements at the inner edge are equal ($u_1 = u_4$) and those at the outer edge are equal ($u_2 = u_3$). In addition, the axial displacements at the outer nodes and inner nodes are equal but opposite in sign ($w_1 = -w_4$ and $w_2 = -w_3$) as a result of the Poisson effect and symmetry. Finally, the axial displacement at the center node is zero ($w_5 = 0$), as it should be because of symmetry.

By using Eq. (9.1.22), we now determine the stresses in each element as

$$\{\sigma\} = [D][\bar{B}]\{d\} \tag{9.2.18}$$

For element 1, we use Eq. (9.2.5) for $[\bar{B}]$, Eq. (9.2.8) for $[D]$, and Eq. (9.2.17) for $\{d\}$ in Eq. (9.2.18) to obtain

$$\sigma_r = -0.1695 \text{ N/m}^2 \qquad \sigma_z = -0.00638 \text{ N/m}^2$$

$$\sigma_\theta = 0.4711 \text{ N/m}^2 \qquad \tau_{rz} = -0.05199 \text{ N/m}^2$$

Similarly, for element 2, we obtain

$$\sigma_r = -0.0525 \text{ N/m}^2 \qquad \sigma_z = -0.0373 \text{ N/m}^2$$

$$\sigma_\theta = 0.345 \text{ N/m}^2 \qquad \tau_{rz} = 0.00 \text{ N/m}^2$$

For element 3, the stresses are

$$\sigma_r = -0.1685 \text{ N/m}^2 \qquad \sigma_z = -0.00625 \text{ N/m}^2$$

$$\sigma_\theta = 0.471 \text{ N/m}^2 \qquad \tau_{rz} = 0.05185 \text{ N/m}^2$$

For element 4, the stresses are

$$\sigma_r = -0.235 \text{ N/m}^2 \qquad \sigma_z = 0.07465 \text{ N/m}^2$$

$$\sigma_\theta = 0.713 \text{ N/m}^2 \qquad \tau_{rz} = 0.00 \text{ N/m}^2$$

Figure 9–13 shows the exact solution [10] along with the results determined here and the results from Reference [5]. Observe that agreement with the exact solution is quite good except for the limited results due to the very coarse mesh used in the longhand example, and in case 1 of Reference [5]. In Reference [5], stresses have been plotted at the center of the quadrilaterals and were obtained by averaging the stresses in the four connecting triangles. ∎

▲ 9.3 Applications of Axisymmetric Elements ▲

Numerous structural (and nonstructural) systems can be classified as axisymmetric. Some typical structural systems whose behavior is modeled accurately using the axisymmetric element developed in this chapter are represented in Figures 9–14, 9–15, and 9–17.

Figure 9–14 illustrates the finite element model of a steel-reinforced concrete pressure vessel. The vessel is a thick-walled cylinder with flat heads. An axis of symmetry (the z axis) exists such that only one-half of the r-z plane passing through the middle of the structure need be modeled. The concrete was modeled by using the axisymmetric triangular element developed in this chapter. The steel elements were laid out along the boundaries of the concrete elements so as to maintain continuity (or perfect bond assumption) between the concrete and the steel. The vessel was then

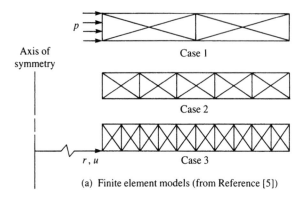

(a) Finite element models (from Reference [5])

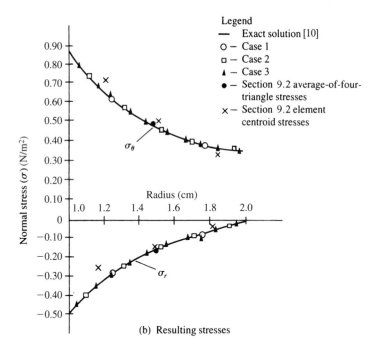

(b) Resulting stresses

Figure 9–13 Finite element analysis of a thick-walled cylinder under internal pressure

subjected to an internal pressure as shown in the figure. Note that the nodes along the axis of symmetry should be supported by rollers preventing motion perpendicular to the axis of symmetry.

Figure 9–15 shows a finite element model of a high-strength steel die used in a thin-plastic-film-making process [7]. The die is an irregularly shaped disk. An axis of symmetry with respect to geometry and loading exists as shown. The die was modeled by using simple quadrilateral axisymmetric elements. The locations of high stress were of primary concern. Figure 9–16 shows a plot of the von Mises stress contours for the

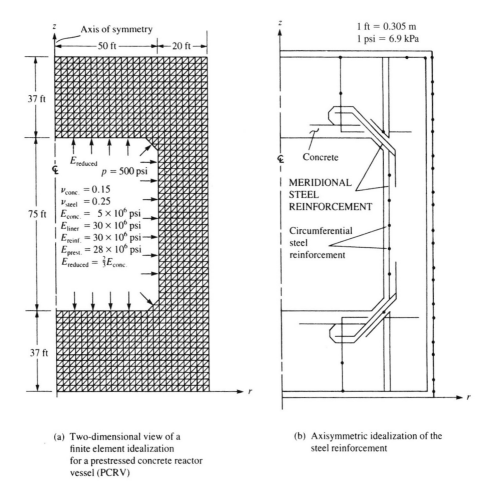

(a) Two-dimensional view of a finite element idealization for a prestressed concrete reactor vessel (PCRV)

(b) Axisymmetric idealization of the steel reinforcement

Figure 9–14 Model of steel-reinforced concrete pressure vessel (Reprinted from Nuclear Engineering and Design Volume 3, Issue 1, Rashid, Yosef R., Analysis of Axisymmetric Composite Structures by the Finite Element Method, Pages No. 163–182, Copyright 1966, with permission from Elsevier.)

die of Figure 9–15. The von Mises (or equivalent, or effective) stress [8] is often used as a failure criterion in design. Notice the artificially high stresses at the location of load F as explained in Section 7.1.

(Recall that the failure criterion based on the maximum distortion energy theory for ductile materials subjected to static loading predicts that a material will fail if the von Mises stress reaches the yield strength of the material.) Also recall from Eqs. (6.5.37) and (6.5.38), the von Mises stress σ_{vm} is related to the principal stresses by the expression

$$\sigma_{vm} = \frac{1}{\sqrt{2}} \sqrt{(\sigma_1 - \sigma_2)^2 + (\sigma_2 - \sigma_3)^2 + (\sigma_3 - \sigma_1)^2} \qquad (9.3.1)$$

where the principal stresses are given by σ_1, σ_2, and σ_3. These results were obtained from the commercial computer code ANSYS [12].

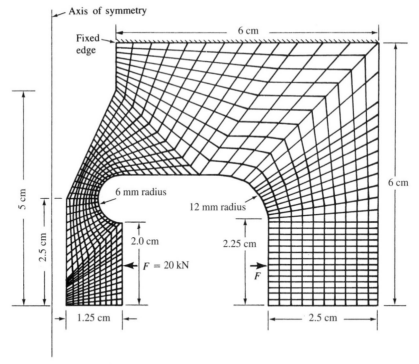

Figure 9–15 Model of a high-strength steel die (924 nodes and 830 elements)

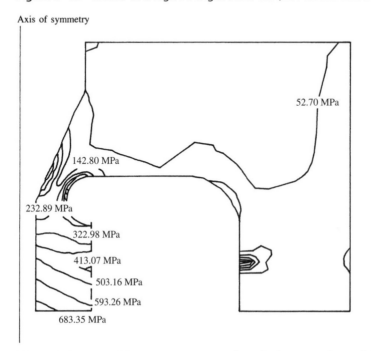

Figure 9–16 von Mises stress contour plot of axisymmetric model of Figure 9–15 (also producing a radial inward deflection of about 0.375 mm)

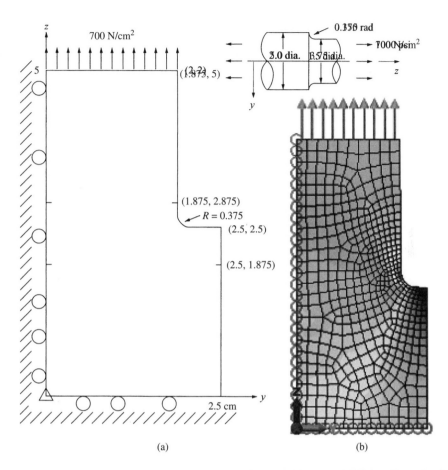

Figure 9-17 (a) Stepped shaft subjected to axial load and (b) the discretized model

Other dies with modifications in geometry were also studied to evaluate the most suitable die before the construction of an expensive prototype. Confidence in the acceptability of the prototype was enhanced by doing these comparison studies. Finally, Figure 9–17 shows a stepped 4130 steel shaft with a fillet radius subjected to an axial pressure of 700 N/cm^2 in tension. Fatigue analysis for reversed axial loading required an accurate stress concentration factor to be applied to the average axial stress of 700 N/cm^2. The stress concentration factor for the geometry shown was to be determined. Therefore, locations of highest stress were necessary. Figure 9–18 shows the resulting maximum principal stress plot using a computer program [11]. The largest principal stress was 1507.4 N/cm^2 at the fillet. Other examples of the use of the axisymmetric element can be found in References [2–6].

In this chapter, we have shown the finite element analysis of axisymmetric systems using a simple three-noded triangular element to be analogous to that of the two-dimensional plane stress problem using three-noded triangular elements as developed in Chapter 6. Therefore, the two-dimensional element in commercial computer programs with the axisymmetric element selected will allow for the analysis of axisymmetric structures.

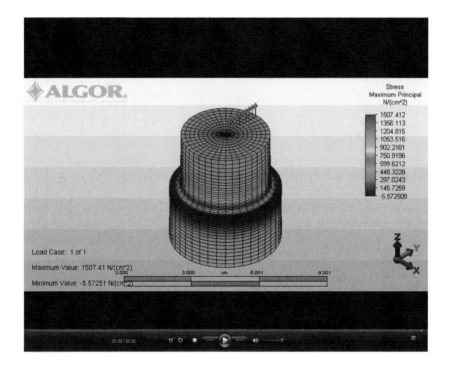

Figure 9-18 Three-dimensional visual of shaft of Figure 9–17 showing principal stress plot (See the full-color insert for a color version of this figure.)

Finally, note that other axisymmetric elements, such as a simple quadrilateral (one with four corner nodes and two degrees of freedom per node, as used in the steel die analysis of Figure 9–15) or higher-order triangular elements, such as in Reference [6], in which a cubic polynomial involving ten terms (ten a's) for both u and w, could be used for axisymmetric analysis. The three-noded triangular element was described here because of its simplicity and ability to describe geometric boundaries rather easily.

▲ Summary Equations

(All pertain to axisymmetric element).

Strain-displacement relationships for axisymmetric behavior:

$$\varepsilon_r = \frac{\partial u}{\partial r} \qquad \varepsilon_\theta = \frac{u}{r} \qquad \varepsilon_z = \frac{\partial w}{\partial z} \qquad \gamma_{rz} = \frac{\partial u}{\partial z} + \frac{\partial w}{\partial r} \tag{9.1.1e}$$

Stress–strain relationships for isotropic material:

$$\begin{Bmatrix} \sigma_r \\ \sigma_z \\ \sigma_\theta \\ \tau_{rz} \end{Bmatrix} = \frac{E}{(1+v)(1-2v)} \begin{bmatrix} 1-v & v & v & 0 \\ v & 1-v & v & 0 \\ v & v & 1-v & 0 \\ 0 & 0 & 0 & \dfrac{1-2v}{2} \end{bmatrix} \begin{Bmatrix} \varepsilon_r \\ \varepsilon_z \\ \varepsilon_\theta \\ \gamma_{rz} \end{Bmatrix} \tag{9.1.2}$$

Displacement functions for axisymmetric triangle element:

$$u(r, z) = a_1 + a_2 r + a_3 z$$
$$w(r, z) = a_4 + a_5 r + a_6 z \tag{9.1.3}$$

Shape functions for axisymmetric triangle element:

$$N_i = \frac{1}{2A}(\alpha_i + \beta_i r + \gamma_i z)$$

$$N_j = \frac{1}{2A}(\alpha_j + \beta_j r + \gamma_j z) \tag{9.1.12}$$

$$N_m = \frac{1}{2A}(\alpha_m + \beta_m r + \gamma_m z)$$

Gradient matrix:

$$[B_i] = \frac{1}{2A}\begin{bmatrix} \beta_i & 0 \\ 0 & \gamma_i \\ \dfrac{\alpha_i}{r} + \beta_i + \dfrac{\gamma_i z}{r} & 0 \\ \gamma_i & \beta_i \end{bmatrix} \tag{9.1.19}$$

and

$$[B] = [[B_i] \ [B_j] \ [B_m]] \tag{9.1.21}$$

Strain–displacement equations in matrix form:

$$\{\varepsilon\} = [B]\{d\} \tag{9.1.20}$$

Stress–displacement equations in matrix form:

$$\{\sigma\} = [D][B]\{d\} \tag{9.1.22}$$

Element stiffness matrix:

$$[k] = 2\pi \iint_A [B]^T [D][B] r \, dr \, dz \tag{9.1.24}$$

First approximation stiffness matrix:

$$[k] = 2\pi \bar{r} A [\bar{B}]^T [D][\bar{B}] \tag{9.1.26}$$

Body force matrix (first approximation):

$$\{f_b\} = \frac{2\pi \bar{r} A}{3} \begin{Bmatrix} \bar{R}_b \\ Z_b \\ \bar{R}_b \\ Z_b \\ \bar{R}_b \\ Z_b \end{Bmatrix} \tag{9.1.31}$$

$$\bar{R}_b = \omega^2 \rho \bar{r} \tag{9.1.32}$$

Surface force matrix on side j–m of element subjected to uniform radial and axial pressure:

$$\{f_s\} = \frac{2\pi r_j(z_m - z_j)}{2} \begin{Bmatrix} 0 \\ 0 \\ p_r \\ p_z \\ p_r \\ p_z \end{Bmatrix} \tag{9.1.36}$$

▲ References

[1] Utku, S., "Explicit Expressions for Triangular Torus Element Stiffness Matrix," *Journal of the American Institute of Aeronautics and Astronautics*, Vol. 6, No. 6, pp. 1174–1176, June 1968.

[2] Zienkiewicz, O. C., *The Finite Element Method*, 3rd ed., McGraw-Hill, London, 1977.

[3] Clough, R., and Rashid, Y., "Finite Element Analysis of Axisymmetric Solids," *Journal of the Engineering Mechanics Division*, American Society of Civil Engineers, Vol. 91, pp. 71–85, Feb. 1965.

[4] Rashid, Y., "Analysis of Axisymmetric Composite Structures by the Finite Element Method," *Nuclear Engineering and Design*, Vol. 3, pp. 163–182, 1966.

[5] Wilson, E., "Structural Analysis of Axisymmetric Solids," *Journal of the American Institute of Aeronautics and Astronautics*, Vol. 3, No. 12, pp. 2269–2274, Dec. 1965.

[6] Chacour, S., "A High Precision Axisymmetric Triangular Element Used in the Analysis of Hydraulic Turbine Components," Transactions of the American Society of Mechanical Engineers, *Journal of Basic Engineering*, Vol. 92, pp. 819–826, 1973.

[7] Greer, R. D., *The Analysis of a Film Tower Die Utilizing the ANSYS Finite Element Package*, M.S. Thesis, Rose-Hulman Institute of Technology, Terre Haute, IN, May 1989.

[8] Gere, J. M., and Goodno, B. J., *Mechanics of Materials*, 7th ed., Cengage Learning, Mason, OH, 2009.

[9] Cook, R. D., Malkus, D. S., Plesha, M. E., and Witt, R. J., *Concepts and Applications of Finite Element Analysis*, 4th ed., Wiley, New York, 2002.

[10] Cook, R. D., and Young, W. C., *Advanced Mechanics of Materials*, Macmillan, New York, 1985.

[11] Algor Interactive Systems, 150 Beta Drive, Pittsburgh, PA 15238.

[12] Swanson, J. A. ANSYS-Engineering Analysis System's User's Manual, Swanson Analysis Systems, Inc., Johnson Rd., P.O. Box 65, Houston, PA 15342.

▲ Problems

9.1 For the elements shown (in cm) in Figure P9–1, evaluate the stiffness matrices using Eq. (9.2.2). The coordinates are shown in the figures. Let $E = 210$ GPa and $v = 0.25$ for each element.

9.2 Evaluate the nodal forces used to replace the linearly varying surface traction shown in Figure P9–2. Hint: Use Eq. (9.1.34).

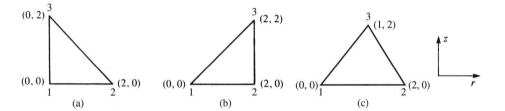

Figure P9–1

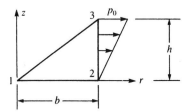

Figure P9–2

9.3 For an element of an axisymmetric body rotating with a constant angular velocity $\omega = 20$ rpm as shown in Figure P9–3, evaluate the body-force matrix. The coordinates (in cm) of the element are shown in the figure. Let the mass density ρ_w be 7850 kg/m^3.

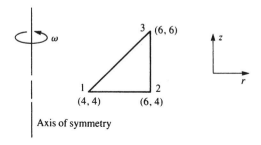

Figure P9–3

9.4 For the axisymmetric elements shown in Figure P9–4, determine the element stresses. Let $E = 210$ GPa and $v = 0.25$. The coordinates (in cm) are shown in the figures, and the nodal displacements for each element are $u_1 = 0.0001$ cm, $w_1 = 0.0002$ cm, $u_2 = 0.0005$ cm, $w_2 = 0.0006$ cm, $u_3 = 0$, and $w_3 = 0$.

9.5 Explicitly show that the integration of Eq. (9.1.35) yields the j surface forces given by Eq. (9.1.36).

9.6 For the elements shown in Figure P9–6, evaluate the stiffness matrices using Eq. (9.2.2). The coordinates (in millimeters) are shown in the figures. Let $E = 210$ GPa and $v = 0.25$ for each element.

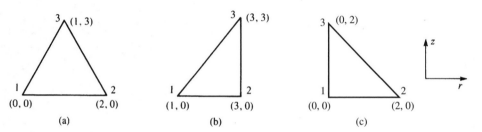

Figure P9–4

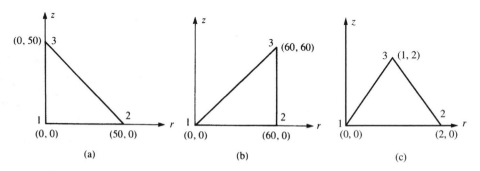

Figure P9–6

9.7 For the axisymmetric elements shown in Figure P9–7, determine the element stresses. Let $E = 210$ GPa and $v = 0.25$. The coordinates (in millimeters) are shown in the figures, and the nodal displacements for each element are

$$u_1 = 0.05 \text{ mm} \qquad w_1 = 0.03 \text{ mm}$$

$$u_2 = 0.02 \text{ mm} \qquad w_2 = 0.02 \text{ mm}$$

$$u_3 = 0.0 \text{ mm} \qquad w_3 = 0.0 \text{ mm}$$

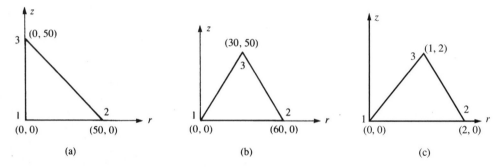

Figure P9–7

9.8 Can we connect plane stress elements with axisymmetric ones? Explain.

9.9 Is the three-noded triangular element considered in Section 9.1 a constant strain element? Why or why not?

9.10 How should one model the boundary conditions of nodes acting on the axis of symmetry?

9.11 How would you evaluate the circumferential strain, ε_θ, at $r = 0$? What is this strain in terms of the a's given in Eq. (9.1.15). Hint: Elasticity theory tells us that the radial strain must equal the circumferential strain at $r = 0$.

9.12 What will be the stresses σ_r and σ_θ at $r = 0$? Hint: Look at Eq. (9.1.2) after considering Problem 9.11.

Solve the following axisymmetric problems using a computer program.

9.13 The soil mass in Figure P9–13 is loaded by a force transmitted through a circular footing as shown. Determine the stresses in the soil. Compare the values of σ_r using an axisymmetric model with the σ_y values using a plane stress model. Let $E = 20$ MPa and $v = 0.45$ for the soil mass.

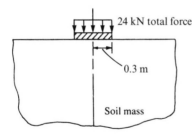

24 kN total force

0.3 m   **Figure P9–13**

Soil mass

9.14 Perform a stress analysis of the pressure vessel shown in Figure P9–14. Let $E = 32$ GPa and $v = 0.15$ for the concrete, and let $E = 200$ GPa and $v = 0.25$ for the steel liner. The steel liner is 50 mm thick. Let the pressure p equal 3.2 MPa. Use a 6 mm radius in the re-entrant corners.

9.15 Perform a stress analysis of the concrete pressure vessel with the steel liner shown in Figure P9–15. Let $E = 30$ GPa and $v = 0.15$ for the concrete, and let $E = 205$ GPa and $v = 0.25$ for the steel liner. The steel liner is 50 mm thick. Let the pressure p equal 700 kPa. Use a 10 mm radius in the re-entrant corners.

9.16 Perform a stress analysis of the disk shown in Figure P9–16 if it rotates with constant angular velocity of $\omega = 50$ rpm. Let $E = 210$ GPa, $v = 0.25$, and the mass density $\rho_w = 72.44$ kN/m³ [mass density, $\rho = \rho_w/(g = 9.81 \text{ m/s}^2)$]. (Use 8 and then 16 elements symmetrically modeled similar to Example 9.4. Compare the finite element solution to the theoretical circumferential and radial stresses given by

$$\sigma_\theta = \frac{3+v}{8}\rho\omega^2 a^2 \left(1 - \frac{1+3vr^2}{3+va^2}\right), \qquad \sigma_r = \frac{3+v}{8}\rho\omega^2 a^2 \left(1 - \frac{r^2}{a^2}\right)$$

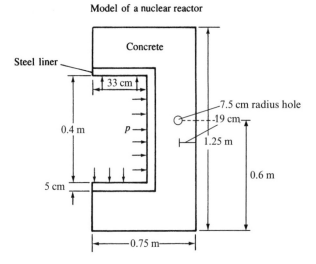

Model of a nuclear reactor

Figure P9-14

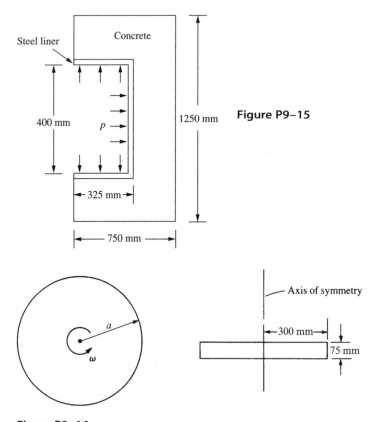

Figure P9-15

Figure P9-16

9.17 For the die casting shown in Figure P9–17, determine the maximum stresses and their locations. Let $E = 210$ GPa and $v = 0.25$. The dimensions are shown in the figure.

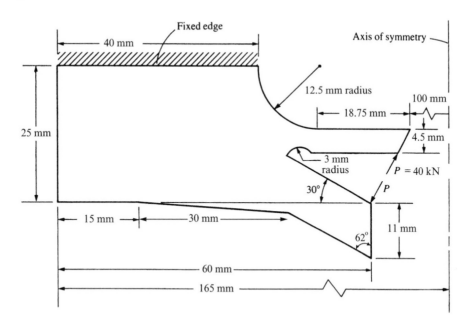

Figure P9–17

9.18 For the axisymmetric connecting rod shown in Figure P9–18, determine the stresses $\sigma_z, \sigma_r, \sigma_\theta,$ and τ_{rz}. Plot stress contours (lines of constant stress) for each of the normal stresses. Let $E = 210$ GPa and $v = 0.25$. The applied loading and boundary conditions are shown in the figure. A typical discretized rod is shown in the figure for illustrative purposes only.

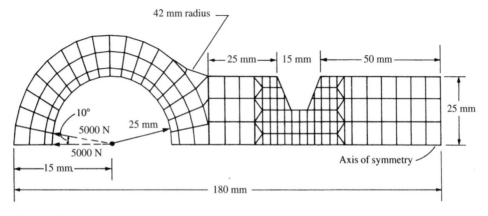

Figure P9–18

9.19 For the thick-walled open-ended cylindrical pipe subjected to internal pressure shown in Figure P9–19, use five layers of elements to obtain the circumferential stress, σ_θ, and the principal stresses and maximum radial displacement. Compare these results to the exact circumferential stress and radial displacement equations given by

$$\sigma_\theta = \left(\frac{P_i a^2 + P_0 b^2}{b^2 - a^2}\right) + \left(\frac{a^2 b^2}{r^2(b^2 - a^2)}\right)(P_i - P_0)$$

$$\Delta r = \frac{r}{E(b^2 - a^2)}\left[(1 - v)(P_i a^2 - P_0 b^2) + \frac{(1 + v)a^2 b^2}{r^2}(P_i - P_0)\right]$$

where

P_i = inner pressure, P_0 = outer pressure (set to zero in this problem)
a = inner radius of vessel, b = outer radius of vessel, r = any radial location

Let $E = 205$ GPa and $v = 0.3$.

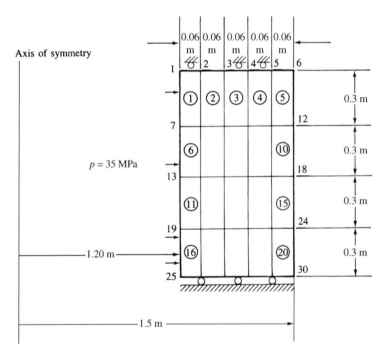

Figure P9–19

9.20 A steel cylindrical pressure vessel with flat plate end caps is shown in Figure P9–20 with vertical axis of symmetry. Addition of thickened sections helps to reduce stress concentrations in the corners. Analyze the design and identify the most critically stressed regions. Note that inside sharp re-entrant corners produce infinite stress concentration zones, so refining the mesh in these regions will not help you get a better answer unless you use an inelastic theory or place small fillet radii there. Recommend any design changes in your report. Let the pressure inside be 1000 kPa.

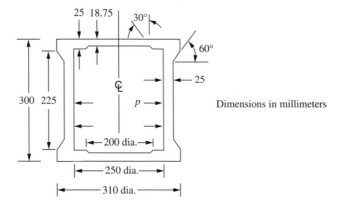

Figure P9–20

9.21 For the cylindrical vessel with hemispherical ends (heads) under uniform internal pressure of intensity $p = 3.2$ MPa shown in Figure P9–21, determine the maximum von Mises stress and where it is located. The material is ASTM—A242 quenched and tempered alloy steel. Use a factor of safety of 3 against yielding. The inner radius is $a = 2.5$ m and the thickness $t = 50$ mm.

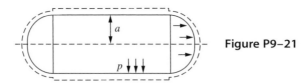

Figure P9–21

9.22 For the cylindrical vessel with *ellipsoidal heads* shown in Figure P9–22a under loading $p = 3.2$ MPa, determine if the vessel is safe against yielding. Use the same material and factor of safety as in Problem 9.21. Now let $a = 2.5$ m and $b = 1.25$ m. Which vessel has the lowest hoop stress? Recommend the preferred head shape of the two based on your answers.

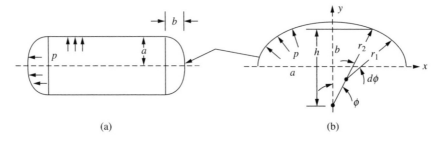

(a)   (b)

Figure P9–22

For modeling purposes, the equation of an ellipse is given by $b^2x^2 + a^2y^2 = a^2b^2$, where a is the major axis and b is the minor axis of the ellipse shown in Figure P9–22(b).

9.23 The syringe with plunger is shown in Figure P9–23. The material of the syringe is glass with $E = 69$ GPa, $v = 0.15$, and tensile strength of 5 MPa. The bottom hole is assumed to be closed under test conditions. Determine the deformation and stresses in the glass. Compare the maximum principal stress in the glass to the ultimate tensile strength. Do you think the syringe is safe? Why?

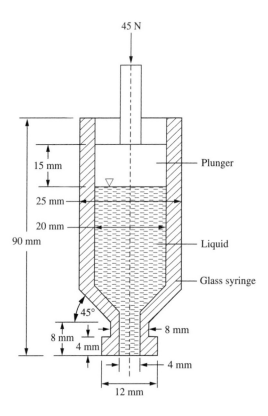

Figure P9–23

9.24 For the tapered solid circular shaft shown in Figure P9–24, a semicircular groove has been machined into the side. The shaft is made of a hot rolled 1040 steel alloy with yield strength of 500 MPa. The shaft is subjected to a uniform axial pressure of 28 MPa. Determine the maximum principal stresses and von Mises stresses at the fillet and at the semicircular groove. Is the shaft safe from failure based on the maximum distortion energy theory?

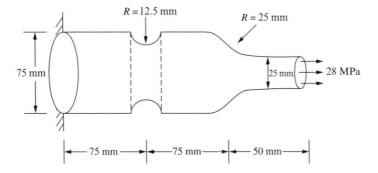

Figure P9–24

9.25 A steel hole punch is shown in Figure P9–25a. Investigate the proper material for a hole punch. Tell me what material you used and why? Model the punch without the side groove (Figure P9–25b) and with the side groove (Figure P9–25c). Determine the von Mises stress distribution throughout the punch for both cases. Are the punches safe under the loading shown?

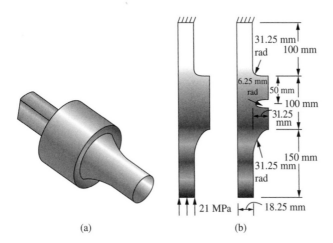

Figure P9–25

ISOPARAMETRIC FORMULATION

CHAPTER OBJECTIVES

- To formulate the isoparametric formulation of the bar element stiffness matrix.
- To present the isoparametric formulation of the plane four-noded quadrilateral (Q4) element stiffness matrix.
- To describe two methods for numerical integration—Newton-Cotes and Gaussian Quadrature—used for numerical evaluation of definite integrals and to demonstrate their application to specific examples.
- To present a flowchart describing how to evaluate the stiffness matrix for the plane quadrilateral element by a four-point Gaussian quadrature rule.
- To solve an explicit example showing the evaluation of the stiffness matrix for the plane quadrilateral element by the four-point Gaussian quadrature rule.
- To illustrate by example how to evaluate the stresses at a given point in a plane quadrilateral element using Gaussian quadrature.
- To describe some higher-order shape functions for the three-noded linear strain bar, the eight-noded quadratic quadrilateral (Q8) element, and the twelve-noded cubic quadrilateral (Q12) element.
- To evaluate the stiffness matrix of the three-noded bar using Gaussian quadrature and compare the result to that found by explicit evaluation of the stiffness matrix for the bar.

Introduction

In this chapter, we introduce the isoparametric formulation of the element stiffness matrices. After considering the linear-strain triangular element in Chapter 8, we can see that the development of element matrices and equations expressed in terms of a global coordinate system becomes an enormously difficult task (if even possible) except for the simplest of elements such as the constant-strain triangle of Chapter 6. Hence, the isoparametric formulation was developed [1]. The isoparametric method may appear somewhat tedious (and confusing initially), but it will lead to a simple computer program formulation, and it is generally applicable for two- and three-dimensional stress analysis and for nonstructural problems. The isoparametric formulation allows elements to be created that are nonrectangular and have curved sides. Furthermore, numerous commercial computer programs (as described in Chapter 1) have adapted this formulation for their various libraries of elements.

We first illustrate the isoparametric formulation to develop the simple bar element stiffness matrix. Use of the bar element makes it relatively easy to understand the method because simple expressions result.

We then consider the development of the isoparametric formulation of the simple quadrilateral element stiffness matrix.

Next, we will introduce numerical integration methods for evaluating the quadrilateral element stiffness matrix and illustrate the adaptability of the isoparametric formulation to common numerical integration methods.

Finally, we will consider some higher-order elements and their associated shape functions.

▲ 10.1 Isoparametric Formulation of the Bar Element Stiffness Matrix ▲

The term *isoparametric* is derived from the use of the same shape functions (or interpolation functions) $[N]$ to define the element's geometric shape as are used to define the displacements within the element. Thus, when the shape function is $u = a_1 + a_2 s$ for the displacement, we use $x = a_1 + a_2 s$ for the description of the nodal coordinate of a point on the bar element and, hence, the physical shape of the element.

Isoparametric element equations are formulated using a **natural** (or **intrinsic**) **coordinate system** s that is defined by element geometry and not by the element orientation in the global-coordinate system. In other words, axial coordinate s is attached to the bar and remains directed along the axial length of the bar, regardless of how the bar is oriented in space. There is a relationship (called a *transformation mapping*) between the natural coordinate system s and the global coordinate system x for each element of a specific structure, and this relationship must be used in the element equation formulations.

We will now develop the isoparametric formulation of the stiffness matrix of a simple linear bar element [with two nodes as shown in Figure 10–1(a)].

Step 1   Select Element Type

First, the natural coordinate s is attached to the element, with the origin located at the center of the element, as shown in Figure 10–1(b). The s axis need not be parallel to the x axis—this is only for convenience.

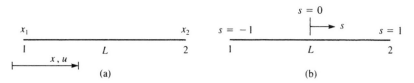

Figure 10–1   Linear bar element in (a) a global coordinate system x and (b) a natural coordinate system s

We consider the bar element to have two degrees of freedom—axial displacements u_1 and u_2 at each node associated with the global x axis.

For the special case when the s and x axes are parallel to each other, the s and x coordinates can be related by

$$x = x_c + \frac{L}{2}s \qquad (10.1.1a)$$

where x_c is the global coordinate of the element centroid.

Using the global coordinates x_1 and x_2 in Eq. (10.1.1a) with $x_c = (x_1 + x_2)/2$, we can express the natural coordinate s in terms of the global coordinates as

$$s = [x - (x_1 + x_2)/2][2/(x_2 - x_1)] \qquad (10.1.1b)$$

The shape functions used to define a position within the bar are found in a manner similar to that used in Chapter 3 to define displacement within a bar (Section 3.1). We begin by relating the natural coordinate to the global coordinate by

$$x = a_1 + a_2 s \qquad (10.1.2)$$

where we note that s is such that $-1 \leqslant s \leqslant 1$. Solving for the a_i's in terms of x_1 and x_2, we obtain

$$x = \frac{1}{2}[(1 - s)x_1 + (1 + s)x_2] \qquad (10.1.3)$$

or, in matrix form, we can express Eq. (10.1.3) as

$$\{x\} = [N_1 \quad N_2]\begin{Bmatrix} x_1 \\ x_2 \end{Bmatrix} \qquad (10.1.4)$$

where the shape functions in Eq. (10.1.4) are

$$N_1 = \frac{1 - s}{2} \qquad N_2 = \frac{1 + s}{2} \qquad (10.1.5)$$

The linear shape functions in Eqs. (10.1.5) map the s coordinate of any point in the element to the x coordinate when used in Eq. (10.1.3). For instance, when we substitute $s = -1$ into Eq. (10.1.3), we obtain $x = x_1$. These shape functions are shown in Figure 10–2, where we can see that they have the same properties as defined for the interpolation functions of Section 3.1. Hence, N_1 represents the physical shape of the coordinate x when plotted over the length of the element for $x_1 = 1$ and $x_2 = 0$, and N_2 represents the coordinate x when plotted over the length of the element for $x_2 = 1$ and $x_1 = 0$. Again, we must have $N_1 + N_2 = 1$.

These shape functions must also be continuous throughout the element domain and have finite first derivatives within the element.

Step 2   Select a Displacement Function

The displacement function within the bar is now defined by the same shape functions, Eqs. (10.1.5), as are used to define the element shape; that is,

$$\{u\} = [N_1 \quad N_2]\begin{Bmatrix} u_1 \\ u_2 \end{Bmatrix} \qquad (10.1.6)$$

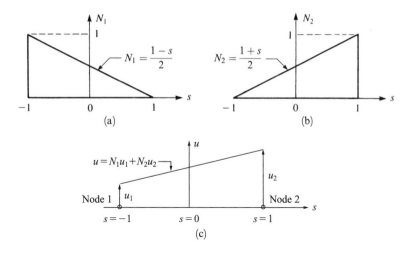

Figure 10–2 Shape function variations with natural coordinates: (a) shape function N_1, (b) shape function N_2, and (c) linear displacement field u plotted over element length

When a particular coordinate s of the point of interest is substituted into $[N]$, Eq. (10.1.6) yields the displacement of a point on the bar in terms of the nodal degrees of freedom u_1 and u_2 as shown in Figure 10–2(c). Since u and x are defined by the same shape functions at the same nodes, comparing Eqs. (10.1.4) and (10.1.6), the element is called *isoparametric*.

Step 3 Define the Strain–Displacement and Stress–Strain Relationships

We now want to formulate element matrix $[B]$ to evaluate $[k]$. We use the isoparametric formulation to illustrate its manipulations. For a simple bar element, no real advantage may appear evident. However, for higher-order elements, the advantage will become clear because relatively simple computer program formulations will result.

To construct the element stiffness matrix, we must determine the strain, which is defined in terms of the derivative of the displacement with respect to x. The displacement u, however, is now a function of s as given by Eq. (10.1.6). Therefore, we must apply the chain rule of differentiation to the function u as follows:

$$\frac{du}{ds} = \frac{du}{dx}\frac{dx}{ds} \tag{10.1.7}$$

We can evaluate (du/ds) and (dx/ds) using Eqs. (10.1.6) and (10.1.3). We seek $(du/dx) = \varepsilon_x$. Therefore, we solve Eq. (10.1.7) for (du/dx) as

$$\frac{du}{dx} = \frac{\left(\dfrac{du}{ds}\right)}{\left(\dfrac{dx}{ds}\right)} \tag{10.1.8}$$

Using Eq. (10.1.6) for u, we obtain

$$\frac{du}{ds} = \frac{u_2 - u_1}{2} \tag{10.1.9a}$$

and using Eq. (10.1.3) for x, we have

$$\frac{dx}{ds} = \frac{x_2 - x_1}{2} = \frac{L}{2} \tag{10.1.9b}$$

because $x_2 - x_1 = L$.

Using Eqs. (10.1.9a) and (10.1.9b) in Eq. (10.1.8), we obtain

$$\{\varepsilon_x\} = \begin{bmatrix} -\dfrac{1}{L} & \dfrac{1}{L} \end{bmatrix} \begin{Bmatrix} u_1 \\ u_2 \end{Bmatrix} \tag{10.1.10}$$

Since $\{\varepsilon\} = [B]\{d\}$, the strain–displacement matrix $[B]$ is then given in Eq. (10.1.10) as

$$[B] = \begin{bmatrix} -\dfrac{1}{L} & \dfrac{1}{L} \end{bmatrix} \tag{10.1.11}$$

We recall that use of linear shape functions results in a constant $[B]$ matrix, and hence, in a constant strain within the element. For higher-order elements, such as the quadratic bar with three nodes, $[B]$ becomes a function of natural coordinate s (see Eq. (10.5.16)).

The stress matrix is again given by Hooke's law as

$$\{\sigma\} = E\{\varepsilon\} = E[B]\{d\}$$

Step 4   Derive the Element Stiffness Matrix and Equations

The stiffness matrix is

$$[k] = \int_0^L [B]^T [D][B] A \, dx \tag{10.1.12}$$

However, in general, we must transform the coordinate x to s because $[B]$ is, in general, a function of s. This general type of transformation is given by References [4] and [5]

$$\int_0^L f(x) \, dx = \int_{-1}^1 f(s)|[J]| \, ds \tag{10.1.13}$$

where $[J]$ is called the *Jacobian* matrix. In the one-dimensional case, we have $|[J]| = J$. For the simple bar element, from Eq. (10.1.9b), we have

$$|[J]| = \frac{dx}{ds} = \frac{L}{2} \tag{10.1.14}$$

Observe that in Eq. (10.1.14), the Jacobian determinant relates an element length (dx) in the global-coordinate system to an element length (ds) in the natural-coordinate system. In general, $|[J]|$ is a function of s and depends on the numerical values of the nodal coordinates. This can be seen by looking at Eq. (10.2.22) for the quadrilateral element. (Section 10.2 further discusses the Jacobian.) Using Eqs. (10.1.13) and (10.1.14) in Eq. (10.1.12), we obtain the stiffness matrix in natural coordinates as

$$[k] = \frac{L}{2} \int_{-1}^{1} [B]^T E [B] A \, ds \tag{10.1.15}$$

where, for the one-dimensional case, we have used the modulus of elasticity $E = [D]$ in Eq. (10.1.15). Substituting Eq. (10.1.11) in Eq. (10.1.15) and performing the simple integration, we obtain

$$[k] = \frac{AE}{L} \begin{bmatrix} 1 & -1 \\ -1 & 1 \end{bmatrix} \tag{10.1.16}$$

which is the same as Eq. (3.1.14). For higher-order one-dimensional elements, the integration in closed form becomes difficult if not impossible (see Example 10.7). Even the simple rectangular element stiffness matrix is difficult to evaluate in closed form (Section 10.2). However, the use of numerical integration, as described in Section 10.3, illustrates the distinct advantage of the isoparametric formulation of the equations.

Body Forces

We will now determine the body-force matrix using the natural coordinate system s. Using Eq. (3.10.20b), the body-force matrix is

$$\{f_b\} = \iiint_V [N]^T \{X_b\} \, dV \tag{10.1.17}$$

Letting $dV = A \, dx$, we have

$$\{f_b\} = A \int_0^L [N]^T \{X_b\} \, dx \tag{10.1.18}$$

Substituting Eqs. (10.1.5) for N_1 and N_2 into $[N]$ and noting that by Eq. (10.1.9b), $dx = (L/2) \, ds$, we obtain

$$\{f_b\} = A \int_{-1}^{1} \left\{ \begin{array}{c} \dfrac{1-s}{2} \\ \dfrac{1+s}{2} \end{array} \right\} \{X_b\} \frac{L}{2} \, ds \tag{10.1.19}$$

On integrating Eq. (10.1.19), we obtain

$$\{f_b\} = \frac{ALX_b}{2} \left\{ \begin{array}{c} 1 \\ 1 \end{array} \right\} \tag{10.1.20}$$

The physical interpretation of the results for $\{f_b\}$ is that since AL represents the volume of the element and X_b the body force per unit volume, then ALX_b is the total body force acting on the element. The factor $\frac{1}{2}$ indicates that this body force is equally distributed to the two nodes of the element.

Surface Forces

Surface forces can be found using Eq. (3.10.20a) as

$$\{f_s\} = \iint_S [N_s]^T \{T_x\} \, dS \tag{10.1.21}$$

Assuming the cross section is constant and the traction is uniform over the perimeter and along the length of the element, we obtain

$$\{f_s\} = \int_0^L [N_s]^T \{T_x\} \, dx \tag{10.1.22}$$

where we now assume T_x is in units of force per unit length. Using the shape functions N_1 and N_2 from Eq. (10.1.5) in Eq. (10.1.22), we obtain

$$\{f_s\} = \int_{-1}^{1} \left\{ \begin{array}{c} \dfrac{1-s}{2} \\ \dfrac{1+s}{2} \end{array} \right\} \{T_x\} \dfrac{L}{2} \, ds \tag{10.1.23}$$

On integrating Eq. (10.1.23), we obtain

$$\{f_s\} = \{T_x\} \dfrac{L}{2} \left\{ \begin{array}{c} 1 \\ 1 \end{array} \right\} \tag{10.1.24}$$

The physical interpretation of Eq. (10.1.24) is that since $\{T_x\}$ is in force-per-unit-length units, $\{T_x\}L$ is now the total force. The $\frac{1}{2}$ indicates that the uniform surface traction is equally distributed to the two nodes of the element. Note that if $\{T_x\}$ were a function of x (or s), then the amounts of force allocated to each node would generally not be equal and would be found through integration as in Example 3.12.

▲ 10.2 Isoparametric Formulation of the Plane Quadrilateral Element Stiffness Matrix ▲

Recall that the term *isoparametric* is derived from the use of the same shape functions to define the element shape as are used to define the displacements within the element. Thus, when the shape function is $u = a_1 + a_2 s + a_3 t + a_4 st$ for the displacement, we use $x = a_1 + a_2 s + a_3 t + a_4 st$ for the description of a coordinate point in the plane element.

The natural-coordinate system s-t is defined by element geometry and not by the element orientation in the global-coordinate system x-y. Much as in the bar element example, there is a transformation mapping between the two coordinate systems for each element of a specific structure, and this relationship must be used in the element formulation.

We will now formulate the isoparametric formulation of the simple linear plane quadrilateral element stiffness matrix. This formulation is general enough to be

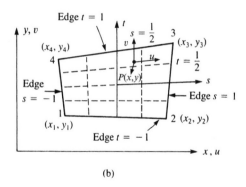

Figure 10–3 (a) Linear square element in *s-t* coordinates and (b) square element mapped into quadrilateral in *x-y* coordinates whose size and shape are determined by the eight nodal coordinates $x_1, y_1, \ldots, y_4$

applied to more complicated (higher-order) elements such as a quadratic plane element with three nodes along an edge, which can have straight or quadratic curved sides. Higher-order elements have additional nodes and use different shape functions as compared to the linear element, but the steps in the development of the stiffness matrices are the same. We will briefly discuss these elements after examining the linear plane element formulation.

Step 1   Select Element Type

First, the natural *s-t* coordinates are attached to the element, with the origin at the center of the element, as shown in Figure 10–3(a). The *s* and *t* axes need not be orthogonal, and neither has to be parallel to the *x* or *y* axis. The orientation of *s-t* coordinates is such that the four corner nodes and the edges of the quadrilateral are bounded by $+1$ or -1. This orientation will later allow us to take advantage more fully of common numerical integration schemes.

We consider the quadrilateral to have eight degrees of freedom, $u_1, v_1, \ldots, u_4$, and v_4 associated with the global *x* and *y* directions. The element then has straight sides but is otherwise of arbitrary shape, as shown in Figure 10–3(b).

For the special case when the distorted element becomes a rectangular element with sides parallel to the global *x-y* coordinates, the *s-t* coordinates can be related to the global element coordinates *x* and *y* by

$$x = x_c + bs \qquad y = y_c + ht \tag{10.2.1}$$

where x_c and y_c are the global coordinates of the element centroid.

We begin by assuming global coordinates *x* and *y* are related to the natural coordinates *s* and *t* as follows:

$$x = a_1 + a_2 s + a_3 t + a_4 st$$
$$y = a_5 + a_6 s + a_7 t + a_8 st \tag{10.2.2}$$

and solving for the a_i's in terms of $x_1, x_2, x_3, x_4, y_1, y_2, y_3,$ and y_4, we obtain

$$x = \frac{1}{4}[(1-s)(1-t)x_1 + (1+s)(1-t)x_2$$
$$+ (1+s)(1+t)x_3 + (1-s)(1+t)x_4]$$

$$y = \frac{1}{4}[(1-s)(1-t)y_1 + (1+s)(1-t)y_2$$
$$+ (1+s)(1+t)y_3 + (1-s)(1+t)y_4]$$

$\qquad(10.2.3)$

Or, in matrix form, we can express Eqs. (10.2.3) as

$$\left\{ \begin{array}{c} x \\ y \end{array} \right\} = \begin{bmatrix} N_1 & 0 & N_2 & 0 & N_3 & 0 & N_4 & 0 \\ 0 & N_1 & 0 & N_2 & 0 & N_3 & 0 & N_4 \end{bmatrix} \left\{ \begin{array}{c} x_1 \\ y_1 \\ x_2 \\ y_2 \\ x_3 \\ y_3 \\ x_4 \\ y_4 \end{array} \right\} \qquad(10.2.4)$$

where the shape functions of Eq. (10.2.4) are now

$$N_1 = \frac{(1-s)(1-t)}{4} \qquad N_2 = \frac{(1+s)(1-t)}{4}$$

$$N_3 = \frac{(1+s)(1+t)}{4} \qquad N_4 = \frac{(1-s)(1+t)}{4}$$

$\qquad(10.2.5)$

The shape functions of Eqs. (10.2.5) are linear. These shape functions are seen to map the s and t coordinates of any point in the square element of Figure 10–3(a) to those x and y coordinates in the quadrilateral element of Figure 10–3(b). For instance, consider square element node 1 coordinates, where $s = -1$ and $t = -1$. Using Eqs. (10.2.4) and (10.2.5), the left side of Eq. (10.2.4) becomes

$$x = x_1 \qquad y = y_1 \qquad\qquad (10.2.6)$$

Similarly, we can map the other local nodal coordinates at nodes 2, 3, and 4 such that the square element in s-t isoparametric coordinates is mapped into a quadrilateral element in global coordinates x_1, y_1 through x_4, y_4. Also observe the property that $N_1 + N_2 + N_3 + N_4 = 1$ for all values of s and t.

We further observe that the shape functions in Eq. (10.2.5) are again such that N_1 through N_4 have the properties that N_i $(i = 1, 2, 3, 4)$ is equal to one at node i and equal to zero at all other nodes. The physical shapes of N_i as they vary over the element with natural coordinates are shown in Figure 10–4. For instance, N_1 represents the geometric shape for $x_1 = 1$, $y_1 = 1$, and $x_2, y_2, x_3, y_3, x_4,$ and y_4 all equal to zero.

Until this point in the discussion, we have always developed the element shape functions either by assuming some relationship between the natural and global coordinates in terms of the generalized coordinates $(a_i$'s) as in Eqs. (10.2.2) or, similarly, by assuming a displacement function in terms of the a_i's. However, physical intuition

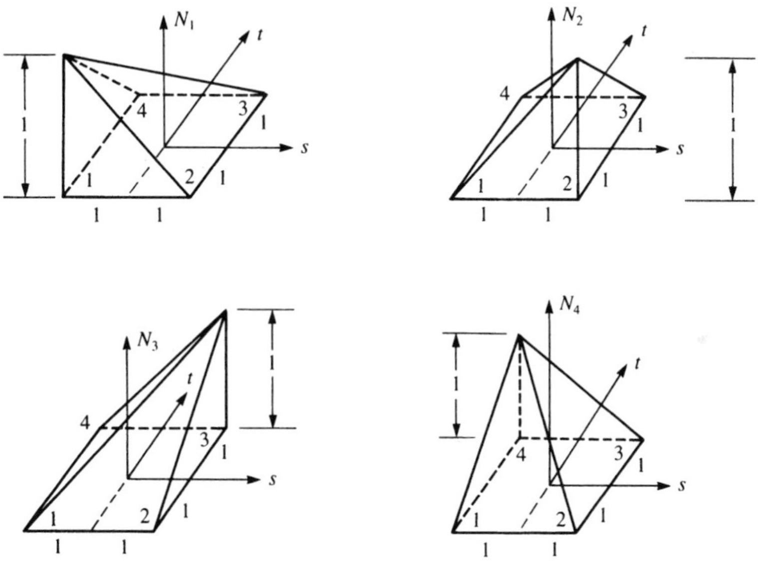

Figure 10–4 Variations of the shape functions over a linear square element

can often guide us in directly expressing shape functions based on the following two criteria set forth in Section 3.2 and used on numerous occasions:

$$\sum_{i=1}^{n} N_i = 1 \qquad (i = 1, 2, \dots, n)$$

where $n =$ the number of shape functions corresponding to displacement shape functions N_i, and $N_i = 1$ at node i and $N_i = 0$ at all nodes other than i. In addition, a third criterion is based on Lagrangian interpolation when displacement continuity is to be satisfied, or on Hermitian interpolation when additional slope continuity needs to be satisfied, as in the beam element of Chapter 4. (For a description of the use of Lagrangian and Hermitian interpolation to develop shape functions, consult References [4] and [6].)

Step 2 Select Displacement Functions

The displacement functions within an element are now similarly defined by the same shape functions as are used to define the element geometric shape; that is,

$$\begin{Bmatrix} u \\ v \end{Bmatrix} = \begin{bmatrix} N_1 & 0 & N_2 & 0 & N_3 & 0 & N_4 & 0 \\ 0 & N_1 & 0 & N_2 & 0 & N_3 & 0 & N_4 \end{bmatrix} \begin{Bmatrix} u_1 \\ v_1 \\ u_2 \\ v_2 \\ u_3 \\ v_3 \\ u_4 \\ v_4 \end{Bmatrix} \qquad (10.2.7)$$

where u and v are displacements parallel to the global x and y coordinates, and the shape functions are given by Eqs. (10.2.5). The displacement of an interior point P located at (x, y) in the element of Figure 10–3(b) is described by u and v in Eq. (10.2.7).

Comparing Eqs. (6.6.6) and (10.2.7), we see similarities between the rectangular element with sides of lengths $2b$ and $2h$ (Figure 6–20) and the square element with sides of length 2. If we let $b = 1$ and $h = 1$, the two sets of shape functions, Eqs. (6.6.5) and (10.2.5), are identical.

Step 3   Define the Strain–Displacement and Stress–Strain Relationships

We now want to formulate element matrix $[B]$ to evaluate $[k]$. However, because it becomes tedious and difficult (if not impossible) to write the shape functions in terms of the x and y coordinates, as seen in Chapter 8, we will carry out the formulation in terms of the isoparametric coordinates s and t. This may appear tedious, but it is easier to use the s- and t-coordinate expressions than to attempt to use the x- and y-coordinate expressions. This approach also leads to a simple computer program formulation.

To construct an element stiffness matrix, we must determine the strains, which are defined in terms of the derivatives of the displacements with respect to the x and y coordinates. The displacements, however, are now functions of the s and t coordinates, as given by Eq. (10.2.7), with the shape functions given by Eqs. (10.2.5). Before, we could determine $(\partial f / \partial x)$ and $(\partial f / \partial y)$, where, in general, f is a function representing the displacement functions u or v. However, u and v are now expressed in terms of s and t. Therefore, we need to apply the chain rule of differentiation because it will not be possible to express s and t as functions of x and y directly. For f as a function of x and y, the chain rule yields

$$\frac{\partial f}{\partial s} = \frac{\partial f}{\partial x}\frac{\partial x}{\partial s} + \frac{\partial f}{\partial y}\frac{\partial y}{\partial s}$$

$$\frac{\partial f}{\partial t} = \frac{\partial f}{\partial x}\frac{\partial x}{\partial t} + \frac{\partial f}{\partial y}\frac{\partial y}{\partial t}$$

(10.2.8)

In Eq. (10.2.8), $(\partial f / \partial s)$, $(\partial f / \partial t)$, $(\partial x / \partial s)$, $(\partial y / \partial s)$, $(\partial x / \partial t)$, and $(\partial y / \partial t)$ are all known using Eqs. (10.2.7) and (10.2.4). We still seek $(\partial f / \partial x)$ and $(\partial f / \partial y)$. The strains can then be found; for example, $\varepsilon_x = (\partial u / \partial x)$. Therefore, we solve Eqs. (10.2.8) for $(\partial f / \partial x)$ and $(\partial f / \partial y)$ using Cramer's rule, which involves evaluation of determinants (Appendix B), as

$$\frac{\partial f}{\partial x} = \frac{\begin{vmatrix} \dfrac{\partial f}{\partial s} & \dfrac{\partial y}{\partial s} \\[2mm] \dfrac{\partial f}{\partial t} & \dfrac{\partial y}{\partial t} \end{vmatrix}}{\begin{vmatrix} \dfrac{\partial x}{\partial s} & \dfrac{\partial y}{\partial s} \\[2mm] \dfrac{\partial x}{\partial t} & \dfrac{\partial y}{\partial t} \end{vmatrix}} \qquad \frac{\partial f}{\partial y} = \frac{\begin{vmatrix} \dfrac{\partial x}{\partial s} & \dfrac{\partial f}{\partial s} \\[2mm] \dfrac{\partial x}{\partial t} & \dfrac{\partial f}{\partial t} \end{vmatrix}}{\begin{vmatrix} \dfrac{\partial x}{\partial s} & \dfrac{\partial y}{\partial s} \\[2mm] \dfrac{\partial x}{\partial t} & \dfrac{\partial y}{\partial t} \end{vmatrix}}$$

(10.2.9)

where the determinant in the denominator is the determinant of the *Jacobian* matrix $[J]$. Hence, the Jacobian matrix is given by

$$[J] = \begin{bmatrix} \dfrac{\partial x}{\partial s} & \dfrac{\partial y}{\partial s} \\[2ex] \dfrac{\partial x}{\partial t} & \dfrac{\partial y}{\partial t} \end{bmatrix} \qquad (10.2.10)$$

We now want to express the element strains as

$$\{\varepsilon\} = [B]\{d\} \qquad (10.2.11)$$

where $[B]$ must now be expressed as a function of s and t. We start with the usual relationship between strains and displacements given in matrix form as

$$\begin{Bmatrix} \varepsilon_x \\ \varepsilon_y \\ \gamma_{xy} \end{Bmatrix} = \begin{bmatrix} \dfrac{\partial(\ )}{\partial x} & 0 \\[2ex] 0 & \dfrac{\partial(\ )}{\partial y} \\[2ex] \dfrac{\partial(\ )}{\partial y} & \dfrac{\partial(\ )}{\partial x} \end{bmatrix} \begin{Bmatrix} u \\ v \end{Bmatrix} \qquad (10.2.12)$$

where the rectangular matrix on the right side of Eq. (10.2.12) is an *operator matrix*; that is, $\partial(\ )/\partial x$ and $\partial(\ )/\partial y$ represent the partial derivatives of any variable we put inside the parentheses.

Using Eqs. (10.2.9) and evaluating the determinant in the numerators, we have

$$\frac{\partial(\ )}{\partial x} = \frac{1}{\|[J]\|}\left[\frac{\partial y}{\partial t}\frac{\partial(\ )}{\partial s} - \frac{\partial y}{\partial s}\frac{\partial(\ )}{\partial t}\right]$$
$$\frac{\partial(\ )}{\partial y} = \frac{1}{\|[J]\|}\left[\frac{\partial x}{\partial s}\frac{\partial(\ )}{\partial t} - \frac{\partial x}{\partial t}\frac{\partial(\ )}{\partial s}\right] \qquad (10.2.13)$$

where $\|[J]\|$ is the determinant of $[J]$ given by Eq. (10.2.10). Using Eq. (10.2.13) in Eq. (10.2.12) we obtain the strains expressed in terms of the natural coordinates $(s\text{-}t)$ as

$$\begin{Bmatrix} \varepsilon_x \\ \varepsilon_y \\ \gamma_{xy} \end{Bmatrix} = \frac{1}{\|[J]\|}\begin{bmatrix} \dfrac{\partial y}{\partial t}\dfrac{\partial(\ )}{\partial s} - \dfrac{\partial y}{\partial s}\dfrac{\partial(\ )}{\partial t} & 0 \\[2ex] 0 & \dfrac{\partial x}{\partial s}\dfrac{\partial(\ )}{\partial t} - \dfrac{\partial x}{\partial t}\dfrac{\partial(\ )}{\partial s} \\[2ex] \dfrac{\partial x}{\partial s}\dfrac{\partial(\ )}{\partial t} - \dfrac{\partial x}{\partial t}\dfrac{\partial(\ )}{\partial s} & \dfrac{\partial y}{\partial t}\dfrac{\partial(\ )}{\partial s} - \dfrac{\partial y}{\partial s}\dfrac{\partial(\ )}{\partial t} \end{bmatrix} \begin{Bmatrix} u \\ v \end{Bmatrix} \qquad (10.2.14)$$

Using Eq. (10.2.7), we can express Eq. (10.2.14) in terms of the shape functions and global coordinates in compact matrix form as

$$\{\varepsilon\} = [D'][N]\{d\} \qquad (10.2.15)$$

where $[D']$ is an operator matrix given by

$$[D'] = \frac{1}{|[J]|} \begin{bmatrix} \dfrac{\partial y}{\partial t}\dfrac{\partial(\ )}{\partial s} - \dfrac{\partial y}{\partial s}\dfrac{\partial(\ )}{\partial t} & 0 \\[2ex] 0 & \dfrac{\partial x}{\partial s}\dfrac{\partial(\ )}{\partial t} - \dfrac{\partial x}{\partial t}\dfrac{\partial(\ )}{\partial s} \\[2ex] \dfrac{\partial x}{\partial s}\dfrac{\partial(\ )}{\partial t} - \dfrac{\partial x}{\partial t}\dfrac{\partial(\ )}{\partial s} & \dfrac{\partial y}{\partial t}\dfrac{\partial(\ )}{\partial s} - \dfrac{\partial y}{\partial s}\dfrac{\partial(\ )}{\partial t} \end{bmatrix} \quad (10.2.16)$$

and $[N]$ is the 2×8 shape function matrix given as the first matrix on the right side of Eq. (10.2.7) and $\{d\}$ is the column matrix on the right side of Eq. (10.2.7).

Defining $[B]$ as

$$\begin{array}{ccc} [B] & = & [D'] \qquad [N] \\ (3 \times 8) & & (3 \times 2) \quad (2 \times 8) \end{array} \quad (10.2.17)$$

we have $[B]$ expressed as a function of s and t and thus have the strains in terms of s and t. Here $[B]$ is of order 3×8, as indicated in Eq. (10.2.17).

The explicit form of $[B]$ can be obtained by substituting Eq. (10.2.16) for $[D']$ and Eqs. (10.2.5) for the shape functions into Eq. (10.2.17). The matrix multiplications yield

$$[B(s, t)] = \frac{1}{|[J]|}[[B_1]\ [B_2]\ [B_3]\ [B_4]] \quad (10.2.18)$$

where the submatrices of $[B]$ are given by

$$[B_i] = \begin{bmatrix} a(N_{i,s}) - b(N_{i,t}) & 0 \\ 0 & c(N_{i,t}) - d(N_{i,s}) \\ c(N_{i,t}) - d(N_{i,s}) & a(N_{i,s}) - b(N_{i,t}) \end{bmatrix} \quad (10.2.19)$$

Here i is a dummy variable equal to 1, 2, 3, and 4, and

$$a = \frac{1}{4}[y_1(s-1) + y_2(-1-s) + y_3(1+s) + y_4(1-s)]$$

$$b = \frac{1}{4}[y_1(t-1) + y_2(1-t) + y_3(1+t) + y_4(-1-t)]$$

$$\qquad\qquad (10.2.20)$$

$$c = \frac{1}{4}[x_1(t-1) + x_2(1-t) + x_3(1+t) + x_4(-1-t)]$$

$$d = \frac{1}{4}[x_1(s-1) + x_2(-1-s) + x_3(1+s) + x_4(1-s)]$$

Using the shape functions defined by Eqs. (10.2.5), we have

$$N_{1,s} = \frac{1}{4}(t-1) \qquad N_{1,t} = \frac{1}{4}(s-1) \qquad \text{(and so on)} \quad (10.2.21)$$

where the comma followed by the variable s or t indicates differentiation with respect to that variable; that is, $N_{1,s} \equiv \partial N_1/\partial s$, and so on. The determinant $|[J]|$ is a polynomial in s and t and is tedious to evaluate even for the simplest case of the linear plane quadrilateral element. However, using Eq. (10.2.10) for $[J]$ and Eqs. (10.2.3) for x and y, we can evaluate $||J||$ as

$$||J|| = \frac{1}{8}\{X_c\}^T \begin{bmatrix} 0 & 1-t & t-s & s-1 \\ t-1 & 0 & s+1 & -s-t \\ s-t & -s-1 & 0 & t+1 \\ 1-s & s+t & -t-1 & 0 \end{bmatrix} \{Y_c\} \qquad (10.2.22)$$

where

$$\{X_c\}^T = [x_1 \quad x_2 \quad x_3 \quad x_4] \qquad (10.2.23)$$

and

$$\{Y_c\} = \begin{Bmatrix} y_1 \\ y_2 \\ y_3 \\ y_4 \end{Bmatrix} \qquad (10.2.24)$$

We observe that $||J||$ is a function of s and t and the known global coordinates $x_1, x_2, \ldots, y_4$. Hence, $[B]$ is a function of s and t in both the numerator and the denominator [because of $||J||$ given by Eq. (10.2.22)] and of the known global coordinates x_1 through y_4.

The stress–strain relationship is again $\{\sigma\} = [D][B]\{d\}$, where because the $[B]$ matrix is a function of s and t, so also is the stress matrix $\{\sigma\}$.

Step 4 Derive the Element Stiffness Matrix and Equations

We now want to express the stiffness matrix in terms of s-t coordinates. For an element with a constant thickness h, we have

$$[k] = \iint_A [B]^T[D][B]h\,dx\,dy \qquad (10.2.25)$$

However, $[B]$ is now a function of s and t, as seen by Eqs. (10.2.18) through (10.2.20), and so it must integrate with respect to s and t. Once again, to transform the variables and the region from x and y to s and t, we must have a standard procedure that involves the determinant of $[J]$. This general type of transformation [4, 5] is given by

$$\iint_A f(x,y)\,dx\,dy = \iint_A f(s,t)||J||\,ds\,dt \qquad (10.2.26)$$

where the inclusion of $||J||$ in the integrand on the right side of Eq. (10.2.26) results from a theorem of integral calculus (see Reference [5] for the complete proof of this theorem). We also observe that the Jacobian (the determinant of the Jacobian matrix) relates an element area ($dx\,dy$) in the global coordinate system to an elemental area ($ds\,dt$) in the natural coordinate system. For rectangles and parallelograms, J is the constant value $J = A/4$, where A represents the physical surface area of the element. Using Eq. (10.2.26) in Eq. (10.2.25), we obtain

$$[k] = \int_{-1}^{1} \int_{-1}^{1} [B]^T[D][B]h||J||\,ds\,dt \qquad (10.2.27)$$

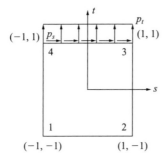

Figure 10–5 Surface traction: p_s and p_t acting at edge $t = 1$

The $\|[J]\|$ and $[B]$ are such as to result in complicated expressions within the integral of Eq. (10.2.27), and so the integration to determine the element stiffness matrix is usually done numerically. A method for numerically integrating Eq. (10.2.27) is given in Section 10.3. The stiffness matrix in Eq. (10.2.27) is of the order 8×8.

Body Forces

The element body-force matrix will now be determined from

$$\underset{(8 \times 1)}{\{f_b\}} = \int_{-1}^{1} \int_{-1}^{1} \underset{(8 \times 2)}{[N]^T} \underset{(2 \times 1)}{\{X\}} \, h\|[J]\| \, ds \, dt \qquad (10.2.28)$$

Like the stiffness matrix, the body-force matrix in Eq. (10.2.28) has to be evaluated by numerical integration.

Surface Forces

The surface-force matrix, say, along edge $t = 1$ (Figure 10–5) with overall length L, is

$$\underset{(4 \times 1)}{\{f_s\}} = \int_{-1}^{1} \underset{(4 \times 2)}{[N_s]^T} \underset{(2 \times 1)}{\{T\}} \, h\frac{L}{2} \, ds \qquad (10.2.29)$$

or

$$\begin{Bmatrix} f_{s3s} \\ f_{s3t} \\ f_{s4s} \\ f_{s4t} \end{Bmatrix} = \int_{-1}^{1} \begin{bmatrix} N_3 & 0 & N_4 & 0 \\ 0 & N_3 & 0 & N_4 \end{bmatrix}^T \Bigg|_{\substack{\text{evaluated} \\ \text{along } t=1}} \begin{Bmatrix} p_s \\ p_t \end{Bmatrix} h\frac{L}{2} \, ds \qquad (10.2.30)$$

because $N_1 = 0$ and $N_2 = 0$ along edge $t = 1$, and hence, no nodal forces exist at nodes 1 and 2. For the case of uniform (constant) p_s and p_t along edge $t = 1$, the total surface-force matrix is

$$\{f_s\} = h\frac{L}{2}[0 \ \ 0 \ \ 0 \ \ 0 \ \ p_s \ \ p_t \ \ p_s \ \ p_t]^T \qquad (10.2.31)$$

Surface forces along other edges can be obtained similar to Eq. (10.2.30) by merely using the proper shape functions associated with the edge where the tractions are applied.

Example 10.1

For the four-noded linear plane quadrilateral element shown in Figure 10–6 with a uniform surface traction along side 2–3, evaluate the force matrix by using the energy equivalent nodal forces obtained from the integral similar to Eq. (10.2.29). Let the thickness of the element be $h = 0.1$ in.

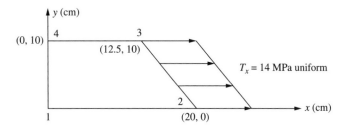

Figure 10–6 Element subjected to uniform surface traction

SOLUTION:

Using Eq. (10.2.29), we have

$$\{f_s\} = \int_{-1}^{1} [N_s]^T \{T\} h \frac{L}{2} dt \tag{10.2.32}$$

With length of side 2–3 given by

$$L = \sqrt{(12.5 - 20)^2 + (10 - 0)^2} = \sqrt{56.25 + 100} = 12.5 \text{ cm} \tag{10.2.33}$$

Shape functions N_2 and N_3 must be used, as we are evaluating the surface traction along side 2–3 (at $s = 1$). Therefore, Eq. (10.2.33) becomes

$$\{f_s\} = \int_{-1}^{1} [N_s]^T \{T\} h \frac{L}{2} dt = \int_{-1}^{1} \begin{bmatrix} N_2 & 0 & N_3 & 0 \\ 0 & N_2 & 0 & N_3 \end{bmatrix}^T \begin{Bmatrix} p_s \\ p_t \end{Bmatrix} h \frac{L}{2} dt \tag{10.2.34}$$

$$\text{evaluated along } s = 1$$

The shape functions for the four-noded linear plane element are taken from Eq. (10.2.5) as

$$N_2 = \frac{(1+s)(1-t)}{4} = \frac{s - t - st + 1}{4} \qquad N_3 = \frac{(1+s)(1+t)}{4} = \frac{s + t + st + 1}{4} \tag{10.2.35}$$

The surface traction matrix is given by

$$\{T\} = \begin{Bmatrix} p_s \\ p_t \end{Bmatrix} = \begin{Bmatrix} 14.0 \\ 0 \end{Bmatrix} \times 10^6 \tag{10.2.36}$$

Substituting Eq. (10.2.33) for L and Eq. (10.2.36) for the surface traction matrix and the thickness $h = 2.5$ mm into Eq. (10.2.32), we obtain

$$\{f_s\} = \int_{-1}^{1} [N_s]^T \{T\} h \frac{L}{2} dt = \int_{-1}^{1} \begin{bmatrix} N_2 & 0 \\ 0 & N_2 \\ N_3 & 0 \\ 0 & N_3 \end{bmatrix} \begin{Bmatrix} 14 \\ 0 \end{Bmatrix} \times 10^6 \times (2.5 \times 10^{-3}) \times \left(\frac{12.5 \times 10^{-2}}{2} \right) dt$$

$$\text{evaluated along } s = 1 \tag{10.2.37}$$

Simplifying Eq. (10.2.37), we obtain

$$\{f_s\} = 156.25 \int_{-1}^{1} \begin{bmatrix} 14N_2 \\ 0 \\ 14N_3 \\ 0 \end{bmatrix} dt = 2.187 \times 10^3 \int_{-1}^{1} \begin{bmatrix} N_2 \\ 0 \\ N_3 \\ 0 \end{bmatrix} dt \tag{10.2.38}$$

$$\text{evaluated along } s = 1$$

Substituting the shape functions from Eq. (10.2.35) into Eq. (10.2.38), we have

$$\{f_s\} = 2.187 \times 10^3 \int_{-1}^{1} \begin{bmatrix} \dfrac{s - t - st + 1}{4} \\ 0 \\ \dfrac{s + t + st + 1}{4} \\ 0 \end{bmatrix} dt \tag{10.2.39}$$

$$\text{evaluated along } s = 1$$

Upon substituting $s = 1$ into the integrand in Eq. (10.2.39) and performing the explicit integration in Eq. (10.2.40), we obtain

$$\{f_s\} = 2.187 \times 10^3 \int_{-1}^{1} \begin{bmatrix} \dfrac{2 - 2t}{4} \\ 0 \\ \dfrac{2t + 2}{4} \\ 0 \end{bmatrix} dt = 2.187 \times 10^3 \begin{bmatrix} 0.50t - \dfrac{t^2}{4} \\ 0 \\ 0.50t + \dfrac{t^2}{4} \\ 0 \end{bmatrix}_{-1}^{1} \tag{10.2.40}$$

Evaluating the resulting integration expression for each limit, we obtain the final expression for the surface traction matrix as

$$\{f_s\} = 2.187 \times 10^3 \begin{bmatrix} 0.50 - 0.25 \\ 0 \\ 0.50 + 0.25 \\ 0 \end{bmatrix} - 2.187 \times 10^3 \begin{bmatrix} -0.50 - 0.25 \\ 0 \\ -0.50 + 0.25 \\ 0 \end{bmatrix} = 2.187 \times 10^3 \begin{bmatrix} 1 \\ 0 \\ 1 \\ 0 \end{bmatrix} \text{N}$$

$$\tag{10.2.41}$$

Or in explicit form the surface tractions at nodes 2 and 3 are

$$\begin{Bmatrix} f_{s2s} \\ f_{s2t} \\ f_{s3s} \\ f_{s3t} \end{Bmatrix} = \begin{bmatrix} 2187 \\ 0 \\ 2187 \\ 0 \end{bmatrix} \text{N} \tag{10.2.42}$$

■

▲ 10.3 Newton-Cotes and Gaussian Quadrature ▲

In this section, we will describe two methods for numerical evaluation of definite integrals, because it has proven most useful for finite element work.

We begin with the simpler more common integration method of Newton-Cotes. The Newton-Cotes methods for one and two intervals of integration are the well-known trapezoid and Simpson's one-third rule, respectively. We will then describe Gauss' method for numerical evaluation of definite integrals. After describing both methods, we will then understand why the Gaussian quadrature method is used in finite element work.

Newton-Cotes Numerical Integration

We first describe the common numerical integration method called the Newton-Cotes method for evaluation of definite integrals. However, the method does not yield as accurate of results as the Gaussian quadrature method and so is not normally used in finite element method evaluations, such as to evaluate the stiffness matrix.

To evaluate the integral

$$I = \int_{-1}^{1} y \, dx$$

we assume the sampling points of $y(x)$ are spaced at equal intervals. Since the limits of integration are from -1 to 1 using the isoparametric formulation, the Newton-Cotes formula is given by

$$I = \int_{-1}^{1} y \, dx = h \sum_{i=0}^{n} C_i y_i = h[C_0 y_0 + C_1 y_1 + C_2 y_2 + C_3 y_3 + \ldots + C_n y_n] \quad (10.3.1)$$

where the C_i are the Newton-Cotes constants for numerical integration with i intervals (the number of intervals will be one less than the number of sampling points, n) and h is the interval between the limits of integration (for limits of integration between -1 and 1 this makes $h = 2$). The Newton-Cotes constants have been published and are summarized in Table 10–1 for $i = 1$ to 6. The case $i = 1$ corresponds to the well-known trapezoid rule illustrated by Figure 10–7. The case $i = 2$ corresponds to the

Table 10–1 Table for Newton-Cotes intervals and points for integration, $\int_{-1}^{1} y(x) \, dx = h \sum_{i=0}^{n} C_i y_i$

Intervals, i	No. of Points, n	C_0	C_1	C_2	C_3	C_4	C_5	C_6
1	2	1/2	1/2			(trapezoid rule)		
2	3	1/6	4/6	1/6		(Simpson's 1/3 rule)		
3	4	1/8	3/8	3/8	1/8	(Simpson's 3/8 rule)		
4	5	7/90	32/90	12/90	32/90	7/90		
5	6	19/288	75/288	50/288	50/288	75/288	19/288	
6	7	41/840	216/840	27/840	272/840	27/840	216/840	41/840

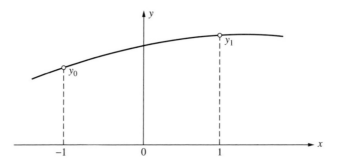

Figure 10–7 Approximation of numerical integration (approximate area under curve) using $i = 1$ interval, $n = 2$ sampling points (trapezoid rule), for

$$I = \int_{-1}^{1} y(x)\, dx = h \sum_{i=0}^{2} C_i y_i$$

well-known Simpson one-third rule. It is shown [9] that the formulas for $i = 3$ and $i = 5$ have the same accuracy as the formulas for $i = 2$ and $i = 4$, respectively. Therefore, it is recommended that the even formulas with $i = 2$ and $i = 4$ be used in practice. To obtain greater accuracy one can then use a smaller interval (include more evaluations of the function to be integrated). This can be accomplished by using a higher-order Newton-Cotes formula, thus increasing the number of intervals i.

It is shown [9] that we need to use n equally spaced sampling points to integrate exactly a polynomial of order at most $n - 1$. On the other hand, using Gaussian quadrature we will show that we use unequally spaced sampling points n and integrate exactly a polynomial of order at most $2n - 1$. For instance, using the Newton-Cotes formula with $n = 2$ sampling points, the highest order polynomial we can integrate exactly is a linear one. However, using Gaussian quadrature, we can integrate a cubic polynomial exactly. Gaussian quadrature is then more accurate with fewer sampling points than Newton-Cotes quadrature. This is because Gaussian quadrature is based on optimizing the position of the sampling points (not making them equally spaced as in the Newton-Cotes method) and also optimizing the weights W_i given in Table 10–2.

Table 10–2 Table for Gauss points for integration from minus one to one,

$$\int_{-1}^{1} y(x)\, dx = \sum_{i=1}^{n} W_i y_i$$

Number of Points	Locations, x_i	Associated Weights, W_i
1	$x_1 = 0.000\ldots$	2.000
2	$x_1, x_2 = \pm 0.57735026918962$	1.000
3	$x_1, x_3 = \pm 0.77459666924148$	$\frac{5}{9} = 0.555\ldots$
	$x_2 = 0.000\ldots$	$\frac{8}{9} = 0.888\ldots$
4	$x_1, x_4 = \pm 0.8611363116$	0.3478548451
	$x_2, x_3 = \pm 0.3399810436$	0.6521451549

After the function is evaluated at the sampling points, the corresponding weights are multiplied by these evaluated functions as is illustrated in Example 10.3.

Example 10.2 is used to illustrate the Newton-Cotes method and compare its accuracy to that of the Gaussian quadrature method subsequently described.

Example 10.2

Using the Newton-Cotes method with $i = 2$ intervals ($n = 3$ sampling points), evaluate the integrals (a) $I = \int_{-1}^{1}[x^2 + \cos(x/2)]dx$ and (b) $I = \int_{-1}^{1}(3^x - x)dx$.

SOLUTION:

Using Table 10–1 with three sampling points means we evaluate the function inside the integrand at $x = -1$, $x = 0$, and $x = 1$, and multiply each evaluated function by the respective Newton-Cotes numbers, 1/6, 4/6, and 1/6. We then add these three products together and finally multiply this sum by the interval of integration ($h = 2$) as follows:

$$I = 2\left[\frac{1}{6}y_0 + \frac{4}{6}y_1 + \frac{1}{6}y_2\right] \tag{10.3.2}$$

(a): Using the integrand in part (a), we obtain

$$y_0 = x^2 + \cos(x/2) \text{ evaluated at } x = -1, \text{ etc. as follows:}$$

$$y_0 = (-1)^2 + \cos(-1/2 \text{ rad}) = 1.8775826$$

$$y_1 = (0)^2 + \cos(0/2) = 1 \tag{10.3.3}$$

$$y_2 = (1)^2 + \cos(1/2 \text{ rad}) = 1.8775826$$

Substituting $y_0 - y_2$ from Eq. (10.3.3) into Eq. (10.3.2), we obtain the evaluation of the integral as

$$I = 2\left[\frac{1}{6}(1.8775826) + \frac{4}{6}(1) + \frac{1}{6}(1.8775826)\right] = 2.585$$

This solution compares exactly to the evaluation performed using Gaussian quadrature subsequently shown in Example 10.3 and to the exact solution. However, for higher-order functions the Gaussian quadrature method yields more accurate results than the Newton-Cotes method as illustrated by part (b) as follows:

(b): Using the integrand in part (b), we obtain

$$y_0 = 3^{(-1)} - (-1) = \frac{4}{3}$$

$$y_1 = 3^0 - 0 = 1$$

$$y_2 = 3^1 - (1) = 2$$

Substituting $y_0 - y_2$ into Eq. (10.3.2) we obtain I as

$$I = 2\left[\frac{1}{6}\left(\frac{4}{3}\right) + \frac{4}{6}(1) + \frac{1}{6}(2)\right] = 2.444$$

The error is $2.444 - 2.427 = 0.017$. This error is larger than that found using Gaussian quadrature (see Example 10.3 (b)). ∎

Gaussian Quadrature

To evaluate the integral

$$I = \int_{-1}^{1} y\,dx \tag{10.3.4}$$

where $y = y(x)$, we might choose (sample or evaluate) y at the midpoint $y(0) = y_1$ and multiply by the length of the interval, as shown in Figure 10–8, to arrive at $I = 2y_1$, a result that is exact if the curve happens to be a straight line. This is an example of what is called **one-point Gaussian quadrature** because only one sampling point was used. Therefore,

$$I = \int_{-1}^{1} y(x)\,dx \cong 2y(0) \tag{10.3.5}$$

which is the familiar midpoint rule. Generalization of the formula [Eq. (10.3.5] leads to

$$I = \int_{-1}^{1} y\,dx = \sum_{i=1}^{n} W_i y_i \tag{10.3.6}$$

That is, to approximate the integral, we evaluate the function at several sampling points n, multiply each value y_i by the appropriate weight W_i, and add the terms. Gauss's method chooses the sampling points so that for a given number of points, the best possible accuracy is obtained. Sampling points are located symmetrically with respect to the center of the interval. Symmetrically paired points are given the same weight W_i. Table 10–2 gives appropriate sampling points and weighting

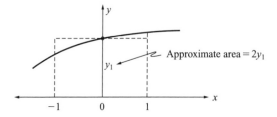

Figure 10–8 Gaussian quadrature using one sampling point

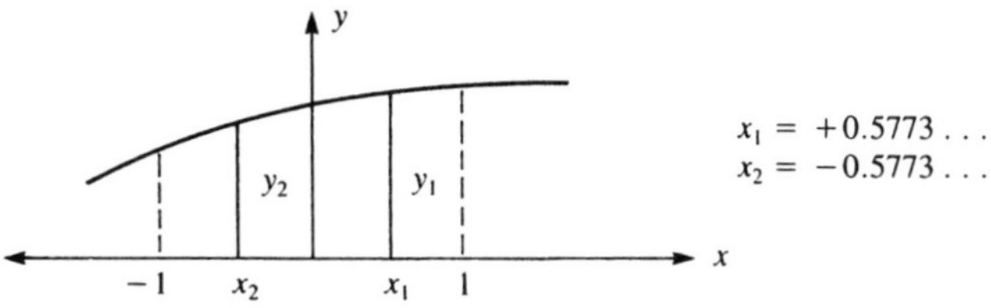

Figure 10-9 Gaussian quadrature using two sampling points

coefficients for the first three orders—that is, one, two, or three sampling points (see Reference [2] for more complete tables). For example, using two points (Figure 10–9), we simply have $I = y_1 + y_2$ because $W_1 = W_2 = 1.000$. This is the exact result if $y = f(x)$ is a polynomial containing terms up to and including x^3. In general, Gaussian quadrature using n points (Gauss points) is exact if the integrand is a polynomial of degree $2n - 1$ or less. In using n points, we effectively replace the given function $y = f(x)$ by a polynomial of degree $2n - 1$. The accuracy of the numerical integration depends on how well the polynomial fits the given curve.

If the function $f(x)$ is not a polynomial, Gaussian quadrature is inexact, but it becomes more accurate as more Gauss points are used. Also, it is important to understand that the ratio of two polynomials is, in general, not a polynomial; therefore, Gaussian quadrature will not yield exact integration of the ratio.

Two-Point Formula

To illustrate the derivation of a two-point ($n = 2$) Gauss formula based on Eq. (10.3.6), we have

$$I = \int_{-1}^{1} y\,dx = W_1 y_1 + W_2 y_2 = W_1 y(x_1) + W_2 y(x_2) \tag{10.3.7}$$

There are four unknown parameters to determine: $W_1, W_2, x_1,$ and x_2. Therefore, we assume a cubic function for y as follows:

$$y = C_0 + C_1 x + C_2 x^2 + C_3 x^3 \tag{10.3.8}$$

In general, with four parameters in the two-point formula, we would expect the Gauss formula to exactly predict the area under the curve. That is,

$$A = \int_{-1}^{1} (C_0 + C_1 x + C_2 x^2 + C_3 x^3)\,dx = 2C_0 + \frac{2C_2}{3} \tag{10.3.9}$$

However, we will assume, based on Gauss's method, that $W_1 = W_2$ and $x_1 = x_2$ as we use two symmetrically located Gauss points at $x = \pm a$ with equal weights. The area predicted by Gauss's formula is

$$A_G = Wy(-a) + Wy(a) = 2W(C_0 + C_2 a^2) \tag{10.3.10}$$

where $y(-a)$ and $y(a)$ are evaluated using Eq. (10.3.8). If the error, $e = A - A_G$, is to vanish for any C_0 and C_2, we must have, using Eqs. (10.3.9) and (10.3.10) in the error expression,

$$\frac{\partial e}{\partial C_0} = 0 = 2 - 2W \quad \text{or} \quad W = 1 \quad (10.3.11)$$

and

$$\frac{\partial e}{\partial C_2} = 0 = \frac{2}{3} - 2a^2 W \quad \text{or} \quad a = \sqrt{\frac{1}{3}} = 0.5773\ldots \quad (10.3.12)$$

Now $W = 1$ and $a = 0.5773\ldots$ are the W_i's and a_i's (x_i's) for the two-point Gaussian quadrature given in Table 10–2.

Example 10.3

Evaluate the integrals (a) $I = \int_{-1}^{1}[x^2 + \cos(x/2)]\,dx$ and (b) $I = \int_{-1}^{1}(3^x - x)dx$ using three-point Gaussian quadrature.

SOLUTION:

(a) Using Table 10–2 for the three Gauss points and weights, we have $x_1 = x_3 = \pm 0.77459\ldots$, $x_2 = 0.000\ldots$, $W_1 = W_3 = \frac{5}{9}$, and $W_2 = \frac{8}{9}$. The integral then becomes

$$I = \left[(-0.77459)^2 + \cos\left(-\frac{0.77459}{2}\text{ rad}\right)\right]\frac{5}{9} + \left[0^2 + \cos\frac{0}{2}\right]\frac{8}{9}$$

$$+ \left[(0.77459)^2 + \cos\left(\frac{0.77459}{2}\text{ rad}\right)\right]\frac{5}{9}$$

$$= 1.918 + 0.667 = 2.585$$

Compared to the exact solution, we have $I_{\text{exact}} = 2.585$.

In this example, three-point Gaussian quadrature yields the exact answer to four significant figures.

(b) Using Table 10–2 for the three Gauss points and weights as in part (a), the integral then becomes

$$I = [3^{(-0.77459)} - (-0.77459)]\frac{5}{9} + [3^0 - 0]\frac{8}{9} + [3^{(0.77459)} - (0.77459)]\frac{5}{9}$$

$$= 0.66755 + 0.88889 + 0.86065 = 2.4229(2.423 \text{ to four significant figures})$$

Compared to the exact solution, we have $I_{\text{exact}} = 2.427$. The error is $2.427 - 2.423 = 0.004$. ■

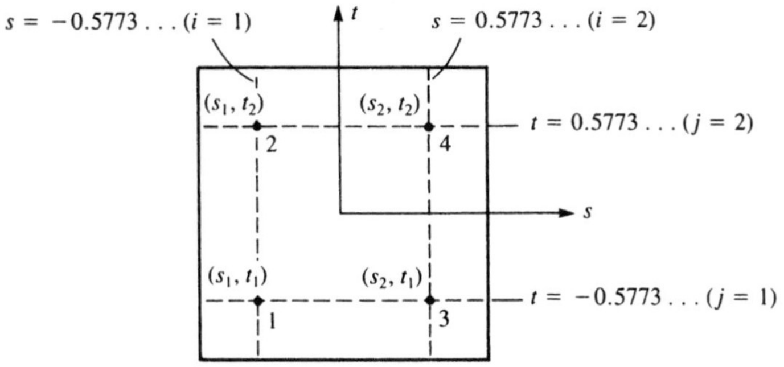

Figure 10–10 Four-point Gaussian quadrature in two dimensions

In two dimensions, we obtain the quadrature formula by integrating first with respect to one coordinate and then with respect to the other as

$$I = \int_{-1}^{1} \int_{-1}^{1} f(s, t) \, ds \, dt = \int_{-1}^{1} \left[\sum_i W_i f(s_i, t) \right] dt$$

$$= \sum_j W_j \left[\sum_i W_i f(s_i, t_j) \right] = \sum_i \sum_j W_i W_j f(s_i, t_j) \qquad (10.3.13)$$

In Eq. (10.3.13), we need not use the same number of Gauss points in each direction (that is, i does not have to equal j), but this is usually done. Thus, for example, a four-point Gauss rule (often described as a 2×2 rule) is shown in Figure 10–10. Equation (10.3.13) with $i = 1, 2$ and $j = 1, 2$ yields

$$I = W_1 W_1 f(s_1, t_1) + W_1 W_2 f(s_1, t_2) + W_2 W_1 f(s_2, t_1) + W_2 W_2 f(s_2, t_2) \quad (10.3.14)$$

where the four sampling points are at s_i, $t_i = \pm 0.5773 \ldots = \pm 1/\sqrt{3}$, and the weights are all 1.000. Hence, the double summation in Eq. (10.3.13) can really be interpreted as a single summation over the four points for the rectangle.

In general, in three dimensions, we have

$$I = \int_{-1}^{1} \int_{-1}^{1} \int_{-1}^{1} f(s, t, z) \, ds \, dt \, dz = \sum_i \sum_j \sum_k W_i W_j W_k f(s_i, t_j, z_k) \qquad (10.3.15)$$

▲ 10.4 Evaluation of the Stiffness Matrix and Stress Matrix by Gaussian Quadrature

Evaluation of the Stiffness Matrix

For the two-dimensional element, we have shown in previous chapters that

$$[k] = \iint_A [B(x, y)]^T [D][B(x, y)] h \, dx \, dy \qquad (10.4.1)$$

where, in general, the integrand is a function of x and y and nodal coordinate values.

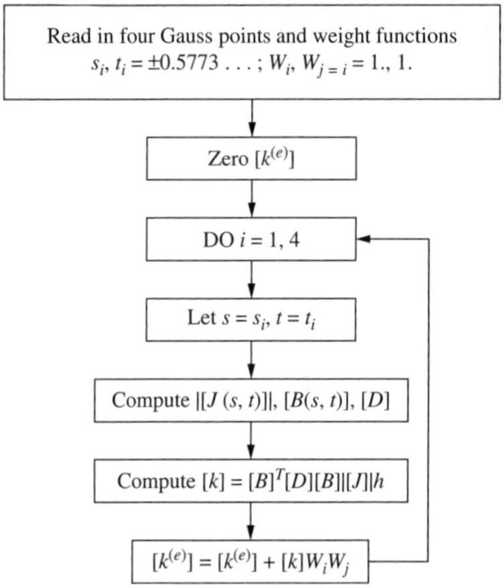

Figure 10–11 Flowchart to evaluate $[k^{(e)}]$ by four-point Gaussian quadrature

We have shown in Section 10.2 that $[k]$ for a quadrilateral element can be evaluated in terms of a local set of coordinates s-t, with limits from minus one to one within the element, and in terms of global nodal coordinates as given by Eq. (10.2.27). We repeat Eq. (10.2.27) here for convenience as

$$[k] = \int_{-1}^{1}\int_{-1}^{1} [B(s,t)]^T[D][B(s,t)]\,\|[J]\|h\,ds\,dt \qquad (10.4.2)$$

where $\|[J]\|$ is defined by Eq. (10.2.22) and $[B]$ is defined by Eq. (10.2.18). In Eq. (10.4.2), each coefficient of the integrand $[B]^T[D][B]\|[J]\|$ must be evaluated by numerical integration in the same manner as $f(s,t)$ was integrated in Eq. (10.3.13).

A flowchart to evaluate $[k]$ of Eq. (10.4.2) for an element using four-point Gaussian quadrature is given in Figure 10–11. The four-point Gaussian quadrature rule is relatively easy to use. Also, it has been shown to yield good results [7]. In Figure 10–11, in explicit form for four-point Gaussian quadrature (now using the single summation notation with $i = 1, 2, 3, 4$), we have

$$[k] = [B(s_1, t_1)]^T[D][B(s_1, t_1)]\|[J(s_1, t_1)]\|hW_1W_1$$

$$+ [B(s_2, t_2)]^T[D][B(s_2, t_2)]\|[J(s_2, t_2)]\|hW_2W_2$$

$$+ [B(s_3, t_3)]^T[D][B(s_3, t_3)]\|[J(s_3, t_3)]\|hW_3W_3$$

$$+ [B(s_4, t_4)]^T[D][B(s_4, t_4)]\|[J(s_4, t_4)]\|hW_4W_4 \qquad (10.4.3)$$

where $s_1 = t_1 = -0.5773$, $s_2 = -0.5773$, $t_2 = 0.5773$, $s_3 = 0.5773$, $t_3 = -0.5773$, and $s_4 = t_4 = 0.5773$ as shown in Figure 10–10, and $W_1 = W_2 = W_3 = W_4 = 1.000$.

Example 10.4

Evaluate the stiffness matrix for the quadrilateral element shown in Figure 10–12 using the four-point Gaussian quadrature rule. Let $E = 30 \times 10^6$ psi and $v = 0.25$. The global coordinates are shown in inches. Assume $h = 1$ in.

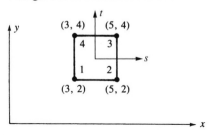

Figure 10–12 Quadrilateral element for stiffness evaluation

SOLUTION:

Using Eq. (10.4.3), we evaluate the $[k]$ matrix. Using the four-point rule, the four points are (also see Figure 10–10).

$$(s_1, t_1) = (-0.5773, -0.5773)$$
$$(s_2, t_2) = (-0.5773, 0.5773)$$
$$(s_3, t_3) = (0.5773, -0.5773)$$
$$(s_4, t_4) = (0.5773, 0.5773)$$

(10.4.4a)

with weights $W_1 = W_2 = W_3 = W_4 = 1.000$.

Therefore, by Eq. (10.4.3), we have

$$[k] = [B(-0.5773, -0.5773)]^T[D][B(-0.5773, -0.5773)]$$
$$\times ||[J(-0.5773, -0.5773)]|(1)(1.000)(1.000)$$
$$+ [B(-0.5773, 0.5773)]^T[D][B(-0.5773, 0.5773)]$$
$$\times ||[J(-0.5773, 0.5773)]|(1)(1.000)(1.000)$$
$$+ [B(0.5773, -0.5773)]^T[D][B(0.5773, -0.5773)]$$
$$\times ||[J(0.5773, -0.5773)]|(1)(1.000)(1.000)$$
$$+ [B(0.5773, 0.5773)]^T[D][B(0.5773, 0.5773)]$$
$$\times ||[J(0.5773, 0.5773)]|(1)(1.000)(1.000)$$

(10.4.4b)

To evaluate $[k]$, we first evaluate $||[J]||$ at each Gauss point by using Eq. (10.2.22). For instance, one part of $||[J]||$ is given by

$$||[J(-0.5773, -0.5773)]|| = \frac{1}{8}[3 \quad 5 \quad 5 \quad 3]$$

$$\times \begin{bmatrix} 0 & 1-(-0.5773) & -0.5773-(-0.5773) & -0.5773-1 \\ -0.5773-1 & 0 & -0.5773+1 & -0.5773-(-0.5773) \\ -0.5773-(-0.5773) & -(-0.5773)-1 & 0 & -0.5773+1 \\ 1-(-0.5773) & -0.5773+(-0.5773) & -0.5773-1 & 0 \end{bmatrix}$$

$$\times \begin{Bmatrix} 2 \\ 2 \\ 4 \\ 4 \end{Bmatrix} = 1.000$$

(10.4.4c)

Similarly,

$$||[J(-0.5773, 0.5773)]|| = 1.000$$

$$||[J(0.5773, -0.5773)]|| = 1.000 \qquad (10.4.4d)$$

$$||[J(0.5773, 0.5773)]|| = 1.000$$

Even though $||[J]|| = 1$ in this example, in general, $||[J]|| \neq 1$ and varies in space.

Then, using Eqs. (10.2.18) and (10.2.19), we evaluate $[B]$. For instance, one part of $[B]$ is

$$[B(-0.5773, -0.5773)] = \frac{1}{||[J(-0.5773, -0.5773)]||} [[B_1] \ [B_2] \ [B_3] \ [B_4]]$$

where, by Eq. (10.2.19),

$$[B_1] = \begin{bmatrix} aN_{1,s} - bN_{1,t} & 0 \\ 0 & cN_{1,t} - dN_{1,s} \\ cN_{1,t} - dN_{1,s} & aN_{1,s} - bN_{1,t} \end{bmatrix} \qquad (10.4.4e)$$

and by Eqs. (10.2.20) and (10.2.21), $a, b, c, d, N_{1,s}$, and $N_{1,t}$ are evaluated. For instance,

$$a = \frac{1}{4}[y_1(s-1) + y_2(-1-s) + y_3(1+s) + y_4(1-s)]$$

$$= \frac{1}{4}\{2(-0.5773 - 1) + 2[-1 - (-0.5773)]\} + 4[1 + (-0.5773)] + 4[1 - (-0.5773)]$$

$$= 1.00 \qquad (10.4.4f)$$

with similar computations used to obtain b, c, and d. Also,

$$N_{1,s} = \frac{1}{4}(t-1) = \frac{1}{4}(-0.5773 - 1) = -0.3943$$

$$\qquad (10.4.4g)$$

$$N_{1,t} = \frac{1}{4}(s-1) = \frac{1}{4}(-0.5773 - 1) = -0.3943$$

Similarly, $[B_2]$, $[B_3]$, and $[B_4]$ must be evaluated like $[B_1]$, at $(-0.5773, -0.5773)$. We then repeat the calculations to evaluate $[B]$ at the other Gauss points [Eq. (10.4.4a)].

Using a computer program written specifically to evaluate $[B]$ at each Gauss point and then $[k]$, we obtain the final form of $[B(-0.5773, -0.5773)]$ as

$$[B(-0.5773, -0.5773)] =$$

$$\begin{bmatrix} -0.1057 & 0 & 0.1057 & 0 & 0 & -0.1057 & 0 & -0.3943 \\ -0.1057 & -0.1057 & -0.3943 & 0.1057 & 0.3943 & 0 & -0.3943 & 0 \\ 0 & 0.3943 & 0 & 0.1057 & 0.3943 & 0.3943 & 0.1057 & -0.3943 \end{bmatrix}$$

$$\qquad (10.4.4h)$$

with similar expressions for $[B(-0.5773, 0.5773)]$, and so on.

From Eq. (6.1.8), the matrix $[D]$ is

$$[D] = \frac{E}{1 - v^2} \begin{bmatrix} 1 & v & 0 \\ v & 1 & 0 \\ 0 & 0 & \dfrac{1-v}{2} \end{bmatrix} = \begin{bmatrix} 32 & 8 & 0 \\ 8 & 32 & 0 \\ 0 & 0 & 12 \end{bmatrix} \times 10^6 \text{ psi} \qquad (10.4.4\text{i})$$

Finally, using Eq. (10.4.4b), the matrix $[k]$ becomes

$$[k] = 10^4 \begin{bmatrix} 1466 & 500 & -866 & -99 & -733 & -500 & 133 & 99 \\ 500 & 1466 & 99 & 133 & -500 & -733 & -99 & -866 \\ -866 & 99 & 1466 & -500 & 133 & -99 & -733 & 500 \\ -99 & 133 & -500 & 1466 & 99 & -866 & 500 & -733 \\ -733 & -500 & 133 & 99 & 1466 & 500 & -866 & -99 \\ -500 & -733 & -99 & -866 & 500 & 1466 & 99 & 133 \\ 133 & -99 & -733 & 500 & -866 & 99 & 1466 & -500 \\ 99 & -866 & 500 & -733 & -99 & 133 & -500 & 1466 \end{bmatrix}$$

$$(10.4.4\text{j}) \quad \blacksquare$$

Evaluation of Element Stresses

The stresses $\{\sigma\} = [D][B]\{d\}$ are not constant within the quadrilateral element. Because $[B]$ is a function of s and t coordinates, $\{\sigma\}$ is also a function of s and t. In practice, the stresses are evaluated at the same Gauss points used to evaluate the stiffness matrix $[k]$. For a quadrilateral using 2×2 integration, we get four sets of stress data. To reduce the data, it is often practical to evaluate $\{\sigma\}$ at $s = 0$, $t = 0$ instead. Another method mentioned in Section 7.4 is to evaluate the stresses in all elements at a shared (common) node and then use an average of these element nodal stresses to represent the stress at the node. Most computer programs use this method. Stress plots obtained in these programs are based on this average nodal stress method. Example 10.5 illustrates the use of Gaussian quadrature to evaluate the stress matrix at the $s = 0$, $t = 0$ location of the element.

Example 10.5

For the rectangular element shown in Figure 10–12 of Example 10.4, assume plane stress conditions with $E = 200$ GPa, $v = 0.3$, and displacements $u_1 = 0$, $v_1 = 0$, $u_2 = 0.02$ mm, $v_2 = 0.03$ mm, $u_3 = 0.06$ mm, $v_3 = 0.032$ mm, $u_4 = 0$, and $v_4 = 0$. Evaluate the stresses, σ_x, σ_y, and τ_{xy} at $s = 0$, $t = 0$.

SOLUTION:

Using Eqs. (10.2.18) through (10.2.20), we evaluate $[B]$ at $s = 0$, $t = 0$.

$$[B] = \frac{1}{|[J]|}[[B_1] \ [B_2] \ [B_3] \ [B_4]]$$ (10.2.18) (repeated)

$$[B(0,0)] = \frac{1}{|[J(0,0)]|}[B_1(0,0)] \ [B_2(0,0)] \ [B_3(0,0)] \ [B_4(0,0)]$$

By Eq. (10.2.22), $|[J]|$ is

$$|[J(0,0)]| = \frac{1}{8}[3 \quad 5 \quad 5 \quad 3]\begin{bmatrix} 0 & 1 & 0 & -1 \\ -1 & 0 & 1 & 0 \\ 0 & -1 & 0 & 1 \\ 1 & 0 & -1 & 0 \end{bmatrix}\begin{Bmatrix} 2 \\ 2 \\ 4 \\ 4 \end{Bmatrix}$$

$$= \frac{1}{8}[-2 \quad -2 \quad 2 \quad 2]\begin{Bmatrix} 2 \\ 2 \\ 4 \\ 4 \end{Bmatrix}$$

$$|[J(0,0)]| = 1$$ (10.4.5a)

Notice that again $|[J]| = 1$ is equal to $A/4$ where $A = 2 \times 2 = 4 \text{ in}^2$ is the physical surface area for the rectangle in Figure 10–12.

By Eq. (10.2.19), we have

$$[B_i] = \begin{bmatrix} aN_{i,s} - bN_{i,t} & 0 \\ 0 & cN_{i,t} - dN_{i,s} \\ cN_{i,t} - dN_{i,s} & aN_{i,s} - bN_{i,t} \end{bmatrix}$$ (10.4.5b)

By Eq. (10.2.20), we obtain

$$a = 1 \qquad b = 0 \qquad c = 1 \qquad d = 0$$

Differentiating the shape functions in Eq. (10.2.5) with respect to s and t and then evaluating at $s = 0$, $t = 0$, we obtain

$$N_{1,s} = -\tfrac{1}{4} \qquad N_{1,t} = -\tfrac{1}{4} \qquad N_{2,s} = \tfrac{1}{4} \qquad N_{2,t} = -\tfrac{1}{4}$$
$$N_{3,s} = \tfrac{1}{4} \qquad N_{3,t} = \tfrac{1}{4} \qquad N_{4,s} = -\tfrac{1}{4} \qquad N_{4,t} = \tfrac{1}{4}$$ (10.4.5c)

Therefore, substituting Eqs. (10.4.5c) into Eq. (10.4.5b), we obtain

$$[B_1] = \begin{bmatrix} -\tfrac{1}{4} & 0 \\ 0 & -\tfrac{1}{4} \\ -\tfrac{1}{4} & -\tfrac{1}{4} \end{bmatrix} \quad [B_2] = \begin{bmatrix} \tfrac{1}{4} & 0 \\ 0 & -\tfrac{1}{4} \\ -\tfrac{1}{4} & \tfrac{1}{4} \end{bmatrix} \quad [B_3] = \begin{bmatrix} \tfrac{1}{4} & 0 \\ 0 & \tfrac{1}{4} \\ \tfrac{1}{4} & \tfrac{1}{4} \end{bmatrix} \quad [B_4] = \begin{bmatrix} -\tfrac{1}{4} & 0 \\ 0 & \tfrac{1}{4} \\ \tfrac{1}{4} & -\tfrac{1}{4} \end{bmatrix}$$ (10.4.5d)

The element stress matrix $\{\sigma\}$ is then obtained by substituting Eqs. (10.4.5a) for $|[J]| = 1$ and (10.4.5d) into Eq. (10.2.18) for $[B]$ and the plane stress $[D]$ matrix from Eq. (6.1.8) into the definition for $\{\sigma\}$ as

$$\{\sigma\} = [D][B]\{d\} = (30)\frac{10^6\begin{bmatrix} 1 & 0.3 & 0 \\ 0.3 & 1 & 0 \\ 0 & 0 & 0.35 \end{bmatrix}}{1-0.09}$$

$$\times \begin{bmatrix} -0.25 & 0 & 0.25 & 0 & 0.25 & 0 & -0.25 & 0 \\ 0 & -0.25 & 0 & -0.25 & 0 & 0.25 & 0 & 0.25 \\ -0.25 & -0.25 & -0.25 & 0.25 & 0.25 & 0.25 & 0.25 & -0.25 \end{bmatrix} \begin{Bmatrix} 0 \\ 0 \\ 0.02 \\ 0.03 \\ 0.06 \\ 0.032 \\ 0 \\ 0 \end{Bmatrix} \times 10^{-3}$$

$$\{\sigma\} = \begin{Bmatrix} 4.40 \\ 1.429 \\ 1.978 \end{Bmatrix} \text{MPa}$$ ∎

▲ 10.5 Higher-Order Shape Functions ▲

In general, higher-order element shape functions can be developed by adding additional nodes to the sides of the linear element. These elements result in higher-order strain variations within each element, and convergence to the exact solution thus occurs at a faster rate using fewer elements. (However, a trade-off exists because a more complicated element takes up so much computation time that even with few elements in the model, the computation time can become larger than for the simple linear element model.) Another advantage of the use of higher-order elements is that curved boundaries of irregularly shaped bodies can be approximated more closely than by the use of simple straight-sided linear elements.

Linear Strain Bar

To illustrate the concept of higher-order elements, we will begin with the three-noded linear strain quadratic displacement (and quadratic shape functions) shown in Figure 10–13. Figure 10–13 shows a quadratic isoparametric bar element (also called a linear strain bar) with three coordinates of nodes, x_1, x_2, and x_3, in the global coordinates.

Example 10.6

For the three-noded linear strain bar isoparametric element shown in Figure 10–13, determine (a) the shape functions, N_1, N_2, and N_3, and (b) the strain–displacement matrix [B]. Assume the general axial displacement function to be a quadratic taken as $u = a_1 + a_2 s + a_3 s^2$.

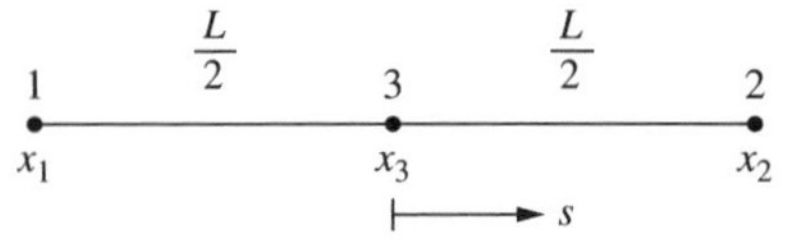

Figure 10–13 Three-noded linear strain bar element

SOLUTION:

(a) As we are formulating shape functions for an isoparametric element, we assume the following axial coordinate function for x as

$$x = a_1 + a_2 s + a_3 s^2 \qquad (10.5.1)$$

Evaluating the a_i's in terms of the nodal coordinates, we obtain

$$x(-1) = a_1 - a_2 + a_3 = x_1 \qquad \text{or} \qquad x_1 = a_1 - a_2 + a_3$$
$$x(0) = a_1 = x_3 \qquad \text{or} \qquad x_3 = a_1$$
$$x(1) = a_1 + a_2 + a_3 = x_2 \qquad \text{or} \qquad x_2 = a_1 - a_2 + a_3 \qquad (10.5.2)$$

Substituting $a_1 = x_3$ from the second of Eqs. (10.5.2), into the first and third of Eqs. (10.5.2), we obtain a_2 and a_3 as follows:

$$x_1 = x_3 - a_2 + a_3 \qquad (10.5.3)$$
$$x_2 = x_3 + a_2 + a_3$$

Adding Eqs. (10.5.3) together and solving for a_3 gives the following:

$$a_3 = (x_1 + x_2 - 2x_3)/2 \qquad (10.5.4)$$
$$x_1 = x_3 - a_2 + ((x_1 + x_2 - 2x_3)/2)$$
$$a_2 = x_3 - x_1 + ((x_1 + x_2 - 2x_3)/2) = (x_2 - x_1)/2 \qquad (10.5.5)$$

Substituting the values for a_1, a_2, and a_3 from Eqs. (10.5.2), (10.5.4), and (10.5.5) into the general equation for x given by Eq. (10.5.1), we obtain

$$x = a_1 + a_2 s + a_3 s^2 = x_3 + \frac{x_2 - x_1}{2} s + \frac{x_1 + x_2 - 2x_3}{2} s^2 \qquad (10.5.6)$$

Combining like terms in x_1, x_2, and x_3, from Eq. (10.5.6), we obtain the final form of x as:

$$x = \left(\frac{s(s-1)}{2}\right) x_1 + \frac{s(s+1)}{2} x_2 + (1 - s^2) x_3 \qquad (10.5.7)$$

Recall that the function x can be expressed in terms of the shape function matrix and the axial coordinates, we have from Eq. (10.5.7)

$$\{x\} = [N_1 \quad N_2 \quad N_3] \begin{Bmatrix} x_1 \\ x_2 \\ x_3 \end{Bmatrix} = \left[\left(\frac{s(s-1)}{2} \right) \quad \frac{s(s+1)}{2} \quad (1-s^2) \right] \begin{Bmatrix} x_1 \\ x_2 \\ x_3 \end{Bmatrix} \quad (10.5.8)$$

Therefore the shape functions are

$$N_1 = \frac{s(s-1)}{2} \qquad N_2 = \frac{s(s+1)}{2} \qquad N_3 = (1-s^2) \qquad (10.5.9)$$

(b) We now determine the strain–displacement matrix $[B]$ as follows:
From our basic definition of axial strain we have

$$\{\varepsilon_x\} = \frac{du}{dx} = \frac{du}{ds}\frac{ds}{dx} = [B] \begin{Bmatrix} u_1 \\ u_2 \\ u_3 \end{Bmatrix} \quad (10.5.10)$$

Using an isoparametric formulation means the displacement function is of the same form as the axial coordinate function. Therefore, using Eq. (10.5.6), we have

$$u = u_3 + \frac{u_2}{2}s - \frac{u_1}{2}s + \frac{u_1}{2}s^2 + \frac{u_2}{2}s^2 - \frac{2u_3}{2}s^2 \quad (10.5.11)$$

Differentiating u with respect to s, we obtain

$$\frac{du}{ds} = \frac{u_2}{2} - \frac{u_1}{2} + u_1 s + u_2 s - 2u_3 s = \left(s - \frac{1}{2} \right) u_1 + \left(s + \frac{1}{2} \right) u_2 + (-2s)u_3 \quad (10.5.12)$$

We have previously proven that $dx/ds = L/2 = ||[J]||$ (see Eq. (10.1.9b). This relationship holds for the higher-order one-dimensional elements as well as for the two-noded constant strain bar element as long as node 3 is at the geometry center of the bar. Using this relationship and Eq. (10.5.12) in Eq. (10.5.10), we obtain

$$\frac{du}{dx} = \frac{du}{ds}\frac{ds}{dx} = \left(\frac{2}{L} \right) \left(\left(s - \frac{1}{2} \right) u_1 + \left(s + \frac{1}{2} \right) u_2 + (-2s)u_3 \right)$$

$$= \left(\frac{2s-1}{L} \right) u_1 + \left(\frac{2s+1}{L} \right) u_2 + \left(\frac{-4s}{L} \right) u_3 \quad (10.5.13)$$

In matrix form, Eq. (10.5.13) becomes

$$\frac{du}{dx} = \left[\frac{2s-1}{L} \quad \frac{2s+1}{L} \quad \frac{-4s}{L} \right] \begin{Bmatrix} u_1 \\ u_2 \\ u_3 \end{Bmatrix} \quad (10.5.14)$$

As Eq. (10.5.14) represents the axial strain, we have

$$\{\varepsilon_x\} = \frac{du}{dx} = \begin{bmatrix} \dfrac{2s-1}{L} & \dfrac{2s+1}{L} & \dfrac{-4s}{L} \end{bmatrix} \begin{Bmatrix} u_1 \\ u_2 \\ u_3 \end{Bmatrix} = [B] \begin{Bmatrix} u_1 \\ u_2 \\ u_3 \end{Bmatrix} \qquad (10.5.15)$$

Therefore the gradient matrix $[B]$ is given by

$$[B] = \begin{bmatrix} \dfrac{2s-1}{L} & \dfrac{2s+1}{L} & \dfrac{-4s}{L} \end{bmatrix} \qquad (10.5.16)$$

∎

Example 10.7

For the three-noded bar element shown previously in Figure 10–13, evaluate the stiffness matrix analytically. Use the $[B]$ from Example 10.6.

SOLUTION:

From Example 10.6, Eq. (10.5.16), we have

$$[B] = \begin{bmatrix} \dfrac{2s-1}{L} & \dfrac{2s+1}{L} & \dfrac{-4s}{L} \end{bmatrix}, \quad [J] = \frac{L}{2} \quad \text{(see Eq. (10.1.9b))} \qquad (10.5.17)$$

Substituting the expression for $[B]$ into Eq. (10.1.15) for the stiffness matrix, we obtain

$$[k] = \frac{L}{2} \int_{-1}^{1} [B]^T E[B] A \, ds = \frac{AEL}{2} \int_{-1}^{1} \begin{bmatrix} \dfrac{(2s-1)^2}{L^2} & \dfrac{(2s-1)(2s+1)}{L^2} & \dfrac{(2s-1)(-4s)}{L^2} \\[2mm] \dfrac{(2s+1)(2s-1)}{L^2} & \dfrac{(2s+1)^2}{L^2} & \dfrac{(2s+1)(-4s)}{L^2} \\[2mm] \dfrac{(-4s)(2s-1)}{L^2} & \dfrac{(-4s)(2s+1)}{L^2} & \dfrac{(-4s)^2}{L^2} \end{bmatrix} ds$$

$$(10.5.18)$$

Simplifying the terms in Eq. (10.5.18) for easier integration, we have

$$[k] = \frac{AE}{2L} \int_{-1}^{1} \begin{bmatrix} 4s^2 - 4s + 1 & 4s^2 - 1 & -8s^2 + 4s \\ 4s^2 - 1 & 4s^2 + 4s + 1 & -8s^2 - 4s \\ -8s^2 + 4s & -8s^2 - 4s & 16s^2 \end{bmatrix} ds \qquad (10.5.19)$$

Upon explicit integration of Eq. (10.5.19), we obtain

$$[k] = \frac{AE}{2L} \begin{bmatrix} \dfrac{4}{3}s^3 - 2s^2 + s & \dfrac{4}{3}s^3 - s & -\dfrac{8}{3}s^3 + 2s^2 \\[2mm] \dfrac{4}{3}s^3 - s & \dfrac{4}{3}s^3 + 2s^2 + s & -\dfrac{8}{3}s^3 - 2s^2 \\[2mm] -\dfrac{8}{3}s^3 + 2s^2 & -\dfrac{8}{3}s^3 - 2s^2 & \dfrac{16}{3}s^3 \end{bmatrix} \Bigg|_{-1}^{1} \qquad (10.5.20)$$

Evaluating Eq. (10.5.20) at the limits 1 and −1, we have

$$[k] = \frac{AE}{2L} \left(\begin{bmatrix} \frac{4}{3} - 2 + 1 & \frac{4}{3} - 1 & -\frac{8}{3} + 2 \\ \frac{4}{3} - 1 & \frac{4}{3} + 2 + 1 & -\frac{8}{3} - 2 \\ -\frac{8}{3} + 2 & -\frac{8}{3} - 2 & \frac{16}{3} \end{bmatrix} - \begin{bmatrix} -\frac{4}{3} - 2 - 1 & -\frac{4}{3} + 1 & \frac{8}{3} + 2 \\ -\frac{4}{3} + 1 & -\frac{4}{3} + 2 - 1 & \frac{8}{3} - 2 \\ \frac{8}{3} + 2 & \frac{8}{3} - 2 & -\frac{16}{3} \end{bmatrix} \right)$$

(10.5.21)

Simplifying Eq. (10.5.21), we obtain the final stiffness matrix as

$$[k] = \frac{AE}{2L} \begin{bmatrix} 4.67 & 0.667 & -5.33 \\ 0.667 & 4.67 & -5.33 \\ -5.33 & -5.33 & 10.67 \end{bmatrix}$$

(10.5.22)

■

Example 10.8

We now illustrate how to evaluate the stiffness matrix for the three-noded bar element shown in Figure 10–14 by using two-point Gaussian quadrature. We can then compare this result to that obtained by the explicit integration performed in Example 10.7.

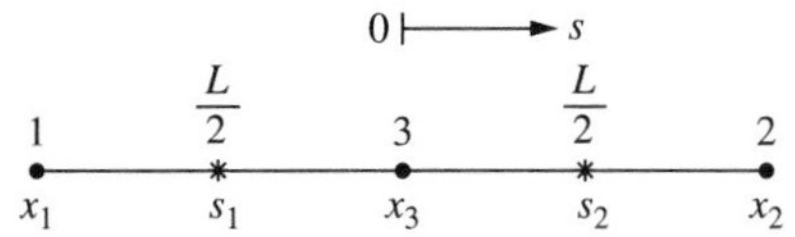

Figure 10–14 Three-noded bar with two Gauss points

SOLUTION:

Starting with Eq. (10.5.18), we have for the stiffness matrix

$$[k] = \frac{L}{2} \int_{-1}^{1} [B]^T E [B] A \, ds = \frac{AEL}{2} \int_{-1}^{1} \begin{bmatrix} \frac{(2s-1)^2}{L^2} & \frac{(2s-1)(2s+1)}{L^2} & \frac{(2s-1)(-4s)}{L^2} \\ \frac{(2s+1)(2s-1)}{L^2} & \frac{(2s+1)^2}{L^2} & \frac{(2s+1)(-4s)}{L^2} \\ \frac{(-4s)(2s-1)}{L^2} & \frac{(-4s)(2s+1)}{L^2} & \frac{(-4s)^2}{L^2} \end{bmatrix} ds$$

(10.5.23)

Using two-point Gaussian quadrature, we evaluate the stiffness matrix at the two points shown in Figure 10–14 (also based on Table 10–2):

$$s_1 = -0.57735, \quad s_2 = 0.57735$$

(10.5.24)

with weights given by

$$W_1 = 1, \quad W_2 = 1 \tag{10.5.25}$$

We then evaluate each term in the integrand of Eq. (10.5.23) at each Gauss point and multiply each term by its weight (here each weight is 1). We then add those Gauss point evaluations together to obtain the final term for each element of the stiffness matrix. For two-point evaluation, there will be two terms added together to obtain each element of the stiffness matrix. We proceed to evaluate the stiffness matrix term by term as follows:

The one–one element:

$$\sum_{i=1}^{2} W_i(2s_i - 1)^2 = (1)[2(-0.57735) - 1]^2 + (1)[2(0.57735) - 1]^2 = 4.6667$$

The one–two element:

$$\sum_{i=1}^{2} W_i(2s_i - 1)(2s_i + 1) = (1)[(2)(-0.57735) - 1][(2)(-0.57735) + 1]$$
$$+ (1)[(2)(0.57735) - 1][(2)(0.57735) + 1] = 0.6667$$

The one–three element:

$$\sum_{i=1}^{2} W_i(-4s_i(2s_i - 1)) = (1)(-4)(-0.57735)[(2)(-0.57735) - 1]$$
$$+ (1)(-4)(0.57735)[(2)(0.57735) - 1] = -5.3333$$

The two–two element:

$$\sum_{i=1}^{2} W_i(2s_i + 1)^2 = (1)[(2)(-0.57735) + 1]^2 + (1)[(2)(0.57735) + 1]^2 = 4.6667$$

The two–three element:

$$\sum_{i=1}^{2} W_i[-4s_i(2s_i + 1)] = (1)(-4)(-0.57735)[(2)(-0.57735) + 1]$$
$$+ (1)(-4)(0.57735)[(2)(0.57735) + 1] = -5.3333$$

The three–three element:

$$\sum_{i=1}^{2} W_i(16s_i^2) = (1)(16)(-0.57735)^2 + (1)(16)(0.57735)^2 = 10.6667$$

By symmetry, the two–one element equals the one–two element, etc. Therefore, from the evaluations of the terms above, the final stiffness matrix is

$$[k] = \frac{AE}{2L} \begin{bmatrix} 4.67 & 0.667 & -5.33 \\ 0.667 & 4.67 & -5.33 \\ -5.33 & -5.33 & 10.67 \end{bmatrix} \tag{10.5.26}$$

Equation (10.5.26) is identical to Eq. (10.5.22) obtained analytically by direct explicit integration of each term in the stiffness matrix. ∎

To further illustrate the concept of higher-order elements, we will consider the quadratic and cubic element shape functions as described in Reference [3].

Quadratic Rectangle (Q8 and Q9)

Figure 10–15 shows a quadratic isoparametric element with four corner nodes and four additional midside nodes. This eight-noded element is often called a "Q8" element.

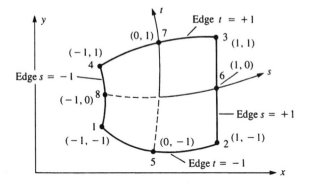

Figure 10–15 Quadratic (Q8) isoparametric element

The shape functions of the quadratic element are based on the incomplete cubic polynomial such that coordinates x and y are

$$x = a_1 + a_2s + a_3t + a_4st + a_5s^2 + a_6t^2 + a_7s^2t + a_8st^2$$
$$y = a_9 + a_{10}s + a_{11}t + a_{12}st + a_{13}s^2 + a_{14}t^2 + a_{15}s^2t + a_{16}st^2 \tag{10.5.27}$$

These functions have been chosen so that the number of generalized degrees of freedom (2 per node times 8 nodes equals 16) are identical to the total number of a_i's. The literature also refers to this eight-noded element as a "serendipity" element as it is based on an incomplete cubic, but it yields good results in such cases as beam bending. We are also reminded that because we are considering an isoparametric formulation, displacements u and v are of identical form as x and y, respectively, in Eq. (10.5.27).

To describe the shape functions, two forms are required—one for corner nodes and one for midside nodes, as given in Reference [3]. For the corner nodes ($i = 1, 2, 3, 4$),

$$N_1 = \frac{1}{4}(1-s)(1-t)(-s-t-1)$$

$$N_2 = \frac{1}{4}(1+s)(1-t)(s-t-1)$$

$$N_3 = \frac{1}{4}(1+s)(1+t)(s+t-1) \qquad (10.5.28)$$

$$N_4 = \frac{1}{4}(1-s)(1+t)(-s+t-1)$$

or, in compact index notation, we express Eqs. (10.5.28) as

$$N_i = \frac{1}{4}(1+ss_i)(1+tt_i)(ss_i + tt_i - 1) \qquad (10.5.29)$$

where i is the number of the shape function and

$$\begin{aligned} s_i &= -1, 1, 1, -1 \qquad (i = 1, 2, 3, 4) \\ t_i &= -1, -1, 1, 1 \qquad (i = 1, 2, 3, 4) \end{aligned} \qquad (10.5.30)$$

For the midside nodes ($i = 5, 6, 7, 8$),

$$N_5 = \frac{1}{2}(1-t)(1+s)(1-s)$$

$$N_6 = \frac{1}{2}(1+s)(1+t)(1-t)$$

$$N_7 = \frac{1}{2}(1+t)(1+s)(1-s) \qquad (10.5.31)$$

$$N_8 = \frac{1}{2}(1-s)(1+t)(1-t)$$

or, in index notation,

$$N_i = \frac{1}{2}(1-s^2)(1+tt_i) \qquad t_i = -1, 1 \qquad (i = 5, 7)$$

$$N_i = \frac{1}{2}(1+ss_i)(1-t^2) \qquad s_i = 1, -1 \qquad (i = 6, 8) \qquad (10.5.32)$$

We can observe from Eqs. (10.5.28) and (10.5.31) that an edge (and displacement) can vary with s^2 (along t constant) or with t^2 (along s constant). Furthermore, $N_i = 1$ at node i and $N_i = 0$ at the other nodes, as it must be according to our usual definition of shape functions.

The displacement functions are given by

$$
\left\{ \begin{matrix} u \\ v \end{matrix} \right\} = \begin{bmatrix} N_1 & 0 & N_2 & 0 & N_3 & 0 & N_4 & 0 & N_5 & 0 & N_6 & 0 & N_7 & 0 & N_8 & 0 \\ 0 & N_1 & 0 & N_2 & 0 & N_3 & 0 & N_4 & 0 & N_5 & 0 & N_6 & 0 & N_7 & 0 & N_8 \end{bmatrix}
$$

$$
\times \left\{ \begin{matrix} u_1 \\ v_1 \\ u_2 \\ v_2 \\ \vdots \\ v_8 \end{matrix} \right\} \tag{10.5.33}
$$

and the strain matrix is now

$$
\{\varepsilon\} = [D'][N]\{d\}
$$

with

$$
[B] = [D'][N]
$$

We can develop the matrix $[B]$ using Eq. (10.2.17) with $[D']$ from Eq. (10.2.16) and with $[N]$ now the 2×16 matrix given in Eq. (10.5.33), where the N's are defined in explicit form by Eq. (10.5.28) and (10.5.31).

To evaluate the matrix $[B]$ and the matrix $[k]$ for the eight-noded quadratic isoparametric element, we now use the nine-point Gauss rule (often described as a 3×3 rule). Results using 2×2 and 3×3 rules have shown significant differences, and the 3×3 rule is recommended by Bathe and Wilson [7]. Table 10–2 indicates the locations of points and the associated weights. The 3×3 rule is shown in Figure 10–16.

By adding a ninth node at $s = 0$, $t = 0$ in Figure 10–15, we can create an element called a "Q9." This is an internal node that is not connected to any other nodes. We then add the $a_{17}s^2t^2$ and $a_{18}s^2t^2$ terms to x and y, respectively in Eq. (10.5.27) and to u and v. The element is then called a Lagrange element as the shape functions can be derived using Lagrange interpolation formulas. For more on this subject consult [8].

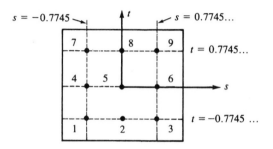

Figure 10–16 3×3 rule in two dimensions

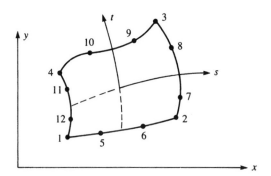

Figure 10–17 Cubic isoparametric element

Cubic Rectangle (Q12)

The cubic (Q12) element in Figure 10–17 has four corner nodes and additional nodes taken to be at one-third and two-thirds of the length along each side. The shape functions of the cubic element (as derived in Reference [3]) are based on the incomplete quartic polynomial such that

$$x = a_1 + a_2 s + a_3 t + a_4 s^2 + a_5 st + a_6 t^2 + a_7 s^2 t + a_8 st^2$$
$$+ a_9 s^3 + a_{10} t^3 + a_{11} s^3 t + a_{12} st^3 \tag{10.5.34}$$

with a similar polynomial for y. For the corner nodes ($i = 1, 2, 3, 4$),

$$N_i = \frac{1}{32}(1 + ss_i)(1 + tt_i)[9(s^2 + t^2) - 10] \tag{10.5.35}$$

with s_i and t_i given by Eqs. (10.5.30). For the nodes on sides $s = \pm 1$ ($i = 7, 8, 11, 12$),

$$N_i = \frac{9}{32}(1 + ss_i)(1 + 9tt_i)(1 - t^2) \tag{10.5.36}$$

with $s_i = \pm 1$ and $t_i = \pm \frac{1}{3}$. For the nodes on sides $t = \pm 1$ ($i = 5, 6, 9, 10$),

$$N_i = \frac{9}{32}(1 + tt_i)(1 + 9ss_i)(1 - s^2) \tag{10.5.37}$$

with $t_i = \pm 1$ and $s_i = \pm \frac{1}{3}$.

Having the shape functions for the quadratic element given by Eqs. (10.5.28) and (10.5.31) or for the cubic element given by Eqs. (10.5.35) through (10.5.37), we can again use Eq. (10.2.17) to obtain $[B]$ and then Eq. (10.2.27) to set up $[k]$ for numerical integration for the plane element. The cubic element requires a 3×3 rule (nine points) to evaluate the matrix $[k]$ exactly. We then conclude that what is really desired is a library of shape functions that can be used in the general equations developed for stiffness matrices, distributed load, and body force and can be applied not only to stress analysis but to nonstructural problems as well.

Since in this discussion the element shape functions N_i relating x and y to nodal coordinates x_i and y_i are of the same form as the shape functions relating u and v to nodal displacements u_i and v_i, this is said to be an *isoparametric formulation*. For instance, for the linear element $x = \sum_{i=1}^{4} N_i x_i$ and the displacement function $u = \sum_{i=1}^{4} N_i u_i$, use the same shape functions N_i given by Eq. (10.2.5). If instead the shape functions for the coordinates are of lower order (say, linear for x) than the shape functions used for displacements (say, quadratic for u), this is called a *subparametric formulation*.

Finally, referring to Figure 10–17, note that an element can have a linear shape along, say, one edge (1–2), a quadratic along, say, two edges (2–3 and 1–4), and a cubic along the other edge (3–4). Hence, the simple linear element can be mixed with different higher-order elements in regions of a model where rapid stress variation is expected. The advantage of the use of higher-order elements is further illustrated in Reference [3].

▲ Summary Equations

Natural coordinates related to global for a two-noded bar element:

$$x = a_1 + a_2 s \qquad (10.1.2)$$

Shape functions in natural coordinate s for two-noded bar:

$$N_1 = \frac{1-s}{2} \qquad N_2 = \frac{1+s}{2} \qquad (10.1.5)$$

Displacement function for two-noded bar:

$$\{u\} = [N_1 \quad N_2]\begin{Bmatrix} u_1 \\ u_2 \end{Bmatrix} \qquad (10.1.6)$$

Gradient matrix for two-noded bar:

$$[B] = \left[-\frac{1}{L} \quad \frac{1}{L}\right] \qquad (10.1.11)$$

Determinant of Jacobian matrix:

$$\|[J]\| = \frac{dx}{ds} = \frac{L}{2} \qquad (10.1.14)$$

Stiffness matrix for two-noded bar:

$$[k] = \frac{L}{2}\int_{-1}^{1} [B]^T E[B] A \, ds \qquad (10.1.15)$$

$$[k] = \frac{AE}{L}\begin{bmatrix} 1 & -1 \\ -1 & 1 \end{bmatrix} \qquad (10.1.16)$$

Body force matrix for two-noded bar:

$$\{f_b\} = \frac{ALX_b}{2}\begin{Bmatrix} 1 \\ 1 \end{Bmatrix} \qquad (10.1.20)$$

Surface force matrix for two-noded bar:

$$\{f_s\} = \{T_x\}\frac{L}{2}\begin{Bmatrix} 1 \\ 1 \end{Bmatrix} \qquad (10.1.24)$$

Relation between global and natural coordinates for quadrilateral element:

$$x = a_1 + a_2 s + a_3 t + a_4 st$$
$$y = a_5 + a_6 s + a_7 t + a_8 st \tag{10.2.2}$$

and

$$x = \frac{1}{4}[(1-s)(1-t)x_1 + (1+s)(1-t)x_2$$
$$+ (1+s)(1+t)x_3 + (1-s)(1+t)x_4]$$
$$y = \frac{1}{4}[(1-s)(1-t)y_1 + (1+s)(1-t)y_2$$
$$+ (1+s)(1+t)y_3 + (1-s)(1+t)y_4] \tag{10.2.3}$$

Shape functions for four-noded quadrilateral element expressed in natural coordinates:

$$N_1 = \frac{(1-s)(1-t)}{4} \qquad N_2 = \frac{(1+s)(1-t)}{4}$$
$$N_3 = \frac{(1+s)(1+t)}{4} \qquad N_4 = \frac{(1-s)(1+t)}{4} \tag{10.2.5}$$

Strain–displacement equations in natural coordinates:

$$\{\varepsilon\} = [D'][N]\{d\} \tag{10.2.15}$$

Determinant of Jacobian matrix for four-noded quadrilateral element:

$$|[J]| = \frac{1}{8}\{X_c\}^T \begin{bmatrix} 0 & 1-t & t-s & s-1 \\ t-1 & 0 & s+1 & -s-t \\ s-t & -s-1 & 0 & t+1 \\ 1-s & s+t & -t-1 & 0 \end{bmatrix} \{Y_c\} \tag{10.2.22}$$

$$\{X_c\}^T = [x_1 \ x_2 \ x_3 \ x_4]$$

where

$$\{Y_c\} = \begin{Bmatrix} y_1 \\ y_2 \\ y_3 \\ y_4 \end{Bmatrix} \tag{10.2.24}$$

Stiffness matrix for four-noded quadrilateral expressed in natural coordinates:

$$[k] = \int_{-1}^{1}\int_{-1}^{1}[B]^T[D][B]h|[J]| \, ds \, dt \tag{10.2.27}$$

Body force matrix for four-noded quadrilateral expressed in natural coordinates:

$$\{f_b\} = \int_{-1}^{1}\int_{-1}^{1} [N]^T \quad \{X\} \ h|[J]| \, ds \, dt \tag{10.2.28}$$
$$(8 \times 1) \qquad \qquad (8 \times 2) \ (2 \times 1)$$

Surface force matrix along an edge $t = 1$:

$$\{f_s\} = \int_{-1}^{1} [N_s]^T \ \{T\} \ h\frac{L}{2}ds \qquad (10.2.29)$$
$$(4 \times 1) \qquad (4 \times 2) \ (2 \times 1)$$

For the case of uniform (constant) p_s and p_t along edge $t = 1$, the total surface-force matrix is

$$\{f_s\} = h\frac{L}{2}[0 \quad 0 \quad 0 \quad 0 \quad p_s \quad p_t \quad p_s \quad p_t]^T \qquad (10.2.31)$$

Newton-Cotes formula for numerical integration:

$$I = \int_{-1}^{1} y \, dx = h\sum_{i=0}^{n} C_i y_i = h[C_0 y_0 + C_1 y_1 + C_2 y_2 + C_3 y_3 + \ldots + C_n y_n] \qquad (10.3.1)$$

See Table 10–1 for Newton-Cotes intervals and points for integration.

Gaussian Quadrature formula for numerical integration:

$$I = \int_{-1}^{1} y \, dx = \sum_{i=1}^{n} W_i y_i \qquad (10.3.6)$$

See Table 10–2 for Gauss points for integration from minus 1 to 1.

Four-point Gaussian quadrature formula to evaluate stiffness matrix of four-noded quadrilateral element:

$$[k] = [B(s_1, t_1)]^T[D][B(s_1, t_1)]|[J(s_1, t_1)]|hW_1 W_1$$
$$+ [B(s_2, t_2)]^T[D][B(s_2, t_2)]|[J(s_2, t_2)]|hW_2 W_2$$
$$+ [B(s_3, t_3)]^T[D][B(s_3, t_3)]|[J(s_3, t_3)]|hW_3 W_3$$
$$+ [B(s_4, t_4)]^T[D][B(s_4, t_4)]|[J(s_4, t_4)]|hW_4 W_4 \qquad (10.4.3)$$

Axial coordinate function for three-noded bar element:

$$x = a_1 + a_2 s + a_3 s^2 \qquad (10.5.1)$$

Shape functions for three-noded bar:

$$N_1 = \frac{s(s-1)}{2} \qquad N_2 = \frac{s(s+1)}{2} \qquad N_3 = (1 - s^2) \qquad (10.5.9)$$

Gradient matrix for three-noded bar:

$$[B] = \begin{bmatrix} \dfrac{2s-1}{L} & \dfrac{2s+1}{L} & \dfrac{-4s}{L} \end{bmatrix} \qquad (10.5.16)$$

x and y coordinate functions for eight-noded (Q8) quadrilateral element:

$$x = a_1 + a_2 s + a_3 t + a_4 st + a_5 s^2 + a_6 t^2 + a_7 s^2 t + a_8 st^2$$
$$y = a_9 + a_{10} s + a_{11} t + a_{12} st + a_{13} s^2 + a_{14} t^2 + a_{15} s^2 t + a_{16} st^2 \qquad (10.5.27)$$

Equations (10.5.28) and (10.5.31) give the shape functions for the eight-noded (Q8) quadrilateral element.

x coordinate function for the 12-noded (Q12) quadrilateral element:

$$x = a_1 + a_2 s + a_3 t + a_4 s^2 + a_5 st + a_6 t^2 + a_7 s^2 t + a_8 st^2$$
$$+ a_9 s^3 + a_{10} t^3 + a_{11} s^3 t + a_{12} st^3 \tag{10.5.34}$$

Equations (10.5.35), (10.5.36), and (10.5.37) give the shape functions for the 12-noded (Q12) quadrilateral element.

▲ References

[1] Irons, B. M., "Engineering Applications of Numerical Integration in Stiffness Methods," *Journal of the American Institute of Aeronautics and Astronautics*, Vol. 4, No. 11, pp. 2035–2037, 1966.

[2] Stroud, A. H., and Secrest, D., *Gaussian Quadrature Formulas*, Prentice-Hall, Englewood Cliffs, NJ, 1966.

[3] Ergatoudis, I., Irons, B. M., and Zienkiewicz, O. C., "Curved Isoparametric, Quadrilateral Elements for Finite Element Analysis," *International Journal of Solids and Structures*, Vol. 4, pp. 31–42, 1968.

[4] Zienkiewicz, O. C., *The Finite Element Method*, 3rd ed., McGraw-Hill, London, 1977.

[5] Thomas, B. G., and Finney, R. L., *Calculus and Analytic Geometry*, Addison-Wesley, Reading, MA, 1984.

[6] Gallagher, R., *Finite Element Analysis Fundamentals*, Prentice-Hall, Englewood Cliffs, NJ, 1975.

[7] Bathe, K. J., and Wilson, E. L., *Numerical Methods in Finite Element Analysis*, Prentice-Hall, Englewood Cliffs, NJ, 1976.

[8] Cook, R. D., Malkus, D. S., Plesha, M. E., and Witt, R. J., *Concepts and Applications of Finite Element Analysis*, 4th ed., Wiley, New York, 2002.

[9] Bathe, Klaus-Jurgen, *Finite Element Procedures in Engineering Analysis*, Prentice-Hall, Englewood Cliffs, New Jersey, 1982.

▲ Problems

10.1 For the three-noded linear strain bar with three coordinates of nodes x_1, x_2, and x_3, shown in Figure P10–1 in the global-coordinate system show that the Jacobian determinate is $||[J]|| = L/2$.

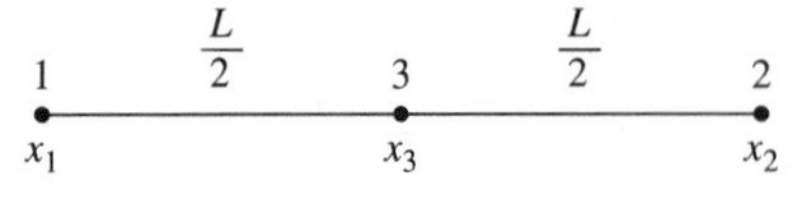

Figure P10–1

10.2 For the two-noded one-dimensional isoparametric element shown in Figure P10–2 (a) and (b), with shape functions given by Eq. (10.1.5), determine (a) intrinsic coordinate s

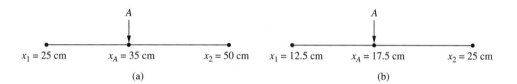

Figure P10–2

at point A and (b) shape functions N_1 and N_2 at point A. If the displacements at nodes one and two are respectively, $u_1 = 0.15$ mm and $u_2 = -0.15$ mm, determine (c) the value of the displacement at point A and (d) the strain in the element.

10.3 Answer the same questions as posed in problem 10.2 with the data listed under the Figure P10–3.

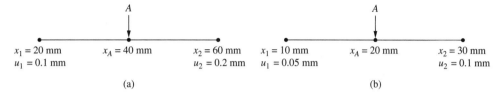

Figure P10–3

10.4 For the four-noded bar element in Figure P10–4, show that the Jacobian determinate is $\|[J]\| = L/2$. Also determine the shape functions $N_1 - N_4$ and the strain/displacement matrix $[B]$. Assume $u = a_1 + a_2 s + a_3 s^2 + a_4 s^3$.

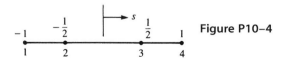   **Figure P10–4**

10.5 Using the three-noded bar element shown in Figure P10–5 (a) and (b), with shape functions given by Eq. (10.5.9), determine (a) the intrinsic coordinate s at point A and (b) the shape functions, N_1, N_2, and N_3 at A. For the displacements of the nodes shown in Figure P10–5, determine (c) the displacement at A and (d) the axial strain expression in the element.

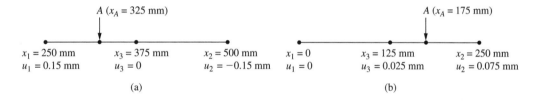

Figure P10–5

10.6 Using the three-noded bar element shown in Figure P10–5 (a) and (b), with shape functions given by Eq. (10.5.9), determine (a) the intrinsic coordinate s at point A and (b) the shape functions, N_1, N_2, and N_3 at point A. For the displacements of the nodes shown in Figure P10–6, determine (c) the displacement at A and (d) the axial strain expression in the element.

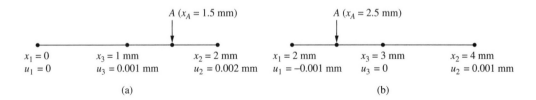

$A (x_A = 1.5 \text{ mm})$ $\qquad\qquad$ $A (x_A = 2.5 \text{ mm})$

$x_1 = 0$	$x_3 = 1$ mm	$x_2 = 2$ mm	$x_1 = 2$ mm	$x_3 = 3$ mm	$x_2 = 4$ mm
$u_1 = 0$	$u_3 = 0.001$ mm	$u_2 = 0.002$ mm	$u_1 = -0.001$ mm	$u_3 = 0$	$u_2 = 0.001$ mm

(a) $\qquad\qquad\qquad\qquad\qquad\qquad$ (b)

Figure P10–6

10.7 For the bar subjected to the linearly varying axial line load shown in Figure P10–7, use the linear strain (three-noded element) with two elements in the model, to determine the nodal displacements and nodal stresses. Compare your answer with that in Figure 3–31 and Eqs. (3.11.6) and (3.11.7). Let $A = 12.5 \times 10^{-4}$ m^2 and $E = 210$ GPa. Hint: Use Eq. (10.5.22) for the element stiffness matrix.

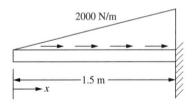

2000 N/m

1.5 m

x

Figure P10–7

10.8 Use the three-noded bar element and find the axial displacement at the end of the rod shown in Figure P10–8. Determine the stress at $x = 0$, $x = L/2$ and $x = L$. Let $A = 2 \times 10^{-4}$ m^2, $E = 205$ GPa, and $L = 4$ m. Hint: Use Eq. (10.5.22) for the element stiffness matrix.

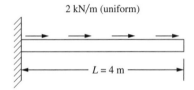

2 kN/m (uniform)

$L = 4$ m

Figure P10–8

10.9   Derive $\|[J]\|$ given by Eq. (10.2.22) for a four-noded isoparametric quadrilateral element.

10.10   Show that for the quadrilateral element described in Section 10.2, $[J]$ can be expressed as

$$[J] = \begin{bmatrix} N_{1,s} & N_{2,s} & N_{3,s} & N_{4,s} \\ N_{1,t} & N_{2,t} & N_{3,t} & N_{4,t} \end{bmatrix} \begin{bmatrix} x_1 & y_1 \\ x_2 & y_2 \\ x_3 & y_3 \\ x_4 & y_4 \end{bmatrix}$$

10.11   Determine the Jacobian matrix $[J]$ and its determinant for the elements shown in Figure P10–11. Show that the determinant of $[J]$ for rectangular and parallelogram shaped elements is equal to $A/4$, where A is the physical area of the element and 4 actually represents the area of the rectangle of sides 2×2 when $b = 1$ cm and $h = 1$ cm Figure 6–20.

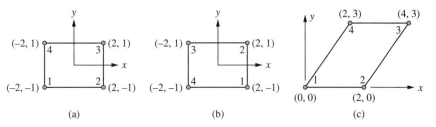

(a)   (b)   (c)

Figure P10–11

10.12   Derive Eq. (10.2.18) with $[B_i]$ given by Eq. (10.2.19) by substituting Eq. (10.2.16) for $[D']$ and Eqs. (10.2.5) for the shape functions into Eq. (10.2.17).

10.13   Use Eq. (10.2.30) with $p_s = 0$ and $p_t = p$ (a constant) alongside 3–4 of the element shown in Figure 10–5 on page 500 to obtain the nodal forces.

10.14   For the element shown in Figure P10–14, replace the distributed load with the energy equivalent nodal forces by evaluating a force matrix similar to Eq. (10.2.29). Let $h = 2.5$ mm thick. The global coordinates (in cm) are shown in Figure P10–14.

10.15   Use Gaussian quadature with two and three Gauss points and Table 10–2 to evaluate the following integrals:

(a) $\displaystyle\int_{-1}^{1} \cos\frac{s}{2}\, ds$ (b) $\displaystyle\int_{-1}^{1} s^2\, ds$ (c) $\displaystyle\int_{-1}^{1} s^4\, ds$

(d) $\displaystyle\int_{-1}^{1} \frac{\cos s}{1 - s^2}\, ds$ (e) $\displaystyle\int_{-1}^{1} s^3\, ds$ (f) $\displaystyle\int_{-1}^{1} s\cos s\, ds$ (g) $\displaystyle\int_{-1}^{1} (4^s - 2s)\, ds$

Then use the Newton-Cotes quadrature with two and three sampling points and Table 10–1 to evaluate the same integrals. Compare your results.

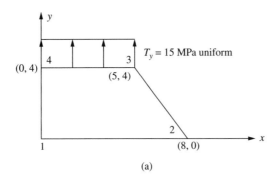

(a)

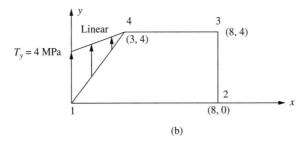

(b)

Figure P10–14

10.16 For the quadrilateral elements shown in Figure P10–16, write a computer program to evaluate the stiffness matrices using four-point Gaussian quadrature as outlined in Section 10.4. Let $E = 210$ GPa and $v = 0.25$. The global coordinates (in cm) are shown in the figures.

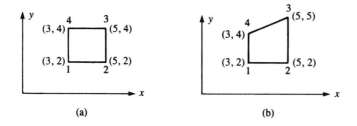

Figure P10–16

10.17 For the quadrilateral elements shown in Figure P10–17, evaluate the stiffness matrices using four-point Gaussian quadrature as outlined in Section 10.4. Let $E = 210$ GPa and $v = 0.25$. The global coordinates (in millimeters) are shown in the figures.

10.18 Evaluate the matrix $[B]$ for the quadratic quadrilateral element shown in Figure 10–15 on page 521 (Section 10.5).

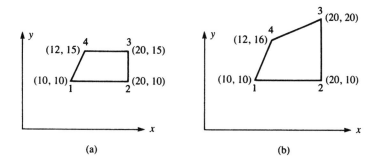

Figure P10–17

10.19 Evaluate the stiffness matrix for the four-noded bar of Problem 10.4 using three-point Gaussian quadrature.

10.20 For the rectangular element of Figure 10–20 with the nodal displacements given by

$$u_1 = 0 \qquad v_1 = 0 \qquad u_2 = 0.125 \text{ mm}$$

$$v_2 = 0.625 \text{ mm} \qquad u_3 = 0.625 \text{ mm} \qquad v_3 = -0.625 \text{ mm}$$

$$u_4 = 0 \qquad v_4 = 0$$

determine the $\{\sigma\}$ matrix at $s = 0$, $t = 0$ using the isoparametric formulation described in Section 10.4. (Also see Example 10.5.) Let $E = 210$ GPa and $v = 0.3$.

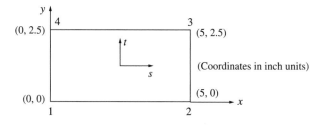

Figure P10–20

10.21 For the three-noded bar (Figure P10–1), what Gaussian quadrature rule (how many Gauss points) would you recommend to evaluate the stiffness matrix? Why?

THREE-DIMENSIONAL
STRESS ANALYSIS ▲

CHAPTER OBJECTIVES

- To introduce concepts of three-dimensional stress and strain.
- To develop the tetrahedral solid-element stiffness matrix.
- To describe how body and surface tractions are treated.
- To illustrate a numerical example of the tetrahedral element stiffness matrix.
- To describe the isoparametric formulation of the stiffness matrix for three-dimensional hexahedral (brick) elements, including the linear (eight-noded) brick, and the quadratic (20 noded) brick.
- To present some commercial computer program examples of three-dimensional solid models and results for real-world applications.
- To present a comparison of the four-noded tetrahedral, the ten-noded tetrahedral, the eight-noded brick, and the twenty-noded brick.

Introduction

In this chapter, we consider the three-dimensional, or solid, element. This element is useful for the stress analysis of general three-dimensional bodies that require more precise analysis than is possible through two-dimensional and/or axisymmetric analysis. Examples of three-dimensional problems are arch dams, thick-walled pressure vessels, and solid forging parts as used, for instance, in the heavy equipment and automotive industries. Figure 11–1 shows finite element models of some typical automobile parts and a subsoiter used in agricultural equipment. Also see Figure 1–7 for a model of a swing casting for a backhoe frame, Figure 1–9 for a model of a pelvis bone with an implant, and Figures 11–7 through 11–10 of a forging part, a foot pedal, a trailer hitch, and an alternator bracket, respectively.

The tetrahedron is the basic three-dimensional element, and it is used in the development of the shape functions, stiffness matrix, and force matrices in terms of a global coordinate system. We follow this development with the isoparametric formulation of the stiffness matrix for the hexahedron, or brick element. Finally, we will provide some typical three-dimensional applications.

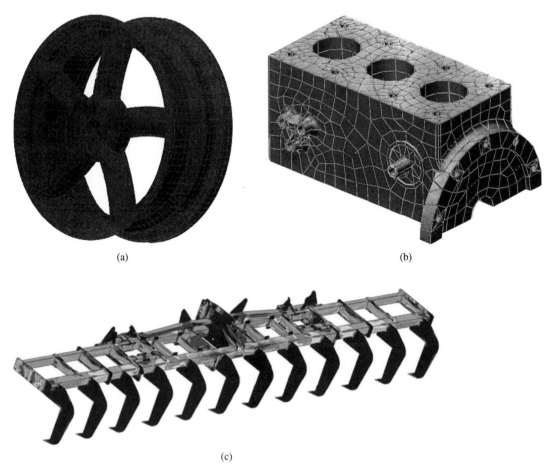

(a)

(b)

(c)

Figure 11–1 (a) wheel rim (Courtesy of Mark Blair); (b) engine block (Courtesy of Mark Guard); and (c) *Subsoiler*—12-row subsoiler used in agricultural equipment (Courtesy of Algor, Inc.) (See the full-color insert for a color version of this figure.)

In the last section of this chapter, we show some three-dimensional problems solved using a computer program.

▲ 11.1 Three-Dimensional Stress and Strain ▲

We begin by considering the three-dimensional infinitesimal element in Cartesian coordinates with dimensions dx, dy, and dz and normal and shear stresses as shown in Figure 11–2. This element conveniently represents the state of stress on three mutually perpendicular planes of a body in a state of three-dimensional stress. As usual, normal stresses are perpendicular to the faces of the element and are represented by σ_x, σ_y, and σ_z. Shear stresses act in the faces (planes) of the element and are represented by $\tau_{xy}, \tau_{yz}, \tau_{zx}$, and so on.

From moment equilibrium of the element, we show in Appendix C that

$$\tau_{xy} = \tau_{yx} \qquad \tau_{yz} = \tau_{zy} \qquad \tau_{zx} = \tau_{xz}$$

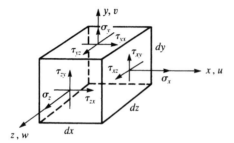

Figure 11–2 Three-dimensional stresses on an element

Hence, there are only three independent shear stresses, along with the three normal stresses.

The element strain–displacement relationships are obtained in Appendix C. They are repeated here, for convenience, as

$$\varepsilon_x = \frac{\partial u}{\partial x} \qquad \varepsilon_y = \frac{\partial v}{\partial y} \qquad \varepsilon_z = \frac{\partial w}{\partial z} \tag{11.1.1}$$

where $u, v,$ and w are the displacements associated with the $x, y,$ and z directions. The shear strains γ are now given by

$$\gamma_{xy} = \frac{\partial u}{\partial y} + \frac{\partial v}{\partial x} = \gamma_{yx}$$

$$\gamma_{yz} = \frac{\partial v}{\partial z} + \frac{\partial w}{\partial y} = \gamma_{zy} \tag{11.1.2}$$

$$\gamma_{zx} = \frac{\partial w}{\partial x} + \frac{\partial u}{\partial z} = \gamma_{xz}$$

where, as for shear stresses, only three independent shear strains exist.

We again represent the stresses and strains by column matrices as

$$\{\sigma\} = \begin{Bmatrix} \sigma_x \\ \sigma_y \\ \sigma_z \\ \tau_{xy} \\ \tau_{yz} \\ \tau_{zx} \end{Bmatrix} \qquad \{\varepsilon\} = \begin{Bmatrix} \varepsilon_x \\ \varepsilon_y \\ \varepsilon_z \\ \gamma_{xy} \\ \gamma_{yz} \\ \gamma_{zx} \end{Bmatrix} \tag{11.1.3}$$

The stress–strain relationships for an isotropic material are again given by

$$\{\sigma\} = [D]\{\varepsilon\} \tag{11.1.4}$$

where $\{\sigma\}$ and $\{\varepsilon\}$ are defined by Eqs. (11.1.3), and the constitutive matrix $[D]$ (see also Appendix C) is now given by

$$[D] = \frac{E}{(1+v)(1-2v)}\begin{bmatrix} 1-v & v & v & 0 & 0 & 0 \\ & 1-v & v & 0 & 0 & 0 \\ & & 1-v & 0 & 0 & 0 \\ & & & \dfrac{1-2v}{2} & 0 & 0 \\ & & & & \dfrac{1-2v}{2} & 0 \\ \text{Symmetry} & & & & & \dfrac{1-2v}{2} \end{bmatrix} \quad (11.1.5)$$

▲ 11.2 Tetrahedral Element ▲

We now develop the tetrahedral stress element stiffness matrix by again using the steps outlined in Chapter 1. The development is seen to be an extension of the plane element previously described in Chapter 6. This extension was suggested in References [1] and [2].

Step 1 Select Element Type

Consider the tetrahedral element shown in Figure 11–3 with corner nodes 1–4. This element is a four-noded solid. The nodes of the element must be numbered such that when viewed from the last node (say, node 4), the first three nodes are numbered in a counterclockwise manner, such as 1, 2, 3, 4 or 2, 3, 1, 4. This ordering of nodes avoids the calculation of negative volumes and is consistent with the counterclockwise node numbering associated with the CST element in Chapter 6. (Using an isoparametric formulation to evaluate the $[k]$ matrix for the tetrahedral element enables us to use the element node numbering in any order. The isoparametric formulation of $[k]$ is left

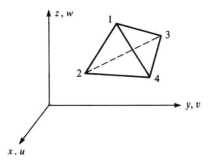

Figure 11–3 Tetrahedral solid element

to Section 11.3.) The unknown nodal displacements are now given by

$$\{d\} = \begin{Bmatrix} u_1 \\ v_1 \\ w_1 \\ \vdots \\ u_4 \\ v_4 \\ w_4 \end{Bmatrix} \tag{11.2.1}$$

Hence, there are 3 degrees of freedom per node, or 12 total degrees of freedom per element.

Step 2 Select Displacement Functions

For a compatible displacement field, the element displacement functions u, v, and w must be linear along each edge because only two points (the corner nodes) exist along each edge, and the functions must be linear in each plane side of the tetrahedron. We then select the linear displacement functions as

$$u(x, y, z) = a_1 + a_2 x + a_3 y + a_4 z$$

$$v(x, y, z) = a_5 + a_6 x + a_7 y + a_8 z \tag{11.2.2}$$

$$w(x, y, z) = a_9 + a_{10} x + a_{11} y + a_{12} z$$

In the same manner as in Chapter 6, we can express the a_i's in terms of the known nodal coordinates $(x_1, y_1, z_1, \ldots, z_4)$ and the unknown nodal displacements $(u_1, v_1, w_1, \ldots, w_4)$ of the element. Skipping the straightforward but tedious details, we obtain

$$\begin{aligned} u(x, y, z) = \frac{1}{6V} \{ & (\alpha_1 + \beta_1 x + \gamma_1 y + \delta_1 z) u_1 \\ & + (\alpha_2 + \beta_2 x + \gamma_2 y + \delta_2 z) u_2 \\ & + (\alpha_3 + \beta_3 x + \gamma_3 y + \delta_3 z) u_3 \\ & + (\alpha_4 + \beta_4 x + \gamma_4 y + \delta_4 z) u_4 \} \end{aligned} \tag{11.2.3}$$

where $6V$ is obtained by evaluating the determinant

$$6V = \begin{vmatrix} 1 & x_1 & y_1 & z_1 \\ 1 & x_2 & y_2 & z_2 \\ 1 & x_3 & y_3 & z_3 \\ 1 & x_4 & y_4 & z_4 \end{vmatrix} \tag{11.2.4}$$

and V represents the volume of the tetrahedron. The coefficients $\alpha_i, \beta_i, \gamma_i,$ and δ_i $(i = 1, 2, 3, 4)$ in Eq. (11.2.3) are given by

$$
\alpha_1 = \begin{vmatrix} x_2 & y_2 & z_2 \\ x_3 & y_3 & z_3 \\ x_4 & y_4 & z_4 \end{vmatrix} \qquad \beta_1 = - \begin{vmatrix} 1 & y_2 & z_2 \\ 1 & y_3 & z_3 \\ 1 & y_4 & z_4 \end{vmatrix}
$$

$$ \tag{11.2.5} $$

$$
\gamma_1 = \begin{vmatrix} 1 & x_2 & z_2 \\ 1 & x_3 & z_3 \\ 1 & x_4 & z_4 \end{vmatrix} \qquad \delta_1 = - \begin{vmatrix} 1 & x_2 & y_2 \\ 1 & x_3 & y_3 \\ 1 & x_4 & y_4 \end{vmatrix}
$$

$$
\alpha_2 = - \begin{vmatrix} x_1 & y_1 & z_1 \\ x_3 & y_3 & z_3 \\ x_4 & y_4 & z_4 \end{vmatrix} \qquad \beta_2 = \begin{vmatrix} 1 & y_1 & z_1 \\ 1 & y_3 & z_3 \\ 1 & y_4 & z_4 \end{vmatrix}
$$

and

$$ \tag{11.2.6} $$

$$
\gamma_2 = - \begin{vmatrix} 1 & x_1 & z_1 \\ 1 & x_3 & z_3 \\ 1 & x_4 & z_4 \end{vmatrix} \qquad \delta_2 = \begin{vmatrix} 1 & x_1 & y_1 \\ 1 & x_3 & y_3 \\ 1 & x_4 & y_4 \end{vmatrix}
$$

$$
\alpha_3 = \begin{vmatrix} x_1 & y_1 & z_1 \\ x_2 & y_2 & z_2 \\ x_4 & y_4 & z_4 \end{vmatrix} \qquad \beta_3 = - \begin{vmatrix} 1 & y_1 & z_1 \\ 1 & y_2 & z_2 \\ 1 & y_4 & z_4 \end{vmatrix}
$$

and

$$ \tag{11.2.7} $$

$$
\gamma_3 = \begin{vmatrix} 1 & x_1 & z_1 \\ 1 & x_2 & z_2 \\ 1 & x_4 & z_4 \end{vmatrix} \qquad \delta_3 = - \begin{vmatrix} 1 & x_1 & y_1 \\ 1 & x_2 & y_2 \\ 1 & x_4 & y_4 \end{vmatrix}
$$

$$
\alpha_4 = - \begin{vmatrix} x_1 & y_1 & z_1 \\ x_2 & y_2 & z_2 \\ x_3 & y_3 & z_3 \end{vmatrix} \qquad \beta_4 = \begin{vmatrix} 1 & y_1 & z_1 \\ 1 & y_2 & z_2 \\ 1 & y_3 & z_3 \end{vmatrix}
$$

and

$$ \tag{11.2.8} $$

$$
\gamma_4 = - \begin{vmatrix} 1 & x_1 & z_1 \\ 1 & x_2 & z_2 \\ 1 & x_3 & z_3 \end{vmatrix} \qquad \delta_4 = \begin{vmatrix} 1 & x_1 & y_1 \\ 1 & x_2 & y_2 \\ 1 & x_3 & y_3 \end{vmatrix}
$$

Expressions for v and w are obtained by simply substituting v_i's for all u_i's and then w_i's for all u_i's in Eq. (11.2.3).

The displacement expression for u given by Eq. (11.2.3), with similar expressions for v and w, can be written equivalently in expanded form in terms of the shape functions and unknown nodal displacements as

$$
\left\{ \begin{array}{c} u \\ v \\ w \end{array} \right\} = \left[\begin{array}{cccccccccccc} N_1 & 0 & 0 & N_2 & 0 & 0 & N_3 & 0 & 0 & N_4 & 0 & 0 \\ 0 & N_1 & 0 & 0 & N_2 & 0 & 0 & N_3 & 0 & 0 & N_4 & 0 \\ 0 & 0 & N_1 & 0 & 0 & N_2 & 0 & 0 & N_3 & 0 & 0 & N_4 \end{array} \right] \left\{ \begin{array}{c} u_1 \\ v_1 \\ w_1 \\ \vdots \\ u_4 \\ v_4 \\ w_4 \end{array} \right\}
$$

(11.2.9)

where the shape functions are given by

$$
N_1 = \frac{(\alpha_1 + \beta_1 x + \gamma_1 y + \delta_1 z)}{6V} \qquad N_2 = \frac{(\alpha_2 + \beta_2 x + \gamma_2 y + \delta_2 z)}{6V}
$$

$$
N_3 = \frac{(\alpha_3 + \beta_3 x + \gamma_3 y + \delta_3 z)}{6V} \qquad N_4 = \frac{(\alpha_4 + \beta_4 x + \gamma_4 y + \delta_4 z)}{6V}
$$

(11.2.10)

and the rectangular matrix on the right side of Eq. (11.2.9) is the shape function matrix $[N]$.

Step 3 Define the Strain–Displacement and Stress–Strain Relationships

The element strains for the three-dimensional stress state are given by

$$
\{\varepsilon\} = \left\{ \begin{array}{c} \varepsilon_x \\ \varepsilon_y \\ \varepsilon_z \\ \gamma_{xy} \\ \gamma_{yz} \\ \gamma_{zx} \end{array} \right\} = \left\{ \begin{array}{c} \dfrac{\partial u}{\partial x} \\[2mm] \dfrac{\partial v}{\partial y} \\[2mm] \dfrac{\partial w}{\partial z} \\[2mm] \dfrac{\partial u}{\partial y} + \dfrac{\partial v}{\partial x} \\[2mm] \dfrac{\partial v}{\partial z} + \dfrac{\partial w}{\partial y} \\[2mm] \dfrac{\partial w}{\partial x} + \dfrac{\partial u}{\partial z} \end{array} \right\}
$$

(11.2.11)

Using Eq. (11.2.9) in Eq. (11.2.11), we obtain

$$
\{\varepsilon\} = [B]\{d\}
$$

(11.2.12)

where

$$
[B] = [[B_1] \ [B_2] \ [B_3] \ [B_4]]
$$

(11.2.13)

The submatrix $[B_1]$ in Eq. (11.2.13) is defined by

$$[B_1] = \begin{bmatrix} N_{1,x} & 0 & 0 \\ 0 & N_{1,y} & 0 \\ 0 & 0 & N_{1,z} \\ N_{1,y} & N_{1,x} & 0 \\ 0 & N_{1,z} & N_{1,y} \\ N_{1,z} & 0 & N_{1,x} \end{bmatrix} \qquad (11.2.14)$$

where, again, the comma after the subscript indicates differentation with respect to the variable that follows. Submatrices $[B_2], [B_3]$, and $[B_4]$ are defined by simply indexing the subscript in Eq. (11.2.14) from 1 to 2, 3, and then 4, respectively. Substituting the shape functions from Eqs. (11.2.10) into Eq. (11.2.14), $[B_1]$ is expressed as

$$[B_1] = \frac{1}{6V} \begin{bmatrix} \beta_1 & 0 & 0 \\ 0 & \gamma_1 & 0 \\ 0 & 0 & \delta_1 \\ \gamma_1 & \beta_1 & 0 \\ 0 & \delta_1 & \gamma_1 \\ \delta_1 & 0 & \beta_1 \end{bmatrix} \qquad (11.2.15)$$

with similar expressions for $[B_2], [B_3]$, and $[B_4]$.

The element stresses are related to the element strains by

$$\{\sigma\} = [D]\{\varepsilon\} \qquad (11.2.16)$$

where the constitutive matrix for an elastic material is now given by Eq. (11.1.5).

Step 4   Derive the Element Stiffness Matrix and Equations

The element stiffness matrix is given by

$$[k] = \iiint_V [B]^T [D][B] \, dV \qquad (11.2.17)$$

Because both matrices $[B]$ and $[D]$ are constant for the simple tetrahedral element, Eq. (11.2.17) can be simplified to

$$[k] = [B]^T [D][B] V \qquad (11.2.18)$$

where, again, V is the volume of the element. The element stiffness matrix is now of order 12×12.

Body Forces

The element body force matrix is given by

$$\{f_b\} = \iiint_V [N]^T \{X\} \, dV \qquad (11.2.19)$$

where $[N]$ is given by the 3×12 matrix in Eq. (11.2.9), and

$$\{X\} = \begin{Bmatrix} X_b \\ Y_b \\ Z_b \end{Bmatrix} \qquad (11.2.20a)$$

For constant body forces, the nodal components of the total resultant body forces can be shown to be distributed to the nodes in four equal parts. That is,

$$\{f_b\} = \frac{1}{4}[X_b \ Y_b \ Z_b \ X_b \ Y_b \ Z_b \ X_b \ Y_b \ Z_b \ X_b \ Y_b \ Z_b]^T \qquad (11.2.20b)$$

The element body force is then a 12×1 matrix.

Surface Forces

Again, the surface forces are given by

$$\{f_s\} = \iint_S [N_s]^T \{T\} \, dS \qquad (11.2.21)$$

where $[N_s]$ is the shape function matrix evaluated on the surface where the surface traction occurs.

For example, consider the case of uniform pressure p acting on the face with nodes 1–3 of the element shown in Figure 11–3 or 11–4. The resulting nodal forces become

$$\{f_s\} = \iint_S [N]^T \Big|_{\substack{\text{evaluated on} \\ \text{surface } 1,2,3}} \begin{Bmatrix} p_x \\ p_y \\ p_z \end{Bmatrix} dS \qquad (11.2.22)$$

where p_x, p_y, and p_z are the x, y, and z components, respectively, of p. Simplifying and integrating Eq. (11.2.22), we can show that

$$\{f_s\} = \frac{S_{123}}{3} \begin{Bmatrix} p_x \\ p_y \\ p_z \\ p_x \\ p_y \\ p_z \\ p_x \\ p_y \\ p_z \\ 0 \\ 0 \\ 0 \end{Bmatrix} \qquad (11.2.23)$$

where S_{123} is the area of the surface associated with nodes 1–3. The use of volume coordinates, as explained in Reference [8], facilitates the integration of Eq. (11.2.22).

Example 11.1

Evaluate the matrices necessary to determine the stiffness matrix for the tetrahedral element shown in Figure 11–4. Let $E = 200$ GPa and $v = 0.30$. The coordinates are shown in the figure in units of centimeters.

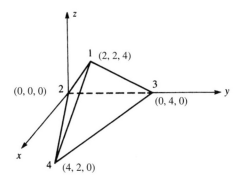

Figure 11–4 Tetrahedral element

SOLUTION:

To evaluate the element stiffness matrix, we first determine the element volume V and all α's, β's, γ's, and δ's from Eqs. (11.2.4) through (11.2.8). From Eq. (11.2.4), we have

$$
6V = \begin{vmatrix} 1 & 2 & 2 & 4 \\ 1 & 0 & 0 & 0 \\ 1 & 0 & 4 & 0 \\ 1 & 4 & 2 & 0 \end{vmatrix} = 64 \text{ cm}^3 \tag{11.2.24}
$$

From Eqs. (11.2.5), we obtain

$$
\alpha_1 = \begin{vmatrix} 0 & 0 & 0 \\ 0 & 4 & 0 \\ 4 & 2 & 0 \end{vmatrix} = 0 \qquad \beta_1 = -\begin{vmatrix} 1 & 0 & 0 \\ 1 & 4 & 0 \\ 1 & 2 & 0 \end{vmatrix} = 0 \tag{11.2.25}
$$

and similarly,

$$
\gamma_1 = 0 \qquad \delta_1 = 16
$$

From Eqs. (11.2.6) through (11.2.8), we obtain

$$
\begin{aligned}
\alpha_2 &= 64 & \beta_2 &= -8 & \gamma_2 &= -16 & \delta_2 &= -4 \\
\alpha_3 &= 0 & \beta_3 &= -8 & \gamma_3 &= 16 & \delta_3 &= -4 \\
\alpha_4 &= 0 & \beta_4 &= 16 & \gamma_4 &= 0 & \delta_4 &= -8
\end{aligned} \tag{11.2.26}
$$

Note that α's typically have units of cubic centimeters or cubic meters, where β's, γ's, and δ's have units of square centimeters or square meters.

Next, the shape functions are determined using Eqs. (11.2.10) and the results from Eqs. (11.2.25) and (11.2.26) as

$$N_1 = \frac{4z}{8} \qquad N_2 = \frac{8 - 2x - 4y - z}{8}$$

$$N_3 = \frac{-2x + 4y - z}{8} \qquad N_4 = \frac{4x - 2z}{8} \tag{11.2.27}$$

Note that $N_1 + N_2 + N_3 + N_4 = 1$ is again satisfied.

The 6×3 submatrices of the matrix $[B]$, Eq. (11.2.13), are now evaluated using Eqs. (11.2.14) and (11.2.27) as

$$[B_1] = (10^4) \begin{bmatrix} 0 & 0 & 0 \\ 0 & 0 & 0 \\ 0 & 0 & \frac{1}{4} \\ 0 & 0 & 0 \\ 0 & \frac{1}{4} & 0 \\ \frac{1}{4} & 0 & 0 \end{bmatrix} \qquad [B_2] = (10^4) \begin{bmatrix} -\frac{1}{8} & 0 & 0 \\ 0 & -\frac{1}{4} & 0 \\ 0 & 0 & -\frac{1}{16} \\ -\frac{1}{4} & -\frac{1}{8} & 0 \\ 0 & -\frac{1}{16} & -\frac{1}{4} \\ -\frac{1}{16} & 0 & -\frac{1}{8} \end{bmatrix} \tag{11.2.28}$$

$$[B_3] = (10^4) \begin{bmatrix} -\frac{1}{8} & 0 & 0 \\ 0 & \frac{1}{4} & 0 \\ 0 & 0 & -\frac{1}{16} \\ \frac{1}{4} & -\frac{1}{8} & 0 \\ 0 & -\frac{1}{16} & \frac{1}{4} \\ -\frac{1}{16} & 0 & -\frac{1}{8} \end{bmatrix} \qquad [B_4] = (10^4) \begin{bmatrix} \frac{1}{4} & 0 & 0 \\ 0 & 0 & 0 \\ 0 & 0 & -\frac{1}{8} \\ 0 & \frac{1}{4} & 0 \\ 0 & -\frac{1}{8} & 0 \\ -\frac{1}{8} & 0 & \frac{1}{4} \end{bmatrix}$$

Next, the matrix $[D]$ is evaluated using Eq. (11.1.5) as

$$[D] = \frac{200 \text{ GPa}}{(1 + 0.3)(1 - 0.6)} \begin{bmatrix} 0.7 & 0.3 & 0.3 & 0 & 0 & 0 \\ & 0.7 & 0.3 & 0 & 0 & 0 \\ & & 0.7 & 0 & 0 & 0 \\ & & & 0.2 & 0 & 0 \\ & & & & 0.2 & 0 \\ \text{Symmetry} & & & & & 0.2 \end{bmatrix} \tag{11.2.29}$$

Finally, substituting the results from Eqs. (11.2.24) for V, (11.2.28) for $[B]$, and (11.2.29) for $[D]$ into Eq. (11.2.18), we obtain the element stiffness matrix. The resulting 12×12 matrix is shown as

$$[k] = [B]^T[D][B]V$$

$$10^6 \times \begin{bmatrix}
3.846 & 0 & 0 & -0.962 & 0 & -1.923 & -0.962 & 0 & -1.923 & -1.923 & 0 & 3.846 \\
0 & 3.846 & 0 & 0 & -0.962 & -3.846 & 0 & -0.962 & 3.846 & 0 & -1.923 & 0 \\
0 & 0 & 13.462 & -2.885 & -5.769 & -3.365 & -2.885 & 5.769 & -3.365 & 5.769 & 0 & -6.731 \\
-0.962 & 0 & -2.885 & 7.452 & 4.808 & 1.202 & -0.24 & -0.962 & 1.202 & -6.25 & -3.846 & 0.481 \\
0 & -0.962 & -5.769 & 4.808 & 14.663 & 2.404 & 0.962 & -12.26 & 0.481 & -5.769 & -1.442 & 2.885 \\
-1.923 & -3.846 & -3.365 & 1.202 & 2.404 & 5.649 & 1.202 & -0.481 & -2.043 & -0.481 & 1.923 & -0.24 \\
-0.962 & 0 & -2.885 & -0.24 & 0.962 & 1.202 & 7.452 & -4.808 & 1.202 & -6.25 & 3.846 & 0.481 \\
0 & -0.962 & 5.769 & -0.962 & -12.26 & -0.481 & -4.808 & 14.663 & -2.404 & 5.769 & -1.442 & -2.885 \\
-1.923 & 3.846 & -3.365 & 1.202 & 0.481 & -2.043 & 1.202 & -2.404 & 5.649 & -0.481 & -1.923 & -0.24 \\
-1.923 & 0 & 5.769 & -6.25 & -5.769 & -0.481 & -6.25 & 5.769 & -0.481 & 14.423 & 0 & -4.808 \\
0 & -1.923 & 0 & -3.846 & -1.442 & 1.923 & 3.846 & -1.442 & -1.923 & 0 & 4.808 & 0 \\
3.846 & 0 & -6.731 & 0.481 & 2.885 & -0.24 & 0.481 & -2.885 & -0.24 & -4.808 & 0 & 7.212
\end{bmatrix}$$

■

▲ 11.3 Isoparametric Formulation

We now describe the isoparametric formulation of the stiffness matrix for some three-dimensional hexahedral (brick) elements.

Linear Hexahedral Element

The basic (linear) hexahedral element [Figure 11–5(a)] now has eight corner nodes with isoparametric natural coordinates given by s, t, and z' as shown in Figure 11–5(b). The element faces are now defined by $s, t, z' = \pm 1$. (We use s, t, and z' for the coordinate axes because they are probably simpler to use than Greek letters ξ, η, and ζ).

The formulation of the stiffness matrix follows steps analogous to the isoparametric formulation of the stiffness matrix for the plane element in Chapter 10.

The function used to describe the element geometry for x in terms of the generalized degrees of freedom a_i's is

$$x = a_1 + a_2 s + a_3 t + a_4 z' + a_5 st + a_6 tz' + a_7 z's + a_8 stz' \tag{11.3.1}$$

The same form as Eq. (11.3.1) is used for y and z as well. Just start with a_9 through a_{16} for y and a_{17} through a_{24} for z.

First, we expand Eq. (10.2.4) to include the z coordinate as follows:

$$\begin{Bmatrix} x \\ y \\ z \end{Bmatrix} = \sum_{i=1}^{8} \left(\begin{bmatrix} N_i & 0 & 0 \\ 0 & N_i & 0 \\ 0 & 0 & N_i \end{bmatrix} \begin{Bmatrix} x_i \\ y_i \\ z_i \end{Bmatrix} \right) \tag{11.3.2}$$

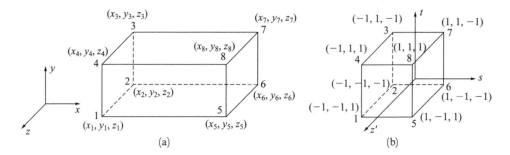

Figure 11–5 Linear hexahedral element (a) in a global-coordinate system and (b) element mapped into a cube of two unit sides placed symmetrically with natural or intrinsic coordinates s, t, and z'

where the shape functions are now given by

$$N_i = \frac{(1 + ss_i)(1 + tt_i)(1 + z'z_i')}{8} \tag{11.3.3}$$

with $s_i, t_i, z_i' = \pm 1$ and $i = 1, 2, \ldots, 8$. For instance,

$$N_1 = \frac{(1 + ss_1)(1 + tt_1)(1 + z'z_1')}{8} \tag{11.3.4}$$

and when, from Figure 11–5, $s_1 = -1$, $t_1 = -1$, and $z_1' = +1$ are used in Eq. (11.3.4), we obtain

$$N_1 = \frac{(1 - s)(1 - t)(1 + z')}{8} \tag{11.3.5a}$$

Explicit forms of the other shape functions follow similarly. The shape functions in Eq. (11.3.3) map the natural coordinates (s, t, z') of any point in the element to any point in the global coordinates (x, y, z) when used in Eq. (11.3.2). For instance, when we let $i = 8$ and substitute $s_8 = 1$, $t_8 = 1$, $z_8' = 1$ into Eq. (11.3.3) for N_8, we obtain

$$N_8 = \frac{(1 + s)(1 + t)(1 + z')}{8} \tag{11.3.5b}$$

Similar expressions are obtained for the other shape functions. Then evaluating all shape functions at node 8, we obtain $N_8 = 1$, and all other shape functions equal zero at node 8. [From Eq. (11.3.5a), we see that $N_1 = 0$ when $s = 1$ or when $t = 1$.] Therefore, using Eq. (11.3.2), we obtain

$$x = x_8 \qquad y = y_8 \qquad z = z_8$$

We see that indeed Eq. (11.3.2) maps any point in the natural-coordinate system to one in the global-coordinate system.

The displacement functions in terms of the generalized degrees of freedom are of the same form as used to describe the element geometry given by Eq. (11.3.1).
That is,

$$u = a_1 + a_2s + a_3t + a_4z' + a_5st + a_6tz' + a_7z's + a_8stz' \qquad (11.3.6a)$$

with similar expressions used for displacements v and w. There are now a total of 24 degrees of freedom in the linear hexahedral element. Therefore, we use the same shape functions as used to describe the geometry (Eq. 11.3.3)). The displacement functions now include w such that

$$\left\{ \begin{array}{c} u \\ v \\ w \end{array} \right\} = \sum_{i=1}^{8} \left(\left[\begin{array}{ccc} N_i & 0 & 0 \\ 0 & N_i & 0 \\ 0 & 0 & N_i \end{array} \right] \left\{ \begin{array}{c} u_i \\ v_i \\ w_i \end{array} \right\} \right) \qquad (11.3.6b)$$

with the same shape functions as defined by Eq. (11.3.3) and the size of the shape function matrix now 3×24.
The Jacobian matrix [Eq. (10.2.10)] is now expanded to

$$[J] = \left[\begin{array}{ccc} \dfrac{\partial x}{\partial s} & \dfrac{\partial y}{\partial s} & \dfrac{\partial z}{\partial s} \\ \dfrac{\partial x}{\partial t} & \dfrac{\partial y}{\partial t} & \dfrac{\partial z}{\partial t} \\ \dfrac{\partial x}{\partial z'} & \dfrac{\partial y}{\partial z'} & \dfrac{\partial z}{\partial z'} \end{array} \right] \qquad (11.3.7)$$

Because the strain–displacement relationships, given by Eq. (11.2.11) in terms of global coordinates, include differentiation with respect to z, we expand Eq. (10.2.9) as follows:

$$\frac{\partial f}{\partial x} = \frac{\left| \begin{array}{ccc} \dfrac{\partial f}{\partial s} & \dfrac{\partial y}{\partial s} & \dfrac{\partial z}{\partial s} \\ \dfrac{\partial f}{\partial t} & \dfrac{\partial y}{\partial t} & \dfrac{\partial z}{\partial t} \\ \dfrac{\partial f}{\partial z'} & \dfrac{\partial y}{\partial z'} & \dfrac{\partial z}{\partial z'} \end{array} \right|}{|[J]|} \qquad \frac{\partial f}{\partial y} = \frac{\left| \begin{array}{ccc} \dfrac{\partial x}{\partial s} & \dfrac{\partial f}{\partial s} & \dfrac{\partial z}{\partial s} \\ \dfrac{\partial x}{\partial t} & \dfrac{\partial f}{\partial t} & \dfrac{\partial z}{\partial t} \\ \dfrac{\partial x}{\partial z'} & \dfrac{\partial f}{\partial z'} & \dfrac{\partial z}{\partial z'} \end{array} \right|}{|[J]|}$$

$$\frac{\partial f}{\partial z} = \frac{\left| \begin{array}{ccc} \dfrac{\partial x}{\partial s} & \dfrac{\partial y}{\partial s} & \dfrac{\partial f}{\partial s} \\ \dfrac{\partial x}{\partial t} & \dfrac{\partial y}{\partial t} & \dfrac{\partial f}{\partial t} \\ \dfrac{\partial x}{\partial z'} & \dfrac{\partial y}{\partial z'} & \dfrac{\partial f}{\partial z'} \end{array} \right|}{|[J]|} \qquad (11.3.8)$$

Table 11–1 Table of Gauss points for linear hexahedral element with associated weights^a

Points, i	s_i	t_i	z_i'	Weight, W_i
1	$-1/\sqrt{3}$	$-1/\sqrt{3}$	$1/\sqrt{3}$	1
2	$1/\sqrt{3}$	$-1/\sqrt{3}$	$1/\sqrt{3}$	1
3	$1/\sqrt{3}$	$1/\sqrt{3}$	$1/\sqrt{3}$	1
4	$-1/\sqrt{3}$	$1/\sqrt{3}$	$1/\sqrt{3}$	1
5	$-1/\sqrt{3}$	$-1/\sqrt{3}$	$-1/\sqrt{3}$	1
6	$1/\sqrt{3}$	$-1/\sqrt{3}$	$-1/\sqrt{3}$	1
7	$1/\sqrt{3}$	$1/\sqrt{3}$	$-1/\sqrt{3}$	1
8	$-1/\sqrt{3}$	$1/\sqrt{3}$	$-1/\sqrt{3}$	1

^{a} $1/\sqrt{3} = 0.57735.$

Using Eqs. (11.3.8) by substituting u, v, and then w for f and using the definitions of the strains, we can express the strains in terms of natural coordinates (s, t, z') to obtain an equation similar to Eq. (10.2.14). In compact form, we can again express the strains in terms of the shape functions and global nodal coordinates similar to Eq. (10.2.15). The matrix $[B]$, given by a form similar to Eq. (10.2.17), is now a function of $s, t,$ and z' and is of order 6×24.

The 24×24 stiffness matrix is now given by

$$[k] = \int_{-1}^{1} \int_{-1}^{1} \int_{-1}^{1} [B]^T [D][B] |[J]| \, ds \, dt \, dz' \tag{11.3.9a}$$

Again, it is best to evaluate $[k]$ by numerical integration (also see Section 10.3); that is, we evaluate (integrate) the eight-node hexahedral element stiffness matrix using a $2 \times 2 \times 2$ rule (or two-point rule). Actually, eight points defined in Table 11–1 are used to evaluate $[k]$ as

$$[k] = \sum_{i=1}^{8} [B(s_i, t_i, z_i')]^T [D][B(s_i, t_i, z_i')] |[J(s_i, t_i, z_i')]| W_i W_j W_k \tag{11.3.9b}$$

where $W_i = W_j = W_k$ for the two-point rule.

As is true with the bilinear quadrilateral element described in Section 10.2, the eight-noded linear hexahedral element cannot model beam-bending action well because the element sides remain straight during the element deformation. During the bending process, the elements will be stretched and can shear lock. This concept of shear locking is described in more detail in [12] along with ways to remedy it. However, the quadratic hexahedral element described subsequently remedies the shear locking problem.

Quadratic Hexahedral Element

For the quadratic hexahedral element shown in Figure 11–6, we have a total of 20 nodes with the inclusion of a total of 12 midside nodes.

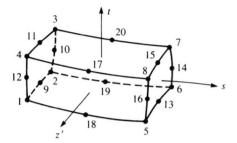

Figure 11–6 Quadratic hexahedral isoparametric element

The function describing the element geometry for x in terms of the 20 a_i's is

$$x = a_1 + a_2 s + a_3 t + a_4 z' + a_5 st + a_6 tz' + a_7 z's + a_8 s^2 + a_9 t^2$$
$$+ a_{10} z'^2 + a_{11} s^2 t + a_{12} st^2 + a_{13} t^2 z' + a_{14} tz'^2 + a_{15} z'^2 s$$
$$+ a_{16} z's^2 + a_{17} stz' + a_{18} s^2 tz' + a_{19} st^2 z' + a_{20} stz'^2 \qquad (11.3.10)$$

Similar expressions describe the y and z coordinates.

The x-displacement function u is described by the same polynomial used for the x element geometry in Eq. (11.3.10). Similar expressions are used for displacement functions v and w. In order to satisfy interelement compatibility, the three cubic terms s^3, t^3, and z'^3 are not included. Instead the three quartic terms $s^2 tz'$, $st^2 z'$, and stz'^2 are used.

The development of the stiffness matrix follows the same steps we outlined before for the linear hexahedral element, where the shape functions now take on new forms. Again, letting $s_i, t_i, z'_i = \pm 1$, we have for the corner nodes $(i = 1, 2, \ldots, 8)$,

$$N_i = \frac{(1 + ss_i)(1 + tt_i)(1 + z'z'_i)}{8}(ss_i + tt_i + z'z'_i - 2) \qquad (11.3.11)$$

For the midside nodes at $s_i = 0$, $t_i = \pm 1$, $z'_i = \pm 1$ $(i = 17, 18, 19, 20)$, we have

$$N_i = \frac{(1 - s^2)(1 + tt_i)(1 + z'z'_i)}{4} \qquad (11.3.12)$$

For the midside nodes at $s_i = \pm 1$, $t_i = 0$, $z'_i = \pm 1$ $(i = 10, 12, 14, 16)$, we have

$$N_i = \frac{(1 + ss_i)(1 - t^2)(1 + z'z'_i)}{4} \qquad (11.3.13)$$

Finally, for the midside nodes at $s_i = \pm 1$, $t_i = \pm 1$, $z'_i = 0$ $(i = 9, 11, 13, 15)$, we have

$$N_i = \frac{(1 + ss_i)(1 + tt_i)(1 - z'^2)}{4} \qquad (11.3.14)$$

The $[B]$ matrix is now a 60 × 60 matrix. Therefore, using Eq. (11.3.9a), the stiffness matrix of the quadratic hexahedral element is of order 60 × 60. This is consistent with the fact that the element has 20 nodes and 3 degrees of freedom $(u_i, v_i, \text{ and } w_i)$ per node.

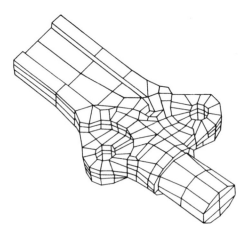

Figure 11–7   Finite element model of a forging using linear and quadratic solid elements

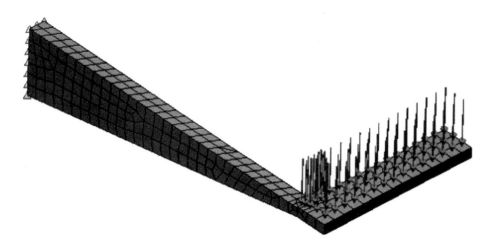

Figure 11–8   Meshed model of steel foot pedal (fixed along left back face and total downward acting surface force of 100 N applied uniformly over front pedal surface) (Courtesy of Justin Hronek)

The stiffness matrix for this 20-node quadratic solid element can be evaluated using a $3 \times 3 \times 3$ rule (27 points). However, a special 14-point rule may be a better choice [9, 10].

As with the eight-noded plane element of Section 10.5 (Figure 10–15), the 20-node solid element is also called a serendipity element.

Figures 1–7 and 11–7 show applications of the use of linear and quadratic (curved sides) solid elements to model three-dimensional solids.

Finally, commercial computer programs, such as [11] (also see references [46–56] of Chapter 1), are available to solve three-dimensional problems. Figures 11–8, 11–9,

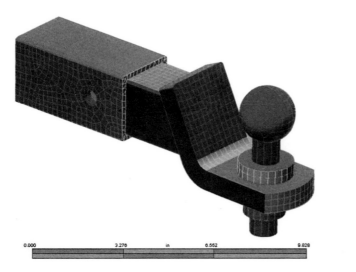

Figure 11–9 Meshed model of a trailer hitch (Courtesy of David Anderson) (See the full-color insert for a color version of this figure.)

Figure 11–10 Meshed model of an alternator bracket (Courtesy of Andrew Heckman, Design Engineer, Seagrave Fire Apparatus, LLC)

and 11–10 show a steel foot pedal, a trailer hitch, and an alternator bracket solved using a computer program [11]. We emphasize that these problems have been solved using the three-dimensional element as opposed to using a two-dimensional element, such as described in Chapters 6 and 8, as these problems have a three-dimensional stress state occurring in them. That is, the three normal and three shear stresses are of similar order of magnitude in some parts of the foot pedal, the trailer hitch, and the alternator bracket. The most accurate results will then occur when modeling these problems using the three-dimensional brick or tetrahedral elements (or a combination of both).

For the foot pedal, modeled with brick elements, the largest von Mises stress was 71.1 MPa located at the interior corner of the elbow. The maximum displacement was

Table 11–2 Table comparing results for cantilever beam modeled using 4-noded-tetrahedral, 10-noded tetrahedral, 8-noded brick, and 20-noded brick element

Solid Element Used	Number of Nodes	Number of Degrees of Freedom	Number of Elements	Free End Displ., mm	Principal Stress, MPa
4-noded tet	30	90	61	0.1346	3.87
4-noded tet	415	1245	1549	0.7163	16.25
4-noded tet	896	2688	3729	1.067	22.64
4-noded tet	1658	4974	7268	1.392	27.97
10-noded tet	144	432	61	2.997	45.51
10-noded tet	2584	7752	1549	3.244	54.95
8-noded brick	64	192	27	3.023	40.63
8-noded brick	343	1029	216	3.183	44.87
8-noded brick	1331	3993	1000	3.243	47.86
20-noded brick	208	624	27	3.175	54.46
20-noded brick	1225	3675	216	3.264	57.57
20-noded brick	4961	14,883	1000	3.294	57.39
Classical solution				3.266	47.85

(Mr. William Gobeli for creating the results for Table 11–2)

0.439 mm down at the front free end corner. (See Problem 11.14 for detailed dimensions and material properties used.)

For the steel trailer hitch shown in Figure 11–9, subjected to both a lateral and downward load of 12.59 kN each on the ball, the largest von Mises stress away from the unrealistic high stress located at the point load that was applied at the base of the ball is 406.75 MPa located at the inside re-entrant curve of the hitch. The largest displacement magnitude was 1.52 mm located at the top of the ball. This displacement magnitude also matches the value obtained through experimental testing of the hitch under the same load conditions used in the finite element analysis.

For the alternator bracket made of ASTM-A36 hot-rolled steel, the model consisted of 13,298 solid brick elements and 10,425 nodes. A total load of 4.45 kN was applied downward to the flat front face piece. The bracket back side was constrained against displacement. The largest von Mises stress was 79.54 MPa located at the top surface near the center (narrowest) section of the bracket. The largest vertical deflection was 0.412 mm at the front tip of the outer edge of the alternator bracket.

It has been shown [3] that use of the simple eight-noded hexahedral element yields better results than use of the constant-strain tetrahedral discussed in Section 11.1. Table 11.2 also illustrates the comparison between the corner-noded (constant-strain) tetrahedral, the linear-strain tetrahedral (mid-edge nodes added), the 8-noded brick, and the 20-noded brick models for a three-dimensional cantilever beam of length 2.54 m, base 15.24 cm, and height 30.48 cm. The beam has an end load of 44.48 kN acting upward and is made of steel ($E = 210$ GPa). A typical 8-noded brick model with the principal stress plot is shown in Figure 11–11. The classical beam theory solution for the vertical displacement and bending stress is also included for comparison. We can observe that the constant-strain tetrahedral gives very poor results, whereas the

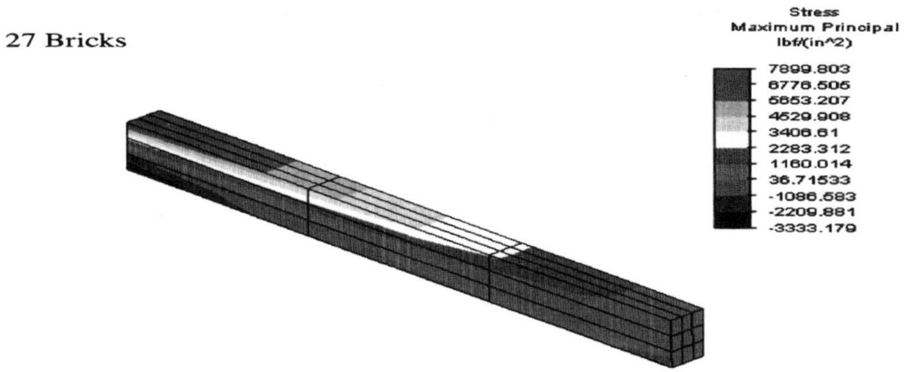

27 Bricks

Stress
Maximum Principal
lbf/(in^2)

7899.803
6776.505
5653.207
4529.908
3406.61
2283.312
1160.014
36.71533
-1086.583
-2209.881
-3333.179

Figure 11–11 Eight-noded brick model (27 Bricks) showing principal stress plot

linear tetrahedral gives much better results. This is because the linear-strain model predicts the beam-bending behavior much better. The 8-noded and 20-noded brick models yield similar but accurate results compared to the classical beam theory results.

In summary, the use of the three-dimensional elements results in a large number of equations to be solved simultaneously. For instance, a model using a simple cube with, say, 20 by 20 by 20 nodes ($= 8000$ total nodes) for a region requires 8000 times 3 degrees of freedom per node ($= 24,000$) simultaneous equations.

References [4–7] report on early three-dimensional programs and analysis procedures using solid elements such as a family of subparametric curvilinear elements, linear tetrahedral elements, and 8-noded linear and 20-noded quadratic isoparametric elements.

▲ Summary Equations

Strain–displacement equations:

$$\varepsilon_x = \frac{\partial u}{\partial x} \qquad \varepsilon_y = \frac{\partial v}{\partial y} \qquad \varepsilon_z = \frac{\partial w}{\partial z} \tag{11.1.1}$$

$$\gamma_{xy} = \frac{\partial u}{\partial y} + \frac{\partial v}{\partial x} = \gamma_{yx} \qquad \gamma_{yx} = \frac{\partial v}{\partial z} + \frac{\partial w}{\partial y} = \gamma_{zy} \qquad \gamma_{zx} = \frac{\partial w}{\partial x} + \frac{\partial u}{\partial z} = \gamma_{xz} \tag{11.1.2}$$

Stress and strain matrices:

$$\{\sigma\} = \begin{Bmatrix} \sigma_x \\ \sigma_y \\ \sigma_z \\ \tau_{xy} \\ \tau_{yz} \\ \tau_{zx} \end{Bmatrix} \qquad \{\varepsilon\} = \begin{Bmatrix} \varepsilon_x \\ \varepsilon_y \\ \varepsilon_z \\ \gamma_{xy} \\ \gamma_{yz} \\ \gamma_{zx} \end{Bmatrix} \tag{11.1.3}$$

Constitutive matrix:

$$[D] = \frac{E}{(1+v)(1-2v)} \begin{bmatrix} 1-v & v & v & 0 & 0 & 0 \\ & 1-v & v & 0 & 0 & 0 \\ & & 1-v & 0 & 0 & 0 \\ & & & \dfrac{1-2v}{2} & 0 & 0 \\ & & & & \dfrac{1-2v}{2} & 0 \\ \text{Symmetry} & & & & & \dfrac{1-2v}{2} \end{bmatrix} \quad (11.1.5)$$

Displacement functions:

$$u(x, y, z) = a_1 + a_2 x + a_3 y + a_4 z$$

$$v(x, y, z) = a_5 + a_6 x + a_7 y + a_8 z \quad (11.2.2)$$

$$w(x, y, z) = a_9 + a_{10} x + a_{11} y + a_{12} z$$

Shape functions for tetrahedral element:

$$N_1 = \frac{(\alpha_1 + \beta_1 x + \gamma_1 y + \delta_1 z)}{6V} \qquad N_2 = \frac{(\alpha_2 + \beta_2 x + \gamma_2 y + \delta_2 z)}{6V}$$

$$N_3 = \frac{(\alpha_3 + \beta_3 x + \gamma_3 y + \delta_3 z)}{6V} \qquad N_4 = \frac{(\alpha_4 + \beta_4 x + \gamma_4 y + \delta_4 z)}{6V} \quad (11.2.10)$$

and

$$6V = \begin{vmatrix} 1 & x_1 & y_1 & z_1 \\ 1 & x_2 & y_2 & z_2 \\ 1 & x_3 & y_3 & z_3 \\ 1 & x_4 & y_4 & z_4 \end{vmatrix} \quad (11.2.4)$$

Gradient matrix:

$$[B_1] = \frac{1}{6V} \begin{bmatrix} \beta_1 & 0 & 0 \\ 0 & \gamma_1 & 0 \\ 0 & 0 & \delta_1 \\ \gamma_1 & \beta_1 & 0 \\ 0 & \delta_1 & \gamma_1 \\ \delta_1 & 0 & \beta_1 \end{bmatrix} \quad (11.2.15)$$

Stiffness matrix for tetrahedral element:

$$[k] = [B]^T [D] [B] V \quad (11.2.18)$$

Body-force matrix for tetrahedral element:

$$\{f_b\} = \frac{1}{4}[X_b \ Y_b \ Z_b \ X_b \ Y_b \ Z_b \ X_b \ Y_b \ Z_b \ X_b \ Y_b \ Z_b]^T \qquad (11.2.20b)$$

Surface-force matrix along face with nodes 1 through 3 for tetrahedral element:

$$\{f_s\} = \frac{S_{123}}{3} \begin{Bmatrix} p_x \\ p_y \\ p_z \\ p_x \\ p_y \\ p_z \\ p_x \\ p_y \\ p_z \\ 0 \\ 0 \\ 0 \end{Bmatrix} \qquad (11.2.23)$$

Function to define the geometry for eight-noded linear hexahedral element:

$$x = a_1 + a_2 s + a_3 t + a_4 z' + a_5 st + a_6 tz' + a_7 z' s + a_8 stz' \qquad (11.3.1)$$

Shape functions for isoparametric 8-noded brick element:

$$N_i = \frac{(1 + ss_i)(1 + tt_i)(1 + z'z_i')}{8} \qquad (11.3.3)$$

x direction displacement function for eight-noded brick element:

$$u = a_1 + a_2 s + a_3 t + a_4 z' + a_5 st + a_6 tz' + a_7 z' s + a_8 stz' \qquad (11.3.6a)$$

Stiffness matrix for eight-noded brick element:

$$[k] = \int_{-1}^{1} \int_{-1}^{1} \int_{-1}^{1} [B]^T [D][B]|[J]| \, ds \, dt \, dz' \qquad (11.3.9a)$$

$2 \times 2 \times 2$ rule (8 point rule) for evaluating stiffness matrix of eight-noded brick element:

$$[k] = \sum_{i=1}^{8} [B(s_i, t_i, z_i')]^T [D][B(s_i, t_i, z_i')]|[J(s_i, t_i, z_i')]| W_i W_j W_k \qquad (11.3.9b)$$

Table 11–1 lists the Gauss points for a linear brick element.

Function describing the element geometry for 20-noded quadratic brick element:

$$x = a_1 + a_2 s + a_3 t + a_4 z' + a_5 st + a_6 tz' + a_7 z' s + a_8 s^2 + a_9 t^2$$
$$+ a_{10} z'^2 + a_{11} s^2 t + a_{12} st^2 + a_{13} t^2 z' + a_{14} tz'^2 + a_{15} z'^2 s$$
$$+ a_{16} z' s^2 + a_{17} stz' + a_{18} s^2 tz' + a_{19} st^2 z' + a_{20} stz'^2 \qquad (11.3.10)$$

Similar expressions describe the y and z coordinates.

Shape functions for 20-noded brick element:

$$N_i = \frac{(1 + ss_i)(1 + tt_i)(1 + z'z'_i)}{8}(ss_i + tt_i + z'z'_i - 2) \quad (i = 1, 2, ..., 8) \quad (11.3.11)$$

$$N_i = \frac{(1 - s^2)(1 + tt_i)(1 + z'z'_i)}{4} \quad (i = 17, 18, 19, 20) \quad (11.3.12)$$

$$N_i = \frac{(1 + ss_i)(1 - t^2)(1 + z'z'_i)}{4} \quad (i = 10, 12, 14, 16) \quad (11.3.13)$$

$$N_i = \frac{(1 + ss_i)(1 + tt_i)(1 - z'^2)}{4} \quad (i = 9, 11, 13, 15) \quad (11.3.14)$$

▲ References

[1] Martin, H. C., "Plane Elasticity Problems and the Direct Stiffness Method." *The Trend in Engineering*, Vol. 13, pp. 5–19, Jan. 1961.

[2] Gallagher, R. H., Padlog, J., and Bijlaard, P. P., "Stress Analysis of Heated Complex Shapes," *Journal of the American Rocket Society*, pp. 700–707, May 1962.

[3] Melosh, R. J., "Structural Analysis of Solids," *Journal of the Structural Division*, American Society of Civil Engineers, pp. 205–223, Aug. 1963.

[4] Chacour, S., "DANUTA, a Three-Dimensional Finite Element Program Used in the Analysis of Turbo-Machinery," Transactions of the American Society of Mechanical Engineers, *Journal of Basic Engineering*, March 1972.

[5] Rashid, Y. R., "Three-Dimensional Analysis of Elastic Solids-I: Analysis Procedure," *International Journal of Solids and Structures*, Vol. 5, pp. 1311–1331, 1969.

[6] Rashid, Y. R., "Three-Dimensional Analysis of Elastic Solids-II: The Computational Problem," *International Journal of Solids and Structures*, Vol. 6, pp. 195–207, 1970.

[7] *Three-Dimensional Continuum Computer Programs for Structural Analysis*, Cruse, T. A., and Griffin, D. S., eds., American Society of Mechanical Engineers, 1972.

[8] Zienkiewicz, O. C., *The Finite Element Method*, 3rd ed., McGraw-Hill, London, 1977.

[9] Irons, B. M., "Quadrature Rules for Brick Based Finite Elements," *International Journal for Numerical Methods in Engineering*, Vol. 3, No. 2, pp. 293–294, 1971.

[10] Hellen, T. K., "Effective Quadrature Rules for Quadratic Solid Isoparametric Finite Elements," *International Journal for Numerical Methods in Engineering*, Vol. 4, No. 4, pp. 597–599, 1972.

[11] Linear Stress and Dynamics Reference Division, Docutech On-line Documentation, Algor, Inc., Pittsburgh, PA.

[12] Cook, R. D., Malkus, D. S., Plesha, M. E., and Witt, R. J., *Concepts and Applications of Finite Element Analysis*, 4th ed., Wiley, New York, 2002.

▲ Problems

11.1 Evaluate the matrix $[B]$ for the tetrahedral solid element shown in Figure P11–1.

11.2 Evaluate the stiffness matrix for the elements shown in Figure P11–1. Let $E = 206$ GPa and $v = 0.3$.

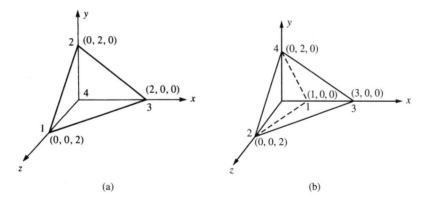

Figure P11–1

11.3 For the elements shown in Figure P11–1, assume the nodal displacements have been determined to be

$$u_1 = 0.125 \text{ mm} \qquad v_1 = 0.0 \qquad w_1 = 0.0$$

$$u_2 = 0.025 \text{ mm} \qquad v_2 = 0.0 \qquad w_2 = 0.025 \text{ mm}$$

$$u_3 = 0.125 \text{ mm} \qquad v_3 = 0.0 \qquad w_3 = 0.0$$

$$u_4 = -0.025 \text{ mm} \qquad v_4 = 0.0 \qquad w_4 = 0.125 \text{ mm}$$

Determine the strains and then the stresses in the elements. Let $E = 206$ GPa and $v = 0.3$.

11.4 What is special about the strains and stresses in the tetrahedral element?

11.5 Show that for constant body force Z_b acting on an element ($X_b = 0$ and $Y_b = 0$),

$$\{f_{bi}\} = \frac{V}{4} \left\{ \begin{matrix} 0 \\ 0 \\ Z_b \end{matrix} \right\}$$

where $\{f_{bi}\}$ represents the body forces at node i of the element with volume V.

11.6 Evaluate the $[B]$ matrix for the tetrahedral solid element shown in Figure P11–6. The coordinates are in units of millimeters.

11.7 For the elements shown in Figure P11–6, assume the nodal displacements have been determined to be

$$u_1 = 0.0 \qquad v_1 = 0.0 \qquad w_1 = 0.0$$

$$u_2 = 0.01 \text{ mm} \qquad v_2 = 0.02 \text{ mm} \qquad w_2 = 0.01 \text{ mm}$$

$$u_3 = 0.02 \text{ mm} \qquad v_3 = 0.01 \text{ mm} \qquad w_3 = 0.005 \text{ mm}$$

$$u_4 = 0.0 \qquad v_4 = 0.01 \text{ mm} \qquad w_4 = 0.01 \text{ mm}$$

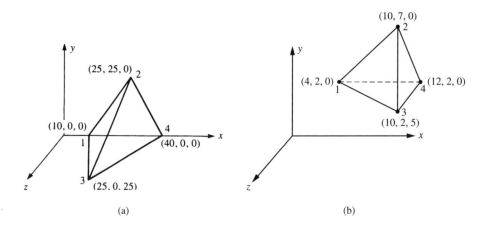

(a)   (b)

Figure P11–6

Determine the strains and then the stresses in the elements. Let $E = 210$ GPa and $v = 0.3$.

11.8 For the linear strain tetrahedral element shown in Figure P11–8, (a) express the displacement fields u, v, and w in the x, y and z directions, respectively. Hint: There are 10 nodes each with three translational degrees of freedom, u_i, v_i, and w_i. Also look at the linear strain triangle given by Eq. (8.1.2) or the expansion of Eqs. (11.2.2).

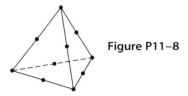

Figure P11–8

11.9 Figure P11–9 shows how solid and plane elements may be connected. What restriction must be placed on the externally applied loads for this connection to be acceptable?

11.10 Express the explicit shape functions N_2 through N_8, similar to N_1 given by Eq. (11.3.4), for the linear hexahedral element shown in Figure 11–5 on page 546.

11.11 Express the explicit shape functions for the corner nodes of the quadratic hexahedral element shown in Figure 11–6 on page 549.

11.12 Write a computer program to evaluate $[k]$ of Eq. (11.3.9a) using a $2 \times 2 \times 2$ Gaussian quadrature rule.

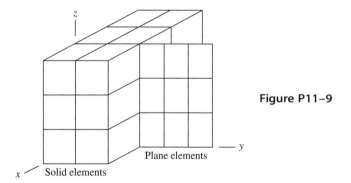

Figure P11-9

Solve the following problems using a computer program.

11.13 Determine the deflections at the four corners of the free end of the structural steel cantilever beam shown in Figure P11–13. Also determine the maximum principal stress. Compare your answer for deflections to the classical beam theory equation $(\delta = PL^3/(3EI))$.

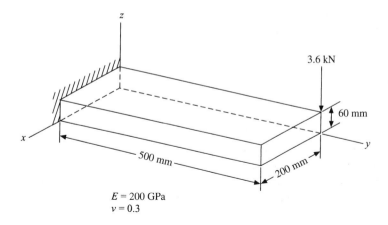

Figure P11-13

11.14 A portion of a structural steel brake pedal in a vehicle is modeled as shown in Figure P11–14. Determine the maximum deflection at the pedal under a uniform pressure acting over the pedal totaling 100 N.

11.15 For the compressor flap valve shown in Figure P11–15, determine the maximum operating pressure such that the material yield stress is not exceeded with a factor of safety of two. The valve is made of hardened 1020 steel with a modulus of elasticity of 210 GPa and a yield strength of 400 MPa. The valve thickness is a uniform 0.457 mm. The valve clip ears support the valve at opposite diameters. The pressure load is applied uniformly around the annular region.

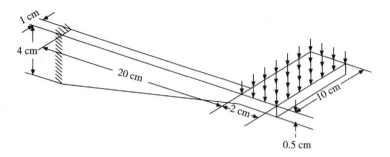

Figure P11–14

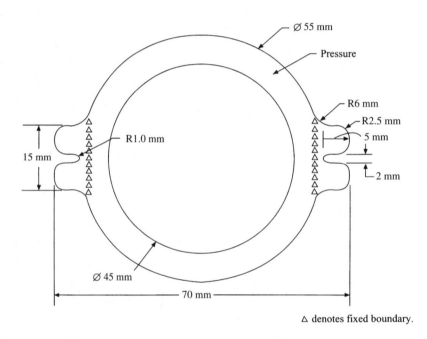

△ denotes fixed boundary.

Figure P11–15

11.16 An S-shaped block used in force measurement as shown in Figure P11–16 is to be designed for a pressure of 7 MPa psi applied uniformly to the top surface. Determine the uniform thickness of the block needed such that the sensor is compressed no more than 1 mm. Also make sure that the maximum stress from the maximum distortion energy failure theory is less than the yield strength of the material. Use a factor of safety of 1.5 on the stress only. The overall size of the block must fit in a 30 mm-high, 20 mm-wide, 20 mm-deep volume. The block should be made of steel.

11.17 A device is to be hydraulically loaded to resist an upward force $P = 27$ kN as shown in Figure P11–17. Determine the thickness of the device such that the maximum deflection is 2 mm vertically and the maximum stress is less than the yield strength

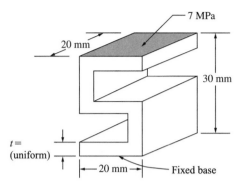

Figure P11–16 S-shaped block

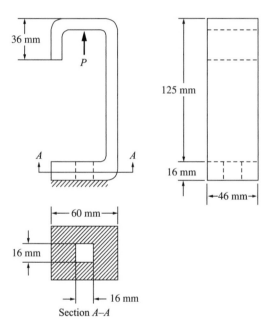

Section A–A

Figure P11–17 Hydraulically loaded device

using a factor of safety of 2 (only on the stress). The device must fit in a space 140 mm high, 60 mm wide, and 46 mm deep. The top flange is bent vertically as shown, and the device is clamped to the floor. Use steel for the material.

11.18 An "Allen" wrench is used to loosen a bolt that has a hex-head cross section. As shown in Figure P11–18. This wrench is a 5 mm size and is made of quenched and tempered carbon steel with modulus of elasticity of 200 GPa, Poisson's ratio of 0.29, and yield strength of 615 MPa. The wrench is used to loosen a rusty bolt. To simulate the fixity a

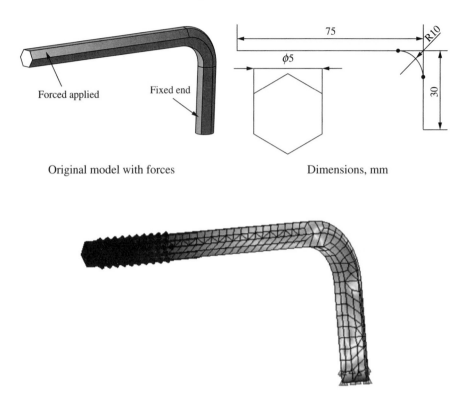

Figure P11–18 Allen wrench showing dimensions, loads, and typical finite element model (Compliments of Justin Hronek)

surface 2.5 mm in depth from the bottom is held fixed. A total force of 125 N is applied uniformly over 25 mm at the end of the horizontal section of the wrench. Determine the maximum von Mises stress in the wrench. Also determine the maximum displacement. Comment on the safety of the wrench based on whether it will yield or not. (This problem is compliments of Justin Hronek.)

11.19 A blacksmith desires to forge a work piece using the anvil shown in Figure P11–19. The anvil is bolted to a workbench with 114 mm diameter bolts. The anvil is made of gray cast iron with $E = 100$ GPa. The tensile and compressive strengths are 214 MPa and 751 MPa, respectively. A surface pressure of 6.9 MPa is applied to the horn of the anvil during the forging process. Determine the maximum principal stress and its location on the anvil. (This problem is compliments of Dan Baxter.)

11.20 A fork from a forklift is constrained by two bars (not shown) that fit into each L-shaped appendage on the vertical part of the fork as shown in Figure P11–20. The fork is made of AISI 4130 steel with $E = 206.84$ GPa, $v = 0.30$, and a yield strength of 360 MPa. The fork is loaded with 46,189 N/m^2 of surface traction on the top surface.

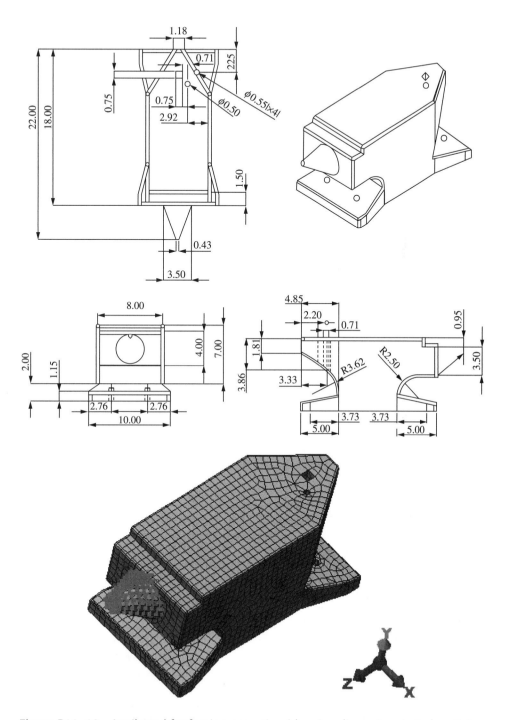

Figure P11–19 Anvil used for forging operation (showing dimensions in inch units) and typical finite element model (Compliments of Dan Baxter) (See the full-color insert for a color version of this figure.)

How much will the fork deflect, and what is the maximum von Mises stress? What is the factor of safety against yielding of the material? (This problem is compliments of Jay Emmerich.)

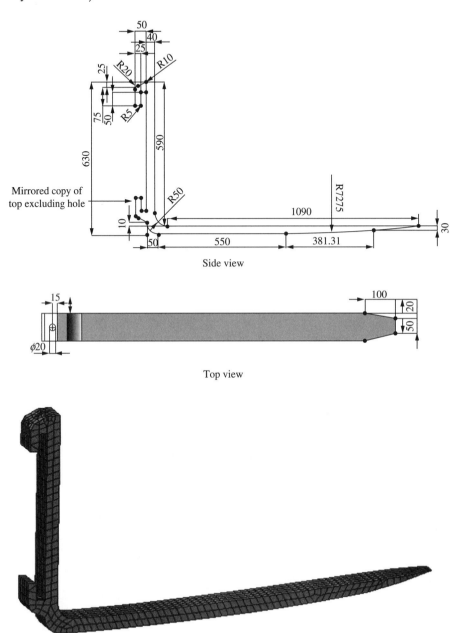

Figure P11–20 Fork from forklift showing dimensions (all dimensions in mm) and typical finite element model (Compliments of Jay Emmerich)

11.21 A radio-control car front steering unit is shown in Figure P11–21. The arm is made of molded ABS plastic with a modulus of elasticity of 2.5 GPa and a tensile strength of 41 MPa. The base of the steering unit is attached to the frame of the car by bolts, so the three holes that the bolts pass through are assumed fixed around their cylindrical surface (as shown in the finite element mode). A force of 13 N is applied circumferentially around the upper finger (as shown in the finite element model). This force represents the typical weight of a remote-control car. Determine the maximum von Mises stress and largest displacement of the control arm. (This problem is compliments of Phillip Grommes.)

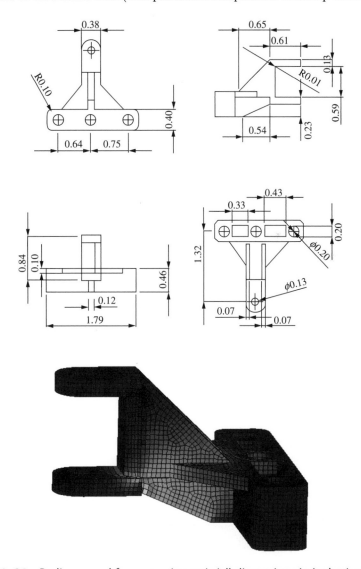

Figure P11–21 Radio-control front steering unit (all dimensions in inches) and finite element model (Compliments of Phillip Grommes) (See the full-color insert for a color version of this figure.)

11.22 The hitch shown in Figure P11–22 is used on an International 496 disk. The hitch is made of cold drawn 1018 steel with $E = 200$ GPa. The yield strength of the material is 370 MPa. The disk requires 200 hp to pull at 10 kmph. The total force of 55.6 kN in the hitch is then determined from force equal to power divided by velocity. Determine the maximum von Mises stress and the deflection of the hitch under the load. In the model, use two 27.8 N applied to each side of the hitch and fix the nodes at the ends of the attachment to the disk frame (as shown in the finite element model). (This problem is compliments of Byron Manternach.)

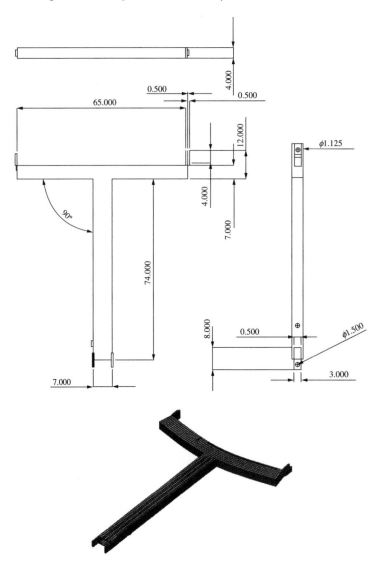

Figure P11–22 Hitch from a 24-foot-wide International 496 disk (dimensions in inches) and typical finite element model (Compliments of Byron Manternach)

11.23 A swivel C bracket shown in Figure P11–23 is mounted to a ceiling of a building and has a speaker (not shown) of 90 N hanging from each mounting hole. The mounting bracket is made of A 36 steel with a modulus of elasticity of 200 GPa, Poisson's ratio of 0.29, and yield strength of 250 MPa. Determine the maximum von Mises stress and deflection in the bracket. (This problem is compliments of Tyler Austin and Kyle Jones.)

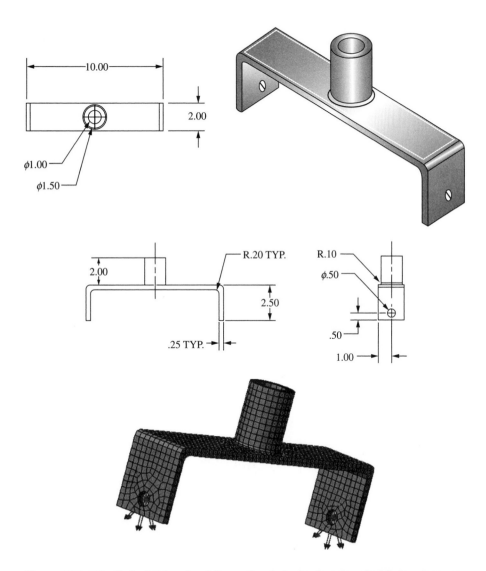

Figure P11–23 Swivel C bracket (dimension in inches) and typical finite element model (Compliments of Tyler Austin and Kyle Jones)

11.24 The lower arms of a front-end loader are shown in Figure P11–24. The loader material is AISI 1010 cold drawn steel with a modulus of elasticity of 205 GPa and Poisson's ratio of 0.29. The yield strength of the material is 305 MPa. In the finite element model, the back faces of the top horizontal members are fixed. Determine the maximum force that can be applied to the bottom of the left arm to cause yielding of the arm. You may want to try loads that are vertical (y-directed) and lateral (z-directed). (This problem is compliments of Quentin Moller.)

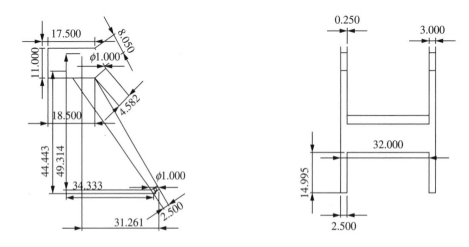

The depth of the 32-in. cross member 5 in. into the paper

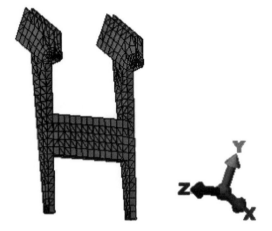

Figure P11–24 Lower arms of front end loader (dimensions in inches) and a typical finite element model (Compliments of Quentin Moller)

11.25 A bicycle stem is shown in Figure P11–25. The stem attaches the handlebars to the steerer tube of the fork. The stem is made of 7075-T6 aluminum alloy with yield strength of 504 MPa. The load of 1200 N is spread over the mounting surface to the handlebar in the *x-y* plane and acts at a 45° angle from the axis of the stem. The inner surface of the stem that attaches to the steerer tube is fixed in translation in the *y*-direction and in rotation about the *y*-axis. Determine the largest von Mises stress and its location on the stem. (This problem is courtesy of Stephen Wilson.)

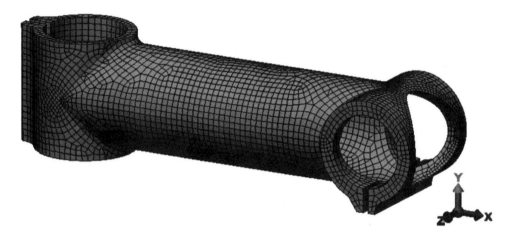

Figure P11–25 Bicycle stem (dimensions in mm) and typical finite element model (Compliments of Stephen Wilson)

11.26 The piston head shown in Figure P11–26 is made of aluminum alloy A356.0-T6, sand cast. The modulus of elasticity is 72.5 GPa. The Poisson's ratio is 0.33. The yield strength is 105 MPa. The pressure on the head is 1.275 MPa. Determine if the piston head is safe based on a factor of safety of 2.5 against yielding. The wrist pin hole is fixed on the top half to represent resistance on the piston head by the connecting rod. (This problem is compliments of Robert Jablonsky.)

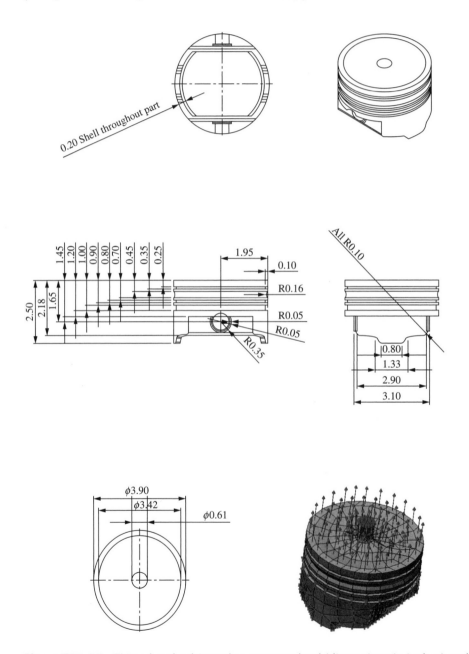

Figure P11–26 Piston head subjected to pressure load (dimensions in inches) and typical finite element model (Compliments of Robert Jablonsky)

11.27 A solid part shown in Figure P11–27 is made to locate parts into proper position. The material is AISI 1005 steel with $E = 200$ GPa and $v = 0.29$. The front faces are fixed, and a pressure P of 100 MPa is applied to the semi-circular face of the inside slot, as shown in the figure. Determine the largest von Mises stress and its location on the locator device.

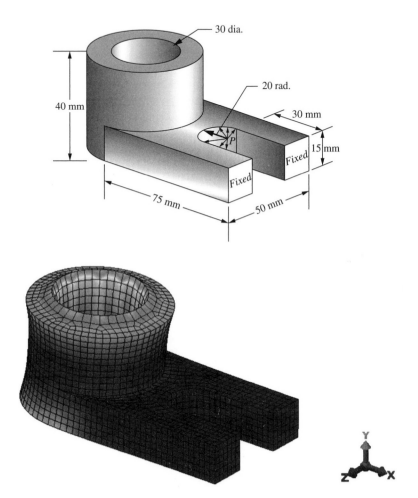

Figure P11–27 Locator part (dimensions in mm) with typical finite element model (See the full-color insert for a color version of this figure.)

CHAPTER OBJECTIVES

- To introduce basic concepts of plate bending.
- To derive a common plate bending element stiffness matrix.
- To present some plate element numerical comparisons.
- To demonstrate some computer solutions for plate bending problems.

Introduction

In this chapter, we will begin by describing elementary concepts of plate bending behavior and theory. The plate element is one of the more important structural elements and is used to model and analyze such structures as pressure vessels, chimney stacks (Figure 1–5), and automobile parts. Figure 12–1 shows finite element models of a computer case and a water tank modeled using the plate bending element described in this chapter. This description of plate bending is followed by a discussion of some commonly used plate finite elements. A large number of plate bending element formulations exist that would require a lengthy chapter to cover. Our purpose in this chapter is to present the derivation of the stiffness matrix for one of the most common plate bending finite elements and then to compare solutions to some classical problems from a variety of bending elements in the literature.

We finish the chapter with a solution to a plate bending problem using a computer program.

▲ 12.1 Basic Concepts of Plate Bending ▲

A plate can be considered the two-dimensional extension of a beam in simple bending. Both beams and plates support loads transverse or perpendicular to their plane and through bending action. A plate is flat (if it were curved, it would become a shell). A beam has a single bending moment resistance, while a plate resists bending about two axes and has a twisting moment.

We will consider the classical thin-plate theory or Kirchhoff plate theory [1]. Many of the assumptions of this theory are analogous to the classical beam theory or Euler–Bernoulli beam theory described in Chapter 4 and in Reference [2].

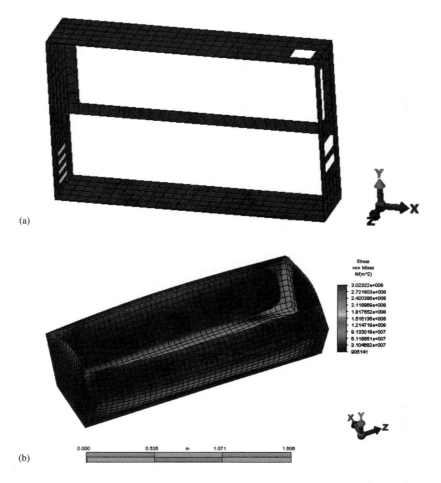

(a)

(b)

Figure 12–1 (a) Computer case and (b) water tank (See the full-color insert for a color version of this figure.)

Basic Behavior of Geometry and Deformation

We begin the derivation of the basic thin-plate equations by considering the thin plate in the x-y plane and of thickness t measured in the z direction shown in Figure 12–2. The plate surfaces are at $z = \pm t/2$, and its midsurface is at $z = 0$. The assumed basic geometry of the plate is as follows: (1) The plate thickness is much smaller than its in-plane dimensions b and c (that is, $t \ll b$ or c). (If t is more than about one-tenth the span of the plate, then transverse shear deformation must be accounted for and the plate is then said to be thick.) (2) The deflection w is much less than the thickness t (that is, $w/t \ll 1$).

Kirchhoff Assumptions

Consider a differential slice cut from the plate by planes perpendicular to the x axis as shown in Figure 12–3(a). Loading q causes the plate to deform laterally or upward in

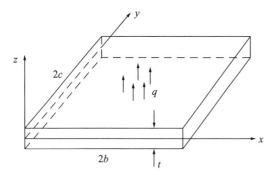

Figure 12–2 Basic thin plate showing transverse loading and dimensions

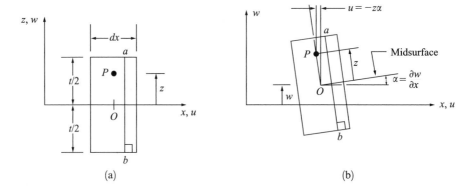

Figure 12–3 Differential slice of plate of thickness t (a) before loading and (b) displacements of point P after loading, based on Kirchhoff theory. Transverse shear deformation is neglected, and so right angles in the cross section remain right angles. Displacements in the y-z plane are similar

the z direction, and the deflection w of point P is assumed to be a function of x and y only; that is, $w = w(x, y)$ and the plate does not stretch in the z direction. A line a-b drawn perpendicular to the plate surfaces before loading remains perpendicular to the surfaces after loading [Figure 12–3(b)]. This is consistent with the Kirchhoff assumptions as follows:

1. **Normals remain normal.** This implies that transverse shear strains $\gamma_{yz} = 0$ and similarly $\gamma_{xz} = 0$. However, γ_{xy} does not equal 0; right angles in the plane of the plate may not remain right angles after loading. The plate may twist in the plane.
2. **Thickness changes can be neglected and normals undergo no extension.** This means normal strain, $\varepsilon_z = 0$.
3. **Normal stress σ_z has no effect on in-plane strains ε_x and ε_y in the stress–strain equations and is considered negligible.**

4. Membrane or in-plane forces are neglected here, and the plane stress resistance can be superimposed later (that is, the constant-strain triangle behavior of Chapter 6 can be superimposed with the basic plate bending element resistance). That is, the in-plane deformations in the x and y directions at the midsurface are assumed to be zero; $u(x, y, 0) = 0$ and $v(x, y, 0) = 0$.

Based on the Kirchhoff assumptions, any point P in Figure 12–3 has displacement in the x direction due to a small rotation α of

$$u = -z\alpha = -z\left(\frac{\partial w}{\partial x}\right) \tag{12.1.1}$$

and similarly the same point has displacement in the y direction of

$$v = -z\left(\frac{\partial w}{\partial y}\right) \tag{12.1.2}$$

The curvatures of the plate are then given as the rate of change of the angular displacements of the normals and are defined as

$$\kappa_x = -\frac{\partial^2 w}{\partial x^2} \qquad \kappa_y = -\frac{\partial^2 w}{\partial y^2} \qquad \kappa_{xy} = -\frac{2\partial^2 w}{\partial x \partial y} \tag{12.1.3}$$

The first of Eqs. (12.1.3) is used in beam theory [Eq. (4.1.1e)].

Using the definitions for the in-plane strains from Eq. (6.1.4), along with Eq. (12.1.3), the in-plane strain–displacement equations become

$$\varepsilon_x = -z\frac{\partial^2 w}{\partial x^2} \qquad \varepsilon_y = -z\frac{\partial^2 w}{\partial y^2} \qquad \gamma_{xy} = -2z\frac{\partial^2 w}{\partial x \partial y} \tag{12.1.4a}$$

or using Eq. (12.1.3) in Eq.(12.1.4a), we have

$$\varepsilon_x = -z\kappa_x \qquad \varepsilon_y = -z\kappa_y \qquad \gamma_{xy} = -z\kappa_{xy} \tag{12.1.4b}$$

The first of Eqs. (12.1.4a) is used in beam theory [see Eq. (4.1.10)]. The others are new to plate theory.

Stress–Strain Relations

Based on the third assumption above, the plane stress equations can be used to relate the in-plane stresses to the in-plane strains for an isotropic material as

$$\sigma_x = \frac{E}{1 - v^2}(\varepsilon_x + v\varepsilon_y)$$

$$\sigma_y = \frac{E}{1 - v^2}(\varepsilon_y + v\varepsilon_x) \tag{12.1.5}$$

$$\tau_{xy} = G\gamma_{xy}$$

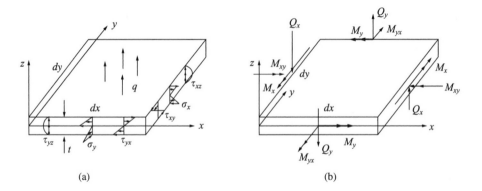

Figure 12–4 Differential element of a plate with (a) stresses shown on the edges of the plate and (b) differential moments and forces

The in-plane normal stresses and shear stress are shown acting on the edges of the plate in Figure 12–4(a). Similar to the stress variation in a beam, these stresses vary linearly in the z direction from the midsurface of the plate. The transverse shear stresses τ_{yz} and τ_{xz} are also present, even though transverse shear deformation is neglected. As in beam theory, these transverse stresses vary quadratically through the plate thickness. The stresses of Eq. (12.1.5) can be related to the bending moments M_x and M_y and to the twisting moment M_{xy} acting along the edges of the plate as shown in Figure 12–4(b).

The moments are actually functions of x and y and are computed per unit length in the plane of the plate so have units of lb–in./in. Therefore, the moments are

$$M_x = \int_{-t/2}^{t/2} z\sigma_x \, dz \qquad M_y = \int_{-t/2}^{t/2} z\sigma_y \, dz \qquad M_{xy} = \int_{-t/2}^{t/2} z\tau_{xy} \, dz \qquad (12.1.6)$$

The moments can be related to the curvatures by substituting Eqs. (12.1.4b) into Eqs. (12.1.5) and then using those stresses in Eq. (12.1.6) to obtain

$$M_x = D(\kappa_x + v\kappa_y) \qquad M_y = D(\kappa_y + v\kappa_x) \qquad M_{xy} = \frac{D(1-v)}{2}\kappa_{xy} \qquad (12.1.7)$$

where $D = Et^3/[12(1-v^2)]$ is called the bending rigidity of the plate (in units of lb–in.).

The maximum magnitudes of the normal stresses on each edge of the plate are located at the top or bottom at $z = t/2$. For instance, it can be shown that

$$\sigma_x = \frac{6M_x}{t^2} \qquad (12.1.8)$$

This formula is similar to the flexure formula $\sigma_x = M_x c/I$ when applied to a unit width of plate and when $c = t/2$.

The governing equilibrium differential equation of plate bending is important in selecting the element displacement fields. The basis for this relationship is the equilibrium differential equations derived by the equilibrium of forces with respect to the

z direction and by the equilibrium of moments about the x and y axes, respectively. These equilibrium equations result in the following differential equations:

$$\frac{\partial Q_x}{\partial x} + \frac{\partial Q_y}{\partial y} + q = 0$$

$$\frac{\partial M_x}{\partial x} + \frac{\partial M_{xy}}{\partial y} - Q_x = 0 \tag{12.1.9}$$

$$\frac{\partial M_y}{\partial y} + \frac{\partial M_{xy}}{\partial x} - Q_y = 0$$

where q is the transverse distributed loading (in units of psi) and Q_x and Q_y are the transverse shear line loads (in units of lb/in.) shown in Figure 12–4(b).

Now substituting the moment/curvature relations from Eq. (12.1.7) into the second and third of Eqs. (12.1.9), then solving those equations for Q_x and Q_y, and finally substituting the resulting expressions into the first of Eqs. (12.1.9), we obtain the governing partial differential equation for an isotropic, thin-plate bending behavior as

$$D\left(\frac{\partial^4 w}{\partial x^4} + \frac{2\partial^4 w}{\partial x^2 \partial y^2} + \frac{\partial^4 w}{\partial y^4}\right) = q \tag{12.1.10}$$

From Eq. (12.1.10), we observe that the solution of thin-plate bending using a displacement point of view depends on selection of the single-displacement component w, the transverse displacement.

If we neglect the differentiation with respect to the y coordinate, Eq. (12.1.10) simplifies to Eq. (4.1.1g) for a beam (where the flexural rigidity D of the plate reduces to EI of the beam when the Poisson effect is set to zero and the plate width becomes unity).

Potential Energy of a Plate

The total potential energy of a plate is given by

$$U = \frac{1}{2}\int (\sigma_x \varepsilon_x + \sigma_y \varepsilon_y + \tau_{xy}\gamma_{xy})\,dV \tag{12.1.11}$$

The potential energy can be expressed in terms of the moments and curvatures by substituting Eqs. (12.1.4b) and (12.1.6) in Eq. (12.1.11) as

$$U = \frac{1}{2}\int (M_x \kappa_x + M_y \kappa_y + M_{xy}\kappa_{xy})\,dA \tag{12.1.12}$$

▲ ## 12.2 Derivation of a Plate Bending Element Stiffness Matrix and Equations ▲

Numerous finite elements for plate bending have been developed over the years, and Reference [3] cites 88 different elements. In this section we will introduce only one element formulation, the basic 12-degrees-of-freedom rectangular element shown in

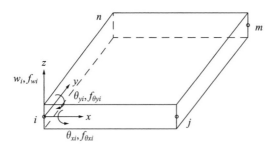

Figure 12–5 Basic rectangular plate element with nodal degrees of freedom

Figure 12–5. For more details of this formulation and of various other formulations including triangular elements, see References [4–18].

The formulation will be developed consistently with the stiffness matrix and equations for the bar, beam, plane stress–strain, axisymmetric, and solid elements of previous chapters.

Step 1 Select Element Type

We will consider the 12-degrees-of-freedom flat-plate bending element shown in Figure 12–5. Each node has 3 degrees of freedom—a transverse displacement w in the z direction, a rotation θ_x about the x axis, and a rotation θ_y about the y axis.

The nodal displacement matrix at node i is given by

$$\{d_i\} = \begin{Bmatrix} w_i \\ \theta_{xi} \\ \theta_{yi} \end{Bmatrix} \tag{12.2.1}$$

where the rotations are related to the transverse displacement by

$$\theta_x = +\frac{\partial w}{\partial y} \qquad \theta_y = -\frac{\partial w}{\partial x} \tag{12.2.2}$$

The negative sign on θ_y is due to the fact that a negative displacement w is required to produce a positive rotation about the y axis.

The total element displacement matrix is now given by

$$\{d\} = \{[d_i] \quad [d_j] \quad [d_m] \quad [d_n]\}^T \tag{12.2.3}$$

Step 2 Select the Displacement Function

Because there are 12 total degrees of freedom for the element, we select a 12-term polynomial in x and y as follows:

$$w = a_1 + a_2x + a_3y + a_4x^2 + a_5xy + a_6y^2 + a_7x^3 + a_8x^2y$$
$$+ a_9xy^2 + a_{10}y^3 + a_{11}x^3y + a_{12}xy^3 \tag{12.2.4}$$

Equation (12.2.4) is an incomplete quartic in the context of the Pascal triangle (Figure 8–2). The function is complete up to the third order (ten terms), and a choice of two more terms from the remaining five terms of a complete quartic must be made. The best choice is the x^3y and xy^3 terms as they ensure that we will have continuity in displacement among the interelement boundaries. (The x^4 and y^4 terms would yield discontinuities of displacement along interelement boundaries and so must be rejected. The x^2y^2 term is alone and cannot be paired with any other terms and so is also rejected.) The function [Eq. (12.2.4)] also satisfies the basic differential equation [Eq. (12.1.10)] over the unloaded part of the plate, although not a requirement in a minimum potential energy approximation.

Furthermore, the function allows for rigid-body motion and constant strain, as terms are present to account for these phenomena in a structure. However, interelement slope discontinuities along common boundaries of elements are not ensured.

To observe this discontinuity in slope, we evaluate the polynomial and its slopes along a side or edge (say, along side *i-j*, the x axis of Figure 12–5). We then obtain

$$w = a_1 + a_2x + a_4x^2 + a_7x^3$$
$$\frac{\partial w}{\partial x} = a_2 + 2a_4x + 3a_7x^2 \tag{12.2.5}$$
$$\frac{\partial w}{\partial y} = a_3 + a_5x + a_8x^2 + a_{12}x^3$$

The displacement w is a cubic as used for the beam element, while the slope $\partial w/\partial x$ is the same as in beam bending. Based on the beam element, we recall that the four constants $a_1, a_2, a_4,$ and a_7 can be defined by invoking the endpoint conditions of $(w_i, w_j, \theta_{yi}, \theta_{yj})$. Therefore, w and $\partial w/\partial x$ are completely defined along this edge. The normal slope $\partial w/\partial y$ is a cubic in x. However, only two degrees of freedom remain for definition of this slope, while four constants $(a_3, a_5, a_8,$ and $a_{12})$ exist. This slope is then not uniquely defined, and a slope discontinuity occurs. Thus, the function for w is said to be nonconforming. The solution obtained from the finite element analysis using this element will not be a minimum potential energy solution. However, this element has proven to give acceptable results, and proofs of its convergence have been shown [8].

The constants a_1 through a_{12} can be determined by expressing the 12 simultaneous equations linking the values of w and its slopes at the nodes when the coordinates take up their appropriate values. First, we write

$$\begin{Bmatrix} w \\ +\dfrac{\partial w}{\partial y} \\ -\dfrac{\partial w}{\partial x} \end{Bmatrix} = \begin{bmatrix} 1 & x & y & x^2 & xy & y^2 & x^3 & x^2y & xy^2 & y^3 & x^3y & xy^3 \\ 0 & 0 & +1 & 0 & +x & +2y & 0 & +x^2 & +2xy & +3y^2 & +x^3 & +3xy^2 \\ 0 & -1 & 0 & -2x & -y & 0 & -3x^2 & -2xy & -y^2 & 0 & -3x^2y & -y^3 \end{bmatrix}$$

$$\times \begin{Bmatrix} a_1 \\ a_2 \\ a_3 \\ \vdots \\ a_{12} \end{Bmatrix} \tag{12.2.6}$$

or in simple matrix form the degrees of freedom matrix is

$$\{\psi\} = [P]\{a\} \tag{12.2.7}$$

where $[P]$ is the 3×12 first matrix on the right side of Eq. (12.2.6).
Next, we evaluate Eq. (12.2.6) at each node point as follows

$$\{d\} = \begin{Bmatrix} w_i \\ \theta_{xi} \\ \theta_{yi} \\ w_j \\ \vdots \end{Bmatrix} = \begin{bmatrix} 1 & x_i & y_i & x_i^2 & x_iy_i & y_i^2 & x_i^3 & x_i^2y_i & x_iy_i^3 & y_i^3 & x_i^3y_i & x_iy_i^3 \\ 0 & 0 & +1 & 0 & +x_i & +2y_i & 0 & +x_i^2 & +2x_iy_i & +3y_i^2 & +x_i^3 & +3x_iy_i^2 \\ \vdots & & & & & & & & & & & \\ \vdots & & & & & & & & & & & \\ \cdots & \cdots & \cdots & \cdots & \cdots & \cdots & \cdots & \cdots & \cdots & \cdots & \cdots & \cdots \end{bmatrix}$$

$$\times \begin{Bmatrix} a_1 \\ a_2 \\ \vdots \\ a_{12} \end{Bmatrix} \tag{12.2.8}$$

In compact matrix form, we express Eq. (12.2.8) as

$$\{d\} = [C]\{a\} \tag{12.2.9}$$

where $[C]$ is the 12×12 matrix on the right side of Eq. (12.2.8).
Therefore, the constants (a's) can be solved for by

$$\{a\} = [C]^{-1}\{d\} \tag{12.2.10}$$

Equation (12.2.7) can now be expressed as

$$\{\psi\} = [P][C]^{-1}\{d\} \tag{12.2.11}$$

or $$\{\psi\} = [N]\{d\} \tag{12.2.12}$$

where $[N] = [P][C]^{-1}$ is the 3×12 shape function matrix. A specific form of the shape functions $N_i, N_j, N_m,$ and N_n is given in Reference [9].

Step 3 Define the Strain (Curvature)-Displacement and Stress (Moment)-Curvature Relationships

The curvature matrix, based on the curvatures of Eq.(12.1.3), is

$$\{\kappa\} = \begin{Bmatrix} \kappa_x \\ \kappa_y \\ \kappa_{xy} \end{Bmatrix} = \begin{Bmatrix} -2a_4 - 6a_7x - 2a_8\,y - 6a_{11}xy \\ -2a_6 - 2a_9x - 6a_{10}y - 6a_{12}xy \\ -2a_5 - 4a_8x - 4a_9y - 6a_{11}x^2 - 6a_{12}y^2 \end{Bmatrix} \tag{12.2.13}$$

or expressing Eq. (12.2.13) in matrix form, we have

$$\{\kappa\} = [Q]\{a\} \tag{12.2.14}$$

where $[Q]$ is the 3×12 coefficient matrix multiplied by the a's in Eq. (12.2.13). Using Eq. (12.2.10) for $\{a\}$, we express the curvature matrix as

$$\{\kappa\} = [B]\{d\} \tag{12.2.15}$$

where

$$[B] = [Q][C]^{-1} \tag{12.2.16}$$

is the 3×12 gradient matrix.

The moment-curvature matrix for a plate is given by

$$\{M\} = \left\{ \begin{array}{c} M_x \\ M_y \\ M_{xy} \end{array} \right\} = [D] \left\{ \begin{array}{c} \kappa_x \\ \kappa_y \\ \kappa_{xy} \end{array} \right\} = [D][B]\{d\} \tag{12.2.17}$$

where the $[D]$ matrix is the constitutive matrix given for isotropic materials by

$$[D] = \frac{Et^3}{12(1 - v^2)} \begin{bmatrix} 1 & v & 0 \\ v & 1 & 0 \\ 0 & 0 & \dfrac{1-v}{2} \end{bmatrix} \tag{12.2.18}$$

and Eq. (12.2.15) has been used in the final expression for Eq. (12.2.17).

Step 4   Derive the Element Stiffness Matrix and Equations

The stiffness matrix is given by the usual form of the stiffness matrix as

$$[k] = \iint [B]^T [D][B] \, dx \, dy \tag{12.2.19}$$

where $[B]$ is defined by Eq. (12.2.16) and $[D]$ is defined by Eq. (12.2.18). The stiffness matrix for the four-noded rectangular element is of order 12×12. A specific expression for $[k]$ is given in References [4] and [5].

The surface force matrix due to distributed loading q acting per unit area in the z direction is obtained using the standard equation

$$\{F_s\} = \iint [N_s]^T q \, dx \, dy \tag{12.2.20}$$

For a uniform load q acting over the surface of an element of dimensions $2b \times 2c$, Eq. (12.2.20) yields the forces and moments at node i as

$$\begin{Bmatrix} f_{wi} \\ f_{\theta xi} \\ f_{\theta yi} \end{Bmatrix} = 4qcb \begin{Bmatrix} 1/4 \\ -c/12 \\ b/12 \end{Bmatrix} \tag{12.2.21}$$

with similar expressions at nodes $j, m,$ and n. We should note that a uniform load yields applied couples at the nodes as part of the work-equivalent load replacement, just as was the case for the beam element (Section 4.4).

The element equations are given by

$$\begin{Bmatrix} f_{wi} \\ f_{\theta xi} \\ f_{\theta yi} \\ \vdots \\ f_{\theta yn} \end{Bmatrix} = \begin{bmatrix} k_{11} & k_{12} & \cdots & k_{1,12} \\ k_{21} & k_{22} & \cdots & k_{2,12} \\ k_{31} & k_{32} & \cdots & k_{3,12} \\ \vdots & \vdots & \cdots & \cdots \\ & & \cdots & \cdots \\ k_{12,1} & & \cdots & k_{12,12} \end{bmatrix} \begin{Bmatrix} w_i \\ \theta_{xi} \\ \theta_{yi} \\ \vdots \\ \theta_{yn} \end{Bmatrix} \tag{12.2.22}$$

The rest of the steps, including assembling the global equations, applying boundary conditions (now boundary conditions on w, θ_x, θ_y), and solving the equations for the nodal displacements and slopes (note three degrees of freedom per node), follow the standard procedures introduced in previous chapters.

▲ ## 12.3 Some Plate Element Numerical Comparisons ▲

We now present some numerical comparisons of quadrilateral plate element formulations. Remember there are numerous plate element formulations in the literature. Figure 12–6 shows a number of plate element formulation results for a square plate simply supported all around and subjected to a concentrated vertical load applied at the center of the plate. The results are shown to illustrate the upper and lower bound solution behavior and demonstrate the convergence of solution for various plate element formulations. Included in these results is the 12-term polynomial described in Section 12.2. We note that the 12-term polynomial converges to the exact solution from above. It yields an upper bound solution. Because the interelement continuity of slopes is not ensured by the 12-term polynomial, the lower bound classical characteristic of a minimum potential energy formulation is not obtained. However, as more elements are used, the solution converges to the exact solution [1].

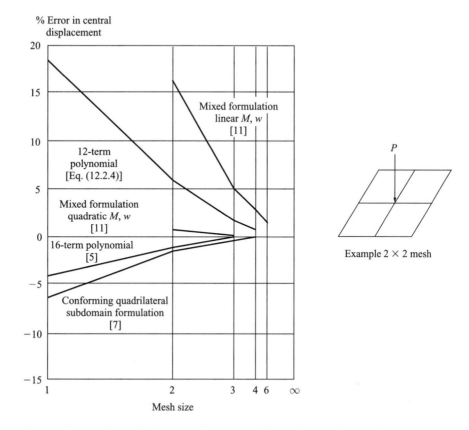

Figure 12–6 Numerical comparisons: quadrilateral plate element formulations
(*Gallagher, Richard H., Finite Element Analysis: Fundamentals, 1st, © 1975*. Printed and
Electronically reproduced by permission of Pearson Education, Inc., Upper Saddle
River, New Jersey.)

Figure 12–7 shows comparisons of triangular plate formulations for the same
centrally loaded simply supported plate used to compare quadrilateral element formulations in Figure 12–6. We can observe from Figures 12–6 and 12–7 a number of different formulations with results that converge from above and below. Some of these
elements produce better results than others.

The Algor program [19] uses, among others, the Veubeke (after Baudoin
Fraeijs de Veubeke) 16-degrees-of-freedom "subdomain" formulation [7], which
converges from below, as it is based on a compatible displacement formulation.
For more information on some of these formulations, consult the references at the
end of the chapter.

Finally, Figure 12–8 shows results for some selected Mindlin plate theory elements. Mindlin plate elements account for bending deformation and for transverse
shear deformation. For more on Mindlin plate theory, see Reference [6]. The "heterosis"
element [10] is the best performing element in Figure 12–8.

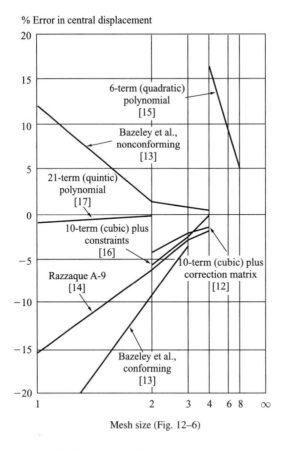

% Error in central displacement

Mesh size (Fig. 12–6)

Figure 12–7 Numerical comparisons for a simply supported square plate subjected to center load triangular element formulations (*Gallagher, Richard H., Finite Element Analysis: Fundamentals, 1st,* © *1975.* Printed and Electronically reproduced by permission of Pearson Education, Inc., Upper Saddle River, New Jersey.)

▲ **12.4 Computer Solutions** ▲
for Plate Bending Problems

A computer program solution for plant bending problems [19] is now illustrated in Example 12.2. The plate element is a three- or four-noded element formulated in three-dimensional space. The element degrees of freedom allowed are all three translations (u, v, and w) and in-plane rotations (θ_x and θ_y). The rotational degrees of freedom normal to the plate are undefined and must be constrained. The element formulated in the computer program is the 16-term polynomial described in References [5] and [7]. This element is known as the Veubeke plate in the program. The 16-node formulation converges from below for the displacement analysis, as it is based on a compatible displacement formulation. This is also shown in Figure 12–6 for the clamped plate subjected to a concentrated center load.

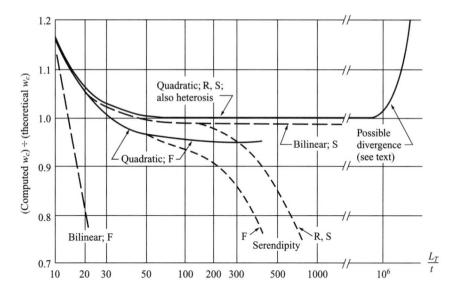

Figure 12–8 Center deflection of a uniformly loaded clamped square plate of side length L_T and thickness t. An 8×8 mesh is used in all cases. Thin plates correspond to large L_T/t. Transverse shear deformation becomes significant for small L_T/t. Integration rules are reduced (R), selective (S), and full (F) [18], based on Mindlin plate element formulations (Cook, R., Malkus, D., and Plesha, M. Concepts and Applications of Finite Element Analysis, 3rd ed., 1989, p. 326. Reprinted by permission of John Wiley & Sons, Inc., New York)

Example 12.1

The problem of a square steel plate fixed along all four edges and subjected to a concentrated load at its center is shown in Figure 12–9. Determine the maximum vertical deflection of the plate.

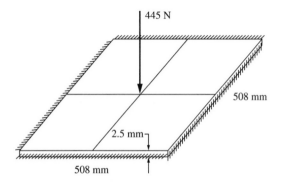

Figure 12–9 A 2×2 mesh model of the clamped plate of Example 12.1

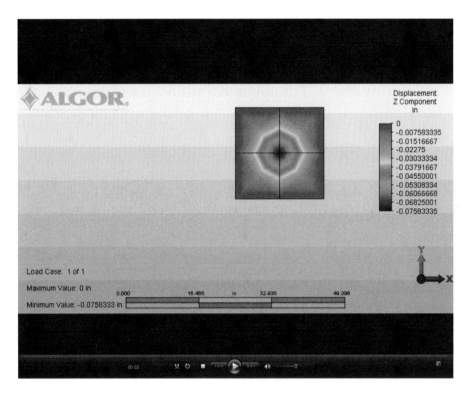

Figure 12–10 Displacement plot of the clamped plate of Example 12.1 (See the full-color insert for a color version of this figure.)

SOLUTION:

A 2 × 2 mesh was created to model the plate. The resulting vertical displacement plot is shown in Figure 12–10. The maximum displacement located at the center of the plate is −1.926 mm.

The classical plate bending solution for the maximum displacement (which occurs under the concentrated center load) is given in Reference [1] as

$$w = 0.0056PL^2/D = 0.0056(-445 \text{ N})(508)^2/(286.17 \text{ N-m}) = -2.247 \times 10^{-3} \text{ m}$$

where

$$D = Et^3/(12(1-v^2)) = (200 \times 10^9 \text{ N/m}^2)(0.0025)^3/[12(1-0.3^2)] = 286.17 \text{ N-m}$$

A mesh refinement to a 4 × 4 mesh would show convergence toward the classical solution. ■

Example 12.2

The clamped plate of Example 12.1 is now reinforced with 5 cm wide × 30 cm deep rectangular cross-section beams spanning the centers in both directions as indicated by the lines dividing the plate into four parts in Figure 12–11(a). (Figure 1–5 also illustrates how a chimney stack was modeled using both beam and plate elements.)

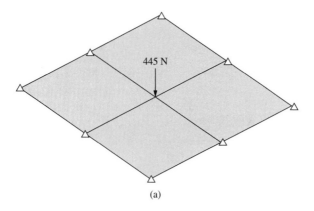

(a)

Fixed Plate with Beam Reinforcement
Concentrated Load of 445 N at center
2.5 mm thick plate
5 × 30 cm beams

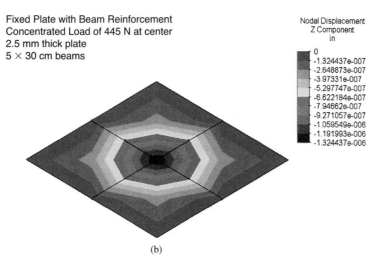

(b)

Figure 12–11   (a) Model of beam and plate elements combined at centerline of elements and (b) vertical deflection plot for model in part (a)

SOLUTION:

The resulting displacement plot is shown in Figure 12–11(b). The maximum displacement is now decreased to -1.324×10^{-6} in. ∎

Example 12.3

A finite element model of a computer case is shown in Figure 12–12(a). The model consists of plate bending elements.

SOLUTION:

Figure 12–12(b) shows the 700-Pa uniform pressure applied to the top surface, the fixed boundary conditions applied to the bottom of the case, and the resulting von Mises stress plot. For more details of the dimensions used, see Problem 12.13. ∎

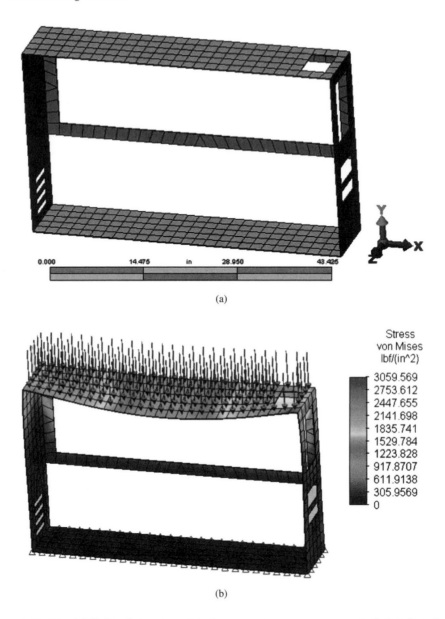

(a)

(b)

Figure 12–12 (a) Finite element model of a computer case composed of plate bending elements and (See the full-color insert for a color version of this figure.) (b) the pressure load, boundary conditions, and resulting von Mises stress (by Nicholas Dachniwskyj)

▲ Summary Equations

Plate curvatures expressions:

$$\kappa_x = -\frac{\partial^2 w}{\partial x^2} \qquad \kappa_y = -\frac{\partial^2 w}{\partial y^2} \qquad \kappa_{xy} = -\frac{2\partial^2 w}{\partial x \partial y} \qquad (12.1.3)$$

Strain–displacement equations:

$$\varepsilon_x = -z\frac{\partial^2 w}{\partial x^2} \qquad \varepsilon_y = -z\frac{\partial^2 w}{\partial y^2} \qquad \gamma_{xy} = -2z\frac{\partial^2 w}{\partial x \partial y} \qquad (12.1.4a)$$

Stress–strain relations:

$$\sigma_x = \frac{E}{1-v^2}(\varepsilon_x + v\varepsilon_y)$$

$$\sigma_y = \frac{E}{1-v^2}(\varepsilon_y + v\varepsilon_x) \qquad (12.1.5)$$

$$\tau_{xy} = G\gamma_{xy}$$

Moment-curvature relations:

$$M_x = D(\kappa_x + v\kappa_y) \qquad M_y = D(\kappa_y + v\kappa_x) \qquad M_{xy} = \frac{D(1-v)}{2}\kappa_{xy} \qquad (12.1.7)$$

where $D = Et^3/[12(1-v^2)]$.

Normal stress on plate due to bending:

$$\sigma_x = \frac{6M_x}{t^2} \qquad (12.1.8)$$

Potential energy in plate:

$$U = \frac{1}{2}\int (M_x\kappa_x + M_y\kappa_y + M_{xy}\kappa_{xy})\,dA \qquad (12.1.12)$$

Transverse displacement function for four-noded rectangular plate:

$$w = a_1 + a_2 x + a_3 y + a_4 x^2 + a_5 xy + a_6 y^2 + a_7 x^3 + a_8 x^2 y$$
$$+ a_9 xy^2 + a_{10} y^3 + a_{11} x^3 y + a_{12} xy^3 \qquad (12.2.4)$$

Gradient matrix:

$$[B] = [Q][C]^{-1} \qquad (12.2.16)$$

Moment-curvature matrix for four-noded rectangular plate:

$$\{M\} = \begin{Bmatrix} M_x \\ M_y \\ M_{xy} \end{Bmatrix} = [D]\begin{Bmatrix} \kappa_x \\ \kappa_y \\ \kappa_{xy} \end{Bmatrix} = [D][B]\{d\} \qquad (12.2.17)$$

Constitutive matrix for plate bending:

$$[D] = \frac{Et^3}{12(1-v^2)}\begin{bmatrix} 1 & v & 0 \\ v & 1 & 0 \\ 0 & 0 & \dfrac{1-v}{2} \end{bmatrix} \qquad (12.2.18)$$

Stiffness matrix:

$$[k] = \int\int [B]^T[D][B]\,dx\,dy \qquad (12.2.19)$$

Surface-force matrix at node i for plate under uniform pressure:

$$\begin{Bmatrix} f_{wi} \\ f_{\theta xi} \\ f_{\theta yi} \end{Bmatrix} = 4qcb\begin{Bmatrix} 1/4 \\ -c/12 \\ b/12 \end{Bmatrix} \qquad (12.2.21)$$

▲ References

[1] Timoshenko, S. and Woinowsky-Krieger, S., *Theory of Plates and Shells*, 2nd ed., McGraw-Hill, New York, 1969.

[2] Gere, J. M., and Goodno, B. J. *Mechanics of Material*, 7th ed., Cengage Learning, Mason, OH, 2009.

[3] Hrabok, M. M., and Hrudley, T. M., "A Review and Catalog of Plate Bending Finite Elements," *Computers and Structures*, Vol. 19, No. 3, 1984, pp. 479–495.

[4] Zienkiewicz, O. C., and Taylor R. L., *The Finite Element Method*, 4th ed., Vol. 2, McGraw-Hill, New York, 1991.

[5] Gallagher, R. H., *Finite Element Analysis Fundamentals*, Prentice-Hall, Englewood Cliffs, NJ, 1975.

[6] Cook, R. D., Malkus, D. S., Plesha, M. E., and Witt, R. J., *Concepts and Applications of Finite Element Analysis*, 4th ed., Wiley, New York, 2002.

[7] Fraeijs De Veubeke, B., "A Conforming Finite Element for Plate Bending," *International Journal of Solids and Structures*, Vol. 4, No. 1, pp. 95–108, 1968.

[8] Walz, J. E., Fulton, R. E., and Cyrus N. J., "Accuracy and Convergence of Finite Element Approximations," Proceedings of the Second Conference on Matrix Method in Structural Mechanics, AFFDL TR 68-150, pp. 995–1027, Oct., 1968.

[9] Melosh, R. J., "Basis of Derivation of Matrices for the Direct Stiffness Method," *Journal of AIAA*, Vol. 1, pp. 1631–1637, 1963.

[10] Hughes, T. J. R., and Cohen, M., "The 'Heterosis' Finite Element for Plate Bending," *Computers and Structures*, Vol. 9, No. 5, 1978, pp. 445–450.

[11] Bron, J., and Dhatt, G., "Mixed Quadrilateral Elements for Bending," *Journal of AIAA*, Vol. 10, No. 10, pp. 1359–1361, Oct., 1972.

[12] Kikuchi, F., and Ando, Y., "Some Finite Element Solutions for Plate Bending Problems by Simplified Hybrid Displacement Method," *Nuclear Engineering Design*, Vol. 23, pp. 155–178, 1972.

[13] Bazeley, G., Cheung, Y., Irons, B., and Zienkiewicz, O., "Triangular Elements in Plate Bending—Conforming and Non-Conforming Solutions," Proceedings of the First Conference on Matrix Methods on Structural Mechanics, AFFDL TR 66-80, pp. 547–576, Oct., 1965.

[14] Razzaque, A. Q., "Program for Triangular Elements with Derivative Smoothing," *International Journal for Numerical Methods in Engineering*, Vol. 6, No. 3, pp. 333–344, 1973.

[15] Morley, L. S. D., "The Constant-Moment Plate Bending Element," *Journal of Strain Analysis*, Vol. 6, No. 1, pp. 20–24, 1971.

[16] Harvey, J. W., and Kelsey, S., "Triangular Plate Bending Elements with Enforced Compatibility," *AIAA Journal*, Vol. 9, pp. 1023–1026, 1971.

[17] Cowper, G. R., Kosko, E., Lindberg, G., and Olson M., "Static and Dynamic Applications of a High Precision Triangular Plate Bending Element", *AIAA Journal*, Vol. 7, No. 10, pp. 1957–1965, 1969.

[18] Hinton, E., and Huang, H. C., "A Family of Quadrilateral Mindlin Plate Elements with Substitute Shear Strain Fields," *Computers and Structures*, Vol. 23, No. 3, pp. 409–431, 1986.

[19] *Linear Stress and Dynamics Reference Division*, Docutech On-line Documentation, Algor, Inc., Pittsburgh, PA, 1999.

▲ Problems

Solve these problems using the plate element from a computer program.

12.1 A square steel plate (Figure P12–1) of dimensions 0.5 m × 0.5 m with thickness of 2.5 mm is clamped all around. The plate is subjected to a uniformly distributed loading

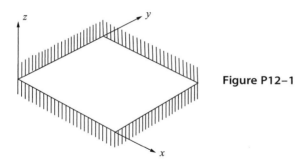

Figure P12–1

of 7 kPa. Using a 2 × 2 mesh and then a 4 × 4 mesh, determine the maximum deflection and maximum stress in the plate. Compare the finite element solution to the classical one in [1].

12.2 An *L*-shaped plate (Figure P12–2) with thickness 2.5 mm is made of ASTM A-36 steel. Determine the deflection under the load and the maximum principal stress and its location using the plate element. Then model the plate as a grid with two beam elements with each beam having the stiffness of each *L*-portion of the plate and compare your answer.

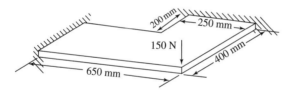

Figure P12–2

12.3 A square (Figure P12–3) simply supported 0.5 m × 0.5 m steel plate with thickness 3.8 mm has a round hole of 100 mm diameter drilled through its center. The plate is uniformly loaded with a load of 15 kN/m^2. Determine the maximum principal stress in the plate.

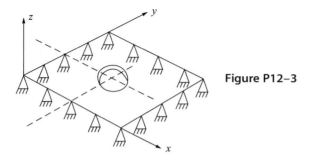

Figure P12–3

12.4 A *C*-channel section (Figure P12–4) structural steel beam of 50 mm wide flanges, 75 mm depth and thickness of both flanges and web of 6 mm is loaded as shown with 500 N acting in the *y* direction on the free end. Determine the free end deflection and angle of twist. Now move the load in the *z* direction until the rotation (angle of twist) becomes zero.

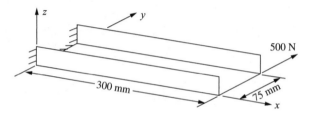

Figure P12–4

This distance is called the shear center (the location where the force can be placed so that the cross section will bend but not twist). You will need to add a beam or plate element to the center of the web extended into the negative z direction and place the load at the end of this proper length beam. (See Table 5–1 for the equation for the shear center location.)

12.5 For the simply supported structural steel $W\ 14 \times 61$ wide flange beam shown in Figure P12–5, compare the plate element model results with the classical beam bending results for deflection and bending stress. The beam is subjected to a central vertical load of 110 kN. The cross-sectional area is 115 cm^2, depth is 352 mm, flange width is 253 mm, flange thickness is 16 mm, web thickness is 9.5 mm, and moment of inertia about the strong axis is 2.65×10^{-4} m^4.

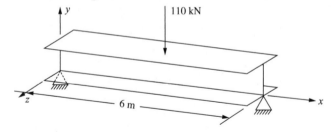

Figure P12–5

12.6 For the structural steel plate structure shown in Figure P12–6 (all dimensions in mm), determine the maximum principal stress and its location. If the stresses are unacceptably

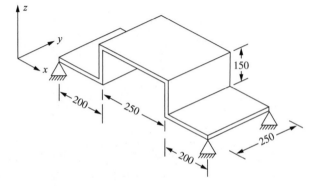

Figure P12–6

high, recommend any design changes. The initial thickness of each plate is 6.25 mm. The left and right edges are simply supported. The load is a uniformly applied pressure of 70 kPa over the top plate.

12.7 Design a steel box structure (Figure P12–7) 1.2 m wide × 2.4 m long made of plates to be used to protect construction workers while working in a trench. That is, determine a recommended thickness of each plate. The depth of the structure must be 2.4 m. Assume the loading is from a side load acting along the long sides due to a wet soil (density of 1000 kg/m^3) and varies linearly with the depth. The allowable deflection of the plate type structure is 1 in. and the allowable stress is 140 MPa.

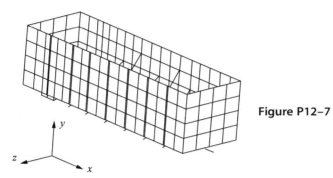

Figure P12–7

12.8 Determine the maximum deflection and maximum principal stress of the circular plate shown in Figure P12–8. The plate is subjected to a uniform pressure $p = 700$ kPa and fixed along its outer edge. Let $E = 200$ GPa, $v = 0.3$, radius $r = 500$ mm, and thickness $t = 12$ mm.

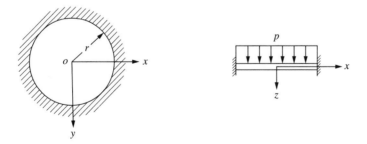

Figure P12–8

12.9 Determine the maximum deflection and maximum principal stress for the plate shown in Figure P12–9. The plate is fixed along all three sides. A uniform pressure of 70 MPa is applied to the surface. The plate is made of steel with $E = 200$ GPa, $v = 0.3$, and thickness $t = 6$ mm, $a = 0.75$ m and $b = 1$ m.

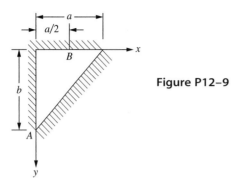

Figure P12–9

12.10 An aircraft cabin window of circular cross section and simple supports all around as shown in Figure P12–10 is made of polycarbonate with $E = 2.5$ GPa, $v = 0.36$, radius $= 0.5$ m, and thickness $t = 18$ mm. The safety of the material is tested at a uniform pressure of 70 kPa. Determine the maximum deflection and maximum principal stress in the material. The yield strength of the material is 63 MPa. Comment on the potential use of this material in regard to strength and deflection.

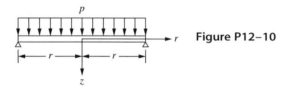

Figure P12–10

12.11 A square steel plate 2 m × 2 m and 10 mm thick at the bottom of a tank must support salt water at a height of 3 m, as shown in Figure P12–11. Assume the plate to be built in (fixed all around). The plate allowable stress is 100 MPa. Let $E = 200$ GPa, $v = 0.3$ for the steel properties. The weight density of salt water is 10.054 kN/m³. Determine the maximum principal stress in the plate and compare to the yield strength.

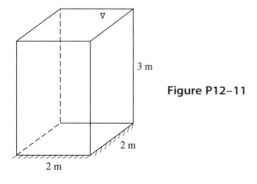

Figure P12–11

12.12 A stockroom floor carries a uniform load of $p = 4$ kN/m^2 over half the floor as shown in Figure P12–12. The floor has opposite edges clamped and remaining edges and mid-span simply supported. The dimensions are 3 m by 6 m. The floor thickness is 15 cm. The floor is made of reinforced concrete with $E = 21$ GPa and $v = 0.25$. Determine the maximum deflection and maximum principal stress in the floor.

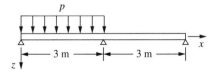

Figure P12–12

12.13 A computer case shown in Figure P12–13 is made of AISI 4130 steel. The top surface is subjected to a uniform pressure load of 700 N/m^2. The thickness of the case is uniformly 3.2 mm. The bottom surface is fully constrained. Model the case using plate bending elements. Determine the maximum von Mises stress and largest deflection of the top face of the case.

12.14 The hopper shown in Figure P12–14 is to be made of plate steel with 6-mm thick walls. Apply a surface traction or pressure load to the walls to simulate a grain loading. Use plate bending elements to model the hopper. Determine through research typical values to be used for the loading. Determine the von Mises stress throughout the vessel.

12.15 A manure spreader tank is shown in Figure P12–15. The tank is 2.25 m long. The bottom axle is 30 cm long measured along the tank axis direction and located in the middle. The single front end coupling is 15 cm in length measured along the axis of the tank. The pressure is a variable surface pressure extending from the top edge to bottom and given by the function shown with a maximum pressure of 80 kPa at the center of the tank. (The density of manure is taken as 980 kg/m^3). Other dimensions are shown in the figures. Assume the tank is made of plate steel with modulus of elasticity of 200 GPa and Poisson's ratio of 0.29. Determine the von Mises stress

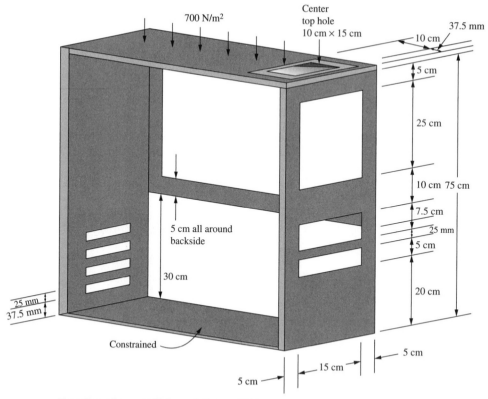

700 N/m²

Center
top hole
10 cm × 15 cm

37.5 mm

10 cm

5 cm

25 cm

10 cm 75 cm

7.5 cm

25 mm

5 cm

20 cm

5 cm all around
backside

30 cm

25 mm
37.5 mm

Constrained

5 cm

15 cm

5 cm

Note: Center four vents 12.5 cm × 2.5 cm w. 18.25 mm space between

Figure P12–13 Computer case

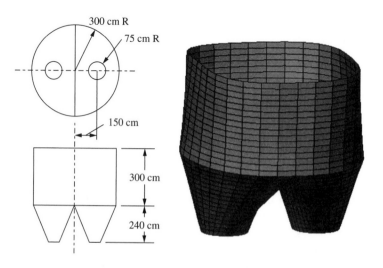

300 cm R

75 cm R

150 cm

300 cm

240 cm

Figure P12–14 Hopper

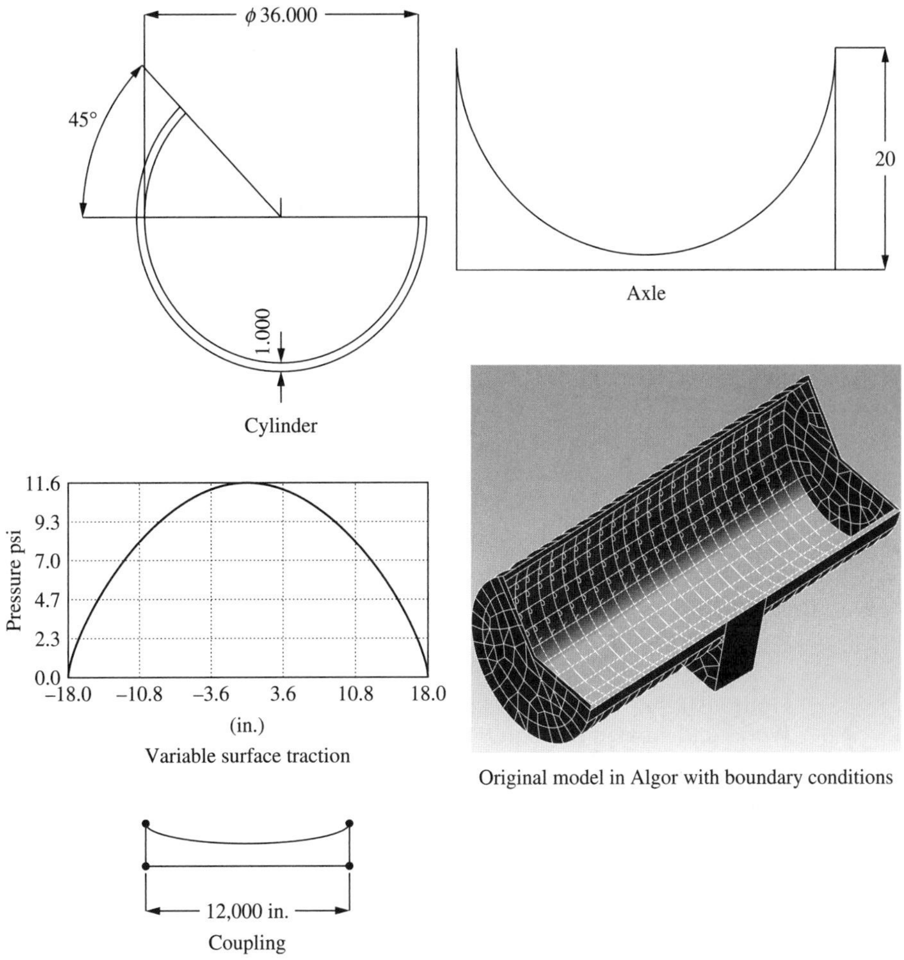

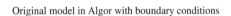

ϕ 36.000
45°
1.000
Cylinder
20
Axle
11.6
9.3
7.0
4.7
2.3
0.0
Pressure psi
−18.0
−10.8
−3.6
3.6
10.8
18.0
(in.)
Variable surface traction
Original model in Algor with boundary conditions
12,000 in.
Coupling

Figure P12–15 Manure spreader tank showing dimensions (in.) and pressure load variation (Compliments of Justin Hronek)

throughout the tank and the maximum displacement. (This problem compliments of Justin Hronek.)

12.16 A tractor bucket with dimensions is shown in Figure P12–16. The load on the bucket is 3 kN spread uniformly over the inside surface. Three split surfaces are fully constrained. The bucket material is A36 structural steel. Determine the von Mises stress throughout the bucket. (This problem was done in Cosmos Works and created by John Mirth and Brian Niggemann.)

Draw the typical bucket using your own dimensions or use the scaled drawing with dimensions shown in Figure P12–16.

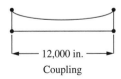

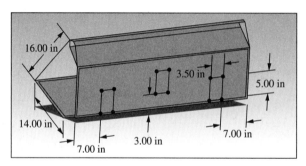

Split surfaces on the bucket

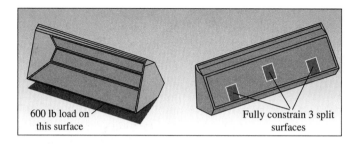

Figure P12–16 Tractor bucket with dimensions shown in inches (Done in Cosmos Works and created by John Mirth and Brian Niggeman)

CHAPTER 13

HEAT TRANSFER AND
MASS TRANSPORT ▲

CHAPTER OBJECTIVES

- To derive the basic differential equation for one-dimensional heat conduction.
- To include heat transfer by convection in the one-dimensional heat transfer model.
- To introduce typical units used for heat transfer.
- To list typical thermal conductivities of materials and heat transfer coefficients based on common modes of free air convection through condensation of water vapor.
- To derive the one-dimensional finite element formulation for heat transfer by conduction and convection.
- To introduce the steps for solving a heat transfer problem by the finite element method.
- To illustrate by examples how to solve one-dimensional heat transfer problems.
- To develop the two-dimensional heat transfer finite element formulation and illustrate an example of a two-dimensional solution.
- To describe how to deal with point or line sources of heat generation.
- To demonstrate when three-dimensional finite element models must be used.
- To introduce the one-dimensional heat transfer with mass transport of the fluid.
- To derive the finite element formulation of heat transfer with mass transport by using Galerkin's method.
- To present a flowchart of two- and three-dimensional heat transfer process.
- To show examples of two- and three-dimensional problems that have been solved using a computer program.

Introduction

In this chapter, we present the first use in this text of the finite element method for solution of nonstructural problems. We first consider the heat-transfer problem, although many similar problems, such as seepage through porous media, torsion of shafts, and magnetostatics [3], can also be treated by the same form of equations (but with different physical characteristics) as that for heat transfer.

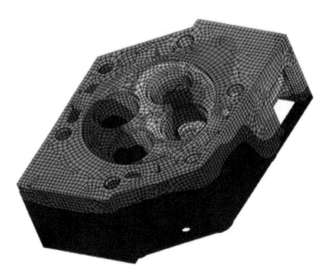

Figure 13–1 Finite element results of cylinder head showing temperature distribution (brick elements were used in the model) (Courtesy of Algor, Inc.) (See the full-color insert for a color version of this figure.)

Familiarity with the heat-transfer problem makes possible determination of the temperature distribution within a body. We can then determine the amount of heat moving into or out of the body and the thermal stresses. Figure 13–1 is an illustration of a three-dimensional model of a cylinder head with the temperature distribution shown throughout the head. The cylinder head is made of stainless steel AISI 410 and is part of a diesel engine that would provide reduced heat rejection and increased power density. The resulting temperature distribution reveals the high temperature of 815 °C in red color at the interface between the two exhaust ports. These temperatures were then fed into the linear stress analyzer to obtain the thermal stresses ranging from 585 MPa to 1380 MPa. The linear stress analysis confirmed the behavior that the engineers saw in the initial prototype tests. The highest thermal stresses coincided with the part of the cylinder head that had been leaking in the preliminary prototypes.

We begin with a derivation of the basic differential equation for heat conduction in one dimension and then extend this derivation to the two-dimensional case. We will then review the units used for the physical quantities involved in heat transfer.

In preceding chapters dealing with stress analysis, we used the principle of minimum potential energy to derive the element equations, where an assumed displacement function within each element was used as a starting point in the derivation. We will now use a similar procedure for the nonstructural heat-transfer problem. We define an assumed temperature function within each element. Instead of minimizing a potential energy functional, we minimize a similar functional to obtain the element equations. Matrices analogous to the stiffness and force matrices of the structural problem result.

We will consider one-, two-, and three-dimensional finite element formulations of the heat-transfer problem and provide illustrative examples of the determination of the temperature distribution along the length of a rod and within a two-dimensional body and show some three-dimensional heat transfer examples as well.

Next, we will consider the contribution of fluid mass transport. The one-dimensional mass-transport phenomenon is included in the basic heat-transfer differential equation. Because it is not readily apparent that a variational formulation is possible for this problem, we will apply Galerkin's residual method directly to the differential equation to obtain the finite element equations. (You should note that the mass transport stiffness matrix is asymmetric.) We will compare an analytical solution to the finite element solution for a heat exchanger design/analysis problem to show the excellent agreement.

Finally, we will present some computer program results for both two- and three-dimensional heat transfer.

▲ 13.1 Derivation of the Basic Differential Equation ▲

One-Dimensional Heat Conduction (without Convection)

We now consider the derivation of the basic differential equation for the one-dimensional problem of heat conduction without convection. The purpose of this derivation is to present a physical insight into the heat-transfer phenomena, which must be understood so that the finite element formulation of the problem can be fully understood. (For additional information on heat transfer, consult texts such as References [1] and [2].) We begin with the control volume shown in Figure 13–2. By conservation of energy, we have

$$E_{in} + E_{generated} = \Delta U + E_{out} \tag{13.1.1}$$

or
$$q_x A \, dt + Q A \, dx \, dt = \Delta U + q_{x+dx} A \, dt \tag{13.1.2}$$
where

E_{in} is the energy entering the control volume, in units of joules (J) or kW · h.

ΔU is the change in stored energy, in units of kW · h (kWh).

q_x is the heat conducted (heat flux) into the control volume at surface edge x, in units of kW/m^2.

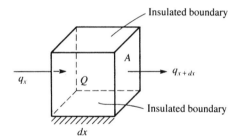

Figure 13–2 Control volume for one-dimensional heat conduction

q_{x+dx} is the heat conducted out of the control volume at the surface edge $x + dx$.

t is time, in h or s.

Q is the internal heat source (heat generated per unit time per unit volume is positive), in kW/m^3 (a heat sink, heat drawn out of the volume, is negative).

A is the cross-sectional area perpendicular to heat flow q, in m^2.

By Fourier's law of heat conduction,

$$q_x = -K_{xx}\frac{dT}{dx} \tag{13.1.3}$$

where

K_{xx} is the thermal conductivity in the x direction, in $kW/(m \cdot °C)$.

T is the temperature, in $°C$.

dT/dx is the temperature gradient, in $°C/m$.

Equation (13.1.3) states that the heat flux in the x direction is proportional to the gradient of temperature in the x direction. The minus sign in Eq. (13.1.3) implies that, by convention, heat flow is positive in the direction opposite the direction of temperature increase. Equation (13.1.3) is analogous to the one-dimensional stress–strain law for the stress analysis problem—that is, to $\sigma_x = E(du/dx)$. Similarly,

$$q_{x+dx} = -K_{xx}\frac{dT}{dx}\bigg|_{x+dx} \tag{13.1.4}$$

where the gradient in Eq. (13.1.4) is evaluated at $x + dx$. By Taylor series expansion, for any general function $f(x)$, we have

$$f_{x+dx} = f_x + \frac{df}{dx}dx + \frac{d^2f}{dx^2}\frac{dx^2}{2} + \cdots$$

Therefore, using a two-term Taylor series, Eq. (13.1.4) becomes

$$q_{x+dx} = -\left[K_{xx}\frac{dT}{dx} + \frac{d}{dx}\left(K_{xx}\frac{dT}{dx}\right)dx\right] \tag{13.1.5}$$

The change in stored energy can be expressed by

$$\Delta U = \text{specific heat} \times \text{mass} \times \text{change in temperature}$$

$$= c(\rho A\, dx)\, dT \tag{13.1.6}$$

where c is the specific heat in $kW \cdot h/(kg \cdot °C)$, and ρ is the mass density in kg/m^3. On substituting Eqs. (13.1.3), (13.1.5), and (13.1.6) into Eq. (13.1.2), dividing Eq. (13.1.2) by $A\, dx\, dt$, and simplifying, we have the one-dimensional heat conduction equation as

$$\frac{\partial}{\partial x}\left(K_{xx}\frac{\partial T}{\partial x}\right) + Q = \rho c\frac{\partial T}{\partial t} \tag{13.1.7}$$

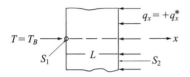 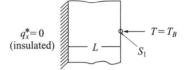

Figure 13–3 Examples of boundary conditions in one-dimensional heat conduction

For steady state, any differentiation with respect to time is equal to zero, so Eq. (13.1.7) becomes

$$\frac{d}{dx}\left(K_{xx}\frac{dT}{dx}\right) + Q = 0 \qquad (13.1.8)$$

For constant thermal conductivity and steady state, Eq. (13.1.7) becomes

$$K_{xx}\frac{d^2T}{dx^2} + Q = 0 \qquad (13.1.9)$$

The boundary conditions are of the form

$$T = T_B \qquad \text{on } S_1 \qquad (13.1.10)$$

where T_B represents a known boundary temperature and S_1 is a surface where the temperature is known, and

$$q_x^* = -K_{xx}\frac{dT}{dx} = \text{constant} \qquad \text{on } S_2 \qquad (13.1.11)$$

where S_2 is a surface where the prescribed heat flux q_x^* or temperature gradient is known. On an insulated boundary, $q_x^* = 0$. These different boundary conditions are shown in Figure 13–3, where by sign convention, positive q_x^* occurs when heat is flowing into the body, and negative q_x^* when heat is flowing out of the body.

Two-Dimensional Heat Conduction (Without Convection)

Consider the two-dimensional heat conduction problem in Figure 13–4. In a manner similar to the one-dimensional case, for steady-state conditions, we can show that for material properties coinciding with the global x and y directions,

$$\frac{\partial}{\partial x}\left(K_{xx}\frac{\partial T}{\partial x}\right) + \frac{\partial}{\partial y}\left(K_{yy}\frac{\partial T}{\partial y}\right) + Q = 0 \qquad (13.1.12)$$

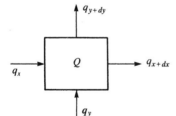

Figure 13–4 Control volume for two-dimensional heat conduction

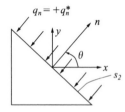

Figure 13–5 Unit vector normal to surface S_2

with boundary conditions

$$T = T_B \qquad \text{on } S_1 \tag{13.1.13}$$

$$q_n = q_n^* = K_{xx}\frac{\partial T}{\partial x}C_x + K_{yy}\frac{\partial T}{\partial y}C_y = \text{constant} \qquad \text{on } S_2 \tag{13.1.14}$$

where C_x and C_y are the direction cosines of the unit vector $\boldsymbol{n}$ normal to the surface S_2 shown in Figure 13–5. Again, q_n^* is by sign convention, positive if heat is flowing into the edge of the body.

▲ 13.2 Heat Transfer with Convection ▲

For a conducting solid in contact with a fluid, there will be a heat transfer taking place between the fluid and solid surface when a temperature difference occurs.

The fluid will be in motion either through external pumping action (**forced convection**) or through the buoyancy forces created within the fluid by the temperature differences within it (**natural** or **free convection**).

We will now consider the derivation of the basic differential equation for one-dimensional heat conduction with convection. Again we assume the temperature change is much greater in the x direction than in the y and z directions. Figure 13–6 shows the control volume used in the derivation. Again, by Eq. (13.1.1) for conservation of energy, we have

$$q_x A \, dt + QA \, dx \, dt = c(\rho A \, dx)\, dT + q_{x+dx}A \, dt + q_h P \, dx \, dt \tag{13.2.1}$$

In Eq. (13.2.1), all terms have the same meaning as in Section 13.1, except the heat flow by convective heat transfer is given by Newton's law of cooling

$$q_h = h(T - T_\infty) \tag{13.2.2}$$

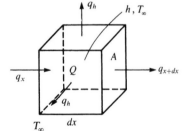

Figure 13–6 Control volume for one-dimensional heat conduction with convection

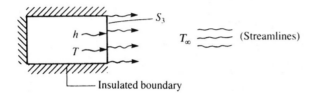

Figure 13–7 Model illustrating convective heat transfer (arrows on surface S_3 indicate heat transfer by convection)

where

h is the heat-transfer or convection coefficient, in kW/(m² · °C).

T is the temperature of the solid surface at the solid/fluid interface.

T_∞ is the temperature of the fluid (here the free-stream fluid temperature).

P in Eq. (13.2.1) denotes the perimeter around the constant cross-sectional area A.

Again, using Eqs. (13.1.3) through (13.1.6) and (13.2.2) in Eq. (13.2.1), dividing by $A\, dx\, dt$, and simplifying, we obtain the differential equation for one-dimensional heat conduction with convection as

$$\frac{\partial}{\partial x}\left(K_{xx}\frac{\partial T}{\partial x}\right) + Q = \rho c \frac{\partial T}{\partial t} + \frac{hP}{A}(T - T_\infty) \qquad (13.2.3)$$

with possible boundary conditions on (1) temperature, given by Eq. (13.1.10), and/or (2) temperature gradient, given by Eq. (13.1.11), and/or (3) loss of heat by convection from the ends of the one-dimensional body, as shown in Figure 13–7. Equating the heat flow in the solid wall to the heat flow in the fluid at the solid/fluid interface, we have

$$-K_{xx}\frac{dT}{dx} = h(T - T_\infty) \qquad \text{on } S_3 \qquad (13.2.4)$$

as a boundary condition for the problem of heat conduction with convection.

▲ 13.3 Typical Units; Thermal Conductivities, K; and Heat-Transfer Coefficients, h ▲

Table 13–1 lists some typical units used for the heat-transfer problem.

Table 13–2 lists some typical thermal conductivities of various solids and liquids. The thermal conductivity K, in W/(m · °C), measures the amount of heat energy (W · h) that will flow through a unit length (ft or m) of a given substance in a unit time (h) to raise the temperature one degree (°C).

Table 13–3 lists approximate ranges of values of convection coefficients for various conditions of convection. The heat transfer coefficient h, in W/(m² · °C),

Table 13–1 Typical units for heat transfer

Variable	SI
Thermal conductivity, K	kW/(m · °C)
Temperature, T	°C or K
Internal heat source, Q	kW/m^3
Heat flux, q	kW/m^2
Heat flow, $\bar{q}$	kW
Convection coefficient, h	kW/(m^2 · °C)
Energy, E	kW · h
Specific heat, c	(kW · h)/(kg · °C)
Mass density, ρ	kg/m^3

Table 13–2 Typical thermal conductivities of some solids and fluids

Material	K [W/(m · °C)]
Solids	
Aluminum, 0 °C (32 °F)	202
Steel (1% carbon), 0 °C	35
Fiberglass, 20 °C (68 °F)	0.035
Concrete, 0 °C	0.81–1.40
Earth, coarse gravelly, 20 °C	0.520
Wood, oak, radial direction, 20 °C	0.17
Fluids	
Engine oil, 20 °C	0.145
Dry air, atmospheric pressure, 20 °C	0.0243

Table 13–3 Approximate values of convection heat-transfer coefficients (from Reference [1])

Mode	h [W/(m^2 · °C)]
Free convection, air	5–25
Forced convection, air	10–500
Forced convection, water	100–15,000
Boiling water	2,500–25,000
Condensation of water vapor	5,000–100,000

measures the amount of heat energy (W · h) that will flow across a unit area (m^2) of a given substance in a unit time (h) to raise the temperature one degree (°C).

    Natural or **free convection** occurs when, for instance, a heated plate is exposed to ambient room air without an external source of motion. This movement of the air, experienced as a result of the density gradients near the plate, is called *natural* or *free convection*. **Forced convection** is experienced, for instance, in the case of a fan blowing air over a plate.

▲ 13.4 One-Dimensional Finite Element Formulation Using a Variational Method ▲

The temperature distribution influences the amount of heat moving into or out of a body and also influences the stresses in a body. Thermal stresses occur in all bodies that experience a temperature gradient from some equilibrium state but are not free to expand in all directions. To evaluate thermal stresses, we need to know the temperature distribution in the body. The finite element method is a realistic method for predicting quantities such as temperature distribution and thermal stresses in a body. In this section, we formulate the one-dimensional heat-transfer equations using a variational method. Examples are included to illustrate the solution of this type of problem.

Step 1   Select Element Type

The basic element with nodes 1 and 2 is shown in Figure 13–8(a).

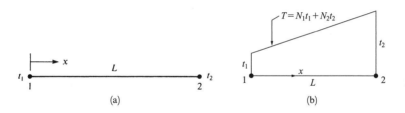

Figure 13–8 (a) Basic one-dimensional temperature element and (b) temperature variation along length of element

Step 2   Choose a Temperature Function

We choose the temperature function T [Figure 13–8(b)] within each element similar to the displacement function of Chapter 3, as

$$T(x) = N_1 t_1 + N_2 t_2 \tag{13.4.1}$$

where t_1 and t_2 are the nodal temperatures to be determined, and

$$N_1 = 1 - \frac{x}{L} \qquad N_2 = \frac{x}{L} \tag{13.4.2}$$

are again the same shape functions as used for the bar element. The $[N]$ matrix is then given by

$$[N] = \left[1 - \frac{x}{L} \quad \frac{x}{L}\right] \tag{13.4.3}$$

and the nodal temperature matrix is

$$\{t\} = \begin{Bmatrix} t_1 \\ t_2 \end{Bmatrix} \tag{13.4.4}$$

In matrix form, we express Eq. (13.4.1) as

$$\{T\} = [N]\{t\} \tag{13.4.5}$$

Step 3 Define the Temperature Gradient/Temperature and Heat Flux/Temperature Gradient Relationships

The temperature gradient matrix $\{g\}$, analogous to the strain matrix $\{\varepsilon\}$, is given by

$$\{g\} = \begin{Bmatrix} \dfrac{dT}{dx} \end{Bmatrix} = [B]\{t\} \tag{13.4.6}$$

where $[B]$ is obtained by substituting Eq. (13.4.1) for $T(x)$ into Eq. (13.4.6) and differentiating with respect to x, that is,

$$[B] = \begin{bmatrix} \dfrac{dN_1}{dx} & \dfrac{dN_2}{dx} \end{bmatrix}$$

Using Eqs. (13.4.2) in the definition for $[B]$, we have

$$[B] = \begin{bmatrix} -\dfrac{1}{L} & \dfrac{1}{L} \end{bmatrix} \tag{13.4.7}$$

The heat flux/temperature gradient relationship is given by

$$q_x = -[D]\{g\} \tag{13.4.8}$$

where the material property matrix is now given by

$$[D] = [K_{xx}] \tag{13.4.9}$$

Step 4 Derive the Element Conduction Matrix and Equations

Equations (13.1.9) through (13.1.11) and (13.2.3) can be shown to be derivable (as shown, for instance, in References [4–6]) by the minimization of the following functional (analogous to the potential energy functional π_p):

$$\pi_h = U + \Omega_Q + \Omega_q + \Omega_h \tag{13.4.10}$$

where

$$U = \frac{1}{2}\iiint_V \left[K_{xx}\left(\frac{dT}{dx}\right)^2 \right] dV$$

$$\Omega_Q = -\iiint_V QT\, dV \qquad \Omega_q = -\iint_{S_2} q^*T\, dS \qquad \Omega_h = \frac{1}{2}\iint_{S_3} h(T - T_\infty)^2\, dS \tag{13.4.11}$$

and where S_2 and S_3 are separate surface areas over which heat flow (flux) q^* (q^* is positive into the surface) and convection loss $h(T - T_\infty)$ are specified. We cannot specify q^* and h on the same surface because they cannot occur simultaneously on the same surface, as indicated by Eqs. (13.4.11).

Using Eqs. (13.4.5), (13.4.6), and (13.4.9) in Eq. (13.4.11) and then using Eq. (13.4.10), we can write π_h in matrix form as

$$\pi_h = \frac{1}{2}\iiint_V [\{g\}^T [D]\{g\}]\, dV - \iiint_V \{t\}^T [N]^T Q\, dV$$

$$- \iint_{S_2} \{t\}^T [N]^T q^*\, dS + \frac{1}{2}\iint_{S_3} h[(\{t\}^T [N]^T - T_\infty)^2]\, dS \qquad (13.4.12)$$

On substituting Eq. (13.4.6) into Eq. (13.4.12) and using the fact that the nodal temperatures $\{t\}$ are independent of the general coordinates x and y and can therefore be taken outside the integrals, we have

$$\pi_h = \frac{1}{2}\{t\}^T \iiint_V [B]^T [D][B]\, dV\{t\} - \{t\}^T \iiint_V [N]^T Q\, dV$$

$$- \{t\}^T \iint_{S_2} [N]^T q^*\, dS + \frac{1}{2}\iint_{S_3} h[\{t\}^T [N]^T [N]\{t\}$$

$$- (\{t\}^T [N]^T + [N]\{t\})T_\infty + T_\infty^2]\, dS \qquad (13.4.13)$$

In Eq. (13.4.13), the minimization is most easily accomplished by explicitly writing the surface integral S_3 with $\{t\}$ left inside the integral as shown. On minimizing Eq. (13.4.13) with respect to $\{t\}$, we obtain

$$\frac{\partial \pi_h}{\partial \{t\}} = \iiint_V [B]^T [D][B]\, dV\{t\} - \iiint_V [N]^T Q\, dV$$

$$- \iint_{S_2} [N]^T q^*\, dS + \iint_{S_3} h[N]^T [N]\, dS\{t\}$$

$$- \iint_{S_3} [N]^T h T_\infty\, dS = 0 \qquad (13.4.14)$$

where the last term $h T_\infty^2$ in Eq. (13.4.13) is a constant that drops out while minimizing π_h. Simplifying Eq. (13.4.14), we obtain

$$\left[\iiint_V [B]^T [D][B]\, dV + \iint_{S_3} h[N]^T [N]\, ds\right]\{t\} = \{f_Q\} + \{f_q\} + \{f_h\} \qquad (13.4.15)$$

where the force matrices have been defined by

$$\{f_Q\} = \iiint_V [N]^T Q \, dV \qquad \{f_q\} = \iint_{S_2} [N]^T q^* \, dS$$

$$\{f_h\} = \iint_{S_3} [N]^T h T_\infty \, dS$$

(13.4.16)

In Eq. (13.4.16), the first term $\{f_Q\}$ (heat source positive, sink negative) is of the same form as the body-force term, and the second term $\{f_q\}$ (heat flux, positive into the surface) and third term $\{f_h\}$ (heat transfer or convection) are similar to surface tractions (distributed loading) in the stress analysis problem. You can observe this fact by comparing Eq. (13.4.16) with Eq. (6.2.46). Because we are formulating element equations of the form $\{f\} = [k]\{t\}$, we have the element conduction matrix[1] for the heat-transfer problem given in Eq. (13.4.15) by

$$[k] = \iiint_V [B]^T [D][B] \, dV + \iint_{S_3} h[N]^T [N] \, dS$$

(13.4.17)

where the first and second integrals in Eq. (13.4.17) are the contributions of conduction and convection, respectively. Using Eq. (13.4.17) in Eq. (13.4.15), for each element, we have

$$\{f\} = [k]\{t\}$$

(13.4.18)

Using the first term of Eq. (13.4.17), along with Eqs. (13.4.7) and (13.4.9), the conduction part of the $[k]$ matrix for the one-dimensional element becomes

$$[k_c] = \iiint_V [B]^T [D][B] \, dV = \int_0^L \left\{ \begin{matrix} -\dfrac{1}{L} \\ \dfrac{1}{L} \end{matrix} \right\} [K_{xx}] \left[-\dfrac{1}{L} \quad \dfrac{1}{L} \right] A \, dx$$

$$= \frac{AK_{xx}}{L^2} \int_0^L \begin{bmatrix} 1 & -1 \\ -1 & 1 \end{bmatrix} dx$$

(13.4.19)

or, finally,

$$[k_c] = \frac{AK_{xx}}{L} \begin{bmatrix} 1 & -1 \\ -1 & 1 \end{bmatrix}$$

(13.4.20)

---

[1] The element conduction matrix is often called the *stiffness matrix* because *stiffness matrix* is becoming a generally accepted term used to describe the matrix of known coefficients multiplied by the unknown degrees of freedom, such as temperatures, displacements, and so on.

The convection part of the $[k]$ matrix becomes

$$[k_h] = \iint_{S_3} h[N]^T[N]\,dS = hP\int_0^L \left\{ \begin{array}{c} 1 - \dfrac{x}{L} \\ \dfrac{x}{L} \end{array} \right\} \left[1 - \dfrac{x}{L} \quad \dfrac{x}{L} \right] dx$$

or, on integrating,

$$[k_h] = \frac{hPL}{6} \begin{bmatrix} 2 & 1 \\ 1 & 2 \end{bmatrix} \tag{13.4.21}$$

where

$$dS = P\,dx$$

and P is the perimeter of the element (assumed to be constant). Therefore, adding Eqs. (13.4.20) and (13.4.21), we find that the $[k]$ matrix is

$$[k] = \frac{AK_{xx}}{L} \begin{bmatrix} 1 & -1 \\ -1 & 1 \end{bmatrix} + \frac{hPL}{6} \begin{bmatrix} 2 & 1 \\ 1 & 2 \end{bmatrix} \tag{13.4.22}$$

When h is zero on the boundary of an element, the second term on the right side of Eq. (13.4.22) (convection portion of $[k]$) is zero. This corresponds, for instance, to an insulated boundary.

The force matrix terms, on simplifying Eq. (13.4.16) and assuming Q, q^*, and product hT_∞ to be constant are

$$\{f_Q\} = \iiint_V [N]^T Q\,dV = QA\int_0^L \left\{ \begin{array}{c} 1 - \dfrac{x}{L} \\ \dfrac{x}{L} \end{array} \right\} dx = \frac{QAL}{2} \left\{ \begin{array}{c} 1 \\ 1 \end{array} \right\} \tag{13.4.23}$$

and

$$\{f_q\} = \iint_{S_2} q^*[N]^T\,dS = q^*P\int_0^L \left\{ \begin{array}{c} 1 - \dfrac{x}{L} \\ \dfrac{x}{L} \end{array} \right\} dx = \frac{q^*PL}{2} \left\{ \begin{array}{c} 1 \\ 1 \end{array} \right\} \tag{13.4.24}$$

and

$$\{f_h\} = \iint_{S_3} hT_\infty[N]^T\,dS = \frac{hT_\infty PL}{2} \left\{ \begin{array}{c} 1 \\ 1 \end{array} \right\} \tag{13.4.25}$$

Therefore, adding Eqs. (13.4.23) through (13.4.25), we obtain

$$\{f\} = \frac{QAL + q^*PL + hT_\infty PL}{2} \left\{ \begin{array}{c} 1 \\ 1 \end{array} \right\} \tag{13.4.26}$$

Equation (13.4.26) indicates that one-half of the assumed uniform heat source Q goes to each node, one-half of the prescribed uniform heat flux q^* (positive q^* enters the body) goes to each node, and one-half of the convection from the perimeter surface hT_∞ goes to each node of an element.

Finally, we must consider the convection from the free end of an element. For simplicity's sake, we will assume convection occurs only from the right end of the

Figure 13–9 Convection force from the end of an element

element, as shown in Figure 13–9. The additional convection term contribution to the stiffness matrix is given by

$$[k_h]_{\text{end}} = \iint\limits_{S_{\text{end}}} h[N]^T[N] \, dS \tag{13.4.27}$$

Now $N_1 = 0$ and $N_2 = 1$ at the right end of the element. Substituting the N's into Eq. (13.4.27), we obtain

$$[k_h]_{\text{end}} = \iint\limits_{S_{\text{end}}} h \begin{Bmatrix} 0 \\ 1 \end{Bmatrix} [0 \quad 1] \, dS = hA \begin{bmatrix} 0 & 0 \\ 0 & 1 \end{bmatrix} \tag{13.4.28}$$

The convection force from the free end of the element is obtained from the application of Eq. (13.4.25) with the shape functions now evaluated at the right end (where convection occurs) and with S_3 (the surface over which convection occurs) now equal to the cross-sectional area A of the rod. Hence,

$$\{f_h\}_{\text{end}} = hT_\infty A \begin{Bmatrix} N_1(x=L) \\ N_2(x=L) \end{Bmatrix} = hT_\infty A \begin{Bmatrix} 0 \\ 1 \end{Bmatrix} \tag{13.4.29}$$

represents the convective force from the right end of an element where $N_1(x=L)$ represents N_1 *evaluated at* $x = L$, and so on.

Step 5   Assemble the Element Equations to Obtain the Global Equations and Introduce Boundary Conditions

We obtain the global or total structure conduction matrix using the same procedure as for the structural problem (called the *direct stiffness method* as described in Section 2.4); that is,

$$[K] = \sum_{e=1}^{N} [k^{(e)}] \tag{13.4.30}$$

typically in units of kW/°C or Btu/(h-°F). The global force matrix is the sum of all element heat sources and is given by

$$\{F\} = \sum_{e=1}^{N} \{f^{(e)}\} \tag{13.4.31}$$

typically in units of kW or Btu/h. The global equations are then

$$\{F\} = [K]\{t\} \tag{13.4.32}$$

with the prescribed nodal temperature boundary conditions given by Eq. (13.1.13). Note that the boundary conditions on heat flux, Eq. (13.1.11), and convection, Eq. (13.2.4), are actually accounted for in the same manner as distributed loading was accounted for in the stress analysis problem; that is, they are included in the column of force matrices through a consistent approach (using the same shape functions used to derive [k]), as given by Eqs. (13.4.2).

The heat-transfer problem is now amenable to solution by the finite element method. The procedure used for solution is similar to that for the stress analysis problem. In Section 13.5, we will derive the specific equations used to solve the two-dimensional heat-transfer problem.

Step 6 Solve for the Nodal Temperatures

We now solve for the global nodal temperature, $\{t\}$, where the appropriate nodal temperature boundary conditions, Eq. (13.1.13), are specified.

Step 7 Solve for the Element Temperature Gradients and Heat Fluxes

Finally, we calculate the element temperature gradients from Eq. (13.4.6), and the heat fluxes, typically from Eq. (13.4.8).

To illustrate the use of the equations developed in this section, we will now solve some one-dimensional heat-transfer problems.

Example 13.1

Determine the temperature distribution along the length of the rod shown in Figure 13–10 with an insulated perimeter. The temperature at the left end is a constant 40°C and the free-stream temperature is -10°C. Let $h = 55$ W/(m² · °C) and $K_{xx} = 35$ W/(m · °C). The value of h is typical for forced air convection and the value of K_{xx} is a typical conductivity for carbon steel (Tables 13–2 and 13–3).

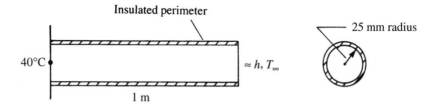

Figure 13–10 One-dimensional rod subjected to temperature variation

SOLUTION:

The finite element discretization is shown in Figure 13–11. For simplicity's sake, we will use four elements, each 0.25 m long. There will be convective heat loss only over the right end of the rod because we consider the left end to have a known

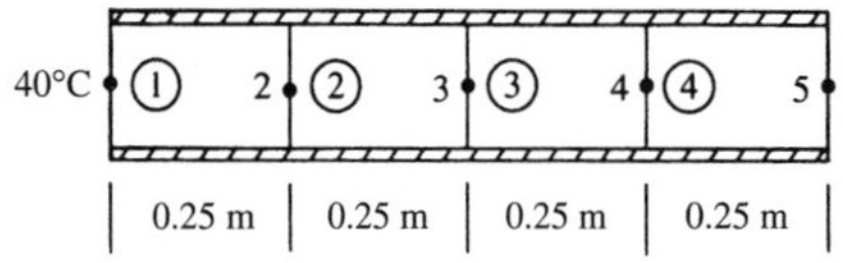

Figure 13-11 Finite element discretized rod

temperature and the perimeter to be insulated. We calculate the stiffness matrices for each element as follows:

$$\frac{AK_{xx}}{L} = \frac{\pi(0.025 \text{ m})^2[35 \text{ W}/(\text{m} \cdot {}^\circ\text{C})]}{(0.25 \text{ m})}$$

$$= 0.275 \text{ W}/{}^\circ\text{C}$$

$$\frac{hPL}{6} = \frac{[55 \text{ W}/(\text{m} \cdot {}^\circ\text{C})](2\pi)}{6}(0.025 \text{ m})(0.25 \text{ m}) \tag{13.4.33}$$

$$= 0.36 \text{ W}/{}^\circ\text{C}$$

$$hT_\infty PL = [55 \text{ W}/(\text{m} \cdot {}^\circ\text{C})](-10 \,{}^\circ\text{C})(2\pi)(0.025 \text{ m})(0.25 \text{ m})$$

$$= -21.6 \text{ W}/{}^\circ\text{C}$$

In general, from Eqs. (13.4.22) and (13.4.27), we have

$$[k] = \frac{AK_{xx}}{L}\begin{bmatrix} 1 & -1 \\ -1 & 1 \end{bmatrix} + \frac{hPL}{6}\begin{bmatrix} 2 & 1 \\ 1 & 2 \end{bmatrix} + \iint\limits_{S_{\text{end}}} h[N]^T[N]\, dS \tag{13.4.34}$$

Substituting Eqs. (13.4.33) into Eq. (13.4.34) for element 1, we have

$$[k^{(1)}] = 0.275 \begin{bmatrix} 1 & -1 \\ -1 & 1 \end{bmatrix} \text{ W}/{}^\circ\text{C} \tag{13.4.35}$$

where the second and third terms on the right side of Eq. (13.4.34) are zero because there are no convection terms associated with element 1. Similarly, for elements 2 and 3, we have

$$[k^{(2)}] = [k^{(3)}] = [k^{(1)}] \tag{13.4.36}$$

However, element 4 has an additional (convection) term owing to heat loss from the flat surface at its right end. Hence, using Eq. (13.4.28), we have

$$[k^{(4)}] = [k^{(1)}] + hA\begin{bmatrix} 0 & 0 \\ 0 & 1 \end{bmatrix}$$

$$= 0.275 \begin{bmatrix} 1 & -1 \\ -1 & 1 \end{bmatrix} + [55 \text{ W}/(\text{m}^2 \cdot {}^\circ\text{C})]\pi(0.025 \text{ m})^2 \begin{bmatrix} 0 & 0 \\ 0 & 1 \end{bmatrix}$$

$$= \begin{bmatrix} 0.275 & -0.275 \\ -0.275 & 0.386 \end{bmatrix} \text{ W}/{}^\circ\text{C} \tag{13.4.37}$$

In general, we would use Eqs. (13.4.23) through (13.4.25), and (13.4.29) to obtain the element force matrices. However, in this example, $Q = 0$ (no heat source), $q^* = 0$ (no heat flux), and there is no convection except from the right end. Therefore,

$$\{f^{(1)}\} = \{f^{(2)}\} = \{f^{(3)}\} = 0 \tag{13.4.38}$$

and
$$\{f^{(4)}\} = hT_\infty A \left\{ \begin{array}{c} 0 \\ 1 \end{array} \right\}$$

$$= [55 \text{ W}/(\text{m}^2 \cdot {}^\circ\text{C})](-10\,{}^\circ\text{C})\pi(0.025 \text{ m})^2 \left\{ \begin{array}{c} 0 \\ 1 \end{array} \right\}$$

$$= -1.8 \left\{ \begin{array}{c} 0 \\ 1 \end{array} \right\} \text{ W}/{}^\circ\text{C} \tag{13.4.39}$$

The assembly of the element stiffness matrices [Eqs. (13.4.35) through (13.4.37)] and the element force matrices [Eqs. (13.4.38) and (13.4.39)], using the direct stiffness method, produces the following system of equations:

$$\begin{bmatrix} 0.275 & -0.275 & 0 & 0 & 0 \\ -0.275 & 0.55 & -0.275 & 0 & 0 \\ 0 & -0.275 & 0.55 & -0.275 & 0 \\ 0 & 0 & -0.275 & 0.55 & -0.275 \\ 0 & 0 & 0 & -0.275 & 0.383 \end{bmatrix} \left\{ \begin{array}{c} t_1 \\ t_2 \\ t_3 \\ t_4 \\ t_5 \end{array} \right\} = \left\{ \begin{array}{c} F_1 \\ 0 \\ 0 \\ 0 \\ -1.8 \end{array} \right\} \tag{13.4.40}$$

where F_1 corresponds to an unknown rate of heat flow at node 1 (analogous to an unknown support force in the stress analysis problem). We have a known nodal temperature boundary condition of $t_1 = 40\,{}^\circ\text{C}$. This nonhomogeneous boundary condition must be treated in the same manner as was described for the stress analysis problem (see Section 2.5 and Appendix B.4). We modify the stiffness (conduction) matrix and force matrix as follows:

$$\begin{bmatrix} 1 & 0 & 0 & 0 & 0 \\ 0 & 0.55 & -0.275 & 0 & 0 \\ 0 & -0.275 & 0.55 & -0.275 & 0 \\ 0 & 0 & -0.275 & 0.55 & -0.275 \\ 0 & 0 & 0 & -0.275 & 0.383 \end{bmatrix} \left\{ \begin{array}{c} t_1 \\ t_2 \\ t_3 \\ t_4 \\ t_5 \end{array} \right\} = \left\{ \begin{array}{c} 40 \\ 11 \\ 0 \\ 0 \\ -1.8 \end{array} \right\} \tag{13.4.41}$$

where the terms in the first row and column of the stiffness matrix corresponding to the known temperature condition, $t_1 = 40\,{}^\circ\text{C}$, have been set equal to 0 except for the main diagonal, which has been set equal to 1, and the first row of the force matrix has been set equal to the known nodal temperature at node 1. Also, the term $(-0.275) \times (40\,{}^\circ\text{C}) = -11$ on the left side of the second equation of Eq. (13.4.40) has been transposed to the right side in the second row (as $+11$) of Eq. (13.4.41). The second through fifth equations of Eq. (13.4.41) corresponding to the rows of unknown nodal temperatures can now be solved (typically by Gaussian elimination). The resulting solution is given by

$$t_2 = 32.36\,°C \qquad t_3 = 24.72\,°C \qquad t_4 = 17.09\,°C \qquad t_5 = 9.45\,°C \qquad (13.4.42)$$

For this elementary problem, the closed-form solution of the differential equation for conduction, Eq. (13.1.9), with the left-end boundary condition given by Eq. (13.1.10) and the right-end boundary condition given by Eq. (13.2.4) yields a linear temperature distribution through the length of the rod. The evaluation of this linear temperature function at 10-in. intervals (corresponding to the nodal points used in the finite element model) yields the same temperatures as obtained in this example by the finite element method. Because the temperature function was assumed to be linear in each finite element, this comparison is as expected. Note that F_1 could be determined by the first of Eqs. (13.4.40). ∎

Example 13.2

To illustrate more fully the use of the equations developed in Section 13.4, we will now solve the heat-transfer problem shown in Figure 13–12. For the one-dimensional rod, determine the temperatures at 75 mm increments along the length of the rod and the rate of heat flow through element 1. Let $K_{xx} = 60$ W/(m · °C), $h = 800$ W/(m² · °C), and $T_\infty = 10\,°C$. The temperature at the left end of the rod is constant at $100\,°C$.

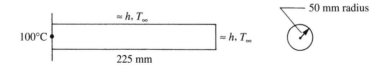

Figure 13–12 One-dimensional rod subjected to temperature variation

SOLUTION:

The finite element discretization is shown in Figure 13–13. Three elements are sufficient to enable us to determine temperatures at the four points along the rod, although more elements would yield answers more closely approximating the analytical solution obtained by solving the differential equation such as Eq. (13.2.3) with the partial derivative with respect to time equal to zero. There will be convective heat loss over the perimeter and the right end of the rod. The left end will not have convective

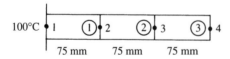

Figure 13–13 Finite element discretized rod of Figure 13–12

heat loss. Using Eqs. (13.4.22) and (13.4.28), we calculate the stiffness matrices for the elements as follows:

$$\frac{AK_{xx}}{L} = \frac{(\pi \times 0.05^2)(60)}{0.075} = 2\pi \text{ W}/^\circ\text{C}$$

$$\frac{hPL}{6} = \frac{(800)(2\pi \times 0.05)(0.075)}{6} = \pi \text{ W}/^\circ\text{C} \tag{13.4.43}$$

$$hA = (800)(\pi \times 0.05^2) = 2\pi \text{ W}/^\circ\text{C}$$

Substituting the results of Eqs. (13.4.43) into Eq. (13.4.22), we obtain the stiffness matrix for element 1 as

$$[k^{(1)}] = 2\pi \begin{bmatrix} 1 & -1 \\ -1 & 1 \end{bmatrix} + \pi \begin{bmatrix} 2 & 1 \\ 1 & 2 \end{bmatrix}$$

$$= 2\pi \begin{bmatrix} 2 & -\frac{1}{2} \\ -\frac{1}{2} & 2 \end{bmatrix} \text{ W}/^\circ\text{C} \tag{13.4.44}$$

Because there is no convection across the ends of element 1 (its left end has a known temperature and its right end is inside the whole rod and thus not exposed to fluid motion), the contribution to the stiffness matrix owing to convection from an end of the element, such as given by Eq. (13.4.28), is zero. Similarly,

$$[k^{(2)}] = [k^{(1)}] = 2\pi \begin{bmatrix} 2 & -\frac{1}{2} \\ -\frac{1}{2} & 2 \end{bmatrix} \text{ W}/^\circ\text{C} \tag{13.4.45}$$

However, element 3 has an additional (convection) term owing to heat loss from the exposed surface at its right end. Therefore, Eq. (13.4.28) yields a contribution to the element 3 stiffness matrix, which is then given by

$$[k^{(3)}] = [k^{(1)}] + hA \begin{bmatrix} 0 & 0 \\ 0 & 1 \end{bmatrix} = 2\pi \begin{bmatrix} 2 & -\frac{1}{2} \\ -\frac{1}{2} & 2 \end{bmatrix} + 2\pi \begin{bmatrix} 0 & 0 \\ 0 & 1 \end{bmatrix}$$

$$= 2\pi \begin{bmatrix} 2 & -\frac{1}{2} \\ -\frac{1}{2} & 3 \end{bmatrix} \text{ W}/^\circ\text{C} \tag{13.4.46}$$

In general, we calculate the force matrices by using Eqs. (13.4.26) and (13.4.29). Because $Q = 0$ and $q^* = 0$, we only have force terms from hT_∞ as given by Eq. (13.4.25). Therefore,

$$\{f^{(1)}\} = \{f^{(2)}\} = \frac{hT_\infty PL}{2} \begin{Bmatrix} 1 \\ 1 \end{Bmatrix} = \frac{(800 \text{ W}/(\text{m}^2 \cdot {}^\circ\text{C}) \, (10\,^\circ\text{C}) \, (2\pi \times 0.05) \, (0.075)}{2} \begin{Bmatrix} 1 \\ 1 \end{Bmatrix}$$

$$= 2\pi \begin{Bmatrix} 15 \\ 15 \end{Bmatrix} \tag{13.4.47a}$$

Element 3 has convection from both the perimeter and the right end. Therefore,

$$\{f^{(3)}\} = \{f^{(1)}\} + hT_\infty A \begin{Bmatrix} 0 \\ 1 \end{Bmatrix} = 2\pi \begin{Bmatrix} 15 \\ 15 \end{Bmatrix} + (800\,\text{W/m}^2 \cdot {}^\circ\text{C})\,(10\,{}^\circ\text{C})\pi(0.05\,\text{m})^2 \begin{Bmatrix} 0 \\ 1 \end{Bmatrix}$$

$$= 2\pi \begin{Bmatrix} 15 \\ 25 \end{Bmatrix} \tag{13.4.47b}$$

The assembly of the element stiffness matrices, Eqs. (13.4.44) through (13.4.46), and the force matrices, Eqs. (13.4.47a) and (13.4.47b), using the direct stiffness method, produces the following system of equations (where the 2π term has been divided out of both sides of Eq. (13.4.48)):

$$\begin{bmatrix} 2 & -\frac{1}{2} & 0 & 0 \\ -\frac{1}{2} & 4 & -\frac{1}{2} & 0 \\ 0 & -\frac{1}{2} & 4 & -\frac{1}{2} \\ 0 & 0 & -\frac{1}{2} & 3 \end{bmatrix} \begin{Bmatrix} t_1 = 100 \\ t_2 \\ t_3 \\ t_4 \end{Bmatrix} = \begin{Bmatrix} F_1' + 15 \\ 15 + 15 \\ 15 + 15 \\ 25 \end{Bmatrix} \tag{13.4.48}$$

Where $F_1' = F_1/2\pi$.

Expressing the second through fourth of Eqs. (13.4.48) in explicit form, we have

$$4t_2 - 0.5t_3 + 0t_4 = 50 + 30$$
$$-0.5t_2 + 4t_3 - 0.5t_4 = 30 \tag{13.4.49}$$
$$0t_2 - 0.5t_3 + 3t_4 = 25$$

Solving for the nodal temperatures $t_2 - t_4$ we obtain

$$t_2 = 21.43\,{}^\circ\text{C} \qquad t_3 = 11.46\,{}^\circ\text{C} \qquad t_4 = 10.24\,{}^\circ\text{C} \tag{13.4.50}$$

Next, we determine the heat flux for element 1 by using Eqs. (13.4.6) and (13.4.9) in Eq. (13.4.8) as

$$q^{(1)} = -K_{xx}[B]\{t\} \tag{13.4.51}$$

Using Eq. (13.4.7) in Eq. (13.4.51), we have

$$q^{(1)} = -K_{xx} \begin{bmatrix} -\dfrac{1}{L} & \dfrac{1}{L} \end{bmatrix} \begin{Bmatrix} t_1 \\ t_2 \end{Bmatrix} \tag{13.4.52}$$

Substituting the numerical values for t_1 and t_2 into Eq. (13.4.52), we obtain

$$q^{(1)} = -60 \begin{bmatrix} -\dfrac{1}{0.075} & \dfrac{1}{0.075} \end{bmatrix} \begin{Bmatrix} 100 \\ 21.43 \end{Bmatrix}$$

or

$$q^{(1)} = 62856\,\text{W/m}^2 \tag{13.4.53}$$

We then determine the rate of heat flow $\bar{q}$ by multiplying Eq. (13.4.53) by the cross-sectional area over which q acts. Therefore,

$$\bar{q}^{(1)} = 62856(\pi \times 0.05^2) = 493.7\,\text{W}$$

Here positive heat flow indicates heat flow from node 1 to node 2 (to the right). ■

Example 13.3

The plane wall shown in Figure 13–14 is 1 m thick. The left surface of the wall $(x = 0)$ is maintained at a constant temperature of 200°C, and the right surface $(x = L = 1 \text{ m})$ is insulated. The thermal conductivity is $K_{xx} = 25 \text{ W/(m} \cdot °\text{C)}$ and there is a uniform generation of heat inside the wall of $Q = 400 \text{ W/m}^3$. Determine the temperature distribution through the wall thickness.

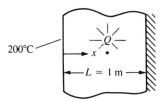

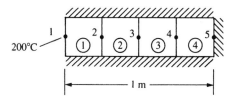

Figure 13–14 Conduction in a plane wall subjected to uniform heat generation

Figure 13–15 Discretized model of Figure 13–14

SOLUTION:

This problem is assumed to be approximated as a one-dimensional heat-transfer problem. The discretized model of the wall is shown in Figure 13–15. For simplicity, we use four equal-length elements all with unit cross-sectional area $(A = 1 \text{ m}^2)$. The unit area represents a typical cross section of the wall. The perimeter of the wall model is then insulated to obtain the correct conditions.

Using Eqs. (13.4.22) and (13.4.28), we calculate the element stiffness matrices as follows:

$$\frac{AK_{xx}}{L} = \frac{(1 \text{ m}^2)[25 \text{ W/(m} \cdot °\text{C)}]}{0.25 \text{ m}} = 100 \text{ W/}°\text{C}$$

For each identical element, we have

$$[k] = 100 \begin{bmatrix} 1 & -1 \\ -1 & 1 \end{bmatrix} \text{ W/}°\text{C} \qquad (13.4.54)$$

Because no convection occurs, h is equal to zero; therefore, there is no convection contribution to $[k]$.

The element force matrices are given by Eq. (13.4.26). With $Q = 400 \text{ W/m}^3$, $q = 0$, and $h = 0$, Eq. (13.4.26) becomes

$$\{f\} = \frac{QAL}{2} \begin{Bmatrix} 1 \\ 1 \end{Bmatrix} \qquad (13.4.55)$$

Evaluating Eq. (13.4.55) for a typical element, such as element 1, we obtain

$$\begin{Bmatrix} f_{1x} \\ f_{2x} \end{Bmatrix} = \frac{(400 \text{ W/m}^3)(1 \text{ m}^2)(0.25 \text{ m})}{2} \begin{Bmatrix} 1 \\ 1 \end{Bmatrix} = \begin{Bmatrix} 50 \\ 50 \end{Bmatrix} \text{ W} \qquad (13.4.56)$$

The force matrices for all other elements are equal to Eq. (13.4.56).

The assemblage of the element matrices, Eqs. (13.4.54) and (13.4.56) and the other force matrices similar to Eq. (13.4.56), yields

$$100 \begin{bmatrix} 1 & -1 & 0 & 0 & 0 \\ -1 & 2 & -1 & 0 & 0 \\ 0 & -1 & 2 & -1 & 0 \\ 0 & 0 & -1 & 2 & -1 \\ 0 & 0 & 0 & -1 & 1 \end{bmatrix} \begin{Bmatrix} t_1 \\ t_2 \\ t_3 \\ t_4 \\ t_5 \end{Bmatrix} = \begin{Bmatrix} F_1 + 50 \\ 100 \\ 100 \\ 100 \\ 50 \end{Bmatrix} \tag{13.4.57}$$

Substituting the known temperature $t_1 = 200\,°C$ into Eq. (13.4.57), dividing both sides of Eq. (13.4.57) by 100, and transposing known terms to the right side, we have

$$\begin{bmatrix} 1 & 0 & 0 & 0 & 0 \\ 0 & 2 & -1 & 0 & 0 \\ 0 & -1 & 2 & -1 & 0 \\ 0 & 0 & -1 & 2 & -1 \\ 0 & 0 & 0 & -1 & 1 \end{bmatrix} \begin{Bmatrix} t_1 \\ t_2 \\ t_3 \\ t_4 \\ t_5 \end{Bmatrix} = \begin{Bmatrix} 200\,°C \\ 201 \\ 1 \\ 1 \\ 0.5 \end{Bmatrix} \tag{13.4.58}$$

The second through fifth equations of Eq. (13.4.58) can now be solved simultaneously to yield

$$t_2 = 203.5\,°C \qquad t_3 = 206\,°C \qquad t_4 = 207.5\,°C \qquad t_5 = 208\,°C \tag{13.4.59}$$

Using the first of Eqs. (13.4.57) yields the rate of heat flow out the left end:

$$F_1 = 100(t_1 - t_2) - 50$$
$$F_1 = 100(200 - 203.5) - 50$$
$$F_1 = -400 \text{ W}$$

The closed-form solution of the differential equation for conduction, Eq. (13.1.9), with the left-end boundary condition given by Eq. (13.1.10) and the right-end boundary condition given by Eq. (13.1.11), and with $q_x^* = 0$, is shown in Reference [2] to yield a parabolic temperature distribution through the wall. Evaluating the expression for the temperature function given in Reference [2] for values of x corresponding to the node points of the finite element model, we obtain

$$t_2 = 203.5\,°C \qquad t_3 = 206\,°C \qquad t_4 = 207.5\,°C \qquad t_5 = 208\,°C \tag{13.4.60}$$

Figure 13–16 is a plot of the closed-form solution and the finite element solution for the temperature variation through the wall. The finite element nodal values and the closed-form values are equal, because the consistent equivalent force matrix has been used. (This was also discussed in Sections 3.10 and 3.11 for the axial bar subjected to distributed loading, and in Section 4.5 for the beam subjected to distributed

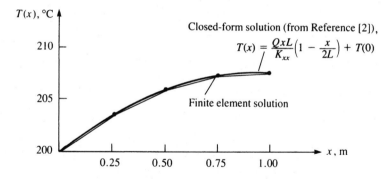

Closed-form solution (from Reference [2]),

$$T(x) = \frac{QxL}{K_{xx}}\left(1 - \frac{x}{2L}\right) + T(0)$$

Finite element solution

Figure 13–16 Comparison of the finite element and closed-form solutions for Example 13.3

loading.) However, recall that the finite element model predicts a linear temperature distribution within each element as indicated by the straight lines connecting the nodal temperature values in Figure 13–16. ∎

Example 13.4

The fin shown in Figure 13–17 is insulated on the perimeter. The left end has a constant temperature of $100\,°\text{C}$. A positive heat flux of $q = 5000\,\text{W/m}^2$ acts on the right end. Let $K_{xx} = 6\,\text{W/(m-°C)}$ and cross-sectional area $A = 0.1\,\text{m}^2$. Determine the temperatures at $L/4$, $L/2$, $3L/4$, and L, where $L = 0.4\,\text{m}$.

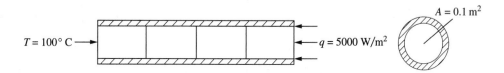

Figure 13–17 Insulated fin subjected to end heat flux

SOLUTION:

Using Eq. (13.4.22) with the second term set to zero as there is no heat transfer by convection from any surfaces due to the insulated perimeter and constant temperature on the left end and constant heat flux on the right end, we obtain

$$[k^{(1)}] = [k^{(2)}] = [k^{(3)}] = \frac{AK_{xx}}{L}\begin{bmatrix} 1 & -1 \\ -1 & 1 \end{bmatrix}$$

$$= \frac{(0.1\,\text{m}^2)(6\,\text{W/(m-°C)})}{0.1\,\text{m}}\begin{bmatrix} 1 & -1 \\ -1 & 1 \end{bmatrix} = \begin{bmatrix} 6 & -6 \\ -6 & 6 \end{bmatrix}\text{W/°C} \qquad (13.4.61)$$

$[k^{(4)}] = [k^{(1)}]$ also

$$\{f^{(1)}\} = \{f^{(2)}\} = \{f^{(3)}\} = \left\{ \begin{array}{c} 0 \\ 0 \end{array} \right\} \text{ as } Q = 0 \text{ (no internal heat source) and } q^* = 0$$

(no surface heat flux)

$$\{f^{(4)}\} = qA \left\{ \begin{array}{c} 0 \\ 1 \end{array} \right\} = (5000 \text{ W/m}^2)(0.1 \text{ m}^2) \left\{ \begin{array}{c} 0 \\ 1 \end{array} \right\} = \left\{ \begin{array}{c} 0 \\ 500 \end{array} \right\} \text{W} \qquad (13.4.62)$$

Assembling the global stiffness matrix from Eq. (13.4.61), and the global force matrix from Eq. (13.4.62), we obtain the global equations as

$$\begin{bmatrix} 6 & -6 & 0 & 0 & 0 \\ & 12 & -6 & 0 & 0 \\ & & 12 & -6 & 0 \\ & & & 12 & -6 \\ \text{Symmetry} & & & & 6 \end{bmatrix} \left\{ \begin{array}{c} t_1 \\ t_2 \\ t_3 \\ t_4 \\ t_5 \end{array} \right\} = \left\{ \begin{array}{c} F_{1x} \\ 0 \\ 0 \\ 0 \\ 500 \end{array} \right\} \qquad (13.4.63)$$

Now applying the boundary condition on temperature, we have

$$t_1 = 100\,°C \qquad (13.4.64)$$

Substituting Eq. (13.4.64) for t_1 into Eq. (13.4.63), we then solve the second through fourth equations (associated with the unknown temperatures $t_2 - t_5$) simultaneously to obtain

$$t_2 = 183.33\,°C, \quad t_3 = 266.67\,°C, \quad t_4 = 350\,°C, \quad t_5 = 433.33\,°C \qquad (13.4.65)$$

Substituting the nodal temperatures from Eq. (13.4.65) into the first of Eqs. (13.4.63), we obtain the nodal heat source at node 1 as

$$F_{1x} = 6(100\,°C - 183.33\,°C) = -500 \text{ W} \qquad (13.4.66)$$

The nodal heat source given by Eq. (13.4.66) has a negative value, which means the heat is leaving the left end. This source is the same as the source coming into the fin at the right end given by $qA = (5000)(0.1) = 500$ W. ∎

To further demonstrate explicit concepts of Fourier's law of heat conduction and Newton's law of cooling, along with heat balance, we solve the following problem.

Example 13.5

A composite furnace wall shown in Figure 13–18 is composed of two homogeneous slabs in contact. Let thermal conductivities be $k_1 = 1$ W/(m-°C) for firebrick slab 1 and $k_2 = 0.3$ W/(m-°C) for insulating slab 2. The left side is exposed to an ambient temperature of $T_{\infty L} = 1000\,°C$ inside the furnace with a heat transfer coefficient of $h_L = 10$ W/(m²-°C). The right-side ambient temperature is 25 °C outside of the furnace with a heat transfer coefficient of $h_R = 3$ W/(m²-°C). The thermal resistance of the

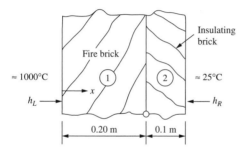

Figure 13–18 Composite furnace wall

interface between the firebrick and insulating brick can be neglected. The thicknesses of the slabs are $L_1 = 0.20$ m and $L_2 = 0.10$ m. Determine the temperatures at the left edge, middle, and right edge of the composite wall and the heat transferred through the wall.

SOLUTION:

We assume the furnace wall is tall enough, such that the heat flux along the vertical direction can be neglected, and therefore, that the heat flow is one-dimensional in the direction of the furnace wall thickness. We can then assume the cross-sectional area to be a unit slice ($A = 1\,\text{m}^2$) in the finite element model. The model will be made of two one-dimensional finite elements as shown in Figure 13–19.

We will develop the equations in two ways. First, using the heat flow balance at each of the three nodes of the model shown in Figure 13–19 and then by using the direct stiffness method.

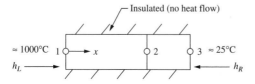

Figure 13–19 Finite element model of the furnace wall

Method 1: Heat Balance Equations at Nodes

By Newton's law of cooling, Eq. (13.2.2), we have the heat flow entering node 1 by convection as

$$\bar{q}_1 = Ah_L(T_{\infty L} - t_1) \tag{13.4.67}$$

By Fourier's law of heat conduction, Eq. (13.1.30), we have heat flow through elements 1 and 2 as

$$\bar{q}_2 = \frac{Ak_1}{L_1}(t_1 - t_2) \tag{13.4.68}$$

and

$$\bar{q}_3 = \frac{Ak_2}{L_2}(t_2 - t_3) \qquad (13.4.69)$$

By Newton's Law of cooling, we have the heat flow exiting node 3 as

$$\bar{q}_4 = Ah_R(t_3 - T_{\infty R}) \qquad (13.4.70)$$

Now realizing the heat flow through the wall is constant ($\bar{q}_1 = \bar{q}_2 = \bar{q}_3 = \bar{q}_4$) and applying the heat balance equations at nodes 1, 2, and 3, we obtain

$$\bar{q}_1 = \bar{q}_2$$

or $\qquad Ah_L(T_{\infty L} - t_1) = \dfrac{Ak_1}{L_1}(t_1 - t_2) \quad$ at node 1 $\qquad (13.4.71)$

$$\bar{q}_2 = \bar{q}_3$$

or $\qquad \dfrac{Ak_1}{L_1}(t_1 - t_2) = \dfrac{Ak_2}{L_2}(t_2 - t_3) \quad$ at node 2 $\qquad (13.4.72)$

$$\bar{q}_3 = \bar{q}_4$$

or $\qquad \dfrac{Ak_2}{L_2}(t_2 - t_3) = Ah_R(t_3 - T_{\infty R}) \quad$ at node 3 $\qquad (13.4.73)$

In matrix form with rearrangement so that the equations are in the form $[K]\{t\} = \{F\}$, we have

$$\begin{bmatrix} \dfrac{Ak_1}{L_1} + Ah_L & \dfrac{-Ak_1}{L_1} & 0 \\ \dfrac{-Ak_1}{L_1} & \dfrac{Ak_1}{L_1} + \dfrac{Ak_2}{L_2} & \dfrac{-Ak_2}{L_2} \\ 0 & \dfrac{-Ak_2}{L_2} & \dfrac{Ak_2}{L_2} + Ah_R \end{bmatrix} \begin{Bmatrix} t_1 \\ t_2 \\ t_3 \end{Bmatrix} = \begin{Bmatrix} Ah_L T_{\infty L} \\ 0 \\ Ah_R T_{\infty R} \end{Bmatrix} \qquad (13.4.74)$$

Method 2: Direct Stiffness Method

By the direct stiffness method, the typical element stiffness matrix given by Eq. (13.4.2) is

$$[k] = \frac{Ak}{L} \begin{bmatrix} 1 & -1 \\ -1 & 1 \end{bmatrix} \qquad (13.4.75)$$

and from the right end due to convection by Eq. (13.4.28), we have

$$[k_h]_{\text{rt end}} = hA \begin{bmatrix} 0 & 0 \\ 0 & 1 \end{bmatrix} \qquad (13.4.76)$$

Similarly, from the left end due to convection, we have

$$[k_h]_{\text{lt end}} = hA \begin{bmatrix} 1 & 0 \\ 0 & 0 \end{bmatrix} \qquad (13.4.77)$$

The right end force terms are given by Eq. (13.4.29) as

$$\{f_h\}_{\text{rt end}} = h_R T_{\infty R} A \begin{Bmatrix} 0 \\ 1 \end{Bmatrix} \tag{13.4.78}$$

Similarly, at the left end, we have

$$\{f_h\}_{\text{lt end}} = h_L T_{\infty L} A \begin{Bmatrix} 1 \\ 0 \end{Bmatrix} \tag{13.4.79}$$

By the direct stiffness method of assembly, we obtain the identical equations, Eq. (13.4.74), as obtained using heat balance at each node.
Substituting the numerical values into Eq. (13.4.74), we obtain

$$\begin{bmatrix} 15 & -5 & 0 \\ -5 & 8 & -3 \\ 0 & -3 & 6 \end{bmatrix} \begin{Bmatrix} t_1 \\ t_2 \\ t_3 \end{Bmatrix} = \begin{Bmatrix} 1 \times 10^4 \\ 0 \\ 75 \end{Bmatrix} \tag{13.4.80}$$

Solving Eq. (13.4.80) simultaneously, we obtain the resulting nodal temperatures as

$$t_1 = 899.14\,°\text{C}, \qquad t_2 = 697.41°\text{C}, \qquad t_3 = 361.21°\text{C} \tag{13.4.81}$$

The heat flow through the wall is determined by using the heat flow equation from Fourier's law as

$$\bar{q} = \frac{-k_1 A}{L_1}(t_2 - t_1) = \frac{(-1\,\text{W/m-}°\text{C})\,(1\,\text{m}^2)\,(697.41 - 899.14)°\text{C}}{0.20\,\text{m}} = 1009\,\text{W}$$

$$\tag{13.4.82}$$

The heat loss through the wall is obtained from the convection boundary equation as

$$\bar{q}_4 = A h_R\,(t_3 - T_{\infty R}) = (+1\,\text{m}^2)\,(3\,\text{W/m}^2\text{-}°\text{C})\,(361.21 - 25)°\text{C} = 1009\,\text{W}$$

$$\tag{13.4.83}$$

The exact solution to this problem is obtained by solving the basic differential equation, Eq. (13.1.8), with $Q = 0$ (as there is no internal heat source). The solution requires integrating the differential equation once to obtain the expression for heat flux and a second time to obtain the expression for temperature function, $T(x)$. You must also realize that we have two expressions for $T(x)$, one for the $0 \le x \le L_1$ (label it $T_1(x)$), and one for $L_1 \le x \le L_2$ (label it $T_2(x)$). There will be four constants of integration from the two temperature functions. The boundary conditions involve setting the convection heat transfer to the conduction heat transfer at nodes 1 and 3. Setting $T_1(0) = t_1$ and $T_2(L) = t_3$. Finally, you must introduce temperature and heat flow conditions at the interface between the two wall surfaces (at $x = L_1$). These conditions are $T_1(x = L_1) = T_2(x = L_1)$ and $q_1 = k_1 dT_1/dx = k_2 dT_2/dx$ at $x = L_1$. The resulting system of equations has six total equations for six unknowns, constants

$c_1 - c_4$, and nodal temperatures t_1 and t_3. After solving for the constants and t_1 and t_3, the resulting temperature functions are

$$T_1(x) = -1008.62x + 899.138 \qquad 0 \le x \le L_1 = 0.20 \text{ m}$$

and

$$T_2(x) = -3362.07x + 1369.83 \qquad L_1 = 0.2 \text{ m} \le x \le L = 0.30 \text{ m}$$

The finite element solution is identical to the analytical solution upon evaluating the temperature functions ($T_1(x)$ at $x = 0$ and $x = 0.2$ m, and $T_2(x)$ at $x = 0.2$ m and $x = 0.3$ m). ∎

Finally, remember that the most important advantage of the finite element method is that it enables us to approximate, with high confidence, more complicated problems, such as those with more then one thermal conductivity, for which closed-form solutions are difficult (if not impossible) to obtain. The automation of the finite element method through general computer programs makes the method extremely powerful.

▲ 13.5 Two-Dimensional Finite Element Formulation ▲

Because many bodies can be modeled as two-dimensional heat-transfer problems, we now develop the equations for an element appropriate for these problems. Examples using this element then follow.

Step 1 Select Element Type

The three-noded triangular element with nodal temperatures shown in Figure 13–20 is the basic element for solution of the two-dimensional heat-transfer problem.

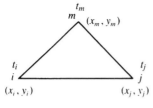

Figure 13–20 Basic triangular element with nodal temperatures

Step 2 Select a Temperature Function

The temperature function is given by

$$\{T\} = [N_i \quad N_j \quad N_m] \begin{Bmatrix} t_i \\ t_j \\ t_m \end{Bmatrix} \tag{13.5.1}$$

where t_i, t_j, and t_m are the nodal temperatures, and the shape functions are again given by Eqs. (6.2.18); that is,

$$N_i = \frac{1}{2A}(\alpha_i + \beta_i x + \gamma_i y) \tag{13.5.2}$$

with similar expressions for N_j and N_m. Here the α's, β's, and γ's are defined by Eqs. (6.2.10).

Unlike the CST element of Chapter 6 where there are 2 degrees of freedom per node (an x and a y displacement), in the heat transfer three-noded triangular element only a single scalar value (nodal temperature) is the primary unknown at each node, as shown by Eq. (13.5.1). This holds true for the three-dimensional elements as well, as shown in Section 13.7. Hence, the heat-transfer problem is sometimes known as a scalar-valued boundary value problem.

Step 3 Define the Temperature Gradient/Temperature and Heat Flux/Temperature Gradient Relationships

We define the gradient matrix analogous to the strain matrix used in the stress analysis problem as

$$\{g\} = \left\{ \begin{array}{c} \dfrac{\partial T}{\partial x} \\[2mm] \dfrac{\partial T}{\partial y} \end{array} \right\} \tag{13.5.3}$$

Using Eq. (13.5.1) in Eq. (13.5.3), we have

$$\{g\} = \begin{bmatrix} \dfrac{\partial N_i}{\partial x} & \dfrac{\partial N_j}{\partial x} & \dfrac{\partial N_m}{\partial x} \\[2mm] \dfrac{\partial N_i}{\partial y} & \dfrac{\partial N_j}{\partial y} & \dfrac{\partial N_m}{\partial y} \end{bmatrix} \left\{ \begin{array}{c} t_i \\ t_j \\ t_m \end{array} \right\} \tag{13.5.4}$$

The gradient matrix $\{g\}$, written in compact matrix form analogously to the strain matrix $\{\varepsilon\}$ of the stress analysis problem, is given by

$$\{g\} = [B]\{t\} \tag{13.5.5}$$

where the $[B]$ matrix is obtained by substituting the three equations suggested by Eq. (13.5.2) in the rectangular matrix on the right side of Eq. (13.5.4) as

$$[B] = \frac{1}{2A} \begin{bmatrix} \beta_i & \beta_j & \beta_m \\ \gamma_i & \gamma_j & \gamma_m \end{bmatrix} \tag{13.5.6}$$

The heat flux/temperature gradient relationship is now

$$\left\{ \begin{array}{c} q_x \\ q_y \end{array} \right\} = -[D]\{g\} \tag{13.5.7}$$

where the material property matrix is

$$[D] = \begin{bmatrix} K_{xx} & 0 \\ 0 & K_{yy} \end{bmatrix} \tag{13.5.8}$$

Step 4 Derive the Element Conduction Matrix and Equations

The element stiffness matrix from Eq. (13.4.17) is

$$[k] = \iiint_V [B]^T [D][B] \, dV + \iint_{S_3} h[N]^T [N] \, dS \qquad (13.5.9)$$

where

$$[k_c] = \iiint_V [B]^T [D][B] \, dV$$

$$= \iiint_V \frac{1}{4A^2} \begin{bmatrix} \beta_i & \gamma_i \\ \beta_j & \gamma_j \\ \beta_m & \gamma_m \end{bmatrix} \begin{bmatrix} K_{xx} & 0 \\ 0 & K_{yy} \end{bmatrix} \begin{bmatrix} \beta_i & \beta_j & \beta_m \\ \gamma_i & \gamma_j & \gamma_m \end{bmatrix} dV \qquad (13.5.10)$$

Assuming constant thickness in the element and noting that all terms of the integrand of Eq. (13.5.10) are constant, we have

$$[k_c] = \iiint_V [B]^T [D][B] \, dV = tA[B]^T [D][B] \qquad (13.5.11)$$

Equation (13.5.11) is the true conduction portion of the total stiffness matrix Eq. (13.5.9). The second integral of Eq. (13.5.9) (the convection portion of the total stiffness matrix) is defined by

$$[k_h] = \iint_{S_3} h[N]^T [N] \, dS \qquad (13.5.12)$$

We can explicitly multiply the matrices in Eq. (13.5.12) to obtain

$$[k_h] = h \iint_{S_3} \begin{bmatrix} N_i N_i & N_i N_j & N_i N_m \\ N_j N_i & N_j N_j & N_j N_m \\ N_m N_i & N_m N_j & N_m N_m \end{bmatrix} dS \qquad (13.5.13)$$

To illustrate the use of Eq. (13.5.13), consider the side between nodes i and j of the triangular element to be subjected to convection (Figure 13–21). Then $N_m = 0$ along side i-j, and we obtain

$$[k_h] = \frac{hL_{i\text{-}j}t}{6} \begin{bmatrix} 2 & 1 & 0 \\ 1 & 2 & 0 \\ 0 & 0 & 0 \end{bmatrix} \qquad (13.5.14)$$

where $L_{i\text{-}j}$ is the length of side i-j.

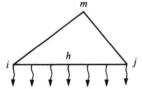

Figure 13–21 Heat loss by convection from side i-j

The evaluation of the force matrix integrals in Eq. (13.4.16) is as follows:

$$\{f_Q\} = \iiint_V Q[N]^T \, dV = Q \iiint_V [N]^T \, dV \tag{13.5.15}$$

for constant heat source Q. Thus it can be shown (left to your discretion) that this integral is equal to

$$\{f_Q\} = \frac{QV}{3} \begin{Bmatrix} 1 \\ 1 \\ 1 \end{Bmatrix} \tag{13.5.16}$$

where $V = At$ is the volume of the element. Equation (13.5.16) indicates that heat is generated by the body in three equal parts to the nodes (like body forces in the elasticity problem). The second force matrix in Eq. (13.4.16) is

$$\{f_q\} = \iint_{S_2} q^*[N]^T \, dS = \iint_{S_2} q^* \begin{Bmatrix} N_i \\ N_j \\ N_m \end{Bmatrix} dS \tag{13.5.17}$$

This reduces to

$$\frac{q^* L_{i\text{-}j} t}{2} \begin{Bmatrix} 1 \\ 1 \\ 0 \end{Bmatrix} \qquad \text{on side } i\text{-}j \tag{13.5.18}$$

$$\frac{q^* L_{j\text{-}m} t}{2} \begin{Bmatrix} 0 \\ 1 \\ 1 \end{Bmatrix} \qquad \text{on side } j\text{-}m \tag{13.5.19}$$

$$\frac{q^* L_{m\text{-}i} t}{2} \begin{Bmatrix} 1 \\ 0 \\ 1 \end{Bmatrix} \qquad \text{on side } m\text{-}i \tag{13.5.20}$$

where $L_{i\text{-}j}$, $L_{j\text{-}m}$, and $L_{m\text{-}i}$ are the lengths of the sides of the element, and the heat flux q^* is assumed constant over each edge. The integral $\iint_{S_3} hT_\infty[N]^T \, dS$ can be found in a manner similar to Eq. (13.5.17) by simply replacing q^* with hT_∞ in Eqs. (13.5.18) through (13.5.20).

Steps 5 through 7

Steps 5 through 7 are identical to those described in Section 13.4.

To illustrate the use of the equations presented in Section 13.5, we will now solve some two-dimensional heat-transfer problems.

Example 13.6

For the two-dimensional body shown in Figure 13–22, determine the temperature distribution. The temperature at the left side of the body is maintained at 40 °C. The edges on the top and bottom of the body are insulated. There is heat convection

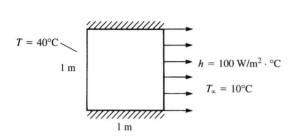

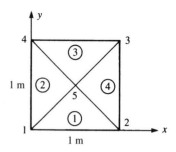

Figure 13–22 Two-dimensional body subjected to temperature variation and convection

Figure 13–23 Discretized two-dimensional body of Figure 13–22

from the right side with convection coefficient $h = 100$ W/(m$^2 \cdot$°C). The free-stream temperature is $T_\infty = 10$°C. The coefficients of thermal conductivity are $K_{xx} = K_{yy} = 40$ W/(m $\cdot$ °C). The dimensions are shown in the figure. Assume the thickness to be 1 m.

SOLUTION:

The finite element discretization is shown in Figure 13–23. We will use four triangular elements of equal size for simplicity of the longhand solution. There will be convective heat loss only over the right side of the body because the other faces are insulated. We now calculate the element stiffness matrices using Eq. (13.5.11) applied for all elements and using Eq. (13.5.14) applied for element 4 only, because convection is occurring only across one edge of element 4.

Element 1

The coordinates of the element 1 nodes are $x_1 = 0$, $y_1 = 0$, $x_2 = 1$, $y_2 = 0$, $x_5 = 0.5$, and $y_5 = 0.5$. Using these coordinates and Eqs. (6.2.10), we obtain

$$\beta_1 = 0 - 0.5 = -0.5 \qquad \beta_2 = 0.5 - 0 = 0.5 \qquad \beta_5 = 0 - 0 = 0$$
$$\gamma_1 = 0.5 - 1 = -0.5 \qquad \gamma_2 = 0 - 0.5 = -0.5 \qquad \gamma_5 = 1 - 0 = 1$$
(13.5.21)

Using Eqs. (13.5.21) in Eq. (13.5.11) with $t = 1$ m and $A = \frac{1}{2}(1$ m $)(0.5$ m$) = 0.25$ m^2, we have

$$[k_c^{(1)}] = tA[B]^T[D][B] = \frac{(1)(0.25)}{(2 \times 0.25)(2 \times 0.25)} \begin{bmatrix} -0.5 & -0.5 \\ 0.5 & -0.5 \\ 0 & 1 \end{bmatrix} \begin{bmatrix} 40 & 0 \\ 0 & 40 \end{bmatrix} \begin{bmatrix} -0.5 & 0.5 & 0 \\ -0.5 & -0.5 & 1 \end{bmatrix}$$
(13.5.22)

where $[B]$ is given by Eq. (13.5.6) and $[D]$ is given by Eq. (13.5.8).

Simplifying Eq. (13.5.22), we obtain

$$[k_c^{(1)}] = \begin{matrix} & 1 & 2 & 5 \\ & \begin{bmatrix} 20 & 0 & -20 \\ 0 & 20 & -20 \\ -20 & -20 & 40 \end{bmatrix} \end{matrix} \text{ W}/(\text{m} \cdot \text{°C})$$
(13.5.23)

where the numbers above the columns indicate the node numbers associated with the matrix.

Element 2

The coordinates of the element 2 nodes are $x_1 = 0$, $y_1 = 0$, $x_5 = 0.5$, $y_5 = 0.5$, $x_4 = 0$, and $y_4 = 1$. Using these coordinates, we obtain

$$\beta_1 = 0.5 - 1 = -0.5 \qquad \beta_5 = 1 - 0 = 1 \qquad \beta_4 = 0 - 0.5 = -0.5$$
$$\gamma_1 = 0 - 0.5 = -0.5 \qquad \gamma_5 = 0 - 0 = 0 \qquad \gamma_4 = 0.5 - 0 = 0.5 \tag{13.5.24}$$

Using Eqs. (13.5.24) in Eq. (13.5.11), we have

$$[k_c^{(2)}] = 1.0 \begin{bmatrix} -0.5 & -0.5 \\ 1 & 0 \\ -0.5 & 0.5 \end{bmatrix} \begin{bmatrix} 40 & 0 \\ 0 & 40 \end{bmatrix} \begin{bmatrix} -0.5 & 1 & -0.5 \\ -0.5 & 0 & 0.5 \end{bmatrix} \tag{13.5.25}$$

Simplifying Eq. (13.5.25), we obtain

$$[k_c^{(2)}] = \begin{matrix} 1 & 5 & 4 \\ \begin{bmatrix} 20 & -20 & 0 \\ -20 & 40 & -20 \\ 0 & -20 & 20 \end{bmatrix} \end{matrix} \; \text{W/}°\text{C} \tag{13.5.26}$$

Element 3

The coordinates of the element 3 nodes are $x_4 = 0$, $y_4 = 1$, $x_5 = 0.5$, $y_5 = 0.5$, $x_3 = 1$, and $y_3 = 1$. Using these coordinates, we obtain

$$\beta_4 = 0.5 - 1 = -0.5 \qquad \beta_5 = 1 - 1 = 0 \qquad \beta_3 = 1 - 0.5 = 0.5$$
$$\gamma_4 = 1 - 0.5 = 0.5 \qquad \gamma_5 = 0 - 1 = -1 \qquad \gamma_3 = 0.5 - 0 = 0.5 \tag{13.5.27}$$

Using Eqs. (13.5.27) in Eq. (13.5.11), we obtain

$$[k_c^{(3)}] = \begin{matrix} 4 & 5 & 3 \\ \begin{bmatrix} 20 & -20 & 0 \\ -20 & 40 & -20 \\ 0 & -20 & 20 \end{bmatrix} \end{matrix} \; \text{W/}°\text{C} \tag{13.5.28}$$

Element 4

The coordinates of the element 4 nodes are $x_2 = 1$, $y_2 = 0$, $x_3 = 1$, $y_3 = 1$, $x_5 = 0.5$, and $y_5 = 0.5$. Using these coordinates, we obtain

$$\beta_2 = 1 - 0.5 = 0.5 \qquad \beta_3 = 0.5 - 0 = 0.5 \qquad \beta_5 = 0 - 1 = -1$$
$$\gamma_2 = 0.5 - 1 = -0.5 \qquad \gamma_3 = 1 - 0.5 = 0.5 \qquad \gamma_5 = 1 - 1 = 0 \tag{13.5.29}$$

Using Eqs. (13.5.29) in Eq. (13.5.11), we obtain

$$[k_c^{(4)}] = \begin{matrix} 2 & 3 & 5 \\ \begin{bmatrix} 20 & 0 & -20 \\ 0 & 20 & -20 \\ -20 & -20 & 40 \end{bmatrix} \end{matrix} \; \text{W/}°\text{C} \tag{13.5.30}$$

For element 4, we have a convection contribution to the total stiffness matrix because side 2–3 is exposed to the free-stream temperature. Using Eq. (13.5.14) with $i = 2$ and $j = 3$, we obtain

$$[k_h^{(4)}] = \frac{(100)(1)(1)}{6} \begin{bmatrix} 2 & 1 & 0 \\ 1 & 2 & 0 \\ 0 & 0 & 0 \end{bmatrix} \tag{13.5.31}$$

Simplifying Eq. (13.5.31) yields

$$[k_h^{(4)}] = \begin{matrix} & 2 & 3 & 5 \\ & \begin{bmatrix} 33.3 & 16.67 & 0 \\ 16.67 & 33.3 & 0 \\ 0 & 0 & 0 \end{bmatrix} & W/°C \end{matrix} \tag{13.5.32}$$

Adding Eqs. (13.5.30) and (13.5.32), we obtain the element 4 total stiffness matrix as

$$[k^{(4)}] = \begin{matrix} & 2 & 3 & 5 \\ & \begin{bmatrix} 53.3 & 16.67 & -20 \\ 16.67 & 53.3 & -20 \\ -20 & -20 & 40 \end{bmatrix} & W/°C \end{matrix} \tag{13.5.33}$$

Superimposing the stiffness matrices given by Eqs. (13.5.23), (13.5.26), (13.5.28), and (13.5.33), we obtain the total stiffness matrix for the body as

$$[K] = \begin{bmatrix} 40 & 0 & 0 & 0 & -40 \\ 0 & 53.3 & 16.67 & 0 & -40 \\ 0 & 16.67 & 53.3 & 0 & -40 \\ 0 & 0 & 0 & 40 & -40 \\ -40 & -40 & -40 & -40 & 160 \end{bmatrix} W/°C \tag{13.5.34}$$

Next, we determine the element force matrices by using Eqs. (13.5.18) through (13.5.20) with q^* replaced by hT_∞. Because $Q = 0$, $q^* = 0$, and we have convective heat transfer only from side 2–3, element 4 is the only one that contributes nodal forces. Hence,

$$\{f^{(4)}\} = \begin{Bmatrix} f_2 \\ f_3 \\ f_5 \end{Bmatrix} = \frac{hT_\infty L_{2-3} t}{2} \begin{Bmatrix} 1 \\ 1 \\ 0 \end{Bmatrix} \tag{13.5.35}$$

Substituting the appropriate numerical values into Eq. (13.5.35) yields

$$\{f^{(4)}\} = \frac{(100)(10)(1)(1)}{2} \begin{Bmatrix} 1 \\ 1 \\ 0 \end{Bmatrix} = \begin{Bmatrix} 500 \\ 500 \\ 0 \end{Bmatrix} W \tag{13.5.36}$$

Using Eqs. (13.5.34) and (13.5.36), we find that the total assembled system of equations is

$$
\begin{bmatrix}
40 & 0 & 0 & 0 & -40 \\
0 & 53.3 & 16.67 & 0 & -40 \\
0 & 16.67 & 53.3 & 0 & -40 \\
0 & 0 & 0 & 40 & -40 \\
-40 & -40 & -40 & -40 & 160
\end{bmatrix}
\begin{Bmatrix}
t_1 \\ t_2 \\ t_3 \\ t_4 \\ t_5
\end{Bmatrix}
=
\begin{Bmatrix}
F_1 \\ 500 \\ 500 \\ F_4 \\ 0
\end{Bmatrix}
\tag{13.5.37}
$$

We have known nodal temperature boundary conditions of $t_1 = 100\,°F$ and $t_4 = 100\,°F$. We again modify the stiffness and force matrices as follows:

$$
\begin{bmatrix}
1 & 0 & 0 & 0 & 0 \\
0 & 53.3 & 16.67 & 0 & -40 \\
0 & 16.67 & 53.3 & 0 & -40 \\
0 & 0 & 0 & 1 & 0 \\
0 & -40 & -40 & 0 & 160
\end{bmatrix}
\begin{Bmatrix}
t_1 \\ t_2 \\ t_3 \\ t_4 \\ t_5
\end{Bmatrix}
=
\begin{Bmatrix}
40 \\ 500 \\ 500 \\ 40 \\ 3200
\end{Bmatrix}
\tag{13.5.38}
$$

The terms in the first and fourth rows and columns corresponding to the known temperature conditions $t_1 = 40\,°C$ and $t_4 = 40\,°C$ have been set equal to zero except for the main diagonal, which has been set equal to one, and the first and fourth rows of the force matrix have been set equal to the known nodal temperatures. Also, the term $(-40)(40\,°C) + (-40) \times (40\,°C) = -3200$ on the left side of the fifth equation of Eq. (13.5.37) has been transposed to the right side in the fifth row (as $+3200$) of Eq. (13.5.38). The second, third and fifth equations of Eq. (13.5.38), corresponding to the rows of unknown nodal temperatures, can now be solved in the usual manner. The resulting solution is given by

$$
t_2 = 26.02\,°C \qquad t_3 = 26.02\,°C \qquad t_5 = 33.0\,°C \tag{13.5.39}
$$

■

Example 13.7

For the two-dimensional body shown in Figure 13–24, determine the temperature distribution. The temperature of the top side of the body is maintained at $100\,°C$. The body is insulated on the other edges. A uniform heat source of $Q = 1000$ W/m^3 acts over the whole plate, as shown in the figure. Assume a constant thickness of 1 m. Let $K_{xx} = K_{yy} = 25$ W/(m · °C).

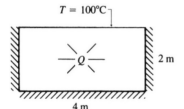

T = 100°C

2 m

4 m

Figure 13–24 Two-dimensional body subjected to a heat source

SOLUTION:

We need consider only the left half of the body, because we have a vertical plane of symmetry passing through the body 2 m from both the left and right edges. This vertical plane can be considered to be an insulated boundary. The finite element model is shown in Figure 13–25.

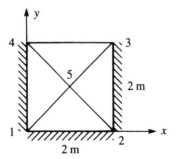

Figure 13–25 Discretized body of Figure 13–24

We will now calculate the element stiffness matrices. Because the magnitudes of the coordinates and conductivities are the same as in Example 13.6, the element stiffness matrices are the same as Eqs. (13.5.23), (13.5.26), (13.5.28), and (13.5.30). Remember that there is no convection from any side of an element, so the convection contribution $[k_h]$ to the stiffness matrix is zero. Superimposing the element stiffness matrices, we obtain the total stiffness matrix as

$$[K] = \begin{bmatrix} 25 & 0 & 0 & 0 & -25 \\ 0 & 25 & 0 & 0 & -25 \\ 0 & 0 & 25 & 0 & -25 \\ 0 & 0 & 0 & 25 & -25 \\ -25 & -25 & -25 & -25 & 100 \end{bmatrix} \text{W}/^\circ\text{C} \qquad (13.5.40)$$

Because the heat source Q is acting uniformly over each element, we use Eq. (13.5.16) to evaluate the nodal forces for each element as

$$\{f^{(e)}\} = \frac{QV}{3} \begin{Bmatrix} 1 \\ 1 \\ 1 \end{Bmatrix} = \frac{1000(1 \text{ m}^3)}{3} \begin{Bmatrix} 1 \\ 1 \\ 1 \end{Bmatrix} = \begin{Bmatrix} 333 \\ 333 \\ 333 \end{Bmatrix} \text{W} \qquad (13.5.41)$$

We then use Eqs. (13.5.40) along with (13.5.41) is applied to each element, to assemble the total system of equations as

$$\begin{bmatrix} 25 & 0 & 0 & 0 & -25 \\ 0 & 25 & 0 & 0 & -25 \\ 0 & 0 & 25 & 0 & -25 \\ 0 & 0 & 0 & 25 & -25 \\ -25 & -25 & -25 & -25 & 100 \end{bmatrix} \begin{Bmatrix} t_1 \\ t_2 \\ t_3 \\ t_4 \\ t_5 \end{Bmatrix} = \begin{Bmatrix} 666 \\ 666 \\ 666 + F_3 \\ 666 + F_4 \\ 1333 \end{Bmatrix} \qquad (13.5.42)$$

We have known nodal temperature boundary conditions of $t_3 = 100\,^\circ\text{C}$ and $t_4 = 100\,^\circ\text{C}$. In the usual manner, as was shown in Example 13.4, we modify the stiffness and force matrices of Eq. (13.5.42) to obtain

$$
\begin{bmatrix}
25 & 0 & 0 & 0 & -25 \\
0 & 25 & 0 & 0 & -25 \\
0 & 0 & 1 & 0 & 0 \\
0 & 0 & 0 & 1 & 0 \\
-25 & -25 & 0 & 0 & 100
\end{bmatrix}
\begin{Bmatrix}
t_1 \\ t_2 \\ t_3 \\ t_4 \\ t_5
\end{Bmatrix}
=
\begin{Bmatrix}
666 \\ 666 \\ 100 \\ 100 \\ 6333
\end{Bmatrix}
\tag{13.5.43}
$$

Equation (13.5.43) satisfies the boundary temperature conditions and is equivalent to Eq. (13.5.42); that is, the first, second, and fifth equations of Eq. (13.5.43) are the same as the first, second, and fifth equations of Eq. (13.5.42), and the third and fourth equations of Eq. (13.5.43) identically satisfy the boundary temperature conditions at nodes 3 and 4. The first, second, and fifth equations of Eq. (13.5.43) corresponding to the rows of unknown nodal temperatures, can now be solved simultaneously. The resulting solution is given by

$$
t_1 = 180\,^\circ\text{C} \qquad t_2 = 180\,^\circ\text{C} \qquad t_5 = 153\,^\circ\text{C}
\tag{13.5.44}
$$

We then use the results from Eq. (13.5.44) in Eq. (13.5.42) to obtain the rates of heat flow at nodes 3 and 4 (that is, F_3 and F_4) as follows:

By Eq. (13.5.42),

$$
25t_3 - 25t_5 = 666 + F_3
$$

Substituting the numerical values for t_3 and t_5, we obtain

$$
25(100) - 25(153) = 666 + F_3
$$

or

$$
F_3 = -1991 \text{ W}
$$

Similarly,

$$
F_4 = -1991 \text{ W}
$$

The negative signs on F_3 and F_4 indicate heat flow out of the body at nodes 3 and 4.

∎

▲ 13.6 Line or Point Sources ▲

A common practical heat-transfer problem is that of a source of heat generation present within a very small volume or area of some larger medium. When such heat sources exist within small volumes or areas, they may be idealized as **line** or **point sources**. Practical examples that can be modeled as line sources include hot-water pipes embedded within a medium such as concrete or earth, and conducting electrical wires embedded within a material.

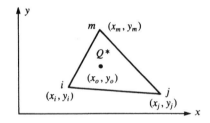

Figure 13–26 Line source located within a typical triangular element

A line or point source can be considered by simply including a node at the location of the source when the discretized finite element model is created. The value of the line source can then be added to the row of the global force matrix corresponding to the global degree of freedom assigned to the node. However, another procedure can be used to treat the line source when it is more convenient to leave the source within an element.

We now consider the line source of magnitude Q^*, with typical units of Btu/(h-ft), located at (x_o, y_o) within the two-dimensional element shown in Figure 13–26. The heat source Q is no longer constant over the element volume.

Using Eq. (13.4.16), we can express the heat source matrix as

$$\{f_Q\} = \iiint_V \left\{ \begin{array}{c} N_i \\ N_j \\ N_m \end{array} \right\} \Bigg|_{x=x_o, y=y_o} \frac{Q^*}{A^*} \, dV \tag{13.6.1}$$

where A^* is the cross-sectional area over which Q^* acts, and the N's are evaluated at $x = x_o$ and $y = y_o$. Equation (13.6.1) can be rewritten as

$$\{f_Q\} = \iint_{A^*} \int_0^t \left\{ \begin{array}{c} N_i \\ N_j \\ N_m \end{array} \right\} \Bigg|_{x=x_o, y=y_o} \frac{Q^*}{A^*} \, dA \, dz \tag{13.6.2}$$

Because the N's are evaluated at $x = x_o$ and $y = y_o$, they are no longer functions of x and y. Thus, we can simplify Eq. (13.6.2) to

$$\{f_Q\} = \left\{ \begin{array}{c} N_i \\ N_j \\ N_m \end{array} \right\} \Bigg|_{x=x_o, y=y_o} Q^* t \text{ W} \tag{13.6.3}$$

From Eq. (13.6.3), we can see that the portion of the line source Q^* distributed to each node is based on the values of N_i, N_j, and N_m, which are evaluated using the coordinates (x_o, y_o) of the line source. Recalling that the sum of the N's at any point within an element is equal to one [that is, $N_i(x_o, y_o) + N_j(x_o, y_o) + N_m(x_o, y_o) = 1$], we see that no more than the total amount of Q^* is distributed and that

$$Q_i^* + Q_j^* + Q_m^* = Q^* \tag{13.6.4}$$

Example 13.8

A line source $Q^* = 65$ W/cm is located at coordinates $(5, 2)$ in the element shown in Figure 13–27. Determine the amount of Q^* allocated to each node. All nodal coordinates are in units of centimeters. Assume an element thickness of $t = 1$ cm.

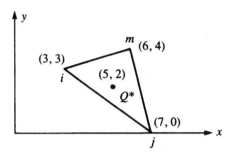

Figure 13–27 Line source located within a triangular element

SOLUTION:

We first evaluate the α's, β's, and γ's, defined by Eqs. (6.2.10), associated with each shape function as follows:

$$\alpha_i = x_j y_m - x_m y_j = 7(4) - 6(0) = 28$$

$$\alpha_j = x_m y_i - x_i y_m = 6(3) - 3(4) = 6$$

$$\alpha_m = x_i y_j - x_j y_i = 3(0) - 7(3) = -21$$

$$\beta_i = y_j - y_m = 0 - 4 = -4$$

$$\beta_j = y_m - y_i = 4 - 3 = 1 \tag{13.6.5}$$

$$\beta_m = y_i - y_j = 3 - 0 = 3$$

$$\gamma_i = x_m - x_j = 6 - 7 = -1$$

$$\gamma_j = x_i - x_m = 3 - 6 = -3$$

$$\gamma_m = x_j - x_i = 7 - 3 = 4$$

Also,

$$2A = \begin{vmatrix} 1 & x_i & y_i \\ 1 & x_j & y_j \\ 1 & x_m & y_m \end{vmatrix} = \begin{vmatrix} 1 & 3 & 3 \\ 1 & 7 & 0 \\ 1 & 6 & 4 \end{vmatrix} = 13 \tag{13.6.6}$$

Substituting the results of Eqs. (13.6.5) and (13.6.6) into Eq. (13.5.2) yields

$$N_i = \frac{1}{13}[28 - 4x - 1y]$$

$$N_j = \frac{1}{13}[6 + x - 3y] \tag{13.6.7}$$

$$N_m = \frac{1}{13}[-21 + 3x + 4y]$$

Equations (13.6.7) for N_i, N_j, and N_m evaluated at $x = 5$ and $y = 2$ are

$$N_i = \frac{1}{13}[28 - 4(5) - 1(2)] = \frac{6}{13}$$

$$N_j = \frac{1}{13}[6 + 5 - 3(2)] = \frac{5}{13} \qquad (13.6.8)$$

$$N_m = \frac{1}{13}[-21 + 3(5) + 4(2)] = \frac{2}{13}$$

Therefore, using Eq. (13.6.3), we obtain

$$\left\{ \begin{matrix} f_{Qi} \\ f_{Qj} \\ f_{Qm} \end{matrix} \right\} = Q^* t \left\{ \begin{matrix} N_i \\ N_j \\ N_m \end{matrix} \right\}_{\substack{x=x_0=5 \\ y=y_0=2}} = \frac{65(1)}{13} \left\{ \begin{matrix} 6 \\ 5 \\ 2 \end{matrix} \right\} = \left\{ \begin{matrix} 30 \\ 25 \\ 10 \end{matrix} \right\} \text{ W} \qquad (13.6.9)$$

▲ 13.7 Three-Dimensional Heat Transfer by the Finite Element Method ▲

When the heat transfer is in all three directions (indicated by q_x, q_y and q_z in Figure 13–28), then we must model the system using three-dimensional elements to account for the heat transfer. The basic partial differential equation for three-dimensional heat transfer by conduction, including the volumetric heat source, Q, is given by Eq. (13.7.1). It is an extension of the one-dimensional heat flow Eq. (13.1.7). It is interpreted as follows: At any point in a body the net heat by conduction into a unit volume plus the volumetric heat source generated must equal the change of thermal energy stored within the volume.

$$\frac{\partial}{\partial x}\left(K_{xx}\frac{\partial T}{\partial x} \right) + \frac{\partial}{\partial y}\left(K_{yy}\frac{\partial T}{\partial y} \right) + \frac{\partial}{\partial z}\left(K_{zz}\frac{\partial T}{\partial z} \right) + Q = \rho c \frac{\partial T}{\partial t} \qquad (13.7.1)$$

Examples of heat transfer that often is three-dimensional are shown in Figure 13–29. Here we see in Figure 13–29(a) and (b) an electronic component soldered to a printed wiring board [11]. The model includes a silicon chip, silver-eutectic die, alumina carrier, solder joints, copper pads, and the printed wiring board. The model actually consisted of 965 8-noded brick elements with 1395 nodes and 216 thermal elements and was modeled in Algor [10]. One-quarter of the actual device was modeled. Figure 13–29(c)

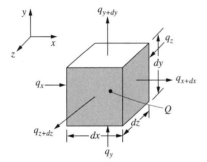

Figure 13–28 Three-dimensional heat transfer

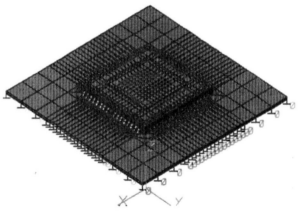

FEA model of 68-pin SMT component

(a) Electronic component soldered to printed circuit board

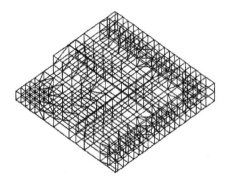

(b1) Carrier of the FEA model

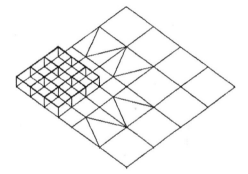

(b2) Silicon chip (left side portion) and Au-Eutectic of FEA model

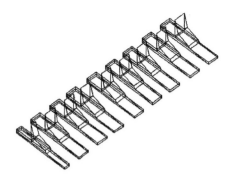

(b3) Solder joints and copper pads of FEA model

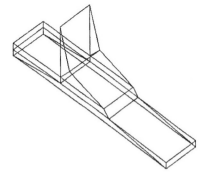

(b4) Close-up of solder and copper pad

(b) Finite element model (quarter thermal model) showing the separate components

Figure 13–29 Examples of three-dimensional heat transfer

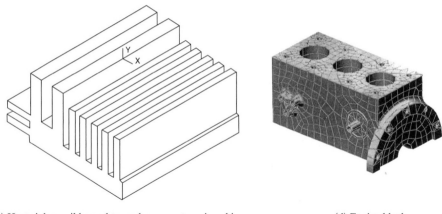

(c) Heat sink possibly used to cool a computer microchip   (d) Engine block

Figure 13–29 (*continued*)

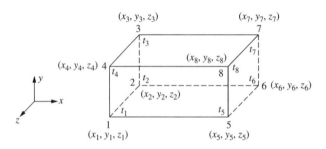

Figure 13–30 Eight-noded brick element showing nodal temperatures for the heat transfer

shows a heat sink used to cool a personal computer microprocessor chip (a two-dimensional model might possibly be used with good results as well). Finally, Figure 13–29(d) shows an engine block, which is an irregularly shaped three-dimensional body requiring a three-dimensional heat transfer analysis.

The elements often included in commercial computer programs to analyze three-dimensional heat transfer are the same as those used in Chapter 11 for three-dimensional stress analysis. These include the four-noded tetrahedral (Figure 11–3), the eight-noded hexahedral (brick) (Figure 11–5), and the twenty-noded hexahedral (Figure 11–6), the difference being that we now have only one degree of freedom at each node, namely a temperature. The temperature functions in the x, y, and z directions can now be expressed by expanding Eq. (13.5.2) to the third dimension or by using shape functions given by Eq. (11.2.10) for a four-noded tetrahedral element or by Eqs. (11.3.3) for the eight-noded brick or the Eqs. (11.3.11) through (11.3.14) for the twenty-noded brick. The typical eight-noded brick element is shown in Figure 13–30 with the nodal temperatures included. ∎

▲ 13.8 One-Dimensional Heat Transfer with Mass Transport ▲

We now consider the derivation of the basic differential equation for one-dimensional heat flow where the flow is due to conduction, convection, and **mass transport** (or **transfer**) of the fluid. The purpose of this derivation including mass transport is to show how Galerkin's residual method can be directly applied to a problem for which the variational method is not applicable. That is, the differential equation will have an odd-numbered derivative and hence does not have an associated functional of the form of Eq. (1.4.3).

The control volume used in the derivation is shown in Figure 13–31. Again, from Eq. (13.1.1) for conservation of energy, we obtain

$$q_x A\, dt + QA\, dx\, dt = c\rho A\, dx\, dT + q_{x+dx} A\, dt + q_h P\, dx\, dt + q_m\, dt \quad (13.8.1)$$

All of the terms in Eq. (13.8.1) have the same meaning as in Sections 13.1 and 13.2, except the additional mass-transport term is given by [1]

$$q_m = \dot{m} c T \quad (13.8.2)$$

where the additional variable $\dot{m}$ is the *mass flow rate* in typical units of kg/h or slug/h.

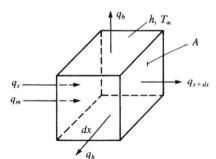

Figure 13–31 Control volume for one-dimensional heat conduction with convection and mass transport

Again, using Eqs. (13.1.3) through (13.1.6), (13.2.2), and (13.8.2) in Eq. (13.8.1) and differentiating with respect to x and t, we obtain

$$\frac{\partial}{\partial x}\left(K_{xx}\frac{\partial T}{\partial x}\right) + Q = \frac{\dot{m} c}{A}\frac{\partial T}{\partial x} + \frac{hP}{A}(T - T_\infty) + \rho c\frac{\partial T}{\partial t} \quad (13.8.3)$$

Equation (13.8.3) is the basic one-dimensional differential equation for heat transfer with mass transport.

▲ 13.9 Finite Element Formulation of Heat Transfer with Mass Transport by Galerkin's Method ▲

Having obtained the differential equation for heat transfer with mass transport, Eq. (13.8.3), we now derive the finite element equations by applying Galerkin's residual method, as outlined in Section 3.12, directly to the differential equation.

We assume here that $Q = 0$ and that we have steady-state conditions so that differentiation with respect to time is zero.

The residual R is now given by

$$R(T) = -\frac{d}{dx}\left(K_{xx}\frac{dT}{dx}\right) + \frac{\dot{m}c}{A}\frac{dT}{dx} + \frac{hP}{A}(T - T_\infty) \qquad (13.9.1)$$

Applying Galerkin's criterion, Eq. (3.12.3), to Eq. (13.9.1), we have

$$\int_0^L \left[-\frac{d}{dx}\left(K_{xx}\frac{dT}{dx}\right) + \frac{\dot{m}c}{A}\frac{dT}{dx} + \frac{hP}{A}(T - T_\infty)\right]N_i\,dx = 0 \qquad (i = 1, 2) \quad (13.9.2)$$

where the shape functions are given by Eqs. (13.4.2). Applying integration by parts to the first term of Eq. (13.9.2), we obtain

$$u = N_i \qquad du = \frac{dN_i}{dx}dx$$

$$dv = -\frac{d}{dx}\left(K_{xx}\frac{dT}{dx}\right)dx \qquad v = -K_{xx}\frac{dT}{dx} \qquad (13.9.3)$$

Using Eqs. (13.9.3) in the general formula for integration by parts [see Eq. (3.12.6)], we obtain

$$\int_0^L \left[-\frac{d}{dx}\left(K_{xx}\frac{dT}{dx}\right)\right]N_i\,dx = -K_{xx}\frac{dT}{dx}N_i\bigg|_0^L + \int_0^L K_{xx}\frac{dT}{dx}\frac{dN_i}{dx}dx \qquad (13.9.4)$$

Substituting Eq. (13.9.4) into Eq. (13.9.2), we obtain

$$\int_0^L \left(K_{xx}\frac{dT}{dx}\frac{dN_i}{dx}\right)dx + \int_0^L \left[\frac{\dot{m}c}{A}\frac{dT}{dx} + \frac{hP}{A}(T - T_\infty)\right]N_i\,dx = K_{xx}\frac{dT}{dx}N_i\bigg|_0^L \qquad (13.9.5)$$

Using Eq. (13.4.2) in (13.4.1) for T, we obtain

$$\frac{dT}{dx} = -\frac{t_1}{L} + \frac{t_2}{L} \qquad (13.9.6)$$

From Eq. (13.4.2), we obtain

$$\frac{dN_1}{dx} = -\frac{1}{L} \qquad \frac{dN_2}{dx} = \frac{1}{L} \qquad (13.9.7)$$

By letting $N_i = N_1 = 1 - (x/L)$ and substituting Eqs. (13.9.6) and (13.9.7) into Eq. (13.9.5), along with Eq. (13.4.1) for T, we obtain the first finite element equation

$$\int_0^L K_{xx}\left(-\frac{t_1}{L} + \frac{t_2}{L}\right)\left(-\frac{1}{L}\right)dx + \int_0^L \frac{\dot{m}c}{A}\left(-\frac{t_1}{L} + \frac{t_2}{L}\right)\left(1 - \frac{x}{L}\right)dx$$

$$+ \int_0^L \frac{hP}{A}\left[\left(1 - \frac{x}{L}\right)t_1 + \left(\frac{x}{L}\right)t_2 - T_\infty\right]\left(\frac{1 - x}{L}\right)dx = q_{x1}^* \qquad (13.9.8)$$

where the definition for q_x given by Eq. (13.1.3) has been used in Eq. (13.9.8). Equation (13.9.8) has a boundary condition q_{x1}^* at $x = 0$ only because $N_1 = 1$ at $x = 0$ and

$N_1 = 0$ at $x = L$. Integrating Eq. (13.9.8), we obtain

$$\left(\frac{K_{xx}A}{L} - \frac{\dot{m}c}{2} + \frac{hPL}{3}\right)t_1 + \left(-\frac{K_{xx}A}{L} + \frac{\dot{m}c}{2} + \frac{hPL}{6}\right)t_2 = q_{x1}^* + \frac{hPL}{2}T_\infty \quad (13.9.9)$$

where q_{x1}^* is defined to be q_x evaluated at node 1.

To obtain the second finite element equation, we let $N_i = N_2 = x/L$ in Eq. (13.9.5) and again use Eqs. (13.9.6), (13.9.7), and (13.4.1) in Eq. (13.9.5) to obtain

$$\left(-\frac{K_{xx}A}{L} - \frac{\dot{m}c}{2} + \frac{hPL}{6}\right)t_1 + \left(\frac{K_{xx}A}{L} + \frac{\dot{m}c}{2} + \frac{hPL}{3}\right)t_2 = q_{x2}^* + \frac{hPL}{2}T_\infty \quad (13.9.10)$$

where q_{x2}^* is defined to be q_x evaluated at node 2. Rewriting Eqs. (13.9.9) and (13.9.10) in matrix form yields

$$\left[\frac{K_{xx}A}{L}\begin{bmatrix} 1 & -1 \\ -1 & 1 \end{bmatrix} + \frac{\dot{m}c}{2}\begin{bmatrix} -1 & 1 \\ -1 & 1 \end{bmatrix} + \frac{hPL}{6}\begin{bmatrix} 2 & 1 \\ 1 & 2 \end{bmatrix}\right]\begin{Bmatrix} t_1 \\ t_2 \end{Bmatrix}$$

$$= \frac{hPLT_\infty}{2}\begin{Bmatrix} 1 \\ 1 \end{Bmatrix} + \begin{Bmatrix} q_{x1}^* \\ q_{x2}^* \end{Bmatrix} \quad (13.9.11)$$

Applying the element equation $\{f\} = [k]\{t\}$ to Eq. (13.9.11), we see that the element stiffness (conduction) matrix is now composed of three parts:

$$[k] = [k_c] + [k_h] + [k_m] \quad (13.9.12)$$

where

$$[k_c] = \frac{K_{xx}A}{L}\begin{bmatrix} 1 & -1 \\ -1 & 1 \end{bmatrix} \qquad [k_h] = \frac{hPL}{6}\begin{bmatrix} 2 & 1 \\ 1 & 2 \end{bmatrix} \qquad [k_m] = \frac{\dot{m}c}{2}\begin{bmatrix} -1 & 1 \\ -1 & 1 \end{bmatrix} \quad (13.9.13)$$

and the element nodal force and unknown nodal temperature matrices are

$$\{f\} = \frac{hPLT_\infty}{2}\begin{Bmatrix} 1 \\ 1 \end{Bmatrix} + \begin{Bmatrix} q_{x1}^* \\ q_{x2}^* \end{Bmatrix} \qquad \{t\} = \begin{Bmatrix} t_1 \\ t_2 \end{Bmatrix} \quad (13.9.14)$$

We observe from Eq. (13.9.13) that the mass transport stiffness matrix $[k_m]$ is asymmetric and, hence, $[k]$ is asymmetric. Also, if heat flux exists, it usually occurs across the free ends of a system. Therefore, q_{x1} and q_{x2} usually occur only at the free ends of a system modeled by this element. When the elements are assembled, the heat fluxes q_{x1} and q_{x2} are usually equal but opposite at the node common to two elements, unless there is an internal concentrated heat flux in the system. Furthermore, for insulated ends, the q_x^*'s also go to zero.

To illustrate the use of the finite element equations developed in this section for heat transfer with mass transport, we will now solve the following problem.

Example 13.9

Air is flowing at a rate of 2.16 kg/h inside a round tube with a diameter of 20 mm and length of 10 cm, as shown in Figure 13–32. The initial temperature of the air entering the tube is 40 °C. The wall of the tube has a uniform constant temperature of 100 °C. The specific heat of the air is 1.005 kJ/(kg · °C), the convection coefficient between

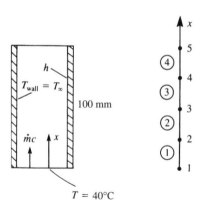

Figure 13–32 Air flowing through a tube, and the finite element model

the air and the inner wall of the tube is 15 W/(m² · °C), and the thermal conductivity is 0.03 W/(m · °C). Determine the temperature of the air along the length of the tube and the heat flow at the inlet and outlet of the tube. Here the flow rate and specific heat are given in force units (Newtons) instead of mass units (kg). This is not a problem because the units cancel in the $\dot{m}c$ product in the formulation of the equations.

We first determine the element stiffness and force matrices using Eqs. (13.9.13) and (13.9.14). To do this, we evaluate the following factors:

$$\frac{K_{xx}A}{L} = \frac{(0.03)\left[\dfrac{\pi(0.02)^2}{4}\right]}{10/100} = 9.42 \times 10^{-5} \text{ W/°C}$$

$$\dot{m}c = \left(\frac{2.16}{3600}\right)(1.005 \times 10^3) = 0.603 \text{ W/°C}$$

$$\frac{hPL}{6} = \frac{(15)(\pi \times 0.02)(0.1)}{6} = 0.01575 \text{ W/°C}$$ (13.9.15)

$$hPLT_\infty = (15)(\pi \times 0.02)(0.1)(100) = 9.45 \text{ W}$$

We can see from Eqs. (13.9.15) that the conduction portion of the stiffness matrix is negligible. Therefore, we neglect this contribution to the total stiffness matrix and obtain

$$[k^{(1)}] = \frac{0.603}{2}\begin{bmatrix} -1 & 1 \\ -1 & 1 \end{bmatrix} + 0.01575\begin{bmatrix} 2 & 1 \\ 1 & 2 \end{bmatrix} = \begin{bmatrix} -0.27 & 0.317 \\ -0.286 & 0.333 \end{bmatrix}$$ (13.9.16)

Similarly, because all elements have the same properties,

$$[k^{(2)}] = [k^{(3)}] = [k^{(4)}] = [k^{(1)}]$$ (13.9.17)

Using Eqs. (13.9.14) and (13.9.15), we obtain the element force matrices as

$$\{f^{(1)}\} = \{f^{(2)}\} = \{f^{(3)}\} = \{f^{(4)}\} = \begin{Bmatrix} 4.73 \\ 4.73 \end{Bmatrix}$$ (13.9.18)

Assembling the global stiffness matrix using Eqs. (13.9.16) and (13.9.17) and the global force matrix using Eq. (13.9.18), we obtain the global equations as

$$
\begin{bmatrix}
-0.27 & 0.317 & 0 & 0 & 0 \\
-0.286 & 0.333 - 0.27 & 0.317 & 0 & 0 \\
0 & -0.286 & 0.333 - 0.27 & 0.317 & 0 \\
0 & 0 & -0.286 & 0.333 - 0.27 & 0.317 \\
0 & 0 & 0 & -0.286 & 0.333
\end{bmatrix}
\begin{Bmatrix}
t_1 \\ t_2 \\ t_3 \\ t_4 \\ t_5
\end{Bmatrix}
$$

$$
= \begin{Bmatrix}
F_1 + 4.73 \\
9.45 \\
9.45 \\
9.45 \\
4.73
\end{Bmatrix}
\tag{13.9.19}
$$

Applying the boundary condition $t_1 = 40\,°C$, we rewrite Eq. (13.9.19) as

$$
\begin{bmatrix}
1 & 0 & 0 & 0 & 0 \\
0 & 0.063 & 0.317 & 0 & 0 \\
0 & -0.286 & 0.063 & 0.317 & 0 \\
0 & 0 & -0.286 & 0.063 & 0.317 \\
0 & 0 & 0 & -0.286 & 0.333
\end{bmatrix}
\begin{Bmatrix}
t_1 \\ t_2 \\ t_3 \\ t_4 \\ t_5
\end{Bmatrix}
= \begin{Bmatrix}
40 \\
9.45 + 11.44 \\
9.45 \\
9.45 \\
4.73
\end{Bmatrix}
\tag{13.9.20}
$$

Solving the second through fifth equations of Eq. (13.9.20) for the unknown temperatures, we obtain

$$
t_2 = 48.83\,°C \qquad t_3 = 56.19\,°C \qquad t_4 = 62.7\,°C \qquad t_5 = 68.05\,°C \tag{13.9.21}
$$

Using Eq. (13.8.2), we obtain the heat flow into and out of the tube as

$$
q_{\text{in}} = \dot{m}ct_1 = \left(\frac{2.16}{3600}\right)(1.005 \times 10^3)(40) = 24.12 \text{ W}
$$

$$
\tag{13.9.22}
$$

$$
q_{\text{out}} = \dot{m}ct_5 = \left(\frac{2.16}{3600}\right)(1.005 \times 10^3)(68.05) = 41.03 \text{ W}
$$

where, again, the conduction contribution to q is negligible; that is, $-kA\Delta T$ is negligible. The analytical solution in Reference [7] yields

$$
t_5 = 68.25\,°C \qquad q_{\text{out}} = 41.15 \text{ W} \tag{13.9.23}
$$

The finite element solution is then seen to compare quite favorably with the analytical solution. ■

The element with the stiffness matrix given by Eq. (13.9.13) has been used in Reference [8] to analyze heat exchangers. Both double-pipe and shell-and-tube heat exchangers were modeled to predict the length of tube needed to perform the task of proper heat exchange between two counterflowing fluids. Excellent agreement was found between the finite element solution and the analytical solutions described in Reference [9].

Finally, remember that when the variational formulation of a problem is difficult to obtain but the differential equation describing the problem is available, a residual method such as Galerkin's method can be used to solve the problem.

▲ 13.10 Flowchart and Examples of a Heat-Transfer Program ▲

Figure 13–33 is a flowchart of the finite element process used for the analysis of two- and three-dimensional heat-transfer problems.

Figures 13–34 and 13–35 show examples of two-dimensional temperature distribution using the two-dimensional heat-transfer element of this chapter (results obtained from Algor [10]).

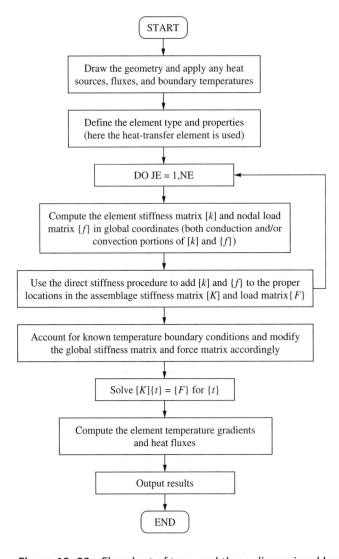

Figure 13–33 Flowchart of two- and three-dimensional heat-transfer process

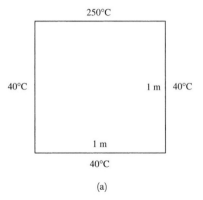

(a)

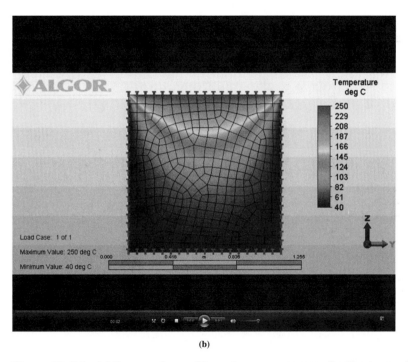

(b)

Figure 13–34 (a) Square plate subjected to temperature distribution and (b) finite element model with resulting temperature variation throughout the plate (Courtesy of David Walgrave) (See the full-color insert for a color version of this figure.)

Figure 13–34(a) shows a square plate subjected to boundary temperatures. Figure 13–34(b) shows the finite element model, along with the temperature distribution throughout the plate.

Figure 13–35(a) shows a square duct that carries hot gases such that its surface temperature is 380°C. The duct is wrapped by a layer of circular fiberglass. The finite element model, along with the temperature distribution throughout the fiberglass is shown in Figure 13–35(b).

Figures 13–36 and 13–37 illustrate the use of the three-dimensional solid element described in Section 13.7 for determining temperature distribution and heat

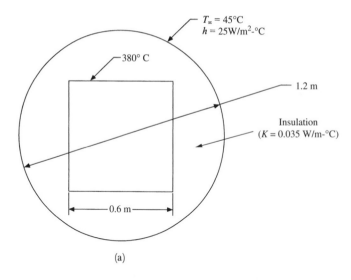

(a)

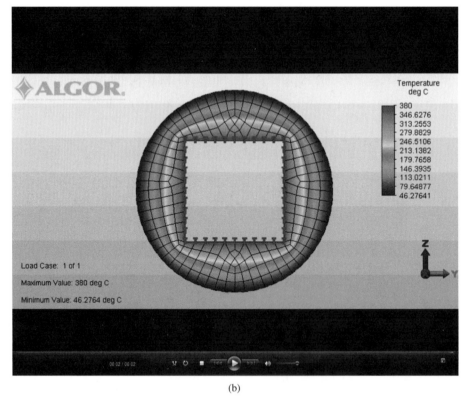

(b)

Figure 13–35 (a) Square duct wrapped by insulation and (b) the finite element model with resulting temperature variation through the insulation (See the full-color insert for a color version of this figure.)

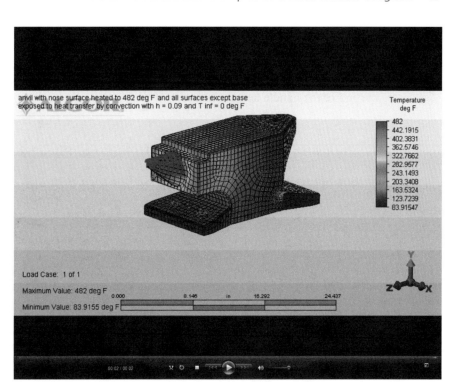

anvil with nose surface heated to 482 deg F and all surfaces except base
exposed to heat transfer by convection with h = 0.09 and T inf = 0 deg F

Temperature
deg F

482
442.1915
402.3831
362.5746
322.7662
282.9577
243.1493
203.3408
163.5324
123.7239
83.91547

Load Case: 1 of 1

Maximum Value: 482 deg F

Minimum Value: 83.9155 deg F

Figure 13–36   Temperature distribution in an anvil (Dan Baxter) (See the full-color insert for a color version of this figure.)

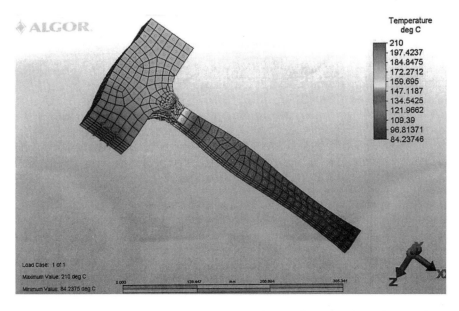

Temperature
deg C

210
197.4237
184.8475
172.2712
159.695
147.1187
134.5425
121.9662
109.39
96.81371
84.23746

Load Case: 1 of 1

Maximum Value: 210 deg C

Minimum Value: 84.2375 deg C

Figure 13–37   Temperature distribution in a forging hammer (by Wilson Arifin)

flux in solid bodies subjected to temperature change. Figure 13–36 is an anvil with the nose heated to 250 °C. The temperature distribution throughout the anvil is shown. Figure 13–37 is a solid model of a steel forging hammer with the flat end of the hammer subjected to a 210 °C surface temperature. Notice that the temperature plot indicates the end of the handle temperature is 84.2 °C.

▲ Summary Equations

Conservation of energy principle (conduction heat transfer):

$$E_{in} + E_{generated} = \Delta U + E_{out} \tag{13.1.1}$$

$$q_x A\,dt + QA\,dx\,dt = \Delta U + q_{x+dx}A\,dt \tag{13.1.2}$$

Fourier's law for heat conduction:

$$q_x = -K_{xx}\frac{dT}{dx} \tag{13.1.3}$$

Basic differential equation for one-dimensional steady-state heat transfer by conduction:

$$\frac{d}{dx}\left(K_{xx}\frac{dT}{dx}\right) + Q = 0 \tag{13.1.8}$$

Basic differential equation for two-dimensional heat conduction:

$$\frac{\partial}{\partial x}\left(K_{xx}\frac{\partial T}{\partial x}\right) + \frac{\partial}{\partial y}\left(K_{yy}\frac{\partial T}{\partial y}\right) + Q = 0 \tag{13.1.12}$$

Conservation of energy principle (with convection):

$$q_x A\,dt + QA\,dx\,dt = c(\rho A\,dx)\,dT + q_{x+dx}A\,dt + q_h P\,dx\,dt \tag{13.2.1}$$

Newton's law of cooling:

$$q_h = h(T - T_\infty) \tag{13.2.2}$$

Basic differential equation for one-dimensional heat conduction with convection:

$$\frac{\partial}{\partial x}\left(K_{xx}\frac{\partial T}{\partial x}\right) + Q = \rho c\frac{\partial T}{\partial t} + \frac{hP}{A}(T - T_\infty) \tag{13.2.3}$$

Temperature function for basic one-dimensional (two-noded) temperature element:

$$T(x) = N_1 t_1 + N_2 t_2 \tag{13.4.1}$$

Shape functions for one-dimensional temperature element:

$$N_1 = 1 - \frac{x}{L} \qquad N_2 = \frac{x}{L} \tag{13.4.2}$$

Temperature gradient matrix:

$$\{g\} = \left\{\frac{dT}{dx}\right\} = [B]\{t\} \tag{13.4.6}$$

Gradient matrix:

$$[B] = \left[\frac{dN_1}{dx} \quad \frac{dN_2}{dx}\right] \tag{13.4.7}$$

Heat flux/temperature gradient relationship:

$$q_x = -[D]\{g\} \tag{13.4.8}$$

Material properties matrix:

$$[D] = [K_{xx}] \tag{13.4.9}$$

Functional for heat transfer:

$$\pi_h = U + \Omega_Q + \Omega_q + \Omega_h \tag{13.4.10}$$

Stiffness matrix for heat transfer due to conduction and convection:

$$[k] = \iiint_V [B]^T [D][B]\, dV + \iint_{S_3} h[N]^T [N]\, dS \tag{13.4.17}$$

Conduction part of stiffness matrix for one-dimensional element:

$$[k_c] = \frac{AK_{xx}}{L}\begin{bmatrix} 1 & -1 \\ -1 & 1 \end{bmatrix} \tag{13.4.20}$$

Convection part of stiffness matrix for one-dimensional elements:

$$[k_h] = \frac{hPL}{6}\begin{bmatrix} 2 & 1 \\ 1 & 2 \end{bmatrix} \tag{13.4.21}$$

Force matrix terms:
    Due to uniform heat source:

$$\{f_Q\} = \iiint_V [N]^T Q\, dV = QA\int_0^L \left\{\begin{matrix} 1 - \dfrac{x}{L} \\ \dfrac{x}{L} \end{matrix}\right\} dx = \frac{QAL}{2}\left\{\begin{matrix} 1 \\ 1 \end{matrix}\right\} \tag{13.4.23}$$

Due to uniform heat flux over perimeter surface of element:

$$\{f_q\} = \frac{q^* PL}{2}\left\{\begin{matrix} 1 \\ 1 \end{matrix}\right\} \tag{13.4.24}$$

Due to uniform convection around perimeter surface of element:

$$\{f_h\} = \iint_{S_3} hT_\infty [N]^T\, dS = \frac{hT_\infty PL}{2}\left\{\begin{matrix} 1 \\ 1 \end{matrix}\right\} \tag{13.4.25}$$

Stiffness matrix contribution for convection from right end of element:

$$[k_h]_{end} = \iint\limits_{S_{end}} h \begin{Bmatrix} 0 \\ 1 \end{Bmatrix} [0 \quad 1] \, dS = hA \begin{bmatrix} 0 & 0 \\ 0 & 1 \end{bmatrix} \tag{13.4.28}$$

Force term due to convection from right end of element:

$$\{f_h\}_{end} = hT_\infty A \begin{Bmatrix} N_1(x = L) \\ N_2(x = L) \end{Bmatrix} = hT_\infty A \begin{Bmatrix} 0 \\ 1 \end{Bmatrix} \tag{13.4.29}$$

Global equations:

$$\{F\} = [K]\{t\} \tag{13.4.32}$$

Temperature function for two-dimensional triangle element:

$$\{T\} = [N_i \quad N_j \quad N_m] \begin{Bmatrix} t_i \\ t_j \\ t_m \end{Bmatrix} \tag{13.5.1}$$

Shape function for two-dimensional triangle element:

$$N_i = \frac{1}{2A}(\alpha_i + \beta_i x + \gamma_i y) \tag{13.5.2}$$

Temperature gradient for two-dimensional triangle element:

$$\{g\} = \begin{Bmatrix} \dfrac{\partial T}{\partial x} \\[2mm] \dfrac{\partial T}{\partial y} \end{Bmatrix} \tag{13.5.3}$$

$$\{g\} = [B]\{t\} \tag{13.5.5}$$

Gradient matrix for two-dimensional triangle element:

$$[B] = \frac{1}{2A} \begin{bmatrix} \beta_i & \beta_j & \beta_m \\ \gamma_i & \gamma_j & \gamma_m \end{bmatrix} \tag{13.5.6}$$

Heat flux/temperature gradient relationship for two-dimensional triangle element:

$$\begin{Bmatrix} q_x \\ q_y \end{Bmatrix} = -[D]\{g\} \tag{13.5.7}$$

Material property matrix for two-dimensional triangle element:

$$[D] = \begin{bmatrix} K_{xx} & 0 \\ 0 & K_{yy} \end{bmatrix} \tag{13.5.8}$$

Stiffness matrix due to conduction for two-dimensional triangle element:

$$[k_c] = \iiint\limits_V [B]^T [D][B] \, dV = tA[B]^T [D][B] \tag{13.5.11}$$

Stiffness matrix due to convection from side *i-j* of two-dimensional triangle element:

$$[k_h] = \frac{hL_{i-j}t}{6} \begin{bmatrix} 2 & 1 & 0 \\ 1 & 2 & 0 \\ 0 & 0 & 0 \end{bmatrix} \tag{13.5.14}$$

Force terms for two-dimensional triangle element:
   Due to uniform heat source:

$$\{f_Q\} = \frac{QV}{3} \begin{Bmatrix} 1 \\ 1 \\ 1 \end{Bmatrix} \tag{13.5.16}$$

Due to uniform heat flux over side *i-j*:

$$\{f_q\} = \frac{q^*L_{i-j}t}{2} \begin{Bmatrix} 1 \\ 1 \\ 0 \end{Bmatrix} \qquad \text{on side } i\text{-}j \tag{13.5.18}$$

Force matrix for line or point source:

$$\{f_Q\} = \begin{Bmatrix} N_i \\ N_j \\ N_m \end{Bmatrix} \Bigg|_{x=x_o, y=y_o} Q^* t \text{ W} \tag{13.6.3}$$

Basic differential equation for three-dimensional heat transfer by conduction:

$$\frac{\partial}{\partial x}\left(K_{xx}\frac{\partial T}{\partial x}\right) + \frac{\partial}{\partial y}\left(K_{yy}\frac{\partial T}{\partial y}\right) + \frac{\partial}{\partial z}\left(K_{zz}\frac{\partial T}{\partial z}\right) + Q = \rho c\frac{\partial T}{\partial t} \tag{13.7.1}$$

Mass transport term:

$$q_m = \dot{m}cT \tag{13.8.2}$$

Flow chart for heat transfer program (See Figure 13–33.)

▲ References

[1] Holman, J. P., *Heat Transfer*, 9th ed., McGraw-Hill, New York, 2002.

[2] Kreith, F., and Black, W. Z., *Basic Heat Transfer*, Harper & Row, New York, 1980.

[3] Lyness, J. F., Owen, D. R. J., and Zienkiewicz, O. C., "The Finite Element Analysis of Engineering Systems Governed by a Non-Linear Quasi-Harmonic Equation," *Computers and Structures*, Vol. 5, pp. 65–79, 1975.

[4] Zienkiewicz, O. C., and Cheung, Y. K., "Finite Elements in the Solution of Field Problems," *The Engineer*, pp. 507–510, Sept. 24, 1965.

[5] Wilson, E. L., and Nickell, R. E., "Application of the Finite Element Method to Heat Conduction Analysis," *Nuclear Engineering and Design*, Vol. 4, pp. 276–286, 1966.

[6] Emery, A. F., and Carson, W. W., "An Evaluation of the Use of the Finite Element Method in the Computation of Temperature," *Journal of Heat Transfer*, American Society of Mechanical Engineers, pp. 136–145, May 1971.

[7] Rohsenow, W. M., and Choi, H. Y., *Heat, Mass, and Momentum Transfer*, Prentice-Hall, Englewood Cliffs, NJ, 1963.

[8] Goncalves, L., *Finite Element Analysis of Heat Exchangers*, M. S. Thesis, Rose-Hulman Institute of Technology, Terre Haute, IN, 1984.

[9] Kern, D. Q., and Kraus, A. D., *Extended and Surface Heat Transfer*, McGraw-Hill, New York, 1972.

[10] *Heat Transfer Reference Division*, Docutech On-line Documentation, Algor, Inc., Pittsburgh, PA.

[11] Beasley, K. G., "Finite Element Analysis Model Development of Leadless Chip Carrier and Printed Wiring Board," M. S., Thesis, Rose-Hulman Institute of Technology, Terre Haute, IN, Nov. 1992.

▲ Problems

13.1 For the one-dimensional composite bar shown in Figure P13–1, determine the interface temperatures. For element 1, let $K_{xx} = 200$ W/(m · °C); for element 2, let $K_{xx} = 100$ W/(m · °C); and for element 3, let $K_{xx} = 50$ W/(m · °C). Let $A = 0.1$ m². The left end has a constant temperature of 100 °C, and the right end has a constant temperature of 300 °C.

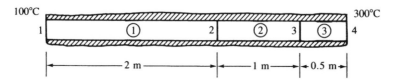

Figure P13–1

13.2 For the one-dimensional rod shown in Figure P13–2 (insulated except at the ends), determine the temperatures at $L/3$, $2L/3$, and L. Let $K_{xx} = 60$ W/(m · °C), $h = 800$ W/(m² · °C), and $T_\infty = 0$ °F. The temperature at the left end is 95 °C.

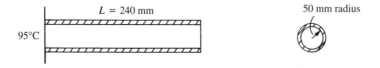

Figure P13–2

13.3 A rod with uniform cross-sectional area of 6 cm² and thermal conductivity of 62.5 W/(m · °C) has heat flow in the x direction only (Figure P13–3). The right end is insulated. The left end is maintained at 10 °C, and the system has the linearly distributed heat flux shown.

Use a two-element model and estimate the temperature at the node points and the heat flow at the left boundary.

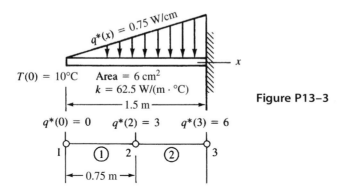

Figure P13–3

13.4 The rod of 25 mm radius shown in Figure P13–4 generates heat internally at the rate of uniform $Q = 10^5$ W/m^3 throughout the rod. The left edge and perimeter of the rod are insulated, and the right edge is exposed to an environment of $T_\infty = 40\,°C$. The convection heat-transfer coefficient between the wall and the environment is $h = 600$ W/(m$^2 \cdot °C$). The thermal conductivity of the rod is $K_{xx} = 12$ Btu/(h-ft-°F). The length of the rod is 75 mm. Calculate the temperature distribution in the rod. Use at least three elements in your finite element model.

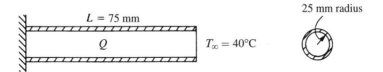

Figure P13–4

13.5 The fin shown in Figure P13–5 is insulated on the perimeter. The left end has a constant temperature of $100\,°C$. A positive heat flux of $q^* = 1000$ W/m^2 acts on the right end. Let $K_{xx} = 6$ W/(m $\cdot °C$) and cross-sectional area $A = 0.1$ m^2. Determine the temperatures at $L/4$, $L/2$, $3L/4$, and L, where $L = 0.4$ m.

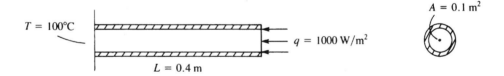

Figure P13–5

13.6 For the composite wall shown in Figure P13–6, determine the interface temperatures. What is the heat flux through the 8-cm portion? Use the finite element method. Use three elements with the nodes shown. 1 cm = 0.01 m.

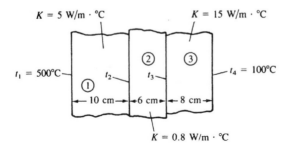

Figure P13–6

13.7 For the composite wall idealized by the one-dimensional model shown in Figure P13–7, determine the interface temperatures. For element 1, let $K_{xx} = 5$ W/(m · °C); for element 2, $K_{xx} = 10$ W/(m · °C); and for element 3, $K_{xx} = 15$ W/(m · °C). The left end has a constant temperature of 200 °C and the right end has a constant temperature of 600 °C.

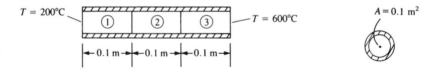

Figure P13–7

13.8 A composite wall is shown in Figure P13–8. For element 1, let $K_{xx} = 5$ W/(m-°C), for element 2 let $K_{xx} = 10$ W/(m-°C), for element 3 let $K_{xx} = 15$ W/(m-°C). The left end has a heat source of 600 W applied to it. The right end is held at 10°C. Determine the left end temperature and the interface temperatures and the heat flux through element 3.

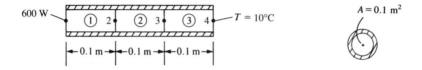

Figure P13–8

13.9 A double-pane glass window shown in Figure P13–9, consists of two 4-mm thick layers of glass with $k = 0.80$ W/m-°C separated by a 10 mm thick stagnant air space with $k = 0.025$ W/m-°C. Determine (a) the temperature at both surfaces of the inside

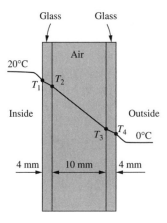

Figure P13–9

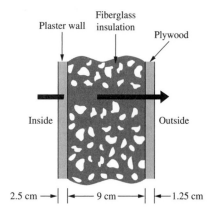

Figure P13–10

layer of glass and the temperature at the outside surfaces of glass, and (b) the steady rate of heat transfer in Watts through the double pane. Assume the inside room temperature $T_{i\infty} = 20\,°C$ with $h_i = 10\ W/m^2\text{-}°C$ and the outside temperature $T_{0\infty} = 0\,°C$ with $h_0 = 30\ W/m^2\text{-}°C$. Assume one-dimensional heat flow through the glass.

13.10 For the composite wall of a house, shown in Figure P13–10, determine the temperatures at the inner and outer surfaces and at the interfaces. The wall is composed of 2.5 cm thick plaster wall $(k = 0.20\ W/m\text{-}°C)$ on the inside, a 9 cm thick layer of fiberglass insulation $(k = 0.038\ W/m\text{-}°C)$, and a 1.25 cm plywood layer $(k = 0.12\ W/m\text{-}°C)$ on the outside. Assume the inside room air is $20\,°C$ with convection coefficient of $10\ W/m^2\text{-}°C$ and the outside air at $-10\,°C$ with convection coefficient of $20\ W/m^2\text{-}°C$. Also, determine the rate of heat transfer through the wall in Watts. Assume one-dimensional heat flow through the wall thickness.

13.11 Condensing steam is used to maintain a room at $20\,°C$. The steam flows through pipes that keep the pipe surface at $100\,°C$. To increase heat transfer from the pipes, stainless steel fins $(k = 15\ W/m\text{-}°C)$, 20 cm long and 0.5 cm in diameter, are welded to the pipe surface as shown in Figure P13–11. A fan forces the room air over the pipe and fins,

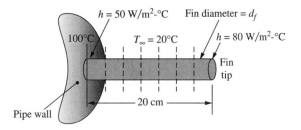

Figure P13–11

resulting in a heat transfer coefficient of $50\,\text{W/m}^2\text{-}^{\circ}\text{C}$ at the base surface of the fin where it is welded to the pipe. However, the air flow distribution increases the heat transfer coefficient to $80\,\text{W/m}^2\text{-}^{\circ}\text{C}$ at the fin tip. Assume the variation in heat transfer coefficient to then vary linearly from left end to right end of the fin surface. Determine the temperature distribution at L/4 locations along the fin. Also determine the rate of heat loss from each fin.

13.12 A tapered aluminum fin $(k = 200\,\text{W/m-}^{\circ}\text{C})$, shown in Figure P13–12, has a circular cross section with base diameter of 1 cm and tip diameter of 0.5 cm. The base is maintained at $200\,^{\circ}\text{C}$ and looses heat by convection to the surroundings at $T_{\infty} = 10\,^{\circ}\text{C}$, $h = 150\,\text{W/m}^2\text{-}^{\circ}\text{C}$. The tip of the fin is insulated. Assume one-dimensional heat flow and determine the temperatures at the quarter points along the fin. What is the rate of heat loss in Watts through each element? Use four elements with an average cross-sectional area for each element.

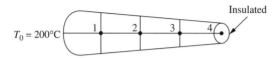

$T_0 = 200\,^{\circ}\text{C}$

Insulated

Figure P13–12

13.13 A wall is constructed of an outer layer of 10 mm thick plywood $(k = 1.5\,\text{W/(m} \cdot {}^{\circ}\text{C}))$, an inner core of 100 mm thick fiberglass insulation $(k = 0.35\,\text{W/(m} \cdot {}^{\circ}\text{C}))$, and an inner layer of 10 mm thick sheetrock $(k = 0.175\,\text{W/(m} \cdot {}^{\circ}\text{C}))$ (Figure P13–13). The inside temperature is $20\,^{\circ}\text{C}$ with $h = 10\,\text{W/(m}^2 \cdot {}^{\circ}\text{C})$, while the outside temperature is $-15\,^{\circ}\text{C}$ with $h = 25\,\text{W/(m}^2 \cdot {}^{\circ}\text{C})$. Determine the temperature at the interfaces of the materials and the rate of heat flow in Watts (W) through the wall.

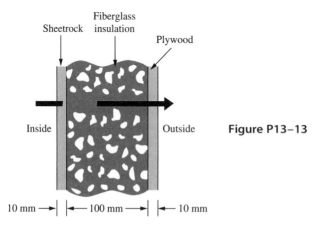

Fiberglass
Sheetrock insulation

Plywood

Inside

Outside   **Figure P13–13**

10 mm →| |← 100 mm →| |← 10 mm

13.14 A large plate of stainless steel with thickness of 5 cm and thermal conductivity of $k = 15\,\text{W/m-}^{\circ}\text{C}$ is subjected to an internal uniform heat generation throughout the

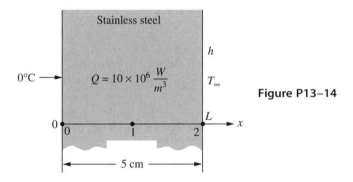

Figure P13–14

plate at constant rate of $Q = 10 \times 10^6 \, \text{W/m}^3$. One side of the plate is maintained at $0°C$ by ice water, and the other side is subjected to convection to an environment at $T_\infty = 35°C$, with heat transfer coefficient $h = 40 \, \text{W/m}^2\text{-}°C$, as shown in Figure P13–14. Use two elements in a finite element model to estimate the temperatures at each surface and in the middle of the plate's thickness. Assume a one-dimensional heat transfer through the plate.

13.15 The base plate of an iron is 0.6 cm thick. The plate is subjected to 100 W of power (provided by resistance heaters inside the iron, as shown in Figure P13–15), over a base plate cross-sectional area of 250 cm^2, resulting in a uniform flux generated on the inside surface. The thermal conductivity of the metal base plate is $k = 20 \, \text{W/m-}°C$. The outside ambient air temperature is $20°C$ with a heat transfer coefficient of $20 \, \text{W/m}^2\text{-}°C$ at steady-state conditions. Assume one-dimensional heat transfer through the plate thickness. Using three elements, model the plate to determine the temperatures at the inner surface and interior one-third points.

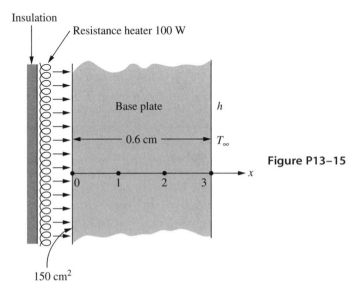

Figure P13–15

13.16 A hot surface is cooled by attaching fins (called pin fins) to it, as shown in Figure P13–16. The surface of the plate (left end of the pin) is 90 °C. The fins are 6 cm long and have a cross-sectional area of 5×10^{-6} m^2 with a perimeter of 0.006 m. The fins are made of copper ($k = 400$ W/m-°C). The temperature of the surrounding air is $T_\infty = 20$ °C with heat transfer coefficient on the surface (including the end surface) of $h = 10$ W/m^2-°C. A model of the typical fin is also shown in Figure P13–16. Use three elements in your finite element model to determine the temperatures along the fin length.

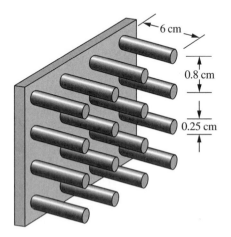

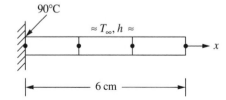

Figure P13–16

13.17 Use the direct method to derive the element equations for the one-dimensional steady-state conduction heat-transfer problem shown in Figure P13–17. The bar is insulated all around and has cross-sectional area A, length L, and thermal conductivity K_{xx}. Determine the relationship between nodal temperatures t_1 and t_2 (°C) and the thermal inputs F_1 and F_2 (in kW-h). Use Fourier's law of heat conduction for this case.

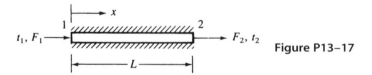

Figure P13–17

13.18 Express the stiffness matrix and the force matrix for convection from the left end of a bar, as shown in Figure P13–18. Let the cross-sectional area of the bar be A, the convection coefficient be h, and the free stream temperature be T_∞.

h, T_∞ Figure P13–18

13.19 For the element shown in Figure P13–19, determine the $[k]$ and $\{f\}$ matrices. The conductivities are $K_{xx} = K_{yy} = 25$ W/(m·°C) and the convection coefficient is $h = 120$ W/(m²·°C). Convection occurs across the i-j surface. The free-stream temperature is $T_\infty = 20\,°C$. The coordinates are expressed in units of feet. Let the line source be $Q^* = 150$ W/m as located in the figure. Take the thickness of the element to be 1 m.

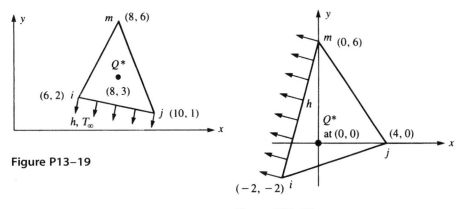

Figure P13–19

Figure P13–20

13.20 Calculate the $[k]$ and $\{f\}$ matrices for the element shown in Figure P13–20. The conductivities are $K_{xx} = K_{yy} = 15$ W/(m·°C) and the convection coefficient is $h = 20$ W/(m²·°C). Convection occurs across the i-m surface. The free-stream temperature is $T_\infty = 15\,°C$. The coordinates are shown expressed in units of meters. Let the line source be $Q^* = 100$ W/m as located in the figure. Take the thickness of the element to be 1 m.

13.21 For the square two-dimensional body shown in Figure P13–21, determine the temperature distribution. Let $K_{xx} = K_{yy} = 45$ W/(m·°C) and $h = 60$ W/(m²·°C). Convection occurs across side 4–5. The free-stream temperature is $T_\infty = 10\,°C$. The temperatures at nodes 1 and 2 are 40°C. The dimensions of the body are shown in the figure. Take the thickness of the body to be 1 m.

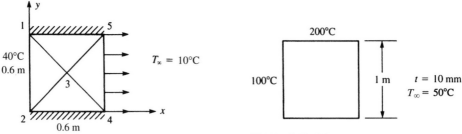

Figure P13–21

Figure P13–22

13.22 For the square plate shown in Figure P13–22, determine the temperature distribution. Let $K_{xx} = K_{yy} = 10$ W/(m · °C) and $h = 20$ W/(m² · °C). The temperature along the left side is maintained at 100 °C and that along the top side is maintained at 200 °C.

Use a computer program to calculate the temperature distribution in the following two-dimensional bodies.

13.23 For the body shown in Figure P13–23, determine the temperature distribution. Surface temperatures are shown in the figure. The body is insulated along the top and bottom edges, and $K_{xx} = K_{yy} = 1.75$ W/(m · °C). No internal heat generation is present.

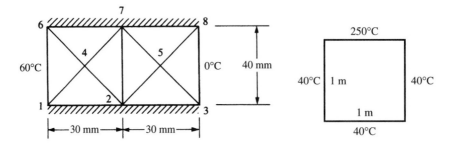

Figure P13–23

Figure P13–24

13.24 For the square two-dimensional body shown in Figure P13–24, determine the temperature distribution. Let $K_{xx} = K_{yy} = 15$ W/(m · °C). The top surface is maintained at 250 °C and the other three sides are maintained at 40 °C. Also, plot the temperature contours on the body.

13.25 For the square two-dimensional body shown in Figure P13–25, determine the temperature distribution. Let $K_{xx} = K_{yy} = 15$ W/(m · °C) and $h = 60$ W/(m² · °C). The top face is maintained at 250 °C, the left face is maintained at 40 °C, and the other two

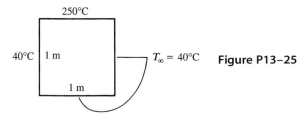

Figure P13–25

faces are exposed to an environmental (free-stream) temperature of 40 °C. Also, plot the temperature contours on the body.

13.26 Hot water pipes are located on 0.6 m centers in a concrete slab with $K_{xx} = K_{yy} = 1.5$ W/(m · °C), as shown in Figure P13–26. If the outside surfaces of the concrete are at 30 °C and the water has an average temperature of 100 °C, determine the temperature distribution in the concrete slab. Plot the temperature contours through the concrete. Use symmetry in your finite element model.

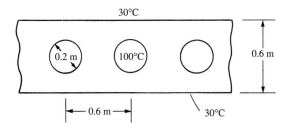

Figure P13–26

13.27 The cross section of a tall chimney shown in Figure P13–27 has an inside surface temperature of 165 °C and an exterior temperature of 55 °C. The thermal conductivity is $K = 0.8$ W/(m · °C). Determine the temperature distribution within the chimney per unit length.

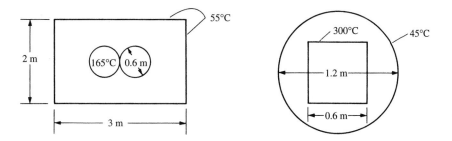

Figure P13–27   Figure P13–28

13.28 The square duct shown in Figure P13–28 carries hot gases such that its surface temperature is 300 °C. The duct is insulated by a layer of circular fiberglass that has a thermal conductivity of $K = 0.04$ W/(m · °C). The outside surface temperature of the fiberglass is maintained at 45 °C. Determine the temperature distribution within the fiberglass.

13.29 The buried pipeline in Figure P13–29 transports oil with an average temperature of 15 °C. The pipe is located 4.5 m below the surface of the earth. The thermal conductivity of the earth is 1.0 W/(m · °C). The surface of the earth is 10 °C. Determine the temperature distribution in the earth.

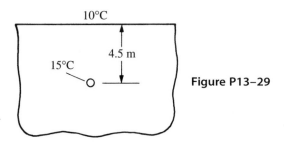

Figure P13–29

13.30 A 250 mm thick concrete bridge deck is embedded with heating cables, as shown in Figure P13–30. If the lower surface is at 0 °C, the rate of heat generation (assumed to be the same in each cable) is 100 W/m and the top surface of the concrete is at 20 °C. The thermal conductivity of the concrete is 1.0 W/(m · °C). What is the temperature distribution in the slab? Use symmetry in your model.

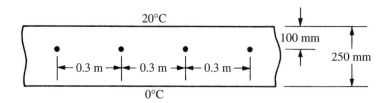

Figure P13–30

13.31 For the circular body with holes shown in Figure P13–31, determine the temperature distribution. The inside surfaces of the holes have temperatures of 150 °C. The outside of the circular body has a temperature of 30 °C. Let $K_{xx} = K_{yy} = 10$ W/(m · °C).

13.32 For the square two-dimensional body shown in Figure P13–32, determine the temperature distribution. Let $K_{xx} = K_{yy} = 10$ W/(m · °C) and $h = 10$ W/(m² · °C). The top face is maintained at 100 °C, the left face is maintained at 0 °C, and the other two faces are exposed to a free-stream temperature of 0 °C. Also, plot the temperature contours on the body.

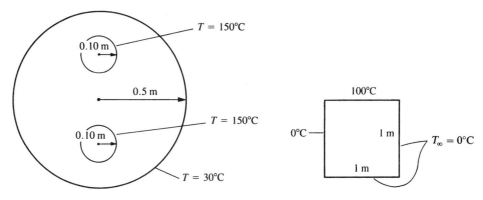

Figure P13–31

Figure P13–32

13.33 A 200-mm-thick concrete bridge deck is embedded with heating cables as shown in Figure P13–33. If the ambient temperature under the deck is $-10\,°C$ with $h = 10$ W/(m²-°C) and the ambient air temperature above the deck is $10\,°C$ with $h = 10$ W/(m²-°C), what is the temperature distribution in the slab? The heating cables are line sources generating heat of $Q^* = 50$ W/m. The thermal conductivity of the concrete is 1.2 W/m-°C. Use symmetry in your model.

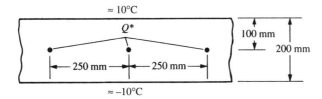

Figure P13–33

13.34 For the two-dimensional body shown in Figure P13–34, determine the temperature distribution. Let the left and right ends have constant temperatures of $200\,°C$ and $100\,°C$, respectively. Let $K_{xx} = K_{yy} = 5$ W/(m · °C). The body is insulated along the top and bottom.

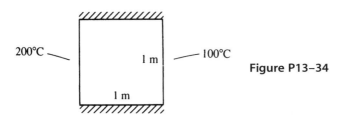

Figure P13–34

13.35 For the two-dimensional body shown in Figure P13–35, determine the temperature distribution. The top and bottom sides are insulated. The right side is subjected to heat transfer by convection. Let $K_{xx} = K_{yy} = 10$ W/(m · °C).

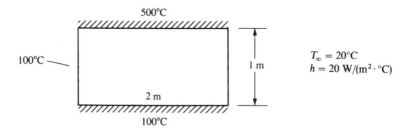

500°C

100°C

1 m

2 m

100°C

$T_\infty = 20°C$
$h = 20$ W/(m² · °C)

Figure P13–35

13.36 For the two-dimensional body shown in Figure P13–36, determine the temperature distribution. The left and right sides are insulated. The top surface is subjected to heat transfer by convection. The bottom and internal portion surfaces are maintained at 300 °C.

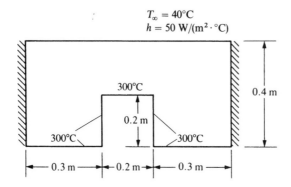

$T_\infty = 40°C$
$h = 50$ W/(m² · °C)

300°C

0.4 m

0.2 m

300°C

300°C

0.3 m — 0.2 m — 0.3 m

Figure P13–36

13.37 Determine the temperature distribution and rate of heat flow through the plain carbon steel ingot shown in Figure P13–37. Let $k = 60$ W/ m-K) for the steel. The top surface is held at 40°C, while the underside surface is held at 0 °C. Assume that no heat is lost from the sides.

13.38 Determine the temperature distribution and rate of heat flow per foot length from a 5 cm outer diameter pipe at 180 °C placed eccentrically within a larger cylinder of insulation ($k = 0.058$ W/ m-°C) as shown in Figure P13–38. The diameter of the outside cylinder is 15 cm, and the surface temperature is 20 °C.

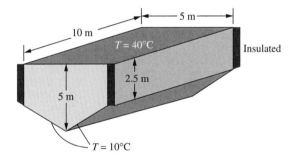

Figure P13–37

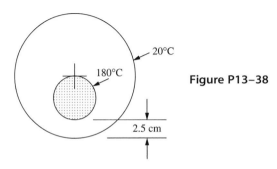

Figure P13–38

13.39 Determine the temperature distribution and rate of heat flow in the molded foam insulation ($k = 0.3$ W/m · °C) shown in Figure P13–39.

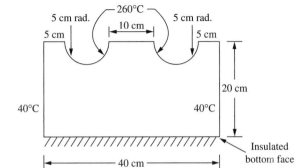

Figure P13–39

13.40 For the basement wall shown in Figure P13–40, determine the temperature distribution and the heat transfer through the wall and soil. The wall is constructed of concrete ($k = 1.75$ W/(m-°C)). The soil has an average thermal conductivity of $k = 1.5$ W/(m-°C). The inside air is maintained at 30 °C with a convection coefficient $h = 10$ W/(m²-°C). The outside air temperature is 0 °C with a heat transfer coefficient of $h = 30$ W/(m²-°C). Assume a reasonable distance from the wall of five feet that the horizontal component of heat transfer becomes negligible. Make sure this assumption is correct.

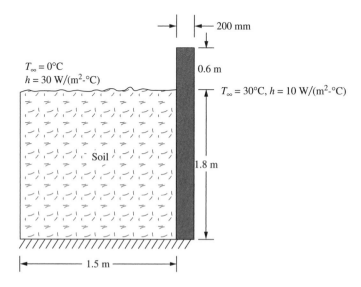

$T_\infty = 0°C$
$h = 30$ W/(m²-°C)

200 mm

0.6 m

$T_\infty = 30°C, h = 10$ W/(m²-°C)

Soil

1.8 m

1.5 m

Figure P13–40

13.41 Now add a 150 mm thick concrete floor to the model of Figure P13–40 (as shown in Figure P13–41). Determine the temperature distribution and the heat transfer through the concrete and soil. Use the same properties as shown in Problem 13–40.

13.42 Aluminum fins ($k = 170$ W/m-K) with triangular profiles shown in Figure P13–42 are used to remove heat from a surface with a temperature of 160 °C. The temperature of the surrounding air is 25 °C. The natural convection coefficient is $h = 25$ W/m²-K. Determine the temperature distribution throughout and the heat loss from a typical fin.

13.43 The forging hammer shown in Figure P13–43 is subjected to a surface temperature of 210 °C acting on the lower flat surface of the steel hammer head. The hammer's thermal conductivity is 20 W/m-K and the room temperature is assumed to be 40 °C with a convection coefficient of 3.216×10^{-6} J/(s-°C-mm²).
    Determine the temperature distribution throughout the hammer.

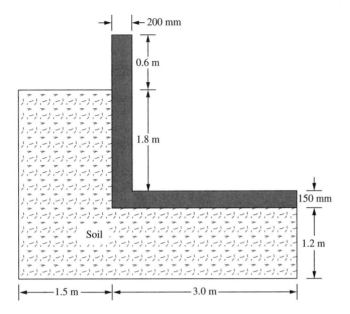

Figure P13–41

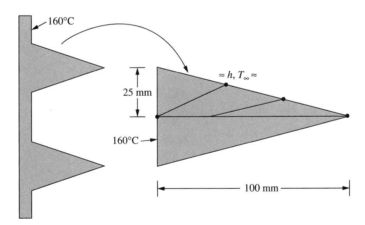

Figure P13–42

 13.44 The Allen wrench shown in Figure P13–44 is unloaded but now exposed to a temperature of 300 K, at its lower end, while the other end has a heat flux of 10 W/m^2 acting over the end surface. Determine the temperature distribution throughout the wrench. The thermal conductivity of the material is 43.6 W/m-K. Assume the wrench is insulated around the perimeter. The dimensions of the wrench are those of Figure P11–18.

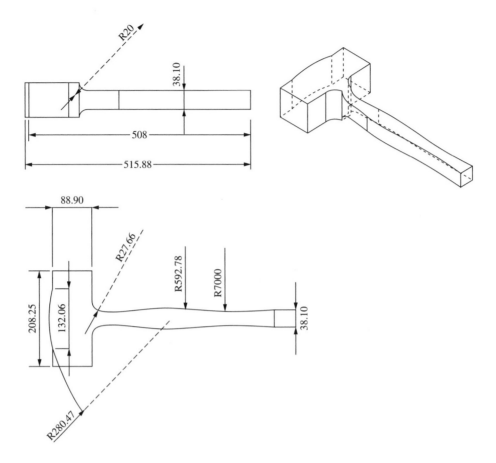

Figure P13-43 Forging hammer (All dimensions are in millimeters)

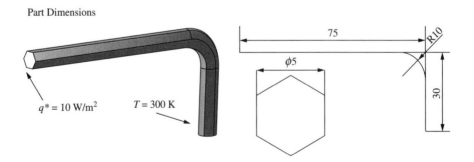

Figure P13-44 Allen wrench

13.45 The forklift from Figure P13–45 (detailed dimensions shown in Figure P11–20) has its load removed. The fork is made of AISI 4130 steel. The thermal conductivity of the steel is 35 W/m°C. The top surface of the fork is at 50 °C. The other surfaces of the L-shaped appendages located at the upper and lower left sides of the forklift are at room temperature (assume 25 °C). Determine the temperature distribution throughout the fork.

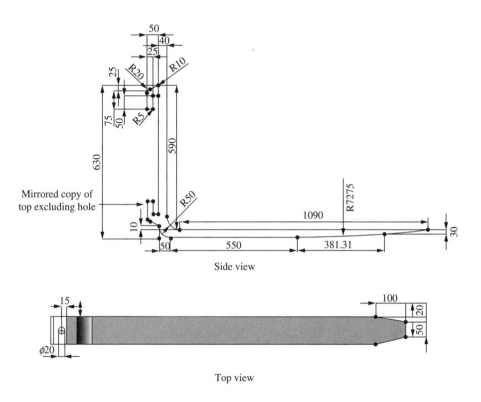

Side view

Top view

Figure P13–45 Fork from forklift (dimensions mm)

13.46 The radio control car front steering unit in Figure P13–46 (detailed dimensions shown in Figure P11–21) is now relieved of stress, but its base has an applied temperature of 40°C. The lower surface of the lower right-side flange has an applied temperature of 10°C. Other surfaces are exposed to $T_\infty = 30$°C and $h = 5$ W/(m²-°C). The unit is made of ABS (acrylonitrile butadine stryrene) with $k = 0.5$ W/(m-°C). Determine the temperature throughout the steering unit.

13.47 The hitch shown in Figure P13–47 (detailed dimensions shown in Figure P11–22) is unloaded but has an applied temperature of 100 °C to the front end and an applied temperature of 0 °C to the rear surface. Determine the temperature distribution throughout the hitch. The rest of the surfaces are exposed to ambient temperature of 30°C with $h = 10$ W/(m²-°C).

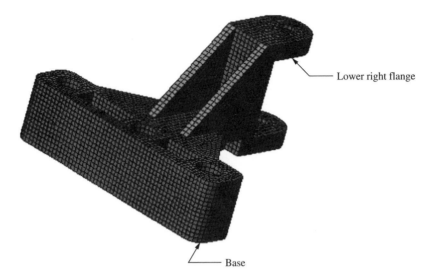

Figure P13–46 Steering unit

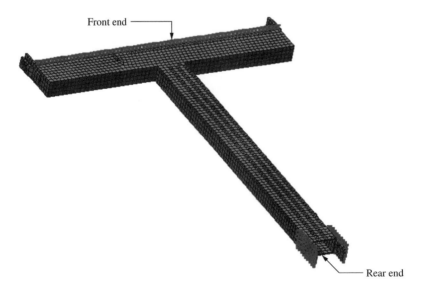

Figure P13–47 Hitch from disk

13.48 Air is flowing at a rate of 5kg/h inside a round tube with a diameter of 30 mm and length of 200 mm, similar to Figure 13–32 on page 644. The initial temperature of the air entering the tube is 10°C. The wall of the tube has a uniform constant temperature of 100°C. The specific heat of the air is 1.004 kJ/(kg-°C), the convection coefficient between the air and the inner wall of the tube is 15 W/(m²-°C), and the thermal conductivity is 0.03 W/(m²-°C). Determine the temperature of the air along the length of the tube and the heat flow at the inlet and outlet of the tube.

14

FLUID FLOW IN POROUS MEDIA AND THROUGH HYDRAULIC NETWORKS; AND ELECTRICAL NETWORKS AND ELECTROSTATICS ▲

CHAPTER OBJECTIVES

- To derive the basic differential equations for steady-state fluid flow through porous media, including Darcy's law.
- To describe the equations used for steady-state, incompressible, and inviscid fluid flow through and around pipes.
- To formulate the one-dimensional finite element fluid flow through porous media and through pipe's stiffness matrix and equations.
- To demonstrate longhand solutions to one-dimensional fluid flow.
- To develop the two-dimensional finite element for fluid flow through porous media and around solid objects or through pipes.
- To derive the stiffness matrix for elements used in hydraulic networks.
- To demonstrate longhand solution to the hydraulic network using the finite-element direct stiffness method.
- To show a flowchart of the fluid flow process.
- To describe electrical network principles, including Ohm's and Kirchhoff's laws, and to introduce the stiffness matrix used to solve electrical network problems.
- To demonstrate the solution of an electrical network by the finite-element direct stiffness method.
- To introduce some basic concepts in electrostatics, including Coulomb's and Gauss's laws and Poisson's equation.
- To present the two-dimensional finite element formulation of the electrostatics problem.
- To perform a longhand finite element solution to an electrostatics problem.
- To show examples of computer program solutions of electrostatics problems.

Introduction

In this chapter, we consider the flow of fluid through porous media, such as the flow of water through an earthen dam, and through pipes or around solid bodies.

We will observe that the form of the equations is the same as that for heat transfer described in Chapter 13.

We begin with a derivation of the basic differential equation in one dimension for an ideal fluid in a steady state, not rotating (that is, the fluid particles are translating only), incompressible (constant mass density), and inviscid (having no viscosity). We then extend this derivation to the two-dimensional case. We also consider the units used for the physical quantities involved in fluid flow. For more advanced topics, such as viscous flow, compressible flow, and three-dimensional problems, consult Reference [1].

We will use the same procedure to develop the element equations as in the heat-transfer problem; that is, we define an assumed fluid head for the flow through porous media (seepage) problem or velocity potential for flow of fluid through pipes and around solid bodies within each element. Then, to obtain the element equations, we use both a direct approach similar to that used in Chapters 2, 3, and 4 to develop the element equations and the minimization of a functional as used in Chapter 13. These equations result in matrices analogous to the stiffness and force matrices of the stress analysis problem or the conduction and associated force matrices of the heat-transfer problem.

Next, we consider both one- and two-dimensional finite element formulations of the fluid-flow problem and provide examples of one-dimensional fluid flow through porous media and through pipes and of flow within a two-dimensional region. We present the results for a two-dimensional fluid-flow problem.

We then consider flow through hydraulic networks and electric networks and show the analogies between these networks and the spring assemblage.

Finally, we describe concepts for electrostatic analysis and develop the two-dimensional finite element formulation for electrostatic analysis, along with computer program examples.

▲ 14.1 Derivation of the Basic Differential Equations ▲

Fluid Flow through a Porous Medium

Let us first consider the derivation of the basic differential equation for the one-dimensional problem of steady-state fluid flow through a porous medium. The purpose of this derivation is to present a physical insight into the fluid-flow phenomena, which must be understood so that the finite element formulation of the problem can be fully comprehended. (For additional information on fluid flow, consult References [2] and [3]). We begin by considering the control volume shown in Figure 14–1. By conservation of mass, we have

$$M_{\text{in}} + M_{\text{generated}} = M_{\text{out}} \tag{14.1.1}$$

or
$$\rho v_x A \, dt + \rho Q \, dt = \rho v_{x+dx} A \, dt \tag{14.1.2}$$

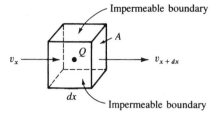

Figure 14–1 Control volume for one-dimensional fluid flow

where

M_{in} is the mass entering the control volume, in units of kilograms.

$M_{generated}$ is the mass generated within the body.

M_{out} is the mass leaving the control volume.

v_x is the velocity of the fluid flow at surface edge x, in units of m/s.

v_{x+dx} is the velocity of the fluid leaving the control volume at surface edge $x + dx$.

t is time, in s.

Q is an internal fluid source (an internal volumetric flow rate), in m³/s.

ρ is the mass density of the fluid, in kg/m³.

A is the cross-sectional area perpendicular to the fluid flow, in m².

By Darcy's law, we relate the velocity of fluid flow to the hydraulic gradient (the change in fluid head with respect to x) as

$$v_x = -K_{xx}\frac{d\phi}{dx} = -K_{xx}g_x \qquad (14.1.3)$$

where

K_{xx} is the permeability coefficient of the porous medium in the x direction, in m/s.

ϕ is the fluid head, in m.

$d\phi/dx = g_x$ is the fluid head gradient or hydraulic gradient, which is a unitless quantity in the seepage problem.

Equation (14.1.3) states that the velocity in the x direction is proportional to the gradient of the fluid head in the x direction. The minus sign in Eq. (14.1.3) implies that fluid flow is positive in the direction opposite the direction of fluid head increase, or that the fluid flows in the direction of lower fluid head. Equation (14.1.3) is analogous to Fourier's law of heat conduction, Eq. (13.1.3).

Similarly,

$$v_{x+dx} = -K_{xx}\frac{d\phi}{dx}\bigg|_{x+dx} \qquad (14.1.4)$$

where the gradient is now evaluated at $x + dx$. By Taylor series expansion, similar to that used in obtaining Eq. (13.1.5), we have

$$v_{x+dx} = -\left[K_{xx} \frac{d\phi}{dx} + \frac{d}{dx}\left(K_{xx} \frac{d\phi}{dx} \right) dx \right] \qquad (14.1.5)$$

where a two-term Taylor series has been used in Eq. (14.1.5). On substituting Eqs. (14.1.3) and (14.1.5) into Eq. (14.1.2), dividing Eq. (14.1.2) by $pA\,dx\,dt$, and simplifying, we have the equation for one-dimensional fluid flow through a porous medium as

$$\frac{d}{dx}\left(K_{xx} \frac{d\phi}{dx} \right) + \bar{Q} = 0 \qquad (14.1.6)$$

where $\bar{Q} = Q/A\,dx$ is the volume flow rate per unit volume in units 1/s. For a constant permeability coefficient, Eq. (14.1.6) becomes

$$K_{xx} \frac{d^2\phi}{dx^2} + \bar{Q} = 0 \qquad (14.1.7)$$

The boundary conditions are of the form

$$\phi = \phi_B \qquad \text{on } S_1 \qquad (14.1.8)$$

where ϕ_B represents a known boundary fluid head and S_1 is a surface where this head is known and

$$v_x^* = -K_{xx} \frac{d\phi}{dx} = \text{constant} \qquad \text{on } S_2 \qquad (14.1.9)$$

where S_2 is a surface where the prescribed velocity v_x^* or gradient is known. On an impermeable boundary, $v_x^* = 0$.

Comparing this derivation to that for the one-dimensional heat conduction problem in Section 13.1, we observe numerous analogies among the variables; that is, ϕ is analogous to the temperature function T, v_x is analogous to heat flux, and K_{xx} is analogous to thermal conductivity.

Now consider the two-dimensional fluid flow through a porous medium, as shown in Figure 14–2. As in the one-dimensional case, we can show that for material properties coinciding with the global x and y directions,

$$\frac{\partial}{\partial x}\left(K_{xx} \frac{\partial\phi}{\partial x} \right) + \frac{\partial}{\partial y}\left(K_{yy} \frac{\partial\phi}{\partial y} \right) + \bar{Q} = 0 \qquad (14.1.10)$$

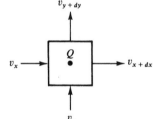

Figure 14–2 Control volume for two-dimensional fluid flow

with boundary conditions

$$\phi = \phi_B \qquad \text{on } S_1 \tag{14.1.11}$$

and

$$K_{xx}\frac{\partial \phi}{\partial x}C_x + K_{yy}\frac{\partial \phi}{\partial y}C_y = \text{constant} \qquad \text{on } S_2 \tag{14.1.12}$$

where C_x and C_y are direction cosines of the unit vector normal to the surface S_2, as previously shown in Figure 13–5.

Fluid Flow in Pipes and Around Solid Bodies

We now consider the steady-state irrotational flow of an incompressible and inviscid fluid. For the ideal fluid, the fluid particles do not rotate; they only translate, and the friction between the fluid and the surfaces is ignored. Also, the fluid does not penetrate into the surrounding body or separate from the surface of the body, which could create voids.

The equations for this fluid motion can be expressed in terms of the stream function or the velocity potential function. We will use the velocity potential analogous to the fluid head that was used for the derivation of the differential equation for flow through a porous medium in the preceding subsection.

The velocity v of the fluid is related to the velocity potential function ϕ by

$$v_x = -\frac{\partial \phi}{\partial x} \qquad v_y = -\frac{\partial \phi}{\partial y} \tag{14.1.13}$$

where v_x and v_y are the velocities in the x and y directions, respectively. In the absence of sources or sinks Q, conservation of mass in two dimensions yields the two-dimensional differential equation as

$$\frac{\partial^2 \phi}{\partial x^2} + \frac{\partial^2 \phi}{\partial y^2} = 0 \tag{14.1.14}$$

Equation (14.1.14) is analogous to Eq. (14.1.10) when we set $K_{xx} = K_{yy} = 1$ and $Q = 0$. Hence, Eq. (14.1.14) is just a special form of Eq. (14.1.10). The boundary conditions are

$$\phi = \phi_B \qquad \text{on } S_1 \tag{14.1.15}$$

and

$$\frac{\partial \phi}{\partial x}C_x + \frac{\partial \phi}{\partial y}C_y = \text{constant} \qquad \text{on } S_2 \tag{14.1.16}$$

where C_x and C_y are again direction cosines of unit vector **n** normal to surface S_2. Also see Figure 14–3. That is, Eq. (14.1.15) states that the velocity potential ϕ_B is known on a boundary surface S_1, whereas Eq. (14.1.16) states that the potential gradient or velocity is known normal to a surface S_2, as indicated for flow out of the pipe shown in Figure 14–3.

To clarify the sign convention on the S_2 boundary condition, consider the case of fluid flowing through a pipe in the positive x direction, as shown in Figure 14–4.

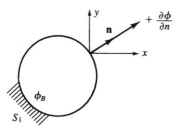

Figure 14–3 Boundary conditions for fluid flow

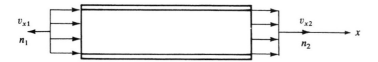

Figure 14–4 Known velocities at left and right edges of a pipe

Assume we know the velocities at the left edge (1) and the right edge (2). By Eq. (14.1.13) the velocity of the fluid is related to the velocity potential by

$$v_x = -\frac{\partial \phi}{\partial x}$$

At the left edge (1) assume we know $v_x = v_{x1}$. Then

$$v_{x1} = -\frac{\partial \phi}{\partial x}$$

But the normal is always positive away, or outward, from the surface. Therefore, positive n_1 is directed to the left, whereas positive x is to the right, resulting in

$$\frac{\partial \phi}{\partial n_1} = -\frac{\partial \phi}{\partial x} = v_{x1} = v_{n1}$$

At the right edge (2) assume we know $v_x = v_{x2}$. Now the normal n_2 is in the same direction as x. Therefore,

$$\frac{\partial \phi}{\partial n_2} = \frac{\partial \phi}{\partial x} = -v_{x2} = -v_{n2}$$

We conclude that the boundary flow velocity is positive if directed into the surface (region), as at the left edge, and is negative if directed away from the surface, as at the right edge.

At an impermeable boundary, the flow velocity and thus the derivative of the velocity potential normal to the boundary must be zero. At a boundary of uniform or constant velocity, any convenient magnitude of velocity potential ϕ may be specified as the gradient of the potential function; see, for instance, Eq. (14.1.13). This idea is also illustrated by Example 14.3.

▲ 14.2 One-Dimensional Finite Element Formulation ▲

We can proceed directly to the one-dimensional finite element formulation of the fluid-flow problem by now realizing that the fluid-flow problem is analogous to the heat-conduction problem of Chapter 13. We merely substitute the fluid velocity potential function ϕ for the temperature function T, the vector of nodal potentials denoted by $\{p\}$ for the nodal temperature vector $\{t\}$, fluid velocity v for heat flux q, and permeability coefficient K for flow through a porous medium instead of the conduction coefficient K. If fluid flow through a pipe or around a solid body is considered, then K is taken as unity. The steps are as follows.

Step 1 Select Element Type

The basic two-node element is again used, as shown in Figure 14–5, with nodal fluid heads, or potentials, denoted by p_1 and p_2.

Step 2 Choose a Potential Function

We choose the potential function ϕ similarly to the way we chose the temperature function of Section 13.4, as

$$\phi = N_1 p_1 + N_2 p_2 \tag{14.2.1}$$

where p_1 and p_2 are the nodal potentials (or fluid heads in the case of the seepage problem) to be determined, and

$$N_1 = 1 - \frac{x}{L} \qquad N_2 = \frac{x}{L} \tag{14.2.2}$$

are again the same shape functions used for the temperature element. The matrix $[N]$ is then

$$[N] = \begin{bmatrix} 1 - \dfrac{x}{L} & \dfrac{x}{L} \end{bmatrix} \tag{14.2.3}$$

Step 3 Define the Gradient/Potential and Velocity/Gradient Relationships

The hydraulic gradient matrix $\{g\}$ is given by

$$\{g\} = \left\{ \frac{d\phi}{dx} \right\} = [B]\{p\} \tag{14.2.4}$$

where $[B]$ is identical to Eq. (13.4.7), given by

$$[B] = \begin{bmatrix} -\dfrac{1}{L} & \dfrac{1}{L} \end{bmatrix} \tag{14.2.5}$$

Figure 14–5 Basic one-dimensional fluid-flow element

Table 14–1 Permeabilities of granular materials

Material	K (cm/s)
Clay	1×10^{-8}
Sandy clay	1×10^{-3}
Ottawa sand	$2–3 \times 10^{-2}$
Coarse gravel	1

and

$$\{p\} = \left\{ \begin{array}{c} p_1 \\ p_2 \end{array} \right\}$$

(14.2.6)

The velocity/gradient relationship based on Darcy's law is given by

$$v_x = -[D]\{g\}$$

(14.2.7)

where the material property matrix is now given by

$$[D] = [K_{xx}]$$

(14.2.8)

with K_{xx} the permeability of the porous medium in the x direction. Typical permeabilities of some granular materials are listed in Table 14–1. High permeabilities occur when $K > 10^{-1}$ cm/s, and when $K < 10^{-7}$ the material is considered to be nearly impermeable. For ideal flow through a pipe or over a solid body, we arbitrarily—but conveniently—let $K = 1$.

Step 4 Derive the Element Stiffness Matrix and Equations

The fluid-flow problem has a stiffness matrix that can be found using the first term on the right side of Eq. (13.4.17). That is, the fluid-flow stiffness matrix is analogous to the conduction part of the stiffness matrix in the heat-transfer problem. There is no comparable convection matrix to be added to the stiffness matrix. However, we will choose to use a direct approach similar to that used initially to develop the stiffness matrix for the bar element in Chapter 3.

Consider the fluid element shown in Figure 14–6 with length L and uniform cross-sectional area A. Recall that the stiffness matrix is defined in the structure problem to relate nodal forces to nodal displacements or in the temperature problem to relate nodal rates of heat flow to nodal temperatures. In the fluid-flow problem, we define

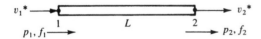

Figure 14–6 Fluid element subjected to nodal velocities

the stiffness matrix to relate nodal volumetric fluid-flow rates to nodal potentials or fluid heads as $\{f\} = [k]\{p\}$. Therefore,

$$f = v^*A \tag{14.2.9}$$

defines the volumetric flow rate f in units of cubic meters or cubic inches per second. Now, using Eqs. (14.2.7) and (14.2.8) in Eq. (14.2.9), we obtain

$$f = -K_{xx}Ag \text{ m}^3/\text{s} \tag{14.2.10}$$

in scalar form; based on Eqs. (14.2.4) and (14.2.5), g is given in explicit form by

$$g = \frac{p_2 - p_1}{L} \tag{14.2.11}$$

Applying Eqs. (14.2.10) and (14.2.11) at nodes 1 and 2, we obtain

$$f_1 = -K_{xx}A\frac{p_2 - p_1}{L} \tag{14.2.12}$$

and

$$f_2 = K_{xx}A\frac{p_2 - p_1}{L} \tag{14.2.13}$$

where f_1 is directed into the element, indicating fluid flowing into the element (p_1 must be greater than p_2 to push the fluid through the element, actually resulting in positive f_1), whereas f_2 is directed away from the element, indicating fluid flowing out of the element; hence the negative sign changes to a positive one in Eq. (14.2.13). Expressing Eqs. (14.2.12) and (14.2.13) together in matrix form, we have

$$\begin{Bmatrix} f_1 \\ f_2 \end{Bmatrix} = \frac{AK_{xx}}{L}\begin{bmatrix} 1 & -1 \\ -1 & 1 \end{bmatrix}\begin{Bmatrix} p_1 \\ p_2 \end{Bmatrix} \tag{14.2.14}$$

The stiffness matrix is then

$$[k] = \frac{AK_{xx}}{L}\begin{bmatrix} 1 & -1 \\ -1 & 1 \end{bmatrix} \text{ m}^2/\text{s} \tag{14.2.15}$$

for flow through a porous medium.

Equation (14.2.15) is analogous to Eq. (13.4.20) for the heat-conduction element or to Eq. (3.1.14) for the one-dimensional (axial stress) bar element. The permeability or stiffness matrix will have units of square meters or square inches per second.

In general, the basic element may be subjected to internal sources or sinks, such as from a pump, or to surface-edge flow rates, such as from a river or stream. To include these or similar effects, consider the element of Figure 14–6 now to include a uniform internal source Q acting over the whole element and a uniform surface flow-rate source q^* acting over the surface, as shown in Figure 14–7. The force matrix terms are

$$\{f_Q\} = \iiint\limits_V [N]^T Q \, dV = \frac{QAL}{2}\begin{Bmatrix} 1 \\ 1 \end{Bmatrix} \text{ m}^3/\text{s} \tag{14.2.16}$$

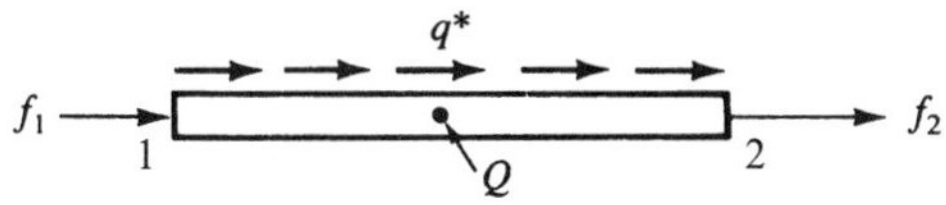

Figure 14-7 Additional sources of volumetric fluid-flow rates

where Q will have units of $m^3/(m^3 \cdot s)$, or $1/s$, and

$$\{f_q\} = \iint_{S_2} q^*[N]^T \, dS = \frac{q^* L t}{2} \begin{Bmatrix} 1 \\ 1 \end{Bmatrix} \, m^3/s \qquad (14.2.17)$$

where q^* will have units of m/s or in./s. Equations (14.2.16) and (14.2.17) indicate that one-half of the uniform volumetric flow rate per unit volume Q (a source being positive and a sink being negative) is allocated to each node and one-half the surface flow rate (again a source is positive) is allocated to each node.

**Step 5   Assemble the Element Equations to Obtain
the Global Equations and Introduce Boundary Conditions**

We assemble the total stiffness matrix $[K]$, total force matrix $\{F\}$, and total set of equations as

$$[K] = \sum [k^{(e)}] \qquad \{F\} = \sum \{f^{(e)}\} \qquad (14.2.18)$$

and $\qquad\qquad\qquad \{F\} = [K]\{p\} \qquad\qquad\qquad\qquad (14.2.19)$

The assemblage procedure is similar to the direct stiffness approach, but it is now based on the requirement that the potentials at a common node between two elements be equal. The boundary conditions on nodal potentials are given by Eq. (14.1.15).

Step 6   Solve for the Nodal Potentials

We now solve for the global nodal potentials, $\{p\}$, where the appropriate nodal potential boundary conditions, Eq. (14.1.15), are specified.

**Step 7   Solve for the Element Velocities and Volumetric
Flow Rates**

Finally, we calculate the element velocities from Eq. (14.2.7) and the volumetric flow rate Q_f as

$$Q_f = (v)(A) \, m^3/s \qquad (14.2.20)$$

Example 14.1

Determine (a) the fluid head distribution along the length of the coarse gravelly medium shown in Figure 14–8, (b) the velocity in the upper part, and (c) the volumetric flow rate in the upper part. The fluid head at the top is 200 mm and that at the bottom is

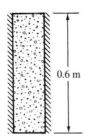

0.6 m   **Figure 14–8** One-dimensional fluid
flow in porous medium

20 mm. Let the permeability coefficient be $K_{xx} = 10$ mm/s. Assume a cross-sectional area of $A = 400$ mm^2.

SOLUTION:

The finite element discretization is shown in Figure 14–9. For simplicity, we will use three elements, each 400 mm long.

We calculate the stiffness matrices for each element as follows:

$$\frac{AK_{xx}}{L} = \frac{(400 \text{ mm}^2)(10 \text{ mm/s})}{200 \text{ mm}} = 20 \text{ mm}^2/\text{s}$$

Using Eq. (14.2.15) for elements 1, 2, and 3, we have

$$[k^{(1)}] = [k^{(2)}] = [k^{(3)}] = 20\begin{bmatrix} 1 & -1 \\ -1 & 1 \end{bmatrix} \text{ mm}^2/\text{s} \qquad (14.2.21)$$

In general, we would use Eqs. (14.2.16) and (14.2.17) to obtain element forces. However, in this example $Q = 0$ (no sources or sinks) and $q^* = 0$ (no applied surface flow rates). Therefore,

$$\{f^{(1)}\} = \{f^{(2)}\} = \{f^{(3)}\} = 0 \qquad (14.2.22)$$

The assembly of the element stiffness matrices from Eq. (14.2.21), via the direct stiffness method, produces the following system of equations:

$$20\begin{bmatrix} 1 & -1 & 0 & 0 \\ -1 & 2 & -1 & 0 \\ 0 & -1 & 2 & -1 \\ 0 & 0 & -1 & 1 \end{bmatrix}\begin{Bmatrix} p_1 \\ p_2 \\ p_3 \\ p_4 \end{Bmatrix} = \begin{Bmatrix} 0 \\ 0 \\ 0 \\ 0 \end{Bmatrix} \qquad (14.2.23)$$

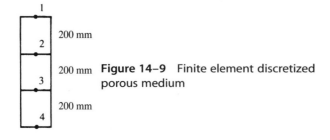

200 mm   **Figure 14–9** Finite element discretized
porous medium

Known nodal fluid head boundary conditions are $p_1 = 200$ mm and $p_4 = 20$ mm. These nonhomogeneous boundary conditions are treated as described for the stress analysis and heat-transfer problems. We modify the stiffness (permeability) matrix and force matrix as follows:

$$
\begin{bmatrix}
1 & 0 & 0 & 0 \\
0 & 40 & -20 & 0 \\
0 & -20 & 40 & 0 \\
0 & 0 & 0 & 1
\end{bmatrix}
\begin{Bmatrix}
p_1 \\
p_2 \\
p_3 \\
p_4
\end{Bmatrix}
=
\begin{Bmatrix}
200 \\
4000 \\
400 \\
20
\end{Bmatrix}
\tag{14.2.24}
$$

where the terms in the first and fourth rows and columns of the stiffness matrix corresponding to the known fluid heads $p_1 = 200$ mm and $p_4 = 20$ mm have been set equal to 0 except for the main diagonal, which has been set equal to 1, and the first and fourth rows of the force matrix have been set equal to the known nodal fluid heads at nodes 1 and 4. Also the terms $(-20) \times (200 \text{ mm}) = -4000$ mm on the left side of the second equation of Eq. (14.2.24) and $(-20) \times (20 \text{ mm}) = -400$ mm on the left side of the third equation of Eq. (14.2.24) have been transposed to the right side in the second and third rows (as +4000 and 400). The second and third equations of Eq. (14.2.24) can now be solved. The resulting solution is given by

$$
p_2 = 140 \text{ mm} \qquad p_3 = 80 \text{ mm} \tag{14.2.25}
$$

Next we use Eq. (14.2.7) to determine the fluid velocity in element 1 as

$$
v_x^{(1)} = -K_{xx}[B]\{p^{(1)}\} \tag{14.2.26}
$$

$$
= -K_{xx}\begin{bmatrix} -\dfrac{1}{L} & \dfrac{1}{L} \end{bmatrix}\begin{Bmatrix} p_1 \\ p_2 \end{Bmatrix} \tag{14.2.27}
$$

or

$$
v_x^{(1)} = 3 \text{ mm/s} \tag{14.2.28}
$$

You can verify that the velocities in the other elements are also 3 mm/s because the cross section is constant and the material properties are uniform. We then determine the volumetric flow rate Q_f in element 1 using Eq. (14.2.20) as

$$
Q_f = vA = (3 \text{ mm/s})(400 \text{ mm}^2) = 1200 \text{ mm}^3/\text{s} \tag{14.2.29}
$$

This volumetric flow rate is constant throughout the length of the medium. ∎

Example 14.2

For the smooth pipe of variable cross section shown in Figure 14–10, determine the potential at the junctions, the velocities in each section of pipe and the volumetric flow rate. The potential at the left end is $p_1 = 10$ m^2/s and that at the right end is $p_4 = 1$ m^2/s.

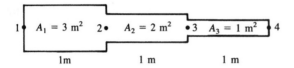

Figure 14–10 Variable-cross-section pipe subjected to fluid flow

SOLUTION:

For the fluid flow through a smooth pipe, $K_{xx} = 1$. The pipe has been discretized into three elements and four nodes, as shown in Figure 14–11. Using Eq. (14.2.15), we find that the element stiffness matrices are

$$[k^{(1)}] = \frac{3}{1}\begin{bmatrix} 1 & -1 \\ -1 & 1 \end{bmatrix}\, m \qquad [k^{(2)}] = \frac{2}{1}\begin{bmatrix} 1 & -1 \\ -1 & 1 \end{bmatrix}\, m \qquad [k^{(3)}] = \frac{1}{1}\begin{bmatrix} 1 & -1 \\ -1 & 1 \end{bmatrix}\, m$$

(14.2.30)

where the units on $[k]$ are now meters for fluid flow through a pipe.

There are no applied fluid sources. Therefore, $\{f^{(1)}\} = \{f^{(2)}\} = \{f^{(3)}\} = 0$. The assembly of the element stiffness matrices produces the following system of equations:

$$\begin{bmatrix} 3 & -3 & 0 & 0 \\ -3 & 5 & -2 & 0 \\ 0 & -2 & 3 & -1 \\ 0 & 0 & -1 & 1 \end{bmatrix}\begin{Bmatrix} 10 \\ p_2 \\ p_3 \\ 1 \end{Bmatrix} = \begin{Bmatrix} 0 \\ 0 \\ 0 \\ 0 \end{Bmatrix}\frac{m^3}{s}$$

(14.2.31)

Solving the second and third of Eqs. (14.2.31) for p_2 and p_3 in the usual manner, we obtain

$$p_2 = 8.365 \text{ m}^2/\text{s} \qquad p_3 = 5.91 \text{ m}^2/\text{s}$$

(14.2.32)

Using Eqs. (14.2.7) and (14.2.20), the velocities and volumetric flow rates in each element are

$$v_x^{(1)} = -[B]\{p^{(1)}\}$$

$$= -\begin{bmatrix} -\dfrac{1}{L} & \dfrac{1}{L} \end{bmatrix}\begin{Bmatrix} 10 \\ 8.365 \end{Bmatrix}$$

$$= 1.635 \text{ m/s}$$

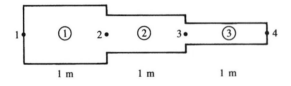

Figure 14–11 Discretized pipe

$$Q_f^{(1)} = Av_x^{(1)} = 3(1.635) = 4.91 \text{ m}^3/\text{s}$$

$$v_x^{(2)} = -(-8.365 + 5.91) = 2.455 \text{ m/s}$$

$$Q_f^{(2)} = 2.455(2) = 4.91 \text{ m}^3/\text{s}$$

$$v_x^{(3)} = -(-5.91 + 1) = 4.91 \text{ m/s}$$

$$Q_f^{(3)} = 4.91(1) = 4.91 \text{ m}^3/\text{s}$$

The potential, being higher at the left and decreasing to the right, indicates that the velocities are to the right. The volumetric flow rate is constant throughout the pipe, as conservation of mass would indicate. ∎

We now illustrate how you can solve a fluid-flow problem where the boundary condition is a known fluid velocity, but none of the p's are initially known.

Example 14.3

For the smooth pipe shown discretized in Figure 14–12 with uniform cross section of 4 cm², determine the flow velocities at the center and right end, knowing the velocity at the left end is $v_x = 4$ cm/s.

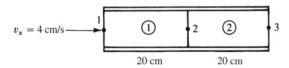

Figure 14–12 Discretized pipe for fluid-flow problem

SOLUTION:

Using Eq. (14.2.15), the element stiffness matrices are

$$[k^{(1)}] = \frac{1}{10} \begin{bmatrix} 1 & -1 \\ -1 & 1 \end{bmatrix} \text{cm} \qquad [k^{(2)}] = \frac{1}{10} \begin{bmatrix} 1 & -1 \\ -1 & 1 \end{bmatrix} \text{cm} \qquad (14.2.33)$$

where now the units on $[k]$ are centimeters for fluid flow through a pipe and $K_{xx} = 0.5$ cm/s.

Assembling the element stiffness matrices produces the following equations:

$$\frac{1}{10} \begin{bmatrix} 1 & -1 & 0 \\ -1 & 2 & -1 \\ 0 & -1 & 1 \end{bmatrix} \begin{Bmatrix} p_1 \\ p_2 \\ p_3 \end{Bmatrix} = \begin{Bmatrix} f_1 \\ f_2 \\ f_3 \end{Bmatrix} \qquad (14.2.34)$$

The specified boundary condition is $v_x = 2$ in./s, so that by Eq. (14.2.9), we have

$$f_1 = v_1 A = (4 \text{ cm/s})(4 \text{ cm}^2) = 16 \text{ cm}^3/\text{s} \qquad (14.2.35)$$

Because p_1, p_2, and p_3 in Eq. (14.2.34) are not known, we cannot determine these potentials directly. The problem is similar to that occurring if we try to solve the structural problem without prescribing displacements sufficient to prevent rigid body motion of the structure. This was discussed in Chapter 2. Because the p's correspond to displacements in the structural problem, it appears that we must specify at least one value of p in order to obtain a solution. We then proceed as follows. Select a convenient value for p_3 (for instance set $p_3 = 0$). (The velocities are functions of the derivatives or differences in p's, so a value of $p_3 = 0$ is acceptable.) Then p_1 and p_2 are the unknowns. The solution will yield p_1 and p_2 relative to $p_3 = 0$. Therefore, from the first two of Eqs. (14.2.34), we have

$$\frac{1}{10} \begin{bmatrix} 1 & -1 \\ -1 & 2 \end{bmatrix} \begin{Bmatrix} p_1 \\ p_2 \end{Bmatrix} = \begin{Bmatrix} 16 \\ 0 \end{Bmatrix} \tag{14.2.36}$$

where $f_1 = 16 \text{ cm}^3/\text{s}$ from Eq. (14.2.35) and $f_2 = 0$, because there is no applied fluid force at node 2.

Solving Eq. (14.2.36), we obtain

$$p_1 = 320 \qquad p_2 = 160 \tag{14.2.37}$$

These are not absolute values for p_1 and p_2; rather, they are relative to p_3. The fluid velocities in each element are absolute values, because velocities depend on the differences in p's. These differences are the same no matter what value for p_3 was chosen. You can verify this by choosing $p_3 = 80$, for instance, and re-solving for the velocities. [You would find $p_1 = 400$ and $p_2 = 240$ and the same v's as in Eq. (14.2.38).]

$$v_x^{(1)} = - \begin{bmatrix} -\dfrac{1}{L} & \dfrac{1}{L} \end{bmatrix} \begin{Bmatrix} 320 \\ 160 \end{Bmatrix} = 4 \text{ cm/s}$$

and

$$v_x^{(2)} = - \begin{bmatrix} -\dfrac{1}{L} & \dfrac{1}{L} \end{bmatrix} \begin{Bmatrix} 160 \\ 0 \end{Bmatrix} = 4 \text{ cm/s} \qquad \blacksquare \tag{14.2.38}$$

Fluid Flow through Hydraulic Networks

Hydraulic or piping networks typically found in buildings, industrial plants, farm irrigation pipe networks, municipal water systems, and power plants also can be analyzed using the finite element method. Pressure flow in these networks can be described by a system of linear equations. In these networks, such as the one shown in Figure 14–13, the fluid flow source (volumetric flow rate) Q (in units of m^3/s) forces fluid through the pipe network. As the fluid flows through each branch, there is resistance in each branch which is typically a function of the fluid viscosity μ (in units of N-s/m^2) (a typical value of μ is $1.002 \times 10^{-6} \text{ N-s/m}^2$ for water at $20\,°\text{C}$), the length of the pipe branch, the diameter of the pipe, the average velocity of the fluid flow in the branch, and the friction factor. These factors cause a pressure drop through the pipe branch. We assume the fluid to be laminar, incompressible, and in a steady state and the pressure drop Δp (in units of N/m^2) in a branch of the network to be proportional to

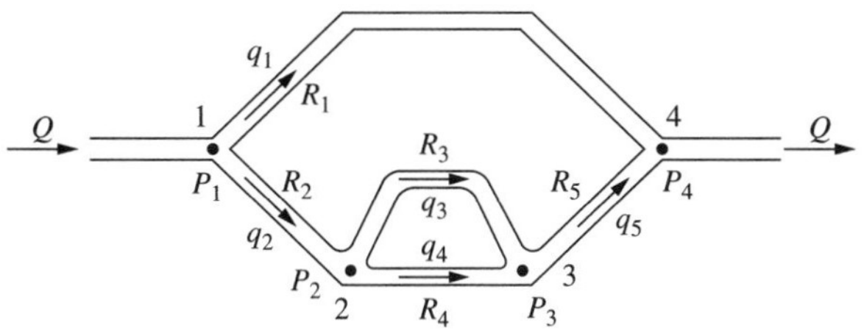

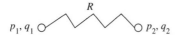

Figure 14–13 Typical pipe network (composed of five branches, 1–4, 1–2, 2–3U, 2–3L, and 3–4 (where U and L stand for upper and lower branches between nodes 2 and 3)

$p_1, q_1 \;\circ\!\!\!\!\!\!\!\raisebox{0pt}{\small$\diagup\!\diagdown\!\diagup\!\diagdown\!\diagup\!\diagdown$}\!\!\!\!\!\!\!\circ\; p_2, q_2$

Figure 14–14 Basic branch resistor element showing nodal pressures and flow rates

the volumetric flow rate q (in units of m^3/s) through that branch, such that by Poiseuille's law

$$\Delta p = Rq \tag{14.2.39}$$

where R is the branch resistance coefficient in units of N-s/m^5. A typical equation to predict R is given by $R = 128\,\mu L/(\pi d^4)$ for flow through long circular pipes, where L is the length of the branch and d the diameter of the pipe both in units consistent with those used for p and q.

Here we consider the basic element as a branch of the network analogous to a spring element, as shown in Figure 14–14.

Using Eq. (14.2.39), we relate the volumetric flow rates to the pressures at each node by the matrix equation as follows:

$$\frac{1}{R}\begin{bmatrix} 1 & -1 \\ -1 & 1 \end{bmatrix}\begin{Bmatrix} p_1 \\ p_2 \end{Bmatrix} = \begin{Bmatrix} q_1 \\ q_2 \end{Bmatrix} \tag{14.2.40}$$

Equation (14.2.40) is considered the element equilibrium relations between the nodal pressures and volumetric flow rates. The stiffness matrix for the pipe resistance is now defined as the matrix relating the nodal volumetric flow rates to the nodal currents. From Eq. (14.2.40), the element stiffness matrix is

$$[k] = \frac{1}{R}\begin{bmatrix} 1 & -1 \\ -1 & 1 \end{bmatrix} \tag{14.2.41}$$

We can draw analogies between the pipe resistor element and the spring element from Chapter 2 as follows: The nodal pressures are analogous to nodal displacements, the nodal volumetric flow rates are analogous to nodal forces, and the resistance R is analogous to the inverse of the spring constant k.

We will now use Eq. (14.2.39) for each branch of Figure 14–13 along with the continuity of flow that states that the mass of fluid passing all sections in a stream of fluid per unit time must be the same. For networks of a single fluid property with constant density, this is equivalent to $Q_1 = Q_2$ at two different cross sections of a pipe. In this network, we assume pressures at nodes 1, 2, and 3 to be unknown and use a baseline pressure of zero for the pressure at node 4.

We first use Eq. (14.2.39) to express the flow rates in each branch (element) as follows:

Branch 1–4: $\qquad q_1 = \dfrac{p_1 - 0}{R_1}$ $\qquad$ Branch 1–2: $\qquad q_2 = \dfrac{p_1 - p_2}{R_2}$

Branch 2–3U: $\qquad q_3 = \dfrac{p_2 - p_3}{R_3}$ $\qquad$ Branch 2–3L: $\qquad q_4 = \dfrac{p_2 - p_3}{R_4}$

Branch 3–4: $\qquad q_5 = \dfrac{p_3 - 0}{R_5}$ $\hfill$ (14.2.42)

We now apply the continuity of flow equations as follows:

$$\text{At node 1:} \qquad Q = q_1 + q_2$$

$$\text{At node 2:} \qquad q_2 = q_3 + q_4 \hfill (14.2.43)$$

$$\text{At node 3:} \qquad q_3 + q_4 = q_5$$

You should note that by continuity of flow, $q_2 = q_5$.

We now use Eq. (14.2.42) in Eq. (14.2.43), to obtain the following set of equations:

$$\text{At node 1:} \qquad Q = \frac{p_1 - 0}{R_1} + \frac{p_1 - p_2}{R_2}$$

$$\text{At node 2:} \qquad \frac{p_1 - p_2}{R_2} = \frac{p_2 - p_3}{R_3} + \frac{p_2 - p_3}{R_4} \hfill (14.2.44)$$

$$\text{At node 3:} \qquad \frac{p_3}{R_5} = \frac{p_2 - p_3}{R_3} + \frac{p_2 - p_3}{R_4}$$

In matrix form, we express Eqs. (14.2.44) as

$$\begin{bmatrix} \dfrac{1}{R_1} + \dfrac{1}{R_2} & -\dfrac{1}{R_2} & 0 \\[2ex] -\dfrac{1}{R_2} & \dfrac{1}{R_3} + \dfrac{1}{R_4} + \dfrac{1}{R_2} & -\dfrac{1}{R_3} - \dfrac{1}{R_4} \\[2ex] 0 & -\dfrac{1}{R_3} - \dfrac{1}{R_4} & \dfrac{1}{R_3} + \dfrac{1}{R_4} + \dfrac{1}{R_5} \end{bmatrix} \begin{Bmatrix} p_1 \\[2ex] p_2 \\[2ex] p_3 \end{Bmatrix} = \begin{Bmatrix} Q \\[2ex] 0 \\[2ex] 0 \end{Bmatrix} \qquad (14.2.45)$$

We will now demonstrate how to use the direct stiffness method to obtain the same system of linear equations as in Eq. (14.2.45). Using Eq. (14.2.41) for the stiffness matrix for each element branch, we have

$$[k^{(1)}] = \frac{1}{R_1} \begin{array}{cc} 1 & 4 \\ \begin{bmatrix} 1 & -1 \\ -1 & 1 \end{bmatrix} \end{array} \quad [k^{(2)}] = \frac{1}{R_2} \begin{array}{cc} 1 & 2 \\ \begin{bmatrix} 1 & -1 \\ -1 & 1 \end{bmatrix} \end{array} \quad [k^{(3)}] = \frac{1}{R_3} \begin{array}{cc} 2 & 3 \\ \begin{bmatrix} 1 & -1 \\ -1 & 1 \end{bmatrix} \end{array}$$

$$[k^{(4)}] = \frac{1}{R_4} \begin{array}{cc} 2 & 3 \\ \begin{bmatrix} 1 & -1 \\ -1 & 1 \end{bmatrix} \end{array} \quad [k^{(5)}] = \frac{1}{R_5} \begin{array}{cc} 3 & 4 \\ \begin{bmatrix} 1 & -1 \\ -1 & 1 \end{bmatrix} \end{array} \quad (14.2.46)$$

where the superscript numbers in Eq. (14.2.46) indicate the element branch. That is, element 1 is from node 1 to node 4, element 2 is from node 1 to node 2, element 3 is from node 2 to 3 in the upper section of pipe from 2 to 3, element 4 is from node 2 to node 3 along the lower section of pipe between nodes 2 and 3, and element 5 is from node 3 to 4, as indicated by the numbers above the matrices in Eq. (14.2.46). Using the stiffness matrices in Eq. (14.2.46) along with the direct stiffness method, we assemble the global stiffness matrix and the global equations in the usual manner as

$$\begin{bmatrix} \dfrac{1}{R_1} + \dfrac{1}{R_2} & -\dfrac{1}{R_2} & 0 \\ -\dfrac{1}{R_2} & \dfrac{1}{R_2} + \dfrac{1}{R_3} + \dfrac{1}{R_4} & -\dfrac{1}{R_3} - \dfrac{1}{R_4} \\ 0 & -\dfrac{1}{R_3} - \dfrac{1}{R_4} & \dfrac{1}{R_3} + \dfrac{1}{R_4} + \dfrac{1}{R_5} \end{bmatrix} \begin{Bmatrix} p_1 \\ p_2 \\ p_3 \end{Bmatrix} = \begin{Bmatrix} Q \\ 0 \\ 0 \end{Bmatrix} \quad (14.2.47)$$

where Q is analogous to an applied global force at node 1.

Comparing Eq. (14.2.47) to Eq. (14.2.45), we observe them to be identical.

Example 14.4

For the piping network shown in Figure 14–13, let $R_1 = 10$, $R_2 = 5$, $R_3 = 2$, $R_4 = 3$, and $R_5 = 5$ all in units of N-s/m^5. Set the pressure at node 4 to zero. Let $Q = 0.5$ m^3/s. Determine the pressures at nodes 1, 2, and 3. Use the direct stiffness method to solve this problem.

SOLUTION:

From Eq. (14.2.46), the element stiffness matrices are

$$[k^{(1)}] = \frac{1}{10} \begin{array}{cc} 1 & 4 \\ \begin{bmatrix} 1 & -1 \\ -1 & 1 \end{bmatrix} \end{array} \quad [k^{(2)}] = \frac{1}{5} \begin{array}{cc} 1 & 2 \\ \begin{bmatrix} 1 & -1 \\ -1 & 1 \end{bmatrix} \end{array} \quad [k^{(3)}] = \frac{1}{2} \begin{array}{cc} 2 & 3 \\ \begin{bmatrix} 1 & -1 \\ -1 & 1 \end{bmatrix} \end{array}$$

$$[k^{(4)}] = \frac{1}{3} \begin{array}{cc} 2 & 3 \\ \begin{bmatrix} 1 & -1 \\ -1 & 1 \end{bmatrix} \end{array} \quad [k^{(5)}] = \frac{1}{5} \begin{array}{cc} 3 & 4 \\ \begin{bmatrix} 1 & -1 \\ -1 & 1 \end{bmatrix} \end{array} \quad (14.2.48)$$

Using the direct stiffness method, we assemble the global stiffness matrix and global equations as

$$
\begin{bmatrix}
\dfrac{1}{10}+\dfrac{1}{5} & \dfrac{-1}{5} & 0 \\[2mm]
\dfrac{-1}{5} & \dfrac{1}{2}+\dfrac{1}{3}+\dfrac{1}{5} & \dfrac{-1}{2}-\dfrac{1}{3} \\[2mm]
0 & \dfrac{-1}{2}-\dfrac{1}{3} & \dfrac{1}{2}+\dfrac{1}{3}+\dfrac{1}{5}
\end{bmatrix}
\begin{Bmatrix} p_1 \\ p_2 \\ p_3 \end{Bmatrix}
=
\begin{Bmatrix} 0.5 \\ 0 \\ 0 \end{Bmatrix}
\qquad (14.2.49)
$$

where the global nodal volumetric flow rate at node 1 is $Q_1 = 0.5 \text{ m}^3/\text{s}$. There are no volumetric flow rates at nodes 2 and 3. Therefore, $Q_2 = Q_3 = 0$.

Solving Eq. (14.2.49) simultaneously, the nodal pressures are

$$
p_1 = 2.642 \text{ N/m}^2 \qquad p_2 = 1.462 \text{ N/m}^2 \qquad p_3 = 1.179 \text{ N/m}^2 \qquad (14.2.50)
$$

■

Finally, we should understand that the assumptions presented in this section do not always apply to real pipe network systems. It should be noted that complex pipe networks often are composed of piping with network fittings such as elbows, tees, contractors, expansions, valves, and pumps. Also, the flow may not always be laminar and steady state. Numerous programs (such as described in References [6, 7, 8]) have been developed to deal with these additional design problems.

▲ 14.3 Two-Dimensional Finite Element Formulation ▲

Because many fluid-flow problems can be modeled as two-dimensional problems, we now develop the equations for an element appropriate for these problems. Examples using this element then follow.

Step 1

The three-node triangular element in Figure 14–15 is the basic element for the solution of the two-dimensional fluid-flow problem.

Step 2

The potential function is

$$
[\phi] = [N_i \quad N_j \quad N_m]
\begin{Bmatrix} p_i \\ p_j \\ p_m \end{Bmatrix}
\qquad (14.3.1)
$$

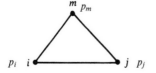

Figure 14–15 Basic triangular element with nodal potentials

where $p_i, p_j,$ and p_m are the nodal potentials (for groundwater flow, ϕ is the piezometric fluid head function, and the p's are the nodal heads), and the shape functions are again given by Eq. (6.2.18) or (13.5.2) as

$$N_i = \frac{1}{2A}(\alpha_i + \beta_i x + \gamma_i y) \tag{14.3.2}$$

with similar expressions for N_j and N_m. The α's, β's, and γ's are defined by Eqs. (6.2.10).

Step 3

The gradient matrix $\{g\}$ is given by

$$\{g\} = [B]\{p\} \tag{14.3.3}$$

where the matrix $[B]$ is again given by

$$[B] = \frac{1}{2A}\begin{bmatrix} \beta_i & \beta_j & \beta_m \\ \gamma_i & \gamma_j & \gamma_m \end{bmatrix} \tag{14.3.4}$$

and

$$\{g\} = \begin{Bmatrix} g_x \\ g_y \end{Bmatrix} \tag{14.3.5}$$

with

$$g_x = \frac{\partial \phi}{\partial x} \qquad g_y = \frac{\partial \phi}{\partial y} \tag{14.3.6}$$

The velocity/gradient matrix relationship is now

$$\begin{Bmatrix} v_x \\ v_y \end{Bmatrix} = -[D]\{g\} \tag{14.3.7}$$

where the material property matrix is

$$[D] = \begin{bmatrix} K_{xx} & 0 \\ 0 & K_{yy} \end{bmatrix} \tag{14.3.8}$$

and the K's are permeabilities (for the seepage problem) of the porous medium in the x and y directions. For fluid flow around a solid object or through a smooth pipe, $K_{xx} = K_{yy} = 1$.

Step 4

The element stiffness matrix is given by

$$[k] = \iiint_V [B]^T [D][B]\, dV \tag{14.3.9}$$

Assuming constant-thickness (t) triangular elements and noting that the integrand terms are constant, we have

$$[k] = tA[B]^T[D][B] \text{ m}^2/\text{s or in}^2/\text{s} \tag{14.3.10}$$

which can be simplified to

$$[k] = \frac{tK_{xx}}{4A}\begin{bmatrix} \beta_i^2 & \beta_i\beta_j & \beta_i\beta_m \\ \beta_i\beta_j & \beta_j^2 & \beta_j\beta_m \\ \beta_i\beta_m & \beta_j\beta_m & \beta_m^2 \end{bmatrix} + \frac{tK_{yy}}{4A}\begin{bmatrix} \gamma_i^2 & \gamma_i\gamma_j & \gamma_i\gamma_m \\ \gamma_i\gamma_j & \gamma_j^2 & \gamma_j\gamma_m \\ \gamma_i\gamma_m & \gamma_j\gamma_m & \gamma_m^2 \end{bmatrix} \tag{14.3.11}$$

The force matrices are

$$\{f_Q\} = \iiint_V Q[N]^T \, dV = Q\iiint_V [N]^T \, dV \tag{14.3.12}$$

for constant volumetric flow rate per unit volume over the whole element. On evaluating Eq. (14.3.12), we obtain

$$\{f_Q\} = \frac{QV}{3}\begin{Bmatrix} 1 \\ 1 \\ 1 \end{Bmatrix} \frac{\text{m}^3}{\text{s}} \text{ or } \frac{\text{in}^3}{\text{s}} \tag{14.3.13}$$

We find that the second force matrix is

$$\{f_q\} = \iint_{S_2} q^*[N]^T \, dS = \iint_{S_2} q^*\begin{Bmatrix} N_i \\ N_j \\ N_m \end{Bmatrix} dS \tag{14.3.14}$$

This reduces to

$$\{f_q\} = \frac{q^*L_{i\text{-}j}t}{2}\begin{Bmatrix} 1 \\ 1 \\ 0 \end{Bmatrix} \frac{\text{m}^3}{\text{s}} \text{ or } \frac{\text{in}^3}{\text{s}} \text{ on side } i\text{-}j \tag{14.3.15}$$

with similar terms on sides j-m and m-i [see Eqs. (13.5.19) and (13.5.20)]. Here $L_{i\text{-}j}$ is the length of side i-j of the element and q^* is the assumed constant surface flow rate. Both Q and q^* are positive quantities if fluid is being added to the element. The units on Q and q^* are m^3/(m$^3 \cdot$ s) and m/s. The total force matrix is then the sum of $\{f_Q\}$ and $\{f_q\}$.

Example 14.5

For the two-dimensional sandy soil region shown in Figure 14–16, determine the potential distribution. The potential (fluid head) on the left side is a constant 10.0 m and that on the right side is 0.0. The upper and lower edges are impermeable. The permeabilities are $K_{xx} = K_{yy} = 25 \times 10^{-5}$ m/s. Assume unit thickness.

The finite element model is shown in Figure 14–16. We use only the four triangular elements of equal size for simplicity of the longhand solution. For increased accuracy

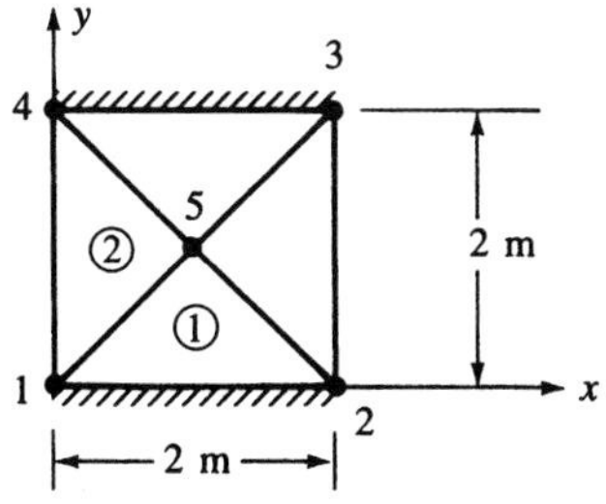

Figure 14–16 Two-dimensional porous medium

in results, we would need to refine the mesh. This body has the same magnitude of coordinates as Figure 13–25. Therefore, the total stiffness matrix is given by Eq. (13.5.40) as

$$[K] = \begin{bmatrix} 25 & 0 & 0 & 0 & -25 \\ 0 & 25 & 0 & 0 & -25 \\ 0 & 0 & 25 & 0 & -25 \\ 0 & 0 & 0 & 25 & -25 \\ -25 & -25 & -25 & -25 & 100 \end{bmatrix} \times 10^{-5} \frac{m^2}{s} \qquad (14.3.16)$$

The force matrices are zero, because $Q = 0$ and $q^* = 0$. Applying the boundary conditions, we have

$$p_1 = p_4 = 10.0 \text{ m} \qquad p_2 = p_3 = 0$$

The assembled total system of equations is then

$$10^{-5} \begin{bmatrix} 25 & 0 & 0 & 0 & -25 \\ 0 & 25 & 0 & 0 & -25 \\ 0 & 0 & 25 & 0 & -25 \\ 0 & 0 & 0 & 25 & -25 \\ -25 & -25 & -25 & -25 & 100 \end{bmatrix} \begin{Bmatrix} 10 \\ 0 \\ 0 \\ 10 \\ p_5 \end{Bmatrix} = \begin{Bmatrix} 0 \\ 0 \\ 0 \\ 0 \\ 0 \end{Bmatrix} \qquad (14.3.17)$$

Solving the fifth of Eqs. (14.3.17) for p_5, we obtain

$$p_5 = 5 \text{ m}$$

Using Eqs. (14.3.7) and (14.3.3) we obtain the velocity in element 2 as

$$\begin{Bmatrix} v_x^{(2)} \\ v_y^{(2)} \end{Bmatrix} = - \begin{bmatrix} 25 & 0 \\ 0 & +25 \end{bmatrix} \times 10^{-5} \frac{1}{2A} \begin{bmatrix} -1 & 2 & -1 \\ -1 & 0 & 1 \end{bmatrix} \begin{Bmatrix} p_1 \\ p_5 \\ p_4 \end{Bmatrix} \qquad (14.3.18)$$

where $\beta_1 = -1$, $\beta_5 = 2$, $\beta_4 = -1$, $\gamma_1 = -1$, $\gamma_5 = 0$, and $\gamma_4 = 1$ were obtained from Eq. (13.5.24). Simplifying Eq. (14.3.18), we obtain

$$v_x^{(2)} = 125 \times 10^{-5} \text{ m/s} \qquad v_y^{(2)} = 0 \qquad \blacksquare$$

A line or point fluid source from a pump, for instance, can be handled in the same manner as described in Section 13.6 for heat sources. If the source is at a node when the discretized finite element model is created, then the source can be added to the row of the global force matrix corresponding to the global degree of freedom assigned to the node. If the source is within an element, we can use Section 13.6 to allocate the source to the proper nodes, as illustrated by the following example.

Example 14.6

A pump, pumping fluid at $Q^* = 6500$ m²/h, is located at coordinates (5, 2) in the element shown in Figure 14–17. Determine the amount of Q^* allocated to each node. All nodal coordinates are in units of meters. Assume unit thickness of $t = 1$ mm.

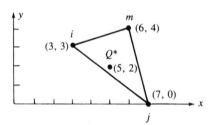

Figure 14–17 Triangular element with pump located within element

The magnitudes of the numbers are the same as in Example 13.8. Therefore, the shape functions are identical to Eq. (13.6.7); when evaluated at the source $x = 5$ m, $y = 2$ m, they are equal to Eq. (13.6.8). Using Eq. (13.6.3), we obtain the amount of Q^* allocated to each node or equivalently the force matrix as

$$\left\{ \begin{array}{c} f_{Qi} \\ f_{Qj} \\ f_{Qm} \end{array} \right\} = Q^* t \left\{ \begin{array}{c} N_i \\ N_j \\ N_m \end{array} \right\} \Bigg|_{\substack{x = x_0 = 5 \text{ m} \\ y = y_0 = 2 \text{ m}}} \qquad (14.3.19)$$

$$= \frac{(6500 \text{ m}^2/\text{h})(1 \text{ mm})}{(13)\left(\dfrac{1000 \text{ mm}}{1 \text{ m}}\right)} \left\{ \begin{array}{c} 6 \\ 5 \\ 2 \end{array} \right\} = \left\{ \begin{array}{c} 3.0 \\ 2.5 \\ 1.0 \end{array} \right\} \frac{\text{m}^3}{\text{h}} \qquad \blacksquare$$

▲ 14.4 Flowchart and Example of a Fluid-Flow Program ▲

Figure 14–18 is a flowchart of a finite element process used for the analysis of two-dimensional steady-state fluid flow through a porous medium or through a pipe. Recall that flow through a porous medium is analogous to heat transfer by conduction. For more complicated fluid flows, see Reference [6].

We now present computer program results for a two-dimensional steady-state, incompressible fluid flow. The program is based on the flowchart of Figure 14–18.

For flow through a porous medium, we recall the analogies between conductive heat transfer and flow through a porous medium and use the heat transfer processor from Reference [4] to solve the problem shown in Figure 14–19. The fluid flow

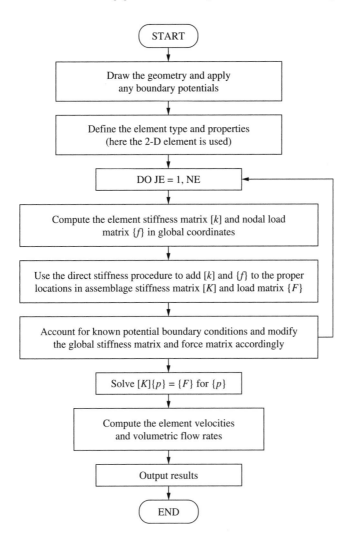

Figure 14–18 Flowchart of two-dimensional fluid-flow process

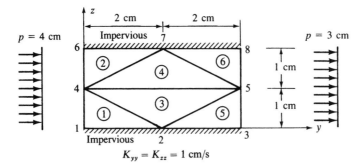

Figure 14–19 Two-dimensional fluid-flow problem

Table 14–2 Nodal potentials

Node Number	Potential
1	4.0000D+00
2	3.5000D+00
3	3.0000D+00
4	4.0000D+00
5	3.0000D+00
6	4.0000D+00
7	3.5000D+00
8	3.0000D+00

problem shown discretized in Figure 14–19 has the top and bottom sides impervious, whereas the right side has a constant head of 3 cm and the left side has a constant head of 4 cm.

Results for the nodal potentials obtained using [6] are shown in Table 14–2. They compare exactly with solutions obtained using another computer program (see Reference [5]).

▲ 14.5 Electrical Networks ▲

Current flow in electrical networks or circuits can be described by a system of linear equations developed using the direct stiffness method. In an electrical network such as the one shown in Figure 14–20, a voltage source (such as from a battery) forces a current of electrons to flow through the network. When the current passes through a resistor (such as a lightbulb or motor), some of the voltage is absorbed by the resistor. By Ohm's law, the voltage drop ΔV across the resistor is given by

$$\Delta V = RI \qquad (14.5.1)$$

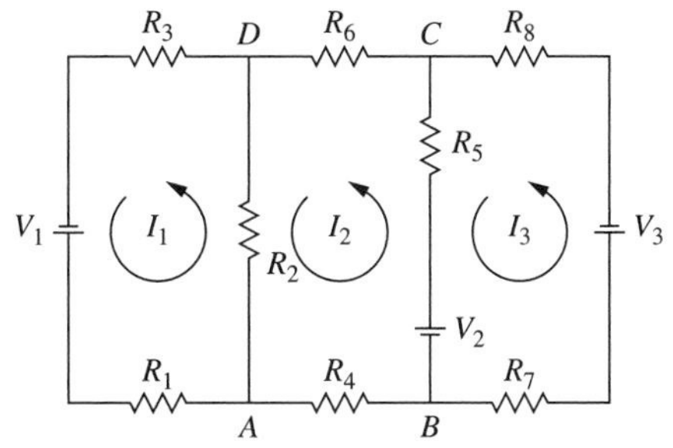

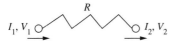

Figure 14–20 Typical electrical network (this one composed of three loops)

Figure 14–21 Basic resistor element showing nodal currents and voltages

where the voltage V is measured in units of volts, the resistance R is measured in ohms (the Greek omega symbol Ω or R are used to designate ohms), and the current flow I is measured in units of amperes or amps.

Here we consider the basic element as the resistor element analogous to a spring element shown in Figure 14–21.

Using Ohm's law, Eq. (14.5.1), we relate the voltage difference across the resistor to the current flow through each end of the resistor by the matrix equation as

$$\left\{ \begin{array}{c} V_1 \\ V_2 \end{array} \right\} = R \begin{bmatrix} 1 & -1 \\ -1 & 1 \end{bmatrix} \left\{ \begin{array}{c} I_1 \\ I_2 \end{array} \right\} \qquad (14.5.2)$$

Equation (14.5.2) represents the element equilibrium relations between the nodal currents and nodal voltages.

The stiffness matrix for the resistor is now defined as the matrix relating the nodal voltages to the nodal currents. From Eq. (14.5.2), the stiffness matrix is then

$$[k] = R \begin{bmatrix} 1 & -1 \\ -1 & 1 \end{bmatrix} \qquad (14.5.3)$$

We can draw analogies between the resistor element and the spring element from Chapter 2 as follows: The nodal currents are analogous to nodal displacements, the nodal voltages are analogous to nodal forces, and the resistance R is analogous to the spring constant k.

We now describe some sign conventions associated with the solution of the network in Figure 14–20. The network contains three closed loops. We designate these loops one (left loop), two (center loop), and three (right loop) and by standard circuit analysis indicate the currents flowing in each loop with current I_1, I_2, and I_3, respectively, shown by the curved arrows inside each loop. The designated directions of the loop currents are arbitrary, but we will use counterclockwise notation as positive in this description. Upon solving the equations for the currents, if a current is negative, then the actual direction of current flow is opposite to that chosen or clockwise. We must be careful regarding the positive direction of the voltages provided by the battery. If the current-direction curved arrowhead shown points from the negative side (shorter side) within the battery to the positive (longer side), the voltage is taken as positive; otherwise the voltage is negative. For example, V_1 is positive as I_1 goes around in a counterclockwise manner into the short side of battery V_1.

Current flow in a loop is described by Kirchhoff's law as follows: The algebraic sum of the product of resistance times current (RI) voltage drops in one direction around a loop equals the algebraic sum of the voltage sources (from batteries for instance) in the same direction around the loop.

We will now use both Ohm's and Kirchhoff's laws, along with afore described sign conventions to set up the three loop equations used to solve for the currents in each loop in Figure 14–20. That is, from Figure 14–20 for the three loops, we have

$$\text{Loop 1:} \qquad V_{R1} + V_{R2} + V_{R3} = V_1 \qquad\qquad (14.5.4)$$

where the voltage drops through the resistors associated with loop 1 are

$$V_{R1} = R_1 I_1, \quad V_{R2} = R_2(I_1 - I_2), \quad V_{R3} = R_3 I_1 \qquad (14.5.5)$$

Note that current I_2 from loop 2 also flows through branch AD of loop 1 with associated RI drop of $R_2 I_2$ $(-R_2 I_2)$ due to I_2 flowing down the branch into R_2 from D towards A, whereas current I_1 flows up the branch into R_2 from A to D. Also, battery voltage V_1 is taken as positive in Eq. (14.5.4) as I_1 goes from the negative (short) side to the positive (long) side of this battery.

Similarly,

$$\text{Loop 2:} \qquad -R_2 I_1 + (R_2 + R_4 + R_5 + R_6)I_2 - R_5 I_3 = V_2 \qquad (14.5.6)$$

where $-R_2 I_1$ is due to the flow I_1 through branch AD with a negative drop as I_1 is flowing through R_2 in the opposite direction from I_2 flowing into R_2 for loop 2.

$$\text{Loop 3:} \qquad -R_5 I_2 + (R_5 + R_7 + R_8)I_3 = -V_2 - V_3 \qquad (14.5.7)$$

Note that the battery voltages V_2 and V_3 are now negative in loop 3 as I_3 passes through the positive (long) side to the negative (short) side of both V_2 and V_3.

Equations (14.5.5) through (14.5.7) could now be expressed in matrix form as $[K]\{I\} = \{V\}$. We will leave this exercise to your discretion and instead illustrate how to use the direct stiffness method to solve this electrical network with numerical values in Example 14.7.

Example 14.7

For the three-loop electrical network shown in Figure 14–22, determine the currents in each loop. The resistances from each resistor and the voltages provided by each battery are shown in the figure. Use the stiffness matrix for the resistors and the direct stiffness method.

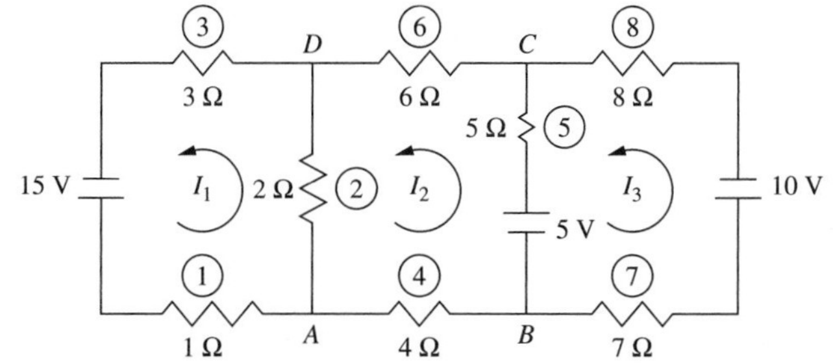

Figure 14–22 Three-loop electrical network

SOLUTION:

The resistor-element stiffness matrices are given by Eq. (14.5.3) as follows:

$$
[k^{(1)}] = \begin{bmatrix} \overset{I_1}{1} & \overset{0}{-1} \\ -1 & 1 \end{bmatrix} \quad
[k^{(2)}] = \begin{bmatrix} \overset{I_1}{2} & \overset{I_2}{-2} \\ -2 & 2 \end{bmatrix} \quad
[k^{(3)}] = \begin{bmatrix} \overset{I_1}{3} & \overset{0}{-3} \\ -3 & 3 \end{bmatrix} \quad
[k^{(4)}] = \begin{bmatrix} \overset{I_2}{4} & \overset{0}{-4} \\ -4 & 4 \end{bmatrix}
$$

$$
[k^{(5)}] = \begin{bmatrix} \overset{I_2}{5} & \overset{I_3}{-5} \\ -5 & 5 \end{bmatrix} \quad
[k^{(6)}] = \begin{bmatrix} \overset{I_2}{6} & \overset{0}{-6} \\ -6 & 6 \end{bmatrix} \quad
[k^{(7)}] = \begin{bmatrix} \overset{I_3}{7} & \overset{0}{-7} \\ -7 & 7 \end{bmatrix} \quad
[k^{(8)}] = \begin{bmatrix} \overset{I_3}{8} & \overset{0}{-8} \\ -8 & 8 \end{bmatrix}
$$

$$(14.5.8)$$

where the labels above the matrices indicate the currents from each loop going through that resistor. A zero means only one current going through that resistor.

Assembling the global equations using the direct stiffness method yields:

$$
\begin{bmatrix} 1+2+3 & -2 & 0 \\ -2 & 2+4+5+6 & -5 \\ 0 & -5 & 5+7+8 \end{bmatrix} \begin{Bmatrix} I_1 \\ I_2 \\ I_3 \end{Bmatrix} = \begin{Bmatrix} V_1 = 15 \\ V_2 = 5 \\ V_3 = -10-5 = -15 \end{Bmatrix} \quad (14.5.9)
$$

We should note that in assembling the equations, the stiffness matrices account for the currents going through each resistor. For instance, by Kirchhoff's law, element 1 only has current I_1 from loop 1 current passing through it, while element 2 has positive current I_1 going through it from the counterclockwise direction around loop 1 (upward current), and negative current I_2 going through it from the counterclockwise flow around loop 2 heading down through element 2, as shown in Figure 14–22. Also, the voltage in loop 1 is +15 V by convention, as the current through loop 1 passes down through the negative (short) side of the battery to the positive (long) side, as shown in Figure 14–22. Similarly, the voltage in loop 2 is considered positive in Eq. (14.5.9), as the current I_2 passes through the negative side of the battery to the positive side. Finally, the voltage in loop 3 is from both the 10 V and 5 V batteries. As the current in loop 3 goes from the positive side of the 10 V battery to the negative side, the voltage is considered -10 V through this battery. Similarly, the current I_3

goes through the 5 V battery through the positive side to the negative side, so this voltage is also considered -5 V. The total voltage for V_3 is then $V_3 = -10 - 5 = -15$ V.

Notice also that by convention we have chosen all loop currents to be positive in the counterclockwise direction resulting in all off-diagonal terms of the stiffness matrix to be negative.

Solving Eq. (14.5.9) simultaneously for the currents, we obtain

$$I_1 = 2.638 \text{ amps} \quad I_2 = 0.414 \text{ amps} \quad I_3 = -0.646 \text{ amps} \quad (14.5.10)$$

The negative sign on I_3 indicates that I_3 is really in a clockwise direction in Figure 14–22.

The branch current also can be obtained after determining the currents in each loop. If only one loop current passes through a branch, such as from A to B in Figure 14–22, the branch current I_2 equals the loop current. If more than one loop current passes through a branch, such as A to D or B to C, the branch current is the algebraic sum of the loop currents in the branch (based on Kirchoff's law). For example, the current in branch AD is $I_1 - I_2 = 2.638 - 0.414 = 2.224$ amps in the direction of I_1. Similarly, the current in branch BC is $I_2 - I_3 = 0.414 - (-0.646) = 1.060$ amps. ■

▲ 14.6 Electrostatics ▲

Electrostatics describes the forces between charged bodies at known positions. We will first present some basic laws associated with the concept of electrostatics. We will then describe the finite element method for solving electrostatics problems where electric potentials and electric fields are of concern. For more details on electrostatics consult [9–12].

Coulomb's Law

The charge on a body is denoted by symbol q. The units of q are Coulombs ($C = $ Amp–s). The charge of an electron is -1.60219×10^{-19} Coulombs. For two small stationary particles with charges q_1 and q_2, separated by distance r, as shown in Figure 14–23,

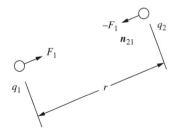

Figure 14–23 Illustration of Coulomb's law for two charged particles

the electric force vector $\mathbf{F}_1$ with magnitude F_1 on charged particle 1 from charged particle 2 is given by Coulomb's law as

$$\mathbf{F}_1 = \frac{q_1 q_2 \mathbf{n}_{21}}{4\pi\varepsilon_0 r^2} \ (N) \tag{14.6.1}$$

where $\mathbf{n}_{21}$ is the unit vector pointing from particle 2 to particle 1 and ε_0 is the permittivity of free space (electric constant) given by value 8.854×10^{-12} farads/m (or Amp–s/V–m) for force in Newtons, charge in Coulombs, and distance r in meters.

For an electrostatic field created by multiple charges, we extend Coulomb's law and use superposition that states the total force on a test or base charge q_0 located at (x_0, y_0, z_0) from several charges N (called a charge distribution) is the vector sum of individual forces. This force can be expressed as follows:

$$\mathbf{F}(x_0, y_0, z_0) = q_0 \sum_{i=1}^{N} \frac{q_i}{4\pi\varepsilon_0 r_i^2} \mathbf{n}_i \tag{14.6.2}$$

where r_i denotes the distance from q_0 to each individual charge q_i.

We now define the total electric field at position (x_0, y_0, z_0) due to a quantity of point charges N as force $F(x_0, y_0, z_0)$ per charge q_0 by dividing both sides of Eq. (14.6.2) by q_0 as follows:

$$\mathbf{E}(x_0, y_0, z_0) = \sum_{I} \frac{q_i}{4\pi\varepsilon_0 r_i^2} \mathbf{n}_i \ (\text{V/m}) \tag{14.6.3}$$

The electric field is expressed in units of volts per meter. Given the electric field for a charge distribution, the force on a test charge q at any location (x, y, z), can be expressed by

$$\mathbf{F}(x, y, z) = q\mathbf{E}(x, y, z) \tag{14.6.4}$$

Electric fields are often illustrated by plots of field lines as shown in Figure 14–24. These will also be shown by the computer program model results given by Example 14.9 and 10.

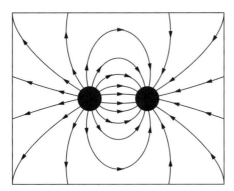

Figure 14–24 Electric field lines surrounding a positive and a negative charge

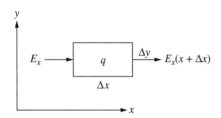

Figure 14–25 Differential control volume showing typical electric field lines entering and exiting the volume along sides x and $x + \Delta x$ and charge density within element

We will subsequently derive the finite element equations for two-dimensional electrostatic problems. The basis for the finite element equations is the Poisson equation which is derived from Gauss's Law. Therefore, we now derive Gauss's law in differential form.

Gauss's Law

Gauss's law relates the relationship between the distribution of electric charge in space, called the charge density ρ (in Coulomb's per cubic meter units (C/m^3)) and the resulting electric field. If the electric field from a point charge is spherically symmetric, Coulomb's law can be derived from Gauss's law.

Consider the three-dimensional differential elemental volume shown in Figure 14–25 with a charge density ρ acting over it where we define charge density as the sum of charge in a differential element divided by the element volume given by

$$\rho(x, y, z) \cong \frac{\sum\limits_{i} q_i}{\Delta x \Delta y \Delta z} \quad (C/m^3) \tag{14.6.5}$$

By Gauss's law, the total flux of electric field lines out of the surface in free space is equal to the total charge enclosed within the differential volume divided by ε_0. If there is no enclosed charge ($\rho = 0$), every field line that enters the volume must exit it.

First consider the two faces normal to the x axis, as shown in Figure 14–25, with areas Δy times Δz. The contribution of the flux out of the control volume surface is

$$-E_x \Delta y \Delta z + E_{x+\Delta x} \Delta y \Delta z \tag{14.6.6}$$

For small variations in E in the y and z directions and keeping the first two terms of the Taylor series expansion of the second term in Eq. (14.6.6), we have

$$E_{x+\Delta x} = E_x + \frac{\partial E_x}{\partial x} \Delta x \tag{14.6.7}$$

Substituting Eq. (14.6.7) into Eq. (14.6.6), yields

$$\frac{\partial E_x}{\partial x} \Delta x \Delta y \Delta z \qquad (14.6.8)$$

Contributions to the other faces normal to the y and z axes have similar forms to Eq. (14.6.8). Setting the contributions from the surface fluxes equal to the total charge within the control volume divided by ε_0, we obtain

$$\frac{\partial E_x}{\partial x} + \frac{\partial E_y}{\partial y} + \frac{\partial E_z}{\partial z} = \frac{\rho}{\varepsilon_0} \; (\text{V/m}^2) \qquad (14.6.9)$$

The left side of Equation (14.6.9) is called the *divergence of the electric field vector*. Therefore, the differential form of Gauss's law is

$$\nabla \cdot E = \frac{\rho}{\varepsilon_0} \qquad (14.6.10)$$

where ∇ is the del operator (nabla symbol) defined by

$$\nabla = \frac{\partial}{\partial x}\mathbf{i} + \frac{\partial}{\partial y}\mathbf{j} + \frac{\partial}{\partial z}\mathbf{k} \qquad (14.6.11)$$

and the dot represents the dot product.

Poisson's Equation

We can express the electric field as a vector E as

$$E = -\nabla\phi \, (\text{V/m}) \qquad (14.6.12)$$

where ϕ is the scalar electrostatic potential in units of volts (the negative sign is due to the fact that the E field is directed from positive to negative charges, while the potential increases in the opposite direction).

Substituting Eq. (14.6.12) into Eq. (14.6.10), we obtain

$$\nabla \cdot (\nabla\phi) = -\frac{\rho}{\varepsilon_0} \qquad (14.6.13)$$

or in explicit form, we write Eq. (14.6.13) as

$$\frac{\partial^2 \phi}{\partial x^2} + \frac{\partial^2 \phi}{\partial y^2} + \frac{\partial^2 \phi}{\partial z^2} = \frac{-\rho}{\varepsilon_0} \qquad (14.6.14)$$

Equation (14.6.14) is called the Poisson equation in cartesian coordinates.

Dielectric constants

For linear isotropic dielectric charges present, (isotropic dielectric constant meaning the dielectric constant is the same in all directions), the total electric field vector points in the same direction as the applied electric field, such that

$$E = E_0/\varepsilon_r$$

where ε_r is called the relative dielectric constant. Relative dielectric constant is also defined as the absolute electric constant of the material ε divided by the electric constant ε_0. Relative dielectric constants (also called relative permittivity or in electrostatics relative static permittivity) are used in solving electrostatics problems and are listed in textbooks on electrostatics [9–10]. Typical values for some materials are:

Air at standard temperature and pressure = 1.00058986,
Polyethylene = 2.25, paper = 3.5,
Silicon dioxide = 3.9, rubber = 7,
Graphite = 10–15, Water = 80 at 20°C.

Dielectric materials are used in transmission lines. In a coaxial cable, polyethylene is often used between the center conductor and outside shield. The relative static permittivity of a solvent is a relative measure of its polarity. For example, water which is a very polar material has a dielectric constant of 80 at 20°C as listed above and hexane used as a spot remover has dielectric constant of 2.

The generalized Poisson's equation, Eq. (14.6.13), with both free space charge and dielectric charge present in any medium is now given by (see [10–12])

$$\nabla \cdot (\varepsilon_r \nabla \phi) = -\frac{\rho}{\varepsilon_0} \tag{14.6.15}$$

Finite Element Formulation of a Two-Dimensional Triangle Element

We will now present the finite element formulation of the electrostatics problem.

The basic differential equation (Poisson's equation) governing the two-dimensional electrostatics problem in any isotropic medium based on Eq. (14.6.15) is given by

$$\frac{\partial}{\partial x}\left(\varepsilon \frac{\partial V}{\partial x}\right) + \frac{\partial}{\partial y}\left(\varepsilon \frac{\partial V}{\partial y}\right) = -\rho \tag{14.6.16}$$

where absolute permittivity $\varepsilon = \varepsilon_r \varepsilon_0$ and $V = \phi$ have been used in Eq. (14.6.16).

The steps defined in Chapter 6 will be followed to derive the stiffness matrix and equations for solving the electrostatics problem.

Step 1 Select Element Type

In the finite element method of solution, the basic triangle as described in Section 6.2 for stress analysis or Section 13.5 for heat transfer is the basis for two-dimensional finite element solutions, although the rectangular element (Section 6.6) and general quadrilateral element (Section 10.2) also can be used. For simplicity sake, we consider the so-called first order triangle with corner nodes only as shown in Figure 14–26. The potentials $v_i, v_j,$ and v_m at nodes $i, j,$ and m are analogous to the nodal temperatures of the heat transfer problem described in Chapter 13.

Step 2 Select a Potential Function

The potential function is described by the bilinear equation

$$V(x, y) = a_1 + a_2 x + a_3 y \tag{14.6.17}$$

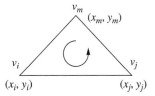

Figure 14–26 Basic triangular element with nodal potentials

Or in a manner as specifically shown in Section 6.2,

$$\{V\} = [N_i \quad N_j \quad N_m]\begin{Bmatrix} v_i \\ v_j \\ v_m \end{Bmatrix} = [N]\{v\} \tag{14.6.18}$$

where v_i, v_j, and v_m are the nodal voltages or potentials, and the shape functions are again given by Eqs. (6.2.18), that is,

$$N_i = \frac{1}{2A}(\alpha_i + \beta_i x + \gamma_i y) \tag{14.6.19}$$

With similar expressions for N_j and N_m. Here the α's, β's, and γ's are defined by Eqs. (6.2.10).

As in the heat transfer problem, only a single scalar value (nodal potential) is the primary unknown at each node, as shown by Eq. (14.6.18). Again, we then have a scalar-valued boundary value problem.

Step 3 Define the Potential Gradient/Potential and Electric Field/ Potential Gradient Relationships

We define the voltage or potential gradient matrix analogous to the temperature gradient material as

$$\{g\} = \begin{Bmatrix} \dfrac{\partial V}{\partial x} \\ \dfrac{\partial V}{\partial y} \end{Bmatrix} \tag{14.6.20}$$

Using Eq. (14.6.18) in Eq. (14.6.20), we have

$$\{g\} = \begin{bmatrix} \dfrac{\partial N_i}{\partial x} & \dfrac{\partial N_j}{\partial x} & \dfrac{\partial N_m}{\partial x} \\ \dfrac{\partial N_i}{\partial y} & \dfrac{\partial N_j}{\partial y} & \dfrac{\partial N_m}{\partial y} \end{bmatrix} \begin{Bmatrix} v_i \\ v_j \\ v_m \end{Bmatrix} \tag{14.6.21}$$

In compact matrix form, we express $\{g\}$ as

$$\{g\} = [B]\{v\} \tag{14.6.22}$$

where gradient matrix $[B]$ is again

$$[B] = \frac{1}{2A}\begin{bmatrix} \beta_i & \beta_j & \beta_m \\ \gamma_i & \gamma_j & \gamma_m \end{bmatrix} \tag{14.6.23}$$

The electric field vector is given by

$$E = -\text{grad } V = -\nabla V = -\mathbf{i}\frac{\partial V}{\partial x} - \mathbf{j}\frac{\partial V}{\partial y} \tag{14.6.24}$$

Using Eq. (14.6.18) with Eq. (14.6.19) in Eq. (14.6.24), the electric field vector can be expressed as

$$E = -\mathbf{i}\frac{1}{2A}(\beta_i v_i + \beta_j v_j + \beta_m v_m) - \mathbf{j}\frac{1}{2A}(\gamma_i v_i + \gamma_j v_j + \gamma_m v_m) \tag{14.6.25}$$

Using Eq. (14.6.20) and Eq. (14.6.22), we can express E in matrix form as

$$\begin{Bmatrix} E_x \\ E_y \end{Bmatrix} = -\{g\} = -[B]\{v\} \tag{14.6.26}$$

The electric displacement field vector for linear and isotropic media $\mathbf{D}$ in C/m^2 is related to the electric field vector $\mathbf{E}$ for linear and isotropic media by

$$\mathbf{D} = \varepsilon\,\mathbf{E} \text{ (C/m}^2 \text{ or A}-\text{s/m}^2) \tag{14.6.27}$$

Using Eq. (14.6.26) in Eq. (14.6.27), we express the electric field displacement/voltage gradient relationship analogous to the heat flux/temperature gradient relationship in Eq. (13.5.7) as

$$\begin{Bmatrix} D_{ex} \\ D_{ey} \end{Bmatrix} = -\begin{bmatrix} \varepsilon & 0 \\ 0 & \varepsilon \end{bmatrix}\begin{Bmatrix} E_x \\ E_y \end{Bmatrix} = -[D]\{g\} \tag{14.6.28}$$

Step 4 Derive the Element Stiffness Matrix and Equations

The element stiffness matrix is based on using the minimization of a functional similar to Eq. (13.4.10) with the functional called the electrostatic energy functional given by

$$\pi_e = U + \Omega \tag{14.6.29}$$

where the potential from internal energy stored in the electric field (field energy) over the volume V' of the element is given by

$$U = \iiint\limits_{V'} \frac{1}{2}\varepsilon E^2 dV' \tag{14.6.30}$$

and the potential energy of the charge density ρ (analogous to weight density as in Eq. (6.2.41) or an internal heat source as in Eq. (13.4.11) is given by

$$\Omega_\rho = -\iiint\limits_{V'} \rho V dV' \tag{14.6.31}$$

Noting that $E^2 = \mathbf{E} \cdot \mathbf{E}$ and using Eq. (14.6.24) for $\mathbf{E}$ in Eq. (14.6.30), we express U as

$$U = \iiint\limits_{V'} \frac{1}{2} \varepsilon \left[\left(\frac{\partial V}{\partial x} \right)^2 + \left(\frac{\partial V}{\partial y} \right)^2 \right] dV' \qquad (14.6.32)$$

We want to express π_e as a function of the nodal voltages so that we can minimize π_e with respect to these voltages to obtain the stiffness matrix and equations for the basic element. It is then most convenient to express the total energy in matrix form as a function of the nodal voltages, by using Eqs. (14.6.18), (14.6.19), (14.6.20), and (14.6.28) in Eqs. (14.6.31) and (14.6.32) as follows:

$$\pi_e = \frac{1}{2} \iiint\limits_{V'} \{g\}^T [D] \{g\} dV' - \iiint\limits_{V'} \{\rho\} \{v\}^T [N]^T dV' \qquad (14.6.33)$$

In general, for proper multiplication of column matrices $\{V\}$ and $\{\rho\}$ in Eq. {14.6.33}, (similar to Eq. (6.2.41) for body forces), the voltage function $\{V\}$ must have a transpose on it. Hence, by the property of matrix transpose multiplication as illustrated in Eq. (A.2.10), $\{V\}^T = \{v\}^T [N]^T$ has been used in Eq. (14.6.33).

Now substituting Eq. (14.6.22) for $\{g\}$ into Eq. (14.6.33), we obtain

$$\pi_e = \frac{1}{2} \{v\}^T \iiint\limits_{V'} \{B\}^T [D][B] dV' \{v\} - \{v\}^T \iiint\limits_{V'} [N]^T \{\rho\} dV' \qquad (14.6.34)$$

where the nodal voltages are independent of the general x-y coordinates, so $\{v\}$ has been taken out of the integrals.

Now minimizing the total energy with respect to the nodal voltage matrix, we obtain

$$\frac{\partial \pi_e}{\partial \{v\}} = \left[\iiint\limits_{V'} [B]^T [D][B] dV' \right] \{v\} - \iiint\limits_{V'} [N]^T \{\rho\} dV' \qquad (14.6.35)$$

The first integral multiplied by the nodal voltage matrix in Eq. (14.6.35) represents the stiffness matrix, while the second integral represents the source or force matrix due to the charge density.

The integrand in Eq. (14.6.35) is constant. Therefore, the stiffness matrix in simplified form becomes

$$[k] = tA[B]^T [D][B] \ (\text{C/V}) \qquad (14.6.36)$$

where t is the constant thickness of the element, A is the surface area of the element as determined by Eq. (6.2.9), $[B]$ is given by Eq. (14.6.23), and $[D]$ is given by Eq. (14.6.28). The specific form of Eq. (14.6.36) for the stiffness matrix of the three-noded triangle can be shown to be

$$[k] = \frac{\varepsilon}{4A} \begin{bmatrix} \beta_i \beta_i + \gamma_i \gamma_i & \beta_i \beta_j + \gamma_i \gamma_j & \beta_i \beta_m + \gamma_i \gamma_m \\ & \beta_j \beta_j + \gamma_j \gamma_j & \beta_j \beta_m + \gamma_j \gamma_m \\ & & \beta_m \beta_m + \gamma_m \gamma_m \end{bmatrix} t \ (\text{C/V}) \qquad (14.6.37)$$

where we must go counterclockwise around the element from the initial arbitrarily chosen node i to node j and then to node m.

For constant charge density acting uniformly over the element, the second integral in Eq. (14.6.35) is evaluated as described in Section 6.3 and given by Eq. (6.3.6) for body forces due to uniform weight density or for constant heat source by Eq. (13.5.16). This integral becomes

$$\{f_\rho\} = \begin{Bmatrix} f_{\rho_1} \\ f_{\rho_2} \\ f_{\rho_3} \end{Bmatrix} = \frac{tA\rho}{3} \begin{Bmatrix} 1 \\ 1 \\ 1 \end{Bmatrix} \text{(C)} \tag{14.6.38}$$

Therefore, one-third of the assumed uniform charge density within the element is applied to each node.

**Step 5 Assemble the Element Equations to obtain the Global
         Equations and Introduce Boundary Conditions**

We obtain the global stiffness matrix, source matrix, and equations by using the direct stiffness method as

$$[K] = \sum [k^{(e)}] \qquad \{F\} = \sum \{f^{(e)}\} \tag{14.6.39}$$

and

$$\{F\} = [K]\{v\} \tag{14.6.40}$$

where $\{v\}$ is the total system nodal voltage matrix.

The boundary conditions are of two types.

1. *Dirichlet* or imposed potentials on surface S_1. These are enforced by applying a known voltage at one or more nodes (analogous to applying a known displacement in the stress analysis problem), such as imposing a potential or voltage at node 1 (say $v_1 = 1$ V) as shown in Figure 14–27.
2. *Neumann* or derivative of voltage or potential is known on surface S_2. In this case, the electric field intensity, $\mathbf{E} = -\text{grad } V$ must be tangential to the surface S_2 as shown in Figure 14–28. (These boundary conditions do not need to be specified in the finite element applications.)

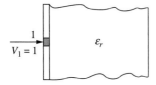

Figure 14–27 Dirichlet or imposed boundary potential or voltage at node 1

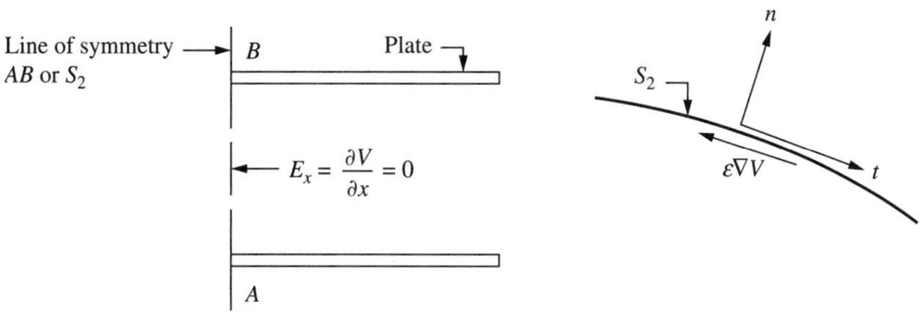

Line of symmetry — B Plate ⌐
AB or S_2

$E_x = \dfrac{\partial V}{\partial x} = 0$

n

S_2

$\varepsilon\nabla V$ t

A

Figure 14–28 Neumann-type derivative boundary conditions on electric field intensity (Line AB is a geometric and potential symmetry line)

Step 6   Solve for the Nodal Potentials or Voltages

We determine the unknown global potentials or voltages by solving the system of algebraic equations given by Eq. (14.6.40) with the boundary conditions invoked.

Step 7   Solve for the Electric Fields and Electric Displacement Fields

We determine the electric fields and electric displacement fields at each node or at element centroids using Eq. (14.6.26) and (14.6.28), respectively.

    We now present a simple hand solution of an electrostatics problem.

Example 14.8

For the two-element model shown in Figure 14–29, determine the nodal voltages at the right end. Node 1 has an applied voltage set to 10 V and node 2 is set to 0 V. The plate is one unit thick ($t = 1$ m) and the permittivity is $\varepsilon = 5$. There is no charge density.

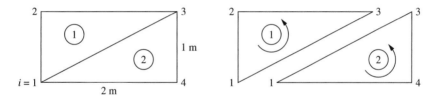

Figure 14–29   Plate subjected to nodal voltages and elements separated

SOLUTION:

Using Eq. (13.5.6) with β's and γ's defined by Eq. (6.2.10), we obtain the following.

For element 1;

$$
\begin{aligned}
\beta_i &= \beta_1 = y_3 - y_2 = 1 - 1 = 0, \\
\beta_j &= \beta_3 = y_2 - y_1 = 1 - 0 = 1, \\
\beta_m &= \beta_2 = y_1 - y_3 = 0 - 1 = -1 \\
\gamma_i &= \gamma_1 = x_2 - x_3 = 0 - 2 = -2, \\
\gamma_j &= \gamma_3 = x_1 - x_2 = 0 - 0 = 0, \\
\gamma_m &= \gamma_2 = x_3 - x_1 = 2 - 0 = 2
\end{aligned}
\tag{14.6.41}
$$

Similarly for element 2;

$$
\begin{array}{lll}
\beta_1 = y_4 - y_3 = 0 - 1 = -1, & \beta_4 = y_3 - y_1 = 1 - 0 = 1, & \beta_3 = y_1 - y_4 = 0 - 0 = 0 \\
\gamma_1 = x_3 - x_4 = 2 - 2 = 0, & \gamma_4 = x_1 - x_3 = 0 - 2 = -2, & \gamma_3 = x_4 - x_1 = 2 - 0 = 2
\end{array}
\tag{14.6.42}
$$

Using Eqs. (14.6.41) and (14.6.42) in Eq. (14.6.23) for $[B]$, Eq. (14.6.28) for $[D]$ and $A = 1 \text{ m}^2$ in Eq. (14.6.36), we obtain the stiffness matrices of the two elements as

$$
[k^{(1)}] = tA[B]^T[D][B] = 1(1)\begin{bmatrix} 0 & 0.5 & -0.5 \\ -1 & 0 & 1 \end{bmatrix}^T \begin{bmatrix} 5 & 0 \\ 0 & 5 \end{bmatrix} \begin{bmatrix} 0 & 0.5 & -0.5 \\ -1 & 0 & 1 \end{bmatrix}
$$

$$
= \begin{bmatrix} 5 & 0 & -5 \\ 0 & 1.25 & -1.25 \\ -5 & -1.25 & 6.25 \end{bmatrix} (\text{C/V})
\tag{14.6.43}
$$

Similarly,

$$
[k^{(2)}] = \begin{bmatrix} 1.25 & -1.25 & 0 \\ -1.25 & 6.25 & -5 \\ 0 & -5 & 5 \end{bmatrix} (\text{C/V})
\tag{14.6.44}
$$

Assembling the total stiffness matrix by the direct stiffness method, we obtain the global equations $\{F\} = [K]\{v\}$ as

$$
\begin{Bmatrix} 0 \\ 0 \\ 0 \\ 0 \end{Bmatrix} = \begin{bmatrix} 6.25 & -5 & 0 & -1.25 \\ -5 & 6.25 & -1.25 & 0 \\ 0 & -1.25 & 6.25 & -5 \\ -1.25 & 0 & -5 & 6.25 \end{bmatrix} \begin{Bmatrix} v_1 = 10 \\ v_2 = 0 \\ v_3 \\ v_4 \end{Bmatrix}
\tag{14.6.45}
$$

Applying the boundary conditions of nodal voltages of $v_1 = 10$, $v_2 = 0$ V, we solve equations 3 and 4 of Eq. (14.6.45) for the nodal voltages v_3 and v_4 as

$$
v_3 = 4.444 \text{ V}, \qquad v_4 = 5.556 \text{ V}
\tag{14.6.46}
$$

The electric field through elements 1 and 2 are determined using Eq. (14.6.26) as

$$\{E^{(1)}\} = -[B^{(1)}]\{v^{(1)}\} = -\begin{bmatrix} 0 & 0.5 & -0.5 \\ -1 & 0 & 1 \end{bmatrix}\begin{pmatrix} 10 \\ 4.444 \\ 0 \end{pmatrix} = \begin{bmatrix} -2.222 \\ 10 \end{bmatrix} \text{V/m}$$

(14.6.47)

and

$$\{E^{(2)}\} = -[B^{(2)}]\{v^{(2)}\} = -\begin{bmatrix} -0.5 & 0.5 & 0 \\ 0 & -1 & 1 \end{bmatrix}\begin{pmatrix} 10 \\ 5.556 \\ 4.444 \end{pmatrix} = \begin{bmatrix} 2.222 \\ 1.112 \end{bmatrix} \text{V/m} \quad (14.6.48)$$

■

We now present two computer model solutions of electrostatics problems.

Example 14.9

An infinitely long enclosed rectangular channel 1 m by 0.4 m is filled with air. The top is insulated from the sides and connected to a potential of 50 V. The sides and bottom are grounded (0 V). Assume the dielectric constant of air to be 1. Determine the voltage variation and the electric field through the channel.

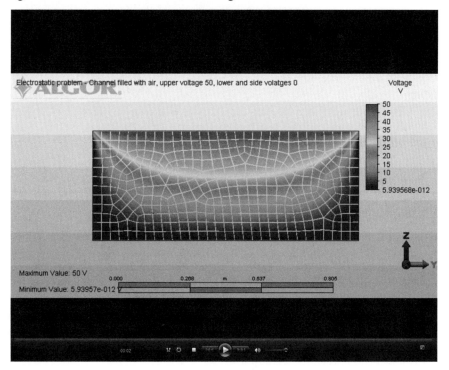

Figure 14–30 Voltage variation throughout channel (See the full-color insert for a color version of this figure.)

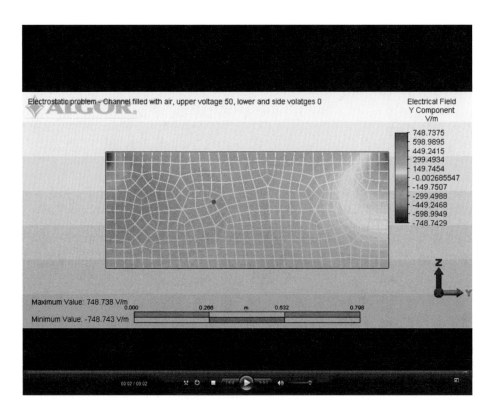

Figure 14–31 Electric field variation throughout channel

SOLUTION:

The finite element model of the channel is shown in Figure 14–30 with the voltage variation throughout the channel. Figure 14–31 shows the electric field variation in the y direction. ■

Example 14.10

A busbar is a rectangular conductor used in the distribution of electric power in a distribution box at 110 V. The sides of the busbar give off 110 V. In the system shown in Figure 14–32, the bottom side is assumed ground at 0 V. Assume the medium around the busbar is air ($\varepsilon = 1$). Also assume symmetry in geometry and potential with respect to a vertical line along the left side of the model. Therefore, the model is really one-half of the whole system. Determine the voltage distribution and maximum electric field intensity. Use a two-dimensional finite element model. Let the thickness be 0.1 m.

SOLUTION:

The results in Figures 14–32 and 14–33 show the voltage distribution through the air around the busbar and the electric field intensity magnitude distribution. As the

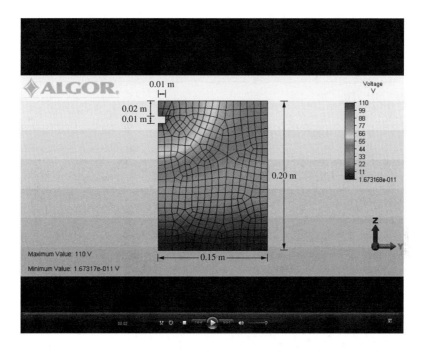

Figure 14–32 Busbar surrounded by air along with the finite element model and the resulting voltage distribution (See the full-color insert for a color version of this figure.)

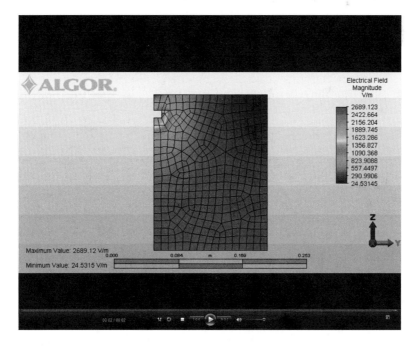

Figure 14–33 Electric field intensity magnitude

voltage has not died out to near zero along the right side of the model, more air should be used in the model. The maximum electric field intensity magnitude is 2689 V/m near the right edge of the busbar. ∎

▲ Summary Equations

Basic Equations for Fluid Flow through Porous Medium and through Pipes

Conservation of mass equation:

$$M_{\text{in}} + M_{\text{generated}} = M_{\text{out}} \tag{14.1.1}$$

or

$$\rho v_x A \, dt + \rho Q \, dt = \rho v_{x+dx} A \, dt \tag{14.1.2}$$

Darcy's law:

$$v_x = -K_{xx}\frac{d\phi}{dx} = -K_{xx} g_x \tag{14.1.3}$$

Basic differential equation for one-dimensional fluid flow through porous medium:

$$\frac{d}{dx}\left(K_{xx}\frac{d\phi}{dx}\right) + \bar{Q} = 0 \tag{14.1.6}$$

Basic differential equation for two-dimensional fluid flow through porous medium:

$$\frac{\partial}{\partial x}\left(K_{xx}\frac{\partial\phi}{\partial x}\right) + \frac{\partial}{\partial y}\left(K_{yy}\frac{\partial\phi}{\partial y}\right) + \bar{Q} = 0 \tag{14.1.10}$$

Velocity/velocity potential relations for flow through pipe and around solid bodies:

$$v_x = -\frac{\partial\phi}{\partial x} \qquad v_y = -\frac{\partial\phi}{\partial y} \tag{14.1.13}$$

Basic differential equation for fluid flow in pipes and around solid bodies:

$$\frac{\partial^2\phi}{\partial x^2} + \frac{\partial^2\phi}{\partial y^2} = 0 \tag{14.1.14}$$

One-Dimensional Fluid Flow Finite Element Equations

Potential function for finite element formulation of one-dimensional fluid flow:

$$\phi = N_1 p_1 + N_2 p_2 \tag{14.2.1}$$

Velocity/gradient relationship based on Darcy's law:

$$v_x = -[D]\{g\} \qquad (14.2.7)$$

Stiffness matrix for one-dimensional fluid flow through porous medium:

$$[k] = \frac{AK_{xx}}{L}\begin{bmatrix} 1 & -1 \\ -1 & 1 \end{bmatrix} \text{ m}^2/\text{s or in}^2/\text{s} \qquad (14.2.15)$$

Force matrix terms for one-dimensional fluid flow through porous medium:

Due to volumetric flow from internal source:

$$\{f_Q\} = \iiint\limits_V [N]^T Q\,dV = \frac{QAL}{2}\begin{Bmatrix} 1 \\ 1 \end{Bmatrix} \text{ m}^3/\text{s or in}^3/\text{s} \qquad (14.2.16)$$

Due to surface flow rate:

$$\{f_q\} = \iint\limits_{S_1} q^*[N]^T\,dS = \frac{q^*Lt}{2}\begin{Bmatrix} 1 \\ 1 \end{Bmatrix} \text{ m}^3/\text{s or in}^3/\text{s} \qquad (14.2.17)$$

Hydraulic Network Equations

Poiseuille's law for flow through hydraulic pipe network:

$$\Delta p = Rq \qquad (14.2.39)$$

Stiffness matrix for flow through pipe with resistance:

$$[k] = \frac{1}{R}\begin{bmatrix} 1 & -1 \\ -1 & 1 \end{bmatrix} \qquad (14.2.41)$$

Two-dimensional Equations for Finite Element Formulation of Fluid Flow

Potential function:

$$[\phi] = [N_i \quad N_j \quad N_m]\begin{Bmatrix} p_i \\ p_j \\ p_m \end{Bmatrix} \qquad (14.3.1)$$

Velocity/gradient matrix relationship:

$$\begin{Bmatrix} v_x \\ v_y \end{Bmatrix} = -[D]\{g\} \qquad (14.3.7)$$

Material property matrix:

$$[D] = \begin{bmatrix} K_{xx} & 0 \\ 0 & K_{yy} \end{bmatrix} \qquad (14.3.8)$$

Stiffness matrix:

$$[k] = \frac{tK_{xx}}{4A} \begin{bmatrix} \beta_i^2 & \beta_i\beta_j & \beta_i\beta_m \\ \beta_i\beta_j & \beta_j^2 & \beta_j\beta_m \\ \beta_i\beta_m & \beta_j\beta_m & \beta_m^2 \end{bmatrix} + \frac{tK_{yy}}{4A} \begin{bmatrix} \gamma_i^2 & \gamma_i\gamma_j & \gamma_i\gamma_m \\ \gamma_i\gamma_j & \gamma_j^2 & \gamma_j\gamma_m \\ \gamma_i\gamma_m & \gamma_j\gamma_m & \gamma_m^2 \end{bmatrix} \tag{14.3.11}$$

Force matrix for uniform volumetric flow rate over element:

$$\{f_Q\} = \frac{QV}{3} \begin{Bmatrix} 1 \\ 1 \\ 1 \end{Bmatrix} \frac{m^3}{s} \text{ or } \frac{in^3}{s} \tag{14.3.13}$$

Force matrix for uniform surface flow over side of element:

$$\{f_q\} = \frac{q^* L_{i\text{-}j} t}{2} \begin{Bmatrix} 1 \\ 1 \\ 0 \end{Bmatrix} \frac{m^3}{s} \text{ or } \frac{in^3}{s} \text{ on side } i\text{-}j \tag{14.3.15}$$

Force matrix for line or point fluid source:

$$\begin{Bmatrix} f_{Qi} \\ f_{Qj} \\ f_{Qm} \end{Bmatrix} = Q^* t \begin{Bmatrix} N_i \\ N_j \\ N_m \end{Bmatrix} \Bigg|_{\substack{x=x_0=5\,m \\ y=y_0=2\,m}} \tag{14.3.19}$$

See Figure 14–18 for flow chart for fluid flow computer program.

Electrical Networks

Ohm's law:

$$\Delta V = RI \tag{14.5.1}$$

Voltage difference/current flow matrix equation:

$$\begin{Bmatrix} V_1 \\ V_2 \end{Bmatrix} = R \begin{bmatrix} 1 & -1 \\ -1 & 1 \end{bmatrix} \begin{Bmatrix} I_1 \\ I_2 \end{Bmatrix} \tag{14.5.2}$$

Stiffness matrix for resistor element:

$$[k] = R \begin{bmatrix} 1 & -1 \\ -1 & 1 \end{bmatrix} \tag{14.5.3}$$

Electrostatics

Basic equations:

Coulomb's law:

$$\mathbf{F}_1 = \frac{q_1 q_2 \mathbf{n}_{21}}{4\pi\varepsilon_0 r^2} \tag{14.6.1}$$

Gauss's law:

$$\nabla \cdot \mathbf{E} = \frac{\rho}{\varepsilon_0} \qquad (14.6.10)$$

Poisson's equation:

$$\frac{\partial^2 \phi}{\partial x^2} + \frac{\partial^2 \phi}{\partial y^2} + \frac{\partial^2 \phi}{\partial z^2} = \frac{-\rho}{\varepsilon_0} \qquad (14.6.14)$$

Dielectric constants, ε, (See page 706).

Finite element equations for two-dimensional electrostatic analysis:
Basic differential equation:

$$\frac{\partial}{\partial x}\left(\varepsilon \frac{\partial V}{\partial x}\right) + \frac{\partial}{\partial y}\left(\varepsilon \frac{\partial V}{\partial y}\right) = -\rho \qquad (14.6.16)$$

Potential function:

$$V(x, y) = a_1 + a_2 x + a_3 y \qquad (14.6.17)$$

Gradient matrix:

$$\{g\} = \begin{Bmatrix} \dfrac{\partial V}{\partial x} \\[2mm] \dfrac{\partial V}{\partial y} \end{Bmatrix} \qquad (14.6.20)$$

Gradient/nodal voltage matrix equation:

$$\{g\} = [B]\{v\} \qquad (14.6.22)$$

where

$$[B] = \frac{1}{2A}\begin{bmatrix} \beta_i & \beta_j & \beta_m \\ \gamma_i & \gamma_j & \gamma_m \end{bmatrix} \qquad (14.6.23)$$

Electric field vector:

$$E = -\text{grad } V = -\nabla V = -\mathbf{i}\frac{\partial V}{\partial x} - \mathbf{j}\frac{\partial V}{\partial y} \qquad (14.6.24)$$

Electric field/nodal voltage matrix equation:

$$\begin{Bmatrix} E_x \\ E_y \end{Bmatrix} = -\{g\} = -[B]\{v\} \qquad (14.6.26)$$

Electric displacement field/voltage gradient matrix equation:

$$\begin{Bmatrix} D_{ex} \\ D_{ey} \end{Bmatrix} = -\begin{bmatrix} \varepsilon & 0 \\ 0 & \varepsilon \end{bmatrix}\begin{Bmatrix} E_x \\ E_y \end{Bmatrix} = -[D]\{g\} \qquad (14.6.28)$$

Electrostatic energy functional:

$$\pi_e = U + \Omega \qquad (14.6.29)$$

Electrostatic functional expressed as function of nodal voltages:

$$\pi_e = \frac{1}{2}\{v\}^T \iiint\limits_{V'} [B]^T[D][B]dV'\{v\} - \{v\}^T \iiint\limits_{V'} [N]^T\{\rho\}dV' \qquad (14.6.34)$$

Stiffness matrix:

$$[k] = \frac{\varepsilon}{4A}\begin{bmatrix} \beta_i\beta_i + \gamma_i\gamma_i & \beta_i\beta_j + \gamma_i\gamma_j & \beta_i\beta_m + \gamma_i\gamma_m \\ & \beta_j\beta_j + \gamma_j\gamma_j & \beta_j\beta_m + \gamma_j\gamma_m \\ & & \beta_m\beta_m + \gamma_m\gamma_m \end{bmatrix} t \ (\text{C/V}) \qquad (14.6.37)$$

Force matrix for uniform charge density within element:

$$\{f_\rho\} = \begin{Bmatrix} f_{\rho_1} \\ f_{\rho_2} \\ f_{\rho_3} \end{Bmatrix} = \frac{tA\rho}{3}\begin{Bmatrix} 1 \\ 1 \\ 1 \end{Bmatrix} \ (\text{C}) \qquad (14.6.38)$$

Global equations:

$$\{F\} = [K]\{v\} \qquad (14.6.40)$$

▲ References

[1] Chung, T. J., *Finite Element Analysis in Fluid Dynamics*, McGraw-Hill, New York, 1978.

[2] John, J. E. A., and Haberman, W. L., *Introduction to Fluid Mechanics*, Prentice-Hall, Englewood Cliffs, NJ, 1988.

[3] Harr, M. E., *Ground Water and Seepage*, McGraw-Hill, New York, 1962.

[4] *Heat Transfer Reference Division*, Algor, Inc., Pittsburgh, PA, 1999.

[5] Logan, D. L., *A First Course in the Finite Element Method*, 2nd ed., PWS-Kent Publishers, Boston, MA, 1992.

[6] *Fluid Flow Reference Division*, Algor, Inc., Pittsburgh, PA, 1999.

[7] Mohtar, R. H., Bralts, V. F., and Shayya, W. H., "*A Finite Element Model for the Analysis and Optimization of Pipe Networks*," Vol. 34(2), 1991, Transactions of ASAE, pp. 393–401.

[8] ANSYS *Engineering Analysis Systems User's Manual*, Swanson Analysis Systems. Inc., Johnson Rd., P.O. Box 65, Houston, PA, 15342.

[9] Jackson, J. D., *Classical Electrodynamics*, 3rd, ed., Wiley, NY, 1998.

[10] Ida, Nathan, *Engineering Electromagnetics*, 2nd ed., Springer-Verlag, NY, 2004.

[11] Bastos, J. P. A. and Sadowski, N., *Electromagnetic Modeling by Finite Element Methods*, Marcel Dekker, NY, 2003.

[12] Humphries, S. Jr., *Field Solutions on Computers*, CRC Press, NY, 1997.

▲ Problems

14.1 For the one-dimensional flow through the porous media shown in Figure P14–1, determine the potentials at one-third and two-thirds of the length. Also determine the velocities in each element. Let $A = 0.2 \text{ m}^2$.

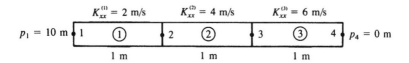

$K_{xx}^{(1)} = 2$ m/s $K_{xx}^{(2)} = 4$ m/s $K_{xx}^{(3)} = 6$ m/s

$p_1 = 10$ m 1   ①   2   ②   3   ③   4 $p_4 = 0$ m

1 m   1 m   1 m

Figure P14–1

14.2 For the one-dimensional flow through the porous medium shown in Figure P14–2 with fluid flux at the right end, determine the potentials at the third points. Also determine the velocities in each element. Let $A = 2$ m^2.

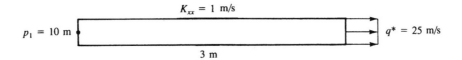

$K_{xx} = 1$ m/s

$p_1 = 10$ m

3 m

$q^* = 25$ m/s

Figure P14–2

14.3 For the one-dimensional fluid flow through the stepped porous medium shown in Figure P14–3, determine the potentials at the junction of each area. Also determine the velocities in each element. Let $K_{xx} = 2$ cm/s.

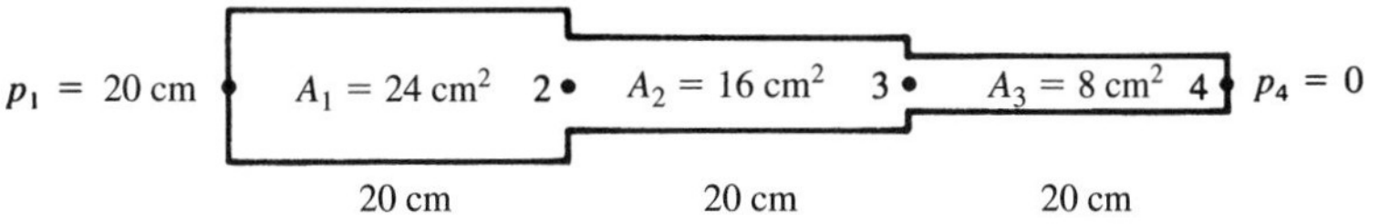

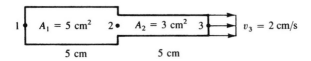

$p_1 = 20$ cm $A_1 = 24$ cm^2 2• $A_2 = 16$ cm^2 3• $A_3 = 8$ cm^2 4 $p_4 = 0$

20 cm   20 cm   20 cm

Figure P14–3

14.4 For the one-dimensional fluid-flow problem (Figure P14–4) with velocity known at the right end, determine the velocities and the volumetric flow rates at nodes 1 and 2. Let $K_{xx} = 2$ cm/s.

1 $A_1 = 5$ cm^2 2• $A_2 = 3$ cm^2 3 $v_3 = 2$ cm/s

5 cm   5 cm

Figure P14–4

14.5 Derive the stiffness matrix, Eq. (14.2.15), using the first term on the right side of Eq. (13.4.17).

14.6 For the one-dimensional fluid-flow problem in Figure P14–6, determine the velocities and volumetric flow rates at nodes 2 and 3. Let $K_{xx} = 0.2$ cm/s.

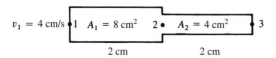

Figure P14–6

14.7 For the simple pipe networks shown in Figure P14–7, determine the pressures at nodes 1, 2, and 3 and the volumetric flow rates through the branches. Assume the pressure at node 4 is zero. In network (a), let $Q = 1$ m³/s. Let the resistances be, $R_1 = 1$, $R_2 = 2$, $R_3 = 3$, $R_4 = 4$ and $R_5 = 5$ all in units of N-s/m⁵. In network (b), let $Q = 1$ m³/s and the resistances be $R_1 = 10$, $R_2 = 20$, $R_3 = 30$, $R_4 = 40$, and $R_5 = 50$ all in units of N-s/m⁵

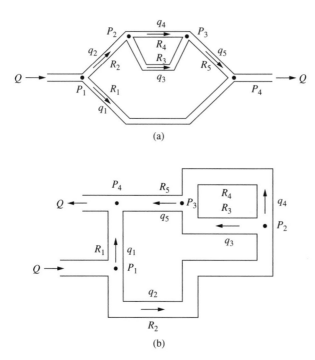

Figure P14–7

14.8 For the triangular element subjected to a fluid source shown in Figure P14–8, determine the amount of Q^* allocated to each node.

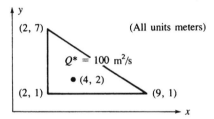

Figure P14–8

14.9 For the triangular element subjected to the surface fluid source shown in Figure P14–9, determine the amount of fluid force at each node.

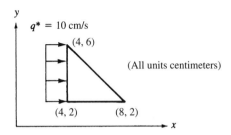

Figure P14–9

14.10 For the two-dimensional fluid flow shown in Figure P14–10, determine the potentials at the center and right edge.

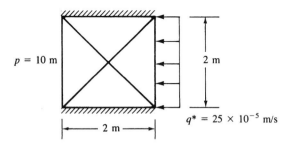

Figure P14–10

14.11– Using a computer program, determine the potential distribution in the two-dimensional
14.16 bodies shown in Figures P14–11–P14–16.

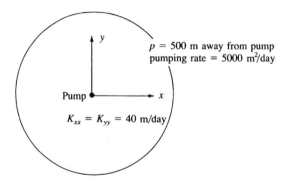

Figure P14–11

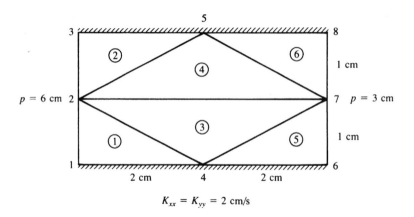

Figure P14–12

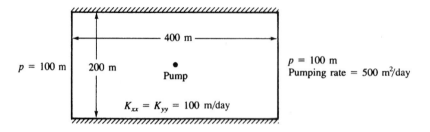

Figure P14–13

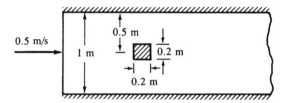

Figure P14–14

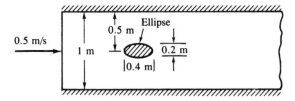

Figure P14–15

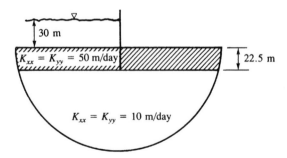

Figure P14–16

14.17–
14.18 For the direct current (dc) electrical networks shown in Figures P14–17 and P14–18, determine the currents through each loop and in branches AD and BC in Figure P14–17 or branches AB and BC in Figure P14–18.

14.19–
14.20 For the direct current (dc) networks consisting of batteries, resistors (shown by the rectangular shapes), and light emitting diodes (LED's) (shown by the triangular shapes) in Figures P14–19 and P14–20, determine the currents through each loop, in branches AD and BC in Figure P14–19, and in branch AD in Figure P14–20. What are the currents in each diode in these figures? If the desired current through the LED's is to be not greater than 0.015 amp, are the standard resistors acceptable?

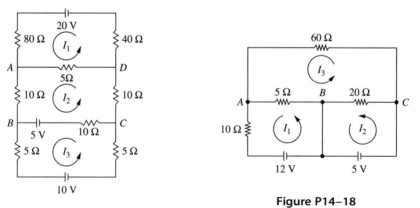

Figure P14–17

Figure P14–18

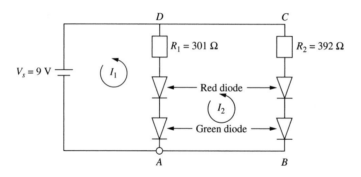

Figure P14–19

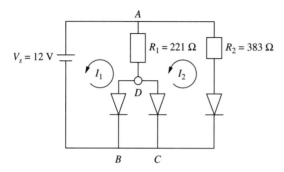

Figure P14–20

14.21   For the thin plate shown in Figure P14–21, determine the voltages at the right end
nodes. The voltages at nodes 1 and 2 are, respectively 20 and 0 V. Let the dielectric
constant of the material be that of silicon dioxide ($\varepsilon = 3.9$). Use a model of two elements as in Example 14.8. Assume a thickness of 0.01 m.

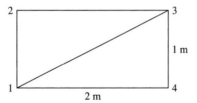

Figure P14–21

For the following problems, use a computer program that solves electrostatics
problems.

14.22   For the infinitely long air-enclosed channel shown in Figure P14–22, determine the
voltage variation through the air ($\varepsilon = 1$) and the largest electric field magnitude and
where it is located.

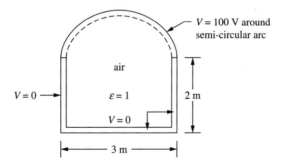

Figure P14–22

14.23   A busbar is a rectangular conductor used in distribution of electric power in a distribution box. The ground and busbar are considered perfect insulators. Assume the potential
of the busbar is 240 V. For the system shown in Figure P14–23, determine the voltage
distribution in the air ($\varepsilon = 1$) around the busbar and the maximum electric field intensity.

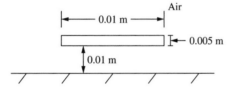

Figure P14–23

THERMAL STRESS ▲

CHAPTER OBJECTIVES

- To formulate the thermal stress problem.
- To derive the thermal force matrix for a one-dimensional bar.
- To derive the thermal force matrix for the three-noded triangle for both plane stress and plane strain.
- To solve examples of bars, trusses, and plane stress problems for thermal stresses due to temperature change.
- To show a finite element computer result for the thermal stress solution of a plate subject to temperature change.
- To demonstrate a finite element computer result for the thermal stress solution of a three-dimensional object subjected to temperature change.

Introduction

In this chapter, we consider the problem of thermal stresses within a body. First, we will discuss the strain energy due to thermal stresses (stresses resulting from the constrained motion of a body or part of a body during a temperature change in the body).

The minimization of the thermal strain energy equation is shown to result in the thermal force matrix. We will then develop this thermal force matrix for the one-dimensional bar element and the two-dimensional plane stress and plane strain elements.

We will outline the procedures for solving both one- and two-dimensional problems and then provide solutions of specific problems, including illustration of a computer program used to solve thermal stress problems for two- and three-dimensional stress problems.

▲ 15.1 Formulation of the Thermal Stress Problem and Examples ▲

In addition to the strains associated with the displacement functions due to mechanical loading, there may be other strains within a body due to temperature variations, swelling (moisture differential), or other causes. We will concern ourselves only with the strains

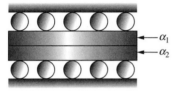

Figure 15–1 Composite member composed of two materials with different coefficients of thermal expansion

due to temperature variation, ε_T, and will consider both one- and two-dimensional problems.

Temperature changes in a structure can result in large stresses if not considered properly in design. In bridges, improper constraint of beams and slabs can result in large compressive stresses and resulting buckling failures due to temperature changes. In statically indeterminate trusses, members subjected to large temperature changes can result in stresses induced in members of the truss. Similarly, machine parts constrained from expanding or contracting may have large stresses induced in them due to temperature changes. Composite members made of two or more different materials may experience large stresses due to temperature change if they are not thermally compatible; that is, if the materials have large differences in their coefficients of thermal expansion, stresses may be induced even under free expansion (Figure 15–1).

When a member undergoes a temperature change the member attempts to change dimensions. For an unconstrained member AB (Figure 15–2) undergoing uniform change in temperature T, the change in the length L is given by

$$\delta_T = \alpha TL \qquad (15.1.1)$$

where α is called the *coefficient of thermal expansion* and T is the change in temperature. The coefficient α is a mechanical property of the material having units of $1/°C$ (where $°C$ is degrees Celsius). In Eq. (15.1.1), δ_T is considered to be positive when expansion occurs and negative when contraction occurs. Typical values of α are: for structural steel $\alpha = 12 \times 10^{-6}/°C$ and for aluminum alloys $\alpha = 23 \times 10^{-6}/°C$.

Based on the definition of normal strain, we can determine the strain due to a uniform temperature change. For the bar subjected to a uniform temperature change T

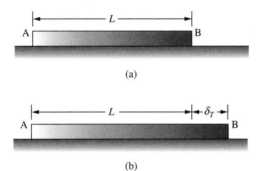

Figure 15–2 (a) Unconstrained member and (b) same member subjected to uniform temperature increase

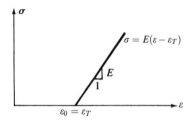

Figure 15–3 Linear stress-strain law with initial thermal strain

(Figure 15–2), the strain is the change in a dimension due to a temperature change divided by the original dimension. Considering the axial direction, we then have

$$\varepsilon_T = \alpha T \tag{15.1.2}$$

Since the bar in Figure 15–2 is free to expand, that is, it is not constrained by other members or supports, the bar will not have any stress in it. In general, for statically determinate structures, a uniform temperature change in one or more members does not result in stress in any of the members. That is, the structure will be stress free. For statically indeterminate structures, a uniform temperature change in one or more members of the structure usually results in stress σ_T in one or more members. We can have strain due to temperature change ε_T without stress due to temperature change, and we can have σ_T without any actual change in member lengths or without strains.

We will now consider the one-dimensional thermal stress problem. The linear stress-strain diagram with initial (thermal) strain ($\varepsilon_0 = \varepsilon_T$) is shown in Figure 15–3.

For the one-dimensional problem, we have, from Figure 15–3,

$$\varepsilon_x = \frac{\sigma_x}{E} + \varepsilon_T \tag{15.1.3}$$

If, in general, we let $1/E = [D]^{-1}$, then in general matrix form Eq. (15.1.3) can be written as

$$\{\varepsilon\} = [D]^{-1}\{\sigma\} + \{\varepsilon_T\} \tag{15.1.4}$$

From Eq. (15.1.4), we solve for $\{\sigma\}$ as

$$\{\sigma\} = D(\{\varepsilon\} - \{\varepsilon_T\}) \tag{15.1.5}$$

The strain energy per unit volume (called strain energy density) is the area under the σ–ε diagram in Figure 15–3 and is given by

$$u_0 = \frac{1}{2}\{\sigma\}(\{\varepsilon\} - \{\varepsilon_T\}) \tag{15.1.6}$$

Using Eq. (15.1.5) in Eq. (15.1.6), we have

$$u_0 = \frac{1}{2}(\{\varepsilon\} - \{\varepsilon_T\})^T [D](\{\varepsilon\} - \{\varepsilon_T\}) \tag{15.1.7}$$

where, in general, the transpose is needed on the strain matrix to multiply the matrices properly.

The total strain energy is then

$$U = \int_V u_0 \, dV \tag{15.1.8}$$

Substituting Eq. (15.1.7) into Eq. (15.1.8), we obtain

$$U = \int_V \frac{1}{2} (\{\varepsilon\} - \{\varepsilon_T\})^T [D] (\{\varepsilon\} - \{\varepsilon_T\}) \, dV \tag{15.1.9}$$

Now, using $\{\varepsilon\} = [B]\{d\}$ in Eq. (15.1.9), we obtain

$$U = \frac{1}{2} \int_V ([B]\{d\} - \{\varepsilon_T\})^T [D] ([B]\{d\} - \{\varepsilon_T\}) \, dV \tag{15.1.10}$$

Simplifying Eq. (15.1.10) yields

$$\begin{aligned} U = \frac{1}{2} \int_V (\{d\}^T [B]^T [D][B]\{d\} - \{d\}^T [B]^T [D]\{\varepsilon_T\} \\ - \{\varepsilon_T\}^T [D][B]\{d\} + \{\varepsilon_T\}^T [D]\{\varepsilon_T\}) \, dV \end{aligned} \tag{15.1.11}$$

The first term in Eq. (15.1.11) is the usual strain energy due to stress produced from mechanical loading—that is,

$$U_L = \frac{1}{2} \int_V \{d\}^T [B]^T [D][B]\{d\} \, dV \tag{15.1.12}$$

Terms 2 and 3 in Eq. (15.1.11) are identical and can be written together as

$$U_T = \int_V \{d\}^T [B]^T [D]\{\varepsilon_T\} \, dV \tag{15.1.13}$$

The last (fourth) term in Eq. (15.1.11) is a constant and drops out when we apply the principle of minimum potential energy by setting

$$\frac{\partial U}{\partial \{d\}} = 0 \tag{15.1.14}$$

Therefore, letting $U = U_L + U_T$ and substituting Eqs. (15.1.12) and (15.1.13) into Eq. (15.1.14), we obtain two contributions as

$$\frac{\partial U_L}{\partial \{d\}} = \int_V [B]^T [D][B] \, dV \{d\} \tag{15.1.15}$$

and
$$\frac{\partial U_T}{\partial \{d\}} = \int_V [B]^T [D]\{\varepsilon_T\} \, dV = \{f_T\} \tag{15.1.16}$$

We recognize the integral term in Eq. (15.1.15) that multiplies by the displacement matrix $\{d\}$ as the general form of the element stiffness matrix $[k]$, whereas Eq. (15.1.16) is the load or force vector due to temperature change in the element.

One-Dimensional Bar

We will now consider the one-dimensional thermal stress problem. We define the **thermal strain matrix** for the one-dimensional bar made of isotropic material with coefficient of thermal expansion α, and subjected to a uniform temperature rise T, as

$$\{\varepsilon_T\} = \{\varepsilon_{xT}\} = \{\alpha T\} \qquad (15.1.17)$$

where the units on α are typically (in./in.)/°F or (mm/mm)/°C.

For the simple one-dimensional bar (with a node at each end), we substitute Eq. (15.1.17) into Eq. (15.1.16) to obtain the thermal force matrix as

$$\{f_T\} = A \int_0^L [B]^T [D] \{\alpha T\}\, dx \qquad (15.1.18)$$

Recall that for the one-dimensional case, from Eqs. (3.10.15) and (3.10.13), we have

$$[D] = [E] \qquad [B] = \left[-\frac{1}{L} \quad \frac{1}{L} \right] \qquad (15.1.19)$$

Substituting Eqs. (15.1.19) into Eq. (15.1.18) and simplifying, we obtain the thermal force matrix as

$$\{f_T\} = \begin{Bmatrix} f_{T1} \\ f_{T2} \end{Bmatrix} = \begin{Bmatrix} -E\alpha T A \\ E\alpha T A \end{Bmatrix} \qquad (15.1.20)$$

Two-Dimensional Plane Stress and Plane Strain

For the two-dimensional thermal stress problem, there will be two normal strains, ε_{xT} and ε_{yT} along with a shear strain γ_{xyT} due to the change in temperature because of the different mechanical properties (such as $E_x \neq E_y$) in the x and y directions for the anisotropic material (see Figure 15–4). The thermal strain matrix for an anisotropic material is then

$$\{\varepsilon_T\} = \begin{Bmatrix} \varepsilon_{xT} \\ \varepsilon_{yT} \\ \gamma_{xyT} \end{Bmatrix} \qquad (15.1.21)$$

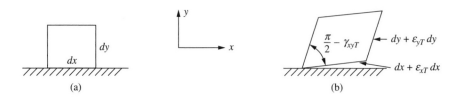

Figure 15–4 Differential two-dimensional element (a) before and (b) after being subjected to uniform temperature change for an anisotropic material

For the case of *plane stress* in an isotropic material ($E_x = E_y$) with coefficient of thermal expansion α subjected to a temperature rise T, the thermal strain matrix is

$$\{\varepsilon_T\} = \begin{Bmatrix} \alpha T \\ \alpha T \\ 0 \end{Bmatrix} \tag{15.1.22}$$

No shear strains are caused by a change in temperature of isotropic materials, only expansion or contraction.

For the case of *plane strain* in an isotropic material, the thermal strain matrix is

$$\{\varepsilon_T\} = (1 + v) \begin{Bmatrix} \alpha T \\ \alpha T \\ 0 \end{Bmatrix} \tag{15.1.23}$$

For a constant-thickness (t), constant-strain triangular element, Eq. (15.1.14) can be simplified to

$$\{f_T\} = [B]^T [D] \{\varepsilon_T\} tA \tag{15.1.24}$$

The forces in Eq. (15.1.24) are contributed to the nodes of an element in an unequal manner and require precise evaluation. It can be shown that substituting Eq. (6.1.8) for $[D]$, Eq. (6.2.34) for $[B]$, and Eq. (15.1.22) for $\{\varepsilon_T\}$ for a plane stress condition into Eq. (15.1.24) reveals the constant-strain triangular element thermal force matrix to be

$$\{f_T\} = \begin{Bmatrix} f_{Tix} \\ f_{Tiy} \\ \vdots \\ f_{Tmy} \end{Bmatrix} = \frac{\alpha E t T}{2(1 - v)} \begin{Bmatrix} \beta_i \\ \gamma_i \\ \beta_j \\ \gamma_j \\ \beta_m \\ \gamma_m \end{Bmatrix} \tag{15.1.25}$$

where the β's and γ's are defined by Eqs. (6.2.10).

Axisymmetric Element

For the case of an *axisymmetric triangular element* of isotropic material subjected to uniform temperature change, the thermal strain matrix is

$$\{\varepsilon_T\} = \begin{Bmatrix} \varepsilon_{rT} \\ \varepsilon_{zT} \\ \varepsilon_{\theta T} \\ \gamma_{rzT} \end{Bmatrix} = \begin{Bmatrix} \alpha T \\ \alpha T \\ \alpha T \\ 0 \end{Bmatrix} \tag{15.1.26}$$

The thermal force matrix for the three-noded triangular element is obtained by substituting the $[B]$ from Eq. (9.1.19) and Eq. (9.1.21) into the following:

$$\{f_T\} = 2\pi \int_A \{\varepsilon\}^T [D] \{\varepsilon_T\} r \, dA \tag{15.1.27}$$

For the element stiffness matrix evaluated at the centroid $(\bar{r}, \bar{z})$, Eq. (15.1.27) becomes

$$\{f_T\} = 2\pi\bar{r}A[\bar{B}]^T[D]\{\varepsilon_T\} \tag{15.1.28}$$

where $[\bar{B}]$ is given by Eq. (9.2.3), A is the surface area of the element which can be found in general from Eq. (6.2.8) when the coordinates of the element are known and $[D]$ is given by Eq. (9.2.6).

We will now describe the solution procedure for both one- and two-dimensional thermal stress problems.

Step 1

Evaluate the thermal force matrix, such as Eq. (15.1.20) or Eq. (15.1.25). Then treat this force matrix as an equivalent (or initial) force matrix $\{F_0\}$ analogous to that obtained when we replace a distributed load acting on an element by equivalent nodal forces (Chapters 4 and 5 and Appendix D).

Step 2

Apply $\{F\} = [K]\{d\} - \{F_0\}$, where if only thermal loading is considered, we solve $\{F_0\} = [K]\{d\}$ for the nodal displacements. Recall that when we formulate the set of simultaneous equations, $\{F\}$ represents the applied nodal forces, which here are assumed to be zero.

Step 3

Back-substitute the now known $\{d\}$ into step 2 to obtain the actual nodal forces, $\{F\}(=[K]\{d\} - \{F_0\})$.

Hence, the thermal stress problem is solved in a manner similar to the distributed load problem discussed for beams and frames in Chapters 4 and 5. We will now solve the following examples to illustrate the general procedure.

Example 15.1

For the one-dimensional bar fixed at both ends and subjected to a uniform temperature rise $T = 30\,°C$ as shown in Figure 15–5, determine the reactions at the fixed ends and the axial stress in the bar. Let $E = 200$ GPa, $A = 24$ cm^2, $L = 1.2$ m, and $\alpha = 12.5 \times 10^{-6}$ (mm/mm)/$°C$.

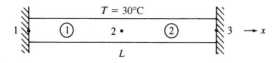

Figure 15–5 Bar subjected to a uniform temperature rise

SOLUTION:

Two elements will be sufficient to represent the bar because internal nodal displacements are not of importance here. To solve $\{F_0\} = [K]\{d\}$, we must determine the global stiffness matrix for the bar. Hence, for each element, we have

$$[k^{(1)}] = \frac{AE}{L/2} \begin{matrix} 1 \\ \\ \end{matrix} \overset{\displaystyle 1 \quad 2}{\begin{bmatrix} 1 & -1 \\ -1 & 1 \end{bmatrix}} \frac{\text{kN}}{\text{m}} \qquad [k^{(2)}] = \frac{AE}{L/2} \overset{\displaystyle 2 \quad 3}{\begin{bmatrix} 1 & -1 \\ -1 & 1 \end{bmatrix}} \frac{\text{kN}}{\text{m}} \qquad (15.1.29)$$

where the numbers above the columns in the $[k]$'s indicate the nodal displacements associated with each element.

Step 1

Using Eq. (15.1.20), the thermal force matrix for each element is given by

$$\{f^{(1)}\} = \begin{Bmatrix} -E\alpha TA \\ E\alpha TA \end{Bmatrix} \qquad \{f^{(2)}\} = \begin{Bmatrix} -E\alpha TA \\ E\alpha TA \end{Bmatrix} \qquad (15.1.30)$$

where these forces are considered to be equivalent nodal forces.

Step 2

Applying the direct stiffness method to Eqs. (15.1.29) and (15.1.30), we assemble the global equations $\{F_0\} = [K]\{d\}$ as

$$\begin{Bmatrix} -E\alpha TA \\ 0 \\ E\alpha TA \end{Bmatrix} = \frac{AE}{L/2} \begin{bmatrix} 1 & -1 & 0 \\ -1 & 1+1 & -1 \\ 0 & -1 & 1 \end{bmatrix} \begin{Bmatrix} u_1 \\ u_2 \\ u_3 \end{Bmatrix} \qquad (15.1.31)$$

Applying the boundary conditions $u_1 = 0$ and $u_3 = 0$ and solving the second part of Eq. (15.1.31), we obtain

$$u_2 = 0 \qquad (15.1.32)$$

Step 3

Back-substituting Eq. (15.1.32) into the global equation $\{F\} = [K]\{d\} - \{F_0\}$ for the nodal forces, we obtain the actual nodal forces as

$$\begin{Bmatrix} F_{1x} \\ F_{2x} \\ F_{3x} \end{Bmatrix} = \begin{Bmatrix} 0 \\ 0 \\ 0 \end{Bmatrix} - \begin{Bmatrix} -E\alpha TA \\ 0 \\ E\alpha TA \end{Bmatrix} = \begin{Bmatrix} E\alpha TA \\ 0 \\ -E\alpha TA \end{Bmatrix} \qquad (15.1.33)$$

Using the numerical quantities for E, α, T, and A in Eq. (15.1.33), we obtain

$$F_{1x} = 180 \text{ kN} \qquad F_{2x} = 0 \qquad F_{3x} = -180 \text{ kN}$$

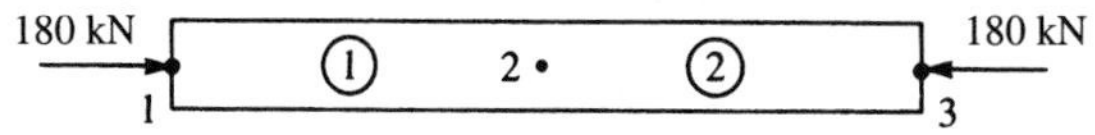

180 kN →   ← 180 kN
① 2 • ②
1   3

Figure 15–6 Free-body diagram of the bar of Figure 15–5

as shown in Figure 15–6. The stress in the bar is then

$$\sigma = \frac{180 \text{ kN}}{24 \times 10^{-4} \text{ m}^2} = 75,000 \text{ kPa} = 75 \text{ MPa} \quad \text{(compressive)} \quad (15.1.34) \quad \blacksquare$$

Example 15.2

For the bar assemblage shown in Figure 15–7, determine the reactions at the fixed ends and the axial stress in each bar. Bar 1 is subjected to a temperature drop of 10 °C. Let bar 1 be aluminum with $E = 70$ GPa, $\alpha = 23 \times 10^{-6}$ (mm/mm)/°C, $A = 12 \times 10^{-4}$ m², and $L = 2$ m. Let bars 2 and 3 be brass with $E = 100$ GPa, $\alpha = 20 \times 10^{-6}$ (mm/mm)/°C, $A = 6 \times 10^{-4}$ m², and $L = 2$ m.

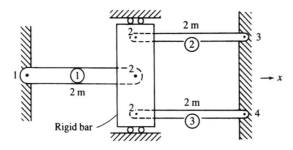

Figure 15–7 Bar assemblage for thermal stress analysis

SOLUTION:

We begin the solution by determining the stiffness matrices for each element.

Element 1

$$[k^{(1)}] = \frac{(12 \times 10^{-4})(70 \times 10^{6})}{2} \begin{bmatrix} 1 & -1 \\ -1 & 1 \end{bmatrix} = 42,000 \begin{matrix} 1 & 2 \\ \begin{bmatrix} 1 & -1 \\ -1 & 1 \end{bmatrix} \end{matrix} \frac{\text{kN}}{\text{m}} \quad (15.1.35)$$

Elements 2 and 3

$$[k^{(2)}] = [k^{(3)}] = \frac{(6 \times 10^{-4})(100 \times 10^{6})}{2} \begin{bmatrix} 1 & -1 \\ -1 & 1 \end{bmatrix} = 30,000 \begin{matrix} 2 & 3 \\ 2 & 4 \\ \begin{bmatrix} 1 & -1 \\ -1 & 1 \end{bmatrix} \end{matrix} \frac{\text{kN}}{\text{m}}$$

$$(15.1.36)$$

Step 1

We obtain the element thermal force matrices by evaluating Eq. (15.1.20). First, evaluating $-E\alpha TA$ for element 1, we have

$$-E\alpha TA = -(70 \times 10^6)(23 \times 10^{-6})(-10)(12 \times 10^{-4}) = 19.32 \text{ kN} \quad (15.1.37)$$

where the -10 term in Eq. (15.1.37) is due to the temperature drop in element 1. Using the result of Eq. (15.1.37) in Eq. (15.1.20), we obtain

$$\{f^{(1)}\} = \begin{Bmatrix} f_{1x} \\ f_{2x} \end{Bmatrix} = \begin{Bmatrix} 19.32 \\ -19.32 \end{Bmatrix} \text{ kN} \quad (15.1.38)$$

There is no temperature change in elements 2 and 3, and so

$$\{f^{(2)}\} = \{f^{(3)}\} = \begin{Bmatrix} 0 \\ 0 \end{Bmatrix} \quad (15.1.39)$$

Step 2

Assembling the global equations using Eqs. (15.1.35), (15.1.36), (15.1.38), and (15.1.39) into $\{F_0\} = [K]\{d\}$, we obtain

$$1000 \begin{array}{cccc} \quad 1 & \quad 2 & \quad 3 & \quad 4 \\ \begin{bmatrix} 42 & -42 & 0 & 0 \\ -42 & 42+30+30 & -30 & -30 \\ 0 & -30 & 30 & 0 \\ 0 & -30 & 0 & 30 \end{bmatrix} \end{array} \begin{Bmatrix} u_1 \\ u_2 \\ u_3 \\ u_4 \end{Bmatrix} = \begin{Bmatrix} +19.32 \\ -19.32 \\ 0 \\ 0 \end{Bmatrix} \quad (15.1.40)$$

where the right-side thermal forces are considered to be equivalent nodal forces. Using the boundary conditions

$$u_1 = 0 \qquad u_3 = 0 \qquad u_4 = 0 \quad (15.1.41)$$

we obtain, from the second equation of Eq. (15.1.40),

$$1000(102)u_2 = -19.32$$

Solving for u_2, we obtain

$$u_2 = -1.89 \times 10^{-4} \text{ m} \quad (15.1.42)$$

Step 3

Back-substituting Eq. (15.1.42) into the global equation for the nodal forces, $\{F\} = [K]\{d\} - \{F_0\}$, we have

$$\begin{Bmatrix} F_{1x} \\ F_{2x} \\ F_{3x} \\ F_{4x} \end{Bmatrix} = 1000 \begin{bmatrix} 42 & -42 & 0 & 0 \\ -42 & 102 & -30 & -30 \\ 0 & -30 & 30 & 0 \\ 0 & -30 & 0 & 30 \end{bmatrix} \begin{Bmatrix} 0 \\ -1.89 \times 10^{-4} \\ 0 \\ 0 \end{Bmatrix} - \begin{Bmatrix} 19.32 \\ -19.32 \\ 0 \\ 0 \end{Bmatrix}$$

$$(15.1.43)$$

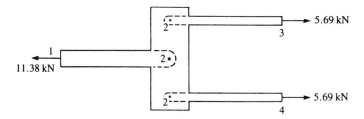

Figure 15–8 Free-body diagram of the bar assemblage of Figure 15–7

Simplifying Eq. (15.1.43), we obtain the actual nodal forces as

$$F_{1x} = -11.38 \text{ kN}$$

$$F_{2x} = 0.0 \text{ kN}$$

$$F_{3x} = 5.69 \text{ kN} \tag{15.1.44}$$

$$F_{4x} = 5.69 \text{ kN}$$

A free-body diagram of the bar assemblage is shown in Figure 15–8. The stresses in each bar are then tensile and given by

$$\sigma^{(1)} = \frac{11.38}{12 \times 10^{-4}} = 9.48 \times 10^3 \text{ kN/m}^2 \quad (9.48 \text{ MPa})$$

$$\sigma^{(2)} = \sigma^{(3)} = \frac{5.69}{6 \times 10^{-4}} = 9.48 \times 10^3 \text{ kN/m}^2 \quad (9.48 \text{ MPa}) \qquad \blacksquare \tag{15.1.45}$$

Example 15.3

For the plane truss shown in Figure 15–9, determine the displacements at node 1 and the axial stresses in each bar. Bar 1 is subjected to a temperature rise of 47.62 °C. Let $E = 210$ GPa, $\alpha = 72.5 \times 10^{-6}$ (mm/mm)/ °C, and $A = 12$ cm² for both bar elements.

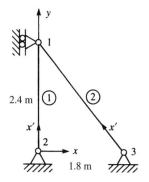

Figure 15–9 Plane truss for thermal stress analysis

SOLUTION:

First, using Eq. (3.4.23), we determine the stiffness matrices for each element.

Element 1

Choosing x' from node 2 to node 1, $\theta = 90°$, and so $\cos \theta = 0$, $\sin \theta = 1$, and

$$[k^{(1)}] = \frac{(12 \times 10^{-4})(210 \times 10^6)}{2.4} \begin{array}{cc} \begin{array}{cccc} \quad 2 & & 1 & \end{array} \\ \begin{bmatrix} 0 & 0 & 0 & 0 \\ & 1 & 0 & -1 \\ & & 0 & 0 \\ \text{Symmetry} & & & 1 \end{bmatrix} \begin{array}{c} \\ \frac{\text{kN}}{\text{m}} \\ \\ \end{array} \end{array}$$
(15.1.46)

Element 2

Choosing x' from node 3 to node 1, $\theta = 180° - 53.13° = 126.87°$, and so $\cos \theta = -0.6$, $\sin \theta = 0.8$, and

$$[k^{(2)}] = \frac{(12 \times 10^{-4})(210 \times 10^6)}{3.0} \begin{array}{cc} \begin{array}{cccc} \quad 3 & & 1 & \end{array} \\ \begin{bmatrix} 0.36 & -0.48 & -0.36 & 0.48 \\ & 0.64 & 0.48 & -0.64 \\ & & 0.36 & -0.48 \\ \text{Symmetry} & & & 0.64 \end{bmatrix} \begin{array}{c} \\ \frac{\text{kN}}{\text{m}} \\ \\ \end{array} \end{array}$$
(15.1.47)

Step 1

We obtain the element thermal force matrices by evaluating Eq. (15.1.20) as follows:

$$-E\alpha TA = -(210 \times 10^{-6})(12.5 \times 10^{-6}) = -150.0 \text{ kN}$$
(15.1.48)

Using the result of Eq. (15.1.48) for element 1, we then have the local thermal force matrix as

$$\{f'^{(1)}\} = \begin{Bmatrix} f'_{2x} \\ f'_{1x} \end{Bmatrix} = \begin{Bmatrix} -150 \\ 150 \end{Bmatrix} \text{ kN}$$
(15.1.49)

There is no temperature change in element 2, so

$$\{f'^{(2)}\} = \begin{Bmatrix} f'_{3x} \\ f'_{1x} \end{Bmatrix} = \begin{Bmatrix} 0 \\ 0 \end{Bmatrix}$$
(15.1.50)

Recall that by Eq. (3.4.16), $\{f'\} = [T]\{f\}$. Since we have shown that $[T]^{-1} = [T]^T$, we can obtain the global forces by premultiplying Eq. (3.4.16) by $[T]^T$ to obtain the element nodal forces in the global reference frame as

$$\{f\} = [T]^T\{f'\}$$
(15.1.51)

Using Eq. (15.1.51), the element 1 global nodal forces are then

$$
\begin{Bmatrix} f_{2x} \\ f_{2y} \\ f_{1x} \\ f_{1y} \end{Bmatrix} = \begin{bmatrix} C & -S & 0 & 0 \\ S & C & 0 & 0 \\ 0 & 0 & C & -S \\ 0 & 0 & S & C \end{bmatrix} \begin{Bmatrix} f'_{2x} \\ f'_{2y} \\ f'_{1x} \\ f'_{1y} \end{Bmatrix}
\tag{15.1.52}
$$

where the order of terms in Eq. (15.1.52) is due to the choice of the x' axis from node 2 to node 1 and where $[T]$, given by Eq. (3.4.15), has been used.

Substituting the numerical quantities $C = 0$ and $S = 1$ (consistent with x' for element 1), and $f'_{1x} = 150$, $f'_{1y} = 0$, $f'_{2x} = -150$, and $f'_{2y} = 0$ into Eq. (15.1.52), we obtain

$$
f_{2x} = 0 \qquad f_{2y} = -150 \text{ kN} \qquad f_{1x} = 0 \qquad f_{1y} = 150 \text{ kN} \tag{15.1.53}
$$

These element forces are now the only equivalent global nodal forces, because element 2 is not subjected to a change in temperature.

Step 2

Assembling the global equations using Eqs. (15.1.46), (15.1.47), and (15.1.53), into $\{F_0\} = [K]\{d\}$, we obtain

$$
84{,}000 \begin{bmatrix} 0.36 & -0.48 & 0 & 0 & 0 & 0 \\ & 1.89 & 0 & -1.25 & 0 & 0 \\ & & 0 & 0 & 0 & 0 \\ & & & 1.25 & 0 & 0 \\ & & & & 0.36 & -0.48 \\ & \text{Symmetry} & & & & 0.64 \end{bmatrix} \begin{Bmatrix} u_1 \\ v_1 \\ u_2 \\ v_2 \\ u_3 \\ v_3 \end{Bmatrix} = \begin{Bmatrix} 0 \\ 150 \\ 0 \\ -150 \\ 0 \\ 0 \end{Bmatrix} \tag{15.1.54}
$$

The boundary conditions are given by

$$
u_1 = 0 \qquad u_2 = 0 \qquad v_2 = 0 \qquad u_3 = 0 \qquad v_3 = 0 \tag{15.1.55}
$$

Using the boundary condition Eqs. (15.1.55) and the second equation of Eq. (15.1.54), we obtain

$$
(1.89 \times 84{,}000)v_1 = 150
$$

or

$$
v_1 = 9.45 \times 10^{-4} \text{ m} = 0.945 \text{ mm} \tag{15.1.56}
$$

Step 3

We now illustrate the procedure used to obtain the local element forces in local coordinates; that is, the local element forces are

$$
\{f'\} = [k']\{d'\} - \{f'_0\} \tag{15.1.57}
$$

We determine the actual local element nodal forces by using the relationship $\{d'\} = [T^*]\{d\}$ in Eq. (15.1.57) the usual bar element $[k']$ matrix [Eq. (3.1.14)], the transformation matrix $[T^*]$ [Eq. (3.4.8)], and the calculated displacements and initial thermal forces applicable for the element under consideration. Substituting the numerical quantities for element 1, from Eq. (15.1.57) into $\{f'\} = [k'][T^*]\{d\} - \{f'_0\}$, we have

$$\left\{ \begin{array}{c} f'_{2x} \\ f'_{1x} \end{array} \right\} = \frac{(12.0 \times 10^{-4})(210 \times 10^6)}{2.4} \begin{bmatrix} 1 & -1 \\ -1 & 1 \end{bmatrix} \begin{bmatrix} 0 & 1 & 0 & 0 \\ 0 & 0 & 0 & 1 \end{bmatrix} \left\{ \begin{array}{l} u_2 = 0 \\ v_2 = 0 \\ u_1 = 0 \\ v_1 = 9.45 \times 10^{-4} \end{array} \right\} - \left\{ \begin{array}{c} -150 \\ 150 \end{array} \right\}$$

$$(15.1.58)$$

Simplifying Eq. (15.1.58), we obtain

$$f'_{2x} = 50.8 \text{ kN} \qquad f'_{1x} = -50.8 \text{ kN} \qquad (15.1.59)$$

Dividing the local element force f'_{1x} (which is the far-end force consistent with the convention used in Section 3.5) by the cross-sectional area, we obtain the stress as

$$\sigma^{(1)} = \frac{-50.8 \text{ kN}}{(12.0 \times 10^{-4} \text{ m}^2)} = -42.3 \times 10^3 \text{ kN/m}^2 = 42.3 \text{ MPa } (C) \qquad (15.1.60)$$

Similarly, for element 2, we have

$$\left\{ \begin{array}{c} f'_{3x} \\ f'_{1x} \end{array} \right\} = \frac{(12.0 \times 10^{-4})(210 \times 10^6)}{3.0} \begin{bmatrix} 1 & -1 \\ -1 & 1 \end{bmatrix} \begin{bmatrix} -0.6 & 0.8 & 0 & 0 \\ 0 & 0 & -0.6 & 0.8 \end{bmatrix} \left\{ \begin{array}{l} 0 \\ 0 \\ 0 \\ 9.45 \times 10^{-4} \end{array} \right\}$$

$$(15.1.61)$$

Simplifying, Eq. (15.1.61), we obtain

$$f'_{3x} = -63.5 \text{ kN} \qquad f'_{1x} = 63.5 \text{ kN} \qquad (15.1.62)$$

where no initial thermal forces were present for element 2 because the element was not subjected to a temperature change. Dividing the far-end force f'_{1x} by the cross-sectional area results in

$$\sigma^{(2)} = 52.9 \text{ MPa } (T) \qquad (15.1.63)$$

For two- and three-dimensional stress problems, this direct division of force by cross-sectional area is not permissible. Hence, the total stress due to both applied loading and temperature change must be determined by

$$\{\sigma\} = \{\sigma_L\} - \{\sigma_T\} \qquad (15.1.64)$$

We now illustrate Eq. (15.1.64) for bar element 1 of the truss of Example 15.3. For the bar, σ_L can be obtained using Eq. (3.5.6) $\{\sigma_L\} = [C']\{d\}$, and σ_T is obtained from

$$\{\sigma_T\} = [D]\{\varepsilon_T\} = E\alpha T \tag{15.1.65}$$

because $[D] = E$ and $\{\varepsilon_T\} = \alpha T$ for the bar element. The stress in bar element 1 is then determined to be

$$\sigma^{(1)} = \frac{E}{L}[-C \quad -S \quad C \quad S]\begin{Bmatrix} u_2 \\ v_2 \\ u_1 \\ v_1 \end{Bmatrix} - E\alpha T \tag{15.1.66}$$

Substituting the numerical quantities for element 1 into Eq. (15.1.66), we obtain

$$\sigma^{(1)} = \frac{210 \times 10^6}{2.4}[0 \quad -1 \quad 0 \quad 1]\begin{Bmatrix} 0 \\ 0 \\ 0 \\ 9.45 \times 10^{-4} \end{Bmatrix} - (210 \times 10^6)(12.5 \times 10^{-6})(47.62)$$

$$\tag{15.1.67}$$

or

$$\sigma^{(1)} = 42.3 \times 10^3 \text{ kN/m}^2 = 42.3 \text{ MPa } (C) \tag{15.1.68} \quad ■$$

We will now illustrate the solutions of two plane thermal stress problems.

Example 15.4

For the plane stress element shown in Figure 15–10, determine the element equations. The element has a 15000 kN/m^2 pressure acting perpendicular to side j-m and is subjected to a 16 °C temperature rise.

SOLUTION:

Recall that the stiffness matrix is given by [Eq. (6.2.52) or (6.4.1)]

$$[k] = [B]^T[D][B]tA \tag{15.1.69}$$

and

$$\beta_i = y_j - y_m = -6 \text{ cm} = 0.06 \text{ m} \qquad \gamma_i = x_m - x_j = -2 \text{ cm} = 0.02 \text{ m}$$
$$\beta_j = y_m - y_i = 6 \text{ cm} = 0.06 \text{ m} \qquad \gamma_j = x_i - x_m = -2 \text{ cm} = 0.02 \text{ m}$$
$$\beta_m = y_i - y_j = 0 \qquad\qquad\quad \gamma_m = x_j - x_i = 4 \text{ cm} = 0.04 \text{ m}$$

and

$$A = \frac{(6)(4)}{2} = 12 \text{ cm}^2 = 12.0 \times 10^{-4} \text{ m}^2 \tag{15.1.70}$$

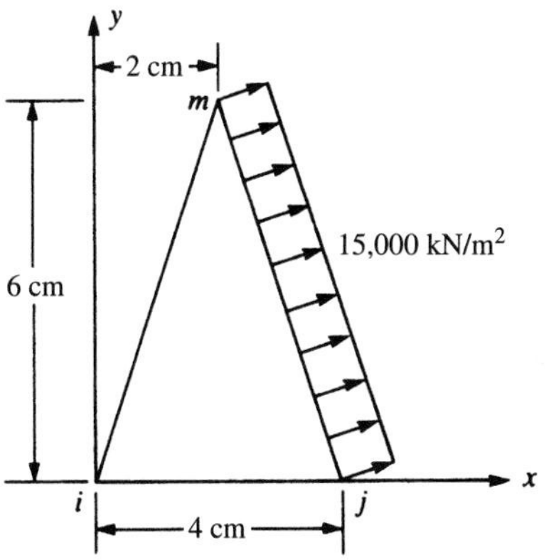

t = 2 cm
E = 210 GPa
α = 12.5 (mm/mm)/°C
ν = 0.25

Figure 15–10 Plane stress element subjected to mechanical loading and a temperature change

Therefore, substituting the results of Eqs. (15.1.70) into Eq. (6.2.34) for $[B]$, we obtain

$$[B] = \frac{10^4}{24}\begin{bmatrix} -6 & 0 & 6 & 0 & 0 & 0 \\ 0 & -2 & 0 & -2 & 0 & 4 \\ -2 & -6 & -2 & 6 & 4 & 0 \end{bmatrix} \times 10^{-2} \qquad (15.1.71)$$

Assuming plane stress conditions to be valid, we have

$$[D] = \frac{E}{1-\nu^2}\begin{bmatrix} 1 & \nu & 0 \\ \nu & 1 & 0 \\ 0 & 0 & \dfrac{1-\nu}{2} \end{bmatrix} = \frac{210 \times 10^6}{1-(0.25)^2}\begin{bmatrix} 1 & 0.25 & 0 \\ 0.25 & 1 & 0 \\ 0 & 0 & 0.375 \end{bmatrix}$$

$$= (28 \times 10^6)\begin{bmatrix} 8 & 2 & 0 \\ 2 & 8 & 0 \\ 0 & 0 & 3 \end{bmatrix} \frac{\text{kN}}{\text{m}^2} \qquad (15.1.72)$$

Also, $$[B]^T[D] = \frac{10^4}{24} \times (2) \times 10^{-2}\begin{bmatrix} -3 & 0 & -1 \\ 0 & -1 & -3 \\ 3 & 0 & -1 \\ 0 & -1 & 3 \\ 0 & 0 & 2 \\ 0 & 2 & 0 \end{bmatrix}(28 \times 10^6)\begin{bmatrix} 8 & 2 & 0 \\ 2 & 8 & 0 \\ 0 & 0 & 3 \end{bmatrix} \qquad (15.1.73)$$

Simplifying Eq. (15.1.73), we obtain

$$[B]^T[D] = \frac{(28)(2) \times 10^8}{24} \begin{bmatrix} -24 & -6 & -3 \\ -2 & -8 & -9 \\ 24 & 6 & -3 \\ -2 & -8 & 9 \\ 0 & 0 & 6 \\ 4 & 16 & 0 \end{bmatrix}$$

(15.1.74)

Therefore, substituting the results of Eqs. (15.1.71) and (15.1.74) into Eq. (15.1.69) yields the element stiffness matrix as

$$[k] = (1 \times 10^{-2} \text{ m})(12 \times 10^{-4} \text{ m}^2)\left(\frac{10^4}{24} \times 2 \times 10^{-2}\right)\left(\frac{56 \times 10^8}{24}\right)$$

$$\times \begin{bmatrix} -24 & -6 & -3 \\ -2 & -8 & -9 \\ 24 & 6 & -3 \\ -2 & -8 & 9 \\ 0 & 0 & 6 \\ 4 & 16 & 0 \end{bmatrix} \begin{bmatrix} -3 & 0 & 3 & 0 & 0 & 0 \\ 0 & -1 & 0 & -1 & 0 & 2 \\ -1 & -3 & -1 & 3 & 2 & 0 \end{bmatrix}$$

(15.1.75)

Simplifying Eq. (15.1.75), we have the element stiffness matrix as

$$[k] = 2.33 \times 10^4 \begin{bmatrix} 75 & 15 & -69 & -3 & -6 & -12 \\ 15 & 35 & 3 & -19 & -18 & -16 \\ -69 & 3 & 75 & -15 & -6 & 12 \\ -3 & -19 & -15 & 35 & 18 & -16 \\ -6 & -18 & -6 & 18 & 12 & 0 \\ -12 & -16 & 12 & -16 & 0 & 32 \end{bmatrix} \frac{\text{kN}}{\text{m}}$$

(15.1.76)

Using Eq. (15.1.25), the thermal force matrix is given by

$$\{f_T\} = \frac{\alpha E t T}{2(1-v)} \begin{Bmatrix} \beta_i \\ \gamma_i \\ \beta_j \\ \gamma_j \\ \beta_m \\ \gamma_m \end{Bmatrix} = \frac{(12.5 \times 10^{-6})(210 \times 10^6)(1 \times 10^{-2})(16)}{2(1-0.25)} \begin{Bmatrix} -3 \\ -1 \\ 3 \\ -1 \\ 0 \\ 2 \end{Bmatrix} 2 \times 10^{-2} = 5.6 \begin{Bmatrix} -3 \\ -1 \\ 3 \\ -1 \\ 0 \\ 2 \end{Bmatrix}$$

or

$$\{f_T\} = \begin{Bmatrix} -16.8 \\ -5.6 \\ 16.8 \\ -5.6 \\ 0 \\ 11.2 \end{Bmatrix} \text{ kN}$$

(15.1.77)

The force matrix due to the pressure applied alongside j-m is determined as follows:

$$L_{j\text{-}m} = [(4-2)^2 + (6-0)^2]^{1/2} = 6.326 \text{ cm}$$

$$p_x = p\cos\theta = 15,000\left(\frac{6}{6.326}\right) = 14,227 \text{ kN/m}^2 \qquad (15.1.78)$$

$$p_y = p\sin\theta = 15,000\left(\frac{1}{6.326}\right) = 4742 \text{ kN/m}^2$$

where θ is the angle measured from the x axis to the normal to surface j-m. Using Eq. (6.3.7) to evaluate the surface forces, we have

$$\{f_L\} = \iint_{S_{j\text{-}m}} [N_s]^T \left\{ \begin{matrix} p_x \\ p_y \end{matrix} \right\} dS$$

$$= \iint_{S_{j\text{-}m}} \begin{bmatrix} N_i & 0 \\ 0 & N_i \\ N_j & 0 \\ 0 & N_j \\ N_m & 0 \\ 0 & N_m \end{bmatrix}_{\substack{\text{evaluated} \\ \text{alongside } j\text{-}m}} \left\{ \begin{matrix} p_x \\ p_y \end{matrix} \right\} dS = \frac{tL_{j\text{-}m}}{2} \begin{bmatrix} 0 & 0 \\ 0 & 0 \\ 1 & 0 \\ 0 & 1 \\ 1 & 0 \\ 0 & 1 \end{bmatrix} \left\{ \begin{matrix} p_x \\ p_y \end{matrix} \right\} \qquad (15.1.79)$$

Evaluating Eq. (15.1.79), we obtain

$$\{f_L\} = \frac{(1 \times 10^{-2} \text{ m})(6.326 \times 10^{-2} \text{ m})}{2} \begin{bmatrix} 0 & 0 \\ 0 & 0 \\ 1 & 0 \\ 0 & 1 \\ 1 & 0 \\ 0 & 1 \end{bmatrix} \left\{ \begin{matrix} 14,227 \\ 4742 \end{matrix} \right\} = \left\{ \begin{matrix} 0 \\ 0 \\ 4.5 \\ 1.5 \\ 4.5 \\ 1.5 \end{matrix} \right\} \text{ kN} \qquad (15.1.80)$$

Using Eqs. (15.1.76), (15.1.77), and (15.1.80), we find that the complete set of element equations is

$$2.33 \times 10^4 \begin{bmatrix} 75 & 15 & -69 & -3 & -6 & -12 \\ & 35 & 3 & -19 & -18 & -16 \\ & & 75 & -15 & -6 & 12 \\ & & & 35 & 18 & -16 \\ & & & & 12 & 0 \\ \text{Symmetry} & & & & & 32 \end{bmatrix} \left\{ \begin{matrix} u_i \\ v_i \\ u_j \\ v_j \\ u_m \\ v_m \end{matrix} \right\} = \left\{ \begin{matrix} -16.8 \\ -5.6 \\ 21.3 \\ -4.8 \\ 4.5 \\ 12.7 \end{matrix} \right\} \qquad (15.1.81)$$

where the force matrix is $\{f_T\} + \{f_L\}$, obtained by adding Eqs. (15.1.77) and (15.1.80). ∎

Example 15.5

For the plane stress plate fixed along one edge and subjected to a uniform temperature rise of $50\,°C$ as shown in Figure 15–11, determine the nodal displacements and the stresses in each element. Let $E = 210$ GPa, $v = 0.30$, $t = 5$ mm, and $\alpha = 12 \times 10^{-6}$ (mm/mm)/$°C$.

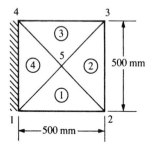

Figure 15–11 Discretized plate subjected to a temperature change

SOLUTION:

The discretized plate is shown in Figure 15–11. We begin by evaluating the stiffness matrix of each element using Eq. (6.2.52).

Element 1

Element 1 has coordinates $x_1 = 0$, $y_1 = 0$, $x_2 = 0.5$, $y_2 = 0$, $x_5 = 0.25$, and $y_5 = 0.25$. From Eqs. (6.2.10), we obtain

$$\beta_1 = y_2 - y_5 = -0.25 \text{ m} \qquad \beta_2 = y_5 - y_1 = 0.25 \text{ m} \qquad \beta_5 = y_1 - y_2 = 0$$

$$\gamma_1 = x_5 - x_2 = -0.25 \text{ m} \qquad \gamma_2 = x_1 - x_5 = -0.25 \text{ m} \qquad \gamma_5 = x_2 - x_1 = 0.5 \text{ m}$$

$$(15.1.82)$$

Using Eqs. (6.2.32) in Eq. (6.2.34), we have

$$[B] = \frac{1}{2A} \begin{bmatrix} \beta_1 & 0 & \beta_2 & 0 & \beta_5 & 0 \\ 0 & \gamma_1 & 0 & \gamma_2 & 0 & \gamma_5 \\ \gamma_1 & \beta_1 & \gamma_2 & \beta_2 & \gamma_5 & \beta_5 \end{bmatrix}$$

$$= \frac{1}{0.125} \begin{bmatrix} -0.25 & 0 & 0.25 & 0 & 0 & 0 \\ 0 & -0.25 & 0 & -0.25 & 0 & 0.5 \\ -0.25 & -0.25 & -0.25 & 0.25 & 0.5 & 0 \end{bmatrix} \frac{1}{\text{m}} \qquad (15.1.83)$$

For plane stress, $[D]$ is given by

$$[D] = \frac{E}{(1 - v^2)} \begin{bmatrix} 1 & v & 0 \\ v & 1 & 0 \\ 0 & 0 & \dfrac{1-v}{2} \end{bmatrix} = \frac{210 \times 10^9}{0.91} \begin{bmatrix} 1 & 0.3 & 0 \\ 0.3 & 1 & 0 \\ 0 & 0 & 0.35 \end{bmatrix} \frac{\text{N}}{\text{m}^2} \qquad (15.1.84)$$

We obtain the element stiffness matrix using

$$[k] = tA[B]^T[D][B] \tag{15.1.85}$$

Substituting the results of Eqs. (15.1.83) and (15.1.84) into Eq. (15.1.85) and carrying out the multiplications, we have

$$[k] = 4.615 \times 10^7 \begin{array}{c} \begin{array}{cccccc} u_1 \quad\;\; & v_1 \quad\;\; & u_2 \quad\;\; & v_2 \quad\;\; & u_5 \quad\;\; & v_5 \end{array} \\ \begin{bmatrix} 8.4375 & 4.0625 & -4.0625 & -0.3125 & -4.375 & -3.75 \\ 4.0625 & 8.4375 & 0.3125 & 4.0625 & -4.375 & -12.5 \\ -4.0625 & 0.3125 & 8.4375 & -4.0625 & -4.375 & 3.75 \\ -0.3125 & 4.0625 & -4.0625 & 8.4375 & 4.375 & -12.5 \\ -4.375 & -4.375 & -4.375 & 4.375 & 8.75 & 0 \\ -3.75 & -12.5 & 3.75 & -12.5 & 0 & 25 \end{bmatrix} \end{array} \dfrac{\text{N}}{\text{m}}$$

$$\tag{15.1.86}$$

Element 2

For element 2, the coordinates are $x_2 = 0.5$, $y_2 = 0$, $x_3 = 0.5$, $y_3 = 0.5$, $x_5 = 0.25$, and $y_5 = 0.25$. Proceeding as for element 1, we obtain

$$\beta_2 = 0.25 \text{ m} \qquad \beta_3 = 0.25 \text{ m} \qquad \beta_5 = -0.5 \text{ m}$$

$$\gamma_2 = -0.25 \text{ m} \qquad \gamma_3 = 0.25 \text{ m} \qquad \gamma_5 = 0$$

The element stiffness matrix then becomes

$$[k] = 4.615 \times 10^7 \begin{array}{c} \begin{array}{cccccc} u_2 \quad\;\; & v_2 \quad\;\; & u_3 \quad\;\; & v_3 \quad\;\; & u_5 \quad\;\; & v_5 \end{array} \\ \begin{bmatrix} 8.4375 & -4.0625 & 4.0625 & -0.3125 & -12.5 & 4.375 \\ -4.0625 & 8.4375 & 0.3125 & -4.0625 & 3.75 & -4.375 \\ 4.0625 & 0.3125 & 8.437 & 4.0625 & -12.5 & -4.375 \\ -0.3125 & -4.0625 & 4.0625 & 8.4375 & -3.75 & -4.375 \\ -12.5 & 3.75 & -12.5 & -3.75 & 25 & 0 \\ 4.375 & -4.375 & -4.375 & -4.375 & 0 & 8.75 \end{bmatrix} \end{array} \dfrac{\text{N}}{\text{m}}$$

$$\tag{15.1.87}$$

Element 3

For element 3, using the same steps as for element 1, we obtain the stiffness matrix as

$$[k] = 4.615 \times 10^7 \begin{array}{c} \begin{array}{cccccc} u_3 \quad\;\; & v_3 \quad\;\; & u_4 \quad\;\; & v_4 \quad\;\; & u_5 \quad\;\; & v_5 \end{array} \\ \begin{bmatrix} 8.437 & 4.0625 & -4.0625 & -0.3125 & -4.375 & -3.75 \\ 4.0625 & 8.437 & 0.3125 & 4.0625 & -4.375 & -12.5 \\ -4.0625 & 0.3125 & 8.437 & -4.0625 & -4.375 & 3.75 \\ -0.3125 & 4.0625 & -4.0625 & 8.4375 & 4.375 & -12.5 \\ -4.375 & -4.375 & -4.375 & 4.375 & 8.75 & 0 \\ -3.75 & -12.5 & 3.75 & -12.5 & 0 & 25 \end{bmatrix} \end{array} \dfrac{\text{N}}{\text{m}}$$

$$\tag{15.1.88}$$

Element 4

Finally, for element 4, we obtain

$$[k] = 4.615 \times 10^7 \begin{array}{c} \\ \\ \\ \\ \\ \\ \end{array} \begin{array}{cccccc} u_4 & v_4 & u_1 & v_1 & u_5 & v_5 \\ \end{array}$$

$$[k] = 4.615 \times 10^7 \begin{bmatrix} 8.437 & -4.0625 & 4.0625 & -0.3125 & -12.5 & 4.375 \\ -4.0625 & 8.4375 & 0.3125 & -4.0625 & 3.75 & -4.375 \\ 4.0625 & 0.3125 & 8.4375 & 4.0625 & -12.5 & -4.375 \\ -0.3125 & -4.0625 & 4.0625 & 8.4375 & -3.75 & -4.375 \\ -12.5 & 3.75 & -12.5 & -3.75 & 25 & 0 \\ 4.375 & -4.375 & -4.375 & -4.375 & 0 & 8.75 \end{bmatrix} \frac{N}{m}$$

$$(15.1.89)$$

Using the direct stiffness method, we assemble the element stiffness matrices, Eqs. (15.1.86) through (15.1.89), to obtain the global stiffness matrix as

$$\begin{array}{ccccc} & u_1 & v_1 & u_2 & v_2 \\ \end{array}$$

$$[K] = 4.615 \times 10^7 \begin{bmatrix} 16.874 & 8.125 & -4.0625 & -0.3125 \\ 8.125 & 16.874 & 0.3125 & 4.0625 \\ -4.0625 & 0.3125 & 16.874 & -8.125 \\ -0.3125 & 4.0625 & -8.125 & 16.875 \\ 0 & 0 & 4.0625 & 0.3125 \\ 0 & 0 & -0.3125 & -4.0625 \\ 4.0625 & -0.3125 & 0 & 0 \\ 0.3125 & -4.0625 & 0 & 0 \\ -16.875 & -8.125 & -16.875 & 8.125 \\ -8.125 & -16.875 & 8.125 & -16.875 \end{bmatrix}$$

$$\begin{array}{cccccc} u_3 & v_3 & u_4 & v_4 & u_5 & v_5 \\ \end{array}$$

$$\begin{bmatrix} 0 & 0 & 4.0625 & 0.3125 & -16.875 & -8.125 \\ 0 & 0 & -0.3125 & -4.0625 & -8.125 & -16.875 \\ 4.0625 & -0.3125 & 0 & 0 & -16.875 & 8.125 \\ 0.3125 & -4.0625 & 0 & 0 & 8.125 & -16.875 \\ 16.875 & 8.125 & -4.0625 & -0.3125 & -16.875 & -8.125 \\ 8.125 & 16.875 & 0.3125 & 4.0625 & -8.125 & -16.875 \\ -4.0625 & 0.3125 & 16.875 & -8.125 & -16.875 & 8.125 \\ -0.3125 & 4.0625 & -8.125 & 16.875 & 8.125 & -16.875 \\ -16.875 & -8.125 & -16.875 & 8.125 & 67.5 & 0 \\ -8.125 & -16.875 & 8.125 & -16.875 & 0 & 67.5 \end{bmatrix} \frac{N}{m}$$

$$(15.1.90)$$

Next, we determine the thermal force matrices for each element by using Eq. (15.1.25) as follows:

Element 1

$$\{f_T\} = \frac{\alpha E t T}{2(1-v)} \begin{Bmatrix} \beta_1 \\ \gamma_1 \\ \beta_2 \\ \gamma_2 \\ \beta_5 \\ \gamma_5 \end{Bmatrix} = \frac{(12 \times 10^{-6})(210 \times 10^9)(0.005 \text{ m})(50)}{2(1-0.3)} \begin{Bmatrix} -0.25 \\ -0.25 \\ 0.25 \\ -0.25 \\ 0 \\ 0.5 \end{Bmatrix}$$

$$= 450,000 \begin{Bmatrix} -0.25 \\ -0.25 \\ 0.25 \\ -0.25 \\ 0 \\ 0.5 \end{Bmatrix} = \begin{Bmatrix} f_{T1x} \\ f_{T1y} \\ f_{T2x} \\ f_{T2y} \\ f_{T5x} \\ f_{T5y} \end{Bmatrix} = \begin{Bmatrix} -112,500 \\ -112,500 \\ 112,500 \\ -112,500 \\ 0 \\ 225,000 \end{Bmatrix} \text{N} \qquad (15.1.91)$$

Element 2

$$\{f_T\} = 450,000 \begin{Bmatrix} 0.25 \\ -0.25 \\ 0.25 \\ 0.25 \\ -0.5 \\ 0 \end{Bmatrix} = \begin{Bmatrix} f_{T2x} \\ f_{T2y} \\ f_{T3x} \\ f_{T3y} \\ f_{T5x} \\ f_{T5y} \end{Bmatrix} = \begin{Bmatrix} 112,500 \\ -112,500 \\ 112,500 \\ 112,500 \\ -225,000 \\ 0 \end{Bmatrix} \text{N} \qquad (15.1.92)$$

Element 3

$$\{f_T\} = 450,000 \begin{Bmatrix} 0.25 \\ 0.25 \\ -0.25 \\ 0.25 \\ 0 \\ -0.5 \end{Bmatrix} = \begin{Bmatrix} f_{T3x} \\ f_{T3y} \\ f_{T4x} \\ f_{T4y} \\ f_{T5x} \\ f_{T5y} \end{Bmatrix} = \begin{Bmatrix} 112,500 \\ 112,500 \\ -112,500 \\ 112,500 \\ 0 \\ -225,000 \end{Bmatrix} \text{N} \qquad (15.1.93)$$

Element 4

$$\{f_T\} = 450,000 \begin{Bmatrix} -0.25 \\ 0.25 \\ -0.25 \\ -0.25 \\ 0.5 \\ 0 \end{Bmatrix} = \begin{Bmatrix} f_{T4x} \\ f_{T4y} \\ f_{T1x} \\ f_{T1y} \\ f_{T5x} \\ f_{T5y} \end{Bmatrix} = \begin{Bmatrix} -112,500 \\ 112,500 \\ -112,500 \\ -112,500 \\ 225,000 \\ 0 \end{Bmatrix} \text{N} \qquad (15.1.94)$$

We then obtain the global thermal force matrix by direct assemblage of the element force matrices [Eqs. (15.1.91) through (15.1.94)]. The resulting matrix is

$$
\begin{Bmatrix} f_{T1x} \\ f_{T1y} \\ f_{T2x} \\ f_{T2y} \\ f_{T3x} \\ f_{T3y} \\ f_{T4x} \\ f_{T4y} \\ f_{T5x} \\ f_{T5y} \end{Bmatrix} = \begin{Bmatrix} -225{,}000 \\ -225{,}000 \\ 225{,}000 \\ -225{,}000 \\ 225{,}000 \\ 225{,}000 \\ -225{,}000 \\ 225{,}000 \\ 0 \\ 0 \end{Bmatrix} \text{N}
\tag{15.1.95}
$$

Using Eqs. (15.1.90) and (15.1.95) and imposing the boundary conditions $u_1 = v_1 = u_4 = v_4 = 0$, we obtain the system of equations for solution as

$$
\begin{Bmatrix} f_{T2x} = 225{,}000 \\ f_{T2y} = -225{,}000 \\ f_{T3x} = 225{,}000 \\ f_{T3y} = 225{,}000 \\ f_{T5x} = 0 \\ f_{T5y} = 0 \end{Bmatrix} = 4.615 \times 10^7
$$

$$
\begin{bmatrix}
16.874 & -8.125 & 4.0625 & -0.3125 & -16.875 & 8.125 \\
-8.125 & 16.875 & 0.3125 & -4.0625 & 8.125 & -16.875 \\
4.0625 & 0.3125 & 16.875 & 8.125 & -16.875 & -8.125 \\
-0.3125 & -4.0625 & 8.125 & 16.875 & -8.125 & -16.875 \\
-16.875 & 8.125 & -16.875 & -8.125 & 67.5 & 0 \\
8.125 & -16.875 & -8.125 & -16.875 & 0 & 67.5
\end{bmatrix}
\begin{Bmatrix} u_2 \\ v_2 \\ u_3 \\ v_3 \\ u_5 \\ v_5 \end{Bmatrix}
$$

$$\tag{15.1.96}$$

Solving Eq. (15.1.96) for the nodal displacements, we have

$$
\begin{Bmatrix} u_2 \\ v_2 \\ u_3 \\ v_3 \\ u_5 \\ v_5 \end{Bmatrix} = \begin{Bmatrix} 3.327 \times 10^{-4} \\ -1.911 \times 10^{-4} \\ 3.327 \times 10^{-4} \\ 1.911 \times 10^{-4} \\ 2.123 \times 10^{-4} \\ 6.654 \times 10^{-9} \end{Bmatrix} \text{m}
\tag{15.1.97}
$$

We now use Eq. (15.1.64) to obtain the stresses in each element. Using Eqs. (6.2.36) and (15.1.65), we write Eq. (15.1.64) as

$$
\{\sigma\} = [D][B]\{d\} - [D]\{\varepsilon_T\}
\tag{15.1.98}
$$

Element 1

$$\left\{\begin{matrix} \sigma_x \\ \sigma_y \\ \tau_{xy} \end{matrix}\right\} = \frac{E}{1-v^2} \begin{bmatrix} 1 & v & 0 \\ v & 1 & 0 \\ 0 & 0 & \frac{1-v}{2} \end{bmatrix} \frac{1}{2A} \begin{Bmatrix} \beta_1 & 0 & \beta_2 & 0 & \beta_5 & 0 \\ 0 & \gamma_1 & 0 & \gamma_2 & 0 & \gamma_5 \\ \gamma_1 & \beta_1 & \gamma_2 & \beta_2 & \gamma_5 & \beta_5 \end{Bmatrix} \left\{\begin{matrix} u_1 \\ v_1 \\ u_2 \\ v_2 \\ u_5 \\ v_5 \end{matrix}\right\}$$

$$-\frac{E}{1-v^2} \begin{bmatrix} 1 & v & 0 \\ v & 1 & 0 \\ 0 & 0 & \frac{1-v}{2} \end{bmatrix} \left\{\begin{matrix} \alpha T \\ \alpha T \\ 0 \end{matrix}\right\} \qquad (15.1.99)$$

Using Eqs. (15.1.82) and (15.1.97) along with the mechanical properties E, v, and α in Eq. (15.1.99), we obtain

$$\left\{\begin{matrix} \sigma_x \\ \sigma_y \\ \tau_{xy} \end{matrix}\right\} = \frac{210 \times 10^9}{0.91} \begin{bmatrix} 1 & 0.3 & 0 \\ 0.3 & 1 & 0 \\ 0 & 0 & 0.35 \end{bmatrix}$$

$$\times \frac{1}{0.125} \begin{bmatrix} -0.25 & 0 & 0.25 & 0 & 0 & 0 \\ 0 & -0.25 & 0 & -0.25 & 0 & 0.5 \\ -0.25 & -0.25 & -0.25 & 0.25 & 0.5 & 0 \end{bmatrix} \left\{\begin{matrix} 0 \\ 0 \\ 3.327 \times 10^{-4} \\ -1.911 \times 10^{-4} \\ 2.123 \times 10^{-4} \\ 6.654 \times 10^{-9} \end{matrix}\right\}$$

$$-\frac{210 \times 10^9}{0.91} \begin{bmatrix} 1 & 0.3 & 0 \\ 0.3 & 1 & 0 \\ 0 & 0 & 0.35 \end{bmatrix} \left\{\begin{matrix} (12 \times 10^{-6})(50) \\ (12 \times 10^{-6})(50) \\ 0 \end{matrix}\right\} \qquad (15.1.100)$$

Simplifying Eq. (15.1.100) yields

$$\left\{\begin{matrix} \sigma_x \\ \sigma_y \\ \tau_{xy} \end{matrix}\right\} = \left\{\begin{matrix} 1.800 \times 10^8 \\ 1.342 \times 10^8 \\ -1.600 \times 10^7 \end{matrix}\right\} - \left\{\begin{matrix} 1.8 \times 10^8 \\ 1.8 \times 10^8 \\ 0 \end{matrix}\right\} = \left\{\begin{matrix} 0 \\ -4.57 \times 10^7 \\ -1.60 \times 10^7 \end{matrix}\right\} \text{Pa}$$

$$(15.1.101)$$

Similarly, we obtain the stresses in element 2 as follows:

Element 2

$$\left\{\begin{matrix} \sigma_x \\ \sigma_y \\ \tau_{xy} \end{matrix}\right\} = \left\{\begin{matrix} 1.640 \times 10^8 \\ 2.097 \times 10^8 \\ -2150 \end{matrix}\right\} - \left\{\begin{matrix} 1.8 \times 10^8 \\ 1.8 \times 10^8 \\ 0 \end{matrix}\right\} = \left\{\begin{matrix} -1.6 \times 10^7 \\ 2.973 \times 10^7 \\ -2150 \end{matrix}\right\} \text{Pa} \qquad (15.1.102)$$

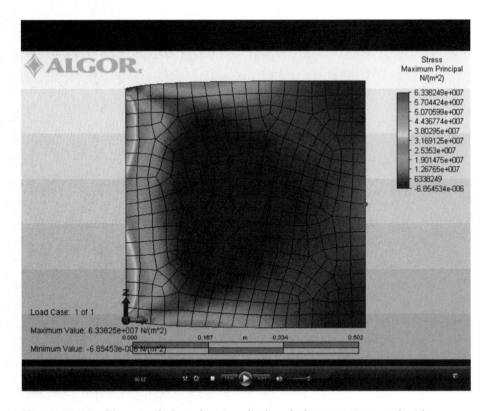

Figure 15–12 Discretized plate showing displaced plate superimposed with maximum principal stress plot in Pa (See the full-color insert for a color version of this figure.)

Stresses in elements 3 and 4 can be determined similarly. The clamped plate subjected to uniform heating (see the longhand solution, Example 15.5) was also solved using the Algor computer program from Reference [1]. The plate was discretized using the "automesh" feature of [1]. These results are similar to those obtained from the longhand solution of Example 15.5 using the very coarse mesh. The computer program solution with 342 elements is naturally more accurate than the longhand solution with only four elements. Figure 15–12 shows the discretized plate with resulting displacement superimposed on the maximum principal stress plot. ∎

▲ Reference

[1] *Linear Stress and Dynamics Reference Division*, Docutech On-line Documentation, Algor, Inc., Pittsburgh, PA.

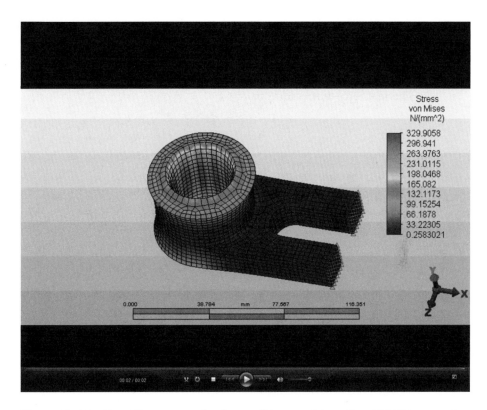

Figure 15–13   Von Mises stress plot for a solid part subjected to 100 °C temperature rise inside the surface of the hole (See the full-color insert for a color version of this figure.)

Finally, Figure 15–13 shows a three-dimensional solid part that is fixed on the small front surfaces and uniformly heated by a temperature increase of 100 °C acting over the entire inside surface of the hole. The resulting von Mises stress plot is shown with the maximum value of 329.9 MPa occurring inside the hole.

▲   Summary Equations

Unconstrained displacement of bar due to uniform temperature change:

$$\delta_T = \alpha T L \tag{15.1.1}$$

Strain due to uniform temperature change for a bar:

$$\varepsilon_T = \alpha T \tag{15.1.2}$$

Thermal strain matrix for a bar:

$$\{\varepsilon_T\} = \{\varepsilon_{xT}\} = \{\alpha T\} \tag{15.1.17}$$

Thermal force matrix for a bar:

$$\{f_T\} = \begin{Bmatrix} f_{T1} \\ f_{T2} \end{Bmatrix} = \begin{Bmatrix} -E\alpha TA \\ E\alpha TA \end{Bmatrix} \tag{15.1.20}$$

Thermal strain matrix for isotropic material in plane stress:

$$\{\varepsilon_T\} = \begin{Bmatrix} \alpha T \\ \alpha T \\ 0 \end{Bmatrix} \tag{15.1.22}$$

Thermal strain matrix for isotropic material in plane strain:

$$\{\varepsilon_T\} = (1+v) \begin{Bmatrix} \alpha T \\ \alpha T \\ 0 \end{Bmatrix} \tag{15.1.23}$$

Thermal force matrix for plane stress triangle:

$$\{f_T\} = \begin{Bmatrix} f_{Tix} \\ f_{Tiy} \\ \vdots \\ f_{Tmy} \end{Bmatrix} = \frac{\alpha E t T}{2(1-v)} \begin{Bmatrix} \beta_i \\ \gamma_i \\ \beta_j \\ \gamma_j \\ \beta_m \\ \gamma_m \end{Bmatrix} \tag{15.1.25}$$

Thermal strain matrix for axisymmetric triangular element:

$$\{\varepsilon_T\} = \begin{Bmatrix} \varepsilon_{rT} \\ \varepsilon_{zT} \\ \varepsilon_{\theta T} \\ \gamma_{rzT} \end{Bmatrix} = \begin{Bmatrix} \alpha T \\ \alpha T \\ \alpha T \\ 0 \end{Bmatrix} \tag{15.1.26}$$

Thermal force matrix for axisymmetric element evaluated at its centroid:

$$\{f_T\} = 2\pi \bar{r} A [\overline{B}]^T [D] \{\varepsilon_T\} \tag{15.1.28}$$

▲ Problems

15.1 For the one-dimensional steel bar fixed at the left end, free at the right end, and subjected to a uniform temperature rise $T = 10\,°C$ as shown in Figure P15–1, determine the free-end displacement, the displacement 60 in. from the fixed end, the reactions at the fixed end, and the axial stress. Let $E = 210$ GPa, $A = 4$ in^2, and $\alpha = 12.0 \times 10^{-6}$ (mm/mm)/$°C$.

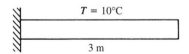

$T = 10°C$

3 m

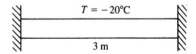

$T = -20°C$

3 m

Figure P15–1   **Figure P15–2**

15.2 For the one-dimensional steel bar fixed at each end and subjected to a uniform temperature drop of $T = 20\,°C$ as shown in Figure P15–2, determine the reactions at the fixed ends and the stress in the bar. Let $E = 210$ GPa, $A = 1 \times 10^{-2}$ m^2, and $\alpha = 11.7 \times 10^{-6}$ (mm/mm)/$°C$.

15.3 For the plane truss shown in Figure P15–3, bar element 2 is subjected to a uniform temperature rise of $T = 10\,°C$. Let $E = 210$ GPa, $A = 12.5$ cm^2, and $\alpha = 12 \times 10^{-6}$ (mm/mm)/$°C$. The lengths of the truss elements are shown in the figure. Determine the stresses in each bar. [*Hint:* See Eqs. (3.6.4) and (3.6.6) in Example 3.5 for the global and reduced $[K]$ matrices.]

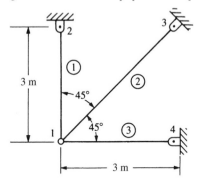

Figure P15–3   **Figure P15–4**

15.4 For the plane truss shown in Figure P15–4, bar element 1 is subjected to a uniform temperature rise of $5\,°C$. Let $E = 210$ GPa, $A = 12$ cm^2, and $\alpha = 12 \times 10^{-6}$ (mm/mm)/$°C$. The lengths of the truss elements are shown in the figure. Determine the stresses in each bar. (*Hint:* Use Problem 3.21 for $[K]$.)

15.5 For the structure shown in Figure P15–5, bar element 1 is subjected to a uniform temperature rise of $T = 20\,°C$. Let $E = 210$ GPa, $A = 2 \times 10^{-2}$ m^2, and $\alpha = 12 \times 10^{-6}$ (mm/mm)/$°C$. Determine the stresses in each bar.

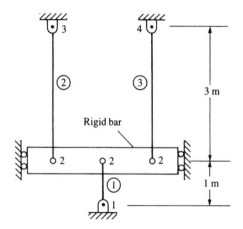

Figure P15–5

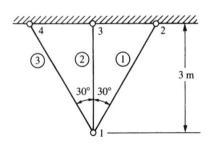

Figure P15–6

15.6 For the plane truss shown in Figure P15–6, bar element 2 is subjected to a uniform temperature drop of $T = 20\,°C$. Let $E = 70$ GPa, $A = 4 \times 10^{-2}$ m^2, and $\alpha = 23 \times 10^{-6}$ (mm/mm)/$°C$. Determine the stresses in each bar and the displacement of node 1.

15.7 For the bar structure shown in Figure P15–7, element 1 is subjected to a uniform temperature rise of $T = 30\,°C$. Let $E = 210$ GPa, $A = 3 \times 10^{-2}$ m^2, and $\alpha = 12 \times 10^{-6}$ (mm/mm)/$°C$. Determine the displacement of node 1 and the stresses in each bar.

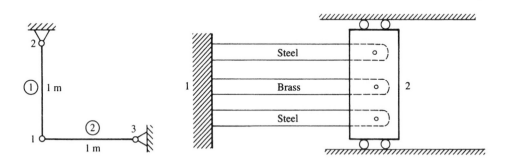

Figure P15–7   Figure P15–8

15.8 A bar assemblage consists of two outer steel bars and an inner brass bar. The three-bar assemblage is then heated to raise the temperature by an amount $T = 20\,°C$. Let all cross-sectional areas be $A = 12.5$ cm^2 and $L = 60$ in., $E_{steel} = 210$ GPa, $E_{brass} = 105$ GPa, $\alpha_{steel} = 12 \times 10^{-6}$ (mm/mm)/$°C$, and $\alpha_{brass} = 18 \times 10^{-6}$ (mm/ mm)/$°C$. Determine (a) the displacement of node 2 and (b) the stress in the steel and brass bars. See Figure P15–8.

15.9 For the plane truss shown in Figure P15–9, bar element 2 is subjected to a uniform temperature rise of $T = 10\,^\circ C$. Let $E = 210$ GPa, $A = 12.5$ cm^2, and $\alpha = 12 \times 10^{-6}/^\circ C$. What temperature change is needed in bars 1 and 3 to remove the stress due to the uniform temperature rise in bar 2? Show enough work to prove your answer. Use a longhand solution.

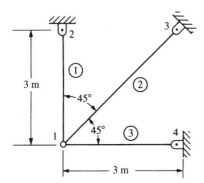

Figure P15–9

15.10 When do stresses occur in a body made of a single material due to uniform temperature change in the body? Consider Problem 15.1 and also compare the solution to Example 15.1 in this chapter.

15.11 Consider two thermally incompatible materials, such as steel and aluminum, attached together as shown in Figure P15–11. Will there be temperature-induced stress in each material upon uniform heating of both materials to the same temperature when the boundary conditions are simple supports (a pin and a roller such that we have a statically determinate system)? Explain. Let there be a uniform temperature rise of $T = 30\,^\circ C$.

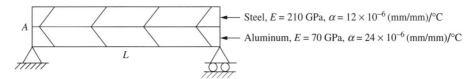

Figure P15–11

15.12 A bimetallic thermal control is made of cold-rolled yellow brass and magnesium alloy bars (Figure P15–12). The bars are arranged with a gap of 0.1 mm between them at 20°C. The brass bar has a length of 20 mm and a cross-sectional area of 0.4 cm^2, and the magnesium bar has a length of 30 mm and a cross-sectional area of 0.6 cm^2. Determine (a) the axial displacement of the end of the brass bar and (b) the stress in each bar after it has closed up due to a temperature increase of 55°C. Use at least one element for each bar in your finite element model.

15.13 For the plane stress element shown in Figure P15–13 subjected to a uniform temperature rise of $T = 30\,^\circ C$, determine the thermal force matrix $\{f_T\}$.

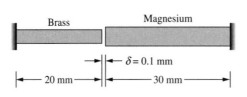

Figure P15–12

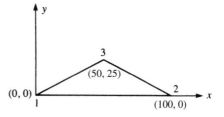

Figure P15–13

Let $E = 70$ GPa, $v = 0.30$, and $\alpha = 23 \times 10^{-6}$ (mm/mm)/°C. The coordinates (in mm) are shown in the figure. The element thickness is $t = 25$ mm.

15.14 For the plane stress element shown in Figure P15–14 subjected to a uniform temperature rise of $T = 30$°C, determine the thermal force matrix $\{f_T\}$. Let $E = 70$ GPa, $v = 0.3$, $\alpha = 23 \times 10^{-6}$ (mm/mm)/°C, and $t = 5$ mm. The coordinates (in millimeters) are shown in the figure.

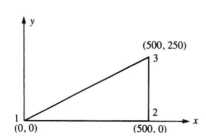

Figure P15–14

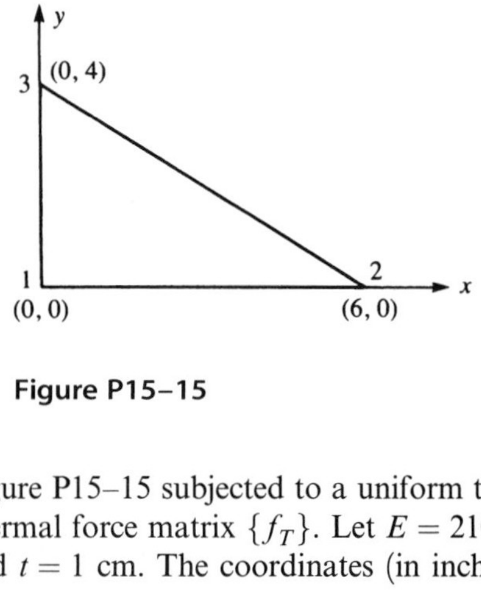

Figure P15–15

15.15 For the plane stress element shown in Figure P15–15 subjected to a uniform temperature rise of $T = 55$°C, determine the thermal force matrix $\{f_T\}$. Let $E = 210$ GPa, $v = 0.3$, $\alpha = 12 \times 10^{-6}$ (mm/mm)/°C, and $t = 1$ cm. The coordinates (in inches) are shown in the figure.

15.16 For the plane stress element shown in Figure P15–16 subjected to a uniform temperature drop of $T = 20$°C, determine the thermal force matrix $\{f_T\}$. Let $E = 210$ GPa, $v = 0.25$, and $\alpha = 12 \times 10^{-6}$ (mm/mm)/°C. The coordinates (in millimeters) are shown in the figure. The element thickness is 10 mm.

15.17 For the plane stress plate fixed along the left and right sides and subjected to a uniform temperature rise of 30°C as shown in Figure P15–17, determine the stresses in each element. Let $E = 70$ GPa, $v = 0.30$, $\alpha = 22.5 \times 10^{-6}$ (mm/mm)/°C, and $t = 10$ mm. The coordinates (in mm) are shown in the figure. (*Hint:* The nodal displacements are all equal to zero. Therefore, the stresses can be determined from $\{\sigma\} = -[D]\{\varepsilon_T\}$.)

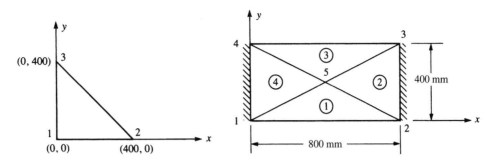

Figure P15–16   Figure P15–17

15.18 For the plane stress plate fixed along all edges and subjected to a uniform temperature decrease of $20\,°\mathrm{C}$ as shown in Figure P15–18, determine the stresses in each element. Let $E = 210$ GPa, $v = 0.25$, and $\alpha = 12 \times 10^{-6}$ (mm/mm)/$°\mathrm{C}$. The coordinates of the plate are shown in the figure. The plate thickness is 10 mm. (*Hint:* The nodal displacements are all equal to zero. Therefore, the stresses can be determined from $\{\sigma\} = -[D]\{\varepsilon_T\}$.)

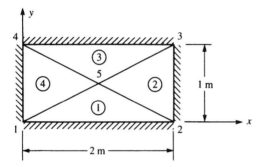

Figure P15–18

15.19 If the thermal expansion coefficient of a bar is given by $\alpha = \alpha_0(1 + x/L)$, determine the thermal force matrix. Let the bar have length L, modulus of elasticity E, and cross-sectional area A.

15.20 Assume the temperature function to vary linearly over the length of a bar as $T = t_1 + t_2 x$; that is, express the temperature function as $\{T\} = [N]\{t\}$, where $[N]$ is the shape function matrix for the two-node bar element. In other words, $[N] = [1 - x/L \quad x/L]$. Determine the force matrix in terms of E, A, α, L, t_1, and t_2. [*Hint:* Use Eq. (15.1.18).]

15.21 Derive the thermal force matrix for the axisymmetric element of Chapter 9. [Also see Eq. (15.1.27).]

Using a computer program, solve the following problems.

15.22 The square plate in Figure P15–22 is subjected to uniform heating of 40 °C. Determine the nodal displacements and element stresses. Let the element thickness be $t = 2$ mm, $E = 210$ GPa, $v = 0.33$, and $\alpha = 18 \times 10^{-6}/°C$.

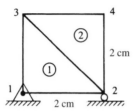   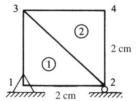

Figure P15–22   **Figure P15–23**

15.23 The square plate in Figure P15–23 has element 1 made of steel with $E = 210$ GPa, $v = 0.33$, and $\alpha = 18 \times 10^{-6}/°C$ and element 2 made of a material with $E = 105$ GPa, $v = 0.25$, and $\alpha = 90 \times 10^{-6}/°C$. Let the plate thickness be $t = 2$ mm. Determine the nodal displacements and element stresses for element 1 subjected to an 80 °F temperature increase and element 2 subjected to a 10 °C temperature increase.

15.24 Solve Problem 15.3 using a computer program.

15.25 Solve Problem 15.6 using a computer program.

15.26 The aluminum tube shown in Figure P15–26 fits snugly into a hole (with surrounding material aluminum) at room temperature. If the temperature of the tube is then increased by 40°C, determine the deformed configuration and the stress distribution of the tube. Let $E = 70$ GPa, $v = 0.33$, and $\alpha = 23 \times 10^{-6}/°C$ for the tube.

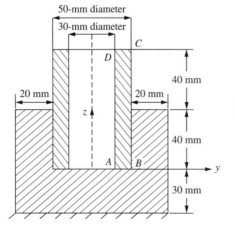

Figure P15–26

15.27 For the solid model of a fixture shown in Figure P15–27, the inside surface of the hole is subjected to a temperature increase of 80 °C. The right end surfaces are fixed. Determine the von Mises stresses throughout the fixture due to this temperature increase. What is the largest von Mises stress? Is it a concern against yielding of the material? Assume the material is AISI 1020 cold-rolled steel.

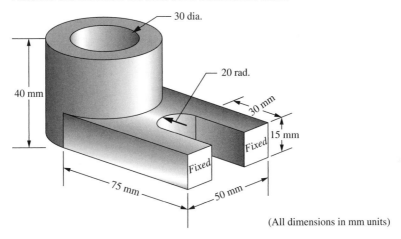

Figure P15–27

15.28 For the fixture shown in Figure P15–28, the inside surfaces of the eight holes are increased in temperature by 50 °C. Determine the von Mises stresses throughout the fixture. What is the largest von Mises stress in the fixture? Is there concern for failure due to yielding of the material? Assume the material is aluminum alloy 6061-O (annealed). Fix the inside surface of the upper hole.

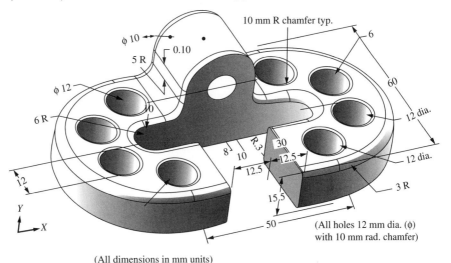

Figure P15–28

15.29 A rectangular slab of concrete in a highway as shown in Figure P15–29 can be considered to have sides parallel to the road free and sides perpendicular to the road simply supported. The critical buckling stress is shown to be

$$\sigma_{cr} = \frac{\pi^2 E}{12(1 - v^2)} \left(\frac{t}{a}\right)^2$$

where for medium strength concrete we assume $E = 25$ GPa and $v = 0.20$. Let the slab dimensions be length $a = 3$ m, width $b = 2.4$ m, and thickness $t = 250$ mm. Determine the uniform temperature increase in the slab that will buckle it. Use a finite element code to solve this problem.

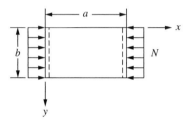

Figure P15–29

CHAPTER 16

STRUCTURAL DYNAMICS AND TIME-DEPENDENT HEAT TRANSFER ▲

CHAPTER OBJECTIVES

- To discuss the dynamics of a single-degree-of freedom spring-mass system.
- To derive the finite element equations for the time-dependent stress analysis of the one-dimensional bar, including derivation of the lumped and consistent mass matrices.
- To introduce procedures for numerical integration in time, including the central difference method, Newmark's method, and Wilson's method.
- To describe how to determine the natural frequencies of bars by the finite element method.
- To illustrate the finite element solution of a time-dependent bar problem.
- To develop the beam element lumped and consistent mass matrices.
- To illustrate the determination of natural frequencies for beams by the finite element method.
- To develop the mass matrices for truss, plane frame, plane stress, plane strain, axisymmetric, and solid elements.
- To derive the time-dependent heat transfer equations, including the consistent and lumped mass matrices in one dimension.
- To describe numerical time integration methods which originate from the generalized trapezoid rule. These include the forward difference, Crank-Nicolson, Galerkin, and backward difference methods.
- To report some results of structural dynamics problems solved using a computer program, including a fixed-fixed beam for natural frequencies, a bar, a fixed-fixed beam, a rigid frame, and a gantry crane—all subjected to time-dependent forcing functions.

Introduction

This chapter provides an elementary introduction to time-dependent problems. We will introduce the basic concepts using the single-degree-of-freedom spring-mass system. We will include discussion of the stress analysis of the one-dimensional bar, beam, truss, and plane frame. This is followed by the analysis of one-dimensional heat transfer.

We will provide the basic equations necessary for structural dynamics analysis and develop both the lumped- and the consistent-mass matrices involved in the analyses of the bar, beam, truss, and plane frame. We will describe the assembly of the global mass matrix for truss and plane frame analysis and then present numerical integration methods for handling the time derivative. We also present the mass matrices for the constant strain triangle and quadrilateral plane elements, for the axisymmetric element, and for the tetrahedral solid element.

We will provide longhand solutions for the determination of the natural frequencies for bars and beams and then illustrate the time-step integration process involved with the stress analysis of a bar subjected to a time-dependent forcing function.

We will next derive the basic equations for the time-dependent one-dimensional heat-transfer problem and discuss their applications. This chapter provides the basic concepts necessary for the solution of time-dependent problems. We conclude with a section on some computer program results for structural dynamics and time-dependent heat-transfer problems.

▲ 16.1 Dynamics of a Spring-Mass System ▲

In this section, we discuss the motion of a single-degree-of-freedom spring-mass system to introduce the important concepts necessary for the later study of continuous systems such as bars, beams, and plane frames. In Figure 16–1, we show the single-degree-of-freedom spring-mass system subjected to a time-dependent force $F(t)$. Here k represents the spring stiffness or constant, and m represents the mass of the system.

The free-body diagram of the mass is shown in Figure 16–2. The spring force $T = kx$ and the applied force $F(t)$ act on the mass, and the mass-times-acceleration term is shown separately.

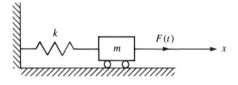

Figure 16–1 Spring-mass system subjected to a time-dependent force

Figure 16–2 Free-body diagram of the mass of Figure 16–1

Applying Newton's second law of motion, $\boldsymbol{f} = \boldsymbol{ma}$, to the mass, we obtain the equation of motion in the x direction as

$$F(t) - kx = m\ddot{x} \qquad (16.1.1)$$

where a dot over a variable denotes differentiation with respect to time; that is, $(\dot{}) = d(\,)/dt$. Rewriting Eq. (16.1.1) in standard form, we have

$$m\ddot{x} + kx = F(t) \qquad (16.1.2)$$

Equation (16.1.2) is a linear differential equation of the second order whose standard solution for the displacement x consists of a homogeneous solution and a particular solution. Standard analytical solutions for this forced vibration can be found in texts on dynamics or vibrations such as Reference [1]. The analytical solution will not be presented here as our intent is to introduce basic concepts in vibration behavior. However, we will solve the problem defined by Eq. (16.1.2) by an approximate numerical technique in Section 16.3 (see Examples 16.1 and 16.2).

The homogeneous solution to Eq. (16.1.2) is the solution obtained when the right side is set equal to zero. A number of useful concepts regarding vibrations are obtained by considering this free vibration of the mass—that is, when $F(t) = 0$. Hence, defining

$$\omega^2 = \frac{k}{m} \qquad (16.1.3)$$

and setting the right side of Eq. (16.1.2) equal to zero, we have

$$\ddot{x} + \omega^2 x = 0 \qquad (16.1.4)$$

where ω is called the **natural circular frequency** of the free vibration of the mass, expressed in units of radians per second or revolutions per minute (rpm). Hence, the natural circular frequency defines the number of cycles per unit time of the mass vibration. We observe from Eq. (16.1.3) that ω depends only on the spring stiffness k and the mass m of the body.

The motion defined by Eq. (16.1.4) is called **simple harmonic motion**. The displacement and acceleration are seen to be proportional but of opposite direction. Again, a standard solution to Eq. (16.1.4) can be found in Reference [1]. A typical displacement/time curve is represented by the sine curve shown in Figure 16–3, where x_m denotes the maximum displacement (called the **amplitude** of the vibration). The time interval required for the mass to complete one full cycle of motion is called the **period** of the vibration τ and is given by

$$\tau = \frac{2\pi}{\omega} \qquad (16.1.5)$$

where τ is measured in seconds. Also the frequency in hertz (Hz = 1/s) is $f = 1/\tau = \omega/(2\pi)$.

Finally, note that all vibrations are damped to some degree by friction forces. These forces may be caused by dry or Coulomb friction between rigid bodies, by

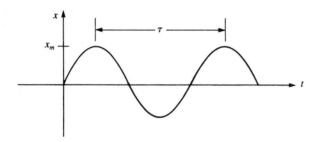

Figure 16–3 Displacement/time curve for simple harmonic motion

internal friction between molecules within a deformable body, or by fluid friction when a body moves in a fluid. Damping results in natural circular frequencies that are smaller than those for undamped systems; maximum displacements also are smaller when damping occurs. A basic treatment of damping can be found in Reference [1] and additional discussion is included in Example 16.12.

▲ 16.2 Direct Derivation of the Bar Element Equations ▲

We will now derive the finite element equations for the time-dependent (dynamic) stress analysis of the one-dimensional bar. Recall that the time-independent (static) stress analysis of the bar was considered in Chapter 3. The steps used in deriving the dynamic equations are the same as those used for the derivation of the static equations.

Step 1   Select Element Type

Figure 16–4 shows the typical bar element of length L, cross-sectional area A, and mass density ρ (with typical units of kg/m³), with nodes 1 and 2 subjected to external time-dependent loads $f_x^e(t)$.

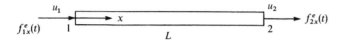

Figure 16–4 Bar element subjected to time-dependent loads

Step 2   Select a Displacement Function

Again, we assume a linear displacement function along the x axis of the bar [see Eq. (3.1.1)]; that is, we let

$$u = a_1 + a_2 x \qquad (16.2.1)$$

As was shown in Chapter 3, Eq. (16.2.1) can be expressed in terms of the shape functions as

$$u = N_1 u_1 + N_2 u_2 \tag{16.2.2}$$

where
$$N_1 = 1 - \frac{x}{L} \qquad N_2 = \frac{x}{L} \tag{16.2.3}$$

Step 3 Define the Strain/Displacement and Stress/Strain Relationships

Again, the strain/displacement relationship is given by

$$\{\varepsilon_x\} = \frac{\partial u}{\partial x} = [B]\{d\} \tag{16.2.4}$$

where
$$[B] = \begin{bmatrix} -\dfrac{1}{L} & \dfrac{1}{L} \end{bmatrix} \qquad \{d\} = \begin{Bmatrix} u_1 \\ u_2 \end{Bmatrix} \tag{16.2.5}$$

and the stress/strain relationship is given by

$$\{\sigma_x\} = [D]\{\varepsilon_x\} = [D][B]\{d\} \tag{16.2.6}$$

Step 4 Derive the Element Stiffness and Mass Matrices and Equations

The bar is generally not in equilibrium under a time-dependent force; hence, $f_{1x} \neq f_{2x}$. Therefore, we again apply Newton's second law of motion, $f = ma$, to each node. In general, the law can be written for each node as "the external (applied) force f_x^e minus the internal force is equal to the nodal mass times acceleration." Equivalently, adding the internal force to the ma term, we have

$$f_{1x}^e = f_{1x} + m_1 \frac{\partial^2 u_1}{\partial t^2} \qquad f_{2x}^e = f_{2x} + m_2 \frac{\partial^2 u_2}{\partial t^2} \tag{16.2.7}$$

where the masses m_1 and m_2 are obtained by lumping the total mass of the bar equally at the two nodes such that

$$m_1 = \frac{\rho A L}{2} \qquad m_2 = \frac{\rho A L}{2} \tag{16.2.8}$$

In matrix form, we express Eqs. (16.2.7) as

$$\begin{Bmatrix} f_{1x}^e \\ f_{2x}^e \end{Bmatrix} = \begin{Bmatrix} f_{1x} \\ f_{2x} \end{Bmatrix} + \begin{bmatrix} m_1 & 0 \\ 0 & m_2 \end{bmatrix} \begin{Bmatrix} \dfrac{\partial^2 u_1}{\partial t^2} \\ \dfrac{\partial^2 u_2}{\partial t^2} \end{Bmatrix} \tag{16.2.9}$$

Using Eqs. (3.1.13) and (3.1.14), we replace $\{f\}$ with $[k]\{d\}$ in Eq. (16.2.9) to obtain the element equations

$$\{f^e(t)\} = [k]\{d\} + [m]\{\ddot{d}\} \qquad (16.2.10)$$

where

$$[k] = \frac{AE}{L} \begin{bmatrix} 1 & -1 \\ -1 & 1 \end{bmatrix} \qquad (16.2.11)$$

is the bar element stiffness matrix, and

$$[m] = \frac{\rho AL}{2} \begin{bmatrix} 1 & 0 \\ 0 & 1 \end{bmatrix} \qquad (16.2.12)$$

is called the **lumped-mass matrix**. Also,

$$\{\ddot{d}\} = \frac{\partial^2\{d\}}{\partial t^2} \qquad (16.2.13)$$

Observe that the lumped-mass matrix has diagonal terms only. This facilitates the computation of the global equations. However, solution accuracy is usually not as good as when a consistent-mass matrix is used [2].

We will now develop the **consistent-mass matrix** for the bar element. Numerous methods are available to obtain the consistent-mass matrix. The generally applicable virtual work principle (which is the basis of many energy principles, such as the principle of minimum potential energy for elastic bodies previously used in this text) provides a relatively simple method for derivation of the element equations and is included in Appendix E. However, an even simpler approach is to use D'Alembert's principle; thus, we introduce an effective body force X^e as

$$\{X^e\} = -\rho\{\ddot{u}\} \qquad (16.2.14)$$

where the minus sign is due to the fact that the acceleration produces D'Alembert's body forces in the direction opposite the acceleration. The nodal forces associated with $\{X^e\}$ are then found by using Eq. (6.3.1), repeated here as

$$\{f_b\} = \iiint_V [N]^T\{X\}\,dV \qquad (16.2.15)$$

Substituting $\{X^e\}$ given by Eq. (16.2.14) into Eq. (16.2.15) for $\{X\}$, we obtain

$$\{f_b\} = -\iiint_V \rho[N]^T\{\ddot{u}\}\,dV \qquad (16.2.16)$$

Recalling from Eq. (16.2.2) that $\{u\} = [N]\{d\}$, we find that the first and second derivatives with respect to time are

$$\{\dot{u}\} = [N]\{\dot{d}\} \qquad \{\ddot{u}\} = [N]\{\ddot{d}\} \qquad (16.2.17)$$

where $\{\dot{d}\}$ and $\{\ddot{d}\}$ are the nodal velocities and accelerations, respectively. Substituting Eqs. (16.2.17) into Eq. (16.2.16), we obtain

$$\{f_b\} = -\iiint_V \rho[N]^T[N]\,dV\{\ddot{d}\} = -[m]\{\ddot{d}\} \tag{16.2.18}$$

where the element mass matrix is defined as

$$[m] = \iiint_V \rho[N]^T[N]\,dV \tag{16.2.19}$$

This mass matrix is called the *consistent-mass matrix* because it is derived from the same shape functions $[N]$ that are used to obtain the stiffness matrix $[k]$. In general, $[m]$ given by Eq. (16.2.19) will be a full but symmetric matrix. Equation (16.2.19) is a general form of the consistent-mass matrix; that is, substituting the appropriate shape functions, we can generate the mass matrix for such elements as the bar, beam, and plane stress.

We will now develop the consistent-mass matrix for the bar element of Figure 16–4 by substituting the shape function Eqs. (16.2.3) into Eq. (16.2.19) as follows:

$$[m] = \iiint_V \rho \left\{ \begin{array}{c} 1-\dfrac{x}{L} \\[2mm] \dfrac{x}{L} \end{array} \right\} \left[1-\dfrac{x}{L} \quad \dfrac{x}{L} \right] dV \tag{16.2.20}$$

Simplifying Eq. (16.2.20), we obtain

$$[m] = \rho A \int_0^L \left\{ \begin{array}{c} 1-\dfrac{x}{L} \\[2mm] \dfrac{x}{L} \end{array} \right\} \left[1-\dfrac{x}{L} \quad \dfrac{x}{L} \right] dx \tag{16.2.21}$$

or, on multiplying the matrices of Eq. (16.2.21),

$$[m] = \rho A \int_0^L \left[\begin{array}{cc} \left(1-\dfrac{x}{L}\right)^2 & \left(1-\dfrac{x}{L}\right)\dfrac{x}{L} \\[3mm] \left(1-\dfrac{x}{L}\right)\dfrac{x}{L} & \left(\dfrac{x}{L}\right)^2 \end{array} \right] dx \tag{16.2.22}$$

On integrating Eq. (16.2.22) term by term, we obtain the consistent-mass matrix for a bar element as

$$[m] = \frac{\rho A L}{6} \begin{bmatrix} 2 & 1 \\ 1 & 2 \end{bmatrix} \tag{16.2.23}$$

Step 5 Assemble the Element Equations to Obtain
the Global Equations and Introduce Boundary Conditions

We assemble the element equations using the direct stiffness method such that interelement continuity of displacements is again satisfied at common nodes and, in addition, interelement continuity of accelerations is also satisfied; that is, we obtain the global equations

$$\{F(t)\} = [K]\{d\} + [M]\{\ddot{d}\} \tag{16.2.24}$$

where
$$[K] = \sum_{e=1}^{N} [k^{(e)}] \qquad [M] = \sum_{e=1}^{N} [m^{(e)}] \qquad \{F\} = \sum_{e=1}^{N} \{f^{(e)}\} \tag{16.2.25}$$

are the global stiffness, mass, and force matrices, respectively. Note that the global mass matrix is assembled in the same manner as the global stiffness matrix. Equation (16.2.24) represents a set of matrix equations discretized with respect to space. To obtain the solution of the equations, discretization in time is also necessary. We will describe this process in Section 16.3 and will later present representative solutions illustrating these equations.

▲ **16.3 Numerical Integration in Time** ▲

We now introduce procedures for the discretization of Eq. (16.2.24) with respect to time. These procedures will enable us to determine the nodal displacements at different time increments for a given dynamic system. The general method used is called *direct integration*. There are two classifications of direct integration: explicit and implicit. We will formulate the equations for three direct integration methods. The first, and simplest, is an explicit method known as the *central difference method* [3, 4]. The second and third, more complicated but more versatile than the central difference method, are implicit methods known as the Newmark-Beta (or Newmark's) method [5] and the Wilson-Theta (or Wilson's) method [7, 8]. The versatility of both Newmark's and Wilson's methods is evidenced by their adaptation in many commercially available computer programs. Wilson's method is used in the Algor computer program [16]. Numerous other integration methods are available in the literature. Among these are Houbolt's method [8] and the alpha method [13].

Central Difference Method

The central difference method is based on finite difference expressions in time for velocity and acceleration at time t given by

$$\{\dot{d}_i\} = \frac{\{d_{i+1}\} - \{d_{i-1}\}}{2(\Delta t)} \tag{16.3.1}$$

$$\{\ddot{d}_i\} = \frac{\{\dot{d}_{i+1}\} - \{\dot{d}_{i-1}\}}{2(\Delta t)} \tag{16.3.2}$$

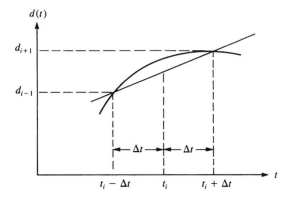

Figure 16–5 Numerical integration (approximation of derivative at t_i)

where the subscripts indicate the time step; that is, for a time increment of Δt, $\{d_i\} = \{d(t)\}$ and $\{d_{i+1}\} = \{d(t + \Delta t)\}$. The procedure used in deriving Eq. (16.3.1) is illustrated by use of the displacement/time curve shown in Figure 16–5. Graphically, Eq. (16.3.1) represents the slope of the line shown in Figure 16–5; that is, given two points at increments $i - 1$ and $i + 1$ on the curve, two Δt increments apart, an approximation of the first derivative at the midpoint i of the increment is given by Eq. (16.3.1). Similarly, using a velocity/time curve, we could obtain Eq. (16.3.2), or we can see that Eq. (16.3.2) is obtained simply by differentiating Eq. (16.3.1) with respect to time.

It has been shown using, for instance, Taylor series expansions [3] that the acceleration can also be expressed in terms of the displacements by

$$\{\ddot{d}_i\} = \frac{\{d_{i+1}\} - 2\{d_i\} + \{d_{i-1}\}}{(\Delta t)^2} \tag{16.3.3}$$

Because we want to evaluate the nodal displacements, it is most suitable to use Eq. (16.3.3) in the form

$$\{d_{i+1}\} = 2\{d_i\} - \{d_{i-1}\} + \{\ddot{d}_i\}(\Delta t)^2 \tag{16.3.4}$$

Equation (16.3.4) will be used to determine the nodal displacements in the next time step $i + 1$ knowing the displacements at time steps i and $i - 1$ and the acceleration at time i.

From Eq. (16.2.24), we express the acceleration as

$$\{\ddot{d}_i\} = [M]^{-1}(\{F_i\} - [K]\{d_i\}) \tag{16.3.5}$$

To obtain an expression for $\{d_{i+1}\}$, we first multiply Eq. (16.3.4) by the mass matrix $[M]$ and then substitute Eq. (16.3.5) for $\{\ddot{d}_i\}$ into this equation to obtain

$$[M]\{d_{i+1}\} = 2[M]\{d_i\} - [M]\{d_{i-1}\} + (\{F_i\} - [K]\{d_i\})(\Delta t)^2 \tag{16.3.6}$$

Combining like terms of Eq. (16.3.6), we obtain

$$[M]\{d_{i+1}\} = (\Delta t)^2 \{F_i\} + [2[M] - (\Delta t)^2 [K]]\{d_i\} - [M]\{d_{i-1}\} \qquad (16.3.7)$$

To start the computations to determine $\{d_{i+1}\}$, $\{\dot{d}_{i+1}\}$, and $\{\ddot{d}_{i+1}\}$, we need the displacement $\{d_{i-1}\}$ initially, as indicated by Eq. (16.3.7). Using Eqs. (16.3.1) and (16.3.4), we solve for $\{d_{i-1}\}$ as

$$\{d_{i-1}\} = \{d_i\} - (\Delta t)\{\dot{d}_i\} + \frac{(\Delta t)^2}{2}\{\ddot{d}_i\} \qquad (16.3.8)$$

The procedure for solution is then as follows:

Step 1 Given: $\{d_0\}$, $\{\dot{d}_0\}$, and $\{F_i(t)\}$.

Step 2 If $\{\ddot{d}_0\}$ is not initially given, solve Eq. (16.3.5) at $t = 0$ for $\{\ddot{d}_0\}$; that is,

$$\{\ddot{d}_0\} = [M]^{-1}(\{F_0\} - [K]\{d_0\})$$

Step 3 Solve Eq. (16.3.8) at $t = -\Delta t$ for $\{d_{-1}\}$; that is,

$$\{d_{-1}\} = \{d_0\} - (\Delta t)\{\dot{d}_0\} + \frac{(\Delta t)^2}{2}\{\ddot{d}_0\}$$

Step 4 Having solved for $\{d_{-1}\}$ in step 3, now solve for $\{d_1\}$ using Eq. (16.3.7) as

$$\{d_1\} = [M]^{-1}\{(\Delta t)^2\{F_0\} + [2[M] - (\Delta t)^2 [K]]\{d_0\} - [M]\{d_{-1}\}\}$$

Step 5 With $\{d_0\}$ initially given, and $\{d_1\}$ determined from step 4, use Eq. (16.3.7) to obtain

$$\{d_2\} = [M]^{-1}\{(\Delta t)^2\{F_1\} + [2[M] - (\Delta t)^2 [K]]\{d_1\} - [M]\{d_0\}\}$$

Step 6 Using Eq. (16.3.5), solve for $\{\ddot{d}_1\}$ as

$$\{\ddot{d}_1\} = [M]^{-1}(\{F_1\} - [K]\{d_1\})$$

Step 7 Using the result of step 5 and the boundary condition for $\{d_0\}$ given in step 1, determine the velocity at the first time step by Eq. (16.3.1) as

$$\{\dot{d}_1\} = \frac{\{d_2\} - \{d_0\}}{2(\Delta t)}$$

Step 8 Use steps 5 through 7 repeatedly to obtain the displacement, acceleration, and velocity for all other time steps.

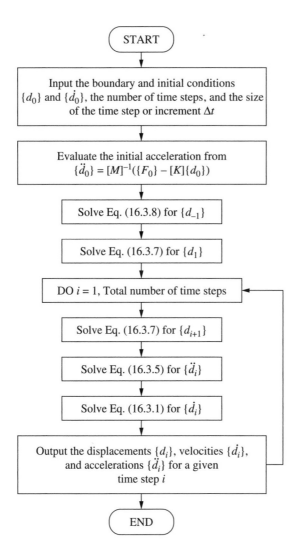

Figure 16–6 Flowchart of the central difference method

Figure 16–6 is a flowchart of the solution procedure using the central difference equations. Note that the recurrence formulas given by equations such as Eqs. (16.3.1) and (16.3.2) are approximate but yield sufficiently accurate results provided the time step Δt is taken small in relation to the variations in acceleration. Methods for determining proper time steps for the numerical integration process are described in Section 16.5.

We will now illustrate the central difference equations as they apply to the following example problem.

Example 16.1

Determine the displacement, velocity, and acceleration at 0.05-s time intervals up to 0.2 s for the one-dimensional spring-mass oscillator subjected to the time-dependent forcing function shown in Figure 16–7. [Guidelines regarding appropriate time intervals (or time steps) are given in Section 16.5.] This forcing function is a typical one assumed for blast loads. The restoring spring force versus displacement curve is also provided. [Note that Figure 16–7 also represents a one-element bar with its left end fixed and right node subjected to $F(t)$ when a lumped mass is used.]

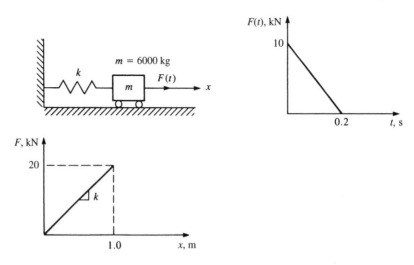

Figure 16–7 Spring-mass oscillator subjected to a time-dependent force

SOLUTION:

Because we are considering the single degree of freedom associated with the mass, the general matrix equations describing the motion reduce to single scalar equations. We will represent this single degree of freedom by d.

The solution procedure follows the steps outlined in this section and in the flow-chart of Figure 16–6.

Step 1

At time $t = 0$, the initial displacement and velocity are zero; therefore,

$$d_0 = 0 \qquad \dot{d}_0 = 0$$

Step 2

The initial acceleration at $t = 0$ is obtained using Eq. (16.3.5) as

$$\ddot{d}_0 = \frac{10 \times 10^3 - (20 \times 10^3)(0)}{6000} = 1.67 \text{ m/s}^2 = 1670 \text{ mm/s}^2$$

where we have used $\{F(0)\} = 10$ kN and $[K] = 20$ kN/m

Step 3

The displacement d_{-1} is obtained using Eq. (16.3.8) as

$$d_{-1} = 0 - 0 + \frac{(0.05)^2}{2}(1.67) = 2.08 \times 10^{-3} \text{ m} = 2.08 \text{ mm}$$

Step 4

The displacement at time $t = 0.05$ s using Eq. (16.3.7) is

$$d_1 = \frac{1}{6000}\left(\{(0.05)^2(10{,}000) + [2(6000) - (0.05)^2(20{,}000)](0) - (6000)(2.08 \times 10^{-3})\}\right)$$

$$= 2.08 \times 10^{-3} \text{ m} = 2.08 \text{ mm}$$

Step 5

Having obtained d_1, we now use Eq. (16.3.7) to determine the displacement at time $t = 0.10$ s as

$$d_2 = \frac{1}{6000}\left((0.05)^2(7500) + [2(6000) - (0.05)^2(20{,}000)](2.08 \times 10^{-3}) - (6000)(0)\right)$$

$$= 7.27 \times 10^{-3} \text{ m} = 7.27 \text{ mm}$$

Step 6

The acceleration at time $t = 0.05$ s is obtained using Eq. (16.3.5) as

$$\ddot{d}_1 = \frac{1}{6000}[7500 - 20{,}000(2.08 \times 10^{-3})] = 1.24 \text{ m/s}^2$$

Step 7

The velocity at time $t = 0.05$ s is obtained using Eq. (16.3.1) as

$$\dot{d}_1 = \frac{7.27 \times 10^{-3} - 0}{2(0.05)} = 0.0727 \text{ m/s} = 72.7 \text{ mm/s}$$

Step 8

Repeated use of steps 5 through 7 will result in the displacement, acceleration, and velocity for additional time steps as desired. We will now perform one more time-step iteration of the procedure.

Repeating step 5 for the next time step, we have displacement d_3 as

$$d_3 = \frac{1}{6000}\left((0.05)^2(5000) + [2(6000) - (0.05)^2(20{,}000)](7.27 \times 10^{-3})\right.$$

$$\left. - (6000)(2.08 \times 10^{-3})\right) = 0.01448 = 14.48 \text{ mm}$$

Repeating step 6 for the next time step, we have acceleration $\ddot{d}_2$ as

$$\ddot{d}_2 = \frac{1}{6000}[(5000) - (20{,}000)(7.27 \times 10^{-3})] = 0.809 \text{ m/s}^2$$

Table 16-1 Results of the analysis of Example 16.1

t (s)	$F(t)$ (kN)	d_i (mm)	Q (kN)	$\ddot{d}_i$ (mm/s²)	$\dot{d}_i$ (mm/s)	d_i (exact) in mm
0	10	0	0	1670	0	0
0.05	7.5	2.08	41.6	1240	72.7	1.795
0.10	5.0	7.27	145.4	810	124.0	6.51
0.15	2.5	14.48	289.6	370	153.4	13.13
0.20	0	22.65	453	−71	160.8	20.63
0.25	0	30.61	612.2	−0.089	156.6	28.3

Finally, repeating step 7 for the next time step, we obtain velocity $\dot{d}_2$ as

$$\dot{d}_2 = \frac{(14.48 \times 10^{-3}) - (2.08 \times 10^{-3})}{2(0.05)} = 124.0$$

Table 16-1 summarizes the results obtained through time $t = 0.25$ s. In Table 16-1, $Q = kd_i$ is the restoring spring force. Also, the exact analytical solution for displacement based on the equation in Reference [14] is given by

$$y = \frac{F_0}{k}(1 - \cos\omega t) + \frac{F_0}{kt_d}\left(\frac{\sin\omega t}{\omega} - t\right)$$

where $F_0 = 2000$ lb, $k = 100$ lb/in., $t_d = 0.2$ s, and

$$\omega = \sqrt{\frac{k}{m}} = \sqrt{\frac{20 \times 10^3}{6000}} = 1.825 \text{ rad/s}$$ ■

Newmark's Method

We will now outline Newmark's numerical method, which, because of its general versatility, has been adopted into numerous commercially available computer programs for purposes of structural dynamics analysis. (Complete development of the equations can be found in Reference [5].) Newmark's equations are given by

$$\{\dot{d}_{i+1}\} = \{\dot{d}_i\} + (\Delta t)[(1 - \gamma)\{\ddot{d}_i\} + \gamma\{\ddot{d}_{i+1}\}] \tag{16.3.9}$$

$$\{d_{i+1}\} = \{d_i\} + (\Delta t)\{\dot{d}_i\} + (\Delta t)^2[(\tfrac{1}{2} - \beta)\{\ddot{d}_i\} + \beta\{\ddot{d}_{i+1}\}] \tag{16.3.10}$$

where β and γ are parameters chosen by the user. The parameter β is generally chosen between 0 and $\frac{1}{4}$, and γ is often taken to be $\frac{1}{2}$. For instance, choosing $\gamma = \frac{1}{2}$ and $\beta = 0$, it can be shown that Eqs. (16.3.9) and (16.3.10) reduce to the central difference Eqs. (16.3.1) and (16.3.2). If $\gamma = \frac{1}{2}$ and $\beta = \frac{1}{6}$ are chosen, Eqs. (16.3.9) and (16.3.10) correspond to those for which a linear acceleration assumption is valid within each time interval. For $\gamma = \frac{1}{2}$ and $\beta = \frac{1}{4}$, it has been shown that the numerical analysis is stable; that is, computed quantities such as displacement and velocities do not become unbounded regardless of the time step chosen. Furthermore, it has been found [5] that a time step of approximately $\frac{1}{10}$ of the shortest natural frequency of the structure being analyzed usually yields the best results.

To find $\{d_{i+1}\}$, we first multiply Eq. (16.3.10) by the mass matrix $[M]$ and then substitute Eq. (16.3.5) for $\{\ddot{d}_{i+1}\}$ into this equation to obtain

$$[M]\{d_{i+1}\} = [M]\{d_i\} + (\Delta t)[M]\{\dot{d}_i\} + (\Delta t)^2[M](\tfrac{1}{2} - \beta)\{\ddot{d}_i\}] + \beta(\Delta t)^2[\{F_{i+1}\} - [K]\{d_{i+1}\}] \tag{16.3.11}$$

Combining like terms of Eq. (16.3.11), we obtain

$$([M] + \beta(\Delta t)^2[K])\{d_{i+1}\} = \beta(\Delta t)^2\{F_{i+1}\} + [M]\{d_i\} + (\Delta t)[M]\{\dot{d}_i\} + (\Delta t)^2[M](\tfrac{1}{2} - \beta)\{\ddot{d}_i\} \tag{16.3.12}$$

Finally, dividing Eq. (16.3.12) by $\beta(\Delta t)^2$, we obtain

$$[K']\{d_{i+1}\} = \{F'_{i+1}\} \tag{16.3.13}$$

where

$$[K'] = [K] + \frac{1}{\beta(\Delta t)^2}[M] \tag{16.3.14}$$

$$\{F'_{i+1}\} = \{F_{i+1}\} + \frac{[M]}{\beta(\Delta t)^2}\left[\{d_i\} + (\Delta t)\{\dot{d}_i\} + \left(\frac{1}{2} - \beta\right)(\Delta t)^2\{\ddot{d}_i\}\right]$$

The solution procedure using Newmark's equations is as follows:

1. Starting at time $t = 0$, $\{d_0\}$ is known from the given boundary conditions on displacement, and $\{\dot{d}_0\}$ is known from the initial velocity conditions.
2. Solve Eq. (16.3.5) at $t = 0$ for $\{\ddot{d}_0\}$ (unless $\{\ddot{d}_0\}$ is known from an initial acceleration condition); that is,

$$\{\ddot{d}_0\} = [M]^{-1}(\{F_0\} - [K]\{d_0\})$$

3. Solve Eq. (16.3.13) for $\{d_1\}$, because $\{F_{i+1}\}$ is known for all time steps and $\{d_0\}, \{\dot{d}_0\}$, and $\{\ddot{d}_0\}$ are now known from steps 1 and 2.
4. Use Eq. (16.3.10) to solve for $\{\ddot{d}_1\}$ as

$$\{\ddot{d}_1\} = \frac{1}{\beta(\Delta t)^2}\left[\{d_1\} - \{d_0\} - (\Delta t)\{\dot{d}_0\} - (\Delta t)^2\left(\frac{1}{2} - \beta\right)\{\ddot{d}_0\}\right]$$

5. Solve Eq. (16.3.9) directly for $\{\dot{d}_1\}$.
6. Using the results of steps 4 and 5, go back to step 3 to solve for $\{d_2\}$ and then to steps 4 and 5 to solve for $\{\ddot{d}_2\}$ and $\{\dot{d}_2\}$. Use steps 3–5 repeatedly to solve for $\{d_{i+1}\}, \{\ddot{d}_{i+1}\}$, and $\{\dot{d}_{i+1}\}$.

Figure 16–8 is a flowchart of the solution procedure using Newmark's equations. The advantages of Newmark's method over the central difference method are that Newmark's method can be made unconditionally stable (for instance, if $\beta = \tfrac{1}{4}$ and $\gamma = \tfrac{1}{2}$) and that larger time steps can be used with better results because, in general, the difference expressions more closely approximate the true acceleration and displacement time behavior [8] to [11]. Other difference formulas, such as Wilson's and Houbolt's, also yield unconditionally stable algorithms.

We will now illustrate the use of Newmark's equations as they apply to the following example problem.

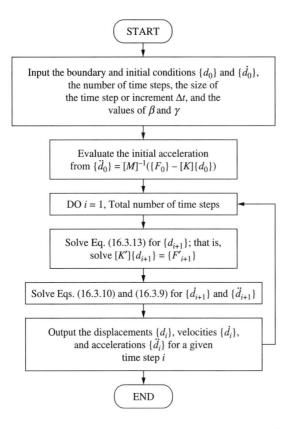

Figure 16–8 Flowchart of numerical integration in time using Newmark's equations

---

Example 16.2

Determine the displacement, velocity, and acceleration at 0.1-s time increments up to a time of 0.5 s for the one-dimensional spring-mass oscillator subjected to the time-dependent forcing function shown in Figure 16–9, along with the restoring spring force versus displacement curve. Assume the oscillator is initially at rest. Let $\beta = \frac{1}{6}$ and $\gamma = \frac{1}{2}$, which corresponds to an assumption of linear acceleration within each time step.

SOLUTION:

Because we are again considering the single degree of freedom associated with the mass, the general matrix equations describing the motion reduce to single scalar equations. Again, we represent this single degree of freedom by d.

The solution procedure follows the steps outlined in this section and in the flow-chart of Figure 16–8.

Step 1

At time $t = 0$, the initial displacement and velocity are zero; therefore,

$$d_0 = 0 \qquad \dot{d}_0 = 0$$

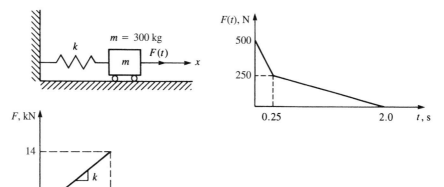

Figure 16–9 Spring-mass oscillator subjected to a time-dependent force

Step 2

The initial acceleration from Eq. (16.3.5) at $t = 0$ is obtained as

$$\ddot{d}_0 = \frac{500 - 14{,}000(0)}{300} = 1.67 \text{ m/s}^2 = 1670 \text{ mm/s}^2$$

where we have used $\{F_0\} = 500$ N and $[K] = 14$ kN/m

Step 3

We now solve for the displacement from Eq. (16.3.13) and Eq. (16.3.14) at time $t = 0.1$ s as

$$K' = 14{,}000 + \frac{1}{\left(\frac{1}{6}\right)(0.1)^2}(300) = 194{,}000 \text{ N/m}$$

$$F_1' = 400 + \frac{300}{\left(\frac{1}{6}\right)(0.1)^2}\left[0 + (0.1)(0) + \left(\frac{1}{2} - \frac{1}{6}\right)(0.1)^2(1.67)\right] = 1402$$

$$d_1 = \frac{1402}{194{,}000} = 0.0072 \text{ m} = 7.2 \text{ mm}$$

Step 4

Solve for the acceleration from Eq. (16.3.10) at time $t = 0.1$ s as

$$\ddot{d}_1 = \frac{1}{\left(\frac{1}{6}\right)(0.1)^2}\left[7.2 \times 10^{-3} - 0 - (0.1)(0) - (0.1)^2\left(\frac{1}{2} - \frac{1}{6}\right)(1.67)\right]$$

$$\ddot{d}_1 = 0.98 \text{ m/s}^2 = 980 \text{ mm/s}^2$$

Step 5

Solve Eq. (16.3.9) for the velocity at time $t = 0.1$ s as

$$\dot{d}_1 = 0 + (0.1)\left[\left(1 - \tfrac{1}{2}\right)(1.67) + \left(\tfrac{1}{2}\right)(0.98)\right]$$

$$\dot{d}_1 = 0.132 \text{ m/s} = 132 \text{ mm/s}$$

Step 6

Repeated use of steps 3 through 5 will result in the displacement, acceleration, and velocity for additional time steps as desired. We will now perform one more time-step iteration.

Repeating step 3 for the next time step $(t = 0.2 \text{ s})$, we have

$$F_2' = 300 + \frac{300}{\left(\tfrac{1}{6}\right)(0.1)^2}\left[7.2 \times 10^{-3} + (0.1)(0.132) + \left(\tfrac{1}{2} - \tfrac{1}{6}\right)(0.1)^2(0.98)\right]$$

$$F_2' = 4560$$

$$d_2 = \frac{4560}{194000} = 0.024 \text{ m} = 24 \text{ mm}$$

Repeating step 4 for time step $t = 0.2$ s, we obtain

$$\ddot{d}_2 = \frac{1}{\left(\tfrac{1}{6}\right)(0.1)^2}\left[0.024 - 7.2 \times 10^{-3} - (0.1)(0.132) - (0.1)^2\left(\tfrac{1}{2} - \tfrac{1}{6}\right)(0.98)\right]$$

$$\ddot{d}_2 = 0.202 \text{ m/s}^2 = 202 \text{ mm/s}^2$$

Finally, repeating step 5 for time step $t = 0.2$ s, we have

$$\dot{d}_2 = 0.132 + (0.1)\left[\left(1 - \tfrac{1}{2}\right)(0.98) + \tfrac{1}{2}(0.202)\right]$$

$$\dot{d}_2 = 0.191 \text{ m/s} = 191 \text{ mm/s}$$

Table 16–2 summarizes the results obtained through time $t = 0.5$ s.

Table 16–2 Results of the analysis of Example 16.2

t (s)	$F(t)$ (N)	d_i (mm)	Q (N)	$\ddot{d}_i$ (mm/s^2)	$\dot{d}_i$ (mm/s)
0.	500	0	0	1670	0
0.1	400	7.2	100.8	980	132
0.2	300	24	336	202	191
0.3	243	39.5	553	−750	149.9
0.4	228.5	50.1	701.4	−1208	50.1
0.5	214.5	48.8	683.2	−1208	−70.5

■

Wilson's Method

We will now outline Wilson's method (also called the Wilson-Theta method). Because of its general versatility, it has been adopted into the Algor computer program for purposes of structural dynamics analysis. Wilson's method is an extension of the linear acceleration method wherein the acceleration is assumed to vary linearly within each time interval now taken from t to $t + \Theta \Delta t$, where $\Theta \geq 1.0$. For $\Theta = 1.0$, the method reduces to the linear acceleration scheme. However, for unconditional stability in the numerical analysis, we must use $\Theta \geq 1.37$ [7, 8]. In practice, $\Theta = 1.40$ is often selected. The Wilson equations are given in a form similar to the previous Newmark's equations, Eqs. (16.3.9) and (16.3.10), as

$$\{\dot{d}_{i+1}\} = \{\dot{d}_i\} + \frac{\Theta \Delta t}{2}(\{\ddot{d}_{i+1}\} + \{\ddot{d}_i\}) \tag{16.3.15}$$

$$\{d_{i+1}\} = \{d_i\} + \Theta \Delta t \{\dot{d}_i\} + \frac{\Theta^2 (\Delta t)^2}{6}(\{\ddot{d}_{i+1}\} + 2\{\ddot{d}_i\}) \tag{16.3.16}$$

where $\{\ddot{d}_{i+1}\}$, $\{\dot{d}_{i+1}\}$, and $\{d_{i+1}\}$ represent the acceleration, velocity, and displacement, respectively, at time $t + \Theta \Delta t$.

We seek a matrix equation of the form of Eq. (16.3.13) that can be solved for displacement $\{d_{i+1}\}$. To obtain this equation, first solve Eqs. (16.3.15) and (16.3.16) for $\{\ddot{d}_{i+1}\}$ and $\{\dot{d}_{i+1}\}$ in terms of $\{d_{i+1}\}$ as follows:

Solve Eq. (16.3.16) for $\{\ddot{d}_{i+1}\}$ to obtain

$$\{\ddot{d}_{i+1}\} = \frac{6}{\Theta^2 (\Delta t)^2}(\{d_{i+1}\} - \{d_i\}) - \frac{6}{\Theta \Delta t}\{\dot{d}_i\} - 2\{\ddot{d}_i\} \tag{16.3.17}$$

Now use Eq. (16.3.17) in Eq. (16.3.15) and solve for $\{\dot{d}_{i+1}\}$ to obtain

$$\{\dot{d}_{i+1}\} = \frac{3}{\Theta \Delta t}(\{d_{i+1}\} - \{d_i\}) - 2\{\dot{d}_i\} - \frac{\Theta \Delta t}{2}\{\ddot{d}_i\} \tag{16.3.18}$$

To obtain the displacement $\{d_{i+1}\}$ (at time $t + \Theta \Delta t$), we use the equation of motion Eq. (16.2.24) rewritten as

$$\{F_{i+1}\} = [M]\{\ddot{d}_{i+1}\} + [K]\{d_{i+1}\} \tag{16.3.19}$$

Now, substituting Eq. (16.3.17) for $\{\ddot{d}_{i+1}\}$ into Eq. (16.3.19), we obtain

$$[M]\left[\frac{6}{\Theta^2 (\Delta t)^2}(\{d_{i+1}\} - \{d_i\}) - \frac{6}{\Theta \Delta t}\{\dot{d}_i\} - 2\{\ddot{d}_i\}\right] + [K]\{d_{i+1}\} = \{F_{i+1}\} \tag{16.3.20}$$

Combining like terms and rewriting in a form similar to Eq. (16.3.13), we obtain

$$[K']\{d_{i+1}\} = \{F'_{i+1}\} \tag{16.3.21}$$

where $\qquad [K'] = [K] + \dfrac{6}{(\Theta \Delta t)^2}[M]$

$$\tag{16.3.22}$$

$$\{F'_{i+1}\} = \{F_{i+1}\} + \frac{[M]}{(\Theta \Delta t)^2}[6\{d_i\} + 6\Theta \Delta t \{\dot{d}_i\} + 2(\Theta \Delta t)^2 \{\ddot{d}_i\}]$$

You will note the similarities between Wilson's Eqs. (16.3.22) and Newmark's Eqs. (16.3.14). Because the acceleration is assumed to vary linearly, the load vector is expressed as

$$\{\bar{F}_{i+1}\} = \{F_i\} + \Theta(\{F_{i+1}\} - \{F_i\}) \tag{16.3.23}$$

where $\{\bar{F}_{i+1}\}$ replaces $\{F_{i+1}\}$ in Eq. (16.3.22). Note that if $\Theta = 1$, $\{\bar{F}_{i+1}\} = \{F_{i+1}\}$.

Also, Wilson's method (like Newmark's) is an implicit integration method, because the displacements show up as multiplied by the stiffness matrix and we implicitly solve for the displacements at time $t + \Theta\Delta t$.

The solution procedure using Wilson's equations is as follows:

1. Starting at time $t = 0$, $\{d_0\}$ is known from the given boundary conditions on displacement, and $\{\dot{d}_0\}$ is known from the initial velocity conditions.
2. Solve Eq. (16.3.5) for $\{\ddot{d}_0\}$ (unless $\{\ddot{d}_0\}$ is known from an initial acceleration condition).
3. Solve Eq. (16.3.21) for $\{d_1\}$, because $\{F'_{i+1}\}$ is known for all time steps, and $\{d_0\}, \{\dot{d}_0\}$, and $\{\ddot{d}_0\}$ are now known from steps 1 and 2.
4. Solve Eq. (16.3.17) for $\{\ddot{d}_1\}$.
5. Solve Eq. (16.3.18) for $\{\dot{d}_1\}$.
6. Using the results of steps 4 and 5, go back to step 3 to solve for $\{d_2\}$, and then return to steps 4 and 5 to solve for $\{\ddot{d}_2\}$ and $\{\dot{d}_2\}$. Use steps 3–5 repeatedly to solve for $\{d_{i+1}\}, \{\dot{d}_{i+1}\}$, and $\{\ddot{d}_{i+1}\}$.

A flowchart similar to Figure 16–8, based on Newmark's equation, is left to your discretion. Again, note that the advantage of Wilson's method is that it can be made unconditionally stable by setting $\Theta \geq 1.37$. Finally, the time step, Δt, recommended is approximately $\frac{1}{10}$ to $\frac{1}{20}$ of the shortest natural period τ_n of the finite element assemblage with n degrees of freedom; that is, $\Delta t \doteq \tau_n/10$. In comparing the Newmark and Wilson methods, we observe little difference in the computational effort, because they both require about the same time step. Wilson's method is very similar to Newmark's, so hand solutions will not be presented. However, we suggest that you rework Example 16.1 by Wilson's method and compare your displacement results with the exact solution listed in Table 16–1.

▲ 16.4 Natural Frequencies of a One-Dimensional Bar ▲

Before solving the structural stress dynamics analysis problem, we will first describe how to determine the natural frequencies of continuous elements (specifically the bar element). The natural frequencies are necessary in a vibration analysis and also are important when choosing a proper time step for a structural dynamics analysis (as will be discussed in Section 16.5).

Natural frequencies are determined by solving Eq. (16.2.24) in the absence of a forcing function $F(t)$. Therefore, we solve the matrix equation

$$[M]\{\ddot{d}\} + [K]\{d\} = 0 \qquad (16.4.1)$$

The standard solution for $\{d(t)\}$ is given by the harmonic equation in time

$$\{d(t)\} = \{d'\}e^{i\omega t} \qquad (16.4.2)$$

where $\{d'\}$ is the part of the nodal displacement matrix called *natural modes* that is assumed to be independent of time, i is the standard imaginary number given by $i = \sqrt{-1}$, and ω is a natural frequency.

Differentiating Eq. (16.4.2) twice with respect to time, we obtain

$$\{\ddot{d}(t)\} = \{d'\}(-\omega^2)e^{i\omega t} \qquad (16.4.3)$$

Substitution of Eqs. (16.4.2) and (16.4.3) into Eq. (16.4.1) yields

$$-[M]\,\omega^2\{d'\}e^{i\omega t} + [K]\{d'\}e^{i\omega t} = 0 \qquad (16.4.4)$$

Combining terms in Eq. (16.4.4), we obtain

$$e^{i\omega t}([K] - \omega^2[M])\{d'\} = 0 \qquad (16.4.5)$$

Because $e^{i\omega t}$ is not zero, from Eq. (16.4.5) we obtain

$$([K] - \omega^2[M])\{d'\} = 0 \qquad (16.4.6)$$

Equation (16.4.6) is a set of linear homogeneous equations in terms of displacement mode $\{d'\}$. Hence, Eq. (16.4.6) has a nontrivial solution if and only if the determinant of the coefficient matrix of $\{d'\}$ is zero; that is, we must have

$$|[K] - \omega^2[M]| = 0 \qquad (16.4.7)$$

In general, Eq. (16.4.7) is a set of n algebraic equations, where n is the number of degrees of freedom associated with the problem.

To illustrate the procedure for determining the natural frequencies, we will solve the following example problem.

Example 16.3

For the bar shown in Figure 16–10 with length $2L$, modulus of elasticity E, mass density ρ, and cross-sectional area A, determine the first two natural frequencies.

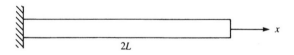

Figure 16–10 One-dimensional bar used for natural frequency determination

SOLUTION:

For simplicity, the bar is discretized into two elements each of length L as shown in Figure 16–11. To solve Eq. (16.4.7), we must develop the total stiffness matrix for

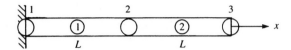

Figure 16–11 Discretized bar of Figure 16–10

the bar by using Eq. (16.2.11). Either the lumped-mass matrix Eq. (16.2.12) or the consistent-mass matrix Eq. (16.2.23) can be used. In general, using the consistent-mass matrix has resulted in solutions that compare more closely to available analytical and experimental results than those found using the lumped-mass matrix. However, the longhand calculations are more tedious using the consistent-mass matrix than using the lumped-mass matrix because the consistent-mass matrix is a full symmetric matrix, whereas the lumped-mass matrix has nonzero terms only along the main diagonal. Hence, the lumped-mass matrix will be used in this analysis.

Using Eq. (16.2.11), the stiffness matrices for each element are given by

$$[k^{(1)}] = \frac{AE}{L} \begin{matrix} 1 & 2 \\ \begin{bmatrix} 1 & -1 \\ -1 & 1 \end{bmatrix} \end{matrix} \qquad [k^{(2)}] = \frac{AE}{L} \begin{matrix} 2 & 3 \\ \begin{bmatrix} 1 & -1 \\ -1 & 1 \end{bmatrix} \end{matrix} \qquad (16.4.8)$$

The usual direct stiffness method for assembling the element matrices, Eqs. (16.4.8), yields the global stiffness matrix for the whole bar as

$$[K] = \frac{AE}{L} \begin{bmatrix} 1 & -1 & 0 \\ -1 & 2 & -1 \\ 0 & -1 & 1 \end{bmatrix} \qquad (16.4.9)$$

Using Eq. (16.2.12), the mass matrices for each element are given by

$$[m^{(1)}] = \frac{\rho A L}{2} \begin{matrix} 1 & 2 \\ \begin{bmatrix} 1 & 0 \\ 0 & 1 \end{bmatrix} \end{matrix} \qquad [m^{(2)}] = \frac{\rho A L}{2} \begin{matrix} 2 & 3 \\ \begin{bmatrix} 1 & 0 \\ 0 & 1 \end{bmatrix} \end{matrix} \qquad (16.4.10)$$

The mass matrices for each element are assembled in the same manner as for the stiffness matrices. Therefore, by assembling Eqs. (16.4.10), we obtain the global mass matrix as

$$[M] = \frac{\rho A L}{2} \begin{bmatrix} 1 & 0 & 0 \\ 0 & 2 & 0 \\ 0 & 0 & 1 \end{bmatrix} \qquad (16.4.11)$$

We observe from the resulting global mass matrix that there are two mass contributions at node 2 because node 2 is common to both elements.

Substituting the global stiffness matrix Eq. (16.4.9) and the global mass matrix Eq. (16.4.11) into Eq. (16.4.6), and using the boundary condition $u_1 = 0$ (or now $d_1' = 0$) to reduce the set of equations in the usual manner, we obtain

$$\left(\frac{AE}{L} \begin{bmatrix} 2 & -1 \\ -1 & 1 \end{bmatrix} - \omega^2 \frac{\rho A L}{2} \begin{bmatrix} 2 & 0 \\ 0 & 1 \end{bmatrix} \right) \begin{Bmatrix} d_2' \\ d_3' \end{Bmatrix} = \begin{Bmatrix} 0 \\ 0 \end{Bmatrix} \qquad (16.4.12)$$

To obtain a solution to the set of homogeneous equations in Eq. (16.4.12), we set the determinant of the coefficient matrix equal to zero as indicated by Eq. (16.4.7). We then have

$$\left| \frac{AE}{L} \begin{bmatrix} 2 & -1 \\ -1 & 1 \end{bmatrix} - \lambda \frac{\rho AL}{2} \begin{bmatrix} 2 & 0 \\ 0 & 1 \end{bmatrix} \right| = 0 \tag{16.4.13}$$

where $\lambda = \omega^2$ has been used in Eq. (16.4.13). Dividing Eq. (16.4.13) by ρAL and letting $\mu = E/(\rho L^2)$, we obtain

$$\begin{vmatrix} 2\mu - \lambda & -\mu \\ -\mu & \mu - \dfrac{\lambda}{2} \end{vmatrix} = 0 \tag{16.4.14}$$

Evaluating the determinant in Eq. (16.4.14), we obtain

$$\lambda = 2\mu \pm \mu\sqrt{2}$$

or
$$\lambda_1 = 0.60\mu \qquad \lambda_2 = 3.41\mu \tag{16.4.15}$$

For comparison, the exact solution is given by $\lambda = 0.616\mu$, whereas the consistent-mass approach yields $\lambda = 0.648\mu$. Therefore, for bar elements, the lumped-mass approach can yield results as good as, or even better than, the results for the consistent-mass approach. However, the consistent-mass approach can be mathematically proved to yield an upper bound on the frequencies, whereas the lumped-mass approach yields results that can be below or above the exact frequencies with no mathematical proof of boundedness. From Eqs. (16.4.15), the first and second natural frequencies are given by

$$\omega_1 = \sqrt{\lambda_1} = 0.77\sqrt{\mu} \qquad \omega_2 = \sqrt{\lambda_2} = 1.85\sqrt{\mu}$$

Letting $E = 210$ GPa, $\rho = 7850$ kg/m^3, and $L = 2.5$ m, we obtain

$$\mu = E/(\rho L^2) = (210 \times 10^9)/[(7850)(2.5)^2] = 4.28 \times 10^{-6} \text{ s}^{-2}$$

Therefore, we obtain the natural circular frequencies as

$$\omega_1 = 1.59 \times 10^3 \text{ rad/s} \qquad \omega_2 = 3.83 \times 10^3 \text{ rad/s} \tag{16.4.16}$$

or in Hertz (1/s) units

$$f_1 = \omega_1/2\pi = 253 \text{ Hz}, \qquad \text{and so on}$$

In conclusion, note that for a bar discretized such that two nodes are free to displace, there are two natural modes and two frequencies. When a system vibrates with a given natural frequency ω_i, that unique shape with arbitrary amplitude corresponding to ω_i is called the *mode*. In general, for an *n*-degrees-of-freedom discrete system, there are *n* natural modes and frequencies. A continuous system actually has an infinite number of natural modes and frequencies. When the system is discretized, only *n* degrees of freedom are created. The lowest modes and frequencies are approximated most often; the higher frequencies are damped out more rapidly and are usually of less importance. A rule of thumb is to use two times as many elements as the number of frequencies desired.

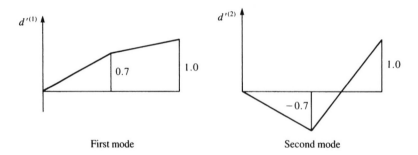

First mode   Second mode

Figure 16–12 First and second modes of longitudinal vibration for the cantilever bar of Figure 16–10

Substituting λ_1 from Eqs. (16.4.15) into Eq. (16.4.12) and simplifying, the first modal equations are given by

$$1.4\mu d_2'^{(1)} - \mu d_3'^{(1)} = 0$$

$$-\mu d_2'^{(1)} + 0.7\mu d_3'^{(1)} = 0$$

(16.4.17)

It is customary to specify the value of one of the natural modes $\{d'\}$ for a given ω_i or λ_i. Letting $d_3'^{(1)} = 1$ and solving Eq. (16.4.17), we find $d_2'^{(1)} = 0.7$. Similarly, substituting λ_2 from Eqs. (16.4.15) into Eq. (16.4.12), we obtain the second modal equations. For brevity's sake, these equations are not presented here. Now letting $d_3'^{(2)} = 1$ results in $d_2'^{(2)} = -0.7$. The modal response for the first and second natural frequencies of longitudinal vibration are plotted in Figure 16–12. The first mode means that the bar is completely in tension or compression, depending on the excitation direction. The second mode means the bar is in compression and tension or in tension and compression. ■

▲ **16.5 Time-Dependent One-Dimensional Bar Analysis** ▲

Example 16.4

To illustrate the finite element solution of a time-dependent problem, we will solve the problem of the one-dimensional bar shown in Figure 16–13(a) subjected to the force shown in Figure 16–13(b). We will assume the boundary condition $u_1 = 0$ and the initial conditions $\{d_0\} = 0$ and $\{\dot{d}_0\} = 0$. For later numerical computation purposes, we let parameters $\rho = 7850$ kg/m^3, $A = 1$ cm^2, $E = 210$ GPa, and $L = 2.5$ m. These parameters are the same values as used in Section 16.4.

SOLUTION:

Because the bar is discretized into two elements of equal length, the global stiffness and mass matrices determined in Section 16.4 and given by Eqs. (16.4.9) and (16.4.11) are applicable. We will again use the lumped-mass matrix because of its

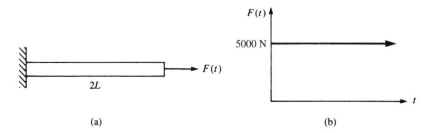

Figure 16–13 (a) Bar subjected to a time-dependent force and (b) the forcing function applied to the end of the bar

Figure 16–14 Discretized bar with lumped masses

resulting computational simplicity. Figure 16–14 shows the discretized bar and the associated lumped masses.

For illustration of the numerical time integration scheme, we will use the central difference method because it is easier to apply for longhand computations (and without loss of generality).

We next select the time step to be used in the integration process. It has been mathematically shown that the time step must be less than or equal to 2 divided by the highest natural frequency when the central difference method is used [7]; that is, $\Delta t \leq 2/\omega_{\max}$. However, for practical results, we must use a time step of less than or equal to three-fourths of this value; that is,

$$\Delta t \leq \frac{3}{4}\left(\frac{2}{\omega_{\max}}\right) \tag{16.5.1}$$

This time step ensures stability of the integration method. This criterion for selecting a time step demonstrates the usefulness of determining the natural frequencies of vibration, as previously described in Section 16.4, before performing the dynamic stress analysis. An alternative guide (used only for a bar) for choosing the approximate time step is

$$\Delta t = \frac{L}{c_x} \tag{16.5.2}$$

where L is the element length, and $c_x = \sqrt{E_x/\rho}$ is called the *longitudinal wave velocity*. Evaluating the time step by using both criteria, Eqs. (16.5.1) and (16.5.2), from Eqs. (16.4.16) for ω, we obtain

$$\Delta t = \frac{3}{4}\left(\frac{2}{\omega_{\max}}\right) = \frac{1.5}{3.8 \times 10^3} = 0.40 \times 10^{-3} \text{ s} \tag{16.5.3}$$

or
$$\Delta t = \frac{L}{c_x} = \frac{100}{\sqrt{2.1 \times 10^{11}/7850}} = 0.48 \times 10^{-3} \text{ s} \tag{16.5.4}$$

Guided by the maximum time steps calculated in Eqs. (16.5.3) and (16.5.4), we choose $\Delta t = 0.25 \times 10^{-3}$ s as a convenient time step for the computations.

Substituting the global stiffness and mass matrices, Eqs. (16.4.9) and (16.4.11), into the global dynamic Eq. (16.2.24), we obtain

$$\frac{AE}{L}\begin{bmatrix} 1 & -1 & 0 \\ -1 & 2 & -1 \\ 0 & -1 & 1 \end{bmatrix}\begin{Bmatrix} u_1 \\ u_2 \\ u_3 \end{Bmatrix} + \frac{\rho AL}{2}\begin{bmatrix} 1 & 0 & 0 \\ 0 & 2 & 0 \\ 0 & 0 & 1 \end{bmatrix}\begin{Bmatrix} \ddot{u}_1 \\ \ddot{u}_2 \\ \ddot{u}_3 \end{Bmatrix} = \begin{Bmatrix} R_1 \\ 0 \\ F_3(t) \end{Bmatrix} \tag{16.5.5}$$

where R_1 denotes the unknown reaction at node 1. Using the procedure for solution outlined in Section 16.3 and in the flowchart of Figure 16–6, we begin as follows:

Step 1

Given: $u_1 = 0$ because of the fixed support at node 1, and all nodal displacements and velocities are zero at time $t = 0$; that is, $\{d_0\} = 0$ and $\{\dot{d}_0\} = 0$. Also, assume $\ddot{u}_1 = 0$ at all times.

Step 2

Solve for $\{\ddot{d}_0\}$ using Eq. (16.3.5) as

$$\{\ddot{d}_0\} = \begin{Bmatrix} \ddot{u}_2 \\ \ddot{u}_3 \end{Bmatrix}_{t=0} = \frac{2}{\rho AL}\begin{bmatrix} \frac{1}{2} & 0 \\ 0 & 1 \end{bmatrix}\begin{bmatrix} \begin{Bmatrix} 0 \\ 1000 \end{Bmatrix} - \frac{AE}{L}\begin{bmatrix} 2 & -1 \\ -1 & 1 \end{bmatrix}\begin{Bmatrix} 0 \\ 0 \end{Bmatrix} \end{bmatrix} \tag{16.5.6}$$

where Eq. (16.5.6) accounts for the conditions $u_1 = 0$ and $\ddot{u}_1 = 0$. Simplifying Eq. (16.5.6), we obtain

$$\{\ddot{d}_0\} = \frac{10000}{\rho AL}\begin{Bmatrix} 0 \\ 1 \end{Bmatrix} = \begin{Bmatrix} 0 \\ 5095 \end{Bmatrix} \text{ m/s}^2 \tag{16.5.7}$$

where the numerical values for ρ, A, and L have been substituted into the final numerical result in Eq. (16.5.7), and

$$[M]^{-1} = \frac{2}{\rho AL}\begin{bmatrix} \frac{1}{2} & 0 \\ 0 & 1 \end{bmatrix} \tag{16.5.8}$$

has been used in Eq. (16.5.6). The computational advantage of using the lumped-mass matrix for longhand calculations is now evident. The inverse of a diagonal matrix, such as the lumped-mass matrix, is obtained simply by inverting the diagonal elements of the matrix.

Step 3

Using Eq. (16.3.8), we solve for $\{d_{-1}\}$ as

$$\{d_{-1}\} = \{d_0\} - (\Delta t)\{\dot{d}_0\} + \frac{(\Delta t)^2}{2}\{\ddot{d}_0\} \tag{16.5.9}$$

Substituting the initial conditions on $\{\dot{d}_0\}$ and $\{d_0\}$ from step 1 and Eq. (16.5.7) for the initial acceleration $\{\ddot{d}_0\}$ from step 2 into Eq. (16.5.9), we obtain

$$\{d_{-1}\} = 0 - (0.25 \times 10^{-3})(0) + \frac{(0.25 \times 10^{-3})^2}{2}(5095)\begin{Bmatrix} 0 \\ 1 \end{Bmatrix}$$

or, on simplification,

$$\left\{ \begin{array}{c} u_2 \\ u_3 \end{array} \right\}_{-1} = \left\{ \begin{array}{c} 0 \\ 1.59 \times 10^{-4} \end{array} \right\} \text{ m} \qquad (16.5.10)$$

Step 4

On premultiplying Eq. (16.3.7) by $[M]^{-1}$, we now solve for $\{d_1\}$ by

$$\{d_1\} = [M]^{-1}\{(\Delta t)^2\{F_0\} + [2[M] - (\Delta t)^2[K]]\{d_0\} - [M]\{d_{-1}\}\} \quad (16.5.11)$$

Substituting the numerical values for $\rho, A, L,$ and E and the results of Eq. (16.5.10) into Eq. (16.5.11), we obtain

$$\left\{ \begin{array}{c} u_2 \\ u_3 \end{array} \right\}_1 = \frac{2}{1.9625} \begin{bmatrix} \frac{1}{2} & 0 \\ 0 & 1 \end{bmatrix} \left\{ (0.25 \times 10^{-3})^2 \left\{ \begin{array}{c} 0 \\ 5000 \end{array} \right\} + \left[\frac{2(1.9625)}{2} \begin{bmatrix} 2 & 0 \\ 0 & 1 \end{bmatrix} \right. \right.$$

$$- (0.25 \times 10^{-3})^2(8.4 \times 10^6) \begin{bmatrix} 2 & -1 \\ -1 & 1 \end{bmatrix} \left] \left\{ \begin{array}{c} 0 \\ 0 \end{array} \right\} \right.$$

$$\left. - \frac{1.9625}{2} \begin{bmatrix} 2 & 0 \\ 0 & 1 \end{bmatrix} \left\{ \begin{array}{c} 0 \\ 1.59 \times 10^{-4} \end{array} \right\} \right\}$$

Simplifying, we obtain

$$\left\{ \begin{array}{c} u_2 \\ u_3 \end{array} \right\}_1 = \frac{2}{1.9625} \begin{bmatrix} \frac{1}{2} & 0 \\ 0 & 1 \end{bmatrix} \left[\left\{ \begin{array}{c} 0 \\ 0.3125 \times 10^{-3} \end{array} \right\} - \left\{ \begin{array}{c} 0 \\ 0.156 \times 10^{-3} \end{array} \right\} \right]$$

Finally, the nodal displacements at time $t = 0.25 \times 10^{-3}$ s become

$$\left\{ \begin{array}{c} u_2 \\ u_3 \end{array} \right\}_1 = \left\{ \begin{array}{c} 0 \\ 1.59 \times 10^{-3} \end{array} \right\} \text{ m} \qquad (\text{at } t = 0.25 \times 10^{-3} \text{ s}) \qquad (16.5.12)$$

Step 5

With $\{d_0\}$ initially given and $\{d_1\}$ determined from step 4, we use Eq. (16.3.7) to obtain

$$\{d_2\} = [M]^{-1}\{(\Delta t)^2\{F_1\} + [2[M] - (\Delta t)^2[K]]\{d_1\} - [M]\{d_0\}\}$$

$$= \frac{2}{1.9625} \begin{bmatrix} \frac{1}{2} & 0 \\ 0 & 1 \end{bmatrix} \left\{ (0.25 \times 10^{-3})^2 \left\{ \begin{array}{c} 0 \\ 5000 \end{array} \right\} + \left[\frac{2(1.9625)}{2} \begin{bmatrix} 2 & 0 \\ 0 & 1 \end{bmatrix} \right. \right.$$

$$- (0.25 \times 10^{-3})^2(8.4 \times 10^6) \begin{bmatrix} 2 & -1 \\ -1 & 1 \end{bmatrix} \right]$$

$$\times \left\{ \begin{array}{c} 0 \\ 1.59 \times 10^{-4} \end{array} \right\} - \frac{1.9625}{2} \begin{bmatrix} 2 & 0 \\ 0 & 1 \end{bmatrix} \left\{ \begin{array}{c} 0 \\ 0 \end{array} \right\} \right\}$$

$$= \frac{2}{1.9625} \begin{bmatrix} \frac{1}{2} & 0 \\ 0 & 1 \end{bmatrix} \left[\left\{ \begin{array}{c} 0 \\ 0.3125 \times 10^{-3} \end{array} \right\} + \left\{ \begin{array}{c} 0.083 \times 10^{-3} \\ 0.2286 \times 10^{-3} \end{array} \right\} \right]$$

Simplifying, we obtain the nodal displacements at time $t = 0.50 \times 10^{-3}$ s as

$$\left\{ \begin{matrix} u_2 \\ u_3 \end{matrix} \right\}_2 = \left\{ \begin{matrix} 0.042 \times 10^{-3} \\ 0.551 \times 10^{-3} \end{matrix} \right\} \text{ m} \qquad (\text{at } t = 0.50 \times 10^{-3} \text{ s}) \qquad (16.5.13)$$

Step 6

Solve for the nodal accelerations $\{\ddot{d}_1\}$ again using Eq. (16.3.5) as

$$\{\ddot{d}_1\} = \frac{2}{1.9625} \begin{bmatrix} \frac{1}{2} & 0 \\ 0 & 1 \end{bmatrix} \left[\left\{ \begin{matrix} 0 \\ 5000 \end{matrix} \right\} - (8.4 \times 10^6) \begin{bmatrix} 2 & -1 \\ -1 & 1 \end{bmatrix} \left\{ \begin{matrix} 0 \\ 0.159 \times 10^{-3} \end{matrix} \right\} \right]$$

Simplifying, we then obtain the nodal accelerations at time $t = 0.25 \times 10^{-3}$ s as

$$\left\{ \begin{matrix} \ddot{u}_2 \\ \ddot{u}_3 \end{matrix} \right\}_1 = \left\{ \begin{matrix} 681 \\ 3734 \end{matrix} \right\} \text{ m/s}^2 \qquad (\text{at } t = 0.25 \times 10^{-3} \text{ s}) \qquad (16.5.14)$$

The reaction R_1 could be found by using the results of Eqs. (16.5.12) and (16.5.14) in Eq. (16.5.5).

Step 7

Using Eq. (16.5.13) from step 5 and the boundary condition for $\{d_0\}$ given in step 1, we obtain $\{\dot{d}_1\}$ as

$$\{\dot{d}_1\} = \frac{\left[\left\{ \begin{matrix} 0.042 \times 10^{-3} \\ 0.551 \times 10^{-3} \end{matrix} \right\} - \left\{ \begin{matrix} 0 \\ 0 \end{matrix} \right\} \right]}{2(0.25 \times 10^{-3})}$$

Simplifying, we obtain

$$\left\{ \begin{matrix} \dot{u}_2 \\ \dot{u}_3 \end{matrix} \right\} = \left\{ \begin{matrix} 0.084 \\ 1.102 \end{matrix} \right\} \text{ m/s} \qquad (\text{at } t = 0.25 \times 10^{-3} \text{ s})$$

Step 8

We now use steps 5 through 7 repeatedly to obtain the displacement, acceleration, and velocity for all other time steps. For simplicity, we calculate the acceleration only.
Repeating step 6 with $t = 0.50 \times 10^{-3}$ s, we obtain the nodal accelerations as

$$\{\ddot{d}_2\} = \frac{2}{1.9625} \begin{bmatrix} \frac{1}{2} & 0 \\ 0 & 1 \end{bmatrix} \left[\left\{ \begin{matrix} 0 \\ 5000 \end{matrix} \right\} - (8.4 \times 10^6) \begin{bmatrix} 2 & -1 \\ -1 & 1 \end{bmatrix} \left\{ \begin{matrix} 0.042 \times 10^{-3} \\ 0.551 \times 10^{-3} \end{matrix} \right\} \right]$$

On simplifying, the nodal accelerations at $t = 0.50 \times 10^{-3}$ s are

$$\left\{ \begin{array}{c} \ddot{u}_2 \\ \ddot{u}_3 \end{array} \right\}_2 = \left\{ \begin{array}{c} 0 \\ 5095 \end{array} \right\} + \left\{ \begin{array}{c} 4000 \\ -4360 \end{array} \right\}$$

$$= \left\{ \begin{array}{c} 4000 \\ 735 \end{array} \right\} \text{m/s}^2 \quad (\text{at } t = 0.5 \times 10^{-3} \text{ s}) \qquad (16.5.15) \quad \blacksquare$$

▲ 16.6 Beam Element Mass Matrices and Natural Frequencies ▲

We now consider the lumped- and consistent-mass matrices appropriate for time-dependent beam analysis. The development of the element equations follows the same general steps as used in Section 16.2 for the bar element.

The beam element with the associated nodal degrees of freedom (transverse displacement and rotation) is shown in Figure 16–15.

The basic element equations are given by the general form, Eq. (16.2.10), with the appropriate nodal force, stiffness, and mass matrices for a beam element. The stiffness matrix for the beam element is that given by Eq. (4.1.14). A lumped-mass matrix is obtained as

$$[m] = \frac{\rho A L}{2} \begin{array}{cccc} v_1 & \phi_1 & v_2 & \phi_2 \\ \begin{bmatrix} 1 & 0 & 0 & 0 \\ 0 & 0 & 0 & 0 \\ 0 & 0 & 1 & 0 \\ 0 & 0 & 0 & 0 \end{bmatrix} \end{array} \qquad (16.6.1)$$

where one-half of the total beam mass has been lumped at each node, corresponding to the translational degrees of freedom. In the lumped mass approach, the inertial effect associated with possible rotational degrees of freedom has been assumed to be zero in obtaining Eq. (16.6.1), although a value may be assigned to these rotational degrees of freedom by calculating the mass moment of inertia of a fraction of the beam segment about the nodal points. For a uniform beam we could then calculate the mass moment of inertia of half of the beam segment about each end node using

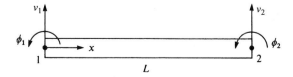

Figure 16–15 Beam element with nodal degrees of freedom

basic dynamics as

$$I = \frac{1}{3}(\rho AL/2)(L/2)^2$$

Again, the lumped-mass matrix given by Eq. (16.6.1) is a diagonal matrix, making matrix numerical calculations easier to perform than when using the consistent-mass matrix. The consistent-mass matrix can be obtained by applying the general Eq. (16.2.19) for the beam element, where the shape functions are now given by Eqs. (4.1.7). Therefore,

$$[m] = \iiint_V \rho [N]^T [N]\, dV \tag{16.6.2}$$

$$[m] = \int_0^L \iint_A \rho \begin{Bmatrix} N_1 \\ N_2 \\ N_3 \\ N_4 \end{Bmatrix} [N_1 \quad N_2 \quad N_3 \quad N_4]\, dA\, dx \tag{16.6.3}$$

with

$$N_1 = \frac{1}{L^3}(2x^3 - 3x^2 L + L^3)$$

$$N_2 = \frac{1}{L^3}(x^3 L - 2x^2 L^2 + xL^3)$$

$$N_3 = \frac{1}{L^3}(-2x^3 + 3x^2 L)$$

$$N_4 = \frac{1}{L^3}(x^3 L - x^2 L^2)$$

(16.6.4)

On substituting the shape function Eqs. (16.6.4) into Eq. (16.6.3) and performing the integration, the consistent-mass matrix becomes

$$[m] = \frac{\rho AL}{420} \begin{bmatrix} 156 & 22L & 54 & -13L \\ 22L & 4L^2 & 13L & -3L^2 \\ 54 & 13L & 156 & -22L \\ -13L & -3L^2 & -22L & 4L^2 \end{bmatrix} \tag{16.6.5}$$

Having obtained the mass matrix for the beam element, we could proceed to formulate the global stiffness and mass matrices and equations of the form given by Eq. (16.2.24) to solve the problem of a beam subjected to a time-dependent load. We will not illustrate the procedure for solution here because it is tedious and similar to that used to solve the one-dimensional bar problem in Section 16.5. However, a computer program can be used for the analysis of beams and frames subjected to time-dependent forces. Section 16.7 provides descriptions of plane frame and other element mass matrices, and Section 16.9 describes some computer program results for dynamics analysis of bars, beams, and frames.

To clarify the procedure for beam analysis, we will now determine the natural frequencies of a beam.

Example 16.5

We now consider the determination of the natural frequencies of vibration for a beam fixed at both ends as shown in Figure 16–16. The beam has mass density ρ, modulus of elasticity E, cross-sectional area A, area moment of inertia I, and length $2L$. For simplicity of the longhand calculations, the beam is discretized into (a) two beam elements of length L (Figure 16–16(a)) and then (b) three beam elements of length L each (Figure 16–16(b)).

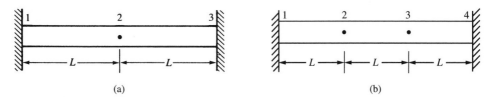

(a)   (b)

Figure 16–16 Beam for determination of natural frequencies

SOLUTION:

(a) Two-Element Solution

We can obtain the natural frequencies by using the general Eq. (16.4.7). First, we assemble the global stiffness and mass matrices (using the boundary conditions $v_1 = 0$, $\phi_1 = 0$, $v_3 = 0$, and $\phi_3 = 0$ to reduce the matrices) as

$$\begin{array}{cc} & v_2 \quad \phi_2 \end{array}$$
$$[K] = \frac{EI}{L^3}\begin{bmatrix} 24 & 0 \\ 0 & 8L^2 \end{bmatrix} \qquad [M] = \frac{\rho AL}{2}\begin{bmatrix} 2 & 0 \\ 0 & 0 \end{bmatrix} \tag{16.6.6}$$

where Eq. (4.1.14) has been used to obtain each element stiffness matrix and Eq. (16.6.1) has been used to calculate the lumped-mass matrix. On substituting Eqs. (16.6.6) into Eq. (16.4.7), we obtain

$$\left| \frac{EI}{L^3}\begin{bmatrix} 24 & 0 \\ 0 & 8L^2 \end{bmatrix} - \omega^2 \rho AL \begin{bmatrix} 1 & 0 \\ 0 & 0 \end{bmatrix} \right| = 0 \tag{16.6.7}$$

Dividing Eq. (16.6.7) by ρAL and simplifying, we obtain

$$\omega^2 = \frac{24EI}{\rho AL^4}$$

or
$$\omega = \frac{4.90}{L^2}\left(\frac{EI}{A\rho}\right)^{1/2} \tag{16.6.8}$$

The exact solution for the first natural frequency, from simple beam theory, is given by References [1] and [6]. It is

$$\omega = \frac{5.59}{L^2}\left(\frac{EI}{A\rho}\right)^{1/2} \tag{16.6.9}$$

(Here L = half the beam length.)

The large discrepancy between the exact solution and the finite element solution is assumed to be accounted for by the coarseness of the finite element model. In Example 16.6, we show for a clamped-free beam that as the number of degrees of freedom increases, convergence to the exact solution results. Furthermore, if we had used the consistent-mass matrix for the beam [Eq. (16.6.5)], the results would have been more accurate than with the lumped-mass matrix as consistent-mass matrices yield more accurate results for flexural elements such as beams.

(b) Three-Element Solution:

Using Eq. (16.6.1), we calculate each element mass matrix as follows:

$$
[m^{(1)}] = \frac{\rho A L}{2}
\begin{matrix}
v_1 & \varphi_1 & v_2 & \varphi_2 \\
\begin{bmatrix}
1 & 0 & 0 & 0 \\
0 & 0 & 0 & 0 \\
0 & 0 & 1 & 0 \\
0 & 0 & 0 & 0
\end{bmatrix}
\end{matrix}
\qquad
[m^{(2)}] = \frac{\rho A L}{2}
\begin{matrix}
v_2 & \varphi_2 & v_3 & \varphi_3 \\
\begin{bmatrix}
1 & 0 & 0 & 0 \\
0 & 0 & 0 & 0 \\
0 & 0 & 1 & 0 \\
0 & 0 & 0 & 0
\end{bmatrix}
\end{matrix}
$$

$$
[m^{(3)}] = \frac{\rho A L}{2}
\begin{matrix}
v_3 & \varphi_3 & v_4 & \varphi_4 \\
\begin{bmatrix}
1 & 0 & 0 & 0 \\
0 & 0 & 0 & 0 \\
0 & 0 & 1 & 0 \\
0 & 0 & 0 & 0
\end{bmatrix}
\end{matrix}
\tag{16.6.10}
$$

Knowing that $v_1 = \varphi_1 = v_4 = \varphi_4$, we obtain the global mass matrix as

$$
[M] = \rho A L
\begin{matrix}
v_2 & \varphi_2 & v_3 & \varphi_3 \\
\begin{bmatrix}
1 & 0 & 0 & 0 \\
0 & 0 & 0 & 0 \\
0 & 0 & 1 & 0 \\
0 & 0 & 0 & 0
\end{bmatrix}
\end{matrix}
\tag{16.6.11}
$$

Using Eq. (4.1.14), we obtain each element stiffness matrix as

$$
[k^{(1)}] = \frac{EI}{L^3}
\begin{matrix}
v_1 & \varphi_1 & v_2 & \varphi_2 \\
\begin{bmatrix}
12 & 6L & -12 & 6L \\
6L & 4L^2 & -6L & 2L^2 \\
-12 & -6L & 12 & -6L \\
6L & 2L^2 & -6L & 4L^2
\end{bmatrix}
\end{matrix}
\qquad
[k^{(2)}] = \frac{EI}{L^3}
\begin{matrix}
v_2 & \varphi_2 & v_3 & \varphi_3 \\
\begin{bmatrix}
12 & 6L & -12 & 6L \\
6L & 4L^2 & -6L & 2L^2 \\
-12 & -6L & 12 & -6L \\
6L & 2L^2 & -6L & 4L^2
\end{bmatrix}
\end{matrix}
$$

$$
[k^{(3)}] = \frac{EI}{L^3}
\begin{matrix}
v_3 & \varphi_3 & v_4 & \varphi_4 \\
\begin{bmatrix}
12 & 6L & -12 & 6L \\
6L & 4L^2 & -6L & 2L^2 \\
-12 & -6L & 12 & -6L \\
6L & 2L^2 & -6L & 4L^2
\end{bmatrix}
\end{matrix}
\tag{16.6.12}
$$

Using Eq. (16.6.12), we asemble the global stiffness matrix as

$$
[K] = \frac{EI}{L^3}
\begin{matrix}
v_2 & \varphi_2 & v_3 & \varphi_3 \\
\begin{bmatrix}
12-12 & 6L+6L & -12 & 6L \\
6L-6L & 4L^2+2L^2 & -6L & 2L^2 \\
-12 & -6L & 12+12 & -6L+6L \\
6L & 2L^2 & -6L+6L & 4L^2+4L^2
\end{bmatrix}
\end{matrix}
= \frac{EI}{L^3}
\begin{matrix}
v_2 & \varphi_2 & v_3 & \varphi_3 \\
\begin{bmatrix}
0 & 12L & -12 & 6L \\
0 & 6L^2 & -6L & 2L^2 \\
-12 & -6L & 24 & 0 \\
6L & 2L^2 & 0 & 8L^2
\end{bmatrix}
\end{matrix}
\tag{16.6.13}
$$

Using the general Eq. (16.4.7), we obtain the frequency equation as

$$
\begin{vmatrix}
\dfrac{EI}{L^3}
\begin{bmatrix}
0 & 12L & -12 & 6L \\
0 & 6L^2 & -6L & 2L^2 \\
-12 & -6L & 24 & 0 \\
6L & 2L^2 & 0 & 8L^2
\end{bmatrix}
- \omega^2 \rho AL
\begin{bmatrix}
1 & 0 & 0 & 0 \\
0 & 0 & 0 & 0 \\
0 & 0 & 1 & 0 \\
0 & 0 & 0 & 0
\end{bmatrix}
\end{vmatrix}
$$

$$
=
\begin{vmatrix}
\omega^2 \rho AL & 12EI/L^2 & -12EI/L^3 & 6EI/L^2 \\
0 & 6EI/L & -6EI/L^2 & 2EI/L \\
-12EI/L^3 & -6EI/L^2 & 24EI/L^3 - \omega^2 \rho AL & 0 \\
6EI/L^2 & 2EI/L & 0 & 8EI/L
\end{vmatrix} = 0 \quad (16.6.14)
$$

Simplifying Eq. (16.6.14), we have

$$
\begin{vmatrix}
-\omega^2 \beta & 12EI/L^2 & -12EI/L^3 & 6EI/L^2 \\
0 & 6EI/L & -6EI/L^2 & 2EI/L \\
-12EI/L^3 & -6EI/L^2 & 24EI/L^3 - \omega^2 \beta & 0 \\
6EI/L^2 & 2EI/L & 0 & 8EI/L
\end{vmatrix} = 0 \quad (16.6.15)
$$

where $\beta = \rho AL$

Upon evaluating the four-by-four determinant in Eq. (16.6.15), we obtain

$$
-\frac{1152\omega^2 E^3 I^3 \beta}{L^5} + \frac{48\omega^4 E^2 I^2 \beta^2}{L^2} + \frac{576 E^4 I^4}{L^8} - \frac{1296 E^4 I^4}{L^8}
$$

$$
+ \frac{96\omega^2 E^3 I^3 \beta}{L^5} - \frac{4\omega^4 \beta^2 E^2 I^2}{L^2} - \frac{6912 E^4 I^4}{L^8} = 0
$$

$$
\frac{44\omega^4 \beta^2 E^2 I^2}{L^2} - \frac{1056\omega^2 \beta E^3 I^3}{L^5} - \frac{7632 E^4 I^4}{L^8} = 0 \quad (16.6.16)
$$

$$
11\omega^4 \beta^2 - \frac{264\omega^2 \beta EI}{L^3} - \frac{1908 E^2 I^2}{L^6} = 0
$$

Dividing Eq. (16.6.16) by $\dfrac{4E^2 I^2}{L^2}$, we obtain two roots for $\omega_1^2 \beta$ as

$$
\omega_1^2 \beta = \frac{-5.817254 EI}{L^3} \qquad \omega_1^2 \beta = \frac{29.817254 EI}{L^3} \quad (16.6.17)
$$

Ignoring the negative root as it is not physically possible and solving explicitly for ω_1, we have

$$
\omega_1^2 = \frac{29.817254 EI}{\beta L^3}
$$

or

$$
\omega_1 = \sqrt{\frac{29.817254 EI}{\beta L^3}} = \frac{5.46}{L^2} \sqrt{\frac{EI}{A\rho}} \quad (16.6.18)
$$

In summary, comparing Eqs (16.6.8) and (16.6.18) with the exact solution, Eq. (16.6.9), for the first natural frequency, we have

$$\text{Two Beam Elements:} \quad \omega = \frac{4.90}{L^2} \sqrt{\frac{EI}{A\rho}}$$

$$\text{Three Beam Elements:} \quad \omega = \frac{5.46}{L^2} \sqrt{\frac{EI}{A\rho}} \qquad (16.6.19)$$

$$\text{Exact solution:} \quad \omega = \frac{5.59}{L^2} \left(\frac{EI}{A\rho}\right)^{1/2}$$

We can observe that with just three elements the accuracy has significantly increased. ∎

Example 16.6

Determine the first natural frequency of vibration of the cantilever beam shown in Figure 16–17 with the following data:

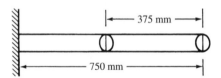

Figure 16–17 Fixed-free beam (two-element model, lumped-mass matrix)

Length of the beam: $L = 750$ mm
Modulus of elasticity: $E = 210$ GPa
Moment of inertia: $I = 3.0 \times 10^4$ mm^4
Cross-sectional area: $A = 6 \times 10^{-2}$ m^2
Mass density: $\rho = 7800$ kg/m^3
Poisson's ratio: $v = 0.3$

SOLUTION:

The finite element longhand solution result for the first natural frequency is obtained similarly to that of Example 16.5 as

$$\omega = \frac{3.148}{L^2} \left(\frac{EI}{A\rho}\right)^{1/2}$$

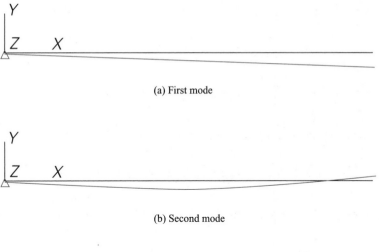

(a) First mode

(b) Second mode

(c) Third mode

Figure 16–18 First, second, and third mode shapes of flexural vibration for a cantilever beam

The exact solution according to beam theory [1] is

$$\omega = \frac{3.516}{L^2} \left(\frac{EI}{\rho A}\right)^{1/2} \qquad (16.6.20)$$

According to vibration theory for a clamped-free beam [1], we relate the second and third natural frequencies to the first natural frequency by

$$\frac{\omega_2}{\omega_1} = 6.2669 \qquad \frac{\omega_3}{\omega_1} = 17.5475$$

Figure 16–18 shows the first, second, and third mode shapes corresponding to the first three natural frequencies for the cantilever beam of Example 16.6 as obtained from a computer program. Note that each mode shape has one fewer node where a node is a point of zero displacement. That is, the first mode has all the elements of the beam of the same sign [Figure 16–18(a)], the second mode has one sign change and at some point along the beam the displacement is zero [Figure 16–18(b)], and

Table 16–3 Finite element computer solution compared to exact solution for Example 16.6

	ω_1 (rad/s)	ω_2 (rad/s)
Exact solution from beam theory	228	1434
Finite element solution		
Using 2 elements	205	1286
Using 6 elements	226	1372
Using 10 elements	227.5	1410
Using 30 elements	228.5	1430
Using 60 elements	228.5	1432

the third mode has two sign changes and at two points along the beam the displacement is zero [Figure 16–18(c)].

Table 16–3 shows the computer solution compared with the exact solution. ■

▲ 16.7 Truss, Plane Frame, Plane Stress, ▲ Plane Strain, Axisymmetric, and Solid Element Mass Matrices

The dynamic analysis of the truss and that of the plane frame are performed by extending the concepts presented in Sections 16.2 and 16.6 to the truss and plane frame, as has previously been done for the static analysis of trusses and frames.

Truss Element

The truss analysis requires the same transformation of the mass matrix from local to global coordinates as in Eq. (3.4.22) for the stiffness matrix; that is, the global mass matrix for a truss element is given by

$$[m] = [T]^T[m'][T] \tag{16.7.1}$$

We are now dealing with motion in two or three dimensions. Therefore, we must reformulate a bar element mass matrix with both axial and transverse inertial properties because mass is included in both the global x and y directions in plane truss analysis (Figure 16–19). Considering two-dimensional motion, we express both local axial

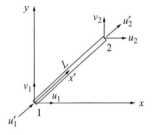

Figure 16–19 Truss element arbitrarily oriented in x-y plane showing nodal degrees of freedom

displacement u and transverse displacement v for the element in terms of the local axial and transverse nodal displacements as

$$\left\{ \begin{array}{c} u' \\ v' \end{array} \right\} = \frac{1}{L} \begin{bmatrix} L - x' & 0 & x' & 0 \\ 0 & L - x' & 0 & x' \end{bmatrix} \left\{ \begin{array}{c} u'_1 \\ v'_1 \\ u'_2 \\ v'_2 \end{array} \right\} \tag{16.7.2}$$

In general, $[\psi'] = [N]\{d'\}$; therefore, the shape function matrix from Eq. (16.7.2) is

$$[N] = \frac{1}{L} \begin{bmatrix} L - x' & 0 & x' & 0 \\ 0 & L - x' & 0 & x' \end{bmatrix} \tag{16.7.3}$$

We can then substitute Eq. (16.7.3) into the general expression given by Eq. (16.2.19) to evaluate the local *truss element consistent-mass matrix* as

$$[m'] = \frac{\rho A L}{6} \begin{bmatrix} 2 & 0 & 1 & 0 \\ 0 & 2 & 0 & 1 \\ 1 & 0 & 2 & 0 \\ 0 & 1 & 0 & 2 \end{bmatrix} \tag{16.7.4}$$

The truss element lumped-mass matrix for two-dimensional motion is obtained by simply lumping mass at each node and remembering that mass is the same in both the x' and y' directions. The local *truss element lumped-mass matrix* is then

$$[m'] = \frac{\rho A L}{2} \begin{bmatrix} 1 & 0 & 0 & 0 \\ 0 & 1 & 0 & 0 \\ 0 & 0 & 1 & 0 \\ 0 & 0 & 0 & 1 \end{bmatrix} \tag{16.7.5}$$

Plane Frame Element

The plane frame analysis requires first expanding and then combining the bar and beam mass matrices to obtain the local mass matrix. Because we recall there are six total degrees of freedom associated with a plane frame element (Figure 16–20), the bar and beam mass matrices are expanded to order 6×6 and superimposed.

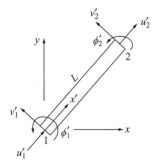

Figure 16–20 Frame element arbitrarily oriented in local coordinate system showing nodal degrees of freedom

On combining the local axes consistent-mass matrices for the bar and beam from Eqs. (16.2.23) and (16.6.5), we obtain

$$[m'] = \rho AL \begin{bmatrix} 2/6 & 0 & 0 & 1/6 & 0 & 0 \\ & 156/420 & 22L/420 & 0 & 54/420 & -13L/420 \\ & & 4L^2/420 & 0 & 13L/420 & -3L^2/420 \\ & & & 2/6 & 0 & 0 \\ & & & & 156/420 & -22L/420 \\ \text{Symmetry} & & & & & 4L^2/420 \end{bmatrix} \quad (16.7.6)$$

On combining the lumped-mass matrices Eqs. (16.2.12) and (16.6.1) for the bar and beam, respectively, the resulting local axes plane frame lumped-mass matrix is

$$[m'] = \frac{\rho AL}{2} \begin{array}{c} \begin{matrix} u_1' & v_1' & \phi_1' & u_2' & v_2' & \phi_2' \end{matrix} \\ \begin{bmatrix} 1 & 0 & 0 & 0 & 0 & 0 \\ 0 & 1 & 0 & 0 & 0 & 0 \\ 0 & 0 & 0 & 0 & 0 & 0 \\ 0 & 0 & 0 & 1 & 0 & 0 \\ 0 & 0 & 0 & 0 & 1 & 0 \\ 0 & 0 & 0 & 0 & 0 & 0 \end{bmatrix} \end{array} \quad (16.7.7)$$

The global mass matrix $[m]$ for a plane frame element arbitrarily oriented in x-y coordinates is transformed according to Eq. (16.7.1), where the transformation matrix $[T]$ is now given by Eq. (5.1.10) and either Eq. (16.7.6) for consistent-mass or (16.7.7) for lumped-mass matrices.

Because a longhand solution of the time-dependent plane frame problem is quite lengthy, only a computer program solution will be presented in Section 16.9.

Plane Stress/Strain Element

The plane stress, plane strain, constant-strain triangle element (Figure 16–21) consistent-mass matrix is obtained by using the shape functions from Eq. (6.2.18) and the shape function matrix given by substituting

$$[N] = \begin{bmatrix} N_1 & 0 & N_2 & 0 & N_3 & 0 \\ 0 & N_1 & 0 & N_2 & 0 & N_3 \end{bmatrix}$$

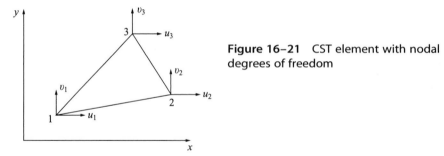

Figure 16–21 CST element with nodal degrees of freedom

into Eq. (16.2.19) to obtain

$$[m] = \rho \int_V [N]^T [N] \, dV \qquad (16.7.8)$$

Letting $dV = t \, dA$ and noting that $\int_A N_1^2 \, dA = \frac{1}{6} A$, $\int_A N_1 N_2 \, dA = \frac{1}{12} A$, and so on, we obtain the *CST* global *consistent-mass matrix* as

$$[m] = \frac{\rho t A}{12}
\begin{bmatrix}
2 & 0 & 1 & 0 & 1 & 0 \\
 & 2 & 0 & 1 & 0 & 1 \\
 & & 2 & 0 & 1 & 0 \\
 & & & 2 & 0 & 1 \\
 & & & & 2 & 0 \\
\text{Symmetry} & & & & & 2
\end{bmatrix}
\qquad (16.7.9)$$

For the isoparametric quadrilateral element for plane stress and plane strain considered in Chapter 10, we use the shape functions given by Eq. (10.2.5) with the shape function matrix given in Eq. (10.2.4) substituted into Eq. (16.7.10). This yields the *quadrilateral element consistent-mass matrix* as

$$[m] = \rho t \int_{-1}^{1} \int_{-1}^{1} [N]^T [N] \, |[J]| \, ds \, dt \qquad (16.7.10)$$

The integral in Eq. (16.7.10) is evaluated best by numerical integration as described in Section 10.4.

Axisymmetric Element

The *axisymmetric triangular element* (considered in Chapter 9 and shown in Figure 16–22) *consistent-mass matrix* is given by

$$[m] = \int_V \rho [N]^T [N] \, dV = \int_A \rho [N]^T [N] 2\pi r \, dA \qquad (16.7.11)$$

Since $r = N_1 r_1 + N_2 r_2 + N_3 r_3$, we have

$$[m] = 2\pi \rho \int_A (N_1 r_1 + N_2 r_2 + N_3 r_3) [N]^T [N] \, dA \qquad (16.7.12)$$

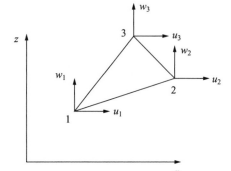

Figure 16–22 Axisymmetric triangular element showing nodal degrees of freedom

Noting that

$$\int_A N_1^3 \, dA = \frac{2A}{20} \qquad \int_A N_1^2 N_2 \, dA = \frac{2A}{60}$$

$$\int_A N_1 N_2 N_3 \, dA = \frac{2A}{120} \qquad \text{and so on} \tag{16.7.13}$$

we obtain

$$[m] = \frac{\pi \rho A}{10}
\begin{bmatrix}
\frac{4}{3}r_1 + 2\bar{r} & 0 & 2\bar{r} - \frac{r_3}{3} & 0 & 2\bar{r} - \frac{r_2}{3} & 0 \\
 & \frac{4}{3}r_1 + 2\bar{r} & 0 & 2\bar{r} - \frac{r_3}{3} & 0 & 2\bar{r} - \frac{r_2}{3} \\
 & & \frac{4}{3}r_2 + 2\bar{r} & 0 & 2\bar{r} - \frac{r_1}{3} & 0 \\
 & & & \frac{4}{3}r_2 + 2\bar{r} & 0 & 2\bar{r} - \frac{r_1}{3} \\
 & & & & \frac{4}{3}r_3 + 2\bar{r} & 0 \\
\text{Symmetry} & & & & & \frac{4}{3}r_3 + 2\bar{r}
\end{bmatrix}$$

$$\tag{16.7.14}$$

where

$$\bar{r} = \frac{r_1 + r_2 + r_3}{3}$$

Tetrahedral Solid Element

Finally, the *tetrahedral solid element* (considered in Chapter 11) *consistent-mass matrix* is obtained by substituting the shape function matrix Eq. (11.2.9) with shape functions defined in Eq. (11.2.10) into Eq. (16.2.19) and performing the integration to obtain

$$[m] = \frac{\rho V}{20}
\begin{bmatrix}
2 & 0 & 0 & 1 & 0 & 0 & 1 & 0 & 0 & 1 & 0 & 0 \\
 & 2 & 0 & 0 & 1 & 0 & 0 & 1 & 0 & 0 & 1 & 0 \\
 & & 2 & 0 & 0 & 1 & 0 & 0 & 1 & 0 & 0 & 1 \\
 & & & 2 & 0 & 0 & 1 & 0 & 0 & 1 & 0 & 0 \\
 & & & & 2 & 0 & 0 & 1 & 0 & 0 & 1 & 0 \\
 & & & & & 2 & 0 & 0 & 1 & 0 & 0 & 1 \\
 & & & & & & 2 & 0 & 0 & 1 & 0 & 0 \\
 & & & & & & & 2 & 0 & 0 & 1 & 0 \\
 & & & & & & & & 2 & 0 & 0 & 1 \\
 & & & & & & & & & 2 & 0 & 0 \\
 & & & & & & & & & & 2 & 0 \\
\text{Symmetry} & & & & & & & & & & & 2
\end{bmatrix}
\tag{16.7.15}$$

▲ 16.8 Time-Dependent Heat Transfer ▲

In this section, we consider the time-dependent heat transfer problem in one dimension only. The basic differential equation for time-dependent heat transfer in one dimension was given previously by Eq. (13.1.7) with the boundary conditions given by Eqs. (13.1.10) and (13.1.11).

The finite element formulation of the equations can be obtained by minimization of the following functional:

$$\pi_h = \frac{1}{2} \iiint_V \left[K_{xx} \left(\frac{\partial T}{\partial x} \right)^2 - 2(Q - c\rho\dot{T})T \right] dV$$

$$- \iint_{S_2} q^* T \, dS + \frac{1}{2} \iint_{S_3} h(T - T_\infty)^2 \, ds \qquad (16.8.1)$$

Equation (16.8.1) is similar to Eq. (13.4.10) with definitions given by Eq. (13.4.11) except that the Q term is now replaced by

$$Q - c\rho\dot{T} \qquad (16.8.2)$$

where, again, c is the specific heat of the material, and the dot over the variable T denotes differentiation with respect to time. Again, Eq. (13.4.22) obtained in Section 13.4 for the conductivity or stiffness matrix and Eqs. (13.4.23) through (13.4.25) for the force matrix terms are applicable here.

The term given by Eq. (16.8.2) yields an additional contribution to the basic element equations previously obtained for the time-independent problem as follows:

$$\Omega_Q = - \iiint_V T(Q - c\rho\dot{T}) \, dV \qquad (16.8.3)$$

Again, the temperature function is given by

$$\{T\} = [N]\{t\} \qquad (16.8.4)$$

where $[N]$ is the shape function matrix given by Eq. (13.4.3) or Eqs. (16.2.3) for the simple one-dimensional element, and $\{t'\}$ is the nodal temperature matrix. Substituting Eq. (16.8.4) into Eq. (16.8.3) and differentiating with respect to time where indicated yields

$$\Omega_Q = - \iiint_V ([N]\{t\}Q - c\rho[N]\{t\}[N]\{\dot{t}\}) \, dV \qquad (16.8.5)$$

where the fact that $[N]$ is a function only of the coordinate system has been taken into account. Equation (16.8.5) must be minimized with respect to the nodal temperatures as follows:

$$\frac{\partial \Omega_Q}{\partial \{t\}} = - \iiint_V [N]^T Q \, dV + \iiint_V c\rho[N]^T [N] \, dV \{\dot{t}\} \qquad (16.8.6)$$

where we have assumed that $\{\dot{t}\}$ remains constant during the differentiation with respect to $\{t\}$. Equation (16.8.6) results in the additional time-dependent term added to Eq. (13.4.18). Hence, using previous definitions for the stiffness and force matrices, we obtain the element equations as

$$\{f\} = [k]\{t\} + [m]\{\dot{t}\}$$ (16.8.7)

where now

$$[m] = \iiint_V c\rho[N]^T[N]\, dV$$ (16.8.8)

For an element with constant cross-sectional area A, the differential volume is $dV = A\,dx$. Substituting the one-dimensional shape function matrix Eq. (13.4.3) into Eq. (16.8.8) yields

$$[m] = c\rho A \int_0^L \left\{ \begin{array}{c} 1 - \dfrac{x}{L} \\ \dfrac{x}{L} \end{array} \right\} \left[1 - \dfrac{x}{L} \quad \dfrac{x}{L} \right] dx$$

or

$$[m] = \frac{c\rho AL}{6} \begin{bmatrix} 2 & 1 \\ 1 & 2 \end{bmatrix}$$ (16.8.9)

Equation (16.8.9) is analogous to the consistent-mass matrix Eq. (16.2.23). The lumped-mass matrix for the heat conduction problem is then

$$[m] = \frac{c\rho AL}{2} \begin{bmatrix} 1 & 0 \\ 0 & 1 \end{bmatrix}$$ (16.8.10)

which is analogous to Eq. (16.2.12) for the one-dimensional stress element.

The time-dependent heat-transfer problem can now be solved in a manner analogous to that for the stress analysis problem. We present the numerical time integration scheme.

Numerical Time Integration

The numerical time integration method described here is similar to Newmark's method used for structural dynamics analysis and can be used to solve time-dependent or transient heat-transfer problems.

We begin by assuming that two temperature states $\{T_i\}$ at time t_i and $\{T_{i+1}\}$ at time t_{i+1} are related by

$$\{T_{i+1}\} = \{T_i\} + [(1 - \beta)\{\dot{T}_i\} + \beta\{\dot{T}_{i+1}\}](\Delta t)$$ (16.8.11)

Equation (16.8.11) is known as the *generalized trapezoid rule*. Much like Newmark's method for numerical time integration of the second-order equations of structural dynamics, Eq. (16.8.11) includes a parameter β that is chosen by the user.

Next we express Eq. (16.8.7) in global form as

$$\{F\} = [K]\{T\} + [M]\{\dot{T}\}$$ (16.8.12)

We now write Eq. (16.8.12) for time t_i and then for time t_{i+1}. We then multiply the first of these two equations by $1 - \beta$ and the second by β to obtain

$$(1 - \beta)([K]\{T_i\} + [M]\{\dot{T}_i\}) = (1 - \beta)\{F_i\} \tag{16.8.13a}$$

$$\beta([K]\{T_{i+1}\} + [M]\{\dot{T}_{i+1}\}) = \beta\{F_{i+1}\} \tag{16.8.13b}$$

Next we add Eqs. (16.8.13a and b) together to obtain

$$[M][(1 - \beta)\{\dot{T}_i\} + \beta\{\dot{T}_{i+1}\}] + [K][(1 - \beta)\{T_i\} + \beta\{T_{i+1}\}]$$
$$= (1 - \beta)\{F_i\} + \beta\{F_{i+1}\} \tag{16.8.14}$$

Now, using Eq. (16.8.11), we can eliminate the time derivative terms from Eq. (16.8.14) to write

$$\frac{[M](\{T_{i+1}\} - \{T_i\})}{\Delta t} + [K][(1 - \beta)\{T_i\} + \beta\{T_{i+1}\}] = (1 - \beta)\{F_i\} + \beta\{F_{i+1}\} \tag{16.8.15}$$

Rewriting Eq. (16.8.15) by grouping the $\{T_{i+1}\}$ terms on the left side, we have

$$\left(\frac{1}{\Delta t}[M] + \beta[K]\right)\{T_{i+1}\}$$

$$= \left[\frac{1}{\Delta t}[M] - (1 - \beta)[K]\right]\{T_i\} + (1 - \beta)\{F_i\} + \beta\{F_{i+1}\} \tag{16.8.16}$$

The time integration to solve for $[T]$ begins as follows. Given a known initial temperature $\{T_0\}$ at time $t = 0$ and a time step Δt, we solve Eq. (16.8.16) for $\{T_1\}$ at $t = \Delta t$. Then, using $\{T_1\}$, we determine $\{T_2\}$ at $t = 2(\Delta t)$, and so on. For a constant Δt, the left-side coefficient of $\{T_{i+1}\}$ need be evaluated only one time (assuming $[M]$ and $[K]$ do not vary with time). The matrix Eq. (16.8.16) can then be solved in the usual manner, such as by Gauss elimination. For a one-dimensional heat-transfer analysis, element $[k]$ is given by Eqs. (13.4.22) and (13.4.28), whereas $\{f\}$ is given by Eqs. (13.4.26) and (13.4.29).

It has been shown that depending on the value of β, the time step Δt may have an upper limit for the numerical analysis to be stable. If $\beta < \frac{1}{2}$, the largest Δt for stability as shown in Reference [12] is

$$\Delta t = \frac{2}{(1 - 2\beta)\lambda_{max}} \tag{16.8.17}$$

where λ_{max} is the largest eigenvalue of

$$([K] - \lambda[M])\{T'\} = 0 \tag{16.8.18}$$

in which, as in Eq. (16.4.2), we have

$$\{T(t)\} = \{T'\}e^{i\lambda t} \tag{16.8.19}$$

with $\{T'\}$ representing the natural modes. If $\beta \geq \frac{1}{2}$, the numerical analysis is unconditionally stable; that is, stability of solution (but not accuracy) is guaranteed for Δt greater than that given by Eq. (16.8.17), or as Δt becomes indefinitely large. Various numerical integration methods result, depending on specific values of β:

$\beta = 0$: Forward difference, or Euler [3], which is said to be conditionally stable (that is, Δt must be no greater than that given by Eq. (16.8.17) to obtain a stable solution).

$\beta = \frac{1}{2}$: Crank-Nicolson, or trapezoid, rule, which is unconditionally stable.

$\beta = \frac{2}{3}$: Galerkin, which is unconditionally stable.

$\beta = 1$: Backward difference, which is unconditionally stable.

If $\beta = 0$, the numerical integration method is called *explicit*; that is, we can solve for $\{T_{i+1}\}$ directly at time Δt knowing only previous information at $t = \{T_i\}$. If $\beta > 0$, the method is called implicit. If a diagonal mass-type matrix $[M]$ exists and $\beta = 0$, the computational effort for each time step is small (see Example 16.4, where a lumped-mass matrix was used), but so must be Δt. The choice of $\beta > \frac{1}{2}$ is often used. However, if $\beta = \frac{1}{2}$ and sharp transients exist, the method generates spurious oscillations in the solution. Using $\beta > \frac{1}{2}$, along with smaller Δt [12], is probably better. Example 16.7 illustrates the solution of a one-dimensional time-dependent heat-transfer problem using the numerical time integration scheme [Eq. (16.8.16)].

Example 16.7

A circular fin (Figure 16–23) is made of pure copper with a thermal conductivity of $K_{xx} = 400$ W/(m · °C), $h = 150$ W/(m² · °C), mass density $\rho = 8900$ kg/m³, and specific heat $c = 375$ J/(kg · °C) (1 J $= 1$ W · s). The initial temperature of the fin is 25 °C. The fin length is 2 cm, and the diameter is 0.4 cm. The right tip of the fin is insulated. The base of the fin is then suddenly increased to a temperature of 85 °C and maintained at this temperature. Use the consistent form of the capacitance matrix, a time step of 0.1 s, and $\beta = \frac{2}{3}$. Use two elements of equal length. Determine the temperature distribution up to 3 s.

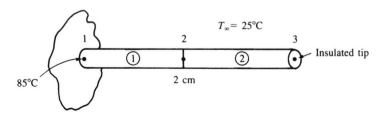

Figure 16–23   Rod subjected to time-dependent temperature

SOLUTION:

Using Eq. (13.4.22), the stiffness matrix is

$$
[k^{(1)}] = [k^{(2)}] = \frac{AK_{xx}}{L} \begin{array}{cc} 1 & 2 \\ \begin{bmatrix} 1 & -1 \\ -1 & 1 \end{bmatrix} \end{array} + \frac{hPL}{6} \begin{array}{cc} 2 & 3 \\ \begin{bmatrix} 2 & 1 \\ 1 & 2 \end{bmatrix} \end{array}
$$

$$
[k^{(1)}] = [k^{(2)}] = \frac{\pi(0.004)^2(400)}{4(0.01)} \begin{bmatrix} 1 & -1 \\ -1 & 1 \end{bmatrix} + \frac{150(2\pi)(0.002)(0.01)}{6} \begin{bmatrix} 2 & 1 \\ 1 & 2 \end{bmatrix}
$$

$$\tag{16.8.20}$$

Assembling the element stiffness matrices, Eq. (16.8.20), we obtain the global stiffness matrix as

$$[K] = \begin{array}{ccc} 1 & 2 & 3 \end{array}$$

$$[K] = \begin{bmatrix} 0.50894 & -0.49951 & 0 \\ -0.49951 & 1.01788 & -0.49951 \\ 0 & -0.49951 & 0.50894 \end{bmatrix} \frac{W}{°C} \qquad (16.8.21)$$

Using Eq. (13.4.25), we obtain each element force matrix as

$$\{f_h^{(1)}\} = \{f_h^{(2)}\} = \frac{hT_\infty PL}{2} \begin{Bmatrix} 1 \\ 1 \end{Bmatrix} = \frac{(150)(25\,°C)(2\pi)(0.002)(0.01)}{2} \begin{Bmatrix} 1 \\ 1 \end{Bmatrix}$$

$$\{f_h^{(1)}\} = \{f_h^{(2)}\} = \begin{Bmatrix} 0.23561 \\ 0.23561 \end{Bmatrix} \qquad (16.8.22)$$

Using Eq. (16.8.22), we find that the assembled global force matrix is

$$\{F\} = \begin{Bmatrix} 0.23561 \\ 0.47122 \\ 0.23561 \end{Bmatrix} W \qquad (16.8.23)$$

Next using Eq. (16.8.9), we obtain each element mass (capacitance) matrix as

$$[m] = \frac{c\rho AL}{6} \begin{bmatrix} 2 & 1 \\ 1 & 2 \end{bmatrix}$$

$$[m^{(1)}] = [m^{(2)}] = \frac{(375)(8900)\dfrac{\pi(0.004)^2}{4}(0.01)}{6} \begin{bmatrix} 2 & 1 \\ 1 & 2 \end{bmatrix}$$

$$= 0.06990 \begin{bmatrix} 2 & 1 \\ 1 & 2 \end{bmatrix} W \cdot s/°C \qquad (16.8.24)$$

Using Eq. (16.8.24), the assembled capacitance matrix is

$$\begin{array}{ccc} 1 & 2 & 3 \end{array}$$

$$[M] = \begin{bmatrix} 0.13980 & 0.06990 & 0 \\ 0.06990 & 0.27960 & 0.06990 \\ 0 & 0.06990 & 0.13980 \end{bmatrix} \frac{W \cdot s}{°C} \qquad (16.8.25)$$

Using Eq. (16.8.16) and Eqs. (16.8.21) and (16.8.25), we obtain

$$\left(\frac{1}{\Delta t}[M] + \beta[K] \right) = \begin{bmatrix} 1.7374 & 0.36603 & 0 \\ 0.36603 & 3.4747 & 0.36603 \\ 0 & 0.36603 & 1.7374 \end{bmatrix} \frac{W}{°C} \qquad (16.8.26)$$

Table 16–4 Nodal temperatures at various times for Example 16.7

Time (s)	Temperature of Node Numbers (°C)		
	1	2	3
0.1	85	18.534	26.371
0.2	85	29.732	21.752
0.3	85	36.404	22.662
0.4	85	41.032	25.655
0.5	85	44.665	29.312
0.6	85	47.749	33.059
0.7	85	50.482	36.669
0.8	85	52.956	40.062
0.9	85	55.218	43.218
1.0	85	57.296	46.139
1.1	85	59.208	48.837
1.2	85	60.969	51.327
1.3	85	62.593	53.623
1.4	85	64.089	55.741
1.5	85	65.469	57.693
1.6	85	66.742	59.493
1.7	85	67.915	61.152
1.8	85	68.996	62.683
1.9	85	69.993	64.094
2.0	85	70.912	65.395
2.1	85	71.760	66.594
2.2	85	72.542	67.700
2.3	85	73.262	68.720
2.4	85	73.926	69.660
2.5	85	74.539	70.527
2.6	85	75.104	71.326
2.7	85	75.624	72.063
2.8	85	76.104	72.742
2.9	85	76.547	73.368
3.0	85	76.955	73.946

and
$$\left[\frac{1}{\Delta t}[M] - (1-\beta)[K]\right] = \begin{bmatrix} 1.2280 & 0.8655 & 0 \\ 0.8655 & 2.457 & 0.8655 \\ 0 & 0.8655 & 1.2280 \end{bmatrix} \frac{W}{°C} \qquad (16.8.27)$$

where $\beta = \frac{2}{3}$ and $\Delta t = 0.1$ s have been used to obtain Eqs. (16.8.26) and (16.8.27). For the first time step, $t = 0.1$ s, we then use Eqs. (16.8.23), (16.8.27), and (16.8.26) in

Eq. (16.8.16) to obtain

$$\begin{bmatrix} 1.7374 & 0.36603 & 0 \\ 0.36603 & 3.4747 & 0.36603 \\ 0 & 0.36603 & 1.7374 \end{bmatrix} \begin{Bmatrix} 85\,^\circ C \\ t_2 \\ t_3 \end{Bmatrix}$$

$$= \begin{bmatrix} 1.2280 & 0.8655 & 0 \\ 0.8655 & 2.457 & 0.8655 \\ 0 & 0.8655 & 1.2280 \end{bmatrix} \begin{Bmatrix} 25\,^\circ C \\ 25\,^\circ C \\ 25\,^\circ C \end{Bmatrix} + \begin{Bmatrix} 0.23561 \\ 0.47122 \\ 0.23561 \end{Bmatrix} \qquad (16.8.28)$$

In Eq. (16.8.28), we should note that because $\{F_i\} = \{F_{i+1}\}$ for all time, the sum of the terms is $(1 - \beta)\{F_i\} + \beta\{F_{i+1}\} = \{F_i\}$ for all time. This is the column matrix on the right side of Eq. (16.8.28). We now solve Eq. (16.8.28) in the usual manner by partitioning the second and third equations of Eq. (16.8.28) from the first equation and solving the second and third equations simultaneously for t_2 and t_3. The results are

$$t_2 = 18.534\,^\circ C \qquad t_3 = 26.371\,^\circ C$$

At time $t = 0.2$ s, Eq. (16.8.28) becomes

$$\begin{bmatrix} 1.7374 & 0.36603 & 0 \\ 0.36603 & 3.4747 & 0.36603 \\ 0 & 0.36603 & 1.7374 \end{bmatrix} \begin{Bmatrix} 85\,^\circ C \\ t_2 \\ t_3 \end{Bmatrix}$$

$$= \begin{bmatrix} 1.2280 & 0.8655 & 0 \\ 0.8655 & 2.457 & 0.8655 \\ 0 & 0.8655 & 1.2280 \end{bmatrix} \begin{Bmatrix} 85\,^\circ C \\ 18.534\,^\circ C \\ 26.371\,^\circ C \end{Bmatrix} + \begin{Bmatrix} 0.23561 \\ 0.47122 \\ 0.23561 \end{Bmatrix} \qquad (16.8.29)$$

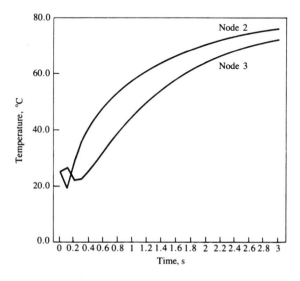

Figure 16–24 Temperature as a function of time for nodes 2 and 3 of Example 16.7

Solving Eq. (16.8.29) for t_2 and t_3, we obtain

$$t_2 = 29.732\,^\circ C \qquad t_3 = 21.752\,^\circ C$$

The results through a time of 3 s are tabulated in Table 16–4 and plotted in Figure 16–24.
∎

▲ 16.9 Computer Program Example Solutions for Structural Dynamics ▲

In this section, we report some results of structural dynamics from a computer program. We report the results of the natural frequencies of a fixed-fixed beam using the plane stress element in Algor [15] and compare how many elements of this type are necessary to obtain correct results. We also report the results of three structural dynamics problems, a bar, a beam, and a frame subjected to time-dependent loadings.

Finally, we show two additional models, one of a time-dependent three-dimensional gantry crane made of beam elements and subjected to an impact loading, and the other of a cab frame that travels along the underside of a crane beam.

Figure 16–25 shows a fixed-fixed steel beam used for natural frequency determination using plane stress elements. Table 16–5 shows the results of the first five natural frequencies using 100 elements and then using 1000 elements. Comparisons to the analytical solutions from beam theory are shown. We observe that it takes a large number of plane stress elements to accurately predict the natural frequencies whereas it

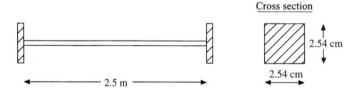

Figure 16–25 Fixed-fixed beam for natural frequency determination modeled using plane stress element

Table 16–5 Results for first five frequencies using 100 and 1000 elements and exact solution

ω (rad/s)	Analytical	100 Elements	1000 Elements
1	130.8	130.7	130.6
2	360.8	359.8	359.7
3	707.3	704.7	704.1
4	1169.2	1163.3	1161.6
5	1746.6	1734.5	1731.0

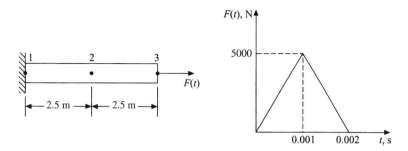

Figure 16–26 Bar subjected to forcing function shown

only took a few beam elements to accurately predict natural frequencies (see Example 16.6 and Table 16–3).

Figure 16–26 shows a steel bar subjected to a time-dependent forcing function. Using two elements in the model, the nodal displacements at nodes 2 and 3 are presented in Table 16–6. A time step of integration of 0.00025 s was used. This time step is based on that recommended by Eq. (16.5.1) and determined in Example 16.4, as the bar has the same properties as that of Example 16.4.

Figure 16–27 shows a plot of the axial displacement of the free end node 3 versus time up to 0.01 s. Notice the oscillatory motion due to damping being neglected.

Table 16–6 lists the largest maximum absolute displacement of 0.1942 mm at time of 0.00425 s. For comparison, the maximum static deflection from $(\delta = PL/AE = (5000 \text{ N})(5 \text{ m})/(6.45 \times 10^{-4} \times 210 \times 10^{9}) = 0.1857 \text{ mm})$.

Figure 16–28 shows a fixed-fixed beam subjected to a forcing function. Here $E = 45$ GPa, $I = 41 \times 10^{-6}$ m^4, mass density of 1065×10^3 kg/m^3 and a time step of integration of 0.001 s were used for the beam. The natural frequencies are shown in Table 16–7.

Table 16–7 lists the first six natural frequencies for the fixed-fixed beam. The natural frequencies 1, 2, 3, and 6 are flexural modes, while mode 5 is an axial mode. These modes are seen by looking at the modes from a frequency analysis. The undamped response of the center node 3 is shown in Figure 16–29 along with a damped response subsequently described. The maximum displacement under the load (at node 3) compares with the solution in Reference [14]. This maximum displacement is at node 3 at a time of 0.086 s with a value of 31 mm. The static deflection for the beam with a concentrated load at mid-span is 15.9 mm as obtained from the classical solution of $y = PL^3/192EI$. The undamped time-dependent response oscillates about zero deflection after the load is removed while the damped response oscillates in a damped manner approaching zero deflection.

A time step of 0.002 was used in the fixed-fixed beam as it meets the recommended time step as suggested in Section 16.3. That is, $\Delta t < T_n/10$ to $T_n/20$ is recommended to provide accurate results for Wilson's direct integration scheme as used in the Algor program. From the frequency analysis (see the output in Table 16–7), the circular frequency $\omega_4 = 170.9$ or the natural frequency is $f_4 = \omega_4/(2\pi) = 27.2$ cycles/s or Hertz (Hz). Now we use $\Delta t = T_n/20 = 1/(20f_4) = 1/[20(27.2)] = 0.002$ s. Therefore, $\Delta t = 0.002$ s is acceptable. Using a time step

Table 16–6 Displacement time history, nodes 2 and 3 of Figure 16–26

	NODE NUMBER – (COMPONENT NUMBER)	
TIME	2–(1)	3–(1)
0.00025	8.496E–07	1.142E–05
0.00050	8.835E–06	8.645E–05
0.00075	4.105E–05	2.630E–04
0.00100	1.232E–04	5.390E–04
0.00125	2.752E–04	8.539E–04
0.00150	4.897E–04	1.104E–03
0.00175	7.177E–04	1.213E–03
0.00200	8.788E–04	1.163E–03
0.00225	8.938E–04	9.933E–04
0.00250	7.260E–04	7.584E–04
0.00275	4.039E–04	4.834E–04
0.00300	1.121E–05	1.565E–04
0.00325	–3.519E–04	–2.343E–04
0.00350	–6.120E–04	–6.603E–04
0.00375	–7.454E–04	–1.039E–03
0.00400	–7.699E–04	–1.263E–03
0.00425	–7.166E–04	–1.251E–03
0.00450	–6.029E–04	–9.932E–04
0.00475	–4.248E–04	–5.568E–04
0.00500	–1.736E–04	–5.643E–05
0.00525	1.393E–04	3.981E–04
0.00550	4.675E–04	7.408E–04
0.00575	7.382E–04	9.565E–04
0.00600	8.807E–04	1.058E–03
0.00625	8.577E–04	1.055E–03
0.00650	6.808E–04	9.370E–04
0.00675	4.025E–04	6.851E–04
0.00700	8.987E–05	3.000E–04
0.00725	–2.025E–04	–1.737E–04
0.00750	–4.448E–04	–6.477E–04
0.00775	–6.270E–04	–1.017E–03
0.00800	–7.423E–04	–1.199E–03
0.00825	–7.759E–04	–1.166E–03
0.00850	–7.066E–04	–9.468E–04
0.00875	–5.221E–04	–6.060E–04
0.00900	–2.361E–04	–2.136E–04
0.00925	1.052E–04	1.782E–04
0.00950	4.335E–04	5.375E–04
0.00975	6.815E–04	8.389E–04
0.01000	8.051E–04	1.050E–03
MAXIMUM ABSOLUTE VALUES		
MAXIMUM	8.938E–04	1.263E–03
TIME	2.250E–03	4.000E–03

greater than $T_n/10$ may result in loss of accuracy as some of the higher mode response contributions to the solution may be missed. Often times a cut-off period or frequency is used to decide what largest natural frequency to use in the analysis. In many applications only a few lower modes contribute significantly to the

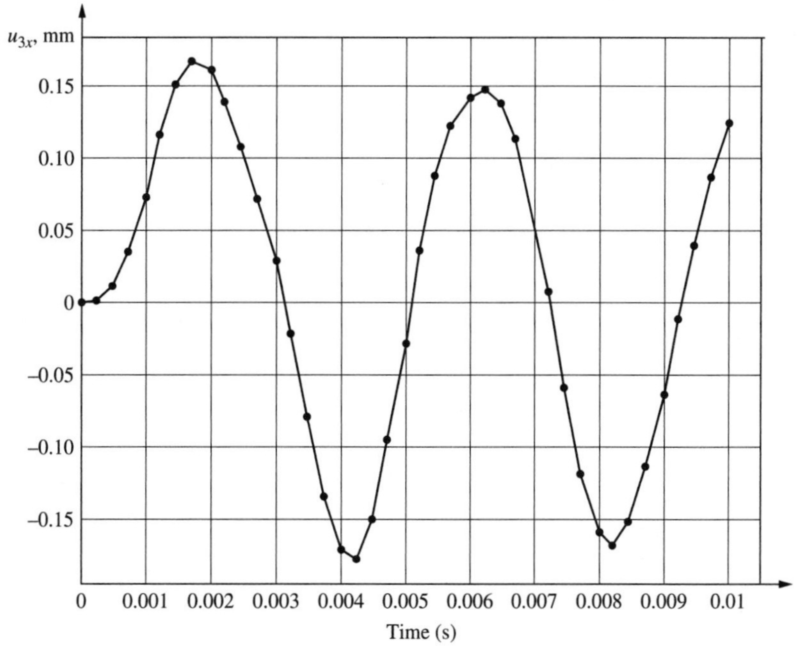

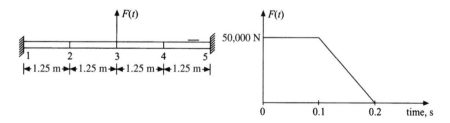

Figure 16–27 Node 3 displacement versus time for bar of Figure 16–26

Figure 16–28 Fixed-fixed beam subjected to forcing function

Table 16–7 Natural frequencies and displacement
time history (nodes 2 and 3, Figure 16–28)

Frequencies =	6
mode number	circular frequency (rad/sec)
1	3.9139E+01
2	1.0398E+02
3	1.3267E+02
4	1.7093E+02
5	2.4514E+02
6	3.2029E+02

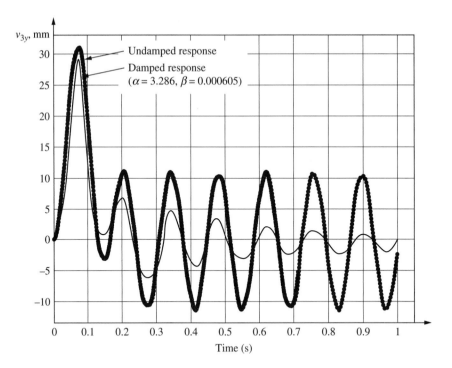

Figure 16–29 Undamped and damped response of center node 3 for fixed-fixed beam of Figure 16–28

response. The higher modes are then not necessary. The highest frequency used in the analysis is called the cut-off frequency. For machinery parts, the cut-off frequency is often taken as high as 250 Hz. In the fixed-fixed beam, we have selected a cut-off frequency of $f_6 = 31.44$ Hz in determining the time step of integration. This frequency is the highest flexural mode frequency computed for the four-element beam model.

Damping

Damping is considered in the fixed-fixed beam example. Computer programs, such as Algor and ANSYS, allow you to consider damping using Rayleigh damping in the direct integration method. For Rayleigh damping, the damping matrix is

$$[C] = \alpha[M] + \beta[K] \tag{16.9.1}$$

where the constants α and β are calculated from the system equations

$$\alpha + \beta\omega_i^2 = 2\omega_i\zeta_i \tag{16.9.2}$$

where ω_i are circular natural frequencies obtained through modal analysis, and ζ_i are damping ratios specified by the analyst. For instance, assuming we assign damping ratios ζ_1 and ζ_2, from the above Eq. (16.9.2), we can show that α and β are

$$\alpha = \frac{2\omega_1\omega_2}{\omega_2^2 - \omega_1^2}(\omega_2\zeta_1 - \omega_1\zeta_2) \qquad \beta = \frac{2}{\omega_2^2 - \omega_1^2}(\omega_2\zeta_2 - \omega_1\zeta_1) \qquad (16.9.3)$$

For $\beta = 0$, $[C] = \alpha[M]$ and the higher modes are only slightly damped, while for $\alpha = 0$, $[C] = \beta[K]$ and higher modes are heavily damped. To obtain α and β, we

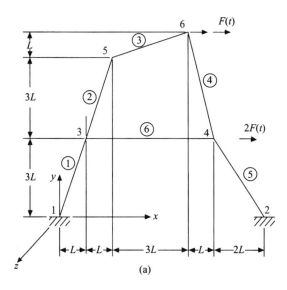

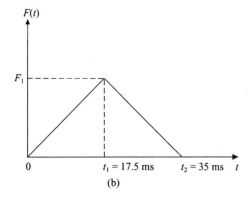

Figure 16–30 (a) Six-member plane frame; (b) dynamic load

then necessarily run the modal analysis program first to obtain the frequencies. For instance, in the fixed-fixed beam, the first two different frequencies are $\omega_1 = 45.23$ rad/s and $\omega_3 = 120.16$ rad/s (ω_2 is the same as ω_3, so use ω_3). Now assume light damping ($\zeta \leq 0.05$). Therefore, let $\zeta_1 = \zeta_2 = 0.05$. Using these ω's and ζ's, in Eqs. (16.9.3), we obtain $\alpha = 3.286$ and $\beta = 0.000605$. These values were used for α and β in the damped response for the fixed-fixed beam to include 5% damping ($\zeta = 0.05$).

Figure 16–30(a) shows a plane frame consisting of six rigidly connected prismatic members with dynamic forces $F(t)$ and $2F(t)$ applied in the x direction at joints 6 and 4, respectively. The time variation of $F(t)$ is shown in Figure 16–30(b). The results are for steel with cross-sectional area of 0.01935 m^2, moment of inertia of 0.4162×10^{-3} m^4, $L = 1.25$ m, and $F_1 = 50,000$ N. Figure 16–31 shows the displaced frame for the worst stress at time of 0.035 s. The largest x displacement of node 6 for the time of 0.035 s is 3.94 mm. This value compares closely with the solution in Reference [16].

Finally, Figures 16–32(a) and 16–33(a) show models of a gantry crane and a cab frame subjected to dynamic loading functions [Figures 16–32(b) and (16–33(b)]. For details of these design solutions consult [17–18].

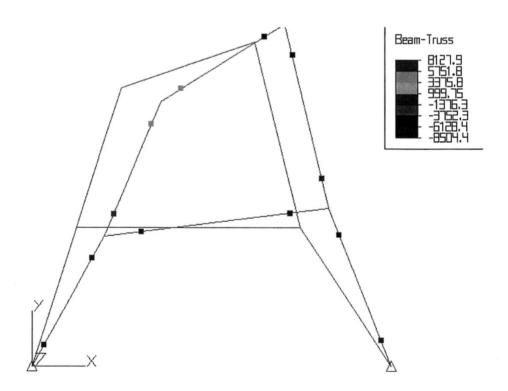

Figure 16–31 Displaced frame with worst stress at time 0.035 s

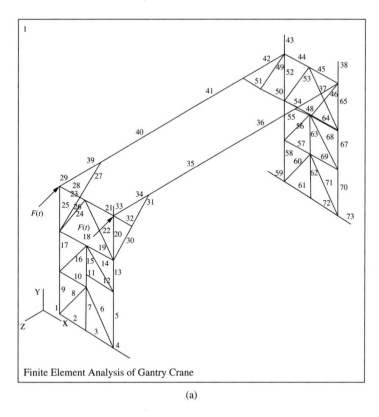

Finite Element Analysis of Gantry Crane

(a)

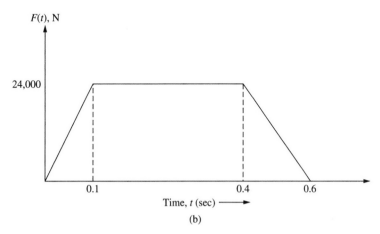

(b)

Figure 16–32 (a) Gantry crane model composed of 73 beam elements and (b) the time-dependent trapezoidal loading function applied to the top edge of the crane [17]

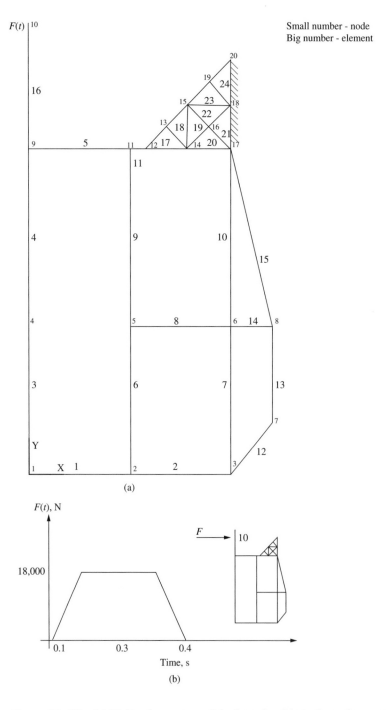

Figure 16–33 (a) Finite element model of a cab with 8 plate elements (upper right triangular elements) and 15 beam elements and (b) the time-dependent trapezoidal loading applied to node 10 [18]

▲ Summary Equations

Equation of motion for spring-mass system:

$$m\ddot{x} + kx = F(t) \tag{16.1.2}$$

Natural circular frequency:

$$\omega^2 = \frac{k}{m} \tag{16.1.3}$$

Period of vibration:

$$\tau = \frac{2\pi}{\omega} \tag{16.1.5}$$

Stiffness matrix for bar:

$$[k] = \frac{AE}{L} \begin{array}{cc} u_1 & u_2 \\ \begin{bmatrix} 1 & -1 \\ -1 & 1 \end{bmatrix} \end{array} \tag{16.2.11}$$

Lumped-mass matrix for bar:

$$[m] = \frac{\rho AL}{2} \begin{array}{cc} u_1 & u_2 \\ \begin{bmatrix} 1 & 0 \\ 0 & 1 \end{bmatrix} \end{array} \tag{16.2.12}$$

Consistent-mass matrix for bar:

$$[m] = \frac{\rho AL}{6} \begin{bmatrix} 2 & 1 \\ 1 & 2 \end{bmatrix} \tag{16.2.23}$$

Global equations of motion:

$$\{F(t)\} = [K]\{d\} + [M]\{\ddot{d}\} \tag{16.2.24}$$

Central difference numerical integration equations for velocity and acceleration:

$$\{\dot{d}_i\} = \frac{\{d_{i+1}\} - \{d_{i-1}\}}{2\Delta t} \tag{16.3.1}$$

$$\{\ddot{d}_i\} = \frac{\{\dot{d}_{i+1}\} - \{\dot{d}_{i-1}\}}{2\Delta t} \tag{16.3.2}$$

For a flowchart for the central difference method, see Figure 16–6.

Newmark's equations for numerical integration:

$$\{\dot{d}_{i+1}\} = \{\dot{d}_i\} + (\Delta t)[(1 - \gamma)\{\ddot{d}_i\} + \gamma\{\ddot{d}_{i+1}\}] \tag{16.3.9}$$

$$\{d_{i+1}\} = \{d_i\} + (\Delta t)\{\dot{d}_i\} + (\Delta t)^2[(\tfrac{1}{2} - \beta)\{\ddot{d}_i\} + \beta\{\ddot{d}_{i+1}\}] \tag{16.3.10}$$

For a flowchart of Newmark's method, see Figure 16–8.

Wilson's equations for numerical integration:

$$\{\dot{d}_{i+1}\} = \{\dot{d}_i\} + \frac{\Theta \Delta t}{2}(\{\ddot{d}_{i+1}\} + \{\ddot{d}_i\}) \tag{16.3.15}$$

$$\{d_{i+1}\} = \{d_i\} + \Theta \Delta t \{\dot{d}_i\} + \frac{\Theta^2 (\Delta t)^2}{6}(\{\ddot{d}_{i+1}\} + 2\{\ddot{d}_i\}) \tag{16.3.16}$$

Determinant to determine natural frequencies:

$$\|[K] - \omega^2 M| = 0 \tag{16.4.7}$$

Time step recommended using central difference method:

$$\Delta t \leq \frac{3}{4}\left(\frac{2}{\omega_{\max}}\right) \tag{16.5.1}$$

Beam element lumped mass matrix:

$$[m] = \frac{\rho A L}{2} \begin{array}{c} \begin{array}{cccc} v_1 & \phi_1 & v_2 & \phi_2 \end{array} \\ \begin{bmatrix} 1 & 0 & 0 & 0 \\ 0 & 0 & 0 & 0 \\ 0 & 0 & 1 & 0 \\ 0 & 0 & 0 & 0 \end{bmatrix} \end{array} \tag{16.6.1}$$

Beam element consistent mass matrix:

$$[m] = \frac{\rho A L}{420} \begin{array}{c} \begin{array}{cccc} v_1 & \phi_1 & v_2 & \phi_2 \end{array} \\ \begin{bmatrix} 156 & 22L & 54 & -13L \\ 22L & 4L^2 & 13L & -3L^2 \\ 54 & 13L & 156 & -22L \\ -13L & -3L^2 & -22L & 4L^2 \end{bmatrix} \end{array} \tag{16.6.5}$$

First natural frequency for beam based on classical beam theory solution:

Fixed-fixed beam:

$$\omega = \frac{5.59}{L^2}\left(\frac{EI}{A\rho}\right)^{1/2} \tag{16.6.9}$$

Fixed-free beam:

$$\omega = \frac{3.516}{L^2}\left(\frac{EI}{\rho A}\right)^{1/2} \tag{16.6.20}$$

Truss element consistent mass matrix:

$$[m'] = \frac{\rho A L}{6} \begin{array}{cccc} u_1 & v_1 & u_2 & v_2 \\ \begin{bmatrix} 2 & 0 & 1 & 0 \\ 0 & 2 & 0 & 1 \\ 1 & 0 & 2 & 0 \\ 0 & 1 & 0 & 2 \end{bmatrix} \end{array} \tag{16.7.4}$$

Truss element lumped mass matrix:

$$[m'] = \frac{\rho A L}{2} \begin{array}{cccc} u_1 & v_1 & u_2 & v_2 \\ \begin{bmatrix} 1 & 0 & 0 & 0 \\ 0 & 1 & 0 & 0 \\ 0 & 0 & 1 & 0 \\ 0 & 0 & 0 & 1 \end{bmatrix} \end{array} \tag{16.7.5}$$

Plane frame consistent element mass matrix:

$$[m'] = \rho A L \begin{array}{cccccc} u_1 & v_1 & \phi_1 & u_2 & v_2 & \phi_2 \\ \begin{bmatrix} 2/6 & 0 & 0 & 1/6 & 0 & 0 \\ & 156/420 & 22L/420 & 0 & 54/420 & -13L/420 \\ & & 4L^2/420 & 0 & 13L/420 & -3L^2/420 \\ & & & 2/6 & 0 & 0 \\ & & & & 156/420 & -22L/420 \\ \text{Symmetry} & & & & & 4L^2/420 \end{bmatrix} \end{array} \tag{16.7.6}$$

Plane frame lumped mass matrix:

$$[m'] = \frac{\rho A L}{2} \begin{array}{cccccc} u_1' & v_1' & \phi_1' & u_2' & v_2' & \phi_2' \\ \begin{bmatrix} 1 & 0 & 0 & 0 & 0 & 0 \\ 0 & 1 & 0 & 0 & 0 & 0 \\ 0 & 0 & 0 & 0 & 0 & 0 \\ 0 & 0 & 0 & 1 & 0 & 0 \\ 0 & 0 & 0 & 0 & 1 & 0 \\ 0 & 0 & 0 & 0 & 0 & 0 \end{bmatrix} \end{array} \tag{16.7.7}$$

Constant strain triangle consistent mass matrix:

$$[m] = \frac{\rho t A}{12} \begin{array}{cccccc} u_1 & v_1 & u_2 & v_2 & u_3 & v_3 \\ \begin{bmatrix} 2 & 0 & 1 & 0 & 1 & 0 \\ & 2 & 0 & 1 & 0 & 1 \\ & & 2 & 0 & 1 & 0 \\ & & & 2 & 0 & 1 \\ & & & & 2 & 0 \\ \text{Symmetry} & & & & & 2 \end{bmatrix} \end{array} \tag{16.7.9}$$

Axisymmetric element consistent mass matrix:

$$[m] = \frac{\pi \rho A}{10} \begin{bmatrix}
\frac{4}{3}r_1 + 2\bar{r} & 0 & 2\bar{r} - \frac{r_3}{3} & 0 & 2\bar{r} - \frac{r_2}{3} & 0 \\
& \frac{4}{3}r_1 + 2\bar{r} & 0 & 2\bar{r} - \frac{r_3}{3} & 0 & 2\bar{r} - \frac{r_2}{3} \\
& & \frac{4}{3}r_2 + 2\bar{r} & 0 & 2\bar{r} - \frac{r_1}{3} & 0 \\
& & & \frac{4}{3}r_2 + 2\bar{r} & 0 & 2\bar{r} - \frac{r_1}{3} \\
& & & & \frac{4}{3}r_3 + 2\bar{r} & 0 \\
\text{Symmetry} & & & & & \frac{4}{3}r_3 + 2\bar{r}
\end{bmatrix}$$

columns labeled $u_1 \quad v_1 \quad u_2 \quad v_2 \quad u_3 \quad v_3$

$$(16.7.14)$$

where
$$\bar{r} = \frac{r_1 + r_2 + r_3}{3}$$

Tetrahedral solid element consistent mass matrix:

$$[m] = \frac{\rho V}{20} \begin{bmatrix}
2 & 0 & 0 & 1 & 0 & 0 & 1 & 0 & 0 & 1 & 0 & 0 \\
& 2 & 0 & 0 & 1 & 0 & 0 & 1 & 0 & 0 & 1 & 0 \\
& & 2 & 0 & 0 & 1 & 0 & 0 & 1 & 0 & 0 & 1 \\
& & & 2 & 0 & 0 & 1 & 0 & 0 & 1 & 0 & 0 \\
& & & & 2 & 0 & 0 & 1 & 0 & 0 & 1 & 0 \\
& & & & & 2 & 0 & 0 & 1 & 0 & 0 & 1 \\
& & & & & & 2 & 0 & 0 & 1 & 0 & 0 \\
& & & & & & & 2 & 0 & 0 & 1 & 0 \\
& & & & & & & & 2 & 0 & 0 & 1 \\
& & & & & & & & & 2 & 0 & 0 \\
& & & & & & & & & & 2 & 0 \\
\text{Symmetry} & & & & & & & & & & & 2
\end{bmatrix}$$

columns labeled $u_1 \ v_1 \ w_1 \ u_2 \ v_2 \ w_2 \ u_3 \ v_3 \ w_3 \ u_4 \ v_4 \ w_4$

$$(16.7.15)$$

One-dimensional bar element consistent mass matrix for heat transfer:

$$[m] = \frac{c\rho A L}{6} \begin{bmatrix} 2 & 1 \\ 1 & 2 \end{bmatrix} \quad \begin{matrix} t_1 & t_2 \end{matrix}$$

$$(16.8.9)$$

One-dimensional bar element lumped mass matrix for heat transfer:

$$[m] = \frac{c\rho A L}{2} \begin{bmatrix} 1 & 0 \\ 0 & 1 \end{bmatrix} \quad \begin{matrix} t_1 & t_2 \end{matrix}$$

$$(16.8.10)$$

Global form of time-dependent heat transfer equation:

$$\{F\} = [K]\{T\} + [M]\{\dot{T}\} \qquad (16.8.12)$$

Upper limit time step for numerical analysis to be stable for heat transfer problem:

$$\Delta t = \frac{2}{(1 - 2\beta)\lambda_{\max}} \qquad (16.8.17)$$

▲ References

[1] Thompson, W. T., and Dahleh, M. D., *Theory of Vibrations with Applications*, 5th ed., Prentice-Hall, Englewood Cliffs, NJ, 1998.
[2] Archer, J. S., "Consistent Matrix Formulations for Structural Analysis Using Finite Element Techniques," *Journal of the American Institute of Aeronautics and Astronautics*, Vol. 3, No. 10, pp. 1910–1918, 1965.
[3] James, M. L., Smith, G. M., and Wolford, J. C., *Applied Numerical Methods for Digital Computation*, 3rd ed., Harper & Row, New York, 1985.
[4] Biggs, J. M., *Introduction to Structural Dynamics*, McGraw-Hill, New York, 1964.
[5] Newmark, N. M., "A Method of Computation for Structural Dynamics," *Journal of the Engineering Mechanics Division*, American Society of Civil Engineers, Vol. 85, No. EM3, pp. 67–94, 1959.
[6] Clark, S. K., *Dynamics of Continuous Elements*, Prentice-Hall, Englewood Cliffs, NJ, 1972.
[7] Bathe, K. J., *Finite Element Procedures in Engineering Analysis*, Prentice-Hall, Englewood Cliffs, NJ, 1982.
[8] Bathe, K. J., and Wilson, E. L., *Numerical Methods in Finite Element Analysis*, Prentice-Hall, Englewood Cliffs, NJ, 1976.
[9] Fujii, H., "Finite Element Schemes: Stability and Convergence," *Advances in Computational Methods in Structural Mechanics and Design*, J. T. Oden, R. W. Clough, and Y. Yamamoto, Eds., University of Alabama Press, Tuscaloosa, AL, pp. 201–218, 1972.
[10] Krieg, R. D., and Key, S. W., "Transient Shell Response by Numerical Time Integration," *International Journal of Numerical Methods in Engineering*, Vol. 17, pp. 273–286, 1973.
[11] Belytschko, T., "Transient Analysis," *Structural Mechanics Computer Programs, Surveys, Assessments, and Availability*, W. Pilkey, K. Saczalski, and H. Schaeffer, Eds., University of Virginia Press, Charlottesville, VA, pp. 255–276, 1974.
[12] Hughes, T. J. R., "Unconditionally Stable Algorithms for Nonlinear Heat Conduction," *Computational Methods in Applied Mechanical Engineering*, Vol. 10, No. 2, pp. 135–139, 1977.
[13] Hilber, H. M., Hughes, T. J. R., and Taylor, R. L., "Improved Numerical Dissipation for Time Integration Algorithms in Structural Dynamics," *Earthquake Engineering in Structural Dynamics*, Vol. 5, No. 3, pp. 283–292, 1977.
[14] Paz, M., *Structural Dynamics Theory and Computation*, 3rd ed., Van Nostrand Reinhold, New York, 1991.
[15] *Linear Stress and Dynamics Reference Division*, Docutech On-line Documentation, Algor, Inc., Pittsburgh, PA, 1999.
[16] Weaver, W., Jr., and Johnston, P. R., *Structural Dynamics by Finite Elements*, Prentice-Hall, Englewood Cliffs, NJ, 1987.

[17] Salemganesan, Hari, *Finite Element Analysis of a Gantry Crane*, M. S. Thesis, Rose-Hulman Institute of Technology, Terre Haute, IN, September 1992.

[18] Leong Cheow Fook, *The Dynamic Analysis of a Cab Using the Finite Element Method*, M. S. Thesis, Rose-Hulman Institute of Technology, Terre Haute, IN, January 1988.

▲ Problems

16.1 Determine the consistent-mass matrix for the one-dimensional bar discretized into two elements as shown in Figure P16–1. Let the bar have modulus of elasticity E, mass density ρ, and cross-sectional area A.

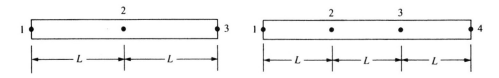

Figure P16–1   **Figure P16–2**

16.2 For the one-dimensional bar discretized into three elements as shown in Figure P16–2, determine the lumped- and consistent-mass matrices. Let the bar properties be $E, \rho,$ and A throughout the bar.

16.3 For the one-dimensional bar shown in Figure P16–3, determine the natural frequencies of vibration, ω's, using two elements of equal length. Use the consistent-mass approach. Let the bar have modulus of elasticity E, mass density ρ, and cross-sectional area A. Compare your answers to those obtained using a lumped-mass matrix in Example 16.3.

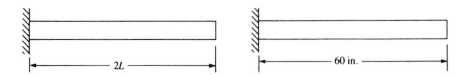

Figure P16–3   **Figure P16–4**

16.4 For the one-dimensional bar shown in Figure P16–4, determine the natural frequencies of longitudinal vibration using first two and then three elements of equal length. Let the bar have $E = 210$ GPa, $\rho = 7800$ kg/m^3, $A = 6$ cm^2, and $L = 1.5$ m.

16.5 For the spring-mass system shown in Figure P16–5, determine the mass displacement, velocity, and acceleration for five time steps using the central difference method. Let $k = 30$ kN/m and $m = 30$ kg. Use a time step of $\Delta t = 0.03$ s. You might want to write a computer program to solve this problem.

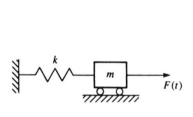
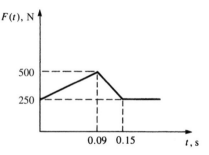

Figure P16–5

16.6 For the spring-mass system shown in Figure P16–6, determine the mass displacement, velocity, and acceleration for five time steps using (a) the central difference method, (b) Newmark's time integration method, and (c) Wilson's method. Let $k = 18$ kN/m and $m = 30$ kg.

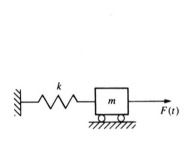
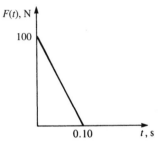

Figure P16–6

16.7 For the bar shown in Figure P16–7, determine the nodal displacements, velocities, and accelerations for five time steps using two finite elements. Let $E = 210$ GPa, $\rho = 7800$ kg/m³, $A = 6$ cm², and $L = 2.5$ m.

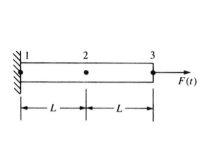
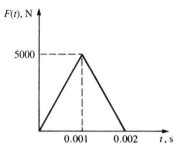

Figure P16–7

16.8 For the bar shown in Figure P16–8, determine the nodal displacements, velocities, and accelerations for five time steps using two finite elements. For simplicity of calculations, let $E = 6.25$ GPa, $\rho = 1 \times 10^7$ kg/m³, $A = 6$ cm², and $L = 2.5$ m. Use Newmark's method and Wilson's method.

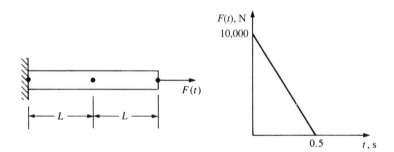

Figure P16–8

16.9, Rework Problems 16.7 and 16.8 using a computer program.
16.10

16.11 For the beams shown in Figure P16–11, determine the natural frequencies using first two and then three elements. Let $E, \rho,$ and A be constant for the beams.

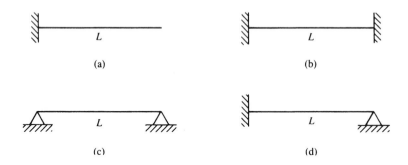

Figure P16–11

16.12 Rework Problem 16.11 using a computer program with $E = 210$ GPa, $\rho = 7800$ kg/m^3, $A = 6.5$ cm^2, $L = 2.5$ m, and $I = 3.5$ cm^4.

16.13, For the beams in Figures P16–13 and P16–14 subjected to the forcing functions
16.14 shown, determine the maximum deflections, velocities, and accelerations. Use a computer program.

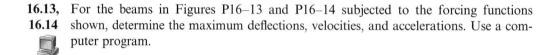

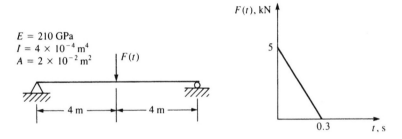

Figure P16–13

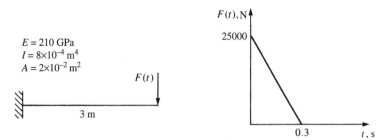

Figure P16–14

16.15, For the rigid frames in Figures P16–15 and P16–16 subjected to the forcing functions
16.16 shown, determine the maximum displacements, velocities, and accelerations. Use a
computer program.

For elements 1 and 9,
$A = 8 \times 10^{-3}\,\mathrm{m}^2, I = 10^{-4}\,\mathrm{m}^4$
For elements 2, 3, 7 and 8,
$A = 3.5 \times 10^{-3}\,\mathrm{m}^2, I = 4 \times 10^{-5}\,\mathrm{m}^4$
For elements 4, 5, and 6,
$A = 9 \times 10^{-3}\,\mathrm{m}^2, I = 3.2 \times 10^{-4}$
For all elements,
$E = 210\,\mathrm{GPa}$

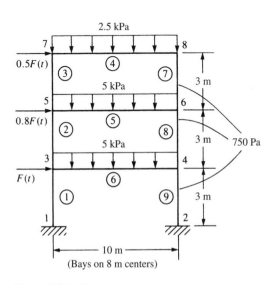

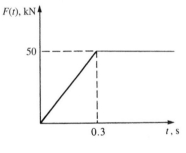

Figure P16–15

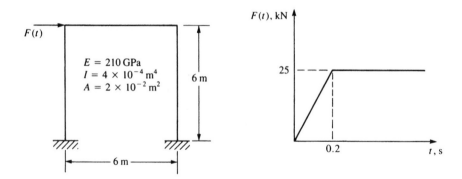

Figure P16–16

16.17 A marble slab with $k = 2$ W/(m · °C), $\rho = 2500$ kg/m³, and $c = 800$ W · s/(kg · °C) is 2 cm thick and at an initial uniform temperature of $T_i = 200$°C. The left surface is suddenly lowered to 0 °C and is maintained at that temperature while the other surface is kept insulated. Determine the temperature distribution in the slab for 40 s. Use $\beta = \frac{2}{3}$ and a time step of 8 s.

16.18 A circular fin is made of pure copper with a thermal conductivity of $k = 400$ W/(m · °C), $h = 150$ W/(m² · °C), mass density $\rho = 8900$ kg/m³, and specific heat $c = 375$ J/(kg · °C). The initial temperature of the fin is 25 °C. The fin length is 2 cm and the diameter is 0.4 cm. The right tip of the fin is insulated. See Figure P16–18. The base of the fin is then suddenly increased to a temperature of 85 °C and maintained at this temperature. Use the lumped form of the capacitance matrix, a time step of 0.1 s, and $\beta = \frac{2}{3}$. Use two elements of equal length. Determine the temperature distribution up to 3 s. Compare your results with Example 16.7, which used the consistent form of the capacitance matrix.

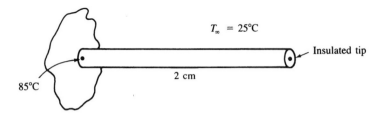

Figure P16–18

16.19, Rework Problems 16.17 and 16.18 using a computer program.
16.20

MATRIX ALGEBRA ▲

Introduction

In this appendix, we provide an introduction to matrix algebra. We will consider the concepts relevant to the finite element method to provide an adequate background for the matrix algebra concepts used in this text.

▲ ## A.1 Definition of a Matrix ▲

A **matrix** *is an m × n array of numbers arranged in m rows and n columns.* The matrix is then described as being of order $m \times n$. Equation (A.1.1) illustrates a matrix with m rows and n columns.

$$[a] = \begin{bmatrix} a_{11} & a_{12} & a_{13} & a_{14} & \cdots & a_{1n} \\ a_{21} & a_{22} & a_{23} & a_{24} & \cdots & a_{2n} \\ a_{31} & a_{32} & a_{33} & a_{34} & \cdots & a_{3n} \\ \vdots & \vdots & \vdots & \vdots & & \vdots \\ a_{m1} & a_{m2} & a_{m3} & a_{m4} & \cdots & a_{mn} \end{bmatrix} \tag{A.1.1}$$

If $m \neq n$ in matrix Eq. (A.1.1), the matrix is called **rectangular**. If $m = 1$ and $n > 1$, the elements of Eq. (A.1.1) form a single row called a **row matrix**. If $m > 1$ and $n = 1$, the elements form a single column called a **column matrix**. If $m = n$, the array is called a **square matrix**. Row matrices and rectangular matrices are denoted by using brackets [ ], and column matrices are denoted by using braces { }. For simplicity, matrices (row, column, or rectangular) are often denoted by using a line under a variable instead of surrounding it with brackets or braces. The order of the matrix should then be apparent from the context of its use. The force and displacement matrices used in structural analysis are column matrices, whereas the stiffness matrix is a square matrix.

To identify an element of matrix [a], we represent the element by a_{ij}, where the subscripts i and j indicate the row number and the column number, respectively, of [a]. Hence, alternative notations for a matrix are given by

$$[a] = [a_{ij}] \tag{A.1.2}$$

Numerical examples of special types of matrices are given by Eqs. (A.1.3) through (A.1.6). A rectangular matrix $[a]$ is given by

$$[a] = \begin{bmatrix} 2 & 1 \\ 3 & 4 \\ 5 & 4 \end{bmatrix} \qquad \text{(A.1.3)}$$

where $[a]$ has three rows and two columns. In matrix $[a]$ of Eq. (A.1.1), if $m = 1$, a row matrix results, such as

$$[a] = \begin{bmatrix} 2 & 3 & 4 & -1 \end{bmatrix} \qquad \text{(A.1.4)}$$

If $n = 1$ in Eq. (A.1.1), a column matrix results, such as

$$[a] = \begin{Bmatrix} 2 \\ 3 \end{Bmatrix} \qquad \text{(A.1.5)}$$

If $m = n$ in Eq. (A.1.1), a square matrix results, such as

$$[a] = \begin{bmatrix} 2 & -1 \\ 3 & -2 \end{bmatrix} \qquad \text{(A.1.6)}$$

Matrices and matrix notation are often used to express algebraic equations in compact form and are frequently used in the finite element formulation of equations. Matrix notation is also used to simplify the solution of a problem.

▲ A.2 Matrix Operations   ▲

We will now present some common matrix operations that will be used in this text.

Multiplication of a Matrix by a Scalar

If we have a scalar k and a matrix $[c]$, then the product $[a] = k[c]$ is given by

$$[a_{ij}] = k[c_{ij}] \qquad \text{(A.2.1)}$$

—that is, every element of the matrix $[c]$ is multiplied by the scalar k. As a numerical example, consider

$$[c] = \begin{bmatrix} 1 & 2 \\ 3 & 1 \end{bmatrix} \qquad k = 4$$

The product $[a] = k[c]$ is

$$[a] = 4\begin{bmatrix} 1 & 2 \\ 3 & 1 \end{bmatrix} = \begin{bmatrix} 4 & 8 \\ 12 & 4 \end{bmatrix}$$

Note that if $[c]$ is of order $m \times n$, then $[a]$ is also of order $m \times n$.

Addition of Matrices

Matrices of the same order can be added together by summing corresponding elements of the matrices. Subtraction is performed in a similar manner. Matrices of unlike order cannot be added or subtracted. Matrices of the same order can be added (or subtracted) in any order (the commutative law for addition applies). That is,

$$[c] = [a] + [b] = [b] + [a] \tag{A.2.2}$$

or, in subscript (index) notation, we have

$$[c_{ij}] = [a_{ij}] + [b_{ij}] = [b_{ij}] + [a_{ij}] \tag{A.2.3}$$

As a numerical example, let

$$[a] = \begin{bmatrix} -1 & 2 \\ -3 & 2 \end{bmatrix} \qquad [b] = \begin{bmatrix} 1 & 2 \\ 3 & 1 \end{bmatrix}$$

The sum $[a] + [b] = [c]$ is given by

$$[c] = \begin{bmatrix} -1 & 2 \\ -3 & 2 \end{bmatrix} + \begin{bmatrix} 1 & 2 \\ 3 & 1 \end{bmatrix} = \begin{bmatrix} 0 & 4 \\ 0 & 3 \end{bmatrix}$$

Again, remember that the matrices $[a]$, $[b]$, and $[c]$ must all be of the same order. For instance, a 2×2 matrix cannot be added to a 3×3 matrix.

Multiplication of Matrices

For two matrices $[a]$ and $[b]$ to be multiplied in the order shown in Eq. (A.2.4), the number of columns in $[a]$ must equal the number of rows in $[b]$. For example, consider

$$[c] = [a][b] \tag{A.2.4}$$

If $[a]$ is an $m \times n$ matrix, then $[b]$ must have n rows. Using subscript notation, we can write the product of matrices $[a]$ and $[b]$ as

$$[c_{ij}] = \sum_{e=1}^{n} a_{ie} b_{ej} \tag{A.2.5}$$

where n is the total number of columns in $[a]$ or of rows in $[b]$. For matrix $[a]$ of order 2×2 and matrix $[b]$ of order 2×2, after multiplying the two matrices, we have

$$[c_{ij}] = \begin{bmatrix} a_{11}b_{11} + a_{12}b_{21} & a_{11}b_{12} + a_{12}b_{22} \\ a_{21}b_{11} + a_{22}b_{21} & a_{21}b_{12} + a_{22}b_{22} \end{bmatrix} \tag{A.2.6}$$

For example, let

$$[a] = \begin{bmatrix} 2 & 1 \\ 3 & 2 \end{bmatrix} \qquad [b] = \begin{bmatrix} 1 & -1 \\ 2 & 0 \end{bmatrix}$$

The product $[a][b]$ is then

$$[a][b] = \begin{bmatrix} 2(1) + 1(2) & 2(-1) + 1(0) \\ 3(1) + 2(2) & 3(-1) + 2(0) \end{bmatrix} = \begin{bmatrix} 4 & -2 \\ 7 & -3 \end{bmatrix}$$

In general, matrix multiplication is *not* commutative; that is,

$$[a][b] \neq [b][a] \tag{A.2.7}$$

The validity of the product of two matrices $[a]$ and $[b]$ is commonly illustrated by

$$\begin{array}{ccc} [a] & [b] & = & [c] \\ (i \times e) & (e \times j) & & (i \times j) \end{array} \tag{A.2.8}$$

where the product matrix $[c]$ will be of order $i \times j$; that is, it will have the same number of rows as matrix $[a]$ and the same number of columns as matrix $[b]$.

Transpose of a Matrix

Any matrix, whether a row, column, or rectangular matrix, can be transposed. This operation is frequently used in finite element equation formulations. The transpose of a matrix $[a]$ is commonly denoted by $[a]^T$. The superscript T is used to denote the transpose of a matrix throughout this text. The transpose of a matrix is obtained by interchanging rows and columns; that is, the first row becomes the first column, the second row becomes the second column, and so on. For the transpose of matrix $[a]$,

$$[a_{ij}] = [a_{ji}]^T \tag{A.2.9}$$

For example, if we let

$$[a] = \begin{bmatrix} 2 & 1 \\ 3 & 2 \\ 4 & 5 \end{bmatrix}$$

then

$$[a]^T = \begin{bmatrix} 2 & 3 & 4 \\ 1 & 2 & 5 \end{bmatrix}$$

where we have interchanged the rows and columns of $[a]$ to obtain its transpose.

Another important relationship that involves the transpose is

$$([a][b])^T = [b]^T [a]^T \tag{A.2.10}$$

That is, the transpose of the product of matrices $[a]$ and $[b]$ is equal to the transpose of the latter matrix $[b]$ multiplied by the transpose of matrix $[a]$ in that order, provided the order of the initial matrices continues to satisfy the rule for matrix multiplication, Eq. (A.2.8). In general, this property holds for any number of matrices; that is,

$$([a][b][c] \ldots [k])^T = [k]^T \ldots [c]^T [b]^T [a]^T \tag{A.2.11}$$

Note that the transpose of a column matrix is a row matrix.

As a numerical example of the use of Eq. (A.2.10), let

$$[a] = \begin{bmatrix} 1 & 2 \\ 3 & 4 \end{bmatrix} \qquad [b] = \begin{Bmatrix} 5 \\ 6 \end{Bmatrix}$$

First,

$$[a][b] = \begin{bmatrix} 1 & 2 \\ 3 & 4 \end{bmatrix} \begin{Bmatrix} 5 \\ 6 \end{Bmatrix} = \begin{Bmatrix} 17 \\ 39 \end{Bmatrix}$$

Then,

$$([a][b])^T = [17 \quad 39] \tag{A.2.12}$$

Because $[b]^T$ and $[a]^T$ can be multiplied according to the rule for matrix multiplication, we have

$$[b]^T [a]^T = \begin{bmatrix} 5 & 6 \end{bmatrix} \begin{bmatrix} 1 & 3 \\ 2 & 4 \end{bmatrix} = \begin{bmatrix} 17 & 39 \end{bmatrix} \qquad (A.2.13)$$

Hence, on comparing Eqs. (A.2.12) and (A.2.13), we have shown (for this case) the validity of Eq. (A.2.10). A simple proof of the general validity of Eq. (A.2.10) is left to your discretion.

Symmetric Matrices

If a square matrix is equal to its transpose, it is called a **symmetric matrix**; that is, if

$$[a] = [a]^T$$

then $[a]$ is a symmetric matrix. As an example,

$$[a] = \begin{bmatrix} 3 & 1 & 2 \\ 1 & 4 & 0 \\ 2 & 0 & 3 \end{bmatrix} \qquad (A.2.14)$$

is a symmetric matrix because each element a_{ij} equals a_{ji} for $i \neq j$. In Eq. (A.2.14), note that the main diagonal running from the upper left corner to the lower right corner is the line of symmetry of the symmetric matrix $[a]$. Remember that only a square matrix can be symmetric.

Unit Matrix

The **unit** (or **identity**) **matrix** $[I]$ is such that

$$[a][I] = [I][a] = [a] \qquad (A.2.15)$$

The unit matrix acts in the same way that the number one acts in conventional multiplication. The unit matrix is always a square matrix of any possible order with each element of the main diagonal equal to one and all other elements equal to zero. For example, the 3×3 unit matrix is given by

$$[I] = \begin{bmatrix} 1 & 0 & 0 \\ 0 & 1 & 0 \\ 0 & 0 & 1 \end{bmatrix}$$

Inverse of a Matrix

The **inverse of a matrix** is a matrix such that

$$[a]^{-1}[a] = [a][a]^{-1} = [I] \qquad (A.2.16)$$

where the superscript, -1, denotes the inverse of $[a]$ as $[a]^{-1}$. Section A.3 provides more information regarding the properties of the inverse of a matrix and gives a method for determining it.

Orthogonal Matrix

A matrix $[T]$ is an orthogonal matrix if

$$[T]^T[T] = [T][T]^T = [I] \tag{A.2.17}$$

Hence, for an orthogonal matrix, we have

$$[T]^{-1} = [T]^T \tag{A.2.18}$$

An orthogonal matrix frequently used is the *transformation* or *rotation* matrix $[T]$. In two-dimensional space, the transformation matrix relates components of a vector in one coordinate system to components in another system. For instance, the displacement (and force as well) vector components of $\bar{d}$ expressed in the x-y system are related to those in the x'-y' system (Figure A–1 and Section 3.3) by

$$\{d'\} = [T]\{d\} \tag{A.2.19}$$

or

$$\begin{Bmatrix} d'_x \\ d'_y \end{Bmatrix} = \begin{bmatrix} \cos\theta & \sin\theta \\ -\sin\theta & \cos\theta \end{bmatrix} \begin{Bmatrix} d_x \\ d_y \end{Bmatrix} \tag{A.2.20}$$

where $[T]$ is the square matrix on the right side of Eq. (A.2.20).

Another use of an orthogonal matrix is to change from the local stiffness matrix to a global stiffness matrix for an element. That is, given a local stiffness matrix $[k']$ for an element, if the element is arbitrarily oriented in the x-y plane, then

$$[k] = [T]^T[k'][T] = [T]^{-1}[k'][T] \tag{A.2.21}$$

Equation (A.2.21) is used throughout this text to express the stiffness matrix $[k]$ in the x-y plane.

By further examination of $[T]$, we see that the trigonometric terms in $[T]$ can be interpreted as the direction cosines of lines Ox' and Oy' with respect to the x-y axes. Thus for Ox' or d'_x, we have from Eq. (A.2.20)

$$\langle t_{11} \quad t_{12} \rangle = \langle \cos\theta \quad \sin\theta \rangle \tag{A.2.22}$$

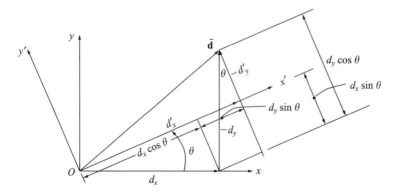

Figure A–1 Components of a vector in x'-y' and x-y coordinates

and for Oy' or d'_y, we have

$$\langle t_{21} \quad t_{22} \rangle = \langle -\sin \theta \quad \cos \theta \rangle \tag{A.2.23}$$

or unit vectors **i** and **j** can be represented in terms of unit vectors **i′** and **j′** [also see Section 3.3 for proof of Eq. (A.2.24)] as

$$\mathbf{i}' = \mathbf{i}\cos \theta + \mathbf{j}\sin \theta$$
$$\mathbf{j}' = -\mathbf{i}\sin \theta + \mathbf{j}\cos \theta \tag{A.2.24}$$

and hence

$$t_{11}^2 + t_{12}^2 = 1 \qquad t_{21}^2 + t_{22}^2 = 1 \tag{A.2.25}$$

and since these vectors (**i′** and **j′**) are orthogonal, by the dot product, we have

$$\langle t_{11}\mathbf{i} + t_{12}\mathbf{j} \rangle \cdot \langle t_{21}\mathbf{i} + t_{22}\mathbf{j} \rangle$$

or

$$t_{11}t_{21} + t_{12}t_{22} = 0 \tag{A.2.26}$$

or we say $[T]$ is orthogonal and therefore $[T]^T[T] = [T][T]^T = [I]$ and that the transpose is its inverse. That is,

$$[T]^T = [T]^{-1} \tag{A.2.27}$$

Differentiating a Matrix

A matrix is differentiated by differentiating every element in the matrix in the conventional manner. For example, if

$$[a] = \begin{bmatrix} x^3 & 2x^2 & 3x \\ 2x^2 & x^4 & x \\ 3x & x & x^5 \end{bmatrix} \tag{A.2.28}$$

the derivative $d[a]/dx$ is given by

$$\frac{d[a]}{dx} = \begin{bmatrix} 3x^2 & 4x & 3 \\ 4x & 4x^3 & 1 \\ 3 & 1 & 5x^4 \end{bmatrix} \tag{A.2.29}$$

Similarly, the partial derivative of a matrix is illustrated as follows:

$$\frac{\partial[b]}{\partial x} = \frac{\partial}{\partial x}\begin{bmatrix} x^2 & xy & xz \\ xy & y^2 & yz \\ xz & yz & z^2 \end{bmatrix} = \begin{bmatrix} 2x & y & z \\ y & 0 & 0 \\ z & 0 & 0 \end{bmatrix} \tag{A.2.30}$$

In structural analysis theory, we sometimes differentiate an expression of the form

$$U = \frac{1}{2}\begin{bmatrix} x & y \end{bmatrix}\begin{bmatrix} a_{11} & a_{12} \\ a_{12} & a_{22} \end{bmatrix}\begin{Bmatrix} x \\ y \end{Bmatrix} \tag{A.2.31}$$

where U might represent the strain energy in a bar. Expression (A.2.31) is known as a quadratic form. By matrix multiplication of Eq. (A.2.31), we obtain

$$U = \frac{1}{2}(a_{11}x^2 + 2a_{12}xy + a_{22}y^2) \qquad \text{(A.2.32)}$$

Differentiating U now yields

$$\frac{\partial U}{\partial x} = a_{11}x + a_{12}y \qquad \text{(A.2.33)}$$

$$\frac{\partial U}{\partial y} = a_{12}x + a_{22}y$$

Equation (A.2.33) in matrix form becomes

$$\begin{Bmatrix} \dfrac{\partial U}{\partial x} \\[2mm] \dfrac{\partial U}{\partial y} \end{Bmatrix} = \begin{bmatrix} a_{11} & a_{12} \\ a_{12} & a_{22} \end{bmatrix} \begin{Bmatrix} x \\ y \end{Bmatrix} \qquad \text{(A.2.34)}$$

A general form of Eq. (A.2.31) is

$$U = \frac{1}{2}\{X\}^T[a]\{X\} \qquad \text{(A.2.35)}$$

Then, by comparing Eq. (A.2.31) and (A.2.35), we obtain

$$\frac{\partial U}{\partial x_i} = [a]\{X\} \qquad \text{(A.2.36)}$$

where x_i denotes x and y. Here Eq. (A.2.36) depends on matrix $[a]$ in Eq. (A.2.35) being symmetric.

Integrating a Matrix

Just as in matrix differentiation, to integrate a matrix, we must integrate every element in the matrix in the conventional manner. For example, if

$$[a] = \begin{bmatrix} 3x^2 & 4x & 3 \\ 4x & 4x^3 & 1 \\ 3 & 1 & 5x^4 \end{bmatrix}$$

we obtain the integration of $[a]$ as

$$\int [a]\,dx = \begin{bmatrix} x^3 & 2x^2 & 3x \\ 2x^2 & x^4 & x \\ 3x & x & x^5 \end{bmatrix}$$

In our finite element formulation of equations, we often integrate an expression of the form

$$\iint [X]^T[A][X]\,dx\,dy \qquad \text{(A.2.37)}$$

The triple product in Eq. (A.2.37) will be symmetric if $[A]$ is symmetric. The form $[X]^T[A][X]$ is also called a *quadratic form.* For example, letting

$$[A] = \begin{bmatrix} 9 & 2 & 3 \\ 2 & 8 & 0 \\ 3 & 0 & 5 \end{bmatrix} \qquad [X] = \begin{Bmatrix} x_1 \\ x_2 \\ x_3 \end{Bmatrix}$$

we obtain

$$\{X\}^T[A]\{X\} = \begin{bmatrix} x_1 & x_2 & x_3 \end{bmatrix} \begin{bmatrix} 9 & 2 & 3 \\ 2 & 8 & 0 \\ 3 & 0 & 5 \end{bmatrix} \begin{Bmatrix} x_1 \\ x_2 \\ x_3 \end{Bmatrix}$$

$$= 9x_1^2 + 4x_1 x_2 + 6x_1 x_3 + 8x_2^2 + 5x_3^2$$

which is in quadratic form.

▲ A.3 Cofactor or Adjoint Method to Determine the Inverse of a Matrix   ▲

We will now introduce a method for finding the inverse of a matrix. This method is useful for longhand determination of the inverse of smaller-order square matrices (preferably of order 4×4 or less). A matrix $[a]$ must be square for us to determine its inverse.

We must first define the determinant of a matrix. This concept is necessary in determining the inverse of a matrix by the cofactor method. *A* **determinant** *is a square array of elements expressed by*

$$||[a]|| = |[a_{ij}]| \tag{A.3.1}$$

where the straight vertical bars, | |, on each side of the array denote the determinant. The resulting determinant of an array will be a single numerical value when the array is evaluated.

To evaluate the determinant of $[a]$, we must first determine the cofactors of $[a_{ij}]$. The cofactors of $[a_{ij}]$ are given by

$$C_{ij} = (-1)^{i+j}|[d]| \tag{A.3.2}$$

where the matrix $[d]$, called the *first minor of* $[a_{ij}]$, is matrix $[a]$ with row i and column j deleted. The inverse of matrix $[a]$ is then given by

$$[a]^{-1} = \frac{[C]^T}{|[a]|} \tag{A.3.3}$$

where $[C]$ is the **cofactor matrix** and $|[a]|$ is the determinant of $[a]$. To illustrate the method of cofactors, we will determine the inverse of a matrix $[a]$ given by

$$[a] = \begin{bmatrix} -1 & 3 & -2 \\ 2 & -4 & 2 \\ 0 & 4 & 1 \end{bmatrix} \tag{A.3.4}$$

Using Eq. (A.3.2), we find that the cofactors of matrix $[a]$ are

$$C_{11} = (-1)^{1+1} \begin{vmatrix} -4 & 2 \\ 4 & 1 \end{vmatrix} = -12$$

$$C_{12} = (-1)^{1+2} \begin{vmatrix} 2 & 2 \\ 0 & 1 \end{vmatrix} = -2$$

$$C_{13} = (-1)^{1+3} \begin{vmatrix} 2 & -4 \\ 0 & 4 \end{vmatrix} = 8$$

$$C_{21} = (-1)^{2+1} \begin{vmatrix} 3 & -2 \\ 4 & 1 \end{vmatrix} = -11 \qquad \text{(A.3.5)}$$

$$C_{22} = (-1)^{2+2} \begin{vmatrix} -1 & -2 \\ 0 & 1 \end{vmatrix} = -1$$

$$C_{23} = (-1)^{2+3} \begin{vmatrix} -1 & 3 \\ 0 & 4 \end{vmatrix} = 4$$

Similarly,
$$C_{31} = -2 \qquad C_{32} = -2 \qquad C_{33} = -2 \qquad \text{(A.3.6)}$$

Therefore, from Eqs. (A.3.5) and (A.3.6), we have

$$[C] = \begin{bmatrix} -12 & -2 & 8 \\ -11 & -1 & 4 \\ -2 & -2 & -2 \end{bmatrix} \qquad \text{(A.3.7)}$$

The determinant of $[a]$ is then

$$|[a]| = \sum_{j=1}^{n} a_{ij} C_{ij} \qquad \text{with } i \text{ any row number } (1 \leqslant i \leqslant n) \qquad \text{(A.3.8)}$$

or
$$|[a]| = \sum_{j=1}^{n} a_{ji} C_{ji} \qquad \text{with } i \text{ any column number } (1 \leqslant i \leqslant n) \qquad \text{(A.3.9)}$$

For instance, if we choose the first rows of $[a]$ and $[C]$, then $i = 1$ in Eq. (A.3.8), and j is summed from 1 to 3 such that

$$|[a]| = a_{11} C_{11} + a_{12} C_{12} + a_{13} C_{13}$$

$$= (-1)(-12) + (3)(-2) + (-2)(8) = -10 \qquad \text{(A.3.10)}$$

Using the definition of the inverse given by Eq. (A.3.3), we have

$$[a]^{-1} = \frac{[C]^T}{|[a]|} = \frac{1}{-10} \begin{bmatrix} -12 & -11 & -2 \\ -2 & -1 & -2 \\ 8 & 4 & -2 \end{bmatrix} \qquad \text{(A.3.11)}$$

We can then check that

$$[a][a]^{-1} = \begin{bmatrix} 1 & 0 & 0 \\ 0 & 1 & 0 \\ 0 & 0 & 1 \end{bmatrix}$$

The transpose of the cofactor matrix is often defined as the **adjoint matrix**; that is,

$$\text{adj}\,[a] = [C]^T$$

Therefore, an alternative equation for the inverse of $[a]$ is

$$[a]^{-1} = \frac{\text{adj}\,[a]}{||[a]||} \tag{A.3.12}$$

An important property associated with the determinant of a matrix is that if the determinant of a matrix is zero—that is, $||[a]|| = 0$—then the matrix is said to be **singular**. A singular matrix does not have an inverse. The stiffness matrices used in the finite element method are singular until sufficient boundary conditions (support conditions) are applied. This characteristic of the stiffness matrix is further discussed in the text.

▲ A.4 Inverse of a Matrix by Row Reduction ▲

The inverse of a nonsingular square matrix $[a]$ can be found by the method of row reduction (sometimes called the *Gauss–Jordan method*) by performing identical simultaneous operations on the matrix $[a]$ and the identity matrix $[I]$ (of the same order as $[a]$) such that the matrix $[a]$ becomes an identity matrix and the original identity matrix becomes the inverse of $[a]$.

A numerical example will best illustrate the procedure. We begin by converting matrix $[a]$ to an upper triangular form by setting all elements below the main diagonal equal to zero, starting with the first column and continuing with succeeding columns. We then proceed from the last column to the first, setting all elements above the main diagonal equal to zero.

We will invert the following matrix by row reduction.

$$[a] = \begin{bmatrix} 2 & 2 & 1 \\ 2 & 1 & 0 \\ 1 & 1 & 1 \end{bmatrix} \tag{A.4.1}$$

To find $[a]^{-1}$, we need to find $[x]$ such that $[a][x] = [I]$, where

$$[x] = \begin{bmatrix} x_{11} & x_{12} & x_{13} \\ x_{21} & x_{22} & x_{23} \\ x_{31} & x_{32} & x_{33} \end{bmatrix}$$

That is, solve

$$\begin{bmatrix} 2 & 2 & 1 \\ 2 & 1 & 0 \\ 1 & 1 & 1 \end{bmatrix}[x] = \begin{bmatrix} 1 & 0 & 0 \\ 0 & 1 & 0 \\ 0 & 0 & 1 \end{bmatrix}$$

We begin by writing $[a]$ and $[I]$ side by side as

$$\left[\begin{array}{ccc|ccc} 2 & 2 & 1 & 1 & 0 & 0 \\ 2 & 1 & 0 & 0 & 1 & 0 \\ 1 & 1 & 1 & 0 & 0 & 1 \end{array}\right] \tag{A.4.2}$$

where the vertical dashed line separates $[a]$ and $[I]$.

1. Divide the first row of Eq. (A.4.2) by 2.

$$\left[\begin{array}{ccc|ccc} 1 & 1 & \frac{1}{2} & \frac{1}{2} & 0 & 0 \\ 2 & 1 & 0 & 0 & 1 & 0 \\ 1 & 1 & 1 & 0 & 0 & 1 \end{array}\right] \tag{A.4.3}$$

2. Multiply the first row of Eq. (A.4.3) by -2 and add the result to the second row.

$$\left[\begin{array}{ccc|ccc} 1 & 1 & \frac{1}{2} & \frac{1}{2} & 0 & 0 \\ 0 & -1 & -1 & -1 & 1 & 0 \\ 1 & 1 & 1 & 0 & 0 & 1 \end{array}\right] \tag{A.4.4}$$

3. Subtract the first row of Eq. (A.4.4) from the third row.

$$\left[\begin{array}{ccc|ccc} 1 & 1 & \frac{1}{2} & \frac{1}{2} & 0 & 0 \\ 0 & -1 & -1 & -1 & 1 & 0 \\ 0 & 0 & \frac{1}{2} & -\frac{1}{2} & 0 & 1 \end{array}\right] \tag{A.4.5}$$

4. Multiply the second row of Eq. (A.4.5) by -1 and the third row by 2.

$$\left[\begin{array}{ccc|ccc} 1 & 1 & \frac{1}{2} & \frac{1}{2} & 0 & 0 \\ 0 & 1 & 1 & 1 & -1 & 0 \\ 0 & 0 & 1 & -1 & 0 & 2 \end{array}\right] \tag{A.4.6}$$

5. Subtract the third row of Eq. (A.4.6) from the second row.

$$\left[\begin{array}{ccc|ccc} 1 & 1 & \frac{1}{2} & \frac{1}{2} & 0 & 0 \\ 0 & 1 & 0 & 2 & -1 & -2 \\ 0 & 0 & 1 & -1 & 0 & 2 \end{array}\right] \tag{A.4.7}$$

6. Multiply the third row of Eq. (A.4.7) by $-\frac{1}{2}$ and add the result to the first row.

$$\left[\begin{array}{ccc|ccc} 1 & 1 & 0 & 1 & 0 & -1 \\ 0 & 1 & 0 & 2 & -1 & -2 \\ 0 & 0 & 1 & -1 & 0 & 2 \end{array}\right] \tag{A.4.8}$$

7. Subtract the second row of Eq. (A.4.8) from the first row.

$$\left[\begin{array}{ccc|ccc} 1 & 0 & 0 & -1 & 1 & 1 \\ 0 & 1 & 0 & 2 & -1 & -2 \\ 0 & 0 & 1 & -1 & 0 & 2 \end{array}\right] \qquad (A.4.9)$$

The replacement of $[a]$ by the inverse matrix is now complete. The inverse of $[a]$ is then the right side of Eq. (A.4.9); that is,

$$[a]^{-1} = \left[\begin{array}{ccc} -1 & 1 & 1 \\ 2 & -1 & -2 \\ -1 & 0 & 2 \end{array}\right] \qquad (A.4.10)$$

For additional information regarding matrix algebra, consult References [1] and [2].

▲ A.5 Properties of Stiffness Matrices ▲

Stiffness matrix $[k]$ is defined in Chapter 2 as relating nodal forces to nodal displacements. The stiffness matrix is also seen (for instance) in the strain energy expressions for springs, Eq. (2.6.20), for bars, Eq. (3.10.28b) and for beams, Eq. (4.7.21). The matrix has the properties of being *square* and *symmetric*, as defined in Sections A.1 and A.2, for nearly all applications in this textbook except for the mass transport problem in Section 13.9.

In the strain energy expression, we see $[k]$ in the quadratic form

$$U = \frac{1}{2}\{d\}^{T}[k]\{d\} \qquad (A.5.1)$$

For most structures, the stiffness matrix is a *positive definite* matrix. That means if arbitrary displacement vectors are chosen, and we calculate U, the result is a positive value. The exception to this is the trivial case where the displacement vector $\{d\}$ is set to zero. Therefore, for any arbitrary displacements of a multi-degree-of-freedom system from its undeformed configuration, the strain energy is positive.

The exception to $[k]$ being positive definite is when a system has rigid-body degrees of freedom. Then the displacement is taken as a rigid-body mode. In this case, $[k]$ is called a *positive semidefinite* matrix. The strain energy U then can be zero for rigid-body modes or greater than zero when we have deformable modes. When $[k]$ is positive semidefinite, $||[k]|| = \det([k]) = 0$. Recall, from Section A.3, a matrix whose determinant is zero is called a *singular* matrix. To physically remove the singularity in a system in static equilibrium, sufficient boundary conditions must be applied. This concept is further described in Chapter 2.

For instance, consider a bar with no supports as shown in Figure A–2. If the bar is discretized into two elements and the 3×3 stiffness matrix of the bar is determined as described in Chapter 2 and as shown by Eq. (A.5.2), the determinant of this stiffness matrix, Eq. (A.5.3), is zero. Now if we fix one end of the bar, making

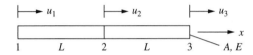

Figure A–2 Two-element bar

$u_1 = 0$, the reduced 2×2 stiffness matrix has a nonzero determinant. (Also see Problem A.12.)

$$[k] = \frac{AE}{L} \begin{bmatrix} 1 & -1 & 0 \\ -1 & 2 & -1 \\ 0 & -1 & 1 \end{bmatrix} \tag{A.5.2}$$

Now the determinant of $[k]$ is

$$\begin{vmatrix} 1 & -1 & 0 \\ -1 & 2 & -1 \\ 0 & -1 & 1 \end{vmatrix} = 1 \begin{vmatrix} 2 & -1 \\ -1 & 1 \end{vmatrix} - (-1) \begin{vmatrix} -1 & -1 \\ 0 & 1 \end{vmatrix} + 0 \tag{A.5.3}$$

$$= 2 - 1 - 1 = 0$$

▲ References

[1] Gere, J. M., *Matrix Algebra for Engineers*, Brooks/Cole Engineering, 2nd Ed, 1983.
[2] Jennings, A., *Matrix Computation for Engineers and Scientists*, Wiley, New York, 1977.

▲ Problems

Solve Problems A.1 through A.6 using matrices $[A]$, $[B]$, $[C]$, $[D]$, and $\{E\}$ given by

$$[A] = \begin{bmatrix} 1 & 0 \\ -1 & 4 \end{bmatrix} \qquad [B] = \begin{bmatrix} 2 & 0 \\ -2 & 8 \end{bmatrix} \qquad [C] = \begin{bmatrix} 3 & 1 & 0 \\ -1 & 0 & 3 \end{bmatrix}$$

$$[D] = \begin{bmatrix} 3 & 1 & 2 \\ 1 & 4 & 0 \\ 2 & 0 & 3 \end{bmatrix} \qquad \{E\} = \begin{Bmatrix} 1 \\ 2 \\ 3 \end{Bmatrix}$$

(Write "nonsense" if the operation cannot be performed.)

A.1 **(a)** $[A] + [B]$ **(b)** $[A] + [C]$
     (c) $[A][C]^T$ **(d)** $[D]\{E\}$
     (e) $[D][C]$ **(f)** $[C][D]$

A.2 Determine $[A]^{-1}$ by the cofactor method.

A.3 Determine $[D]^{-1}$ by the cofactor method.

A.4 Determine $[C]^{-1}$.

A.5 Determine $[B]^{-1}$ by row reduction.

A.6 Determine $[D]^{-1}$ by row reduction.

A.7 Show that $([A][B])^T = [B]^T[A]^T$ by using

$$[A] = \begin{bmatrix} a_{11} & a_{12} \\ a_{21} & a_{22} \end{bmatrix} \qquad [B] = \begin{bmatrix} b_{11} & b_{12} & b_{13} \\ b_{21} & b_{22} & b_{23} \end{bmatrix}$$

A.8 Find $[T]^{-1}$ given that

$$[T] = \begin{bmatrix} \cos\theta & \sin\theta \\ -\sin\theta & \cos\theta \end{bmatrix}$$

and show that $[T]^{-1} = [T]^T$ and hence that $[T]$ is an orthogonal matrix.

A.9 Given the matrices

$$[X] = \begin{bmatrix} x & y \\ 1 & x \end{bmatrix} \qquad [A] = \begin{bmatrix} a & b \\ b & c \end{bmatrix}$$

show that the triple matrix product $[X]^T[A][X]$ is symmetric.

A.10 Evaluate the following integral in explicit form:

$$[k] = \int_0^L [B]^T E[B]\, dx$$

where

$$[B] = \begin{bmatrix} -\dfrac{1}{L} & \dfrac{1}{L} \end{bmatrix}$$

and E is the modulus of elasticity.

[*Note:* This is the step needed to obtain Eq. (10.1.16) from Eq. (10.1.15).]

A.11 The following integral represents the strain energy in a bar of length L and cross-sectional area A:

$$U = \frac{A}{2} \int_0^L \{d\}^T[B]^T[D][B]\{d\}\, dx$$

where

$$\{d\} = \begin{Bmatrix} u_1 \\ u_2 \end{Bmatrix} \qquad [B] = \begin{bmatrix} -\dfrac{1}{L} & \dfrac{1}{L} \end{bmatrix} \qquad [D] = E$$

and E is the modulus of elasticity.

Show that $dU/d\{d\}$ yields $[k]\{d\}$, where $[k]$ is the bar stiffness matrix given by

$$[k] = \frac{AE}{L}\begin{bmatrix} 1 & -1 \\ -1 & 1 \end{bmatrix}$$

A.12 A two-element bar as shown in Figure PA–12 with element lengths L, cross-sectional area A, and Young's modulus E can be shown to have a stiffness matrix of

$$[k] = \frac{AE}{L}\begin{bmatrix} 1 & -1 & 0 \\ -1 & 2 & -1 \\ 0 & -1 & 1 \end{bmatrix}$$

Show that the det $([k]) = 0$ and hence that $[k]$ is positive semidefinite and the matrix is also singular. Now fix the left end (set $u_1 = 0$) and show that the reduced $[k]$ is

$$[k] = \frac{AE}{L}\begin{bmatrix} 2 & -1 \\ -1 & 1 \end{bmatrix}$$

and that the det $([k])$ is no longer 0.

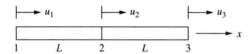

Figure PA–12

METHODS FOR SOLUTION OF SIMULTANEOUS LINEAR EQUATIONS ▲

Introduction

Many problems in engineering and mathematical physics require the solution of a system of simultaneous linear algebraic equations. Stress analysis, heat transfer, and vibration analysis are engineering problems for which the finite element formulation for solution typically involves the solving of simultaneous linear equations. This appendix introduces methods applicable to both longhand and computer solutions of simultaneous linear equations. Many methods are available for the solution of equations; for brevity's sake, we will discuss only some of the more common methods.

▲ B.1 General Form of the Equations ▲

In general, the set of equations will have the form

$$a_{11}x_1 + a_{12}x_2 + \cdots + a_{1n}x_n = c_1$$

$$a_{21}x_1 + a_{22}x_2 + \cdots + a_{2n}x_n = c_2$$

$$\vdots \qquad \vdots \qquad \qquad \vdots \qquad \vdots$$

$$a_{n1}x_1 + a_{n2}x_2 + \cdots + a_{nn}x_n = c_n$$

(B.1.1)

where the a_{ij}'s are the coefficients of the unknown x_j's, and the c_i's are the known right-side terms. In the structural analysis problem, the a_{ij}'s are the stiffness coefficients k_{ij}'s, the x_j's are the unknown nodal displacements d_i's, and the c_i's are the known nodal forces F_i's.

If the c's are not all zero, the set of equations is *nonhomogeneous*, and all equations must be independent to yield a unique solution. Stress analysis problems typically involve solving sets of nonhomogeneous equations.

If the c's are all zero, the set of equations is *homogeneous*, and nontrivial solutions exist only if all equations are not independent. Buckling and vibration problems typically involve homogeneous sets of equations.

▲ **B.2 Uniqueness, Nonuniqueness, and Nonexistence of Solution** ▲

To solve a system of simultaneous linear equations means to determine a unique set of values (if they exist) for the unknowns that satisfy every equation of the set simultaneously. A unique solution exists if and only if the determinant of the square coefficient matrix is not equal to zero. (All of the engineering problems considered in this text result in square coefficient matrices.) The problems in this text usually result in a system of equations that has a unique solution. Here we will briefly illustrate the concepts of uniqueness, nonuniqueness, and nonexistence of solution for systems of equations.

Uniqueness of Solution

$$2x_1 + 1x_2 = 6$$
$$1x_1 + 4x_2 = 17$$

(B.2.1)

For Eqs. (B.2.1), the determinant of the coefficient matrix is not zero, and a unique solution exists, as shown by the single common point of intersection of the two Eqs. (B.2.1) in Figure B–1.

Nonuniqueness of Solution

$$2x_1 + 1x_2 = 6$$
$$4x_1 + 2x_2 = 12$$

(B.2.2)

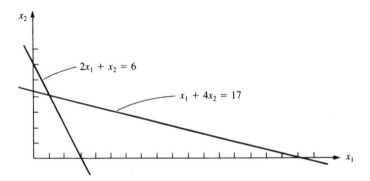

Figure B–1 Uniqueness of solution

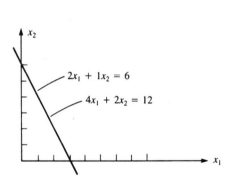

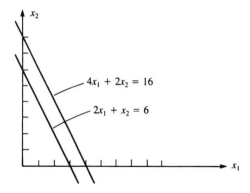

Figure B–2 Nonuniqueness of solution   **Figure B–3** Nonexistence of solution

For Eqs. (B.2.2), the determinant of the coefficient matrix is zero; that is,

$$\begin{vmatrix} 2 & 1 \\ 4 & 2 \end{vmatrix} = 0$$

Hence the equations are called *singular*, and either the solution is not unique or it does not exist. In this case, the solution is not unique, as shown in Figure B–2.

Nonexistence of Solution

$$2x_1 + x_2 = 6$$
$$4x_1 + 2x_2 = 16$$

(B.2.3)

Again, the determinant of the coefficient matrix is zero. In this case, no solution exists because we have parallel lines (no common point of intersection), as shown in Figure B–3.

▲ B.3 Methods for Solving Linear Algebraic Equations ▲

We will now present some common methods for solving systems of linear algebraic equations that have unique solutions. Some of these methods work best for small sets of equations solved longhand, whereas others are well suited for computer application.

Cramer's Rule

We begin by introducing a method known as *Cramer's rule*, which is useful for the longhand solution of small numbers of simultaneous equations. Consider the set of equations

$$[a]\{x\} = [c]$$

(B.3.1)

or, in index notation,

$$\sum_{i=1}^{n} a_{ij}x_j = c_i \tag{B.3.2}$$

We first let $[d^{(i)}]$ be the matrix $[a]$ with column i replaced by the column matrix $[c]$. Then the unknown x_i's are determined by

$$x_i = \frac{|[d^{(i)}]|}{|[a]|} \tag{B.3.3}$$

As an example of Cramer's rule, consider the following equations:

$$-x_1 + 3x_2 - 2x_3 = 2$$
$$2x_1 - 4x_2 + 2x_3 = 1 \tag{B.3.4}$$
$$4x_2 + x_3 = 3$$

In matrix form, Eqs. (B.3.4) become

$$\begin{bmatrix} -1 & 3 & -2 \\ 2 & -4 & 2 \\ 0 & 4 & 1 \end{bmatrix} \begin{Bmatrix} x_1 \\ x_2 \\ x_3 \end{Bmatrix} = \begin{Bmatrix} 2 \\ 1 \\ 3 \end{Bmatrix} \tag{B.3.5}$$

By Eq. (B.3.3), we can solve for the unknown x_i's as

$$x_1 = \frac{|[d^{(1)}]|}{|[a]|} = \frac{\begin{vmatrix} 2 & 3 & -2 \\ 1 & -4 & 2 \\ 3 & 4 & 1 \end{vmatrix}}{\begin{vmatrix} -1 & 3 & -2 \\ 2 & -4 & 2 \\ 0 & 4 & 1 \end{vmatrix}} = \frac{-41}{-10} = 4.1$$

$$x_2 = \frac{|[d^{(2)}]|}{|[a]|} = \frac{\begin{vmatrix} -1 & 2 & -2 \\ 2 & 1 & 2 \\ 0 & 3 & 1 \end{vmatrix}}{-10} = 1.1 \tag{B.3.6}$$

$$x_3 = \frac{|[d^{(3)}]|}{|[a]|} = \frac{\begin{vmatrix} -1 & 3 & 2 \\ 2 & -4 & 1 \\ 0 & 4 & 3 \end{vmatrix}}{-10} = -1.4$$

In general, to find the determinant of an $n \times n$ matrix, we must evaluate the determinants of n matrices of order $(n-1) \times (n-1)$. It has been shown that the solution of n simultaneous equations by Cramer's rule, evaluating determinants by expansion by minors, requires $(n-1)(n+1)!$ multiplications. Hence, this method takes large amounts of computer time and therefore is not used in solving large systems of simultaneous equations either longhand or by computer.

Inversion of the Coefficient Matrix

The set of equations $[a]\{x\} = \{c\}$ can be solved for $\{x\}$ by inverting the coefficient matrix $[a]$ and premultiplying both sides of the original set of equations by $[a]^{-1}$, such that

$$[a]^{-1}[a][x] = [a]^{-1}\{c\}$$

$$[I]\{x\} = [a]^{-1}\{c\} \qquad \text{(B.3.7)}$$

$$\{x\} = [a]^{-1}\{c\}$$

Two methods for determining the inverse of a matrix (the cofactor method and row reduction) were discussed in Appendix A.

The inverse method is much more time-consuming (because much time is required to determine the inverse of $[a]$) than either the elimination method or the iteration method, which are discussed subsequently. Therefore, inversion is practical only for small systems of equations.

However, the concept of inversion is often used during the formulation of the finite element equations, even though elimination or iteration is used in achieving the final solution for the unknowns (such as nodal displacements).

Besides the tedious calculations necessary to obtain the inverse, the method usually involves determining the inverse of sparse, banded matrices (stiffness matrices in structural analysis usually contain many zeros with the nonzero coefficients located in a band around the main diagonal). This sparsity and banded nature can be used to advantage in terms of storage requirements and solution algorithms on the computer. The inverse results in a dense, full matrix with loss of the advantages resulting from the sparse, banded nature of the original coefficient matrix.

To illustrate the solution of a system of equations by the inverse method, consider the same equations that we solved previously by Cramer's rule. For convenience's sake, we repeat the equations here.

$$\begin{bmatrix} -1 & 3 & -2 \\ 2 & -4 & 2 \\ 0 & 4 & 1 \end{bmatrix} \begin{Bmatrix} x_1 \\ x_2 \\ x_3 \end{Bmatrix} = \begin{Bmatrix} 2 \\ 1 \\ 3 \end{Bmatrix} \qquad \text{(B.3.8)}$$

The inverse of this coefficient matrix was found in Eq. (A.3.11) of Appendix A. The unknowns are then determined as

$$\begin{Bmatrix} x_1 \\ x_2 \\ x_3 \end{Bmatrix} = -\frac{1}{10} \begin{bmatrix} -12 & -11 & -2 \\ -2 & -1 & -2 \\ 8 & 4 & -2 \end{bmatrix} \begin{Bmatrix} 2 \\ 1 \\ 3 \end{Bmatrix} = \begin{Bmatrix} 4.1 \\ 1.1 \\ -1.4 \end{Bmatrix} \qquad \text{(B.3.9)}$$

Gaussian Elimination

We will now consider a commonly used method called *Gaussian elimination* that is easily adapted to the computer for solving systems of simultaneous equations. It is based on triangularization of the coefficient matrix and evaluation of the unknowns by back-substitution starting from the last equation.

The general system of n equations with n unknowns given by

$$\begin{bmatrix} a_{11} & a_{12} & \cdots & a_{1n} \\ a_{21} & a_{22} & \cdots & a_{2n} \\ \vdots & \vdots & & \vdots \\ a_{n1} & a_{n2} & \cdots & a_{nn} \end{bmatrix} \begin{Bmatrix} x_1 \\ x_2 \\ \vdots \\ x_n \end{Bmatrix} = \begin{Bmatrix} c_1 \\ c_2 \\ \vdots \\ c_n \end{Bmatrix} \tag{B.3.10}$$

will be used to explain the Gaussian elimination method.

1. Eliminate the coefficient of x_1 in every equation except the first one. To do this, select a_{11} as the pivot, and
   a. Add the multiple $-a_{21}/a_{11}$ of the first row to the second row.
   b. Add the multiple $-a_{31}/a_{11}$ of the first row to the third row.
   c. Continue this procedure through the nth row.
   The system of equations will then be reduced to the following form:

$$\begin{bmatrix} a_{11} & a_{12} & \cdots & a_{1n} \\ 0 & a'_{22} & \cdots & a'_{2n} \\ \vdots & & & \vdots \\ 0 & a'_{n2} & \cdots & a'_{nn} \end{bmatrix} \begin{Bmatrix} x_1 \\ x_2 \\ \vdots \\ x_n \end{Bmatrix} = \begin{Bmatrix} c_1 \\ c'_2 \\ \vdots \\ c'_n \end{Bmatrix} \tag{B.3.11}$$

2. Eliminate the coefficient of x_2 in every equation below the second equation. To do this, select a'_{22} as the pivot, and
   a. Add the multiple $-a'_{32}/a'_{22}$ of the second row to the third row.
   b. Add the multiple $-a'_{42}/a'_{22}$ of the second row to the fourth row.
   c. Continue this procedure through the nth row.
   The system of equations will then be reduced to the following form:

$$\begin{bmatrix} a_{11} & a_{12} & a_{13} & \cdots & a_{1n} \\ 0 & a'_{22} & a'_{23} & \cdots & a'_{2n} \\ 0 & 0 & a''_{33} & \cdots & a''_{3n} \\ \vdots & & & & \vdots \\ 0 & 0 & a''_{n3} & \cdots & a''_{nn} \end{bmatrix} \begin{Bmatrix} x_1 \\ x_2 \\ x_3 \\ \vdots \\ x_n \end{Bmatrix} = \begin{Bmatrix} c_1 \\ c'_2 \\ c''_3 \\ \vdots \\ c''_n \end{Bmatrix} \tag{B.3.12}$$

We repeat this process for the remaining rows until we have the system of equations (called *triangularized*) as

$$\begin{bmatrix} a_{11} & a_{12} & a_{13} & a_{14} & \cdots & a_{1n} \\ 0 & a'_{22} & a'_{23} & a'_{24} & \cdots & a'_{2n} \\ 0 & 0 & a''_{33} & a''_{34} & \cdots & a''_{3n} \\ 0 & 0 & 0 & a'''_{44} & \cdots & a'''_{4n} \\ \vdots & \vdots & \vdots & \vdots & & \vdots \\ 0 & 0 & 0 & 0 & \cdots & a^{n-1}_{nn} \end{bmatrix} \begin{Bmatrix} x_1 \\ x_2 \\ x_3 \\ x_4 \\ \vdots \\ x_n \end{Bmatrix} = \begin{Bmatrix} c_1 \\ c'_2 \\ c''_3 \\ c'''_4 \\ \vdots \\ c^{n-1}_n \end{Bmatrix} \tag{B.3.13}$$

3. Determine x_n from the last equation as

$$x_n = \frac{c^{n-1}_n}{a^{n-1}_{nn}} \tag{B.3.14}$$

and determine the other unknowns by back-substitution. These steps are summarized in general form by

$$k = 1, 2, \ldots, n - 1$$

$$a_{ij} = a_{ij} - a_{kj} \frac{a_{ik}}{a_{kk}} \qquad i = k + 1, \ldots, n$$

$$j = k, \ldots, n + 1 \qquad \text{(B.3.15)}$$

$$x_i = \frac{1}{a_{ii}} \left(a_{i,n+1} - \sum_{r=i+1}^{n} a_{ir} x_r \right)$$

where $a_{i,n+1}$ represent the latest right side c's given by Eq. (B.3.13).

We will solve the following example to illustrate the Gaussian elimination method.

Example B.1

Solve the following set of simultaneous equations using Gauss elimination method.

$$2x_1 + 2x_2 + 1x_3 = 9$$

$$2x_1 + 1x_2 \qquad\quad = 4 \qquad \text{(B.3.16)}$$

$$1x_1 + 1x_2 + 1x_3 = 6$$

SOLUTION:

Step 1

Eliminate the coefficient of x_1 in every equation except the first one. Select $a_{11} = 2$ as the pivot, and

a. Add the multiple $-a_{21}/a_{11} = -2/2$ of the first row to the second row.
b. Add the multiple $-a_{31}/a_{11} = -1/2$ of the first row to the third row.
 We then obtain

$$2x_1 + 2x_2 + 1x_3 = 9$$

$$0x_1 - 1x_2 - 1x_3 = 4 - 9 = -5 \qquad \text{(B.3.17)}$$

$$0x_1 + 0x_2 + \frac{1}{2}x_3 = 6 - \frac{9}{2} = \frac{3}{2}$$

Step 2

Eliminate the coefficient of x_2 in every equation below the second equation. In this case, we accomplished this in Step 1.

Step 3

Solve for x_3 in the third of Eqs. (B.3.17) as

$$x_3 = \frac{\frac{3}{2}}{\frac{1}{2}} = 3$$

Solve for x_2 in the second of Eqs. (B.3.17) as

$$x_2 = \frac{-5+3}{-1} = 2$$

Solve for x_1 in the first of Eqs. (B.3.17) as

$$x_1 = \frac{9 - 2(2) - 3}{2} = 1$$

To illustrate the use of the index Eqs. (B.3.15), we re-solve the same example as follows. The ranges of the indexes in Eqs. (B.3.15) are $k = 1, 2$; $i = 2, 3$; and $j = 1, 2, 3, 4$.

Step 1

For $k = 1$, $i = 2$, and j indexing from 1 to 4,

$$a_{21} = a_{21} - a_{11} \frac{a_{21}}{a_{11}} = 2 - 2\left(\frac{2}{2}\right) = 0$$

$$a_{22} = a_{22} - a_{12} \frac{a_{21}}{a_{11}} = 1 - 2\left(\frac{2}{2}\right) = -1$$

$$a_{23} = a_{23} - a_{13} \frac{a_{21}}{a_{11}} = 0 - 1\left(\frac{2}{2}\right) = -1 \qquad \text{(B.3.18)}$$

$$a_{24} = a_{24} - a_{14} \frac{a_{21}}{a_{11}} = 4 - 9\left(\frac{2}{2}\right) = -5$$

Note that these new coefficients correspond to those of the second of Eqs. (B.3.17), where the right-side a's of Eqs. (B.3.18) are those from the previous step [here from Eqs. (B.3.16)], the right side a_{24} is really $c_2 = 4$, and the left side a_{24} is the new $c_2 = -5$.

For $k = 1$, $i = 3$, and j indexing from 1 to 4,

$$a_{31} = a_{31} - a_{11} \frac{a_{31}}{a_{11}} = 1 - 2\left(\frac{1}{2}\right) = 0$$

$$a_{32} = a_{32} - a_{12} \frac{a_{31}}{a_{11}} = 1 - 2\left(\frac{1}{2}\right) = 0$$

$$a_{33} = a_{33} - a_{13} \frac{a_{31}}{a_{11}} = 1 - 1\left(\frac{1}{2}\right) = \frac{1}{2} \qquad \text{(B.3.19)}$$

$$a_{34} = a_{34} - a_{14} \frac{a_{31}}{a_{11}} = 6 - 9\left(\frac{1}{2}\right) = \frac{3}{2}$$

where these new coefficients correspond to those of the third of Eqs. (B.3.17) as previously explained.

Step 2

For $k = 2$, $i = 3$, and $j\ (= k)$ indexing from 2 to 4,

$$a_{32} = a_{32} - a_{22}\left(\frac{a_{32}}{a_{22}}\right) = 0 - (-1)\left(\frac{0}{-1}\right) = 0$$

$$a_{33} = a_{33} - a_{23}\left(\frac{a_{32}}{a_{22}}\right) = \frac{1}{2} - (-1)\left(\frac{0}{-1}\right) = \frac{1}{2} \qquad \text{(B.3.20)}$$

$$a_{34} = a_{34} - a_{24}\left(\frac{a_{32}}{a_{22}}\right) = \frac{3}{2} - (-5)\left(\frac{0}{-1}\right) = \frac{3}{2}$$

where the new coefficients again correspond to those of the third of Eqs. (B.3.17), because Step 1 already eliminated the coefficients of x_2 as observed in the third of Eqs. (B.3.17), and the a's on the right side of Eqs. (B.3.20) are taken from Eqs. (B.3.18) and (B.3.19).

Step 3

By Eqs. (B.3.15), for x_3, we have

$$x_3 = \frac{1}{a_{33}}(a_{34} - 0)$$

or, using a_{33} and a_{34} from Eqs. (B.3.20),

$$x_3 = \frac{1}{\left(\frac{1}{2}\right)}\left(\frac{3}{2}\right) = 3$$

where the summation is interpreted as zero in the second of Eqs. (B.3.15) when $r > n$ (for x_3, $r = 4$, and $n = 3$). For x_2, we have

$$x_2 = \frac{1}{a_{22}}(a_{24} - a_{23}x_3)$$

or, using the appropriate a's from Eqs. (B.3.18),

$$x_2 = \frac{1}{-1}[-5 - (-1)(3)] = 2$$

and for x_1, we have

$$x_1 = \frac{1}{a_{11}}(a_{14} - a_{12}x_2 - a_{13}x_3)$$

or, using the a's from the first of Eqs. (B.3.16),

$$x_1 = \frac{1}{2}[9 - 2(2) - 1(3)] = 1$$

In summary, the latest a's from the previous steps have been used in Eqs. (B.3.15) to obtain the x's. ∎

Note that the pivot element was the diagonal element in each step. However, the diagonal element must be nonzero because we divide by it in each step. An original matrix with all nonzero diagonal elements does not ensure that the pivots in each step will remain nonzero, because we are adding numbers to equations below the pivot in each following step. Therefore, a test is necessary to determine whether the pivot a_{kk} at each step is zero. If it is zero, the current row (equation) must be interchanged with one of the following rows—usually with the next row unless that row has a zero at the position that would next become the pivot. Remember that the right-side corresponding element in $\{c\}$ must also be interchanged. After making this test and, if necessary, interchanging the equations, continue the procedure in the usual manner.

An example will now illustrate the method for treating the occurrence of a zero pivot element.

Example B.2

Solve the following set of simultaneous equations.

$$2x_1 + 2x_2 + 1x_3 = 9$$
$$1x_1 + 1x_2 + 1x_3 = 6 \qquad \text{(B.3.21)}$$
$$2x_1 + 1x_2 \qquad\;\; = 4$$

SOLUTION:

It will often be convenient to set up the solution procedure by considering the coefficient matrix $[a]$ plus the right-side matrix $\{c\}$ in one matrix without writing down the unknown matrix $\{x\}$. This new matrix is called the *augmented matrix*. For the set of Eqs. (B.3.21), we have the augmented matrix written as

$$\begin{bmatrix} 2 & 2 & 1 & | & 9 \\ 1 & 1 & 1 & | & 6 \\ 2 & 1 & 0 & | & 4 \end{bmatrix} \qquad \text{(B.3.22)}$$

We use the steps previously outlined as follows:

Step 1

We select $a_{11} = 2$ as the pivot and

a. Add the multiple $-a_{21}/a_{11} = -1/2$ of the first row to the second row of Eq. (B.3.22).

b. Add the multiple $-a_{31}/a_{11} = -2/2$ of the first row to the third row of Eq. (B.3.22) to obtain

$$\begin{bmatrix} 2 & 2 & 1 & | & 9 \\ 0 & 0 & \frac{1}{2} & | & \frac{3}{2} \\ 0 & -1 & -1 & | & -5 \end{bmatrix} \qquad \text{(B.3.23)}$$

At the end of Step 1, we would normally choose a_{22} as the next pivot. However, a_{22} is now equal to zero. If we interchange the second and third rows of Eq. (B.3.23), the new a_{22} will be nonzero and can be used as a pivot. Interchanging rows 2 and 3 results in

$$
\begin{bmatrix}
2 & 2 & 1 & \vdots & 9 \\
0 & -1 & -1 & \vdots & -5 \\
0 & 0 & \frac{1}{2} & \vdots & \frac{3}{2}
\end{bmatrix}
\tag{B.3.24}
$$

For this special set of only three equations, the interchange has resulted in an upper-triangular coefficient matrix and concludes the elimination procedure. The back-substitution process of Step 3 now yields

$$x_3 = 3 \qquad x_2 = 2 \qquad x_1 = 1 \qquad \blacksquare$$

A second problem when selecting the pivots in sequential manner without testing for the best possible pivot is that loss of accuracy due to rounding in the results can occur. In general, the pivots should be selected as the largest (in absolute value) of the elements in any column. For example, consider the set of equations given by

$$
\begin{aligned}
0.002x_1 + 2.00x_2 &= 2.00 \\
3.00x_1 + 1.50x_2 &= 4.50
\end{aligned}
\tag{B.3.25}
$$

whose actual solution is given by

$$x_1 = 1.0005 \qquad x_2 = 0.999 \tag{B.3.26}$$

The solution by Gaussian elimination without testing for the largest absolute value of the element in any column is

$$
\begin{aligned}
0.002x_1 + 2.00x_2 &= 2.00 \\
-2998.5x_2 &= -995.5 \\
x_2 &= 0.3320 \\
x_1 &= 668
\end{aligned}
\tag{B.3.27}
$$

This solution does not satisfy the second of Eqs. (B.3.25). The solution by interchanging equations is

$$
\begin{aligned}
3.00x_1 + 1.50x_2 &= 4.50 \\
0.002x_1 + 2.00x_2 &= 2.00
\end{aligned}
$$

or

$$
\begin{aligned}
3.00x_1 + 1.50x_2 &= 4.50 \\
1.999x_2 &= 1.997 \\
x_2 &= 0.999 \\
x_1 &= 1.0005
\end{aligned}
\tag{B.3.28}
$$

Equations (B.3.28) agree with the actual solution [Eqs. (B.3.26)].

Hence, in general, the pivots should be selected as the largest (in absolute value) of the elements in any column. This process is called *partial pivoting*. Even better results can be obtained by choosing the pivot as the largest element in the whole matrix of the remaining equations and performing appropriate interchanging of rows. This is called *complete pivoting*. Complete pivoting requires a large amount of testing, so it is not recommended in general.

The finite element equations generally involve coefficients with different orders of magnitude, so Gaussian elimination with partial pivoting is a useful method for solving the equations.

Finally, it has been shown that for n simultaneous equations, the number of arithmetic operations required in Gaussian elimination is n divisions, $\frac{1}{3}n^3 + n^2$ multiplications, and $\frac{1}{3}n^3 + n$ additions. If partial pivoting is included, the number of comparisons needed to select pivots is $n(n+1)/2$.

Other elimination methods, including the Gauss–Jordan and Cholesky methods, have some advantages over Gaussian elimination and are sometimes used to solve large systems of equations. For descriptions of other methods, see References [1–3].

Gauss–Seidel Iteration

Another general class of methods (other than the elimination methods) used to solve systems of linear algebraic equations is the *iterative methods*. Iterative methods work well when the system of equations is large and sparse (many zero coefficients). The Gauss–Seidel method starts with the original set of equations $[a]\{x\} = \{c\}$ written in the form

$$x_1 = \frac{1}{a_{11}}(c_1 - a_{12}x_2 - a_{13}x_3 - \cdots - a_{1n}x_n)$$

$$x_2 = \frac{1}{a_{22}}(c_2 - a_{21}x_1 - a_{23}x_3 - \cdots - a_{2n}x_n)$$

$$\vdots$$

$$x_n = \frac{1}{a_{nn}}(c_n - a_{n1}x_1 - a_{n2}x_2 - \cdots - a_{n,n-1}x_{n-1})$$

(B.3.29)

The following steps are then applied.

1. Assume a set of initial values for the unknowns $x_1, x_2, \ldots, x_n$, and substitute them into the right side of the first of Eqs. (B.3.29) to solve for the new x_1.
2. Use the latest value for x_1 obtained from Step 1 and the initial values for $x_3, x_4, \ldots, x_n$ in the right side of the second of Eqs. (B.3.29) to solve for the new x_2.
3. Continue using the latest values of the x's obtained in the left side of Eqs. (B.3.29) as the next trial values in the right side for each succeeding step.
4. Iterate until convergence is satisfactory.

A good initial set of values (guesses) is often $x_i = c_i/a_{ii}$. An example will serve to illustrate the method.

Example B.3

Consider the set of linear simultaneous equations given by

$$
\begin{aligned}
4x_1 - x_2 &= 2 \\
-x_1 + 4x_2 - x_3 &= 5 \\
-x_2 + 4x_3 - x_4 &= 6 \\
-x_3 + 2x_4 &= -2
\end{aligned}
\qquad \text{(B.3.30)}
$$

Determine x_1 through x_4.

SOLUTION:

Using the initial guesses given by $x_i = c_i/a_{ii}$, we have

$$
x_1 = \tfrac{2}{4} = \tfrac{1}{2} \qquad x_2 = \tfrac{5}{4} \approx 1 \qquad x_3 = \tfrac{6}{4} \approx 1 \qquad x_4 = -1
$$

Solving the first of Eqs. (B.3.30) for x_1 yields

$$
x_1 = \frac{1}{4}(2 + x_2) = \frac{1}{4}(2 + 1) = \frac{3}{4}
$$

Solving the second of Eqs. (B.3.30) for x_2, we have

$$
x_2 = \frac{1}{4}(5 + x_1 + x_3) = \frac{1}{4}(5 + \tfrac{3}{4} + 1) = 1.68
$$

Solving the third of Eqs. (B.3.30) for x_3, we have

$$
x_3 = \frac{1}{4}(6 + x_2 + x_4) = \frac{1}{4}[6 + 1.68 + (-1)] = 1.672
$$

Solving the fourth of Eqs. (B.3.30) for x_4, we obtain

$$
x_4 = \frac{1}{2}(-2 + x_3) = \frac{1}{2}(-2 + 1.67) = -0.16
$$

The first iteration has now been completed. The second iteration yields

$$
x_1 = \frac{1}{4}(2 + 1.68) = 0.922
$$

$$
x_2 = \frac{1}{4}(5 + 0.922 + 1.672) = 1.899
$$

$$
x_3 = \frac{1}{4}[6 + 1.899 + (-0.16)] = 1.944
$$

$$
x_4 = \frac{1}{2}(-2 + 1.944) = -0.028
$$

Table B–1 lists the results of four iterations of the Gauss–Seidel method and the exact solution. From Table B–1, we observe that convergence to the exact solution

Table B–1 Results of four iterations of the Gauss–Seidel method for Eqs. (B.3.30)

Iteration	x_1	x_2	x_3	x_4
0	0.5	1.0	1.0	−1.0
1	0.75	1.68	1.672	−0.16
2	0.922	1.899	1.944	−0.028
3	0.975	1.979	1.988	−0.006
4	0.9985	1.9945	1.9983	−0.0008
Exact	1.0	2.0	2.00	0

has proceeded rapidly by the fourth iteration, and the accuracy of the solution is dependent on the number of iterations. ■

In general, iteration methods are self-correcting, such that an error made in calculations at one iteration will be corrected by later iterations. However, there are certain systems of equations for which iterative methods are not convergent. The following example illustrates a system of equations for which the Gauss–Seidel iteration method will not converge to the exact solution, as the main diagonal terms are smaller than the off-diagonal terms.

$$1x_1 + 3x_2 = 5$$
$$4x_1 - 1x_2 = 12 \tag{B.3.31}$$

When the equations can be arranged such that the diagonal terms are greater than the off-diagonal terms, which can be done for the previous Eq. (B.3.31), the possibility of convergence is usually enhanced.

Finally, it has been shown that for n simultaneous equations, the number of arithmetic operations required by Gauss–Seidel iteration is n divisions, n^2 multiplications, and $n^2 - n$ additions for each iteration.

▲ B.4 Banded-Symmetric Matrices, Bandwidth, Skyline, and Wavefront Methods ▲

The coefficient matrix (stiffness matrix) for the linear equations that occur in structural analysis is always symmetric and banded. Because a meaningful analysis generally requires the use of a large number of variables, the implementation of compressed storage of the stiffness matrix is desirable both from the standpoint of fitting into memory (immediate access portion of the computer) and for computational efficiency. We will discuss the banded-symmetric format, which is not necessarily the most efficient format but is relatively simple to implement on the computer.

Another method, based on the concept of the skyline of the stiffness matrix, is often used to improve the efficiency in solving the equations. *The* **skyline** *is an envelope that begins with the first nonzero coefficient in each column of the stiffness matrix* (Figure B–5). In skylining, only the coefficients between the main diagonal and the

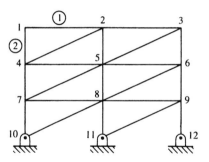

Figure B–4   Plane truss for bandwidth illustration

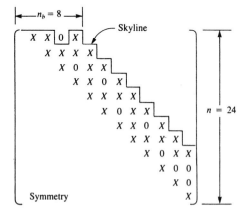

Figure B–5   Stiffness matrix for the plane truss of Figure B–4, where X denotes, in general, blocks of 2×2 submatrices with nonzero coefficients

skyline are stored (normally by successive columns) in a one-dimensional array. In general, this procedure takes even less storage space in the computer and is more efficient in terms of equation solving than the conventional banded format. (For more information on skylining, consult References [3, 10–11].)

A matrix is **banded** if the nonzero terms of the matrix are gathered about the main diagonal. To illustrate this concept, consider the plane truss of Figure B–4.

From Figure B–4, we see that element 2 connects nodes 1 and 4. Therefore, the 2×2 submatrices at positions 1–1, 1–4, 4–1, and 4–4 of Figure B–5 have nonzero coefficients. Figure B–5 represents the total stiffness matrix of the plane truss. The X's denote nonzero coefficients. From Figure B–5, we observe that the nonzero terms are within the band shown. When we use a banded storage format, only the main diagonal and the nonzero upper codiagonals need be stored as shown in Figure B–6. Note that any codiagonal with a nonzero term requires storage of the whole codiagonal *and* any codiagonals between it and the main diagonal. The use of banded storage is efficient for computational purposes. The Scientific Subroutine Package gives a more detailed explanation of banded compressed storage [4].

We now define the semibandwidth n_b as $n_b = n_d(m + 1)$, where n_d is the number of degrees of freedom per node and m is the maximum difference in node numbers

$$\left|\leftarrow n_b = 8 \rightarrow\right|$$

$$\begin{pmatrix} X & X & 0 & X \\ X & X & X & X \\ X & 0 & X & X \\ X & X & 0 & X \\ X & X & X & X \\ X & 0 & X & X \\ X & X & 0 & X \\ X & X & X & X \\ X & 0 & X & X \\ X & 0 & 0 & 0 \\ X & 0 & 0 & 0 \\ X & 0 & 0 & 0 \end{pmatrix}$$

Figure B–6 Banded storage format of the stiffness matrix of Figure B–5

determined by calculating the difference in node numbers for each element of a finite element model. In the example for the plane truss of Figure B–4, $m = 4 - 1 = 3$ and $n_d = 2$, so $n_b = 2(3 + 1) = 8$.

Execution time (primarily equation-solving time) is a function of the number of equations to be solved. It has been shown [5] that when banded storage of global stiffness matrix $[K]$ is not used, execution time is proportional to $(1/3)n^3$, where n is the number of equations to be solved, or, equivalently, the size of $[K]$. When banded storage of $[K]$ is used, the execution time is proportional to $(n)n_b^2$. The ratio of time of execution without banded storage to that with banded storage is then $(1/3)(n/n_b)^2$. For the plane truss example, this ratio is $(1/3)(24/8)^2 = 3$. Therefore, it takes about three times as long to execute the solution of the example truss if banded storage is not used.

Hence, to reduce bandwidth we should number systematically and try to have a minimum difference between adjacent nodes. A small bandwidth is usually achieved by consecutive node numbering across the shorter dimension, as shown in Figure B–4. Some computer programs use the banded-symmetric format for storing the global stiffness matrix, $[K]$.

Several automatic node-renumbering schemes have been computerized [6]. This option is available in most general-purpose computer programs. Alternatively, the wavefront or frontal method is becoming popular for optimizing equation solution time. In the **wavefront method**, elements, instead of nodes, are automatically renumbered.

In the wavefront method, the assembly of the equations alternates with their solution by Gauss elimination. The sequence in which the equations are processed is determined by element numbering rather than by node numbering. The first equations eliminated are those associated with element 1 only. Next, the contributions of stiffness coefficients of the adjacent element, element 2, are added to the system of equations. If any additional degrees of freedom are contributed by elements 1 and 2 only—that is, if no other elements contribute stiffness coefficients to specific degrees of freedom—these equations are eliminated (condensed) from the system of equations. As one or more additional elements make their contributions to the system of equations and additional degrees of freedom are contributed only by these elements, those degrees of freedom are eliminated from the solution. This repetitive alternation between assembly and solution was initially seen as a wavefront that sweeps over the structure in a pattern determined by the element numbering. For greater efficiency of this method, consecutive element numbering should be done across the structure in a direction that spans the smallest number of nodes.

The wavefront method, though somewhat more difficult to understand and to program than the banded-symmetric method, is computationally more efficient. A banded solver stores and processes any blocks of zeros created in assembling the stiffness matrix. In the wavefront method, these blocks of zero coefficients are not stored or processed. Many large-scale computer programs are now using the wavefront method to solve the system of equations. (For additional details of this method, see References [7–9].) Example B.4 illustrates the wavefront method for solution of a truss problem.

Example B.4

For the plane truss shown in Figure B–7, illustrate the wavefront solution procedure.

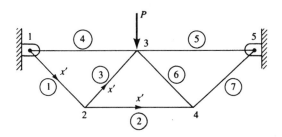

Figure B–7 Truss for wavefront solution

SOLUTION:

We will solve this problem in symbolic form. Merging k's for elements 1, 2, and 3 and enforcing boundary conditions at node 1, we have

$$
\begin{array}{cccccc}
\quad u_2 & \quad v_2 & u_3 & v_3 & u_4 & v_4
\end{array}
$$

$$
\begin{bmatrix}
k_{33}^{(1)}+k_{11}^{(2)}+k_{11}^{(3)} & k_{34}^{(1)}+k_{12}^{(2)}+k_{12}^{(3)} & k_{13}^{(3)} & k_{14}^{(3)} & k_{13}^{(2)} & k_{14}^{(2)} \\
k_{43}^{(1)}+k_{21}^{(2)}+k_{21}^{(3)} & k_{44}^{(1)}+k_{22}^{(2)}+k_{22}^{(3)} & k_{23}^{(3)} & k_{24}^{(3)} & k_{23}^{(2)} & k_{24}^{(2)} \\
\hline
k_{31}^{(2)} & k_{32}^{(3)} & k_{33}^{(3)} & k_{34}^{(3)} & k_{33}^{(2)} & k_{34}^{(2)} \\
k_{41}^{(3)} & k_{42}^{(3)} & k_{43}^{(3)} & k_{44}^{(3)} & k_{43}^{(2)} & k_{44}^{(2)} \\
k_{31}^{(2)} & k_{32}^{(2)} & 0 & 0 & 0 & 0 \\
k_{41}^{(2)} & k_{42}^{(2)} & 0 & 0 & 0 & 0
\end{bmatrix}
\begin{Bmatrix} u_2 \\ v_2 \\ \hline u_3' \\ v_3' \\ u_4' \\ v_4' \end{Bmatrix}
=
\begin{Bmatrix} 0 \\ 0 \\ \hline 0 \\ -P \\ 0 \\ 0 \end{Bmatrix}
$$

$$(B.4.1)$$

Eliminating u_2 and v_2 (all stiffness contributions from node 2 degrees of freedom have been included from these elements; these contributions are from elements 1–3) by static condensation or Gauss elimination yields

$$[k_c'] \begin{Bmatrix} u_3' \\ v_3' \\ u_4' \\ v_4' \end{Bmatrix} = \{F_c'\} \tag{B.4.2}$$

where the condensed stiffness and force matrices are (also see Section 7.5)

$$[k_c'] = [K_{22}'] - [K_{21}'][K_{11}']^{-1}[K_{12}'] \tag{B.4.3}$$

$$\{F_c'\} = \{F_2'\} - [K_{21}'][K_{11}']^{-1}\{F_1'\} \tag{B.4.4}$$

where primes on the degrees of freedom, such as u_3' in Eq. (B.4.1), indicate that all stiffness coefficients associated with that degree of freedom have not yet been included. Now include elements 4–6 for degrees of freedom at node 3. The resulting equations are

$$
\begin{array}{cccc}
u_3 & v_3 & u_4 & v_4
\end{array}
$$

$$
\left[
\begin{array}{cc:cc}
k_{c11}' + k_{33}^{(4)} + k_{11}^{(5)} + k_{11}^{(6)} & k_{34}^{(4)} + k_{12}^{(5)} + k_{12}^{(6)} + k_{c12}' & k_{13}^{(6)} + k_{c13}' & k_{14}^{(6)} + k_{c14}' \\
k_{c21}' + k_{34}^{(4)} + k_{21}^{(5)} + k_{21}^{(6)} & k_{44}^{(4)} + k_{22}^{(5)} + k_{22}^{(6)} + k_{c22}' & k_{23}^{(6)} + k_{c23}' & k_{24}^{(6)} + k_{c24}' \\
\hdashline
k_{c31}' + k_{31}^{(6)} & k_{c32}' + k_{32}^{(6)} & k_{c33}' + k_{33}^{(6)} & k_{c34}' + k_{34}^{(6)} \\
k_{c41}' + k_{41}^{(6)} & k_{c42}' + k_{42}^{(6)} & k_{c43}' + k_{43}^{(6)} & k_{c44}' + k_{44}^{(6)}
\end{array}
\right]
$$

$$
\times \begin{Bmatrix} u_3 \\ \hdashline v_3 \\ \hdashline u_4' \\ v_4' \end{Bmatrix} = \begin{Bmatrix} 0 \\ \hdashline -P \\ \hdashline 0 \\ 0 \end{Bmatrix} \tag{B.4.5}
$$

Using static condensation, we eliminate u_3 and v_3 (all contributions from node 3 degrees of freedom have been included from each element) to obtain

$$[k_c''] \begin{Bmatrix} u_4' \\ v_4' \end{Bmatrix} = \{F_c''\} \tag{B.4.6}$$

where

$$[k_c''] = [K_{22}''] - [K_{21}''][K_{11}'']^{-1}[K_{12}''] \tag{B.4.7}$$

$$\{F_c''\} = \{F_2''\} - [K_{21}''][K_{11}'']^{-1}\{F_1''\} \tag{B.4.8}$$

Next we include element 7 contributions to the stiffness matrix. The condensed set of equations yield

$$[k_c'''] \begin{Bmatrix} u_4 \\ v_4 \end{Bmatrix} = \{F_c'''\} \tag{B.4.9}$$

$$[k_c'''] = [K_{22}'''] - [K_{21}'''][K_{11}''']^{-1}[K_{12}''']$$ (B.4.10)

where $$\{F_c'''\} = \{F_2'''\} - [K_{21}'''][K_{11}''']^{-1}\{F_1'''\}$$ (B.4.11)

The elimination procedure is now complete, and we solve Eq. (B.4.9) for u_4 and v_4. Then we back-substitute u_4 and v_4 into Eq. (B.4.5) to obtain u_3 and v_3. Finally, we back-substitute u_3 through v_4 into Eq. (B.4.1) to obtain u_2 and v_2. Static condensation and Gauss elimination with back-substitution have been used to solve the set of equations for all the degrees of freedom. The solution procedure has then proceeded as though it were a wave sweeping over the structure, starting at node 2, engulfing node 2 and elements with degrees of freedom at node 2, and then sweeping through node 3 and finally node 4. ■

We now describe a practical computer scheme often used in computer programs for the solution of the resulting system of algebraic equations. The significance of this scheme is that it takes advantage of the fact that the stiffness method produces a banded $[K]$ matrix in which the nonzero elements occur about the main diagonal in $[K]$. While the equations are solved, this banded format is maintained.

Example B.5

We will now use a simple example to illustrate this computer scheme. Consider the three-spring assemblage shown in Figure B–8. The assemblage is subjected to forces at node 2 of 100 lb in the x direction and 200 lb in the y direction. Node 1 is completely constrained from displacement in both the x and y directions, whereas node 3 is completely constrained in the y direction but is displaced a known amount δ in the x direction.

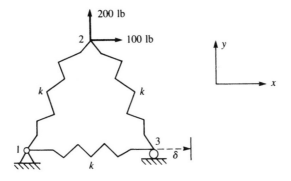

Figure B–8 Three-spring assemblage

SOLUTION:

Our purpose here is not to obtain the actual $[K]$ for the assemblage but rather to illustrate the scheme used for solution. The general solution can be shown to be given by

$$
\begin{bmatrix}
k_{11} & k_{12} & k_{13} & k_{14} & k_{15} & k_{16} \\
 & k_{22} & k_{23} & k_{24} & k_{25} & k_{26} \\
 & & k_{33} & k_{34} & k_{35} & k_{36} \\
 & & & k_{44} & k_{45} & k_{46} \\
 & & & & k_{55} & k_{56} \\
\text{Symmetry} & & & & & k_{66}
\end{bmatrix}
\begin{Bmatrix}
u_1 \\ v_1 \\ u_2 \\ v_2 \\ u_3 \\ v_3
\end{Bmatrix}
=
\begin{Bmatrix}
F_{1x} \\ F_{1y} \\ F_{2x} = 100 \\ F_{2y} = 200 \\ F_{3x} \\ F_{3y}
\end{Bmatrix}
\tag{B.4.12}
$$

where $[K]$ has been left in general form. Upon our imposing the boundary conditions, the computer program transforms Eq. (B.4.12) to:

$$
\begin{bmatrix}
1 & 0 & 0 & 0 & 0 & 0 \\
0 & 1 & 0 & 0 & 0 & 0 \\
0 & 0 & k_{33} & k_{34} & 0 & 0 \\
0 & 0 & k_{43} & k_{44} & 0 & 0 \\
0 & 0 & 0 & 0 & 1 & 0 \\
0 & 0 & 0 & 0 & 0 & 1
\end{bmatrix}
\begin{Bmatrix}
u_1 \\ v_1 \\ u_2 \\ v_2 \\ u_3 \\ v_3
\end{Bmatrix}
=
\begin{Bmatrix}
0 \\ 0 \\ 100 - k_{35}\delta \\ 200 - k_{45}\delta \\ \delta \\ 0
\end{Bmatrix}
\tag{B.4.13}
$$

From Eq. (B.4.13), we can see that $u_1 = 0$, $v_1 = 0$, $v_3 = 0$, and $u_3 = \delta$. These displacements are consistent with the imposed boundary conditions. The unknown displacements, u_2 and v_2, can be determined routinely by solving Eq. (B.4.13).

We will now explain the computer scheme that is generally applicable to transform Eq. (B.4.12) to Eq. (B.4.13). First, the terms associated with the known displacement boundary condition(s) within each equation were transformed to the right side of those equations. In the third and fourth equations of Eq. (B.4.12), $k_{35}\delta$ and $k_{45}\delta$ were transformed to the right side, as shown in Eq. (B.4.13). Then the right-side force term corresponding to the known displacement row was equated to the known displacement. In the fifth equation of Eq. (B.4.12), where $u_3 = \delta$, the right-side, fifth-row force term F_{3x} was equated to the known displacement δ, as shown in Eq. (B.4.13). For the homogeneous boundary conditions, the affected rows of $\{F\}$, corresponding to the zero-displacement rows, were replaced with zeros. Again, this is done in the computer scheme only to obtain the nodal displacements and does not imply that these nodal forces are zero. We obtain the unknown nodal forces by determining the nodal displacements and back-substituting these results into the original Eq. (B.4.12). Because $u_1 = 0$, $v_1 = 0$, and $v_3 = 0$ in Eq. (B.4.12), the first, second, and sixth rows of the force matrix of Eq. (B.4.13) were set to zero. Finally, for both non-homogeneous and homogeneous boundary conditions, the rows and columns of $[K]$ corresponding to these prescribed boundary conditions were set to zero except the main diagonal, which was made unity. That is, the first, second, fifth, and sixth rows and columns of $[K]$ in Eq. (B.4.12) were set to zero, except for the main diagonal terms, which were made unity. Although doing so is not necessary, setting the main diagonal terms equal to 1 facilitates the simultaneous solution of the six equations in Eq. (B.4.13) by an elimination method used in the computer program. This modification is shown in the $[K]$ matrix of Eq. (B.4.13). ∎

▲ References

[1] Chapra, S. C., and Canale, R. P., "*Numerical Methods for Engineers*," 5th ed., McGraw-Hill, NY, 2008.

[2] Rao, S. S., *Applied Numerical Methods for Engineers and Scientists*, Prentice Hall, NY, 2001.

[3] Bathe, K. J., and Wilson, E. L., *Numerical Methods in Finite Element Analysis*, Prentice-Hall, Englewood Cliffs, NJ, 1976.

[4] SYSTEM/360, Scientific Subroutine Package, IBM.

[5] Kardestuncer, H., *Elementary Matrix Analysis of Structures*, McGraw-Hill, New York, 1974.

[6] Collins, R. J., "Bandwidth Reduction by Automatic Renumbering," *International Journal For Numerical Methods in Engineering*, Vol. 6, pp. 345–356, 1973.

[7] Melosh, R. J., and Bamford, R. M., "Efficient Solution of Load-Deflection Equations," *Journal of the Structural Division*, American Society of Civil Engineers, No. ST4, pp. 661–676, April 1969.

[8] Irons, B. M., "A Frontal Solution Program for Finite Element Analysis," *International Journal for Numerical Methods in Engineering*, Vol. 2, No. 1, pp. 5–32, 1970.

[9] Meyer, C., "Solution of Linear Equations-State-of-the-Art," *Journal of the Structural Division*, American Society of Civil Engineers, Vol. 99, No. ST7, pp. 1507–1526, 1973.

[10] Jennings, A., *Matrix Computation for Engineers and Scientists*, Wiley, London, 1977.

[11] Cook, R. D., Malkus, D. S., Plesha, M. E., and Witt, R. J., *Concepts and Applications of Finite Element Analysis*, 4th ed., Wiley, New York, 2002.

▲ Problems

B.1 Determine the solution of the following simultaneous equations by Cramer's rule.

$$1x_1 + 3x_2 = 5$$
$$4x_1 - 1x_2 = 12$$

B.2 Determine the solution of the following simultaneous equations by the inverse method.

$$1x_1 + 3x_2 = 5$$
$$4x_1 - 1x_2 = 12$$

B.3 Solve the following system of simultaneous equations by Gaussian elimination.

$$x_1 - 4x_2 - 5x_3 = 4$$
$$3x_2 + 4x_3 = -1$$
$$-2x_1 - 1x_2 + 2x_3 = -3$$

B.4   Solve the following system of simultaneous equations by Gaussian elimination.

$$2x_1 + 1x_2 - 3x_3 = 11$$
$$4x_1 - 2x_2 + 3x_3 = 8$$
$$-2x_1 + 2x_2 - 1x_3 = -6$$

B.5   Given that

$$x_1 = 2y_1 - y_2 \qquad z_1 = -x_1 - x_2$$
$$x_2 = y_1 - y_2 \qquad z_2 = 2x_1 + x_2$$

a. Write these relationships in matrix form.
b. Express $\{z\}$ in terms of $\{y\}$.
c. Express $\{y\}$ in terms of $\{z\}$.

B.6   Starting with the initial guess $\{X\}^T = \begin{bmatrix} 1 & 1 & 1 & 1 & 1 \end{bmatrix}$, perform five iterations of the Gauss–Seidel method on the following system of equations. On the basis of the results of these five iterations, what is the exact solution?

$$2x_1 - 1x_2 \qquad\qquad\qquad = -1$$
$$-1x_1 + 6x_2 - 1x_3 \qquad\qquad = 4$$
$$-2x_2 + 4x_3 - 1x_4 \qquad = 4$$
$$-1x_3 + 4x_4 - 1x_5 = 6$$
$$-1x_4 + 2x_5 = -2$$

B.7   Solve Problem B.1 by Gauss–Seidel iteration.

B.8   Classify the solutions to the following systems of equations according to Section B.2 as unique, nonunique, or nonexistent.

a. $2x_1 - 4x_2 = 2$ **b.** $10x_1 + 1x_2 = 0$
   $-9x_1 + 12x_2 = -6$ $5x_1 + \frac{1}{2}x_2 = 3$
c. $2x_1 + 1x_2 + 1x_3 = 6$ **d.** $1x_1 + 1x_2 + 1x_3 = 1$
   $3x_1 + 1x_2 - 1x_3 = 4$ $2x_1 + 2x_2 + 2x_3 = 2$
   $5x_1 + 2x_2 + 2x_3 = 8$ $3x_1 + 3x_2 + 3x_3 = 3$

B.9   Determine the bandwidths of the plane trusses shown in Figure PB–9. What conclusions can you draw regarding labeling of nodes?

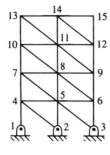
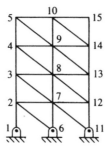

Figure PB–9

EQUATIONS FROM
ELASTICITY THEORY ▲

Introduction

In this appendix, we will develop the basic equations of the theory of elasticity. These equations should be referred to frequently throughout the structural mechanics portions of this text.

There are three basic sets of equations included in theory of elasticity. These equations must be satisfied if an exact solution to a structural mechanics problem is to be obtained. These sets of equations are (1) the differential equations of equilibrium formulated here in terms of the stresses acting on a body, (2) the strain/displacement and compatibility differential equations, and (3) the stress/strain or material constitutive laws.

▲ C.1 Differential Equations of Equilibrium   ▲

For simplicity, we initially consider the equilibrium of a plane element subjected to normal stresses σ_x and σ_y, in-plane shear stress τ_{xy} (in units of force per unit area), and body forces X_b and Y_b (in units of force per unit volume), as shown in Figure C–1. The stresses are assumed to be constant as they act on the width of each face. However, the stresses are assumed to vary from one face to the opposite. For example, we have σ_x acting on the left vertical face, whereas $\sigma_x + (\partial\sigma_x/\partial x)\,dx$ acts on the right vertical face. The element is assumed to have unit thickness.

Summing forces in the x direction, we have

$$\sum F_x = 0 = \left(\sigma_x + \frac{\partial\sigma_x}{\partial x}dx\right)dy(1) - \sigma_x\,dy(1) + X_b\,dx\,dy(1)$$

$$+ \left(\tau_{yx} + \frac{\partial\tau_{yx}}{\partial y}dy\right)dx(1) - \tau_{yx}\,dx(1) = 0 \qquad (C.1.1)$$

After simplifying and canceling terms in Eq. (C.1.1), we obtain

$$\frac{\partial\sigma_x}{\partial x} + \frac{\partial\tau_{yx}}{\partial y} + X_b = 0 \qquad (C.1.2)$$

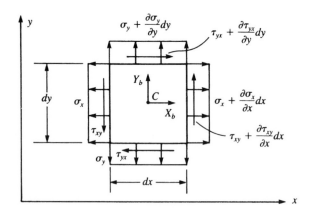

Figure C–1 Plane differential element subjected to stresses

Similarly, summing forces in the y direction, we obtain

$$\frac{\partial \sigma_y}{\partial y} + \frac{\partial \tau_{xy}}{\partial x} + Y_b = 0 \tag{C.1.3}$$

Because we are considering only the planar element, three equilibrium equations must be satisfied. The third equation is equilibrium of moments about an axis normal to the x-y plane; that is, taking moments about point C in Figure C–1, we have

$$\sum M_z = 0 = \tau_{xy}\, dy(1)\frac{dx}{2} + \left(\tau_{xy} + \frac{\partial \tau_{xy}}{\partial x}dx\right)\frac{dx}{2}$$

$$- \tau_{yx}\, dx(1)\frac{dy}{2} - \left(\tau_{yx} + \frac{\partial \tau_{yx}}{\partial y}dy\right)\frac{dy}{2} = 0 \tag{C.1.4}$$

Simplifying Eq. (C.1.4) and neglecting higher-order terms yields

$$\tau_{xy} = \tau_{yx} \tag{C.1.5}$$

We now consider the three-dimensional state of stress shown in Figure C–2, which shows the additional stresses σ_z, τ_{xz}, and τ_{yz}. For clarity, we show only the stresses on three mutually perpendicular planes. With a straightforward procedure, we can extend the two-dimensional equations (C.1.2), (C.1.3), and (C.1.5) to three dimensions. The resulting total set of equilibrium equations is

$$\frac{\partial \sigma_x}{\partial x} + \frac{\partial \tau_{xy}}{\partial y} + \frac{\partial \tau_{xz}}{\partial z} + X_b = 0$$

$$\frac{\partial \tau_{xy}}{\partial x} + \frac{\partial \sigma_y}{\partial y} + \frac{\partial \tau_{yz}}{\partial z} + Y_b = 0 \tag{C.1.6}$$

$$\frac{\partial \tau_{xz}}{\partial x} + \frac{\partial \tau_{yz}}{\partial y} + \frac{\partial \sigma_z}{\partial z} + Z_b = 0$$

and $\qquad\qquad \tau_{xy} = \tau_{yx} \qquad \tau_{xz} = \tau_{zx} \qquad \tau_{yz} = \tau_{zy} \tag{C.1.7}$

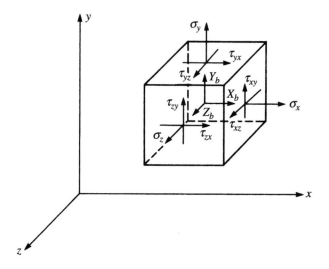

Figure C–2 Three-dimensional stress element

▲ C.2 Strain/Displacement and Compatibility   ▲
Equations

We first obtain the strain/displacement or kinematic differential relationships for the two-dimensional case. We begin by considering the differential element shown in Figure C–3, where the undeformed state is represented by the dashed lines and the deformed shape (after straining takes place) is represented by the solid lines.

Considering line element AB in the x direction, we can see that it becomes $A'B'$ after deformation, where u and v represent the displacements in the x and y directions. By the definition of engineering normal strain (that is, the change in length divided by

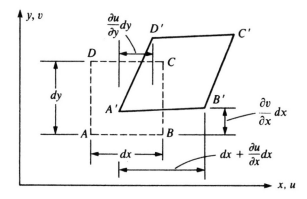

Figure C–3 Differential element before and after deformation

the original length of a line), we have

$$\varepsilon_x = \frac{A'B' - AB}{AB} \tag{C.2.1}$$

Now

$$AB = dx \tag{C.2.2}$$

and

$$(A'B')^2 = \left(dx + \frac{\partial u}{\partial x}dx\right)^2 + \left(\frac{\partial v}{\partial x}dx\right)^2 \tag{C.2.3}$$

Therefore, evaluating $A'B'$ using the binomial theorem and neglecting the higher-order terms $(\partial u/\partial x)^2$ and $(\partial v/\partial x)^2$ (an approach consistent with the assumption of small strains), we have

$$A'B' = dx + \frac{\partial u}{\partial x}dx \tag{C.2.4}$$

Using Eqs. (C.2.2) and (C.2.4) in Eq. (C.2.1), we obtain

$$\varepsilon_x = \frac{\partial u}{\partial x} \tag{C.2.5}$$

Similarly, considering line element AD in the y direction, we have

$$\varepsilon_y = \frac{\partial v}{\partial y} \tag{C.2.6}$$

The shear strain γ_{xy} is defined to be the change in the angle between two lines, such as AB and AD, that originally formed a right angle. Hence, from Figure C–3, we can see that γ_{xy} is the sum of two angles and is given by

$$\gamma_{xy} = \frac{\partial u}{\partial y} + \frac{\partial v}{\partial x} \tag{C.2.7}$$

Equations (C.2.5) through (C.2.7) represent the strain/displacement relationships for in-plane behavior.

For three-dimensional situations, we have a displacement w in the z direction. It then becomes straightforward to extend the two-dimensional derivations to the three-dimensional case to obtain the additional strain/displacement equations as

$$\varepsilon_z = \frac{\partial w}{\partial z} \tag{C.2.8}$$

$$\gamma_{xz} = \frac{\partial u}{\partial z} + \frac{\partial w}{\partial x} \tag{C.2.9}$$

$$\gamma_{yz} = \frac{\partial v}{\partial z} + \frac{\partial w}{\partial y} \tag{C.2.10}$$

Along with the strain/displacement equations, we need compatibility equations to ensure that the displacement components u, v, and w are single-valued continuous functions so that tearing or overlap of elements does not occur. For the planar-elastic case, we obtain the compatibility equation by differentiating γ_{xy} with respect to both x and y and then using the definitions for ε_x and ε_y given by Eqs. (C.2.5) and (C.2.6). Hence,

$$\frac{\partial^2 \gamma_{xy}}{\partial x \partial y} = \frac{\partial^2}{\partial x \partial y} \frac{\partial u}{\partial y} + \frac{\partial^2}{\partial x \partial y} \frac{\partial v}{\partial x} = \frac{\partial^2 \varepsilon_x}{\partial y^2} + \frac{\partial^2 \varepsilon_y}{\partial x^2} \qquad (C.2.11)$$

where the second equation in terms of the strains on the right side is obtained by noting that single-valued continuity of displacements requires that the partial differentiations with respect to x and y be interchangeable in order. Therefore, we have $\partial^2/\partial x \partial y = \partial^2/\partial y \partial x$. Equation (C.2.11) is called the *condition of compatibility*, and it must be satisfied by the strain components in order for us to obtain unique expressions for u and v. Equations (C.2.5), (C.2.6), (C.2.7), and (C.2.11) together are then sufficient to obtain unique single-valued functions for u and v.

In three dimensions, we obtain five additional compatibility equations by differentiating γ_{xz} and γ_{yz} in a manner similar to that described above for γ_{xy}. We need not list these equations here; details of their derivation can be found in Reference [1].

In addition to the compatibility conditions that ensure single-valued continuous functions within the body, we must also satisfy displacement or kinematic boundary conditions. This simply means that the displacement functions must also satisfy prescribed or given displacements on the surface of the body. These conditions often occur as support conditions from rollers and/or pins. In general, we might have

$$u = u_0 \qquad v = v_0 \qquad w = w_0 \qquad (C.2.12)$$

at specified surface locations on the body. We may also have conditions other than displacements prescribed (for example, prescribed rotations).

▲ C.3 Stress-Strain Relationships ▲

We will now develop the three-dimensional stress-strain relationships for an isotropic body only. This is done by considering the response of a body to imposed stresses. We subject the body to the stresses σ_x, σ_y, and σ_z independently as shown in Figure C–4.

We first consider the change in length of the element in the x direction due to the independent stresses σ_x, σ_y, and σ_z. We assume the principle of superposition to hold; that is, we assume that the resultant strain in a system due to several forces is the algebraic sum of their individual effects.

Considering Figure C–4(b), the stress in the x direction produces a positive strain

$$\varepsilon_x' = \frac{\sigma_x}{E} \qquad (C.3.1)$$

where Hooke's law, $\sigma = E\varepsilon$, has been used in writing Eq. (C.3.1), and E is defined as the *modulus of elasticity*. Considering Figure C–4(c), the positive stress in the

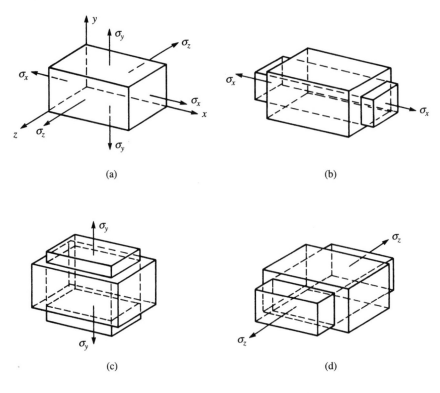

Figure C–4 Element subjected to normal stress acting in three mutually perpendicular directions

y direction produces a negative strain in the x direction as a result of Poisson's effect given by

$$\varepsilon_x'' = -\frac{v\sigma_y}{E} \qquad (C.3.2)$$

where v is Poisson's ratio. Similarly, considering Figure C–4(d), the stress in the z direction produces a negative strain in the x direction given by

$$\varepsilon_x''' = -\frac{v\sigma_z}{E} \qquad (C.3.3)$$

Using superposition of Eqs. (C.3.1) through (C.3.3), we obtain

$$\varepsilon_x = \frac{\sigma_x}{E} - v\frac{\sigma_y}{E} - v\frac{\sigma_z}{E} \qquad (C.3.4)$$

The strains in the y and z directions can be determined in a manner similar to that used to obtain Eq. (C.3.4) for the x direction. They are

$$\varepsilon_y = -v\frac{\sigma_x}{E} + \frac{\sigma_y}{E} - v\frac{\sigma_z}{E}$$
$$\varepsilon_z = -v\frac{\sigma_x}{E} - v\frac{\sigma_y}{E} + \frac{\sigma_z}{E} \qquad (C.3.5)$$

Solving Eqs. (C.3.4) and (C.3.5) for the normal stresses, we obtain

$$\sigma_x = \frac{E}{(1+v)(1-2v)}\left[\varepsilon_x(1-v) + v\varepsilon_y + v\varepsilon_z\right]$$

$$\sigma_y = \frac{E}{(1+v)(1-2v)}\left[v\varepsilon_x + (1-v)\varepsilon_y + v\varepsilon_z\right] \quad\quad (C.3.6)$$

$$\sigma_z = \frac{E}{(1+v)(1-2v)}\left[v\varepsilon_x + v\varepsilon_y + (1-v)\varepsilon_z\right]$$

The Hooke's law relationship, $\sigma = E\varepsilon$, used for normal stress also applies for shear stress and strain; that is,

$$\tau = G\gamma \quad\quad (C.3.7)$$

where G is the *shear modulus*. Hence, the expressions for the three different sets of shear strains are

$$\gamma_{xy} = \frac{\tau_{xy}}{G} \quad\quad \gamma_{yz} = \frac{\tau_{yz}}{G} \quad\quad \gamma_{zx} = \frac{\tau_{zx}}{G} \quad\quad (C.3.8)$$

Solving Eqs. (C.3.8) for the stresses, we have

$$\tau_{xy} = G\gamma_{xy} \quad\quad \tau_{yz} = G\gamma_{yz} \quad\quad \tau_{zx} = G\gamma_{zx} \quad\quad (C.3.9)$$

In matrix form, we can express the stresses in Eqs. (C.3.6) and (C.3.9) as

$$
\begin{Bmatrix} \sigma_x \\ \sigma_y \\ \sigma_z \\ \tau_{xy} \\ \tau_{yz} \\ \tau_{zx} \end{Bmatrix} = \frac{E}{(1+v)(1-2v)}
$$

$$
\times \begin{bmatrix}
1-v & v & v & 0 & 0 & 0 \\
 & 1-v & v & 0 & 0 & 0 \\
 & & 1-v & 0 & 0 & 0 \\
 & & & \dfrac{1-2v}{2} & 0 & 0 \\
 & & & & \dfrac{1-2v}{2} & 0 \\
\text{Symmetry} & & & & & \dfrac{1-2v}{2}
\end{bmatrix}
\begin{Bmatrix} \varepsilon_x \\ \varepsilon_y \\ \varepsilon_z \\ \gamma_{xy} \\ \gamma_{yz} \\ \gamma_{zx} \end{Bmatrix} \quad\quad (C.3.10)
$$

where we note that the relationship

$$G = \frac{E}{2(1+v)}$$

has been used in Eq. (C.3.10). The square matrix on the right side of Eq. (C.3.10) is called the *stress-strain* or *constitutive matrix* and is defined by $[D]$, where $[D]$ is

$$[D] = \frac{E}{(1+v)(1-2v)} \begin{bmatrix} 1-v & v & v & 0 & 0 & 0 \\ & 1-v & v & 0 & 0 & 0 \\ & & 1-v & 0 & 0 & 0 \\ & & & \dfrac{1-2v}{2} & 0 & 0 \\ & & & & \dfrac{1-2v}{2} & 0 \\ \text{Symmetry} & & & & & \dfrac{1-2v}{2} \end{bmatrix} \qquad (C.3.11)$$

▲ Reference

[1] Timoshenko, S., and Goodier, J., *Theory of Elasticity*, 3rd ed., McGraw-Hill, New York, 1970.

EQUIVALENT NODAL FORCES

The equivalent nodal (or joint) forces for different types of loads on beam elements are shown in Table D–1 (on the following page).

▲ Problems

D.1 Determine the equivalent joint or nodal forces for the beam elements shown in Figure PD–1.

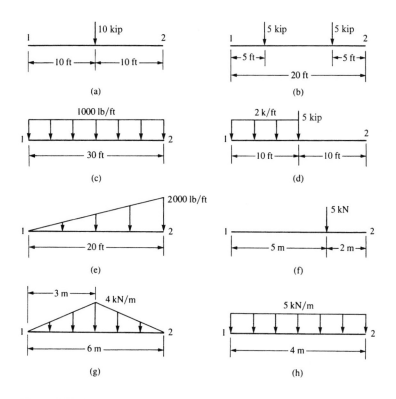

Figure PD–1

Table D–1 Single element equivalent joint forces f_0 for different types of loads

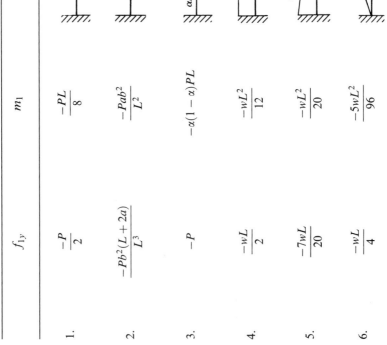

Positive nodal force conventions

	f_{1y}	m_1	Loading case	f_{2y}	m_2
1.	$-\dfrac{P}{2}$	$-\dfrac{PL}{8}$		$-\dfrac{P}{2}$	$\dfrac{PL}{8}$
2.	$-\dfrac{Pb^2(L+2a)}{L^3}$	$-\dfrac{Pab^2}{L^2}$		$-\dfrac{Pa^2(L+2b)}{L^3}$	$\dfrac{Pa^2b}{L^2}$
3.	$-P$	$-\alpha(1-\alpha)PL$		$-P$	$\alpha(1-\alpha)PL$
4.	$-\dfrac{wL}{2}$	$-\dfrac{wL^2}{12}$		$-\dfrac{wL}{2}$	$\dfrac{wL^2}{12}$
5.	$-\dfrac{7wL}{20}$	$-\dfrac{wL^2}{20}$		$-\dfrac{3wL}{20}$	$\dfrac{wL^2}{30}$
6.	$-\dfrac{wL}{4}$	$-\dfrac{5wL^2}{96}$		$-\dfrac{wL}{4}$	$\dfrac{5wL^2}{96}$

(Continued)

Table D–1 *(Continued)*

	f_{1y}	m_1	Loading case	f_{2y}	m_2
7.	$\dfrac{-13wL}{32}$	$\dfrac{-11wL^2}{192}$		$\dfrac{-3wL}{32}$	$\dfrac{5wL^2}{192}$
8.	$\dfrac{-wL}{3}$	$\dfrac{-wL^2}{15}$		$\dfrac{-wL}{3}$	$\dfrac{wL^2}{15}$
9.	$\dfrac{-M(a^2 + b^2 - 4ab - L^2)}{L^3}$	$\dfrac{Mb(2a - b)}{L^2}$		$\dfrac{M(a^2 + b^2 - 4ab - L^2)}{L^3}$	$\dfrac{Ma(2b - a)}{L^2}$

PRINCIPLE OF VIRTUAL WORK ▲

In this appendix, we will use the principle of virtual work to derive the general finite element equations for a dynamic system.

Strictly speaking, the principle of virtual work applies to a static system, but through the introduction of D'Alembert's principle, we will be able to use the principle of virtual work to derive the finite element equations applicable for a dynamic system.

The principle of virtual work is stated as follows:

> If a deformable body in equilibrium is subjected to arbitrary virtual (imaginary) displacements associated with a compatible deformation of the body, the virtual work of external forces on the body is equal to the virtual strain energy of the internal stresses.

In the principle, *compatible displacements* are those that satisfy the boundary conditions and ensure that no discontinuities, such as voids or overlaps, occur within the body. Figure E–1 shows the hypothetical actual displacement, a compatible (admissible) displacement, and an incompatible (inadmissible) displacement for a simply supported beam. Here δv represents the variation in the transverse displacement function v. In the finite element formulation, δv would be replaced by nodal degrees of freedom δd_i. The inadmissible displacements shown in Figure E–1(b) are the result when the support condition at the right end of the beam and the continuity of displacement and slope within the beam are not satisfied. For more details of this principle, consult structural mechanics references such as Reference [1]. Also, for additional descriptions of strain energy and work done by external forces (as applied to a bar), see Section 3.10.

Applying the principle to a finite element, we have

$$\delta U^{(e)} = \delta W^{(e)} \tag{E.1}$$

where $\delta U^{(e)}$ is the virtual strain energy due to internal stresses and $\delta W^{(e)}$ is the virtual work of external forces on the element. We can express the internal virtual strain energy using matrix notation as

$$\delta U^{(e)} = \iiint\limits_V \delta\{\varepsilon\}^T \{\sigma\} \, dV \tag{E.2}$$

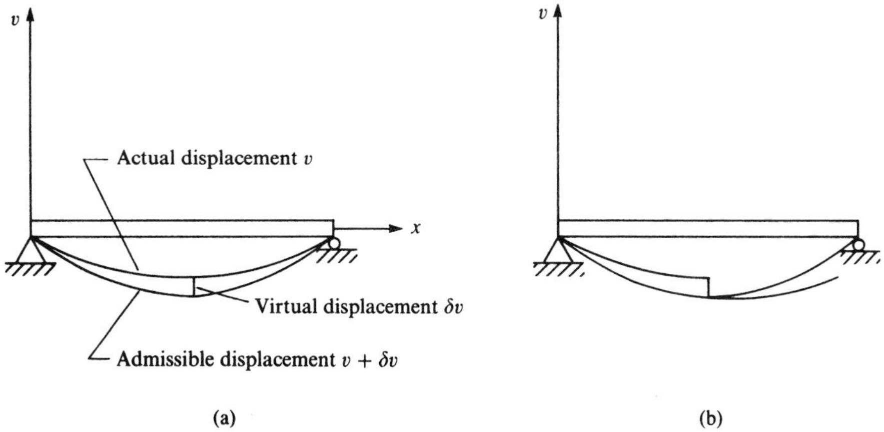

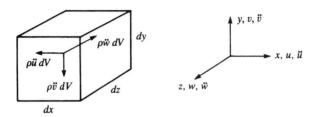

Figure E-1 (a) Admissible and (b) inadmissible virtual displacement functions

Figure E-2 Effective forces acting on an element

From Eq. (E.2), we can observe that internal strain energy is due to internal stresses moving through virtual strains $\delta\varepsilon$. The external virtual work is due to nodal, surface, and body forces. In addition, application of D'Alembert's principle yields effective or inertial forces $\rho\ddot{u}\,dV$, $\rho\ddot{v}\,dV$, and $\rho\ddot{w}\,dV$, where the double dots indicate second derivatives of the translations u, v, and w in the x, y, and z directions, respectively, with respect to time. These forces are shown in Figure E-2. According to D'Alembert's principle, these effective forces act in directions that are opposite to the assumed positive sense of the accelerations. We can now express the external virtual work as

$$\delta W^{(e)} = \delta\{d\}^T\{P\} + \iint_S \delta\{\psi_s\}^T\{T_s\}\,dS + \iiint_V \delta\{\psi\}^T(\{X\} - \rho\{\ddot{\psi}\})\,dV \qquad \text{(E.3)}$$

where $\delta\{d\}$ is the vector of virtual nodal displacements, $\delta\{\psi\}$ is the vector of virtual displacement functions $\delta u, \delta v$, and $\delta w, \delta\{\psi_s\}$ is the vector of virtual displacement functions acting over the surface where surface tractions occur, $\{P\}$ is the nodal load matrix, $\{T_s\}$ is the surface force per unit area matrix, and $\{X\}$ is the body force per unit volume matrix.

Substituting Eqs. (E.2) and (E.3) into Eq. (E.1), we obtain

$$\iiint_V \delta\{\varepsilon\}^T\{\sigma\}\,dV = \delta\{d\}^T\{P\} + \iint_S \delta\{\psi_s\}^T\{T_s\}\,dS + \iiint_V \delta\{\psi\}^T(\{X\} - \rho\{\ddot\psi\})\,dV$$

(E.4)

As shown throughout this text, shape functions are used to relate displacement functions to nodal displacements as

$$\{\psi\} = [N]\{d\} \qquad \{\psi_s\} = [N_s]\{d\}$$

(E.5)

$[N_s]$ is the shape function matrix evaluated on the surface where traction $\{T_s\}$ occurs. Strains are related to nodal displacements as

$$\{\varepsilon\} = [B]\{d\}$$

(E.6)

and stresses are related to strains by

$$\{\sigma\} = [D]\{\varepsilon\}$$

(E.7)

Hence, substituting Eqs. (E.5), (E.6), and (E.7) for $\{\psi\}$, $\{\varepsilon\}$, and $\{\sigma\}$ into Eq. (E.4), we obtain

$$\iiint_V \delta\{d\}^T[B]^T[D][B]\{d\}\,dV = \delta\{d\}^T\{P\} + \iint_S \delta\{d\}^T[N_s]^T\{T_s\}\,dS$$
$$+ \iiint_V \delta\{d\}^T[N]^T(\{X\} - \rho[N]\{\ddot d\})\,dV$$

(E.8)

Note that the shape functions are independent of time. Because $\{d\}$ (or $\{d\}^T$) is the matrix of nodal displacements, which is independent of spatial integration, we can simplify Eq. (E.8) by taking the $\{d\}^T$ terms from the integrals to obtain

$$\delta\{d\}^T \iiint_V [B]^T[D][B]\,dV\{d\} = \delta\{d\}^T\{P\} + \delta\{d\}^T \iint_S [N_s]^T\{T_s\}\,dS$$
$$+ \delta\{d\}^T \iiint_V [N]^T(\{X\} - \rho[N]\{\ddot d\})\,dV$$

(E.9)

Because $\delta\{d\}^T$ is an arbitrary virtual nodal displacement vector common to each term in Eq. (E.9), the following relationship must be true.

$$\iiint_V [B]^T[D][B]\,dV\{d\} = \{P\} + \iint_S [N_s]^T\{T_s\}\,dS + \iiint_V [N]^T\{X\}\,dV - \iiint_V \rho[N]^T[N]\,dV\{\ddot d\}$$

(E.10)

We now define

$$[m] = \iiint_V \rho[N]^T\{N\}\,dV$$

(E.11)

$$[k] = \iiint\limits_V [B]^T [D][B] \, dV \tag{E.12}$$

$$\{f_s\} = \iint\limits_S [N_s]^T \{T_s\} \, dS \tag{E.13}$$

$$\{f_b\} = \iiint\limits_V [N]^T \{X\} \, dV \tag{E.14}$$

Using Eqs. (E.11) through (E.14) in Eq. (E.10) and moving the last term of Eq. (E.10) to the left side, we obtain

$$[m]\{\ddot{d}\} + [k]\{d\} = \{P\} + \{f_s\} + \{f_b\} \tag{E.15}$$

The matrix $[m]$ in Eq. (E.11) is the element consistent-mass matrix [2], $[k]$ in Eq. (E.12) is the element stiffness matrix, $\{f_s\}$ in Eq. (E.13) is the matrix of element equivalent nodal loads due to surface forces, and $\{f_b\}$ in Eq. (E.14) is the matrix of element equivalent nodal loads due to body forces.

Specific applications of Eq. (E.15) are given in Chapter 16 for bars and beams subjected to dynamic (time-dependent) forces. For static problems, we set $\{\ddot{d}\}$ equal to zero in Eq. (E.15) to obtain

$$[k]\{d\} = \{P\} + \{f_s\} + \{f_b\} \tag{E.16}$$

Chapters 3 through 9, 11, and 12 illustrate the use of Eq. (E.16) applied to bars, trusses, beams, frames, and to plane stress, axisymmetric stress, three-dimensional stress, and plate-bending problems.

▲ References

[1] Oden, J. T., and Ripperger, E. A., *Mechanics of Elastic Structures*, 2nd ed., McGraw-Hill, New York, 1981.
[2] Archer, J. S., "Consistent Matrix Formulations for Structural Analysis Using Finite Element Techniques," *Journal of the American Institute of Aeronautics and Astronautics*, Vol. 3, No. 10, pp. 1910–1918, 1965.

GEOMETRIC PROPERTIES OF STRUCTURAL STEEL WIDE-FLANGE SECTIONS (W SHAPES) ▲

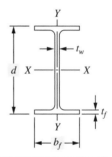

Wide-Flange Sections (W Shapes) SI Units

			Web	Flange							
	Area	Depth	Thickness	Width	Thickness		$x\text{-}x$ axis			$y\text{-}y$ axis	
Designation	A	d	t_w	b_f	t_f	I	S	r_x	I	S	r_y
mm × kg/m	mm^2	mm	mm	mm	mm	10^6 mm^4	10^3 mm^3	mm	10^6 mm^4	10^3 mm^3	mm
W610 × 155	19800	611	12.70	324.0	19.0	1290	4220	255	108.0	667	73.9
W610 × 140	17870	617	13.08	230.0	22.2	1124	3640	251	45.4	394	50.4
W610 × 125	15940	612	11.94	229.0	19.6	985	3220	249	39.3	343	49.7
W610 × 113	14450	608	11.18	228.0	17.3	874	2880	246	34.3	301	48.7
W610 × 101	12970	603	10.54	228.0	14.9	762	2530	242	29.3	257	47.5
W610 × 92	11800	603	10.90	179.0	13.15	646	2140	234	14.4	161	34.9
W610 × 82	10500	599	10.00	178.0	12.8	560	1870	231	12.1	136	33.9
W530 × 138	17610	549	14.73	214.0	23.6	862	3138	221	38.7	362	46.9
W530 × 124	15677	544	13.06	212.0	21.2	762	2799	220	33.9	319	46.5
W530 × 109	13871	539	11.56	211.0	18.8	666	2469	219	29.4	279	46.0
W530 × 101	12903	537	10.92	210.0	17.4	616	2298	218	26.9	258	45.7
W530 × 92	11806	533	10.16	209.0	15.6	554	2077	217	23.9	229	45.0
W530 × 82	10473	528	9.60	209.0	13.2	475	1800	213	20.1	192	43.8

Wide-Flange Sections (W Shapes) SI Units

			Web	Flange							
	Area	Depth	Thickness	Width	Thickness		x-x axis			y-y axis	
Designation	A	d	t_w	b_f	t_f	I	S	r_x	I	S	r_y
mm × kg/m	mm^2	mm	mm	mm	mm	10^6 mm^4	10^3 mm^3	mm	10^6 mm^4	10^3 mm^3	mm
W460 × 177	22645	482	16.64	286.0	26.9	912	3784	201	105.0	736	88.2
W460 × 144	18387	472	13.59	283.0	22.1	728	3085	199	83.7	591	67.5
W460 × 113	14387	463	10.80	280.0	17.3	554	2394	196	63.3	451	66.3
W460 × 97	12320	466	11.43	193.0	19.1	445	1911	190	22.8	237	43.0
W460 × 89	11350	463	10.54	192.0	17.7	410	1766	190	20.9	217	42.9
W460 × 82	10450	460	9.91	191.0	16.0	370	1611	188	18.7	195	42.3
W460 × 74	9484	457	9.02	190.0	14.5	333	1457	187	16.7	175	42.0
W460 × 68	8733	459	9.14	154.0	15.4	296	1293	184	9.4	122	32.8
W460 × 60	7588	455	8.00	153.0	13.3	255	1120	183	8.0	104	32.4
W460 × 52	6640	450	7.62	152.0	10.8	212	944	179	6.4	84	31.0
W410 × 85	10840	417	10.92	181.0	18.2	316	1512	171	17.9	198	40.7
W410 × 74	9484	413	9.65	180.0	16.0	274	1328	170	15.5	172	40.4
W410 × 67	8581	410	8.76	179.0	14.4	244	1200	169	13.7	153	39.9
W410 × 54	6839	403	7.49	177.0	10.9	186	926	165	10.2	115	38.6
W410 × 46	5884	403	6.99	140.0	11.2	156	774	163	5.14	73.6	29.6
W410 × 39	4955	399	6.35	140.0	8.8	125	629	159	3.99	57.1	28.4
W360 × 196	25032	372	16.40	374.0	26.2	636	3421	159	229.0	1222	95.6
W360 × 162	20630	364	13.30	371.0	21.8	515	2832	158	186.0	1001	94.9
W360 × 134	17061	356	11.20	369.0	18.0	415	2332	156	151.0	817	94.0
W360 × 79	10100	354	9.40	205.0	16.8	227	1280	150	24.2	236	48.9
W360 × 64	8150	347	7.75	203.0	13.5	179	1030	148	18.8	185	48.0
W360 × 57	7226	358	7.87	172.0	13.1	160	895	149	11.10	129	39.2
W360 × 51	6452	355	7.24	171.0	11.6	142	797	148	9.70	113	38.8
W360 × 45	5710	352	6.86	171.0	9.8	121	689	146	8.16	95.4	37.8
W360 × 39	4961	353	6.48	128.0	10.7	102	577	143	3.71	58.1	27.3
W360 × 33	4187	349	5.84	127.0	8.5	82.9	475	141	2.91	45.9	26.4
W310 × 253	32212	356	24.40	319.0	39.6	682	3833	146	215.0	1346	81.6
W310 × 202	25744	341	20.10	315.0	31.7	519.0	3042	142	165.0	1050	80.1
W310 × 158	20046	327	15.50	310.0	25.1	386.0	2363	139	125.0	805	78.9
W310 × 129	16500	318	13.10	308.0	20.6	308.0	1935	137	100.0	652	78.0
W310 × 74	9480	310	9.4	205.0	16.3	165.0	1060	132	23.4	228	49.7
W310 × 67	8530	306	8.51	204.0	14.6	145.0	948	130	20.7	203	49.3
W310 × 39	4935	310	5.84	165.0	9.7	84.9	547	131	7.20	87.4	38.2
W310 × 33	4180	313	6.6	102.0	10.8	65.0	415	125	1.92	37.6	21.4
W310 × 24	3040	305	5.59	101.0	6.7	42.8	281	119	1.16	23	19.5
W310 × 21	2680	303	5.08	101.0	5.7	37.0	244	117	0.99	19.5	19.2
W250 × 149	18970	282	17.27	263.0	28.4	259.0	1839	117	86.2	656	67.4
W250 × 80	10200	256	9.4	255.0	15.6	126.0	984	111	42.9	337	64.9
W250 × 67	8560	257	8.89	204.0	15.7	104.0	809	110	22.2	218	50.9
W250 × 58	7400	252	8.0	203.0	13.5	87.3	693	109	18.8	185	50.4
W250 × 45	5700	266	7.62	148.0	13.0	70.8	535	111	6.95	94.2	34.9
W250 × 28	3626	260	6.35	102.0	10.0	40.1	308	105	1.79	34.9	22.2
W250 × 22	2850	254	5.84	102.0	6.9	28.7	226	100	1.20	23.7	20.6
W250 × 18	2284	251	4.83	101.0	5.3	22.4	179	99	0.91	18	19.9
W200 × 100	12700	229	14.50	210.0	23.7	113.0	987	94.3	36.6	349	53.7

(Continued)

Wide-Flange Sections (W Shapes) SI Units (*Continued*)

			Web	Flange							
	Area	Depth	Thickness	Width	Thickness		x-x axis			y-y axis	
Designation	A	d	t_w	b_f	t_f	I	S	r_x	I	S	r_y
mm × kg/m	mm^2	mm	mm	mm	mm	10^6 mm^4	10^3 mm^3	mm	10^6 mm^4	10^3 mm^3	mm
W200 × 86	11000	222	13.00	209.0	20.6	94.7	853	92.8	31.4	300	53.4
W200 × 71	9100	216	10.20	206.0	17.4	76.6	709	91.7	25.4	247	52.8
W200 × 59	7580	210	9.14	205.0	14.2	61.2	583	89.9	20.4	199	51.9
W200 × 46	5890	203	7.24	203.0	11.0	45.5	448	87.9	15.3	151	51.0
W200 × 36	4570	201	6.22	165.0	10.2	34.4	342	86.8	7.64	92.6	40.9
W200 × 22	2860	206	6.22	102.0	8.0	20.0	194	83.6	1.42	27.8	22.3
W150 × 37	4730	162	8.13	154.0	11.6	22.2	274	68.5	7.07	91.8	38.7
W150 × 30	3790	157	6.60	153.0	9.3	17.1	218	67.2	5.54	72.4	38.2
W150 × 22	2860	152	5.84	152.0	6.6	12.1	159	65	3.87	50.9	36.8
W150 × 24	3060	160	6.60	102.0	10.3	13.4	168	66.2	1.83	35.9	24.5
W150 × 18	2290	153	5.84	102.0	7.1	9.19	120	63.3	1.26	24.7	23.5
W150 × 14	1730	150	4.32	100.0	5.5	6.84	91.2	62.9	0.912	18.2	23

I = area moment of inertia, S = Section modulus, r = radius of gyration

Chapter 2

2.1 **a.** $[K] = \begin{bmatrix} k_1 & 0 & -k_1 & 0 \\ 0 & k_3 & 0 & -k_3 \\ -k_1 & 0 & k_1 + k_2 & -k_2 \\ 0 & -k_3 & -k_2 & k_2 + k_3 \end{bmatrix}$

b. $u_3 = \dfrac{k_2 P}{k_1 k_2 + k_1 k_3 + k_2 k_3}, \quad u_4 = \dfrac{(k_1 + k_2) P}{k_1 k_2 + k_1 k_3 + k_2 k_3}$

c. $F_{1x} = \dfrac{-k_1 k_2 P}{k_1 k_2 + k_1 k_3 + k_2 k_3}, \quad F_{2x} = \dfrac{-k_3 (k_1 + k_2) P}{k_1 k_2 + k_1 k_3 + k_2 k_3}$

2.3 **a.** $[K] = \begin{bmatrix} k & -k & 0 & 0 & 0 \\ -k & 2k & -k & 0 & 0 \\ 0 & -k & 2k & -k & 0 \\ 0 & 0 & -k & 2k & -k \\ 0 & 0 & 0 & -k & k \end{bmatrix}$

b. $u_2 = \dfrac{P}{2k}, \quad u_3 = \dfrac{P}{k}, \quad u_4 = \dfrac{P}{2k}$ **c.** $F_{1x} = -\dfrac{P}{2}, \quad F_{5x} = -\dfrac{P}{2}$

2.4 **a.** $[K]$ same as 2.3a. **b.** $u_2 = \dfrac{\delta}{4}, \quad u_3 = \dfrac{\delta}{2}, \quad u_4 = \dfrac{3\delta}{4}$ **c.** $F_{1x} = \dfrac{-k\delta}{4}, \quad F_{5x} = \dfrac{k\delta}{4}$

2.5 $[K] = \begin{bmatrix} 200 & -200 & 0 & 0 \\ -200 & 2000 & 0 & -1800 \\ 0 & 0 & 1000 & -1000 \\ 0 & -1800 & -1000 & 2800 \end{bmatrix}$

2.6 $u_2 = 11.86$ mm

2.7 $[k] = \begin{bmatrix} k & -k \\ -k & k \end{bmatrix}$

2.8 $u_2 = 2.5$ cm, $u_3 = 5$ cm

$f_{1x}^{(1)} = -f_{2x}^{(1)} = -2500$ N, $f_{2x}^{(2)} = -f_{3x}^{(2)} = -2500$ N, $F_{1x} = -2500$ N

2.9 $u_1 = 0$, $u_2 = 7.5$ cm, $u_3 = 17.5$ cm, $u_4 = 27.5$ cm

$f_{1x}^{(1)} = -f_{2x}^{(1)} = -15,000$ N, $f_{2x}^{(2)} = -f_{3x}^{(2)} = -20,000$ N

$f_{3x}^{(3)} = -f_{4x}^{(3)} = -20,000$ N, $F_{1x} = -15,000$ N

2.10 $u_2 = -5$ cm

$f_{1x}^{(1)} = -f_{2x}^{(1)} = 10,000$ N, $f_{2x}^{(2)} = -f_{3x}^{(2)} = -5000$ N

$f_{2x}^{(3)} = -f_{4x}^{(3)} = -5000$ N, $F_{1x} = 10,000$ N, $F_{3x} = F_{4x} = 5000$ N

2.11 $u_2 = 0.01$ m, $f_{1x}^{(1)} = -f_{2x}^{(1)} = -20$ N

$f_{2x}^{(2)} = -f_{3x}^{(2)} = -20$ N, $F_{1x} = -20$ N

2.12 $u_2 = 0.027$ m, $u_3 = 0.018$ m

$f_{1x}^{(1)} = -f_{2x}^{(1)} = -270$ N, $f_{2x}^{(2)} = -f_{3x}^{(2)} = 180$ N

$f_{3x}^{(3)} = -f_{4x}^{(3)} = 180$ N, $F_{1x} = -270$ N, $F_{4x} = -180$ N

2.13 $u_2 = 0.125$ m, $u_3 = 0.25$ m, $u_4 = 0.125$ m

$f_{1x}^{(1)} = -f_{2x}^{(1)} = -2.5$ kN, $f_{2x}^{(2)} = -f_{3x}^{(2)} = -2.5$ kN

$f_{3x}^{(3)} = -f_{4x}^{(3)} = 2.5$ kN, $f_{4x}^{(4)} = -f_{5x}^{(4)} = 2.5$ kN

$F_{1x} = -2.5$ kN, $F_{5x} = -2.5$ kN

2.14 $u_2 = -0.25$ m, $u_3 = -0.75$ m

$f_{1x}^{(1)} = -f_{2x}^{(1)} = 100$ N, $f_{2x}^{(2)} = -f_{3x}^{(2)} = 200$ N

$F_{1x} = 100$ N

2.15 $u_3 = 0.001$ m, $f_{1x}^{(1)} = -f_{3x}^{(1)} = -0.5$ kN

$f_{2x}^{(2)} = -f_{3x}^{(2)} = -0.5$ kN, $f_{3x}^{(3)} = -f_{4x}^{(3)} = 1$ kN

$F_{1x} = -0.5$ kN, $F_{2x} = -0.5$ kN, $F_{4x} = -1$ kN

2.16 $u_2 = 8.33$ mm, $u_3 = -8.33$ mm

2.17 $u_2 = 0.526$ mm, $u_3 = 1.316$ mm, $F_{1x} = -263.2$ N, $F_{4x} = -736.8$ N

2.18 **a.** $x = 1$ cm $\downarrow$, $\pi_{p_{\min}} = -2500$ N · cm

b. $x = 4$ cm $\leftarrow$, $\pi_{p_{\min}} = -10,000$ N · cm

c. $x = 1.962$ mm $\downarrow$, $\pi_{p_{\min}} = -3849$ N · mm

d. $x = 2.4525$ mm $\rightarrow$, $\pi_{p_{\min}} = -1203$ N · mm

2.19 $x = 40.0$ mm ↑

2.20 $x = 1$ cm ←, $\pi_{p_{\min}} = -1666.7$ N · cm

2.21 Same as 2.10

2.22 Same as 2.15

Chapter 3

3.1 a. $[K] = \begin{bmatrix} \dfrac{A_1 E_1}{L_1} & \dfrac{-A_1 E_1}{L_1} & 0 & 0 \\[2ex] \dfrac{-A_1 E_1}{L_1} & \dfrac{A_1 E_1}{L_1} + \dfrac{A_2 E_2}{L_2} & \dfrac{-A_2 E_2}{L_2} & 0 \\[2ex] 0 & \dfrac{-A_2 E_2}{L_2} & \dfrac{A_2 E_2}{L_2} + \dfrac{A_3 E_3}{L_3} & \dfrac{-A_3 E_3}{L_3} \\[2ex] 0 & 0 & \dfrac{-A_3 E_3}{L_3} & \dfrac{A_3 E_3}{L_3} \end{bmatrix}$

b. $u_2 = \dfrac{PL}{3AE}, \quad u_3 = \dfrac{2PL}{3AE}$

c.   **i.** $u_2 = 9.9 \times 10^{-4}$ cm, $u_3 = 19.8 \times 10^{-4}$ cm
  ii. $F_{1x} = -1666.7$ N, $F_{4x} = -3333.3$ N
  iii. $\sigma^{(1)} = 2772$ kPa (T), $\sigma^{(2)} = 2772$ kPa (T), $\sigma^{(3)} = -5544$ kPa (C)

3.2 $u_2 = -0.595 \times 10^{-4}$ m, $u_3 = -1.19 \times 10^{-4}$ m, $F_{1x} = 5$ kN
  $f_{1x}^{(1)} = -f_{2x}^{(1)} = 5$ kN, $f_{2x}^{(2)} = -f_{3x}^{(2)} = 5$ kN

3.3 $u_2 = 8.714 \times 10^{-3}$ cm, $F_{1x} = -28{,}570$ N, $F_{3x} = -11{,}430$ N
  $f_{1x}^{(1)} = -f_{2x}^{(1)} = -28{,}570$ N, $f_{2x}^{(2)} = -f_{3x}^{(2)} = 11{,}430$ N

3.4 $u_2 = -0.5 \times 10^{-5}$ m, $u_3 = -0.4 \times 10^{-5}$ m
  $F_{1x} = 3335$ N, $F_{4x} = 26{,}680$ N
  $f_{1x}^{(1)} = -f_{2x}^{(1)} = 3335$ N, $f_{2x}^{(2)} = -f_{3x}^{(2)} = 23{,}345$ N
  $f_{3x}^{(3)} = -f_{4x}^{(3)} = -26{,}680$ N

3.5 $u_2 = 9.375 \times 10^{-5}$ m, $u_3 = 2.813 \times 10^{-4}$ m, $F_{1x} = -75{,}000$ N
  $f_{1x}^{(1)} = -f_{2x}^{(1)} = f_{2x}^{(2)} = -f_{3x}^{(2)} = -75{,}000$ N

3.6 $u_2 = 9.02 \times 10^{-3}$ cm, $F_{1x} = -18{,}940$ N, $F_{3x} = F_{4x} = -10{,}530$ N
  $f_{1x}^{(1)} = -f_{2x}^{(1)} = -18{,}940$ N, $f_{2x}^{(2)} = -f_{3x}^{(2)} = f_{2x}^{(3)} = -f_{4x}^{(3)} = 10{,}530$ N

3.7 $u_2 = 8.61 \times 10^{-5}$ cm, $u_3 = 20.75 \times 10^{-3}$ cm

$F_{1x} = -206.6$ N, $F_{4x} = -49,800$ N

$f_{1x}^{(1)} = -f_{2x}^{(1)} = f_{2x}^{(2)} = -f_{3x}^{(2)} = -206.6$ N, $f_{3x}^{(3)} = -f_{4x}^{(3)} = 49,800$ N

3.8 $u_2 = -0.250$ mm, $u_3 = -1.678$ mm, $F_{1x} = 20$ kN

3.9 $u_2 = 0.01238$ m, $F_{1x} = -520$ kN, $F_{3x} = 530$ kN

$f_{1x}^{(1)} = -f_{2x}^{(1)} = -520$ kN, $f_{2x}^{(2)} = -f_{3x}^{(2)} = -530$ kN

3.10 $u_2 = 0.935 \times 10^{-3}$ m, $u_3 = 0.727 \times 10^{-3}$ m

$F_{1x} = -6.546$ kN, $F_{4x} = -1.455$ kN

$f_{1x}^{(1)} = -f_{2x}^{(1)} = -6.546$ kN, $f_{2x}^{(2)} = -f_{3x}^{(2)} = 1.455$ kN,

$f_{3x}^{(3)} = -f_{4x}^{(3)} = 1.455$ kN

3.11 $u_2 = 3.572 \times 10^{-4}$ m, $F_{1x} = -7.50$ kN, $F_{3x} = F_{4x} = F_{5x} = -7.50$ kN

$f_{1x}^{(1)} = -f_{2x}^{(1)} = -7.50$ kN,

$f_{2x}^{(2)} = -f_{3x}^{(2)} = f_{2x}^{(3)} = -f_{4x}^{(3)} = f_{2x}^{(4)} = -f_{5x}^{(4)} = 7.50$ kN

3.12 two-element solution, $u_1 = -1.96 \times 10^{-3}$ cm

one-element solution, $u_1 = -1.905 \times 10^{-3}$ cm

3.13 $[B] = \left[-\dfrac{1}{L} + \dfrac{4x}{L^2} \quad \dfrac{-8x}{L^2} \quad \dfrac{1}{L} + \dfrac{4x}{L^2} \right]$, $[k] = A \displaystyle\int_{-L/2}^{L/2} [B]^T E[B]\, dx$

3.15 **a.** $[k] = 4 \times 10^8 \begin{bmatrix} 1 & 1 & -1 & -1 \\ 1 & 1 & -1 & -1 \\ -1 & -1 & 1 & 1 \\ -1 & -1 & 1 & 1 \end{bmatrix} \dfrac{\text{N}}{\text{m}}$

b. $[k] = \dfrac{10^8}{2} \begin{bmatrix} 1 & -\sqrt{3} & -1 & \sqrt{3} \\ -\sqrt{3} & 3 & \sqrt{3} & -3 \\ -1 & \sqrt{3} & 1 & -\sqrt{3} \\ \sqrt{3} & -3 & -\sqrt{3} & 3 \end{bmatrix} \dfrac{\text{N}}{\text{m}}$

c. $[k] = 7000 \begin{bmatrix} 3 & -\sqrt{3} & -3 & \sqrt{3} \\ -\sqrt{3} & 1 & \sqrt{3} & -1 \\ -3 & \sqrt{3} & 3 & -\sqrt{3} \\ \sqrt{3} & -1 & -\sqrt{3} & 1 \end{bmatrix} \dfrac{\text{kN}}{\text{m}}$

d. $[k] = 1.4 \times 10^4 \begin{bmatrix} 0.883 & 0.321 & -0.883 & -0.321 \\ 0.321 & 0.117 & -0.321 & -0.117 \\ -0.883 & -0.321 & 0.883 & 0.321 \\ -0.321 & -0.117 & 0.321 & 0.117 \end{bmatrix} \dfrac{\text{kN}}{\text{m}}$

3.16 **a.** $u_1 = 0.866$ cm, $u_2 = 1.183$ cm

b. $u_1 = 0.866$ cm, $u_2 = -0.317$ cm

3.17 a. $u_1 = 2.165$ mm, $v_1 = -1.25$ mm,

$u_2 = 0.098$ mm, $v_2 = -5.83$ mm

b. $u_1 = -1.25$ mm, $v_1 = 2.165$ mm,

$u_2 = 3.03$ mm, $v_2 = 5.098$ mm

3.18 a. $\sigma = 74.25$ MPa, **b.** 45.47 MPa

3.19 a. $[K] = k \begin{bmatrix} 2 & 0 & -\frac{1}{2} & \frac{1}{2} & -1 & 0 & -\frac{1}{2} & -\frac{1}{2} \\ 0 & 1 & \frac{1}{2} & -\frac{1}{2} & 0 & 0 & -\frac{1}{2} & -\frac{1}{2} \\ -\frac{1}{2} & \frac{1}{2} & \frac{1}{2} & -\frac{1}{2} & 0 & 0 & 0 & 0 \\ \frac{1}{2} & -\frac{1}{2} & -\frac{1}{2} & \frac{1}{2} & 0 & 0 & 0 & 0 \\ -1 & 0 & 0 & 0 & 1 & 0 & 0 & 0 \\ 0 & 0 & 0 & 0 & 0 & 0 & 0 & 0 \\ -\frac{1}{2} & -\frac{1}{2} & 0 & 0 & 0 & 0 & \frac{1}{2} & \frac{1}{2} \\ -\frac{1}{2} & -\frac{1}{2} & 0 & 0 & 0 & 0 & \frac{1}{2} & \frac{1}{2} \end{bmatrix}$

b. $u_1 = 0$, $v_1 = \dfrac{-50}{k}$

3.20 $u_2 = 0$, $v_2 = 0.393$ cm, $\sigma^{(1)} = \sigma^{(2)} = 5892$ kPa (T)

3.21 $u_1 = \dfrac{1155L}{AE}$, $v_1 = \dfrac{217.5L}{AE}$

3.22 $u_1 = \dfrac{2110L}{AE}$, $v_1 = \dfrac{7902L}{AE}$

$\sigma^{(1)} = \dfrac{2893}{A}$ (C), $\sigma^{(2)} = \dfrac{2110}{A}$ (T), $\sigma^{(3)} = \dfrac{5004}{A}$ (T)

3.23 $u_1 = 0.714$ cm, $v_1 = 0$, $\sigma^{(1)} = 100$ MPa

3.24 $u_2 = \dfrac{266,750}{AE}$, $v_2 = \dfrac{1,050,210}{AE}$, $u_3 = \dfrac{-266,750}{AE}$, $v_3 = \dfrac{1,050,210}{AE}$

$f_{1x}^{\prime(1)} = -f_{2x}^{\prime(1)} = -13,333$ N, $f_{1x}^{\prime(2)} = -f_{3x}^{\prime(2)} = -16,667$ N

$f_{2x}^{\prime(3)} = -f_{4x}^{\prime(3)} = 16,667$ N, $f_{2x}^{\prime(4)} = -f_{3x}^{\prime(4)} = 0$

$f_{3x}^{\prime(5)} = -f_{4x}^{\prime(5)} = 13,333$ N, $f_{1x}^{\prime(6)} = -f_{4x}^{\prime(6)} = 0$

3.25 $u_2 = 0$, $v_2 = \dfrac{2,250,000}{AE}$, $u_3 = \dfrac{-533,400}{AE}$, $v_3 = \dfrac{2,100,000}{AE}$

$f_{1x}^{\prime(1)} = -f_{2x}^{\prime(1)} = 0$, $f_{1x}^{\prime(2)} = -f_{3x}^{\prime(2)} = -33,333$ N

$f_{2x}^{\prime(4)} = -f_{3x}^{\prime(4)} = 10,000$ N, $f_{3x}^{\prime(5)} = -f_{4x}^{\prime(5)} = 26,665$ N

$f_{1x}^{\prime(6)} = -f_{4x}^{\prime(6)} = 0$

3.26 No, the truss is unstable, $\|[K]\| = 0$.

3.27 $u_3 = 0.347$ cm, $v_3 = -0.132$ cm

$f_{1x}^{\prime(1)} = -f_{3x}^{\prime(1)} = -62.76$ kN, $f_{2x}^{\prime(2)} = -f_{3x}^{\prime(2)} = 20.52$ kN

$f_{3x}^{\prime(3)} = -f_{4x}^{\prime(3)} = -66$ kN

3.28 $[T]^T = \begin{bmatrix} C & -S & 0 & 0 \\ S & C & 0 & 0 \\ 0 & 0 & C & -S \\ 0 & 0 & S & C \end{bmatrix}$ and $[T][T]^T = \begin{bmatrix} 1 & 0 & 0 & 0 \\ 0 & 1 & 0 & 0 \\ 0 & 0 & 1 & 0 \\ 0 & 0 & 0 & 1 \end{bmatrix}$

$\therefore [T]^T = [T]^{-1}$

3.29 $u_1 = -0.893 \times 10^{-4}$ m, $v_1 = -4.46 \times 10^{-4}$ m

$\sigma^{(1)} = 31.2$ MPa (T), $\sigma^{(2)} = 26.5$ MPa (T), $\sigma^{(3)} = 6.25$ MPa (T)

3.30 $u_1 = 1.71 \times 10^{-4}$ m, $v_1 = -7.55 \times 10^{-4}$ m

$\sigma^{(1)} = 79.28$ MPa (T), $\sigma^{(2)} = 11.97$ MPa (T), $\sigma^{(3)} = -23.87$ MPa (C)

3.31 $u_1 = 8.25 \times 10^{-4}$ m, $v_1 = -3.65 \times 10^{-3}$ m

$\sigma^{(2)} = 57.74$ MPa (T), $\sigma^{(3)} = -115.5$ MPa (C)

3.32 $u_2 = 0.135 \times 10^{-2}$ m, $v_2 = -0.850 \times 10^{-2}$ m,

$v_3 = -0.137 \times 10^{-1}$ m, $v_4 = -0.164 \times 10^{-1}$ m,

$\sigma^{(1)} = -198$ MPa (C), $\sigma^{(2)} = 0$, $\sigma^{(3)} = 44.6$ MPa (T)

$\sigma^{(4)} = -31.6$ MPa (C), $\sigma^{(5)} = -191$ MPa (C),

$\sigma^{(6)} = -63.1$ MPa (C)

3.33 a. $u_1 = -3.448 \times 10^{-3}$ m, $v_1 = -6.896 \times 10^{-3}$ m

$\sigma^{(1)} = 102.4$ MPa (T), $\sigma^{(2)} = -72.4$ MPa (C)

3.34 $u_4 = 39.72 \times 10^{-3}$ cm, $v_4 = -9.86 \times 10^{-3}$ cm

$\sigma^{(1)} = 105$ MPa (T), $\sigma^{(2)} = 11.62$ MPa (T), $\sigma^{(3)} = -5.18$ MPa (C)

$\sigma^{(4)} = -10.42$ MPa (C), $\sigma^{(5)} = 0$

3.35 $v_1 = -1.667 \times 10^{-3}$ cm, $\sigma^{(1)} = 625$ kPa (T)

3.36 $u_1 = 4.24$ mm

3.37 $u_1 = 0.804$ mm

3.38 $u_2 = 16.98$ mm

3.39 $u_2 = 1.71$ mm

3.40 $u_1 = -3.018 \times 10^{-5}$ m, $v_1 = -1.517 \times 10^{-5}$ m,

$w_1 = 2.684 \times 10^{-5}$ m, $\sigma^{(1)} = -338$ kN/m^2 (C),

$\sigma^{(2)} = -1690$ kN/m^2 (C), $\sigma^{(3)} = -7965$ kN/m^2 (C)

$\sigma^{(4)} = -2726$ kN/m^2 (C)

3.41 $u_1 = 1.383 \times 10^{-3}$ m, $v_1 = -5.119 \times 10^{-5}$ m

$w_1 = 6.015 \times 10^{-5}$ m, $\sigma^{(1)} = 20.51$ MPa (T),

$\sigma^{(2)} = 4.21$ MPa (T), $\sigma^{(3)} = -5.29$ MPa (C)

3.42 $u_5 = 0.189$ cm, $v_5 = 0$, $w_5 = -0.0552$ cm

$\sigma^{(1)} = \sigma^{(4)} = 965$ MPa (T), $\sigma^{(2)} = \sigma^{(3)} = 122.6$ MPa (C)

3.43 $u_4 = 1.654$ mm, $v_4 = 0$, $w_4 = -1.463$ mm

$\sigma^{(1)} = -137.3$ MPa (C)

3.46 $v_2 = -1.92$ mm, $v_3 = -1.68$ mm, $u_1 = -0.426$ mm

$\sigma^{(1)} = -27.2$ MPa (C), $\sigma^{(2)} = 21.3$ MPa (T), $\sigma^{(3)} = 8$ MPa (T)

3.48 $v_2 = -0.955 \times 10^{-2}$ m, $v_4 = -1.03 \times 10^{-2}$ m,

$\sigma^{(1)} = 67.1$ MPa (C), $\sigma^{(2)} = 60.0$ MPa (T), $\sigma^{(3)} = 22.4$ MPa (C)

$\sigma^{(4)} = 44.7$ MPa (C), $\sigma^{(5)} = 20.0$ MPa (T)

3.49 $u_1' = 0$, $v_2 = -0.01414$ cm, $F_{2x} = 10{,}000$ N

$\sigma^{(1)} = 0$, $\sigma^{(2)} = 18.85$ MPa (T), $\sigma^{(3)} = 0$

3.50 $v_2 = -0.01414$ cm

3.51 $u_2' = 0.0882$ mm

3.52 **a.** $u_1 = 0.025$ cm ↓, $\pi_{p_{\min}} = -12.5$ N · m

b. $u_1 = 0.002$ cm →, $\pi_{p_{\min}} = -12.0$ N · m

3.53 $[k] = \dfrac{3A_0 E}{2L} \begin{bmatrix} 1 & -1 \\ -1 & 1 \end{bmatrix}$

3.54 two-element solution: $u_2 = 0.2475$ mm, $u_3 = 0.36$ mm, $\sigma^{(1)} = 65.95$ MPa (T),

$\sigma^{(2)} = 30$ MPa (T)

3.56 $u_2 = 2.25 \times 10^{-3}$ cm, $\sigma^{(1)} = 6$ MPa (T)

3.57 $u_1 = \gamma L^2/(2E)$, $u_2 = 3\gamma L^2/(8E)$, $\sigma^{(1)} = \gamma L/8$, $\sigma^{(2)} = 3\gamma L/8$

3.58 **a.** $f_{1x} = 2.87$ N

b. $f_{1x} = 26.7$ kN, $f_{2x} = 80$ kN

Chapter 4

4.3 $v_2 = \dfrac{-7PL^3}{768EI}$, $\phi_1 = \dfrac{-PL^2}{32EI}$, $\phi_2 = \dfrac{PL^2}{128EI}$

$F_{1y} = \dfrac{5P}{16}$, $M_1 = 0$, $F_{3y} = \dfrac{11P}{16}$, $M_3 = \dfrac{-3PL}{16}$

4.4 $v_1 = \dfrac{-PL^3}{3EI}$, $\phi_1 = \dfrac{PL^3}{2EI}$, $F_{2y} = P$, $M_2 = -PL$

4.5 $v_1 = -7.5$ cm, $\phi_1 = 0.0160$ rad, $\phi_2 = 0.00536$ rad

$F_{2y} = 12.5$ kN, $F_{3y} = -7.5$ kN, $M_3 = 15$ kN · m

4.6 $v_3 = -75$ mm

4.7 $v_2 = -13.9$ mm, $\phi_2 = -0.0119$ rad, $v_3 = -45.6$ mm, $\phi_3 = -0.0179$ rad

4.8 $v_2 = -1.34 \times 10^{-4}$ m, $\phi_2 = 8.93 \times 10^{-5}$ rad

$F_{1y} = 10$ kN, $M_1 = 12.5$ kN · m, $F_{3y} = 1.87$ N, $M_3 = -2.5$ kN · m

4.9 $v_3 = -7.619 \times 10^{-4}$ m, $\phi_2 = -3.809 \times 10^{-4}$ rad, $\phi_1 = 1.904 \times 10^{-4}$ rad

$F_{1y} = -0.889$ kN, $F_{2y} = 4.889$ kN

4.11 $v_2 = -7.934 \times 10^{-3}$ m, $\phi_1 = -2.975 \times 10^{-3}$ rad

$F_{1y} = 5.208$ kN, $F_{3y} = 5.208$ kN

$F_{\text{spring}} = 1.587$ kN

4.12 $v_2 = v_4 = \dfrac{-1wL^4}{607.5EI}$, $v_3 = \dfrac{-wL^4}{507EI}$

$\phi_2 = \dfrac{-1wL^3}{270EI}$, $\phi_4 = -\phi_2$

$F_{1y} = \dfrac{wL}{2}$, $M_1 = \dfrac{wL^2}{12}$

4.13 $v_2 = \dfrac{-wL^4}{384EI}$, $F_{1y} = \dfrac{wL}{2}$, $M_1 = \dfrac{wL^2}{12}$

4.14 $v_2 = \dfrac{-5wL^4}{384EI}$, $\phi_1 = -\phi_3 = \dfrac{-wL^3}{24EI}$, $F_{1y} = \dfrac{wL}{2}$

4.15 $v_3 = \dfrac{-wL^4}{4EI}$, $\phi_2 = \dfrac{-wL^3}{8EI}$, $\phi_3 = \dfrac{-7wL^3}{24EI}$

$F_{1y} = \dfrac{-3wL}{4}$, $M_1 = \dfrac{-wL^2}{4}$, $F_{2y} = \dfrac{7wL}{4}$

4.16 $f_{1y} = \dfrac{-3wL}{20}$, $m_1 = \dfrac{-wL^2}{30}$, $f_{2y} = \dfrac{-7wL}{20}$, $m_2 = \dfrac{wL^2}{20}$

4.17 $F_{1y} = \dfrac{wL}{4}$, $M_1 = \dfrac{5wL^2}{96}$, $F_{3y} = \dfrac{wL}{4}$, $M_3 = \dfrac{-5wL^2}{96}$, $v_2 = \dfrac{-7wL^4}{3840EI}$

4.18 $\phi_2 = \dfrac{wL^3}{80EI}$, $F_{1y} = \dfrac{9wL}{40}$, $M_1 = \dfrac{7wL^2}{120}$, $F_{2y} = \dfrac{11wL}{40}$

4.19 $v_3 = -0.0244$ m, $\phi_3 = -0.0071$ rad, $\phi_2 = -0.00305$ rad

$F_{1y} = -24$ kN, $M_1 = -32$ kN · m, $F_{2y} = 56$ kN

$f_{1y}^{(1)} = -f_{2y}^{(1)} = -24$ kN, $m_1^{(1)} = -32$ kN · m, $m_2^{(1)} = -64$ kN · m

$f_{2y}^{(2)} = 32$ kN, $m_2^{(2)} = 64$ kN · m, $f_{3y}^{(2)} = 0$, $m_3^{(2)} = 0$

4.20 $\phi_1 = -0.0032$ rad, $v_2 = -0.0115$ m, $\phi_3 = 0.0032$ rad

$F_{1y} = 29.94$ kN, $F_{2y} = 0.1152$ kN, $F_{3y} = 29.94$ kN

$f_{1y}^{(1)} = 29.94$ kN, $m_1^{(1)} = 0$, $f_{2y}^{(1)} = 0.058$ kN, $m_2^{(1)} = 59.65$ kN · m

4.21 $v_2 = -97.66$ mm, $\phi_2 = -0.009766$ rad, $\phi_3 = 0.03906$ rad

$F_{1y} = 187.5$ kN, $M_1 = 375$ kN $\cdot$ m $F_{3y} = 112.5$ kN

4.24 $\phi_1 = -3.596 \times 10^{-4}$ rad, $\phi_2 = 9.92 \times 10^{-5}$ rad, $\phi_3 = 1.091 \times 10^{-4}$ rad

$F_{1y} = 9875$ N, $F_{2y} = 28,406$ N, $F_{3y} = 6719$ N

4.25 $v_{max} = -0.000756$ m   at midspan of AB and BC

$\sigma_{max} = 34.3$ MPa   at midspan of AB and BC

$\sigma_{min} = -51.0$ MPa   at B

4.26 $v_{max} = -0.1953$ m   at midspan of BC

$\sigma_{min} = -469$ MPa

4.28 $v_{max} = -0.0419$ m   at C

$\sigma_{max} = 66.97$ MPa   at fixed end A

$\sigma_{min} = -133.9$ MPa   at B

4.30 $v_{max} = -0.087$ m   at C

$\sigma_{max} = 257$ MPa   at B

4.37 $v_2 = \dfrac{-PL^3}{192EI} - \dfrac{wL^4}{384EI}$, $F_{1y} = \dfrac{P + wL}{2}$, $M_1 = \dfrac{PL}{8} + \dfrac{wL^2}{12}$

4.38 $v_2 = \dfrac{-5PL^3}{648EI}$

4.39 $v_2 = \dfrac{-(25P + 22wL)L^3}{240EI}$, $\phi_2 = \dfrac{-(PL^2 + wL^3)}{8EI}$

$F_{1y} = P + \dfrac{wL}{2}$, $M_1 = \dfrac{PL}{2} + \dfrac{wL^2}{3}$

4.40 $v_2 = -1.57 \times 10^{-4}$ m, $\phi_2 = 1.19 \times 10^{-4}$ rad

4.41 $v_2 = -3.18 \times 10^{-4}$ m, $\phi_2 = 1.58 \times 10^{-4}$ rad, $\phi_3 = 1.58 \times 10^{-4}$ rad

4.42 $v_3 = -4.26 \times 10^{-5}$ m, $\phi_2 = -2.56 \times 10^{-5}$ rad, $\phi_3 = 5.38 \times 10^{-5}$ rad

4.44 $[k] = \dfrac{GA_W}{L}\begin{bmatrix} 1 & -1 \\ -1 & 1 \end{bmatrix}$

4.47 $[k] = EI \displaystyle\int_0^L [B]^T [B]\, dx + k_f \int_0^L [N]^T [N]\, dx$

4.48 Same answer as 4.47

4.77 For 400 mm span,

$\delta = 1.28$ mm (No shear area effect)

$\delta = 1.34$ mm (Shear area included)

For 100 mm span,

$\delta = 0.02$ mm (No shear area effect)

$\delta = 0.0355$ mm (Shear area effect included)

Chapter 5

5.1 $u_2 = 0.000634$ m, $v_2 = 0$, $\phi_2 = 0$

$f_{1x}^{\prime(1)} = -f_{2x}^{\prime(1)} = -33{,}285$ N, $f_{1y}^{\prime(1)} = -f_{2y}^{\prime(1)} = 73.84$ N

$m_1^{\prime(1)} = 554.57$ N · m, $m_2^{\prime(1)} = 0$

5.2 $u_2 = u_3 = 0.01887$ m, $v_2 = -v_3 = 0.000047$ m

$\phi_2 = -\phi_3 = -0.001897$ rad

$f_{1x}^{\prime(1)} = -f_{2x}^{\prime(1)} = -10286$ N, $f_{1y}^{\prime(1)} = -f_{2y}^{\prime(1)} = 11{,}966.25$ N

$m_1^{\prime(1)} = 41{,}133.7$ N · m, $m_2^{\prime(1)} = 30{,}795.61$ N · m

$f_{2x}^{\prime(2)} = -f_{3x}^{\prime(2)} = 8750$ N, $f_{2y}^{\prime(2)} = -f_{3y}^{\prime(2)} = -11{,}966.25$ N

$m_2^{\prime(2)} = 31{,}079$ N · m, $m_3^{\prime(2)} = 31{,}047.5$ N · m

$f_{3x}^{\prime(3)} = -f_{4x}^{\prime(3)} = 10{,}281$ N, $f_{3y}^{\prime(3)} = -f_{4y}^{\prime(3)} = 8750$ N

$m_3^{\prime(3)} = 30795.6$ N · m, $m_4^{\prime(3)} = 41{,}133.7$ N · m

$F_{1x} = F_{4x} = 10{,}286$ N, $F_{1y} = -F_{4y} = -11{,}966.25$ N

$M_1 = 41{,}133.7$ N · m, $M_4 = 41{,}133.7$ N · m

5.3 Channel section 6×8.2 based on $M_{max} = 10{,}452$ N · m

5.5 $u_2 = 0.001339$ N · m, $v_2 = -0.004266$ m, $\phi_2 = -0.008614$ rad

$f_{1x}^{\prime(1)} = 360.11$ kN, $f_{1y}^{\prime(1)} = 15.34$ kN, $m_1^{\prime(1)} = 36.25$ kN · m

$f_{2x}^{\prime(1)} = -293.55$ kN, $f_{2y}^{\prime(1)} = 29.05$ kN, $m_2^{\prime(1)} = -110.48$ kN · m

$f_{2x}^{\prime(2)} = -f_{3x}^{\prime(2)} = 187.09$ kN, $f_{2y}^{\prime(2)} = 68.14$ kN, $m_2^{\prime(2)} = 110.48$ kN · m

$f_{3y}^{\prime(2)} = 91.45$ kN, $m_3^{\prime(2)} = -217$ kN · m

$F_{1x} = F_{3x} = 187.15$ kN, $F_{1y} = 308.12$ kN, $M_1 = 36.25$ kN · m

$F_{3y} = 91.45$ kN, $M_3 = -217$ kN · m

5.7 $u_2 = 0.4308 \times 10^{-4}$ m, $v_2 = -0.9067 \times 10^{-4}$ m,

$\phi_2 = -0.1403 \times 10^{-2}$ rad

$f_{1x}^{\prime(1)} = -f_{2x}^{\prime(1)} = 23.8$ kN, $f_{1y}^{\prime(1)} = 17.26$ kN, $m_1^{\prime(1)} = 32.77$ kN · m

$f_{2y}^{\prime(1)} = 22.74$ kN, $m_2^{\prime(1)} = -54.64$ kN · m

$f_{2x}^{\prime(2)} = -f_{3x}^{\prime(2)} = 11.31$ kN, $f_{2y}^{\prime(2)} = 37.19$ kN, $m_2^{\prime(2)} = 65.09$ kN · m

$f_{3y}^{\prime(2)} = 42.81$ kN, $m_3^{\prime(2)} = -87.54$ kN · m

$f_{2x}^{\prime(3)} = -f_{4x}^{\prime(3)} = 17.55$ kN, $f_{2y}^{\prime(3)} = -f_{4y}^{\prime(3)} = 1.40$ kN

$m_2^{\prime(3)} = -10.51$ kN · m, $m_4^{\prime(3)} = -5.30$ kN · m

$F_{1x} = -17.26$ kN, $F_{1y} = 23.80$ kN, $M_1 = 32.77$ kN · m

$F_{3x} = -11.31$ kN, $F_{3y} = 42.81$ kN, $M_3 = -87.54$ kN · m

$F_{4x} = -11.42$ kN, $F_{4y} = 13.40$ kN, $M_4 = -5.30$ kN · m

5.9 $u_2 = -4.95 \times 10^{-5}$ m, $v_2 = -2.56 \times 10^{-5}$ m, $\phi_2 = 2.66 \times 10^{-3}$ rad

$f'^{(1)}_{1x} = -f'^{(1)}_{2x} = 26.9$ kN, $f'^{(1)}_{1y} = -f'^{(1)}_{2y} = -42.0$ kN

$m'^{(1)}_1 = 55.9$ kN · m, $m'^{(1)}_2 = 111.7$ kN · m

$f'^{(2)}_{2x} = -f'^{(2)}_{3x} = -42.0$ kN, $f'^{(2)}_{2y} = -f'^{(2)}_{3y} = 26.9$ kN

$M_1 = 55.9$ kN · m, $M_3 = 44.7$ kN · m

5.10 $v_2 = -0.1423 \times 10^{-2}$ m, $\phi_2 = -0.5917 \times 10^{-3}$ rad

$f'^{(1)}_{1x} = 0$, $f'^{(1)}_{1y} = 10$ kN, $m'^{(1)}_1 = 23.3$ kN · m, $f'^{(1)}_{2x} = 0$,

$f'^{(1)}_{2y} = -10$ kN, $m'^{(1)}_2 = 6.7$ kN · m

5.11 $v_2 = -3.712 \times 10^{-5}$ m, $F_{1x} = 5440$ N, $F_{1y} = 10,000$ N, $M_1 = 112$ N · m

5.12 $u_2 = -0.2143$ m, $v_1 = -0.250$ m, $\phi_1 = 0.0893$ rad, $u_2 = -0.2143$ m,

$v_2 = -0.357 \times 10^{-4}$ m, $\phi_2 = 0.0714$ m

5.15 $u_2 = -1.76 \times 10^{-2}$ m, $v_2 = -1.87 \times 10^{-5}$ m, $\phi_2 = 5.00 \times 10^{-3}$ rad

$u_3 = -1.76 \times 10^{-2}$ m, $\phi_3 = -2.49 \times 10^{-3}$ rad

$F_{1x} = 20.0$ kN, $F_{1y} = 13.1$ kN, $M_1 = -57.4$ kN · m, $F_{3y} = -13.1$ kN

5.16 $v_3 = -2.83 \times 10^{-5}$ m, $u_4 = 1.0 \times 10^{-5}$ m, $v_4 = -2.83 \times 10^{-5}$ m

5.17 $v_3 = -0.001324$ m, $\phi_3 = 0$

5.18 $u_2 = v_2 = -0.01 \times 10^{-3}$ m, $\phi_2 = 1.766 \times 10^{-4}$ rad

5.21 Use a W16 × 31 for all sections

5.27 $v_2 = -0.0153$ m, $f'^{(1)}_{1x} = 30$ kN, $f'^{(1)}_{1y} = -6.67$ kN, $m'^{(1)}_1 = 0$

5.28 $u_2 = 5.70$ mm, $v_2 = -0.0244$ mm, $\phi_2 = 0.00523$ rad

5.32 $u_2 = 4.30$ mm, $\phi_2 = -0.241 \times 10^{-3}$ rad

$F_{1x} = -8339$ N, $F_{1y} = -4995$ N, $M_1 = 26,700$ N · m

$F_{4x} = -6661$ N, $F_{4y} = 4995$ N, $M_4 = 23,330$ N · m

5.33 $u_7 = 0.0264$ m, $v_7 = 0.463 \times 10^{-4}$ m, $\phi_7 = 0.171 \times 10^{-2}$ rad

$f'^{(1)}_{1x} = -21.1$ N, $f'^{(1)}_{1y} = 30.4$ N, $m'^{(1)}_1 = 74.95$ N · m

$f'^{(1)}_{3x} = 21.1$ N, $f'^{(1)}_{3y} = -30.4$ N, $m'^{(1)}_3 = 46.65$ N · m

5.35 $u_9 = 0.0174$ m, $f'^{(1)}_{1x} = -22.6$ kN, $f'^{(1)}_{1y} = 16.0$ kN, $m'^{(1)}_1 = 53.6$ kN · m

$f'^{(1)}_{3x} = 22.6$ kN, $f'^{(1)}_{3y} = -16.0$ kN, $m'^{(1)}_3 = 42.4$ kN · m

5.36 $v_5 = -2.80 \times 10^{-7}$ m, $v_7 = -4.87 \times 10^{-7}$ m

5.37 $v_5 = -1.29 \times 10^{-2}$ m

5.38 $u_2 = 1.43 \times 10^{-1}$ m

5.39 Truss: $u_7 = 0.0260$ m, $v_7 = 0.00566$ m

Frame: $u_7 = 0.0180$ m, $v_7 = 0.00424$ m

Truss, element 1: $f_{1x} = -49,730$ N, $f_{1y} = 0$

Frame, element 1: $f_{1x} = -43,060$ N, $f_{1y} = 22,670$ N

5.40 $v_{\max} = -0.0105$ m $\qquad$ at midspan

$M_{\max} = 1.568 \times 10^6$ N $\cdot$ m $\qquad$ at C

5.41 $v_{\max} = 0.0524$ m

$M_{\max} = 6.22 \times 10^4$ N $\cdot$ m

5.46 $[K] = 15\dfrac{GJ_0}{L}\begin{bmatrix} 1 & -1 \\ -1 & 1 \end{bmatrix}$

5.51 $v_1 = -0.690 \times 10^{-2}$ m

5.55 $v_3 = -2.54 \times 10^{-3}$ m

5.57 $v_5 = -2.22 \times 10^{-2}$ m

Chapter 6

6.1 Use Eq. (6.2.10) in Eq. (6.2.18) to show $N_i + N_j + N_m = 1$.

6.3 **a.** $[k] = 2.8 \times 10^8 \begin{bmatrix} 2.5 & 1.25 & -2.0 & -1.5 & -0.5 & 0.25 \\ & 4.375 & -1.0 & -0.75 & -0.25 & -3.625 \\ & & 4.0 & 0 & -2.0 & 1.0 \\ & & & 1.5 & 1.5 & -0.75 \\ & & & & 2.5 & -1.25 \\ \text{Symmetry} & & & & & 4.375 \end{bmatrix} \dfrac{\text{N}}{\text{m}}$

b. $[k] = 93.33 \times 10^9 \begin{bmatrix} 1.54 & 0.75 & -1.0 & -0.45 & -0.54 & -0.3 \\ & 1.815 & -0.3 & -0.375 & -0.45 & -1.44 \\ & & 1.0 & 0 & 0 & 0.3 \\ & & & 0.375 & 0.45 & 0 \\ & & & & 0.54 & 0 \\ \text{Symmetry} & & & & & 1.44 \end{bmatrix} \dfrac{\text{N}}{\text{m}}$

c. $[k] = 10^8 \times \begin{bmatrix} 32.48 & 16.24 & -5.6 & -13.44 & -26.88 & -2.8 \\ 16.24 & 26.12 & -2.8 & -6.72 & -13.44 & -22.4 \\ -5.6 & -2.8 & 5.6 & 0 & 0 & 2.8 \\ -13.44 & -6.72 & 0 & 6.72 & 13.44 & 0 \\ -26.88 & -13.44 & 0 & 13.44 & 26.88 & 0 \\ -2.8 & -22.4 & 2.8 & 0 & 0 & 22.4 \end{bmatrix}$

6.4 **a.** $\sigma_x = 336$ MPa, $\sigma_y = 84$ MPa, $\tau_{xy} = -262.5$ MPa

  $\sigma_1 = 501.17$ MPa, $\sigma_2 = -81.17$ MPa, $\theta_p = -32.2°$

  b. $\sigma_x = 560$ MPa, $\sigma_y = 140$ MPa, $\tau_{xy} = -437.5$ MPa

  $\sigma_1 = 835.28$ MPa, $\sigma_2 = -135.28$ MPa, $\theta_p = -32.2°$

  c. Same answers as Part a.

6.5 **a.** $\sigma_{vM} = 546.29$ MPa, **b.** $\sigma_{vM} = 910.48$ MPa, **c.** $\sigma_{vM} = 546.29$ MPa

6.6 **a.** $[k] = 2.074 \times 10^5 \begin{bmatrix} 8437.5 & 1687.5 & -7762.5 & -337.5 & -675 & -1350 \\ 1687.5 & 3937.5 & 337.5 & -2137.5 & -2025 & -1800 \\ -7762.5 & 337.5 & 8437.5 & -1687.5 & -675 & 1350 \\ -337.5 & -2137.5 & -1687.5 & 3937.5 & 2025 & -1800 \\ -675 & -2025 & -675 & 2025 & 1350 & 0 \\ -1350 & -1800 & 1350 & -1800 & 0 & 3600 \end{bmatrix}$ N/m

  b. $[k] = 4.48 \times 10^7 \begin{bmatrix} 25.0 & 0 & -12.5 & 6.25 & -12.5 & -6.25 \\ & 9.375 & 9.375 & -4.6875 & -9.375 & -4.6875 \\ & & 15.625 & -7.8125 & -3.125 & -1.5625 \\ & & & 27.343 & 1.5625 & -3.125 \\ & & & & 15.625 & 7.8125 \\ \text{Symmetry} & & & & & 27.343 \end{bmatrix}$ N/m

  c. $[k] = \begin{bmatrix} 1.225 \times 10^9 & 3.5 \times 10^8 & -1.015 \times 10^9 & -7 \times 10^7 & -2.1 \times 10^8 & -2.8 \times 10^8 \\ 3.5 \times 10^8 & 7 \times 10^8 & 7 \times 10^7 & -1.4 \times 10^8 & -4.2 \times 10^8 & -5.6 \times 10^8 \\ -1.015 \times 10^9 & 7 \times 10^7 & 1.225 \times 10^9 & -3.5 \times 10^8 & -2.1 \times 10^8 & 2.8 \times 10^8 \\ -7 \times 10^7 & -1.4 \times 10^8 & -3.5 \times 10^8 & 7 \times 10^8 & 4.2 \times 10^8 & -5.6 \times 10^8 \\ -2.1 \times 10^8 & -4.2 \times 10^8 & -2.1 \times 10^8 & 4.2 \times 10^8 & 4.2 \times 10^8 & 0 \\ -2.8 \times 10^8 & -5.6 \times 10^8 & 2.8 \times 10^8 & -5.6 \times 10^8 & 0 & 1.12 \times 10^9 \end{bmatrix}$

6.7 **a.** $\sigma_x = -5.289$ GPa, $\sigma_y = -0.156$ GPa, $\tau_{xy} = 0.233$ GPa

  $\sigma_1 = -0.1459$ GPa, $\sigma_2 = -5.30$ GPa, $\theta_p = -2.59°$

  b. $\sigma_x = 0$, $\sigma_y = 42.0$ MPa, $\tau_{xy} = 33.6$ MPa

  $\sigma_1 = 60.6$ MPa, $\sigma_2 = -18.6$ MPa, $\theta_p = -29°$

  c. $\sigma_1 = 3942$ MPa, $\sigma_2 = -3194$ MPa, $\theta_p = -10.28°$

6.8 **a.** $\sigma_{vM} = 5.231$ GPa, **b.** $\sigma_{vM} = 71.73$ GPa, **c.** $\sigma_{vM} = 68.16$ GPa

6.9 **a.** $\sigma_x = -262.5$ MPa, $\sigma_y = -787.5$ MPa, $\tau_{xy} = -315$ MPa

  $\sigma_1 = -114.96$ MPa, $\sigma_2 = -935.038$ MPa, $\theta_p = -25.1°$

  b. $\sigma_x = -262.08$ MPa, $\sigma_y = -787.5$ MPa, $\tau_{xy} = -367.5$ MPa

  $\sigma_1 = -73.04$ MPa, $\sigma_2 = -976.53$ MPa, $\theta_p = -27.2°$

  c. $\sigma_x = -524.9$ MPa, $\sigma_y = -1574.7$ MPa, $\tau_{xy} = -367.5$ MPa

  $\sigma_1 = -409.03$ MPa, $\sigma_2 = -1690.5$ MPa, $\theta_p = -17.47°$

    d. $\sigma_1 = 85.64$ MPa, $\sigma_2 = -820.6$ MPa, $\theta_p = -40.0°$

    e. $\sigma_1 = -247.76$ MPa, $\sigma_2 = -6892.23$ MPa, $\theta_p = -39.2°$

    f. $\sigma_x = -393.75$ MPa, $\sigma_y = -1181.25$ MPa, $\tau_{xy} = -367.5$ MPa

       $\sigma_1 = -248.89$ MPa, $\sigma_2 = -1326.10$ MPa, $\theta_p = -21.5°$

6.10   **a.** $\sigma_x = -52.5$ MPa, $\sigma_y = -32.8$ MPa, $\tau_{xy} = -5.38$ MPa

       $\sigma_1 = -31.4$ MPa, $\sigma_2 = -53.9$ MPa, $\theta_p = -14.3°$

    b. $\sigma_x = -31.4$ MPa, $\sigma_y = -13.5$ MPa, $\tau_{xy} = 5.38$ MPa

       $\sigma_1 = -12.0$ MPa, $\sigma_2 = -32.9$ MPa, $\theta_p = -15.5°$

    c. $\sigma_x = -27.6$ MPa, $\sigma_y = -19.5$ MPa, $\tau_{xy} = 4.04$ MPa

       $\sigma_1 = -17.9$ MPa, $\sigma_2 = -29.3$ MPa, $\theta_p = -22.5°$

    d. $\sigma_x = -1.05$ MPa, $\sigma_y = 7.0$ MPa, $\tau_{xy} = 3.5$ MPa

       $\sigma_1 = 3.43$ MPa, $\sigma_2 = -3.78$ MPa, $\theta_p = -38°$

6.11   **a.** $f_{s1x} = 0$, $f_{s1y} = 0$, $f_{s2x} = p_0 Lt/6$, $f_{s2y} = 0$

       $f_{s3x} = p_0 Lt/3$, $f_{s3y} = 0$

    b. $f_{s1x} = 0$, $f_{s2x} = p_0 Lt/12$, $f_{s3x} = p_0 Lt/4$

6.12   **a.** $f_{s1y} = p_1 Lt/6$, $f_{s3y} = 1p_2 Lt/3$

    b. $f_{s1y} = f_{s2y} = p_0 Lt/\pi$

6.13 $u_3 = 71.4 \times 10^{-6}$ m, $v_3 = -39.3 \times 10^{-6}$ m

    $u_4 = -86.9 \times 10^{-6}$ m, $v_4 = -41.8 \times 10^{-6}$ m

    $\sigma_x^{(1)} = 25.84$ MPa, $\sigma_y^{(1)} = 7.38$ MPa, $\tau_{xy}^{(1)} = -51.69$ MPa

    $\sigma_1^{(1)} = 69.11$ MPa, $\sigma_2^{(1)} = -35.89$ MPa, $\theta_p^{(1)} = -40°$

    $\sigma_x^{(2)} = -26.4$ MPa, $\sigma_y^{(2)} = 8.67$ MPa, $\tau_{xy}^{(2)} = -12.92$ MPa

    $\sigma_1^{(2)} = 12.91$ MPa, $\sigma_2^{(2)} = -30.64$ MPa, $\theta_p^{(2)} = 18.15°$

6.14   **a.** $u_2 = 0.281 \times 10^{-4}$ m, $v_2 = -0.330 \times 10^{-4}$ m

       $u_5 = 0.115 \times 10^{-4}$ m, $v_5 = -0.103 \times 10^{-4}$ m

       $\sigma_x^{(2)} = 16.4$ MPa, $\sigma_y^{(2)} = 15.2$ MPa

       $\tau_{xy}^{(2)} = -6.99$ MPa, $\sigma_1^{(2)} = 22.8$ MPa

       $\sigma_2^{(2)} = 8.80$ MPa, $\theta_p^{(2)} = -42.7°$

       $\sigma_x^{(1)} = 10.6$ MPa, $\sigma_y^{(1)} = 3.18$ MPa

       $\tau_{xy}^{(1)} = -3.34$ MPa, $\sigma_1^{(1)} = 11.9$ MPa

       $\sigma_2^{(1)} = 1.90$ MPa, $\theta_p^{(1)} = -21.0°$

    c. $u_1 = -u_2 = -0.165 \times 10^{-5}$ m, $v_1 = v_2 = -0.125 \times 10^{-4}$ m

       $u_5 = 0.274 \times 10^{-12}$ m, $v_5 = -0.163 \times 10^{-4}$ m

       $\sigma_x^{(1)} = 5.99 \times 10^5$ N/m^2, $\sigma_y^{(1)} = -3.78 \times 10^6$ N/m^2

$$\tau_{xy}^{(1)} = 4.05 \times 10^{-1} \text{ N/m}^2, \quad \sigma_1^{(1)} = 5.99 \times 10^5 \text{ N/m}^2$$
$$\sigma_2^{(1)} = -3.78 \times 10^6 \text{ N/m}^2, \quad \theta_p^{(1)} = 0°, \quad \sigma_x^{(3)} = 5.64 \times 10^6 \text{ N/m}^2$$
$$\sigma_y^{(3)} = 1.88 \times 10^7 \text{ N/m}^2, \quad \tau_{xy}^{(3)} = -1.11 \times 10^{-1} \text{ N/m}^2$$
$$\sigma_1^{(3)} = 1.88 \times 10^7 \text{ N/m}^2, \quad \sigma_2^{(3)} = 5.64 \times 10^6 \text{ N/m}^2, \quad \theta_p^{(3)} = -90°$$

6.15 All f_{bx}'s are equal to 0.

    a. $f_{b1y} = f_{b2y} = f_{b3y} = f_{b4y} = -10.28$ N, $f_{b5y} = -20.56$ N

    c. $f_{b1y} = f_{b2y} = f_{b3y} = f_{b4y} = -8.03$ N, $f_{b5y} = -16.06$ N

6.18 **b.** Yes,   **c.** Yes,   **e.** Yes,   **g.** No

6.20 **a.** $n_b = 8$,   **b.** $n_b = 12$

6.25 $\varepsilon_x = 0.0009375$ cm/cm, $\varepsilon_y = -0.00125$ cm/cm, $\gamma_{xy} = -0.000625$ rad

    $\sigma_x = 129.8$ MPa, $\sigma_y = -223.6$ MPa, $\tau_{xy} = -50.5$ MPa

Chapter 7

7.13 Stress approaches 25 kPa near edge of hole.

7.16 $\sigma_1 = 133$ MPa at fillet

7.22 $\sigma_1 = 3$ kN/m^2 (round hole model)

    $\sigma_1 = 3.51$ kN/m^2 (square hole with corner radius)

7.24 $\sigma_{VM} = 8.1$ MPa

7.26 $\sigma_1 = 6.6$ MPa at hole, 10.4 MPa at transition

7.28 Largest von Mises stress 35–45 MPa at inside edge at junction of narrow to larger section of wrench

7.30 Largest principal stress $\sigma_1 = 111$ MPa at narrowest width of member

7.38 For a 1 cm thick wrench, $\sigma_{VM} = 502$ MPa

Chapter 8

8.2 $\varepsilon_x = \dfrac{1}{3b}(-u_1 + u_2 + 4u_4 - 4u_5), \quad \varepsilon_y = \dfrac{1}{3h}(-v_1 + v_3 + 4v_4 - 4v_6)$

    $\gamma_{xy} = \dfrac{1}{3h}(-u_1 + u_3 + 4u_4 - 4u_6) + \dfrac{1}{3b}(-v_1 + v_3 + 4v_4 - 4v_6)$

    $\sigma_x = \dfrac{E}{1 - v^2}(\varepsilon_x + v\varepsilon_y), \quad \sigma_y = \dfrac{E}{1 - v^2}(\varepsilon_y + v\varepsilon_x), \quad \tau_{xy} = G\gamma_{xy}$

8.3 $f_{s1x} = f_{s3x} = \dfrac{-pth}{6}, \quad f_{s5x} = \dfrac{-2pth}{3}$

8.4 $f_{s1x} = 0, \quad f_{s3x} = \dfrac{-p_0 th}{6}, \quad f_{s5x} = \dfrac{-p_0 th}{3}$

8.6 $\varepsilon_x = 2.54 \times 10^{-3}$

$\varepsilon_y = -7.62 \times 10^{-3}$

$\gamma_{xy} = -7.04 \times 10^{-3}$

8.7 $N_1 = 1 - \dfrac{x}{20} + \dfrac{x^2}{1800}$, $N_2 = \dfrac{-x+y}{60} + \dfrac{x^2+y^2}{1800} - \dfrac{xy}{900}$

$N_3 = \dfrac{-y}{60} + \dfrac{y^2}{1800}$, $N_4 = \dfrac{xy}{900} - \dfrac{y^2}{900}$, $N_5 = \dfrac{y}{15} - \dfrac{xy}{900}$, etc.

Chapter 9

9.1 **a.** $[K] = 175.84 \times 10^7 \begin{bmatrix} 5 & 1 & 0 & -1 & 1 & 0 \\ 1 & 4 & -2 & -1 & -2 & -3 \\ 0 & -2 & 8 & 0 & 4 & 2 \\ -1 & -1 & 0 & 1 & 1 & 0 \\ 1 & -2 & 4 & 1 & 4 & 1 \\ 0 & -3 & 2 & 0 & 1 & 3 \end{bmatrix} \dfrac{\text{N}}{\text{m}}$

b. $[K] = 151.59 \times 10^7 \begin{bmatrix} 2.75 & 0 & -2.25 & 0.5 & 0.25 & -0.5 \\ 0 & 1 & 1 & -1 & -1 & 0 \\ -2.25 & 1 & 5.75 & -2.5 & 0.25 & 1.5 \\ 0.5 & -1 & -2.5 & 4 & 0.5 & -3 \\ 0.25 & -1 & 0.25 & 0.5 & 1.75 & 0.5 \\ -0.5 & 0 & 1.5 & -3 & 0.5 & 3 \end{bmatrix} \dfrac{\text{N}}{\text{m}}$

9.2 $f_{s2r} = \dfrac{2\pi b p_0 h}{6}$, $f_{s3r} = \dfrac{2\pi b p_0 h}{3}$

9.3 $f_{b1r} = f_{b2r} = f_{b3r} = 0.153 \text{ N}$

$f_{b1z} = f_{b2z} = f_{b3z} = -6.428 \text{ N}$

9.6 **a.** $[k] = 7.037 \begin{bmatrix} 3125 & 625 & 0 & -625 & 625 & 0 \\ & 2500 & -1250 & -625 & -1250 & -1875 \\ & & 5000 & 0 & 2500 & 1250 \\ & & & 625 & 625 & 0 \\ & & & & 2500 & 625 \\ \text{Symmetry} & & & & & 1875 \end{bmatrix} \text{kN/mm}$

b. $[k] = 11.73 \begin{bmatrix} 2475 & 0 & -2025 & 450 & 225 & -450 \\ & 900 & 900 & -900 & -900 & 0 \\ & & 5175 & -2250 & 225 & 1350 \\ & & & 3600 & 450 & -2700 \\ & & & & 1575 & 450 \\ \text{Symmetry} & & & & & 2700 \end{bmatrix} \text{kN/mm}$

$$\textbf{c. } [k] = \begin{bmatrix} 8.577 \times 10^8 & 1.759 \times 10^8 & -3.738 \times 10^8 & -8.796 \times 10^7 & 4.398 \times 10^7 & -8.796 \times 10^7 \\ 1.759 \times 10^8 & 4.618 \times 10^8 & -8.796 \times 10^7 & -6.597 \times 10^7 & -3.519 \times 10^8 & -3.958 \times 10^8 \\ -3.738 \times 10^8 & -8.796 \times 10^7 & 1.561 \times 10^9 & -3.519 \times 10^8 & 3.958 \times 10^8 & 4.398 \times 10^8 \\ -8.796 \times 10^7 & -6.597 \times 10^7 & -3.519 \times 10^8 & 4.618 \times 10^8 & 1.759 \times 10^8 & -3.958 \times 10^8 \\ 4.398 \times 10^7 & -3.519 \times 10^8 & 3.958 \times 10^8 & 1.759 \times 10^8 & 6.158 \times 10^8 & 1.759 \times 10^8 \\ -8.796 \times 10^7 & -3.958 \times 10^8 & 4.398 \times 10^8 & -3.958 \times 10^8 & 1.759 \times 10^8 & 7.917 \times 10^8 \end{bmatrix}$$

9.7 a. $\sigma_r = -84$ MPa, $\sigma_z = -84$ MPa, $\sigma_\theta = 252$ MPa, $\tau_{rz} = -101$ MPa

   b. $\sigma_r = -103$ MPa, $\sigma_z = -103$ MPa, $\sigma_\theta = 112$ MPa, $\tau_{rz} = -73$ MPa

   c. $\sigma_r = -2870$ MPa, $\sigma_z = -2450$ MPa, $\sigma_\theta = 3570$ MPa, $\tau_{rz} = -1890$ MPa

9.12 $\sigma_\theta = \sigma_r$

9.19 $\sigma_\theta = 159$ MPa, $u_r = 23.9$ mm

9.20 $\sigma_1 = 64.1$ MPa, $u = 0.0782$ m at top and bottom center of plates

Chapter 10

10.2 a. $s = -0.2$ **b.** $N_1 = 0.6$, $N_2 = 0.4$ (for both of the 3-noded bars)

10.3 a. $s = 0$,   **b.** $N_1 = 0.5$, $N_2 = 0.5$

10.4 $N_1 = -(2/3)s^3 + (2/3)s^2 + s/6 - 1/6$, $N_2 = (4/3)s^3 - (2/3)s^2 - (4/3)s + 2/3$
    $N_3 = -(4/3)s^3 - (2/3)s^2 + (4/3)s + 2/3$, $N_4 = (2/3)s^3 + (2/3)s^2 - s/6 - 1/6$

10.5 a. $s = -0.4$ **b.** $N_1 = 0.28$, $N_2 = -0.12$, $N_3 = 0.84$

10.6 a. $s = 0.5$ **b.** $N_1 = 0.125$, $N_2 = 0.375$, $N_3 = 0.75$

10.8 $u_2 = 4.859 \times 10^{-4}$ m (right end), $u_3 = 2.793 \times 10^{-4}$ m (center)

10.13 $f_{s3s} = 0$, $f_{s3t} = pLt/2$, $f_{s4s} = 0$, $f_{s4t} = pLt/2$

10.14 a. $f_{s3t} = 500$ N, $f_{s4t} = 500$ N,   **b.** $f_{s1t} = 166.66$ N, $f_{s4t} = 83.33$ N

10.15 a. 1.918,   **b.** 0.667,   **c.** 0.400,   **d.** 2.87,   **f.** 0

Chapter 11

11.1 a. $[B] = \dfrac{1}{8}\begin{bmatrix} 0 & 0 & 0 & 0 & 0 & 0 & 4 & 0 & 0 & -4 & 0 & 0 \\ 0 & 0 & 0 & 0 & 4 & 0 & 0 & 0 & 0 & 0 & -4 & 0 \\ 0 & 0 & 4 & 0 & 0 & 0 & 0 & 0 & 0 & 0 & 0 & -4 \\ 0 & 0 & 0 & 4 & 0 & 0 & 0 & 4 & 0 & -4 & -4 & 0 \\ 0 & 4 & 0 & 0 & 0 & 4 & 0 & 0 & 0 & 0 & -4 & -4 \\ 4 & 0 & 0 & 0 & 0 & 0 & 0 & 0 & 4 & -4 & 0 & -4 \end{bmatrix}$

   b. $[B] = \begin{bmatrix} -0.5 & 0 & 0 & 0 & 0 & 0 & 0.5 & 0 & 0 & 0 & 0 & 0 \\ 0 & -0.75 & 0 & 0 & 0 & 0 & 0 & 0.25 & 0 & 0 & 0.5 & 0 \\ 0 & 0 & -0.75 & 0 & 0 & 0.5 & 0 & 0 & 0.25 & 0 & 0 & 0 \\ -0.75 & -0.5 & 0 & 0 & 0 & 0 & 0.25 & 0.5 & 0 & 0.5 & 0 & 0 \\ 0 & -0.75 & -0.75 & 0 & 0.5 & 0 & 0 & 0.25 & 0.25 & 0 & 0 & 0.5 \\ -0.75 & 0 & -0.5 & 0.5 & 0 & 0 & 0.25 & 0 & 0.5 & 0 & 0 & 0 \end{bmatrix}$

11.3 a. $\sigma_x = 1.3$ GPa, $\sigma_y = 144.16$ MPa, $\sigma_z = -816.66$ MPa

$\quad\quad\tau_{xy} = 191.66$ MPa, $\tau_{yz} = -385$ MPa, $\tau_{zx} = 96.16$ MPa

11.6 a. $[B] = \dfrac{1}{18,750}$

$$\times \begin{bmatrix}
-625 & 0 & 0 & 0 & 0 & 0 & 0 & 0 & 0 & 625 & 0 & 0 \\
0 & -375 & 0 & 0 & 750 & 0 & 0 & 0 & 0 & 0 & -375 & 0 \\
0 & 0 & -375 & 0 & 0 & 0 & 0 & 0 & 750 & 0 & 0 & -375 \\
-375 & -625 & 0 & 750 & 0 & 0 & 0 & 0 & 0 & -375 & 625 & 0 \\
0 & -375 & -375 & 0 & 0 & 750 & 0 & 750 & 0 & 0 & -375 & -375 \\
-375 & 0 & -625 & 0 & 0 & 0 & 750 & 0 & 0 & -375 & 0 & 625
\end{bmatrix}$$

b. $[B] =$

$$\begin{bmatrix}
-0.125 & 0 & 0 & 0 & 0 & 0 & 0 & 0 & 0 & 0.125 & 0 & 0 \\
0 & -0.05 & 0 & 0 & 0.2 & 0 & 0 & 0 & 0 & 0 & -0.15 & 0 \\
0 & 0 & -0.05 & 0 & 0 & 0 & 0 & 0 & 0.2 & 0 & 0 & -0.15 \\
-0.05 & -0.125 & 0 & 0.2 & 0 & 0 & 0 & 0 & 0 & -0.15 & 0.125 & 0 \\
0 & -0.05 & -0.05 & 0 & 0 & 0.2 & 0 & 0.2 & 0 & 0 & -0.15 & -0.15 \\
-0.05 & 0 & -0.125 & 0 & 0 & 0 & 0.2 & 0 & 0 & -0.15 & 0 & 0.125
\end{bmatrix}$$

11.7 a. $\sigma_x = 72.7$ MPa, $\sigma_y = 169.6$ MPa, $\sigma_z = 72.7$ MPa

$\quad\quad\tau_{xy} = 59.2$ MPa, $\tau_{yz} = 32.3$ MPa, $\tau_{zx} = 91.5$ MPa

11.8 $u = a_1 + a_2x + a_3y + a_4z + a_5xy + a_6xz + a_7yz + a_8x^2 + a_9y^2 + a_{10}z^2$

11.9 Loads must be in the y-z plane

11.10 $N_2 = \dfrac{(1-s)(1-t)(1-z')}{8},\quad N_3 = \dfrac{(1-s)(1+t)(1-z')}{8},$

$\quad\quad N_4 = \dfrac{(1-s)(1+t)(1+z')}{8},$

$\quad\quad N_5 = \dfrac{(1+s)(1-t)(1+z')}{8},\quad N_6 = \dfrac{(1+s)(1-t)(1-z')}{8},$

$\quad\quad N_7 = \dfrac{(1+s)(1+t)(1-z')}{8},\quad N_8 = \dfrac{(1+s)(1+t)(1+z')}{8}$

11.11 $N_1 = \dfrac{(1-s)(1-t)(1+z')(-s-t+z'-2)}{8},$

$\quad\quad N_2 = \dfrac{(1-s)(1-t)(1-z')(-s-t-z'-2)}{8}$

11.13 $w = -0.231$ mm under the load, $w = -0.187$ mm at front corner

11.14 $d_{\max} = -0.56$ mm at free end

11.18 Largest $\sigma_{vM} = 758$ MPa at elbow, $\delta_{\max} = 4.13$ mm at free end

11.20 Largest $\sigma_{vM} = 219$ MPa at elbow, $\delta_{\max} = 18.8$ mm at free end

11.25 Largest $\sigma_{vM} = 534$ MPa due to 1200 N

11.27 Largest $\sigma_{vM} = 186$ MPa at inner semi-circular face

Chapter 12

12.2 $\delta_{max} = -0.1808$ mm, $\sigma_{max} = 10.17$ MPa

12.3 $\delta_{max} = -3.716$ mm, $\sigma_{max} = 142$ MPa

12.4 $\delta_{max} = 0.00308$ in.

12.9 $\delta_{max} = 0.214$ mm, $\sigma_1 = 40$ MPa

12.10 $\delta_{max} = 68.9$ mm (too large for small deflection assumption)

Chapter 13

13.1 $t_2 = 166.7\,°C$, $t_3 = 233.3\,°C$

13.2 $t_2 = 66.2\,°C$, $t_3 = 41.8\,°C$, $t_4 = 20.3\,°C$

13.4 $t_1 = 67.54\,°C$, $t_2 = 65.92\,°C$, $t_3 = 61.06\,°C$, $t_4 = 52.96\,°C$

13.5 $t_2 = 117\,°C$, $t_3 = 133\,°C$, $t_4 = 150\,°C$, $t_5 = 167\,°C$

13.6 $t_2 = 421\,°C$, $t_3 = 121\,°C$, $q^{(3)} = 3975$ W/m^2

13.7 $t_2 = 418.2\,°C$, $t_3 = 527.3\,°C$

13.8 $t_1 = 230\,°C$, $t_2 = 110\,°C$, $t_3 = 50\,°C$, $q^{(3)} = 6000$ W/m^2

13.9 $t_1 = 16.3\,°C$, $t_2 = 16.1\,°C$, $t_3 = 1.4\,°C$, $t_4 = 1.2\,°C$

13.10 $t_1 = 18.9\,°C$, $t_2 = 17.5\,°C$, $t_3 = -8.32\,°C$, $t_4 = -9.45\,°C$, $q^{(1)} = 5.46$ W/m^2

13.12 $185\,°C$ at right end, $q_{max} = 439$ W

13.16 $t_2 = 87.95\,°C$, $t_3 = 86.72\,°C$, $t_4 = 86.28\,°C$

13.17 $[k] = \dfrac{AK_{xx}}{L}\begin{bmatrix} 1 & -1 \\ -1 & 1 \end{bmatrix}$

13.18 $[k_h]_{\text{left}} = hA\begin{bmatrix} 1 & 0 \\ 0 & 0 \end{bmatrix}$, $\{f_h\}_{\text{left}} = hT_\infty A\begin{Bmatrix} 1 \\ 0 \end{Bmatrix}$

13.20 $\{f\} = \begin{Bmatrix} 1291 \\ 27.3 \\ 1254 \end{Bmatrix}$ W

13.37 $12\,°C$ at 2.5 cm from top, $25\,°C$ 1.25 cm from top, $\bar{q}_{max} = 1416$ W, $\bar{q}_{min} = -1083$ W

13.44 $T = 323$ K located where q* is applied

Chapter 14

14.1 $p_2 = 4.545$ m, $p_3 = 1.818$ m, $v_x^{(1)} = 10.91$ m/s, $Q_f^{(1)} = 21.82$ m³/s

14.2 $p_2 = -15$ m, $p_3 = -40$ m, $p_4 = -65$ m, $v_x^{(1)} = 25$ m/s, $Q_1 = 50$ m³/s

14.4 $p_2 = -3$ cm, $p_3 = -8$ cm, $v_x^{(1)} = 1.2$ cm/s, $v_x^{(2)} = 2$ cm/s,

$Q_1 = Q_2 = 6$ cm³/s

14.7 **a.** $p_1 = 0.897$ N/m², $p_2 = 0.691$ N/m², $p_3 = 0.515$ N/m², $q_1 = 0.897$ m³/s,

$q_2 = 0.103$ m³/s, $q_3 = 0.059$ m³/s, $q_4 = 0.044$ m³/s, $q_5 = 0.103$ m³/s

14.8 $\{f_Q\} = \begin{Bmatrix} 54.76 \\ 28.57 \\ 16.67 \end{Bmatrix}$ m³/s

14.10 $p_2 = p_3 = 12$ m, $p_5 = 11$ m

14.17 $I_1 = 0.161$ amps, $I_2 = 0.027$ amps, $I_3 = -0.487$ amps, branch amps: $I_{AD} = 0.134$ amps, $I_{BC} = 0.513$ amps

14.18 $I_1 = -0.853$ amps, $I_2 = -0.458$ amps, $I_3 = -0.158$ amps, $I_{AB} = -0.695$ amps, $I_{BC} = -0.30$ amps

14.19 Original resistors too small, standard resistor sizes, $R_1 = 715$ ohms, $R_2 = 806$ ohms make $I_1 = 0.024$ amps, $I_2 = 0.011$ amps and branch amps: $I_{AD} = 0.013$ amps, $I_{BC} = 0.011$ amps

14.20 Original resistors too small, standard resistor sizes, $R_1 = 2000$ ohms, $R_2 = 1270$ ohms, make $I_1 = 0.015$ amps, $I_2 = 0.00945$ amps. So diode amps are less than 0.015 amps.

Chapter 15

15.1 $u_2 = 0.18$ mm, $u_3 = 0.36$ mm, $\sigma_x = 0$

15.2 $u_2 = 0$, $\sigma_x = 49.14$ MPa

15.3 $-u_1 = v_1 = -2.55 \times 10^{-4}$ m, $\sigma^{(1)} = 17.85$ MPa (T)

$\sigma^{(2)} = -25.2$ MPa (C), $\sigma^{(3)} = 17.85$ MPa (T)

15.5 $u_2 = 1.44 \times 10^{-4}$ m, $\sigma^{(1)} = -20.2$ MPa (C), $\sigma^{(2)} = \sigma^{(3)} = -10.1$ MPa (C)

15.6 $u_1 = 0$, $v_1 = 6.0 \times 10^{-4}$ m, $\sigma^{(1)} = \sigma^{(3)} = -10.5$ MPa (C)

$\sigma^{(2)} = 18.2$ MPa (T)

15.7 $u_1 = 0$, $v_1 = -3.6 \times 10^{-4}$ m, $\sigma^{(1)} = \sigma^{(2)} = 0$

15.9 $10\,°C$ increase in elements 1 and 3 also

15.11 Yes, $u = 406.25 \times 10^{-6}$ L, $\sigma_{st} = 2437$ psi (T), $\sigma_{al} = -2437$ psi (C)

15.12 **a.** 7.02×10^{-4} cm **b.** $\sigma_{br} = -78,630$ kPa, $\sigma_{mg} = -52,406$ kPa

15.14 $f_{T1x} = -43.125$ kN, $f_{T1y} = 0$, $f_{T2x} = 43.125$ kN, $f_{T2y} = -86.250$ kN
$f_{T3x} = 0$, $f_{T3y} = 86.250$ kN

15.16 $f_{T1x} = 134$ kN, $f_{T1y} = 134$ kN, $f_{T2x} = -134$ kN, $f_{T2y} = 0$
$f_{T3x} = 0$, $f_{T3y} = -134$ kN

15.18 $\sigma_x = 67.2$ MPa, $\sigma_y = 67.2$ MPa, $\tau_{xy} = 0$

15.19 $\{f_T\} = \dfrac{3AE\alpha_0 T}{2} \begin{Bmatrix} -1 \\ 1 \end{Bmatrix}$

15.20 $\dfrac{AE\alpha}{2} \begin{Bmatrix} -t_1 - t_2 \\ t_1 + t_2 \end{Bmatrix}$

15.21 $\{f_T\} = \dfrac{2\pi \bar{r} AE\alpha(T)[\bar{B}]^T}{1 - 2v} \begin{Bmatrix} 1 \\ 1 \\ 1 \\ 0 \end{Bmatrix}$

Chapter 16

16.1 $[M] = \dfrac{\rho AL}{6} \begin{bmatrix} 2 & 1 & 0 \\ 1 & 4 & 1 \\ 0 & 1 & 2 \end{bmatrix}$

16.2 a. $[M] = \dfrac{\rho AL}{2} \begin{bmatrix} 1 & 0 & 0 & 0 \\ 0 & 2 & 0 & 0 \\ 0 & 0 & 2 & 0 \\ 0 & 0 & 0 & 1 \end{bmatrix}$

b. $[M] = \dfrac{\rho AL}{6} \begin{bmatrix} 2 & 1 & 0 & 0 \\ 1 & 4 & 1 & 0 \\ 0 & 1 & 4 & 1 \\ 0 & 0 & 1 & 2 \end{bmatrix}$

16.3 $\omega_1 = 0.806\sqrt{\mu}$, $\omega_2 = 2.81\sqrt{\mu}$

16.4 $\omega_1 = 5.496 \times 10^3$ rad/s, $\omega_2 = 17.974 \times 10^3$ rad/s

16.5 a.

t (s)	d_i (m)	$\dot{d}_i$ (m/s)	$\ddot{d}_i$ (m/s^2)
0	0	8.33	0
0.03	0.00375	7.36	0.235
0.06	0.01412	−0.231	0.3425
0.09	0.0243	−7.63	0.224
0.12	0.0276	−15.1	−0.1167
0.15	0.0173	−8.967	−0.478

16.6 **a.**

t (s)	d_i (m)	$\dot{d}_i$ (m/s)	$\ddot{d}_i$ (m/s^2)
0	0	0	3.33
0.02	6.67×10^{-4}	0.056	2.267
0.04	2.24×10^{-3}	0.092	0.656
0.06	4.34×10^{-3}	0.092	-1.27
0.08	5.93×10^{-3}	0.0505	-2.89
0.10	6.36×10^{-3}	-0.0168	-3.816

b.

t (s)	d_i (m)	$\dot{d}_i$ (m/s)	$\ddot{d}_i$ (m/s^2)	$F(t)$ N
0.00	0.00000	0.000	3.33	100
0.02	5.98×10^{-4}	0.0564	2.31	80
0.04	2.08×10^{-3}	0.0864	0.69	60
0.06	3.835×10^{-3}	0.0836	-0.975	40
0.08	5.21×10^{-3}	0.048	-2.51	20
0.10	5.61×10^{-3}	-0.0109	-3.38	0

16.8 Using Newmark's method with $\gamma = \frac{1}{2}$, $\beta = \frac{1}{6}$

Node	t (s)	d_i (m)	$\dot{d}_i$ (m/s)	$\ddot{d}_i$ (m/s^2)	$F(t)$ N
2	0	0	0	0	10,000
	0.05	5.74×10^{-5}	3.43×10^{-3}	0.137	9,000
	0.10	5.13×10^{-4}	0.0172	0.414	8,000
	0.15	1.94×10^{-3}	0.041	0.53	7,000
3	0	0	0	1.33	10,000
	0.05	1.49×10^{-3}	0.056	0.91	9,000
	0.10	5.11×10^{-3}	0.0817	0.117	8,000
	0.15	9.09×10^{-3}	0.072	-0.49	7,000

16.11 **a.** $\omega_1 = \dfrac{3.15}{L^2} \left(\dfrac{EI}{\rho A} \right)^{1/2}$, $\omega_2 = \dfrac{16.24}{L^2} \left(\dfrac{EI}{\rho A} \right)^{1/2}$ (2 element model)

b. $\omega_1 = \dfrac{198.4}{L^2} \left(\dfrac{EI}{\rho A} \right)^{1/2}$ (3 element model)

c. $\omega_1 = \dfrac{9.8}{L^2} \left(\dfrac{EI}{\rho A} \right)^{1/2}$ (2 element model), **d.** $\omega = \dfrac{14.8}{L^2} \left(\dfrac{EI}{\rho A} \right)^{1/2}$ (2 element model)

16.17

Node:		1	2	3	4	5	6
i	t (s)			Temperature ($^\circ$C)			
0	0	200	200	200	200	200	200
1	8	0	159.0095	191.4441	198.2110	199.6110	199.8444
2	16	0	135.5852	178.1491	193.6620	198.2112	199.1445
3	24	0	120.2309	165.7003	187.3485	195.5379	197.5152
4	32	0	109.1993	154.9587	180.4038	191.7446	194.8115
5	40	0	100.7600	145.7784	173.4129	187.1268	191.1242
6	48	0	94.00311	137.8529	166.6182	181.9599	186.6590
7	56	0	88.39929	130.9034	160.1012	176.4598	181.6395
8	64	0	83.61745	124.7101	153.8759	170.7856	176.2620
9	72	0	79.43935	119.1075	147.9316	165.0508	170.6822
10	80	0	75.71603	113.9733	142.2502	159.3352	165.0171

16.18

		Node		
Time (s)	1	2	3	(using consistent capacitance matrix)
		Temperature ($^\circ$C)		
0	25	25	25	
0.1	85	18.53611	26.36189	
0.2	85	29.61303	21.63526	
0.3	85	36.18435	22.42717	
0.4	85	40.72491	25.30428	
0.5	85	44.27834	28.85201	
0.6	85	47.29072	32.49614	
0.7	85	49.95809	36.01157	
0.8	85	52.37152	39.31761	
0.9	85	54.57756	42.39278	
1	85	56.60353	45.23933	
1.1	85	58.46814	47.86852	
1.2	85	60.1859	50.29457	
1.3	85	61.76908	52.53218	
1.4	85	63.22852	54.59557	
1.5	85	64.574	56.49814	
1.6	85	65.81448	58.25235	
1.7	85	66.95818	59.86974	
1.8	85	68.01265	61.36096	
1.9	85	68.98485	62.73586	
2	85	69.88121	64.0035	
2.1	85	70.70765	65.17226	
2.2	85	71.46961	66.24984	

(*continued*)

16.18 (*continued*)

Time (s)	Node 1	Node 2	Node 3
		Temperature (°C)	
0	25	25	25
2.3	85	72.17214	67.24336
2.4	85	72.81986	68.15938
2.5	85	73.41705	69.00393
2.6	85	73.96766	69.78261
2.7	85	74.47531	70.50053
2.8	85	74.94336	71.16246
2.9	85	75.3749	71.77274
3	85	75.77277	72.33542

Appendix A

A1. **a.** $\begin{bmatrix} 3 & 0 \\ -3 & 12 \end{bmatrix}$ **b.** Nonsense   **c.** Nonsense

d. $\begin{Bmatrix} 11 \\ 9 \\ 11 \end{Bmatrix}$ **e.** Nonsense   **f.** $\begin{bmatrix} 10 & 7 & 6 \\ 3 & -1 & 7 \end{bmatrix}$

A2. $\begin{bmatrix} 1 & 0 \\ \frac{1}{4} & \frac{1}{4} \end{bmatrix}$

A3. $\frac{1}{17}\begin{bmatrix} 12 & -3 & -8 \\ -3 & 5 & 2 \\ -8 & 2 & 11 \end{bmatrix}$

A4. Nonsense

A5. $\begin{bmatrix} \frac{1}{2} & 0 \\ \frac{1}{8} & \frac{1}{8} \end{bmatrix}$

A6. Same as A3

A8. $\begin{bmatrix} \cos\theta & -\sin\theta \\ \sin\theta & \cos\theta \end{bmatrix}$

A10. $\frac{E}{L}\begin{bmatrix} 1 & -1 \\ -1 & 1 \end{bmatrix}$

Appendix B

B1. $x_1 = 3.15,\quad x_2 = 0.62$

B2. $x_1 = 3.15,\quad x_2 = 0.62$

B3. $x_1 = 2.5, \quad x_2 = -1, \quad x_3 = 0.5$

B4. $x_1 = 3, \quad x_2 = -1, \quad x_3 = -2$

B5. **a.** $\begin{Bmatrix} x_1 \\ x_2 \end{Bmatrix} = \begin{bmatrix} 2 & -1 \\ 1 & -1 \end{bmatrix} \begin{Bmatrix} y_1 \\ y_2 \end{Bmatrix}$ **b.** $\begin{Bmatrix} z_1 \\ z_2 \end{Bmatrix} = \begin{bmatrix} -3 & 2 \\ 5 & -3 \end{bmatrix} \begin{Bmatrix} y_1 \\ y_2 \end{Bmatrix}$

B6. $x_1 = 0, \quad x_2 = 1, \quad x_3 = 2, \quad x_4 = 2, \quad x_5 = 0$

B7. $x_1 = 3.15, \quad x_2 = 0.62$

B8. **a.** Unique   **b.** Nonexistent   **c.** Unique   **d.** Nonunique

Appendix D

D1. **a.** $f_{1y} = f_{2y} = -25$ kN, $\quad m_1 = -m_2 = -37.5$ kN · m
  b. $f_{1y} = f_{2y} = -25$ kN, $\quad m_1 = -m_2 = -28.125$ kN · m
  c. $f_{1y} = f_{2y} = -75$ kN, $\quad m_1 = -m_2 = -125$ kN · m
  d. $f_{1y} = -85.625$ kN, $\quad f_{2y} = -29.375$ kN, $\quad m_1 = -80.625$ kN · m, $\quad m_2 = 46.875$ kN · m
  e. $f_{1y} = -27$ kN, $\quad f_{2y} = -63$ kN, $\quad m_1 = -36$ kN · m, $\quad m_2 = -54$ kN · m
  f. $f_{1y} = -4.0$ kN, $\quad f_{2y} = -0.99$ kN, $\quad m_1 = 5.10$ kN · m, $\quad m_2 = -2.04$ kN · m
  g. $f_{1y} = f_{2y} = -6$ kN, $\quad m_1 = -m_2 = -7.5$ kN · m
  h. $f_{1y} = f_{2y} = -10$ kN, $\quad m_1 = -m_2 = -6.67$ kN · m

PRINCIPAL UNITS USED IN MECHANICS

Quantity	International System (SI)			U.S. Customary System (USCS)		
	Unit	Symbol	Formula	Unit	Symbol	Formula
Acceleration (angular)	radian per second squared		rad/s^2	radian per second squared		rad/s^2
Acceleration (linear)	meter per second squared		m/s^2	foot per second squared		ft/s^2
Area	square meter		m^2	square foot		ft^2
Density (mass) (Specific mass)	kilogram per cubic meter		kg/m^3	slug per cubic foot		$slug/ft^3$
Density (weight) (Specific weight)	newton per cubic meter		N/m^3	pound per cubic foot	pcf	lb/ft^3
Energy; work	joule	J	$N{\cdot}m$	foot-pound		ft-lb
Force	newton	N	$kg{\cdot}m/s^2$	pound	lb	(base unit)
Force per unit length (Intensity of force)	newton per meter		N/m	pound per foot		lb/ft
Frequency	hertz	Hz	s^{-1}	hertz	Hz	s^{-1}
Length	meter	m	(base unit)	foot	ft	(base unit)
Mass	kilogram	kg	(base unit)	slug		$lb\text{-}s^2/ft$
Moment of a force; torque	newton meter		$N{\cdot}m$	pound-foot		lb-ft
Moment of inertia (area)	meter to fourth power		m^4	inch to fourth power		$in.^4$
Moment of inertia (mass)	kilogram meter squared		$kg{\cdot}m^2$	slug foot squared		$slug\text{-}ft^2$
Power	watt	W	J/s $(N{\cdot}m/s)$	foot-pound per second		ft-lb/s
Pressure	pascal	Pa	N/m^2	pound per square foot	psf	lb/ft^2
Section modulus	meter to third power		m^3	inch to third power		$in.^3$
Stress	pascal	Pa	N/m^2	pound per square inch	psi	$lb/in.^2$
Time	second	s	(base unit)	second	s	(base unit)
Velocity (angular)	radian per second		rad/s	radian per second		rad/s
Velocity (linear)	meter per second		m/s	foot per second	fps	ft/s
Volume (liquids)	liter	L	$10^{-3}\ m^3$	gallon	gal.	$231\ in.^3$
Volume (solids)	cubic meter		m^3	cubic foot	cf	ft^3